LEGAL GUIDE TO
AIA DOCUMENTS
FIFTH EDITION

ASPEN PUBLISHERS

LEGAL GUIDE TO AIA DOCUMENTS
FIFTH EDITION

WERNER SABO

Sabo & Zahn
Chicago, Illinois

Wolters Kluwer
Law & Business

AUSTIN   BOSTON   CHICAGO   NEW YORK   THE NETHERLANDS

Printed in the United States of America

1 2 3 4 5 6 7 8 9 0

ISBN 978-0-7355-7454-0

About Wolters Kluwer Law & Business

Wolters Kluwer Law & Business is a leading provider of research information and workflow solutions in key specialty areas. The strengths of the individual brands of Aspen Publishers, CCH, Kluwer Law International and Loislaw are aligned within Wolters Kluwer Law & Business to provide comprehensive, in-depth solutions and expert-authored content for the legal, professional and education markets.

CCH was founded in 1913 and has served more than four generations of business professionals and their clients. The CCH products in the Wolters Kluwer Law & Business group are highly regarded electronic and print resources for legal, securities, antitrust and trade regulation, government contracting, banking, pension, payroll, employment and labor, and healthcare reimbursement and compliance professionals.

Aspen Publishers is a leading information provider for attorneys, business professionals and law students. Written by preeminent authorities, Aspen products offer analytical and practical information in a range of specialty practice areas from securities law and intellectual property to mergers and acquisitions and pension/benefits. Aspen's trusted legal education resources provide professors and students with high-quality, up-to-date and effective resources for successful instruction and study in all areas of the law.

Kluwer Law International supplies the global business community with comprehensive English-language international legal information. Legal practitioners, corporate counsel and business executives around the world rely on the Kluwer Law International journals, loose-leafs, books and electronic products for authoritative information in many areas of international legal practice.

Loislaw is a premier provider of digitized legal content to small law firm practitioners of various specializations. Loislaw provides attorneys with the ability to quickly and efficiently find the necessary legal information they need, when and where they need it, by facilitating access to primary law as well as state-specific law, records, forms and treatises.

Wolters Kluwer Law & Business, a unit of Wolters Kluwer, is headquartered in New York and Riverwoods, Illinois. Wolters Kluwer is a leading multinational publisher and information services company.

PREFACE

This book is written for architects, to use as a reference when drafting the Owner-Architect Agreement, the General Conditions, and the Owner-Contractor Agreement; for owners to understand the work of the architect and the construction process; for contractors who must use these documents in the construction of buildings; and for attorneys who review and amend these documents. It is intended as a reference work and is cross-referenced by cases and topics. The layout is designed to facilitate understanding and revision of the documents, by setting forth the actual language of the documents followed by commentary and description of relevant cases.

Each discussion of a document is followed by a chapter that describes alternate language. This alternate language may favor one party or another and should be used or modified only after consulting with an attorney who is knowledgeable about construction law in the area where the project is to be constructed. Often, the use of such alternate language requires revising language in other parts of the same document or in other documents. This must be done with extreme care. With the widespread use of electronic versions of AIA documents, editing can be done within the body of the document. Users of paper documents can make changes in riders to the document, or directly on the face of the document.

Comments regarding these documents are generally applicable to other AIA documents, because much of the language is similar throughout the family of documents. The language and comments also generally apply to older versions of the documents, particularly because the court cases cited involve cases dating back a number of years.

The AIA language excerpted and highlighted by italicized printing is taken from the standard AIA Documents intended to be used as "consumables," rather than being excerpted as if the language were a model. (Consumables are further defined in Senate Report No. 94-473 on the Copyright Act of 1976.) The consumable nature of this work does not imply permission for further reproduction.

The American Institute of Architects has granted permission to reproduce excerpts from AIA Documents A101, A201, B101, B102, B103, and B104 under license number 28003.

Chicago, Illinois   WERNER SABO
January, 2008   *E-Mail: wsabo@sabozahn.com*
                                   Internet: http://www.sabozahn.com

ACKNOWLEDGMENTS

I would like to thank my wife, Therese, and daughter, Elizabeth, for their support. Thanks also go to my partner, Jim Zahn, for his special insights into these documents and to the staff at Sabo & Zahn for their assistance with this book. Finally, I would like to thank my editor, Pj Iraca, for his enthusiasm and assistance.

W.S.

ABOUT THE AUTHOR

Werner Sabo is a partner in the law firm of Sabo & Zahn in Chicago, Illinois, concentrating in construction law. He is also a licensed architect, having practiced architecture for a number of years before establishing his law practice. He is a Fellow of the American Institute of Architects and the Association of Licensed Architects and has been active with the AIA, CSI, and professional associations for many years. Mr. Sabo also serves as a panel arbitrator and mediator for the American Arbitration Association. He has authored articles for the Chicago AIA *Focus*, the Chicago CSI *Change Order*, the CSI *Specifier*, and other publications, and continues to lecture to students and construction professionals on legal issues. Representing contractors, architects, and other parties in the construction arena, he arbitrates and litigates at the state and federal levels.

SUMMARY CONTENTS

TABLE OF CONTENTS

CHAPTER 1

INTRODUCTION

More than one hundred years ago, the American Institute of Architects ("AIA") began drafting and promoting standard agreements for the construction industry. Over time, these documents have become the gold standard for construction agreements, encompassing the means and methods whereby projects are designed and built, at least in the United States. Each of the participants in the construction process has at least one choice of agreement, based on one or more factors. As of this writing, there are almost 100 different AIA documents to cover a broad range of situations in the construction process.

The AIA documents have achieved success primarily because they are the fairest and most neutral of any document readily available to the construction public. Other constituencies have, over the years, prepared and promulgated their own documents. Engineers, in the form of the Engineers Joint Contract Documents Committee (EJCDC) documents, have achieved some success in drafting documents that are primarily used for projects led by engineers, such as power stations, road projects, and the like. Various owner groups have attempted to draft standard documents without much success. Contractors, through the Associated General Contractors (AGC), have worked with the AIA and, on a parallel track, have drafted their own version of construction documents. In 2007, the AGC led a consortium of construction groups in drafting a series of documents called ConsensusDOCS. It is too early to tell whether or not these documents will be greeted with applause or dismissed as a narrow attempt to promote the interests of general contractors to the detriment of owners and designers.

In late 2007, the AIA released the latest set of documents, including the flagship A201, General Conditions of the Contract for Construction, the A101, Owner-Contractor Agreement, and several new versions of owner-architect agreements. Numerous other documents were released at the same time to make a cohesive and complimentary set of documents for the construction industry. One of the greatest strengths of the AIA documents is that the documents are coordinated. This means that the parties can use a family of documents that include agreements between all of the players in the construction process, general conditions, and related documents, and not worry about inconsistencies in language, style, or intent. This, by itself, makes the AIA documents a bargain.

The AIA documents also reflect the current state of the construction industry. This includes the methodology of how buildings are built, as well as the latest techniques for delivering information to the participants in the project. The AIA documents, particularly A201, reflect a consensus as to how a normal construction project is run, who does what, and what the standards in the construction industry are at this time. Anyone wishing to learn how a construction project in the United States in the early 21st century is done need merely peruse the A201. Anyone wishing to learn what the architect's normal duties and responsibilities on a construction project might be needs to review the latest version of B101. These, and the other new documents, reflect the latest thinking by the AIA and their consultants, advisors, and others involved in this process, as to how best to perform a construction project.

Many people believe that the AIA has as its principal goal the protection of the architects' interests. While there is some truth to that, the real goal of these documents is to further the process of construction, to fairly and equitably allocate the risks and rewards inherent in the construction process, and to create documents that reflect those concerns. If these documents were one-sided, no owner would use them. The reality is that most architects strive to look out for the owners' interests while trying to be fair to the contractor. The documents put into effect a system of checks and balances that is reasonable and works in the interests of the project.

A major advantage of the AIA documents is that they have been tested in the courts. There are literally thousands of court decisions that involve AIA documents. This book is a compendium of these decisions by topic. The format is the documents themselves. The major documents are examined, including several owner-architect agreements, the major owner-contractor agreement, and the general conditions. These documents cover the vast majority of language used in these and all other AIA documents. If the reader has issues with a different AIA document (or, for that matter, any other construction document), it is likely that the language found in this book will be identical to, or closely similar to that found in the other document. Courts have construed these provisions over the last hundred years, providing a vast body of precedent useful in both the drafting of new documents as well as litigation involving such language.

Each section of this book cites the latest AIA language, references the prior language, and gives court citations to the extent that courts have examined this, or similar, language. Not every case is cited, so readers should use the cited cases as a starting point in researching the law about a particular point. Following each section is a chapter with alternative language. This is to assist the transactional attorney in preparing amendments to the documents. Some of this alternative language is taken from, or adopted from, cases that discuss the cited language. Other clauses are taken from, and often modified from, contract forms that the author has drafted or has reviewed. In many cases, there will be alternate clauses that favor one side or the other. It is not the intent of the author to imply that one clause or the other is better. This is left to the architect or attorney involved in a particular situation. Often, the alternative language itself will need to be further modified to suit a situation.

Attorneys that are not familiar with the construction process are warned that one cannot simply take contract provisions from a book such as this and insert them into an agreement. Understanding the legal and practical aspects of construction is essential to drafting a good document. The goal, after all, is to construct the project with a minimum of problems. The secondary goal is to protect the client. If this is properly done, with the risks and rewards properly allocated, a successful project is almost assured.

CHAPTER 2

AIA DOCUMENT B101 STANDARD FORM OF AGREEMENT BETWEEN OWNER AND ARCHITECT

§ 2.1   Introduction

This contract form contains many elements that are not contained in letter forms of agreement.[1] One of these elements is the stipulation that the architect is not responsible for construction means or methods. This protects the architect from liability for many construction accidents. Sometimes, if an architect presents the owner with this contract and it is never executed, a contract can still be found.[2]

---

[1] AIA Document B101 copyright 2007 The American Institute of Architects. All rights reserved. AIA® AIA copyrighted material shown here in italics has been reproduced with permission of the American Institute of Architects under permission number _____ FURTHER REPRODUCTION IS PROHIBITED. Because AIA documents are revised from time to time, users should ascertain from the AIA the current edition of the document reproduced herein. Copies of AIA document may be purchased from the AIA or its local distributors. The AIA language excerpted and highlighted by italicized printing herein is taken from standard AIA Documents intended to be used as "consumables," rather than being excerpted from as if the language was a model. (Consumables are further defined in Senate Report No. 94-473 on the Copyright Act of 1976.) The consumable nature of this work does not imply permission for further reproduction.

[2] Willis v. Russell, 68 N.C. App. 424, 315 S.E.2d 91 (1984) (jury presented with question about contract terms and found that owner's conduct indicated existence of a contract). In Clark & Enersen v. Schimmel Hotels, 194 Neb. 810, 235 N.W.2d 870 (1975), the court found that the architect had an action for quantum meruit but not in contract, although the owner had requested and the architect delivered a standard AIA contract that was never executed. In Matthews v. Neal, Greene & Clark, 177 Ga. App. 26, 338 S.E.2d 496 (1985), the architect was allowed to recover in quantum meruit in the absence of a contract. The owner had abandoned the project after the bids came in too high. The architect attempted to collect based on a percentage of the last estimated construction cost, as would have been permitted under the AIA contract. However, in the absence of a contract, the architect was required to prove the number of hours actually spent on the project. In Supreme Indus. v. Town of Bloomfield, 2007 WL 901805 (Conn. Super., March 8, 2007), there was no signed agreement between the owner and contractor, although the parties had heavily negotiated the terms of A201. The court found that, by conduct, the parties had agreed to the terms of A201.

If no written contract exists, the architect may be found responsible for supervision of the construction when it is not intended.[3] In that case, the architect is under a duty to prevent gross carelessness or imperfect construction. Mere detection of defective workmanship does not relieve the architect of a duty to prevent it.[4]

A contract also defines who the architect is. This is not always clear.[5] The contract can also define the architect's responsibilities for events that occurred before it was executed.[6] Use of the AIA agreement places the obligations and rights of the parties according to modern construction practice. The architect is responsible for the design of the project, whereas the contractor is responsible for the construction. B101 is typically used along with AIA Document A201, General Conditions of the Contract for Construction.

§ 2.2   Prior Editions

The current edition of B101 was issued in November 2007 as part of a major revision of the most significant AIA documents. An extremely important point to remember is that documents from different series cannot be mixed. Thus, the 2007 B101 cannot be used on the same project as the 1997 A201.[7] The text in this chapter uses the most recent version of B101. However, the referenced court opinions are from earlier versions of the document. Because use of the current B101 started in November 2007, it should be expected that no court opinions relating to this edition will appear for at least a year following release. The language of many parts of the prior documents, however, is close enough to provide

---

[3] *See* Kleb v. Wendling, 67 Ill. App. 3d 1016, 385 N.E.2d 346 (1978). However, St. John Pub. Sch. Dist. v. Engineers-Architects, 414 N.W.2d 285 (N.D. 1987), involved an architect who was hired directly by the contractor. The architect was not liable to the owner for failing to supervise the construction because there was no evidence that owner relied on any statement by the architect that he would supervise.

[4] Kleb v. Wendling, 67 Ill. App. 3d 1016, 385 N.E.2d 346 (1978); Lotholz v. Fiedler, 59 Ill. App. 379 (1895).

[5] In Harmon v. Christy Lumber, Inc., 402 N.W.2d 690 (S.D. 1987), the architect was "moonlighting." The owner brought an action against the architect's firm, although the firm's only connection with the project was to turn it down and refer it to the architect. In this situation, the firm should have documented the situation and the individual architect should have had a written contract.

[6] Fruzyna v. Walter C. Carlson Assocs., 78 Ill. App. 3d 1050, 398 N.E.2d 60 (1979). The architect had begun work pursuant to an oral agreement which was later reduced to writing. In the interim, a worker on the construction project was injured and sued the architect under the Illinois Structural Work Act (since repealed). Based on the contract, the architect was not "in charge of the work" and not liable under the Act.

[7] An example of problems caused by mixing documents is found in Eis Group/Cornwall Hill v. Rinaldi Constr., 154 A.D.2d 429, 546 N.Y.S.2d 105 (1989), in which the owner-contractor agreement was a 1987 AIA form that incorporated the 1987 version of A201. However, the 1976 version of A201 was actually attached to the contract, along with various modifications which included striking the arbitration clause. The court found an ambiguity (obviously!) and declined to order arbitration.

insights into judicial interpretation of certain sections of B101. Relevant cases should be closely read to determine if they apply to specific situations.

The previous edition of this document was AIA Document B141, which was issued in 1997. Future revisions are expected every ten years. At the end of each section of AIA text, there is a reference to the prior version of the B141-1997 language in bold text.

CAUTION: The recommended method of amending this document is to use the electronic documents software that is available from the AIA. Using this software, the document is amended on its face, with additions and deletions indicated either in the margin, or by underlines and strikethroughs. Another way is to attach separate written amendments that refer back to B101. This is normally done by filling in Article 12, or additional pages referred to in Article 12. An alternative method is to graphically delete material and insert new material in the margins of the standard form. However, this method may create ambiguity, and it may result in the stricken material's being rendered illegible. It is illegal to make any copies of this document in violation of the AIA copyright or to reproduce this document by computerized means. Refer to the instruction sheet that comes with B101 for additional information about that copyright.

The cited language of this document is as published by AIA in late 2007 in the Electronic Documents. There appear to be some minor inconsistencies between this document and others that are probably attributable to scriveners' errors. AIA often corrects such errors in intervening years without any notice and without any way to determine exactly when a particular version of a document was actually released.

§ 2.3 Title Page

The date of the agreement should be the earliest date that any sort of agreement was entered into, even if that agreement was oral and preliminary. In many states, lien rights are established as of the date of the contract. Thus, a lien claimant can have a priority over a mortgage recorded after the date of the contract if the lien is perfected.[8] This is especially important if the owner becomes insolvent and the lender forecloses on the property. Because architects are often entitled to liens against the property in much the same way that

---

[8] Pittsburgh Plate Glass Co. v. Kransz, 291 Ill. 84, 125 N.E. 730 (1919). However, in Ketchum, Konkel v. Heritage Mountain, 784 P.2d 1217 (Utah Ct. App. 1989), the court held that, under Utah law, the architect was not entitled to priority over the lender because the architect's work was "off-site" and did not constitute "commencement of work" for priority purposes under the mechanic's lien statute.

contractors are, it is important that architects be aware of procedures relating to liens in their jurisdiction.[9]

The owner must be properly identified.[10] The "Owner" might not be the title holder of the property. The client might be a tenant or a beneficiary of the owner. Care should be taken to identify the actual titleholder and, if possible, to list that party and obtain permission from it. It might be advisable to notify (by registered or certified mail) this "real" owner that the architectural work is under way, so that there can be no objection later that the true owner was unaware of the work. Any lien would be filed against the interest of this true owner, even if that owner did not know that any work was going on.

---

[9] A lien is a right asserted against a property. These are frequently called *mechanic's liens* in the context of construction. Perfecting the lien involves following statutory procedures to make sure that the lien is properly filed with the recorder of deeds or other governmental official, adhering to any time limits, and giving proper notice. These requirements can be very strict, and an attorney should be consulted as soon as a bill is overdue to advise whether a lien is possible and how it is to be perfected.

Once a lien has been perfected, it is a matter of public record, and foreclosure proceedings can be used to collect the fee. Priority is important because if a lienholder has priority, he is ahead of another lienholder in obtaining funds from a sale of foreclosed property. For instance, suppose an owner hires an architect, obtains a construction loan, and work is started. The owner then becomes insolvent and the architect cannot collect its fee. The architect can file a lien against the property and start foreclosure proceedings. Other lien holders would probably join in the foreclosure suit. Often, the court orders the property sold at auction to pay the various claimants. Often, too, the sale of the property does not bring enough to pay all the claims. In that case, the party with priority has a greater chance of collecting than those with lesser priority.

[10] In Keller Constr. Co. v. Kashani, 220 Cal. App. 3d 222, 269 Cal. Rptr. 259 (1990), the architect entered into a contract with a limited partnership. Kashani, the sole general partner, signed the agreement on behalf of the partnership. A dispute developed and the partnership filed bankruptcy. The architect sought arbitration against Kashani. The court held that Kashani was bound by the arbitration agreement.

In Cheek v. Uptown Square Wine Merchants, 538 So. 2d 663 (La. Ct. App. 1989), the owner argued that he signed the AIA contract in a representative capacity and not individually. The court held otherwise, based on the plain language of the contract.

In Dunn v. Westlake, 573 N.E.2d 84 (Ohio 1991), the owner formed a corporation after beginning negotiations with the architect but before the contract was executed. The court found the owner personally liable because the owner failed to disclose the representative capacity to the architect. It is important to note that if the parties intend that one party be a corporation or limited partnership, that fact must be apparent from the face of the agreement, or a later modification to the agreement must be signed by all parties. *See also* ¶ 10.3 of AIA Document B101, relating to assignments of the agreement.

In Silver Dollar City v. Kitsmiller Constr., 931 S.W.2d 909 (Mo. Ct. App. 1996), the owner thought it was contracting with a joint venture consisting of two parties. The first page of the contract named both entities the "Contractor," but the signature page listed only one. There was thus an apparent ambiguity and the contract could have been voided by the owner at that time, particularly because the contractors knew of the mistake. However, by failing to take prompt action to rescind or correct the contract, the court found that the owner had ratified the contract.

The status of the owner should also be indicated. The following are some examples:

Smith Partners, an Illinois general partnership

Smith Partners, Ltd., an Illinois limited partnership

Smith Corp., an Illinois corporation

John Smith, authorized agent for the Owner, XYZ

Smith Bank, Trustee under Trust No. 111.

In the last example, the party executing the contract is a trustee for the owner. In Illinois, for example, a bank can hold title to a parcel of property as a trustee for the beneficial owner. This is sometimes referred to as a *secret land trust* because the beneficiary is not disclosed to the public. This beneficiary is the true owner and directs the trustee to deal with the property. Usually, the trustee would execute the owner-architect agreement, which would contain an exculpatory clause saying something to the effect that the parties agree that the trustee is acting merely as a trustee and will not be personally liable in case of any default. In this situation, the architect may find that it cannot enforce its contract against anyone: the trustee who executed the contract has the exculpatory clause, and the beneficiary (true owner) did not sign the contract. Of course, it might be possible to bring in the beneficiary on some other theories, but it would be much cleaner if the beneficiary also executes the contract as an additional party in interest at the outset.

In the case of a corporation, the architect would be wise to try to add an individual as a signatory to the contract. This provides additional protection for the architect. For instance, the party could be identified as "Smith Corp., an Illinois corporation and John Smith, individually."[11]

If the owner is a governmental body, the architect should be sure that the body passed the proper resolution to enable it to enter into the contract. Without such authority, it may be difficult or impossible for the architect to collect a fee.[12]

---

[11] *See, e.g.*, the language in Cheek v. Uptown Square Wine Merchants, 538 So. 2d 663 (La. Ct. App. 1989).

[12] County of Stephenson v. Bradley & Bradley, Inc., 2 Ill. App. 3d 421, 275 N.E.2d 675 (1971). In Blue Ridge Sewer Improvement Dist. v. Lowry & Assocs., 149 Ariz. 373, 718 P.2d 1026 (Ct. App. 1986), an engineer was unable to collect fees. He attempted to collect on the basis of quantum meruit (unjust enrichment or implied contract). The court held that "one who provides services under a contract with a political subdivision may not recover for the value of those services under quantum meruit if the contract was entered into in violation of a state law requiring approval of a majority of property owners in the area affected." *Accord* Scofield Eng'g Co. v. City of Danville, 126 F.2d 942 (4th Cir. 1942); Galion Iron Works & Mfg. Co. v. City of Georgetown, 322 Ill. App. 498, 54 N.E.2d 601 (1944); Dempsey v. City Univ., 106 A.D.2d 486, 483 N.Y.S.2d 24 (1984). In Cuyahoga County Bd. of Comm'rs v. Richard L. Bowen & Assocs., Inc., 2003 WL 21555080 (Ohio App. 8 Dist., July 10, 2003), the county board failed to comply with applicable law in the evaluation and selection of the architect. The contract between the architect and the board was void.

This can occur even when the contract is executed by a board superintendent or other responsible official.[13]

In a Florida case, the architect entered into a contract with developers of a condominium project.[14] The contract listed the owner as "Adalia Court, Inc., Richard Seaman, et al." The "et al." was inserted by the architect because he had been told that a limited partnership would be created to replace the corporation as owner. A limited partnership was later formed and took over the assets of the corporation. The partnership did not, however, sign the contract. Thereafter, the work proceeded, and the limited partnership made several payments to the architect. After a disagreement, the limited partnership refused to pay the balance of the architectural fees. The appellate court found that the limited partnership was not liable under the written contract. The architect was, however, permitted to introduce evidence of an implied contract.

In a situation like this, the parties should sign an amendment to the contract that assigns rights and obligations to the successor. Note that ¶ 10.3 deals with assignments.

In Murphy v. City of Brockton, 364 Mass. 377, 305 N.E.2d 103 (1973), the architect was unable to collect more from the city than the amount of the initial appropriation. The court stated that "it is our opinion that the plaintiff was obligated to proceed no further with its work under the contract than was covered by an appropriation," despite the fact that a contract was signed between the parties to cover all architectural services.

In Jablonsky v. Callaway County, 865 S.W.2d 698 (Mo. Ct. App. 1993), the architect was held not entitled to his fee, when the county accounting officer failed to certify that there was an unencumbered balance to the credit of the appropriation to which the contract was to be charged and a cash balance in the county treasury sufficient to meet the obligation. In Cuyahoga County Bd. of Comm'rs v. Richard L. Bowen & Assocs., Inc., 2003 WL 21555080 (Ohio App. 8 Dist., July 10, 2003), the county board failed to comply with applicable law in the evaluation and selection of the architect. The contract between the architect and the board was void.

[13] In D.C. Consulting Eng'rs v. Batavia Park Dist., 143 Ill. App. 3d 195, 492 N.E.2d 1000 (1986), the park district superintendent contracted with an engineer to inspect a park district building and prepare a report. The court held that the contract was void because the superintendent had no authority to bind the district. The fact that the park district may have benefited did not help the engineer.

In McKee v. City of Cohoes Bd. of Educ., 99 A.D.2d 923, 473 N.Y.S.2d 269 (1984), an architect was not entitled to collect his fee from a school board. The school superintendent had executed the contract, but not in conformance with statute. The contract was thus void.

In Hazelton Area Sch. Dist. v. Krasnoff, 672 A.2d 858 (Pa. Commw. Ct. 1996), a school district hired an architect under a fast-track schedule and for a flat fee. Following the election of a new school board, the architect was terminated. He then submitted a bill for 15 different claims for additional services, totaling more than $464,000. The matter went to arbitration, where the architect submitted additional claims for interest and early termination of the contract. The evidence showed that the president of the school board had verbally agreed to the extras, with the knowledge of a majority of the board. The arbitration concluded with a sizable award in favor of the architect. The courts reversed the arbitrators on the ground that the contract, as well as state law, required written authorization for additional services. All 15 claims were rejected. The court stated that any modification of a contract that increased the indebtedness of a school district must comply with requirements for a formal vote and approval.

[14] Craig W. Sharp, P.A. v. Adalia Bayfront Condominium, Ltd., 547 So. 2d 674 (Fla. Dist. Ct. App. 1989).

On occasion, parties other than the owner may file a contract action against an architect under a third-party beneficiary theory. This theory states that a party who was intended to benefit from a contract may sue to enforce that contract, even if that party is not named in the contract. In one case, the Illinois Housing Development Authority (IHDA) sued an architect for breach of contract.[15] IHDA had loaned money to the developer for the construction of an apartment complex. As a result of numerous problems with the complex, substantial repairs were required. The court rejected IHDA's claim against the architect because it was not named as an owner and no clear intent to benefit IHDA was shown in the contract. If third parties are intended to benefit by the contract, they should be specified in the body of the contract. Paragraph 10.5 states that there are no third-party beneficiaries of this agreement.

The contractor is not a third-party beneficiary of the owner-architect agreement,[16] and neither are construction workers.[17] In one case, the owner was held

---

[15] Illinois Hous. Dev. Auth. v. M-Z Constr. Corp., 110 Ill. App. 3d 129, 441 N.E.2d 1179 (1982).

[16] Harbor Mech., Inc. v. Arizona Elec., 496 F. Supp. 681 (D. Ariz. 1980); Blecick v. Sch. Dist. No. 18, 2 Ariz. App. 115, 406 P.2d 750 (1965); A.R. Moyer, Inc. v. Graham, 285 So. 2d 397 (Fla. 1973).

In Spancrete, Inc. v. Ronald E. Frazier & Assocs., 630 So. 2d 1197 (Fla. Dist. Ct. App. 1994), the architect owed no duty to a subcontractor because the architect had no right to stop the work. In Fleischer v. Hellmuth, Obata & Kassabaum, 870 S.W.2d 832 (Mo. Ct. App. 1993), the court held that an architect does not owe a duty to the construction manager.

In Engineered Refrigeration Sys., Inc. v. Albertson's Inc., 1996 U.S. Dist. LEXIS 9090 (N.D. Fla. May 29, 1996), the court reviewed *A.R. Moyer* and *Spancrete* in determining that an architect and the construction manager had no power to stop the work. Therefore, a negligence action against them was barred by the economic loss rule.

However, in D.I.C. Commercial Constr. Corp. v. Broward County, 668 So. 2d 697, 698 (Fla. Dist. Ct. App. 1996), the appellate court reversed the trial court's dismissal of a complaint against an architect by the contractor, who had alleged that the architect's negligent performance of its contractual duties had caused the contractor economic loss. The architect had won based on *Spancrete*, but the instant contract was not a standard AIA contract. It included terms that sound quite similar to the B 141 provisions.

The architect's contract with the owner, which was attached to the complaint, required the architect to administer the construction contract, to become familiar with the progress and quality of the work, and to determine if the work was proceeding in accordance with the contract documents. Further, the architect was required to notify the contract administrator of the contractor's failure to follow plans and specifications, to interpret the requirements of the contract documents, judge the performance thereunder by the contractor, and decide claims of the owner and the contractor relating to the execution and progress of the work. The architect was also required to review the contractor's requests for payment, and to recommend to the owner when the payment should be made and in what amount, to determine when the project was substantially and finally completed, and to determine when final payment should be made.

Under the contracts, if there were discrepancies in measurements, the contractor was not to proceed until the architect clarified the numbers. The owner could terminate the contractor's work if the owner received written certification from the architect of the contractor's "delay, neglect, or default." The complaint also alleged, without quoting a specific contractual provision, that the architect in fact had the authority to stop the work and had done so.

The use of a non-AIA contract hurt the architect because the architect was unable to use prior law to show that it was not supervising the construction.

[17] Zukowksi v. Howard, Needles, Tammen & Bergendorr, Inc., 657 F. Supp. 926 (D. Colo. 1987).

to be a third-party beneficiary when the architect agreed with a contractor to design a solar heating system.[18] The architect owed a duty to this third party to "exercise ordinary professional skill and diligence and to conform to accepted architectural standards."

The status of the architect should be specified if it is a corporation or partnership. If the architect is not licensed in the particular state, arrangements must be made as soon as possible to legally perform the work, either through a joint venture with a qualified professional or by obtaining that state's license.[19] In some states, an architect may not solicit business for or enter into contracts for any services that require an architectural license unless it is licensed in that state.[20]

---

[18] Keel v. Titan Constr. Corp., 721 P.2d 828 (Okla. 1986).

[19] See Annotation, *Right of Architect or Engineer Licensed in One State to Recover Compensation for Services Rendered in Another State, or in Connection with Construction in Another State, Where He Was Not Licensed in the Latter State*, 32 A.L.R.3d 1151 (1970).

In O'Kon & Co. v. Riedel, 588 So. 2d 1025, 1030 (Fla. Dist. Ct. App. 1991), the court found that a Georgia architectural corporation had violated Florida law and was not entitled to enforce a lien or collect its fee. The firm, apparently anticipating this problem, had associated a licensed Florida architect and made him an officer of the corporation, in charge of the firm's Florida architecture division. Apparently another architect, not licensed in Florida, had prepared most of the plans. The majority of the court, however, determined that this was a violation of Florida law. This is certainly unrealistic and contrary to the practice of architects across the country. The dissenting opinion stated:

> The majority seem to read section 481.219(l)(a) to say that where a corporation contracts to provide architectural services through licensees, every corporate employee performing services coming within the 481.203(6) definition of architecture must be licensed. I disagree. My reading of Chapter 481 and the rules enacted thereunder reveals a regulatory scheme which anticipates that much of the work done in an architect's office can be done by unlicensed persons, so long as the work of the unlicensed persons is performed "under [the licensed architect's] responsible supervising control."

The dissent is the preferred view and is in line with architectural practice. This opinion does, however, reinforce the necessity of carefully reviewing state licensing laws. The dissent in *O'Kon* and the majority view was followed in District Board of Trustees v. Morgan, 890 So. 2d 1155 (Fla. 5th Dist. 2005), in which the architect, a two-member partnership, entered into a contract with a college for the design of a performing arts complex. The two architects were individually licensed in Florida, but they failed to obtain the certificate of authorization required for architectural partnerships in Florida. The issue was whether the contract was void *ab initio* and therefore unenforceable by the architects. The appellate court concluded that the contract was merely voidable, and was enforceable while it was still in existence.

In Carlson v. SALA Architects, Inc., 732 N.W.2d 324 (Minn. App. 2007), the project architect was not licensed, although the firm was properly licensed. When the owners sued the firm, they claimed that they were never told that this individual was unlicensed and asked for all of their fees back. The appellate court reversed the trial court, finding that there were unresolved factual issues as to whether the firm actually held this person out to be licensed.

[20] In Kansas Quality Constr., Inc. v. Chaisson, 112 Ill. App. 2d 277, 250 N.E.2d 785 (1969), the court held that a contractor was not practicing architecture in performing the following: (a) developing plat layout to show building placement, total buildings, recreational area, parking, and so on; (b) causing plans of proposed buildings to be drawn for loan submission; (c) assisting in obtaining financing. Item (b) did not constitute the practice of architecture so long as it was for the limited purpose of obtaining a loan, and other drawings were prepared for construction.

In State ex rel. Love v. Howell, 285 S.C. 53, 328 S.E.2d 77 (1985), South Carolina brought a contempt action against an unlicensed person. The court found him in contempt of an earlier court order prohibiting him from practicing architecture. After the first order, he hired three licensed architects to work for him. Because he retained control, the court found this practice illegal. "These three individuals were hired for the exclusive purpose of signing architectural plans completed and supervised by Howell so as to give them an appearance of legitimacy."

In a Missouri case, Hospital Dev. Corp. v. Park Lane Land Co., 813 S.W.2d 904 (Mo. 1991), an architectural firm was denied fees from an owner because the firm was not properly licensed in Missouri. The unlicensed firm contracted with an owner to perform architectural services and then contracted with a related company, which was licensed to engage in architectural services in Missouri, to perform those architectural services. However, the court held that this was not good enough. The fact that the plaintiff in this case was not licensed meant that it could not enter into a valid contract with an owner to perform architectural services, even though the actual work was done by a properly licensed firm.

In fact, the owner counterclaimed and was able to recover all fees paid to the firm. The firm argued that it had performed nonarchitectural services, such as "coordination of rezoning work," and had incurred expenses, such as printing costs. It wanted at least to be compensated for these items. However, the court found that, under the broad terms of the Missouri licensing act, it could not easily separate out the so-called nonarchitectural services. Therefore, the owner was entitled to recover all fees.

Not every state has adopted this very strict interpretation of licensing acts. However, the lesson for design professionals is to be very careful about complying with the licensing act of any state in which projects are contemplated.

In another Missouri case, Strain-Japan R-16 Sch. Dist. v. Landmark Sys., Inc., 965 S.W.2d 278 (Mo. Ct. App. 1998), the contractor sought to arbitrate a dispute with the owner. The owner sued for an injunction against the arbitration proceeding, based on the fact that the contractor was not licensed as an architectural firm, while Missouri law makes contracts for engineering services rendered by an unlicensed firm unenforceable. The work was for a pre-engineered metal building that was to conform to preliminary plans furnished by the owner. The contract called for plans by the general contractor and specified the particular manufacturer of the pre-engineered structure. The contractor then arranged for plans to be prepared by licensed engineers, which were then approved by the owner. In the lawsuit, the owner argued that its contract with the general contractor is a contract for architectural or engineering services, since it requires the contractor to provide those services. The court rejected these arguments, finding that if that argument were correct, either the owner would be forced to contract directly with architects or engineers, or all corporations engaged in general contracting would need to register as architects or engineers. The court was not willing to read the statutes in that manner. Furthermore, the court found that the owner contracted not for professional engineering services but for a building conforming to its initial sketches.

In Space Planners Architects, Inc. v. Frontier Town-Missouri, Inc., 107 S.W.3d 398 (Mo. App. 2003), an architect sought to foreclose a mechanic's lien. On a motion for summary judgment, the architect presented an uncertified copy of the required architect's license. This, however, was insufficient for the appellate court, which stated that the architect had to plead and prove as part of its case that it was properly licensed. Failing to provide the court with a certified copy of the license required reversal of the summary judgment in the architect's favor.

In Howard v. Usiak, 775 A.2d 909 (Vt. 2001), an out-of-state architect was hired to design a veterinary clinic. The trial court found that the written contract was executed in late January 1994, after the architect was licensed on January 19, 1994. The owner sought return of all fees because the architect had negotiated the contract, accepted a $2,000 retainer, visited Vermont to consult with the client, and completed schematic drawings of the final design prior to obtaining his license. The court ruled in favor of the architect in part because it was the owner seeking return of fees paid, not the architect seeking to obtain unpaid fees. The court held that the Vermont licensing statute does not authorize recovery of fees. Thus, because the statute's penalty provision was a fine, the court refused to create an additional penalty.

Each jurisdiction has different rules defining the practice of architecture. Some states permit an out-of-state architect to obtain a temporary license on a single

---

In Kourafas v. Basic Food Flavors, Inc., 88 P.3d 822 (Nev. 2004), the architect entered into two separate contracts with the owner, one for the design of the project and a second to act as project manager. When the architect was not paid, he filed suit. The owner defended on the grounds that the architect was not licensed as a contractor. Under Nevada law, a contractor is required to be licensed by the state. The appellate court reversed the trial court's finding in favor of the owner and directed the trial court to determine whether the architect performed construction management services that an architect cannot perform.

See also Osherow v. Diao, 2003 WL 21805484 (Conn. Super., July 16, 2003) (unlicensed architect's contract not void because the services fell within an exemption for one- and two-family residences).

For a case involving a criminal conviction of an unlicensed architect, *see* Benitez v. State of Florida, 2003 WL 21976094 (Fla. App. 3 Dist., Aug. 20, 2003).

In Park Ave. & 35th St. Corp. v. Piazza, 170 A.D.2d 410, 566 N.Y.S.2d 297 (1991), a contractor was denied the balance of his fee because he had held himself out as an architect. Work he had done included reviewing bid documents for compliance with specifications and accepted construction standards and drafting of floor plans.

In Kansas City Community Ctr. v. Heritage Indus., Inc., 972 F.2d 185 (8th Cir. 1992), the contract between the owner and the manufacturer of prefabricated modular housing was invalid because the manufacturer failed to obtain a certificate of authority to practice either architecture or engineering in Missouri. The contract was not divisible, because the architectural and engineering services were an integral part of the contract. Quantum meruit was not available.

In American Motorists Ins. Co. v. Republic Ins. Co., 830 P.2d 785 (Alaska 1992), the court was asked to interpret "professional services" in the context of a professional liability insurance policy. The architect prepared a bid for a school district project which consisted of some 160 pages, including drawings, schedules, project approach, and other materials. When the architect was sued by a competitor for misrepresentation, it tendered the defense to its malpractice carrier. The carrier denied coverage, arguing that a bid for services is not the same as professional services, that is, there is a distinction between the preparation to render professional services and the rendering of the services themselves. The court disagreed, holding that only an architect using his or her specialized knowledge, labor, and skills could have prepared the bid.

In Kaplan v. Tabb Assocs., 657 N.E.2d 1065 (Ill. App. Ct. 1995), the owner entered into a contract for architectural services with a corporation. The sole shareholder of the corporation was properly licensed, but the corporation itself failed to comply with Illinois's Architecture Act by failing to be properly licensed. When a dispute over payment arose, the architectural corporation demanded arbitration, and the owner sought an injunction against the arbitration on the grounds that the contract was void because the architect was not properly licensed. The appellate court agreed with the owner, because the Act carries a criminal penalty for noncompliance. In Stein v. Deason, 165 S.W.3d 406 (Tex. App.—Dallas, 2005), a Texas owner brought an action against a California architect for fraud, breach of contract, and other causes relating to the design of a building in California. The California architect was not licensed in California, but, according to the owner, represented that he was a licensed architect. The architect claimed that he was not required by California law to be licensed as an architect to provide architectural services. The court found that Texas courts could exercise jurisdiction over the California architect.

In G.M. Fedorchak v. Chicago Title Land Trust Co., 355 Ill. App. 3d, 822 N.E.2d 905 (3d Dist. 2005), the plaintiff, an Indiana architectural firm, filed an Illinois mechanic's lien against the owner after the owner failed to pay its bills. At the time of contract formation, the firm was not registered in Illinois, the senior member of the firm was not licensed, and the son held an inactive license in Illinois. Based on these facts, the court held that the contract was void and the mechanic's lien was not valid.

project. In most cases, an architect would be advised to enroll with the National Council of Architectural Registration Boards (NCARB) to expedite licensing in other states. Whenever an architect contemplates work in another jurisdiction, the state's licensing laws should be researched immediately.

Also, states differ in qualifying corporations. For instance, in Illinois, a corporation may practice architecture if two-thirds of the directors are licensed architects or engineers. There are no licensing requirements for shareholders. In Ohio, on the other hand, a majority of the shares must be held by one or more Ohio-registered architects. There are no licensing requirements for directors. Thus, an Illinois corporation wishing to practice in Ohio probably must do so indirectly, possibly through a subsidiary corporation in which the primary corporation is a minority shareholder. Partnerships are similarly regulated. In Illinois, for instance, two-thirds of the partners must be licensed in some state to practice architecture or engineering.

The work must be specified by typing in a description in the blank space. Wording can be very important here. If prior design work has been performed, those documents can be referenced to define the project. Otherwise, the scope should be as detailed as possible, with exceptions specified. Possibilities include:

> A new three-bedroom single-family residence to be located on a lot at the southeast corner of Oak and Elm Streets, Oak Park, Illinois. All site work is included.

> Rehabilitation of a six-story loft building at 100 N. Michigan, Chicago, Illinois. The existing parking lot and all exterior work is excluded from this contract, with the exception of new windows and frames, which work is included.

> Construction of a new 100,000-sf factory building as shown on the Architect's drawings SK-1 through SK-5, dated December 24, 1986.

In many cases, the description may be more comprehensive. However, the project description can be too specific, particularly if a maximum price is listed. In one case, the architect contracted for "a multiple purpose building suitable to the needs of the Owner, at an approximate estimated cost of $250,000.00."[21] The evidence at trial showed that the architect and owner subsequently orally agreed that the building was to be of prestressed concrete of at least 40,000 square feet. The architect prepared bid drawings that provided alternative bids for prestressed concrete and for steel. The bids came in at $317,000 for the prestressed concrete and under $250,000 for the steel. Because the parties had agreed on the concrete building at the specific price, the architect was not entitled to any fee because the price was exceeded. The fact that the steel scheme came in under budget made no difference because the later modification required the concrete building.

In another case, an engineer undertook to "analyze the piping, valving, and structural characteristics of the existing" water filtration and distribution plant

---

[21] Stevens v. Fanning, 59 Ill. App. 2d 285, 207 N.E.2d 136 (1965).

and to design changes to the system.[22] He assumed that the intake piping was 18-inch piping. It was later discovered that portions of the existing pipes were only 12 inches in diameter, resulting in an inadequate intake capacity. Based on the duty to analyze the existing system, the engineer was liable for the resulting damages.

In a New York case, the owner hired an engineer to design a single-family house, advising the engineer that he had a budget of $100,000.[23] The contract stated that "all work is predicated upon [the plaintiff's] intention to maintain a budget for the cost of construction in the area of $100,000." When the lowest bid came in at $160,000, the owner sued for breach of contract and fraud. The court found for the engineer because the engineer never represented that he was an expert with respect to construction costs or that the cost of constructing the residence as proposed and approved by the owner would not exceed $100,000. The court found that several changes had been made to the design subsequent to the agreement, that the owner had approved the changes, and the parties had never discussed the potential costs of these changes.

In a Missouri case, the contract described the project as "a new wing and rehabilitating the existing facility totaling 120 beds for FHA Project No. 084-43041-PM-SR, known as Homesdale Convalescent Center, 8034 Holmes, Kansas City, Missouri."[24] The contract was for a flat fee. The project was abandoned after bids came in too high. The architect sued for additional fees for two final drawings and specifications that added 20 private baths to 20 existing rooms and other engineering modifications, claiming he did not plan these at the time the contract was executed. The question for the court was whether the additional work was within the contemplation of the parties at the time of the signing of the contract. The court found that at the time an architect's contract is signed, the plans for the project are necessarily amorphous. The architect is hired to prepare these plans, which make the project definite.

> When an architect contracts with an owner to provide full architectural services for the construction of a given building, it is implicit that during the course of the project, the owner may make changes in plans or specifications for the building, thus causing the architect to prepare additional and revised drawings to be included within the contractual fee.

The changes in this case were, thus, "precisely the sort of changes that should be held to be within the scope of any such owner-architect contract."[25] A more detailed description of the work would have avoided the problem in this case. Also, the architect should obtain written approval from the owner of the initial set of schematic drawings. Then, if the owner wants additional services, the architect can invoke ¶ 4.2.1.1 to obtain additional fees.

---

[22] City of Eveleth v. Ruble, 302 Minn. 249, 225 N.W.2d 521 (1974).

[23] Brenner v. DeBruin, 589 N.Y.S.2d 56, 57 (App. Div. 1992).

[24] Waddington v. Wick, 652 S.W.2d 147 (Mo. Ct. App. 1983).

[25] *See also* Ingram v. State Prop. & Bldg. Comm'n, 309 S.W.2d 169 (Ky. 1957); Osterling v. First Nat'l Bank, 262 Pa. 448, 105 A. 633 (1918).

In another case, the owner refused to pay the architect because the construction was not finished on time.[26] The owner alleged, and the architect denied, that the parties orally agreed that the construction would be finished on a certain date. The contract did not specify any such date. The court found that there was no such modification and that the burden of proving the modification was on the owner. If the architect had conceded that a date had been agreed upon, even if he had specified a different date, the burden might have shifted to the architect and he might have lost. Because the construction is not within the architect's control, the architect should never agree to any definite construction schedule.

In a federal case,[27] the issue was whether or not the contract between the owner and architect had expired. The school had hired the architect for two stages. Stage 1 would involve preparation of designs in order to get a school referendum passed. In Stage 2, after the referendum passed, the architect would perform the remainder of his standard services. The referendum did not pass, but the school asked the architect to prepare a second set of documents for another referendum the following year. When this did not pass, the school hired another architect, who prepared a third set of documents for a referendum that did pass. The first architect then sued the school district, claiming lost profits for Stage 2. The court held that the contract was not ambiguous and the contract automatically terminated when the referendum failed. Even if the contract were resurrected for the second referendum, that also terminated when the second referendum failed. This case illustrates the importance of a clear definition of the services of the architect. When more than one phase is involved, the parties should consider the consequences of an early phase not being approved. Presumably, if that had been done here, the architect would not have brought this action.

In a Louisiana case,[28] the architect argued that the B141 (predecessor to B101) agreement required arbitration with the owner. However, the project at issue was a different project than the one indicated in the description of the B141, which was an earlier project. The description did not indicate that all future projects were to come under this agreement.

§ 2.4 Article 1: Initial Information

This section has been revised from the 1997 version by placing most of the required information into "Exhibit A, Initial Information." It is important that this exhibit actually be filled out at the time this document is prepared, even if some of the information is unknown at that time. The purpose of this information is to enable both the owner and architect to define the parameters of the project as much as

---

[26] Orput-Orput Assocs. v. McCarthy, 12 Ill. App. 3d 88, 298 N.E.2d 225 (1973).

[27] Kimball Assocs. v. Homer Cent. Sch. Dist., 2000 U.S. Dist. LEXIS 16759 (N.D.N.Y. Nov. 9, 2000).

[28] Town of Homer, Inc. v. General Design, Inc., 960 So. 2d 310 (La. Ct. App. 2007).

possible at the outset of the relationship in order to minimize problems. The more that the scope of the project and the relationship among the parties can be clarified up front, the easier the entire design process will be.

1.1 This Agreement is based on the Initial Information set forth in this Article 1 or in optional Exhibit A, Initial Information:

(Complete Exhibit A, Initial Information, and incorporate it into the Agreement at Section 13.2, or state below Initial Information such as details of the Project's site and program, Owner's contractors and consultants, Architect's consultants, owner's budget for the cost of the work, authorized representatives, anticipated procurement method, and other information relevant to the Project.) **(1.1.1; substantial revisions)**

The primary purpose of the following provisions in Exhibit A is to have the parties carefully consider what this project is all about, who will be required to work on the project, and the overall costs of the project to see if it is even feasible as proposed. B141-1997 asked for similar information in the body of the document, as opposed to a separate exhibit. These sections, if properly considered, should be tremendously helpful in avoiding problems that might otherwise occur later during the project. As many of these sections of the exhibit should be filled in as possible. Those sections that are not filled in with details should indicate what is not applicable or will be determined later. Alternatively, this information could be inserted here.

1.2 The Owner's anticipated dates for commencement of construction and Substantial Completion of the Work are set forth below: **(new)**

Commencement of Construction date: **(new)**

Substantial Completion date: **(new)**

This is a new section that deals with the start and end dates of the construction. Often, an owner will not have a clear understanding of such dates. The architect and owner should discuss the process leading up to the start of construction, as well as how long construction will likely take. By filling in these dates, the owner and architect are making assumptions that impact the fee that the architect will charge. If these dates substantially change, the architect's fee will likely be impacted. For instance, if the commencement date is moved up, the architect will have to accelerate its work, while if the completion date is moved back, the architect will be on the project for a longer period of time.

1.3 The Owner and Architect may rely on the Initial Information. Both parties, however, recognize that such information may materially change and, in that event, the Owner and the Architect shall appropriately adjust the schedule, the Architect's services and the Architect's compensation. **(1.1.6; revisions)**

Both the owner and architect are entitled to rely on the information that is furnished in this section and/or Exhibit A. If anything changes, then the price and/or time for the architect's services may change.

§ 2.5   Article 2: Architect's Responsibilities

ARTICLE 2 ARCHITECT'S RESPONSIBILITIES

2.1 The Architect shall provide the professional services as set forth in this Agreement. **(1.2.3.1; substantial revisions)**

This simply means that the architect will perform the services that are listed in this document. This provision implies that the architect will use consultants as well as its own employees to complete the project if the architect does not have the required engineering services in-house. This does not mean that the architect is required to use outside consultants or that the scope of the services is unlimited. It is good practice to specify the services that are and are not included by the architect in order to avoid misunderstandings by the owner. For instance, the owner may assume that the architect is providing electrical engineering services when the architect assumes that the landlord of the building is providing these services, for which the owner is paying separately. To avoid this, the agreement should specify that the owner will be hiring these consultants and paying for them directly.

2.2 The Architect shall perform its services consistent with the professional skill and care ordinarily provided by architects practicing in the same or similar locality under the same or similar circumstances. The Architect shall perform its services as expeditiously as is consistent with such professional skill and care and the orderly progress of the Project. **(1.2.3.2; substantial revisions)**

This is a standard of care provision, and it establishes the architect's standard of care as the "normal" or "average" standard of care. Owners sometimes want to require that the architect conform to a higher standard of care, or the "highest" standard. Such a higher standard of care may actually be uninsurable under the architect's professional liability policy. Claims for breach of the standard of care are arbitrable.[29]

Some owners may want to change "expeditiously" in ¶ 2.2 to some other language, such as a "time is of the essence" clause, thereby giving primary importance to time limits and schedules.[30] Inserting this makes the architect the guarantor of the time schedule and is very unwise from the architect's viewpoint. Most projects have some delays. The owner may want to use even a short delay as an excuse to avoid payment to the architect. Also, it is possible that the architect's insurance would not cover liability in the case of a time is of the essence clause.[31]

---

[29] In Moses Taylor Hosp. v. GSGSB, 2000 WL 1865006 (Pa. Comm. Pl., Jan. 20, 2000) (citing a prior version of this provision), arbitration was proper when the claims included tort claims for negligence, breach of professional responsibility, and breach of warranty. *See also* Southern Oklahoma Health Care Corp. v. JHBR, 900 P.2d 1017 (Okla. 1995).

[30] In Outlaw v. Airtech Air Conditioning and Heating, Inc., 412 F.3d 156 (2005), the architect was not liable to the owner for failing to secure the permits in a timely fashion. The agreement did not stipulate a time for securing the permits.

[31] *See* Gibraltar Cas. Co. v. Sargent & Lundy, 214 Ill. App. 3d 768, 574 N.E.2d 664 (1991) (engineer's insurer had a duty to defend a claim against engineer who alleged delay damages).

One case held an architect responsible for the damages of the owner attributable to the architect's delay, stating, "An architect who fails to deliver plans, specifications, and detailed working drawings in time for the successful continuation and completion of work, must be held accountable for the damages which the owner or the builder suffers because of that delay."[32]

The duties stated in the agreement are not the only ones that must be performed by the architect.[33] In an Illinois case,[34] the architect was hired to design a school. After construction was completed, efflorescence was noted. The evidence at trial showed that the architect had specified type O mortar instead of type N, which was the proper mortar type for that climate. Because of that mistake, the extreme freeze-thaw cycles of the area led to the efflorescence. Thus, the architect failed to specify the use of reasonably good materials (which was a breach of his implied obligation), failed to perform his work in a reasonably workmanlike manner, and failed to reasonably satisfy the work requirements.[35]

The traditional standard of care for architects was stated thus:

> The undertaking of an architect implies that he possesses skill and ability, including taste, sufficient to enable him to perform the required services at least ordinarily and reasonably well; and that he will exercise and apply, in the given case, his skill and ability, his judgment and taste, reasonably and without neglect. But the undertaking does not imply or warrant a satisfactory result.[36]

The duty of architects has also been stated as: "The common law duty of an architect in matters of design is to exercise that degree of skill and care that is

---

[32] Edwards v. Hall, 293 Pa. 97, 141 A. 638, 640 (1928). In Miami Heart Inst., Inc. v. Heery Architects & Eng'rs, Inc., 765 F. Supp. 1083 (S.D. Fla. 1991), the hospital sued the architect for breach of contract. There was a delay in obtaining a certificate of occupancy that the hospital attributed to the architect's failure to design the building to code. The hospital sought the rental value of the property as its damages. The architect argued that this measure of damages applied only to contractors and not to architects. The court rejected this argument, stating that no logic would make delay damages dependent on the particular trade or profession of the party that caused the damages. The court found that the hospital was entitled to the reasonable rental value of the difference between its old building and the new structure.

[33] *See* Taylor v. Cannaday, 230 Mont. 151, 749 P.2d 63 (1988) (Architect's Handbook of Professional Practice, published by AIA, was properly admitted into evidence as evidence of negligence, but deviation from the standards in that handbook is not negligence per se).

[34] Board of Educ. v. Del Bianco & Assocs., 57 Ill. App. 3d 302, 372 N.E.2d 953 (1978).

[35] *See also* Stanley v. Chastek, 34 Ill. App. 2d 220, 180 N.E.2d 512 (1962). In Pearce & Pearce v. Kroh Bros. Dev. Co., 474 So. 2d 369 (Fla. Dist. Ct. App. 1985), the architect failed to include flashings in a masonry building. The building leaked and the owner sued the architect and contractor. The court held against the architect as being solely responsible for the damage, although some evidence of shoddy construction was presented.

[36] Coombs v. Beede, 89 Me. 197, 36 A. 104, 105 (1896).
  Coombs was cited in Howard v. Usiak, 775 A.2d 909 (Vt. 2001). In *Howard*, an out-of-state architect was hired to design a veterinary clinic. During the permit stage, the issue arose as to whether the structure required an elevator. When the municipality insisted that an elevator be added to the design, the architect offered to redraw the plans at no additional cost, but the owner terminated the architect and sued to recover fees previously paid. The court held that the architect had not breached any duty to the owner for a variety of reasons.

customarily used by competent architects in the community."[37] Yet another definition is:

Architects, doctors, engineers, attorneys, and others deal in somewhat inexact sciences and are continually called upon to exercise their skilled judgment in order

---

[37] La Bombarbe v. Phillips Swager Assocs., 130 Ill. App. 3d 891, 474 N.E.2d 942, 945 (1985) (citing Miller v. DeWitt, 37 Ill. 2d 273, 226 N.E.2d 630 (1967); Laukkanen v. Jewel Tea Co., 78 Ill. App. 2d 153, 222 N.E.2d 584 (1966)).

In State v. Wolfengarger & McCulley, PA., 236 Kan. 183, 690 P.2d 380 (1984), the architect was held to be negligent when the contractor substantially followed the drawings. The architect had designed air intake louvers with 40% free air space. The contractor installed some with less free air space, and the building was damaged when snow entered the plenum space. The contractor demonstrated that even the proper louvers had a snow problem. The fault was the design and not the contractor's slight deviation. The court found that the architect had breached his duty to exercise care and skill.

In Duncan v. Missouri Bd. for Architects, 744 S.W.2d 524, 540 (Mo. Ct. App. 1988), the court stated that "the level of care required of a professional engineer is directly proportional to the potential for harm arising from his design."

In Urbandale v. Frevert-Ramsey-Kobes, Architects, 435 N.W.2d 400 (Iowa Ct. App. 1988), the architect was found to be 50% at fault for negligent design of a recreation building when the walls and roof began to deteriorate because they retained too much moisture. The appellate court refused to overturn the award, apparently because the breach of the standard of care was obvious.

In Corcoran v. Sanner, 854 P.2d 1376 (Colo. Ct. App. 1993), the court held that a statewide, rather than a local, standard of care applied to architects.

An interesting case is Murphy v. A.A. Mathews, 841 S.W.2d 671, 674 (Mo. 1992), in which the engineer was hired to prepare and document a subcontractor's claim in an arbitration hearing. The subcontractor later claimed that the engineer was negligent in these duties. The engineer claimed witness immunity. The court rejected this contention and found that the engineer owed its client "a duty of care commensurate with the degree of care, skill and proficiency commonly exercised by ordinarily skillful, careful and prudent professionals."

In Garaman, Inc. v. Danny Williams, A.I.A., 912 P.2d 1121 (Wyo. 1996), one issue involved the architect's standard of care. The owner attempted to use the architect's own testimony to establish the standard of care. The court held that the owner did not elicit proper testimony from the architect to establish a standard of care. The architect had testified that he had a duty to know the applicable building codes and that, in certain instances, his design did not comply with the codes as the local municipality subsequently interpreted them. He also testified that he believed that he had complied with the code and that interpreting code requirements is frequently difficult. The court found this to be too general to establish a breach of the standard of care. It left open the issue of whether, under proper circumstances, an owner can use an architect's own testimony to establish the standard of care and breach thereof. The court also held that noncompliance with a building code is not, by itself, negligence. Architects do not warrant perfect results. They are only required to exercise the appropriate degree of care.

In Logsdon v. Cardinal Indus. Insulation, 2006 WL 2382501 (Ky. App., Aug. 18, 2006), a decedent's estate filed a wrongful death action against various parties, including an architect after the decedent developed cancer and died, allegedly as a result of asbestos exposure in the work-place. There was testimony about asbestos insulation being installed in the 1950s and 1960s. The court held that expert testimony was required to establish not only that the architect specifications fell below the standard of care required of architects or engineers at that time, but also that there were alternatives to asbestos available in the 1950s when the facilities were designed. Failure to establish the standard of care or that the architect breached any such duty arising under the standard of care required summary judgment in favor of the architect. The estate also argued that the architect had a continuing duty to warn of potential hazards from building materials as such

to anticipate and provide for random factors which are incapable of precise measurement. The indeterminate nature of these factors makes it impossible for professional service people to gauge them with complete accuracy in every instance. Thus, doctors cannot promise that every operation will be successful; a lawyer can never be certain that a contract he drafts is without latent ambiguity; and an architect cannot be certain that a structural design will interact with natural forces as anticipated. Because of the inescapable possibility of error which inheres in these services, the law has traditionally required, not perfect results, but rather the exercise of that skill and judgment which can be reasonably expected from similarly situated professionals.[38]

Although some older cases use a community or local standard of care, it would seem the better practice to utilize a national standard of care, particularly when national contract forms, such as the AIA documents, are used. Only a handful of building codes are used throughout the country, with only minor local modifications. Architecture schools gear their programs toward national practice. The licensing examinations are administered by the NCARB, a national body. Thus, there is no reason to think that an architect in one area of the country would be subject to a professional standard any different from that in another area, as long as the project is the same.

The contract defines the scope of the architect's duties.[39] However, the duty of the architect is not necessarily limited by the contract.[40] The architect can assume

---

hazards become known. The court rejected that argument, finding that an architect is liable for damages occasioned by defective plans but he does not undertake that plans will be absolutely perfect and is liable only for a failure to exercise reasonable skill in the preparation of the plans.

[38] City of Mounds View v. Walijarvi, 263 N.W.2d 420, 424 (Minn. 1978), *cited in* Waldor Pump v. Orr-Schelen-Mayeron & Assocs., 386 N.W.2d 375 (Minn. Ct. App. 1986).

[39] Gables CVF, Inc. v. Bahr, Vermeer & Haecker, Architect, Ltd., 244 Neb. 346, 506 N.W.2d 706 (1993) ("Implicit in every contract for architectural services is the duty of the architect to exercise skill and care which are commensurate with requirements of the profession."); Harman v. C.E.&M., Inc., 493 N.E.2d 1319 (Ind. Ct. App. 1986). In Strauss Veal Feeds v. Mead & Hunt, Inc., 538 N.E.2d 299, 303 (Ind. Ct. App. 1989), the architect was not liable when a veal feed processing facility violated state environmental laws. The architect's plans were in compliance with codes; it was the owner's innovative process that caused the problems. The court also held that "an architect does not owe a fiduciary duty to its employer; rather, the architect's duties to its employer depend upon the agreement it has entered into with that employer." In R. G. Nelson, A.I.A. v. M.L. Steer, 797 P.2d 117 (Idaho 1990), the court found no fiduciary relationship. However, the dissent presented a strong argument for the existence of a fiduciary duty on the part of the architect. In Carlson v. SALA Architects, Inc., 732 N.W.2d 324 (Minn. App. 2007), the stated that "the relationship of architect and client is not a fiduciary one. However, whether a fiduciary relationship exists is a fact question." AIA documents do not impose such a fiduciary relationship. In E-Med, Inc., v. Mainstreet Architects & Planners, 2007 WL 1536803 (Cal. App. 2 Dist., June 19, 2007), the court assumed the existence of a fiduciary duty. Apparently, no party argued otherwise.

In Woodbridge Care LLC v. Englebrecht & Griffin, 1997 Conn. Super. LEXIS 828 (Mar. 27, 1997), the court held that a violation of the AIA Code of Ethics does not create a cause of action upon which a civil remedy may be based. The court also found that the architect had not engaged in unfair trade practices because there was no contractual provision that required the architect to work exclusively on the client's work.

[40] Seattle W. Indus., Inc. v. David A. Mowat, 110 Wash. 2d 1, 750 P.2d 245 (1988).

additional duties by affirmative conduct, such as when it undertakes additional duties that are not included in the contract.

The architect must prepare documents that reasonably meet the needs of the owner as stated in the program. In one case, a design-build firm contracted with an owner to furnish a set of plans for a house.[41] The owner decided to be his own general contractor, with the design-build firm constructing only the shell of the building. The drawings included a foundation plan. Shortly after the owners moved in, the house began to settle, causing substantial damage. An expert testified that the foundation was improperly designed because a foundation was partly on rock and partly on new fill, so piers should have been designed to rest on rock to support the footings. Because this was not done, there were some four inches of settlement. The drawings were held to be unsuitable for the purpose for which they were drawn.[42]

One court has held that the designer of a truss was under a duty to furnish the truss so that it would function properly once it was in place, and also to design it such that it could withstand the ordinary stresses to which it would be subjected during erection by reasonably anticipated methods.[43]

The implied agreement to reasonably meet the owner's needs includes the architect's duty to draw plans and specifications that conform to building codes, zoning, and other local ordinances,[44] although the current version of B101 makes

---

[41] O'Dell v. Custom Builders Corp., 560 S.W.2d 862 (Mo. 1978).

[42] In Bowman v. Coursey, 433 So. 2d 251 (La. Ct. App. 1983), the owner wanted to reduce the cost of his warehouse. The engineer revised the drawings, but told the owner that there would be some settlement of the concrete floors. When the settlement occurred, the owner refused to pay the engineer. The court found that, because the plans complied with code and the building was not dangerous, the owner was required to pay the engineer, because the owner had accepted these defects in return for a lower construction cost.

In Nicholson & Loup, Inc. v. Carl E. Woodward, Inc., 596 So. 2d 374 (La. Ct. App. 1992), the design-builder's architect used a subsoil report from the owner's engineer to design a supermarket. After the building settled, the owner sued the design-builder, the architect, and the engineer. The court found all of the defendants liable. The architect was liable because he had placed his seal on the drawings. Under Louisiana law, an architect's seal cannot be placed on any drawings "not prepared either by him or under his responsible supervision." In this case, the architect sealed only the architectural drawings but not the other drawings that were part of the set. The court disregarded this based on the fact that the architect was the only professional named in the contract with the owner, which included the cost of all the drawings, and also the fact that the architect had placed the following note on the first page of the drawings:

These plans and specifications have been prepared by or under my close personal supervision and to the best of my knowledge and belief they comply with all city requirements, that I am supervising the work.

This note did not limit itself to the architectural drawings. Therefore, the court reasoned, the architect unequivocally assumed responsibility for all the documents in the packet, including the defective foundation plans that were not prepared by the architect.

[43] Swaney v. Peden Steel Co., 259 N.C. 531, 131 S.E.2d 601 (1963).

[44] Krestow v. Wooster, 360 So. 2d 32 (Fla. Dist. Ct. App. 1978); Bott v. Moser, 175 Va. 11, 7 S.E.2d 217 (1940); Bebb v. Jordan, 111 Wash. 73, 189 P. 553 (1920) (architect could not recover fees for a building design for which owner directed him to draw a building of a certain size that violated the

zoning restrictions when the owner did not know of violation). In Graulich v. Frederick H. Berlowe & Assocs., 338 So. 2d 1109 (Fla. Dist. Ct. App. 1976), the architect failed to investigate the zoning for a parcel of property. The architect's work fell below the professional standard, and the owner was not obligated to pay the architect's fees.

In St. Joseph Hosp. v. Corbetta Constr. Co., 21 Ill. App. 3d 925, 316 N.E.2d 51 (1974), the architect specified a wall paneling that had a flame spread rating some 17 times the maximum allowed by the building code. There was only one type of paneling that complied with the code, but it was not specified. The court cited Scott v. Potomac Ins. Co., 217 Or. 323, 341 P.2d 1083, 1088 (1959) ("It ill behooves a man professing professional skill to say I know nothing of an article which I am called upon to use in the practice of my profession."). The architect had a duty to know the characteristics of the materials he specified (or, as in this case, approved as an alternative), and to make sure that such materials complied with code.

In Duncan v. Missouri Bd. for Architects, 744 S.W.2d 524 (Mo. Ct. App. 1988), the engineer's license was revoked when his failure to conform to the building code resulted in collapse of the walkways in the Kansas City Hyatt Hotel, with serious loss of life.

The measure of damages when an architect fails to follow codes and the owner is delayed is the difference in the reasonable rental value between the old building and the new structure. Miami Heart Inst., Inc. v. Heery Architects & Eng'rs, Inc., 765 F. Supp. 1083 (S.D. Fla. 1991).

Merely because the architect follows codes or governmental or industry standards does not prevent a court from finding the architect negligent. Francisco v. Manson, Inc., 145 Mich. App. 255, 377 N.W.2d 313 (1985) (student injured on diving platform selected by architect).

In Edward J. Seibert, A.I.A. v. Bayport Beach & Tennis Club Ass'n, 573 So. 2d 889 (Fla. 1990), condominium owners brought suit against the architect on the basis of his statutory responsibility to the association for building code violations. The architect had designed the development so that each second-floor dwelling unit had one independent, unenclosed set of stairs leading from the front door of the unit directly to the ground. It was alleged that this did not comply with the standard building code. The architect had presented the plans to the chief code enforcement officer, who had interpreted the code to allow this design and had issued a permit. The association presented its own expert, who testified that his interpretation of the code was different. The appellate court found that "when an agency with the authority to implement a statute construes the statute in a permissible way, that interpretation must be sustained even though another interpretation may be possible." The architect had a right to rely on the code official's interpretation of the code in fulfilling his duty.

In Garaman, Inc. v. Danny Williams, A.I.A., 912 P.2d 1121 (Wyo. 1996), one issue involved the architect's standard of care. The owner attempted to use the architect's own testimony to establish the standard of care. The court held that the owner did not elicit proper testimony from the architect to establish a standard of care. The architect had testified that he had a duty to know the applicable building codes and that, in certain instances, his design did not comply with the codes as the local municipality subsequently interpreted them. He also testified that he believed that he had complied with the code and that interpreting code requirements is frequently difficult. The court found this to be too general to establish a breach of the standard of care. It left open the issue of whether, under proper circumstances, an owner can use an architect's own testimony to establish the standard of care and breach thereof. The court also held that noncompliance with a building code is not, by itself, negligence. Architects do not warrant perfect results; they are only required to exercise the appropriate degree of care.

In Outlaw v. Airtech Air Conditioning and Heating, Inc., 412 F.3d 156 (2005), the architect was accused of failing to design the HVAC system to code. The court rejected this claim in affirming the trial court's order dismissing the action. A report presented by the plaintiff presented "a litany of shortcomings in the design and installation of the HVAC system but also cited no specific code violations" and concluded that the "cooling differentials all met accepted industry standards as of the time of inspection." Thus, the plaintiff had provided no evidence linking the architect's design plans to the alleged defects in the HVAC system, and that the plaintiff had not identified any code provisions that the design violated. Further, there was no demonstration indicating how the

this an explicit undertaking in ¶ 3.2.1. An exception to this rule is when the owner directs the architect to deviate from such codes or ordinances.

There is a clear distinction between the architect's duties to the owner, with whom the architect has a contract, and the architect's duties to third parties, with whom there is no contract. The architect's duties are defined by the contract insofar as the owner (or other party with whom the architect has a contract) is concerned. This is also true relative to the contractor when the general conditions define the architect's duties towards the contractor and its subcontractors. Additional duties that are implied may apply if they do not conflict with the contract or if they are imposed by public policy. However, when there is no contract (as with members of the public, for instance), the architect's duties may be broader. They are still defined by the contract, but they may be broader. For instance, if the contract states (as does the AIA contract) that the architect does not supervise the work, the architect will not be responsible to third parties for safety issues in most circumstances. On the other hand, indemnification provisions may not protect the architect in the case of third parties.

Architects must also comply with laws such as the Americans with Disabilities Act (ADA), 42 U.S.C. §§ 12,101 *et seq*. This law is actually a civil rights law, not a building code. In a California case,[45] the architect was hired by a tenant to perform interior design services. The owner of the property sued the architect alleging that substantial defects had resulted from the architect's work. The architect defended on the basis of owing no duty to the owner. The court agreed, saying that the owner could have had the tenant insert a provision in the contract stating that the owner was an intended beneficiary. There was also no evidence that the owner was responsible for the repair or maintenance of the tenant's space. Architects face liability under the ADA, however, if they fail to conform to the requirements of that law.[46]

Architects have a duty to the public. A Wisconsin court stated:

> The very essence of a profession is that the services are rendered with the understanding that the duties of the profession cannot be undertaken on behalf of a client without an awareness and a responsibility to the public welfare. The entire ambit of state regulations as they apply to the profession of architecture is intended, not solely for the protection of the person with whom the architect deals, but for the protection of the world at large. Professionalism is the very antithesis to irresponsibility to all interests other than those of an immediate employer.[47]

---

architect's plans were a cause in fact of the alleged deficiencies, let alone a proximate cause. "When there is a construction defect, the architect is one of the usual suspects, but [the architect's] proximity to the problem and [the plaintiff's] accusation alone are not enough to survive summary judgment."

[45] Mar Canyon Torrance LLC v. John Wolcott Associates, Inc., 2006 WL 2724027 (Cal. App. 2 Dist., Sept. 25, 2006).

[46] *See, e.g.*, United States v. Ellerbe Becket, 976 F. Supp. 1262 (Minn. Sept. 30, 1997).

[47] A.E. Inv. Corp. v. Link Builders, Inc., 62 Wis. 2d 479, 214 N.W.2d 764, 769 (1974). Privity was not necessary for a foreseeable tenant to maintain an action for negligence against an architect.

This public duty includes the duty to those who would be likely to use the structure to exercise care that the design is safe for the building's intended use.[48]

---

[48] La Bombarbe v. Phillips Swager Assocs., 130 Ill. App. 3d 891, 474 N.E.2d 942 (1985) (architect did not have a duty to design jail cells without grilles so as to avoid providing anchor points from which inmate might hang himself).

In Tittle v. Giattina, Fisher & Co., Architects, Inc., 597 So. 2d 679 (Ala. 1992), a prisoner had committed suicide. The court followed *La Bombarbe* and found that "an architect designing a prison or jail owes no duty to design the prison or jail to be suicide-proof." Even though such suicides may be foreseeable, it was primarily the ability of the jailers to recognize those inmates who exhibit suicidal tendencies that would act to prevent suicides.

In Easterday v. Masiello, 518 So. 2d 260, 261 (Fla. 1988), a prisoner hung himself from the "yard arm" in an air conditioning duct that did not have a guard grille. The court found this to be a patent defect (an obvious danger). "A contractor, architect or engineer is not insulated from liability if there is a latent defect."

Waldor Pump v. Orr-Schelen-Mayeron & Assocs., 386 N.W.2d 375, 377 (Minn. Ct. App. 1986), involved economic losses resulting from negligent provision of engineering services. "[T]he reasonable skill and judgment expected of professionals must be rendered to those who foreseeably rely upon the services." *Waldor* was followed in Mid-Western Elec., Inc. v. De Wild Grant Reckert & Assocs., 500 N.W.2d 250 (S.D. 1993), wherein the electrical subcontractor was allowed to sue the architect for economic damages. But see the cases arising out of Moorman Mfg. Co. v. Nat'l Tank Co., 91 Ill. 2d 69, 435 N.E.2d 443 (1982), which have held that there is no tort recovery for economic losses. *Moorman* was applied to architects in a case in which the architect's negligence caused only economic damages. 2314 Lincoln Park W. Condominium Ass'n v. Mann, Gin, Ebel & Frazier, Ltd., 136 Ill. 2d 302, 555 N.E.2d 346 (1990).

In Alberti, LaRochelle & Hodson Eng'g Corp. v. FDIC, 844 F. Supp. 832, 843 (D. Me. 1994), the engineering and construction management firm was liable to the lender for negligence and negligent misrepresentation regarding the developer's budget prior to the loan and in issuing certificates for payments. The court was convinced that the engineer knew or should have known that a number of items were missing from the budget presented to the lender. The court stated:

> [W]hen [the engineers] signed the requisitions [certificates for payment], they were giving the Bank assurances about the continued accuracy of the construction budget when they knew the budget was substantially less than the amount of money needed to complete Phase I of the project. These assurances about the accuracy of the budget led the Bank to wrongfully disburse loan proceeds to the Developer.

The court in Resurgence Prop., Inc. v. W.E. O'Neil Constr. Co., No. 92C6618, 1995 U.S. Dist. LEXIS 4939 (N.D. Ill., Apr. 14, 1995), stated:

> The testimony in this case underscores the responsibilities of construction and design professional when giving information to lenders. After careful review of all the evidence in this case, the Court concludes that under these circumstances, [the engineer] owed a duty of care to the Bank.

In Aetna Cas. & Sur. Co. v. Leo A. Daly Co., 870 F. Supp. 925, 937 (S.D. Iowa 1994), the court found that the architect could limit his liability contractually so long as third parties were not injured:

> The court concludes first that the design professionals here could properly limit by contract their liability for the harm alleged, because the harm did not involve physical injury or death to any person, as was the circumstance in the cases cited above rejecting contractual limitations on liability. Thus, the contractual limitations here were not in violation of public policy, and are therefore valid. Furthermore, the harm for which liability is sought to be disclaimed is harm to the contracting party, and not harm to a third party.

The architect, Daly, was contractually relieved of the obligation to review the contractor's proposed changes in the details of the building. When a sprinkler pipe froze and broke, each party blamed the other. The architect's original design for the area had been revised at the suggestion of the contractor.

In Pierce v. ALSC Architects, P.S., 52 Mont. 93, 890 P.2d 1254 (1995), the architect designed a remodeling project for a supermarket, which had an existing walk-in cooler located on the main floor immediately below the observation and storage room. The roof of the cooler was even with the security walkway and provided a floor for the storage room and a place for storing seasonal displays used in the store. During the remodeling, the cooler was removed and replaced with a smaller walk-in freezer. A suspended ceiling was installed in the space between the new freezer and the observation walkway. It was agreed that the walkway was to be sealed off by removing the access door (to be used in another location) and covering the opening with dry-wall. It was understood that if the walkway was not to be sealed off, guardrails, lighting, and walkway improvements would be required by code. During construction, the contractor had an extra door and used it where the access door was to go. The access door was not replaced or sealed. Later the plaintiff, a clerk at the supermarket, was asked by a customer to retrieve a poster from the area of the walkway. The clerk, unaware that the roof of the cooler had been replaced, stepped through the access door. The area was unlighted and appeared the same as it had before the remodeling. The clerk fell through the ceiling and injured himself.

The architect claimed that the accepted work doctrine barred recovery. That doctrine holds that, once the owner accepts the work, the contractor or architect is free from liability to third parties. In rejecting this doctrine, the Supreme Court of Montana stated:

> This defense, as previously applied, has the undesirable effect of shifting responsibility for negligent acts or omissions from the negligent party to an innocent person who paid for the negligent party's services. Furthermore, the shifting of responsibility is based on the legal fiction that by accepting a contractor's work, the owner of property fully appreciates the nature of any defect or dangerous condition and assumes responsibility for it. In reality, the opposite is usually true. Contractors, whether they be building contractors, or architects, are hired for their expertise and knowledge. The reason they are paid for their services is that the average property owner does not have sufficient knowledge or expertise to design or construct real property improvements safely and soundly. The mere fact that expert testimony is required to establish professional negligence makes it clear that nonexperts are incapable of recognizing substandard performance on their own. How then can we logically conclude that simply because the professional has completed his or her services and the contractee has paid for those services, liability for the contractor's negligence should shift to the innocent and uninformed contractee? We cannot.

In Morris v. Attia, 2005 WL 709821 (N.Y. Sup., Mar. 15, 2005), the defendant architect designed a duplex apartment. At the rear of the fourth floor were two 7-foot–5-inch-high sliding glass windows that flanked a picture window. The windows stood upon a riser that ran the length of the back wall and was 18 inches high and 29 inches deep. These windows opened directly to the outside of the building, as there was no balcony, nor were there any protective measures in place to prevent them from opening to their full 29-inch width. In the early morning hours, plaintiff's decedent fell through one of the open windows and plunged to his death on the concrete patio below. Plaintiff contended that opening the windows required an unreasonable amount of force, so that in opening one while standing on the polished wood riser, the decedent slid and accidentally fell out of the window. The architect asserted that he designed the windows to include a stopping mechanism and a safety bar. Later, he asserted that the window subcontractor was responsible for the design of the windows and the stopping mechanism. The architect's drawings were not clear on this issue. The court cited this language: "An architect is required to use that reasonable care and skill in design, or that degree of care that a reasonably prudent architect would use to avoid a defect that would create an unreasonable risk of harm." The court found that the architect was not liable

This duty also extends to workers on the site during construction,[49] although the agreement limits it by placing the duty for safety on the contractor (**see** ¶ 3.6.1.2

---

for injuries to third parties based on his alleged failure to supervise construction. However, factual issues remained, thereby precluding summary judgment.

In Mellon v. Shriners Hosp. for Children, 2007 WL 1977968 (Mass. Super., June 26, 2007), a grate was installed improperly, creating a tripping hazard which caused the plaintiff's injury. The court refused to grant the architect summary judgment based on the AIA provisions in a prior version of ¶ 3.6.1.2. The architect had authority to reject nonconforming work and to guard the owner against defects. This created issues of fact.

In Affholder, Inc. v. North Am. Drillers, Inc., 2006 WL 3192537 (S.D. W.Va., Nov. 1, 2006), the engineer prepared drawings and specifications for a tunnel project. The documents gave the wrong strength for the rock, resulting in a substantial cost increase for the subcontractor. The subcontractor sued the engineer who defended based on the economic loss doctrine. The court held that in West Virginia a design professional may owe a duty of care to a contractor under the right circumstances, such as in this case.

In Logsdon v. Cardinal Indus. Insulation, 2006 WL 2382501 (Ky. App., Aug. 18, 2006), a decedent's estate filed a wrongful death action against various parties, including an architect after the decedent developed cancer and died allegedly as a result of asbestos exposure in the workplace. There was testimony about asbestos insulation being installed in the 1950s and 1960s. The court held that expert testimony was required to establish not only that the architect specifications fell below the standard of care required of architects or engineers at that time, but also that there were alternatives to asbestos available in the 1950s when the facilities were designed. Failure to establish the standard of care or that the architect breached any such duty arising under the standard of care required summary judgment in favor of the architect. The estate also argued that the architect had a continuing duty to warn of potential hazards from building materials as and when such hazards became known. The court rejected that argument, finding that an architect is liable for damages occasioned by defective plans but he does not undertake that plans will be absolutely perfect, and is liable only for a failure to exercise reasonable skill in the preparation of the plans.

The court also found that the architect was guilty of negligence per se because the area in question was in violation of the building code.

In American Towers Owners Ass'n v. CCI Mech., 930 P.2d 1182, (Utah 1996), a condominium association sued the designers and contractors of a large condominium complex, alleging design and construction defects in the plumbing and mechanical systems. The court found that the economic loss rule applied. The association argued that there was damage to other property as a result of the leaking pipes. This included damage to walls, wall coverings, carpeting, wall hangings, curtains, and other furnishings. The court found that the property in this case was the entire complex itself that was constructed under one general contract. What the unit owners purchased was a finished product, their dwellings, and not the individual components of the dwellings.

[49] Evans v. Howard R. Green Co., 231 N.W.2d 907 (Iowa 1975) (architect improperly designed a sewage treatment plant, resulting in the production of poisonous hydrogen sulfide gas; two workers were killed by the gas, and architect was held responsible for negligent design). Unless an architect has undertaken by conduct or contract to supervise a construction project, he is under no duty to notify or warn workers or employees of the contractor or subcontractor about hazardous conditions on the construction site. Jones v. James Reeves Contractors, Inc., 701 So. 2d 774 (Miss. 1997); Hobson v. Waggoner Eng'g, Inc., 2003 WL 21789396 (Miss. App., Aug. 5, 2003).

Hobson was cited in Lyndon Prop. Ins. Co. v. Duke Levy and Assocs., LLC, 475 F.3d 268 (5th Cir. 2007), where the surety for a county sewer project sued the engineer for the amount the surety paid to fix and test work that had previously been approved by the engineer. A non-AIA exculpatory provision was not sufficient to protect the engineer under the surety's equitable subrogation theory.

and Article 10 of AIA Document A201).[50] The duty of the architect is nondelegable, that is, the architect cannot rely on information from other parties, such as engineers or manufacturers, in an attempt to circumvent responsibility for damages to the owner or injured third parties (although the architect might be entitled to indemnification).[51]

In 1981, the second and fourth floor walkways of the Hyatt Hotel in Kansas City collapsed, killing 114 people and injuring 186. The state brought actions against the structural engineers in *Duncan v. Missouri Board for Architects*.[52] The engineers were found guilty of gross negligence and their certificates of registration were revoked. In examining the engineers' argument concerning negligence, the court stated:

> An act which demonstrates a conscious indifference to a professional duty would appear to be a reckless act or more seriously a willful and wanton abrogation of professional responsibility. The very nature of the obligations and responsibility of a professional engineer would appear to make evident to him the probability of harm from his conscious indifference to professional duty and conscious indifference includes indifference to the harm as well as to the duty. The structural engineer's duty is to determine that the structural plans which he designs or approves will provide structural safety because if they do not a strong probability of harm exists. Indifference to the duty is indifference to the harm.

A California court described three different roles of the architect:

1. He is an independent contractor in the preparation of the initial plans and specifications. It is now well settled that in such capacity the architect may be sued for negligence in the preparations of plans and specifications either by his client or by third persons;[53]

2. He is an agent of the owner in supervising[54] the construction work as it progresses; and

3. He is a quasi-judicial officer with certain immunity when he acts as arbiter in resolving disputes between the owner and the contractor.[55]

The architect must be extremely careful when asked to deviate from codes. Because codes are meant to protect the public, any injured third party may well

---

[50] Pugh v. Butler Tel. Co., 512 So. 2d 1317 (Ala. 1987) (engineer had no duty to injured worker because contract was for benefit of owner and was to insure compliance with the plans and specifications, and had nothing to do with safety of workers); Case v. Midwest Mech. Contractors, 876 S.W.2d 51 (Mo. Ct. App. 1994).

[51] Mayor & City Council v. Clark-Dietz, 550 F. Supp. 610 (N.D. Miss. 1982).

[52] 744 S.W.2d 524, 533 (Mo. Ct. App. 1988).

[53] It should be noted that this is not well settled in all jurisdictions. *See, e.g.*, R.H. Macy & Co. v. Williams Tile & Terrazzo, 585 F. Supp. 175 (N.D. Ga. 1984).

[54] This is another unfortunate use of the term "supervising" by the courts.

[55] Huber, Hunt & Nichols, Inc. v. Moore, 67 Cal. App. 3d 278, 299-300, 136 Cal. Rptr. 603, 616 (1977).

have an action against the architect. In *Duncan*, an engineer's license was revoked for gross failure to comply with code.[56] Architectural acts are also meant to protect the public from unqualified practitioners, not to provide work for architects.[57]

On the other hand, an architect does not imply or guarantee a perfect plan or satisfactory result,[58] except when such an agreement is specifically stated. In one case, an architect contractually agreed to receive daily reports from a soils engineer to be hired by the excavator and to hire a soils engineer when the architect deemed necessary.[59] Subsequent cracking in the concrete slabs was caused by improperly compacted soil. The architect breached his duty to the owner by not having a soils engineer present on the job site.

---

[56] Duncan v. Missouri Bd. for Architects, 744 S.W.2d 524, 540 (Mo. Ct. App. 1988) (The Kansas City Building Code "is intended to provide a required level of safety for buildings within the City. It is difficult to conclude that gross failure to comply with that Code can constitute other than conscious indifference to duty by a structural engineer.").

In Prince v. Rescorp Realty, 940 F.2d 1104 (7th Cir. 1991), the court held that the Illinois state Fire Marshal Act and the local village building code created a substantive and clearly mandated public policy with respect to fire prevention systems. Although that case was a retaliatory discharge case, it could apply if design professionals deviate from codes and statutes designed to protect the public.

[57] Duncan v. Missouri Bd. for Architects, 744 S.W.2d 524, 531, 538 (Mo. Ct. App. 1988) ("The purpose behind licensing statutes is to protect the public rather than to punish the licensed professional. The purpose of disciplinary action against licensed professionals is not the infliction of punishment, but rather the protection of the public.").

[58] Mayor & City Council v. Clark-Dietz, 550 F. Supp. 610 (N.D. Miss. 1982); Mississippi Meadows, Inc. v. Hodson, 13 Ill. App. 3d 26, 299 N.E.2d 359 (1973); Weill Constr. Co. v. Thibodeaux, 491 So. 2d 166 (La. 1986); Frischhertz Elec. Co. v. Housing Auth., 534 So. 2d 1310 (La. Ct. App. 1988); Klein v. Catalano, 386 Mass. 701, 437 N.E.2d 514 (1982); Malden Cliffside Apartments, LLC v. Steffian Bradley Assocs., Inc., 844 N.E.2d 1124 (Mass. App. Ct. 2006). Borman's, Inc. v. Lake State Dev. Co., 60 Mich. App. 175, 230 N.W.2d 363 (1975); Diocese of Rochester v. R-Monde Contractors, Inc., 148 Misc. 2d 926, 562 N.Y.S.2d 593, 596 (Sup. Ct. 1989) ("An architect is not subject to an action for breach of warranty."); Three Affiliated Tribes v. Wold Eng'g, 419 N.W.2d 920 (N.D. 1988); Bloomburg Mills, Inc. v. Sordoni Constr. Co., 401 Pa. 358, 164 A.2d 201 (1960); Ryan v. Morgan Spear Assocs., 546 S.W.2d 678 (Tex. 1977).

In Allied Stores v. Gulf Gate Joint Venture, 726 S.W.2d 194, 197 (Tex. Ct. App. 1987), the court differentiated between choosing a design and the design itself. "In the total absence of any evidence that the design was defective or negligent in any engineering or architectural aspect, it is impossible for the choosing of such design to be a proximate cause of any loss."

In Seiler v. Ostarly, 525 So. 2d 1207 (La. Ct. App. 1988), under an oral agreement, the architect was hired to prepare plans for an apartment complex, but the duties did not include complete architectural services. The architect was not liable when expansive soil caused structural damage.

A unique case is McKeen Homeowners Ass'n v. Oliver, 586 So. 2d 679 (La. Ct. App. 1991), in which the architect altered the title block of existing drawings to reflect his own logo and affixed his own seal, even though he prepared only one page of the plans. When the exterior walls failed, the homeowners association sued this architect, who had been hired by the builder. The court found that the architect was not liable because he had not prepared the drawings that related to the failed walls. The court rejected the argument that, by adopting the drawings as his own, the architect took the responsibility. This case should be relied on only with great caution.

[59] Rosos Litho Supply Corp. v. Hansen, 123 Ill. App. 3d 290, 462 N.E.2d 566 (1984).

Recent attempts to impose an implied warranty standard on architects and engineers have thus far met with little success.[60] It might be wise to include a

---

[60] *See, e.g.*, State v. Gathman-Matotan, 98 N.M. 790, 653 P.2d 166 (Ct. App. 1982). An architectural firm was hired to develop plans for a renovation of an existing prison, including the central control area. After the work was completed, a riot broke out, and the prisoners gained control over the central control area. The state argued that the architect had breached an implied warranty of sufficiency of its design to provide an area adequate to serve as a control stronghold in the event of such a riot. The matter was dismissed by the appellate court, which found no such duty by the architect.

In Tamarac Dev. v. Delamater, Freund, 234 Kan. 618, 675 P.2d 361 (1984), a Kansas court found the architect had breached a warranty to accurately supervise the grading of a trailer park. The warranty in question, however, was in an oral contract. *See also* Johnson-Voiland-Archuleta, Inc. v. Roark Assocs., 40 Colo. App. 269, 572 P.2d 1220 (1977); Board of Trustees v. Kennedy, 167 N.J. Super. 311, 400 A.2d 850 (1979), and the cases cited therein; Borman's, Inc. v. Lake State Dev. Co., 60 Mich. App. 175, 230 N.W.2d 363 (1975).

An implied warranty was found in Bloomburg Mills, Inc. v. Sordoni Constr. Co., 401 Pa. 358, 164 A.2d 201 (1960), when an architect designed a roof for a rayon and nylon weaving mill that required constant maintenance of a particular temperature and high humidity. When the roof insulation became soggy and condensation formed on the ceiling within a few years of construction, the roof had to be replaced. The architect was held liable for improper design of the roof system. The architect claimed that he relied on the manufacturer's claims, and that the owner approved the system, but these defenses were rejected. *See also* Hill v. Polar Pantries, 219 S.C. 263, 64 S.E.2d 885 (1951); Nave v. McGrane, 113 P. 82 (Idaho 1910); Prier v. Refrigeration Eng'g Co., 74 Wn. 2d 25, 442 P.2d 621 (1968); Broyles v. Brown Eng'g Co., Inc., 151 So. 2d 767 (Ala. 1963). In SME Indus., Inc. v. Thompson, Ventulett, Stainback & Assocs., Inc. 28 P.3d 669 (Utah 2001), the court surveyed the current state of the law as to whether architects impliedly warrant perfect plans or satisfactory results and found that a "solid majority" of courts refuse to so hold. However, when architects bind themselves by contract to perform a service, they agree by implication to use reasonable care and skill in doing it. A breach of a duty to use reasonable or customary care in the provision of professional services gives rise to an action under contract for negligent services.

In Cruet v. Carroll, 2001 WL 1570228 (Conn. Super., Nov. 27, 2001), the owner counterclaimed against the architect in a fee collection case, alleging breach of the implied warranty of fitness for a particular purpose. Following the majority rule, the court dismissed the warranty count, holding that "the nature of professional services are best measured by a negligence standard rather than an implied warranty."

In Bryles v. Brown Eng'g Co., 275 Ala. 35, 151 So. 2d 767 (1963), the court distinguished between the services provided by engineers and those provided by doctors, dentists, and architects, referring to the services of the latter three as largely experimental and subject to conditions beyond their control and influence. In comparison, an engineering survey of drainage requirements of a tract of land was not so involved with unknown, uncontrollable topographical or landscape conditions as to prevent the imposition of an implied warranty.

In White Budd Van Ness P'ship v. Major-Gladys Drive Joint Venture, 798 S.W.2d 805 (Tex. 1990), the court found an implied warranty of good and workmanlike performance of services to exist under the Texas Deceptive Trade Practices-Consumer Protection Act. This was based, in part, on the jury's findings that the owner was unsophisticated and relied on the architect and that the architect had acted in an "unconscionable manner."

In Beachwalk Villas Condominium Ass'n, Inc. v. Martin, 406 S.E.2d 372 (S.C. 1991), the Supreme Court of South Carolina found that there was an implied warranty by an architect to homebuyers who were not in privity with the architect. That court also declined to apply the economic loss rule to architects.

disclaimer of warranties in the contract.[61]

2.3 The Architect shall identify a representative authorized to act on behalf of the Architect with respect to the Project. **(1.2.3.3; substantial revisions)**

This designated representative has authority to sign documents to the extent permitted in this agreement. If no one is designated in the exhibit, the parties can later agree to designate such a representative.

2.4 Except with the Owner's knowledge and consent, the Architect shall not engage in any activity, or accept any employment, interest or contribution that would reasonably appear to compromise the Architect's professional judgment with respect to this Project. **(1.2.3.5; no change)**

This provision prohibits the architect from getting kickbacks from contractors or suppliers, or doing anything that, even though not illegal, has an appearance of impropriety. If the architect has any doubt about a particular situation, the owner should be informed and asked to approve.

2.5 The Architect shall maintain the following insurance for the duration of this agreement. If any of the requirements set forth below exceed the types and limits the Architect normally maintains, the Owner shall reimburse the Architect for any additional cost: **(new)**

In Barnett v. City of Yonkers, 731 F. Supp. 594 (S.D.N.Y. 1990), the owner asserted a claim against the architect based on implied warranty. The court stated that "New York law is crystal clear that in service-oriented contracts, such as agreements to render architectural services, no action in breach of implied warranty or strict product liability will lie for the negligent performance of professional services."

In Georgetown Steel Corp. v. Union Carbide Corp., 806 F. Supp. 74 (D.S.C. 1992), the court found that there is no action for breach of implied warranty in contracts for professional services. In SME Indus., Inc. v. Thompson, Ventulett, Stainback & Assocs., Inc. 28 P.3d 669 (Utah 2001), the court surveyed the current state of the law as to whether architects impliedly warrant perfect plans or satisfactory results and found that a "solid majority" of courts refuse to so hold. However, when architects bind themselves by contract to perform a service, they agree by implication to use reasonable care and skill in doing it. A breach of a duty to use reasonable or customary care in the provision of professional services gives rise to an action under contract for negligent services.

In Cruet v. Carroll, 2001 WL 1570228 (Conn. Super., Nov. 27, 2001), the owner counterclaimed against the architect in a fee collection case, alleging breach of the implied warranty of fitness for a particular purpose. Following the majority rule, the court dismissed the warranty count, holding that "the nature of professional services are best measured by a negligence standard rather than an implied warranty."

[61] In Harman v. C.E.&M., Inc., 493 N.E.2d 1319 (Ind. Ct. App. 1986), the owner sued an engineer when a floor slab sank six to eight inches. The engineer had been hired by the contractor, and the verbal contract specifically excluded any responsibility for the slab or the foundation and the drawings for each. Because the owners were third-party beneficiaries, their rights were tied to the contract provisions. Because the contract excluded liability for the defective conditions, the owners were not entitled to any recovery.

(Identify types and limits of insurance coverage, and other insurance requirements applicable to the Agreement, if any.)

   .1   *General Liability* **(new)**

   .2   *Automobile Liability* **(new)**

   .3   *Workers' Compensation* **(new)**

   .4   *Professional Liability* **(new)**

Here, the parties will identify the types of insurance coverage that the architect will maintain for the duration of the architect's work for this project, as well as the limits for each. If the owner wants higher limits for professional liability or other coverage than the architect's consultants normally carry, Section 11.8.1.8 allows the architect to be reimbursed for the additional cost of such higher coverage. Here, however, the parties will agree on insurance limits for the architect to be included in Basic Services. If the owner subsequently want the architect to carry higher coverages, that will be an extra.

§ 2.6   Article 3: Scope of Architect's Basic Services

ARTICLE 3 SCOPE OF ARCHITECT'S BASIC SERVICES

***3.1** The Architect's Basic Services consist of those described in Article 3 and include usual and customary structural, mechanical, and electrical engineering services. Services not set forth in Article 3 are Additional Services.* **(2.4.1; modified)**

The phrase "usual and customary . . . engineering services" is rather ambiguous. On most projects, it means whatever engineering services are reasonably required to complete the project. Thought should be given, however, to further defining the meaning of this phrase. For instance, it may not be clear whether the services of a lighting specialist are contemplated. Frequently, civil engineering is done under a separate contract. In the event of any possible ambiguity, the parties should clarify their intent in this agreement. Also, on some projects, the owner may be expected to retain engineers from a developer, such as when the owner is a tenant in a newly constructed building. The developer may insist on the tenant's retaining certain engineers, even though the tenant hires his own architect. In that case, the agreement should clearly specify who will pay the consultants, and how they will interact. These matters should be as clear as possible to avoid later misunderstandings.

Sometimes, the owner may want to delegate some of these services to the contractors.[62] The owner attempts to save money by doing so, but the owner often finds that it costs more in the long run to have the contractor design portions of the

---

[62] In Stevens Constr. Corp. v. Carolina Corp., 63 Wis. 2d 342, 217 N.W.2d 291 (1974), the concrete subcontractor designed the prestressed concrete system. Unfortunately, an error was made and the structure was designed to hold 120 pounds per square foot less than it should have.

building. The design professional strives to give the owner the best project for the least cost, whereas the contractor attempts to maximize profits by increasing the cost of the building. (An architect can usually increase profits but at the same time decrease the cost of the building.) An owner should carefully consider this option.

> *3.1.1 The Architect shall manage the Architect's services, consult with the Owner, research applicable design criteria, attend Project meetings, communicate with members of the Project team and report progress to the Owner.* **2.1.1; substantial revisions)**

This paragraph gives an overview of the architect's services. Several duties are stated, from consulting with the owner; researching design criteria (this includes code issues, zoning restrictions, and various other matters that would affect the design); attending project meetings; communicating with the owner, contractors, and consultants; and reporting to the owner the status of the project (this could mean a periodic status report, reports of project meetings, or other report to keep the team advised of project issues).

> *3.1.2 The Architect shall coordinate its services with those services provided by the Owner and the Owner's consultants. The Architect shall be entitled to rely on the accuracy and completeness of services and information furnished by the Owner and the Owner's consultants. The Architect shall provide prompt written notice to the Owner if the Architect becomes aware of any error, omission or inconsistency in such services or information.* **(1.2.3.7, 2.1.1; revisions)**

The architect has a duty to coordinate the activities of the architect and its consultants with the activities of the owner and its consultants. If the architect has any doubts about the accuracy of information furnished by the owner or the owner's consultants, the architect should query the owner in writing and ask for instructions. If the owner wants the architect to verify such information, it is an additional service.

> *3.1.3 As soon as practicable after the date of this Agreement, the Architect shall submit for the Owner's approval a schedule for the performance of the Architect's services. The schedule initially shall include anticipated dates for the commencement of construction and for Substantial Completion of the Work as set forth in the Initial Information. The schedule shall include allowances for periods of time required for the Owner's review, for the performance of the Owner's consultants, and for approval of submissions by authorities having jurisdiction over the Project. Once approved by the Owner, time limits established by the schedule shall not, except for reasonable cause, be exceeded by the Architect or Owner. With the Owner's approval, the Architect shall adjust the schedule, if necessary, as the Project proceeds until the commencement of construction.* **(1.2.3.2, 2.1.2; substantial revisions)**

This schedule is a change from prior versions of the AIA owner/architect agreements. In the 1997 version of B141, for instance, there were two schedules: one for the architect's work, and a separate schedule for the actual construction work. This schedule combines the two, at least until construction begins when the contractor takes over the duty of preparing construction schedules.

The schedule should be realistic, with each element assigned time. For instance, governmental approvals, such as permits, can delay the project if they are not provided within the times projected. The time schedule should allow for contingencies just as the construction budget does.

3.1.4 The Architect shall not be responsible for an Owner's directive or substitution made without the Architect's approval. (**new**)

This provision covers situations where the owner gives directions to the contractor without consulting with the architect. If the architect finds out about such an event, the architect should document the situation.

3.1.5 The Architect shall, at appropriate times, contact the governmental authorities required to approve the Construction Documents and the entities providing utility services to the Project. In designing the Project, the Architect shall respond to applicable design requirements imposed by such governmental authorities and by such entities providing utility services. (**1.2.3.6; substantial revisions; added reference to utilities**)

This provision was substantially rewritten in 2007. The architect is required to contact governmental authorities, such as zoning and code enforcement officials, at "appropriate times." This likely means more than once for many projects. In addition, the architect is to contact utility providers, such as electric, gas, water, cable, telephone, and the like. Note that these contacts are for purposes of obtaining necessary requirements of the officials and utility providers so that the final construction documents comply with those requirements. It is the owner's responsibility for actually obtaining the permits that may be required.

3.1.6 The Architect shall assist the Owner in connection with the Owner's responsibility for filing documents required for the approval of governmental authorities having jurisdiction over the Project. (**2.1.6; minor modification**)

In most cases, this means preparing drawings for building permits from the local municipality, although other governmental bodies may also have jurisdiction. However, not all governmental bodies require submissions of drawings. All appropriate governmental bodies should be identified during the design phases so that requirements can be ascertained. For instance, the Federal Aviation Administration may have certain height limitations for buildings near airports. If the design exceeds that limit, substantial redesign would result. Obviously, this is a factor that should be known from the beginning. Many times, the contractor, not the owner, obtains a permit using the architect's drawings. In jurisdictions in which the architect may expect to be called on to meet with these governmental authorities on more than one occasion and make changes beyond what is in the code, the architect may consider making that portion of its responsibilities an additional service on an hourly basis. The owner should realize that building codes are subject to interpretation, and these interpretations are not always reasonable or logical. In some cases, these interpretations are subject to political pressure to favor a particular building group. For that reason, the owner cannot expect that the initial submittal to the building officials will go through without alteration or according to the owner's schedule.

It should be noted that it is the owner's and not the architect's responsibility to obtain governmental approvals. The architect cannot and should not guarantee that permits can be obtained. The architect's insurance probably would not cover such a guarantee in the event a permit was not issued.

§ 2.7   Schematic Design Phase Service: ¶ 3.2

3.2 SCHEMATIC DESIGN PHASE SERVICES

This phase is actually composed of several sub-phases. First, there is an evaluation of all of the information provided by the owner and obtained by the architect. Next, the architect presents the owner with the results of this evaluation and an agreement as to the requirements of the project is reached. Then, the architect prepares a preliminary design for the owner's approval. Finally, a schematic design is prepared.

Generally, schematic design involves the first sketches of the project in which the design concept is formulated, site conditions are investigated, and other elements affecting the project are initially examined. This phase should end with the owner's approving a set of schematic design drawings and other documents and the architect's proceeding to the design development phase.

3.2.1 The Architect shall review the program and other information furnished by the Owner and shall review laws, codes, and regulations applicable to the Architect's services. **(1.2.3.6; substantial revisions)**

This states the obvious, that the architect is required to conform to laws, codes, and ordinances. However, this requirement is tempered by language that is limited to a "review" of such laws. The architect will be held to a negligence standard in this respect, so that if the architect fails to conform to a particular ordinance or requirement, the owner will have to prove that it was unreasonable for the architect to have failed to comply with the particular requirement.

3.2.2 The Architect shall prepare a preliminary evaluation of the Owner's program, schedule, budget for the Cost of the Work, Project site, and the proposed procurement or delivery method and other Initial Information, each in terms of the other, to ascertain the requirements of the Project. The Architect shall notify the Owner of (1) any inconsistencies discovered in the information, and (2) other information or consulting services that may be reasonably needed for the Project. **(2.3.1, 2.3.2, 2.3.3; modifications)**

The architect is required to perform an evaluation of the information provided in the Initial Information in conjunction with a review of the site and other information made available to the architect at the outset. On a large project, the architect might consider retaining a cost consultant or recommending to the owner that a contractor be retained to give cost input at the earliest possible stage. All references to project costs or budget can be stricken from the agreement if the architect will not have input in this area.

The program is simply the owner's set of requirements. It should define as specifically as possible what the owner's needs are. Typically, the program should list functions, area requirements for each function, interrelationships among the functions, usage, population, and other data. The architect's design is then based on this program. On simple projects, the program may be only a single page long, but on a complex project, such as a hospital, the program may be a thick volume of requirements. The architect and owner must work out this program, along with a schedule and budget for the project at the outset. At this point, if the owner is unreasonable on any of these points, the architect should consider withdrawing from the project. All of these items should be recorded for the architect's files.

The schedule should be realistic, with each element assigned time. For instance, governmental approvals, such as permits, can delay the project if not provided within the times projected. The time schedule should allow for contingencies just as the construction budget does. Note that the word "schedule" as used in this paragraph means the owner's entire schedule, of which construction is only one part. Note, also, that the architect is to provide a schedule under ¶ 3.1.3, which would normally be prepared after this initial review.

The architect is also to give an evaluation of the delivery method. That is, the architect is to advise the owner of alternatives to the construction of the project. Examples would be design-build, construction management, or traditional construction. The architect should have some knowledge of the cost and other implications of these construction delivery systems.

3.2.3 The Architect shall present its preliminary evaluation to the Owner and shall discuss with the Owner alternative approaches to design and construction of the Project, including the feasibility of incorporating environmentally responsible design approaches. The Architect shall reach an understanding with the Owner regarding the requirements of the Project. (**new**)

The preliminary evaluation prepared pursuant to the preceding paragraph will be presented to the owner. While this can be done orally, for most larger projects this will likely mean a written report followed by a meeting to discuss the report. The "understanding" requires a meeting of the minds. This would be best done with a document that memorializes the agreement so that there is no later misunderstanding of what was agreed to.

3.2.4 Based on the Project's requirements agreed upon with the Owner, the Architect shall prepare and present for the Owner's approval a preliminary design illustrating the scale and relationship of the Project components. **(2.4.2; substantial modifications)**

Following the evaluations of the prior paragraphs, the architect is to prepare a preliminary design. This is likely to be not much more than a concept sketch or series of drawings. Depending on the particular architect, these may be freehand drawings, watercolors, or computerized animations. The purpose is to illustrate a design concept to the owner for approval. On larger projects, there may be several such sketches illustrating several approaches to the design. The AIA language, however, contemplates a single such design. At the meeting with the owner to discuss this sketch, the owner will have input that will result in changes to the design. As long as these changes are not radical, requiring a re-do of this entire step, the changes would be incorporated into the schematic design drawings referenced in the next paragraph. Note that the owner's approval is required to proceed to preparation of the schematic design drawings.

3.2.5 Based on the Owner's approval of the preliminary design, the Architect shall prepare Schematic Design Documents for the Owner's approval. The Schematic Design Documents shall consist of drawings and other documents including a site plan, if appropriate, and preliminary building plans, sections and elevations; and may include some combination of study models, perspective sketches, or digital modeling. Preliminary selections of major building systems and construction materials shall be noted on the drawings or described in writing. **(2.4.2.1; substantial modifications)**

This stage involves more detailed drawings and/or other ways to communicate the design than the prior stage. The list of drawings (site plan, plans, sections, etc.) is meant to suggest what may be provided by the architect, but it is the architect's decision as to what will be provided to the owner in order to properly document this phase of the work.

Under an earlier version of this clause,[63] a court held that when the contract did not specify the size, location, style, material, time of completion, or cost requirement of the project, the owner was permitted to offer testimony of oral statements relating to these requirements.[64] The cost of the project was found to be normally a requirement of the owner, meaning that the owner typically provided the architect with a cost budget.

3.2.5.1 The Architect shall consider environmentally responsible design alternatives, such as material choices and building orientation, together with other considerations based on program and aesthetics, in developing a design that is consistent

---

[63] AIA Document B131, ¶ 1.1.1 (1967).

[64] Williams & Assocs. v. Ramsey Prod. Corp., 19 N.C. App. 1, 198 S.E.2d 67 (1973). *See also* Reynolds v. Long, 115 Ga. App. 182, 154 S.E.2d 299 (1967) (court held that under an earlier version of this provision, parties contemplated that they would "arrive at the construction requirements, including any price limitations thereon, by an extrinsic contemporaneous or subsequent determination.").

with the Owner's program, schedule and budget for the Cost of the Work. The Owner may obtain other environmentally responsible design services under Article 4. (**new**)

This new section for 2007 reflects the trend towards environmentally responsible design. This merely states that the architect will take such design into consideration when doing the project. Article 4 contains references to additional services for "extensive environmentally responsible design" (¶ 4.1.23) and "LEED Certification" (¶ 4.1.24).

3.2.5.2 The Architect shall consider the value of alternative materials, building systems, and equipment, together with other considerations based on program and aesthetics in developing a design for the Project that is consistent with the Owner's program, schedule, and budget for the Cost of the Work. (**2.1.3; revised**)

The architect is to evaluate different types of construction, materials, and so on in arriving at a design. This does not mean that the architect will create several alternate designs, although it could do so. Instead, it means that there is to be an evaluation of these alternatives and a discussion with the owner of various alternatives.

3.2.6 The Architect shall submit to the Owner an estimate of the Cost of the Work prepared in accordance with Section 6.3. (**2.1.7.1; substantial revisions**)

Section 6.3 goes into detail about the architect's responsibility concerning this duty. The "Cost of the Work" is also defined in that section. The architect is required to prepare an estimate of the cost of the construction and to revise it periodically as the project changes. Previously, the architect was required to submit a "preliminary estimate of Construction Cost." Under the 1976 edition, this estimate was termed a Statement of Probable Cost.[65] In reality, it was simply an estimate, and frequently it was a pure guess. Architects generally depend on prior similar projects, reference books, or input from contractors to come up with unit costs. Owners usually do not realize the extent to which these estimates are inexact. Not only is the architect generally unskilled in this area, but construction costs are subject to wide fluctuation. This can lead to serious conflict between the owner and architect. The language in Section 6.3 tells the owner that the architect's estimate is not a guarantee but is simply an educated guess at the cost. The architect is expected to develop a design that falls within the mutually agreed-upon budget for the project. By submitting design and construction

---

[65] In Brown v. Cox, 459 S.W.2d 471 (Tex. 1970), the court held that this estimate did not require a detailed cost estimate because other provisions of the agreement provide for additional fees to the architect for such detailed estimates. Further, the estimate was not required to be in unit costs but could merely be a lump sum estimate.

    In Gunter Hotel Inc. v. Buck, 775 S.W.2d 689 (Tex. App. 1989), one issue revolved around whether a construction budget had been established. The architect argued that this budget included only those items over which the architect had control, but the owner argued that this budget included all costs. A jury found in favor of the architect.

documents to the owner for approval, the architect impliedly states that the work falls within this budget. If the estimate exceeds the budget, the architect is required to revise the design to fall within the budget. This is similar to the concept of a fixed limit of construction cost found in AIA Document B141-1987.

On a large project, the architect might consider retaining a cost consultant or recommending to the owner that a contractor be retained to give cost input at the earliest possible stage. All references to project costs or budget can be stricken from the agreement if the architect will have no input in this area.

3.2.7 The Architect shall submit the Schematic Design Documents to the Owner, and request the Owner's approval. **(new)**

Unless the owner approves the schematic design, the architect is not authorized to proceed to the next phase, design development. It is best to get this approval in writing.

§ 2.8   Design Development Phase Services: ¶ 3.3

3.3 DESIGN DEVELOPMENT PHASE SERVICES

Once the owner has approved the schematic design phase, the project moves into the design development phase. Sometimes, these phases overlap, particularly when there are changes to the project.

3.3.1 Based on the Owner's approval of the Schematic Design Documents and on the Owner's authorization of any adjustments in the Project requirements and the budget for the Cost of the Work, the Architect shall prepare Design Development Documents for the Owner's approval. The Design Development Documents shall illustrate and describe the development of the approved Schematic Design Documents and shall consist of drawings and other documents including plans, sections, elevations, typical construction details, and diagrammatic layouts of building systems to fix and describe the size and character of the Project as to architectural, structural, mechanical and electrical systems, and such other elements as may be appropriate. The Design Development Documents shall also include outline specifications that identify major materials and systems and establish in general their quality levels. **(2.4.3.1; substantial revisions)**

The design development phase is meant to take the previously approved schematic design phase and develop the various details into a more comprehensive set of drawings. Often, a preliminary drawing is found not to work, possibly because of conflicts between the various mechanical, electrical, and structural components. Perhaps an elevator cannot fit into a certain area and a new location must be found. These sorts of issues should be solved in this phase (although it is not uncommon to make such changes in the construction documents phase). Most of the major design elements should be resolved here, with only fine tuning done afterwards. This is not to say that no design will be performed during the construction documents phase (**see ¶** 3.4). Many details will be resolved there, and occasionally a major element will be changed. Once again, the architect should obtain written approval from the owner authorizing the architect to proceed to the next phase.[66]

3.3.2 The Architect shall update the estimate of the Cost of the Work. (**new**)

The architect will review any changes to the design that have occurred since the estimate for the project was last reviewed, taking into account any major changes in market conditions. Owner should, however, be aware of the limitations of an architect's estimate, particularly at such an early stage in the development of a project. These estimates will likely be based on gross cost per square foot of building, taking into account the building type and location. Such an estimate can vary significantly from the true cost of the building.

3.3.3 The Architect shall submit the Design Development documents to the Owner, advise the Owner of any adjustments to the estimate of the Cost of the Work, and request the Owner's approval. (**new**)

At the end of the design development phase, the architect must submit these documents to the owner for approval, along with a review of any changes to the cost of the project. Once the owner has approved this design and the cost, the architect can proceed to the next phase.

§ 2.9   Construction Documents Phase Services: ¶ 3.4

3.4 CONSTRUCTION DOCUMENTS PHASE SERVICES

This is the phase in which the design is translated into the final "blueprints," although real blueprints seldom are used anymore. Drawings and specifications are coordinated to assure that all the various elements, including structural, electrical, and mechanical, fit together in a logical and economical fashion to achieve the design purpose of the project. At the end of this phase, a set of documents, including drawings, specifications, and other documents, is produced from which a general contractor can construct the project (subject to required submissions by the contractor, such as shop drawings).

In a suit to recover his fees, an architect was not required to prove that he delivered the final plans to the owner, because the contract did not require such

---

[66] *See* Brixen & Christopher v. Elton, 777 P.2d 1039 (Utah Ct. App. 1989) (court interpreted "for approval by Owner" clause in ¶ 2.3.1 (1987 version of B141) as not requiring a formal written approval).

delivery.[67] Actions by architects to recover fees are rather common. Of course, the architect should keep the owner advised of the progress of the documents and should send check sets to the owner, retaining copies of transmittals in case proof is required later. This will aid in preventing an owner from later claiming that it was unaware of any progress on the project. A difficult situation arises when the owner balks at paying for the cost of printing these check sets, which can be quite expensive. One answer might be to have periodic meetings in the architect's office to review the original documents, then send minutes of the meetings to all parties.

Regular meetings of the project team are important. On a large project, they are vital. Careful minutes, detailing all decisions and the reasons behind these decisions, can come in quite handy at the litigation phase of the project. It also helps the team preserve a businesslike attitude toward the project. Another valuable tool is the telephone memo. Each call of any importance is transcribed immediately afterward. These can be distributed or merely filed with the project file. Many important decisions are made as a result of telephone calls to the owner, contractor, distributer, or manufacturer. It is almost impossible to later remember that the conversation took place at all, much less what was said.

At least one court has found that an architect could be liable to a contractor for negligent misrepresentation.[68] The architect was aware that certain subsurface debris was located on a site but failed to disclose this on his drawings. In reliance on the drawings, the contractor submitted a low bid. After the debris was discovered by the contractor, additional work was required to remove it, causing extra expense to the contractor.

3.4.1 Based on the Owner's approval of the Design Development Documents, and the Owner's authorization of any adjustments in the Project requirements and the budget for the Cost of the Work, the Architect shall prepare construction Documents for the Owner's approval. The Construction Documents shall illustrate and describe the further development of the approved Design Development Documents and shall consist of Drawings and specifications setting forth in detail the quality levels of materials and systems and other requirements for the construction of the Work. The Owner and Architect acknowledge that in order to construct the Work the Contractor

---

[67] Michalowski v. Richter Spring Corp., 112 Ill. App. 2d 451, 251 N.E.2d 299 (1969).

[68] Gulf Contracting v. Bibb County, 795 F.2d 980 (11th Cir. 1986). Lack of privity did not deter the court. Privity is not required to support an action for negligent misrepresentation, although liability is limited to a foreseeable person for whom the information was intended, such as the contractor. A number of cases have stated that the architect can be liable to a contractor for negligence. *See, e.g.,* Huber, Hunt & Nichols, Inc. v. Moore, 67 Cal. App. 3d 278, 136 Cal. Rptr. 603(1977).

In Clevecon, Inc. v. Northeast Ohio Reg'l Sewer Dist., 90 Ohio App. 3d 215, 628 N.E.2d 143 (1993), the general contractor sued the architect for malpractice to recover delay damages. The jury found the architect negligent in the preparation and drafting of the plans and specifications and in the administration of the project. It also found that the architect had breached warranties in the preparation and drafting of the plans and specifications. In affirming the trial court's judgment, the appellate court found that the architect had exercised a substantial amount of control over the project, giving orders to the contractor during the construction.

will provide additional information, including Shop Drawings, Product Data, Samples and other similar submittals, which the Architect shall review in accordance with Section 3.6.4. (**2.4.4.1; substantial revisions**)

Because the contract documents are based on the approved design development documents, it is important that such approvals be in writing.[69] Then, if the owner makes changes, the architect can be compensated, and the project schedule can be extended.

This provision, under a prior version of the document, was cited as supporting a decision to pay an architect despite the fact that the construction loan was not approved.[70] The contract had language indicating that certain payment was due "if the project is approved and not closed." The owner argued that this phrase meant that the fee was due only if the "project loan" was approved. Using this language, the court found that project approval was a function of an owner and not of a lender.

"Design" should also be differentiated from the "plans" and "specifications." The *design drawings* generally refer to the sketches and other drawings produced during the schematic and design development phases. These indicate the design of the project without necessarily showing how everything fits together. The *plans* are the drawings, usually on large sheets, that depict the floor layouts, details, elevations,[71] various schedules, ceiling layouts,[72] and other pictorial elements of the project. The *specifications* are the written portions that provide the contractor with various items of information, such as acceptable manufacturers of components, standards to be adhered to, cleanup, and many other facets of the project that are inappropriate on the drawings. On some smaller projects, the specifications can be found on the drawings.

The specifications are usually divided into sections, according to industry practice. Most architects and specifiers follow this practice, and contractors are familiar with it. In 2004, the Construction Specifications Institute issued a revised version of MasterFormat™ which is considered the specifications-writing standard for most commercial building design and construction projects in North America. For instance, Section 09 24 23 is for Portland Cement Stucco, while Section 09 24 33 is for Portland Cement Parging.

---

[69] *But see* Brixen & Christopher v. Elton, 777 P.2d 1039 (Utah Ct. App. 1989) (court interpreted "for approval by Owner" clause in ¶ 2.4.1 (1987 version) as not requiring a formal written approval).

[70] *Id.*

[71] "Elevation" in the context of architectural drawings means a drawing depicting a facade, or vertical face, of a building or interior space. It is a "picture" of a wall or vertical area. An elevation also refers to a height, usually "feet above sea level as shown by U.S.G.S. surveys." Robinson v. Powers, 777 S.W.2d 675 n. 1 (Mo. Ct. App. 1989). This type of elevation is shown as a specific number on a drawing.

[72] These ceiling layouts are normally referred to as *reflected ceiling plans*. One can imagine looking down at a floor plan, with the floor as a mirror. That mirror then reflects what the ceiling would look like to the observer above the plan. Thus, the shape of the floor plan and reflected ceiling plan are usually identical, but one depicts the floor and the other depicts the ceiling.

The drawings and specifications should be complimentary and must be coordinated. These drawings and specifications are the contract documents that tell the contractor how the parts go together. Other contract documents, such as AIA Document A201, give legal and procedural information.

It should also be noted that the plans and specifications can never be totally complete.[73] Not every detail can be illustrated and not every problem can be anticipated. For this reason, shop drawings are provided by contractors and manufacturers and the architect is available during the construction phase to resolve problems. The contractor is also required to use reasonable discretion in executing the work.

Although most architects still prepare their own specifications, particularly on small and medium-sized projects, there are a growing number of specifications specialists. The Construction Specifications Institute (CSI) is a national organization that certifies specifiers. Traditionally, specifiers have listed one or more products for a particular application by trade name and have added an "or equal" provision to permit the contractor to submit other products. This was done to keep costs down and promote competition, as well as to allow for other products of which the specifier was not aware. Courts have interpreted such provisions to mean that the listed products define the intended performance and that contractors could submit articles of equal performance capabilities, even if those other articles do not offer identical performance.[74] In other words, the performance of the substitute product need not be identical, but it must be as good as or better than the specified product. This is usually a matter of judgment.

Specifiers have also followed the practice of *proprietary specifications*, whereby only one product is given, without an "or equal" clause. Proprietary specifications have been upheld as not being violative of antitrust legislation.[75] It is improper, however, for proprietary specifications to be unnamed or disguised, thereby allowing the impression that alternative products could be submitted. In one case, a court permitted a contractor to recover damages from an engineer who had given performance specifications for a sludge dewatering system.[76] The contractor based his bid on a particular manufacturer, but the engineer refused to review test results that showed that the equipment met the specifications, stating that only a competitive manufacturer's product was acceptable. In an Alabama case, an architect specified a material called "Boncoat" (although it did not call it

---

[73] However, in Skidmore, Owings & Merrill v. Intrawest I, 1997 Wash. App. LEXIS 1505 (Sept. 8, 1997), the plans prepared by the architect were deficient and breached the contract. The architect was liable for the extra work costs and loan interest incurred as a result of necessary design changes because the architect knew that the contractor would rely on the drawings in arriving at a guaranteed maximum cost and the owner, in turn, would rely on that cost in determining whether or not to proceed with the project.

[74] Sherwin v. United States, 436 F.2d 992 (Ct. Cl. 1971).

[75] Whitten Corp. v. Paddock, Inc., 424 F.2d 25 (D. Mass. 1970), *cert. denied*, 421 U.S. 1004, 95 S. Ct. 2407 (1975).

[76] Waldinger Corp. v. Ashbrook-Simon-Hartley, Inc., 546 F. Supp. 970 (C.D. Ill. 1983).

by name) for the exterior of a building.[77] The material leaked, but the court held that the contractor was responsible pursuant to a specific guarantee provision in the contract.

An architect must be cautious in using manufacturer's specifications. In a New Mexico case, an architect designed a school.[78] The roof was to consist of a 15-inch monolithic slab of concrete over steel joists, measuring 201 feet by 144 feet, all without expansion or other joints. This was to be topped by a waterproof type of roofing which, in turn, was to be topped by three inches of insulation and three more inches of concrete. Flintcote, a roofing manufacturer, was contacted, and its Monofilm Compound CMR/FM 200 and Monoform Glass Roofing Type 1 was specified. Prior to construction, the design of the roof was changed from the monolithic slab to a structure containing 12 separate slabs. No waterstops, expansion joints, or other waterproofing protection was used, and the original Flintcote specifications were not altered. During the construction, Flintcote's agents were present, but they did not warn the architect of any potential problems. The roof subsequently leaked at the joints between the 12 slabs. Investigation showed that the revision in the slabs resulted in active joints, and that the Flintcote system was capable of withstanding minute movements but could not take the movements that occurred in these joints and had ruptured. The court found that the manufacturer's agents had no duty to warn the architect. The architect had changed the system and should have investigated all the ramifications of the change. If the original scheme had been followed, the roof would not have leaked.

The lesson is clear: when any change is made to the contract documents, the architect must go back and carefully review everything affected. It should then contact all manufacturers whose specifications are being used and request a new review of their product, with a written response.[79]

The 2007 version of this document acknowledges that the contractor will be furnishing shop drawings, samples, and similar items that will be necessary to construct the building, since no set of drawings prepared by the architect can be absolutely complete.

---

[77] United States Fidelity & Guar. Co. v. Jacksonville, 357 So. 2d 952 (Ala. 1978). This case was poorly reasoned in rejecting the logic of two cases cited in the opinion involving defective materials specified for those projects and in which the contractor was found not at fault. The *Jacksonville* case involved a specified material that did not perform. There was no inquiry as to whether the material was inherently defective and thus the damage was either the fault of the architect or manufacturer, or whether it was improperly installed. The case states that there were numerous meetings to discuss the problem, suggesting a defective product. In that case, the contractor should not be at fault for simply following the contractual provisions.

[78] Standhardt v. Flintcote Co., 84 N.M. 796, 508 P.2d 1283 (1973).

[79] In St. Joseph Hosp. v. Corbetta Constr. Co., 21 Ill. App. 3d 925, 316 N.E.2d 51 (1974), the manufacturer of a wall paneling system was liable to the owner (and the architect) for not notifying the architect that the panels did not meet code for flame spread rating. The manufacturer had stated that "our material is not flame rated," but it had been tested by Underwriters Laboratory and found to have a flame spread 17 times the maximum allowable by code. This amounted to fraud and deceit.

3.4.2 The Architect shall incorporate into the Construction Documents the design requirements of governmental authorities having jurisdiction over the Project. **(1.2.3.6; revisions)**

The architect is required to conform the construction documents to the requirements of code and building officials. Normally, the architect will perform an analysis of applicable codes and ordinances before starting with the work. Later, the architect may submit a preliminary set of drawings to the code officials for review. Often, the code officials will have their own interpretation of the applicable laws and ordinances, and may require revisions to the architect's plans. This does not mean that the architect was wrong in its interpretation of the codes, but may merely mean that the code official has a different interpretation. If this is a major issue, the owner may need to have its attorney argue the point to the building official, since many of these issues are matters of interpretation.

3.4.3 During the development of the Construction Documents, the Architect shall assist the Owner in the development and preparation of (1) bidding and procurement information that describes the time, place and conditions of bidding, including bidding or proposal forms; (2) the form of agreement between the Owner and Contractor; and (3) the Conditions of the Contract for Construction (General, Supplementary and other Conditions). The Architect shall also compile a project manual that includes the Conditions of the Contract for Construction and Specifications and may include bidding requirements and sample forms. **(2.4.4.2; revisions)**

Note that the architect is only required to "assist" the owner in preparing and organizing these documents. The architect can use standard form documents such as those prepared by the AIA, but the ultimate responsibility for the documents listed in ¶ 3.4.3 is with the owner. The architect's area of expertise is in designing buildings and not in drafting contracts or other legal documents. Thus, it is the owner and its attorney who must review these various legal documents and approve or revise them. These documents should be the standard AIA documents of the same vintage. Because the AIA changes these documents on a regular basis, the architect or attorney should check that all documents are current. The owner should be made aware that the architect is not providing legal advice; only lawyers can do that. The owner should be encouraged to consult its own attorney to review the contract documents, particularly the various agreements. The second sentence includes documents that the architect has prepared, including the specifications. The drawings are not referenced in this paragraph, as the architect is not assisting the owner in their preparation, and they would not be part of the project manual. Note that a project manual was specifically referenced in the 1997 version of A201, but not in the 2007 version.

3.4.4 The Architect shall update the estimates for the Cost of the Work. **(2.4.3; revisions)**

Here, again, the architect will review any changes that have been made to the project since the last review of the costs and budget, as well as any significant changes in the market, and adjust the project estimate accordingly.

> *3.4.5 The Architect shall submit the Construction Documents to the Owner, advise the Owner of any adjustments to the estimate of the Cost of the Work, take any action required under Section 6.5, and request the Owner's approval.* **(new)**

At the completion of the construction documents phase, the architect will submit these documents, including the drawings and specifications, to the owner for approval. The architect must also advise the owner of adjustments to the estimate. Under Section 6.5, the architect may need to revise the drawings to conform to the last estimate, or the owner may elect to change the budget for the project.

§ 2.10   Bidding or Negotiation Phase Services: ¶ 3.5

3.5 BIDDING OR NEGOTIATION PHASE SERVICES
3.5.1 GENERAL

The Architect shall assist the Owner in establishing a list of prospective contractors. Following the Owner's approval of the Construction Documents, the Architect shall assist the Owner in (1) obtaining either competitive bids or negotiated proposals; (2) confirming responsiveness of bids or proposals; (3) determining the successful bid or proposal, if any; and, (4) awarding and preparing contracts for construction. **(2.5.1, 2.5.2, 2.5.3; substantial revisions)**

Generally, the architect acts as the owner's agent during the bidding phase. Thus, when an owner rejects certain bids on the advice of the architect, the contractor cannot maintain a suit against the architect for interference with the contractual rights of the disappointed contractor.[80] Note that the architect "assists" the owner in obtaining bids or negotiating with contractors and with awarding or

---

[80] Commercial Indus. Constr. v. Anderson, 683 P.2d 378 (Colo. Ct. App. 1984). An architect sent invitations to bid to five contractors including the plaintiff. A statement was included that the owner "reserves the right to reject all bids, to waive informalities, and to accept any bid deemed desirable." At the bid opening, one of the contractors informed the architect of an error in its bid. The plaintiff appeared to be the low bidder. The architect met with the owner and informed him that, after correcting the errors, another contractor appeared to be lowest in price. The owner then instructed the architect to prepare a contract with this other contractor. The plaintiff sued the architect, claiming tortious interference with contractual relations. The court found that the architect was acting as an agent for the owner. Because the owner had the authority stated in the invitation, and the owner cannot interfere with his own contract, no such action can be maintained against an agent acting within his authority.

In Kecko Piping Co. v. Town of Monroe, 172 Conn. 197, 374 A.2d 179 (1977), the architect requested information from a subcontractor for the apparent low bidder. The sub failed to respond. Other information indicated that the sub was overextended and did not have sufficient experience with jobs of that size. The architect advised the owner to reject the sub, which was done. The sub

preparing contracts. These functions are the responsibility of the owner and his attorney.[81] The architect can provide standard form contracts such as those furnished by the AIA, but it cannot provide legal services. Some AIA documents that may be helpful to the owner at this stage, or earlier, are AIA Document A501, Recommended Guide for Competitive Bidding Procedures and Contract Awards for Building Construction, and AIA Document G612, Owner's Instructions Regarding the Construction Contract, Insurance and Bonds, and Bidding Procedures.

The owner should be advised in writing of the architect's recommendations regarding the number of bidders and other matters. Sometimes, an owner may

---

sued the owner and architect for interference with its contract. The court held for the defendants. The architect was under an obligation to advise the owner as to the suitability of contractors and subcontractors and accordingly enjoyed a qualified privilege.

In Professional Bldg. Concepts, Inc. v. City of Cent. Falls Hous. Auth., 783 F. Supp. 1558 (D.R.I. 1992), the low bidder on a project challenged the validity of a contract awarded to the second lowest bidder. The original low bidder did not comply with the bidding requirement of a bid guarantee in the form of a certified check, bank draft, or bid bond. Instead, the bidder submitted a corporate check. The issue for the court was whether the submission of the corporate check instead of the required guarantee was a "material noncompliance in the bidding process." The court found that the purpose of a bid guarantee is to provide assurances, in the form of a firm commitment, that the bidder will, if successful, perform the contract. An uncertified company check is not a firm commitment, because payment could be stopped or there may be insufficient funds in the account.

In Hoon v. Pate Constr. Co., 607 So. 2d 423, 428 (Fla. Dist. Ct. App. 1992), the low bidder sued the owner and architect after it was not awarded the contract. The court found that no contract is formed when a bid is made pursuant to an invitation to bid. The contractor also sued for negligent misrepresentation, alleging that it had been informed that the owner intended to award the contract to the lowest bidder and that it relied on this information. The court held that this statement of intention did not constitute negligent misrepresentation, because the bid forms also stated that the owner reserved the right to reject any or all bids "for whatever reason he may deem necessary for his best interest." Finally, the contractor alleged that the architect was guilty of defamation "by innuendo," because the public would assume that there was something wrong with the contractor because it was not awarded the contract when it was the low bidder. The court also rejected this contention.

In Real Estate Conn., Inc. v. Montagno Constr., Inc., No. CV 94 0139344S, 1995 Conn. Super. LEXIS 2010 (July 7, 1995), the court found that there was no agency relationship between the owner and architect. Instead, the relationship was one of independent contractor. This case should be read cautiously, because the facts indicate that the architect was actually performing design-build work for an owner, who was the lessee of the premises. The architect hired the contractor directly. Thus, there was no direct relationship between the owner and contractor, as would be the case in the usual situation.

In Absher Constr. Co. v. Kent Sch. Dist. No. 415, 77 Wash. App. 137, 143, 890 P.2d 1071, 1074 (1995), the court stated that "where the contract is silent, an architect and its sub-consultants are not a general agent of his or her employer and have no implied authority to make a new contract or alter an existing one for the employer." This is in keeping with the AIA language, which gives the architect limited authority and does not permit the architect to change the owner-contractor contract.

[81] In Mularz v. Greater Park City Co., 623 F.2d 139 (10th Cir. 1980), the court held that the architect was entitled to his fee through the bidding phase because he had no definite duties to perform during that phase. The fact that the bidding process was not completed merely delayed the payment of the fee for a reasonable time. There was no condition precedent to payment.

have only one or two bidders, perhaps because the owner has a personal relationship with a particular contractor. When the bids come in high, the owner may blame the architect. If the architect recommends, in writing, that at least three bids be obtained, it will be easier for the architect to obtain its payment. Also, if the architect recommends a particular contractor, it should have the owner thoroughly review the contractor's qualifications, including financial ability, independently of the architect. This is to avoid any suggestion by the owner that the architect is an agent for the contractor or that the architect obtained payments from the contractor for the recommendation.

It should be noted that the architect is generally in no position to analyze a contractor's financial capabilities. The owner has the obligation to investigate whether the contractor is financially sound.[82] The architect can examine the contractor's past work and make a recommendation to the owner as to the contractor's past workmanship.

A Mississippi case held an architect liable to a contractor for misrepresenting the terms of a contract during the bidding period.[83] This case is instructive concerning the duties of the architect towards bidders.

The architect is usually in a better position to know of contractors who do that type of work. However, the architect cannot guarantee that any contractor will

---

[82] In Travelers Indem. Co. v. Ewing, Cole, Erdman & Eubank, 711 F.2d 14 (3d Cir. 1983), the contractor went into bankruptcy and the owner alleged that the architect had a duty to advise the owner about the possibility of the contractor's bankruptcy. The court stated that no court had ever suggested the existence of such a duty. Although an architect can be held liable for failing to furnish certain information to the client, this duty relates only to matters within the special knowledge of a design professional.

[83] Godfrey, Bassett v. Huntington, 584 So. 2d 1254 (Miss. 1991). The architect had been hired by a school to prepare plans for the renovation of certain areas of the school. When the plans went out for bids, they included a requirement in the specifications that the bidders include a $9,000 contingency in their bids. Any unused portion would be credited to the owner. The architect then issued an addendum to all bidders that deleted the contingency. When the bids were opened, all bids exceeded the budget, and the architect was asked to reduce the scope of the work.

When the new plans were reissued, portions of the old plans and specifications were reused, including the section that required the $9,000 contingency. After one of the bidders noticed this, he called the senior partner at the architectural firm (with whom he was friendly) and asked whether the $9,000 contingency was intentionally included. The architect informed the contractor that an addendum was being mailed out to delete this contingency, just as had been done during the prior bidding. In fact, an addendum was sent out, but it did not delete the $9,000 contingency. The contractor never saw the addendum but was the low bidder and was awarded the job. He signed the contract that incorporated the addendum. After the work was started, the contractor learned that the contingency had not been eliminated and that he was expected by the owner to have the contingency in place. At the end of the project, the contractor credited the owner the $9,000 and sued the architect to recover this amount.

The court found that it appeared that all parties had made an honest mistake. However, the contractor relied on the architect's representation that the contingency was not included. In the absence of this misrepresentation, the contractor would have read the addendum and learned the truth before submitting his bid. The negligence of the contractor in failing to carefully read the contract documents was less than the negligence of the architect in failing to make the planned corrections to the specifications. The contractor's negligence resulted directly from the

perform as required or is financially stable. The owner must do its own evaluation of these matters.

The architect should work with the owner in evaluating the bids and proposals, as well as the qualifications of the various contractors who are being considered for the project.

§ 2.11   Competitive Bidding: ¶ 3.5.2

3.5.2 COMPETITIVE BIDDING

3.5.2.1 *Bidding Documents shall consist of bidding requirements and proposed Contract Documents.* **(2.5.4.1; minor revisions)**

Note that the bidding requirements are not a part of the contract between the owner and contractor (**see** AIA Document A201, ¶ 1.1.1).

3.5.2.2 *The Architect shall assist the Owner in bidding the Project by*

   .1   *procuring the reproduction of Bidding Documents for distribution to prospective bidders;*

This is frequently handled by the architect. The owner pays for the costs of reproduction, messenger services, and so on. As part of the basic services, the architect arranges for distribution of the bidding documents, answers questions, and otherwise assists in the bidding process.

   .2   *distributing the Bidding Documents to prospective bidders, requesting their return upon completion of the bidding process, and maintaining a log of distribution and retrieval and of the amounts of deposits, if any, received from and returned to prospective bidders;*

It is normal to charge contractors a deposit for the bidding documents to assure their return. The owner can then give these returned documents to the successful bidder. The log of distribution is necessary to track the distribution of these documents.

---

misrepresentation by the architect. The court held that the contractor was entitled to the $9,000 from the architect.

Although there are some flaws in the legal reasoning of this case, including the fact that the owner received a windfall in the form of a contract price that was $9,000 less than it should have been, the lesson is to take care in the preparation of contract documents and to be careful what is said to bidders. Be especially careful during a rebid when some documents are reused; it is very easy to make a mistake in that situation.

   .3 organizing and conducting a pre-bid conference for prospective bidders;

Pre-bid conferences are extremely valuable to assure that all bidders have the same understanding of the project. They give each bidder an opportunity to ask questions of the architect and the consultants (who should also attend this meeting). The meeting is usually conducted at the job site.

   .4 preparing responses to questions from prospective bidders and providing clarifications and interpretations of the Bidding Documents to all prospective bidders in the form of addenda; and

During the bidding process, the bidders often find discrepancies in the drawings and specifications. These are then called to the architect's attention (**see** AIA Document A201, ¶ 3.2.3), and the architect must then issue an addendum that clarifies or corrects the problem. If the problem is a major one, the bidding period may be extended to allow for reissue of the documents.

   .5 organizing and conducting the opening of the bids, and subsequently documenting and distributing the bidding results, as directed by the Owner. **(2.5.4; modifications)**

When public bodies are involved, the opening of the bids is a formal process that is usually open to the public. On private projects, this may be very informal, with the architect's receiving the bids by mail or messenger. Under this provision, the owner may direct the architect to notify the bidders of the results of the bids.

3.5.2.3 The Architect shall consider requests for substitutions, if the Bidding Documents permit substitutions, and shall prepare and distribute addenda identifying approved substitutions to all prospective bidders. **(2.5.4.4; minor modifications)**

Often, prospective bidders offer suggestions as to better ways to accomplish the design. This might be in the form of a substitution of materials or alternate designs for elements of the project. The architect should evaluate these suggestions and, if they have merit, issue addenda to all of the bidders for inclusion in their bids.

§ 2.12 Negotiated Proposals: ¶ 3.5.3

3.5.3 NEGOTIATED PROPOSALS

Instead of bidding the project, the owner may elect to negotiate directly with one or more pre-selected general contractors. This section deals with that alternative.

3.5.3.1 Proposal Documents shall consist of proposal requirements and proposed Contract Documents. **(2.5.5.1; no change)**

3.5.3.2 The Architect shall assist the Owner in obtaining proposals by

   .1   procuring the reproduction of Proposal Documents for distribution to prospective contractors, and requesting their return upon completion of the negotiation process;

Here, the architect will assist the owner in obtaining proposals. The owner pays for the costs of reproduction, messenger services, and so on. As part of the basic services, the architect arranges for distribution of the proposed contract documents and proposal requirements, answers questions, and otherwise assists in this process.

   .2   organizing and participating in selection interviews with prospective contractors; and

The architect will organize the selection process and participate with the owner in interviews with the various contractors.

   .3   participating in negotiations with prospective contractors, and subsequently preparing a summary report of the negotiations results, as directed by the Owner. **(2.5.5; modifications)**

The architect will work with the owner in negotiations with the various contractors. Ultimately, it will be the owner's choice as to which contractor to select. At the end of this process, the architect will prepare a report of the negotiations if the owner requests such a report.

3.5.3.3 The Architect shall consider requests for substitutions, if the Proposal Documents permit substitutions, and shall prepare and distribute addenda identifying approved substitutions to all prospective contractors. **(2.5.5.4; minor modification)**

Often, prospective bidders offer suggestions as to better ways to accomplish the design. This might be in the form of a substitution of materials or alternate designs for elements of the project. The architect should evaluate these suggestions and, if they have merit, issue addenda to all of the bidders for inclusion in their bids.

§ 2.13   Construction Phase Services: ¶ 3.6

3.6 CONSTRUCTION PHASE SERVICES

3.6.1 GENERAL

3.6.1.1 The Architect shall provide administration of the Contract between the Owner and the Contractor as set forth below and in AIA Document A201-2007, General Conditions of the Contract for Construction. If the Owner and Contractor modify AIA Document A201-2007, those modifications shall not affect the

Architect's services under this Agreement unless the Owner and the Architect amend this Agreement. **(2.6.1.1; substantial modifications)**

This clause provides that the architect can assume that the General Conditions of A201 will apply to the construction. If the owner decides to use other general conditions, the architect may be entitled to additional compensation, because that may affect a number of the documents the architect prepares. There is a potential for conflict in this provision if the owner modifies the General Conditions or uses nonstandard general conditions. Under this language, the architect would presumably be entitled to extra compensation for services that it is asked to perform under nonstandard general conditions. Usually, the architect participates in modifications to the General Conditions (A201) by preparing supplementary general conditions and specifications. Presumably, this participation would constitute a written approval by the architect under this provision in lieu of a written amendment to this agreement.

Of course, it is always a good idea for the architect to immediately notify the owner of any discrepancy between the owner-architect agreement and any other nonstandard agreement.

3.6.1.2 The Architect shall advise and consult with the Owner during the Construction Phase Services. The Architect shall have authority to act on behalf of the Owner only to the extent provided in this Agreement. The Architect shall not have control over, charge of, or responsibility for, the construction means, methods, techniques, sequences or procedures, or for safety precautions and programs in connection with the Work, nor shall the Architect be responsible for the Contractor's failure to perform the Work in accordance with the requirements of the Contract Documents. The Architect shall be responsible for the Architect's negligent acts or omissions, but shall not have control over or charge of and shall not be responsible for acts or omissions of the Contractor or of any other persons or entities performing portions of the Work. **(2.6.1.3, 2.6.2.1; substantial modifications)**

The responsibility under this provision ends only when final payment to the contractor is due. Under some circumstances, however, the architect may be entitled to additional payments if the services extend beyond 60 days after substantial completion.

The architect is the owner's agent[84] during the construction phase, although this agency is a very limited one. This provision was inserted as a result of a number of

---

[84] An "agency relationship" is a fiduciary relationship that results from the consent by one person, the principal, to another, the agent, for the agent to act on the principal's behalf and subject to the principal's control. Tri-City Constr. Co. v. A.C. Kirkwood & Assocs., 738 S.W.2d 925 (Mo. Ct. App. 1987).

See the dissent in R.G. Nelson, A.I.A. v. M.L. Steer, 797 P.2d 117 (Idaho 1990), for a discussion of fiduciary relationship in the context of an owner-architect relationship.

In Tunica-Biloxi Indians of La. v. Pecot, 2006 WL 1228902 (W.D. La. May 3, 2006), the court stated that an architect is an agent of the owner when acting as the owner's supervisor of a project, but is an independent contractor when preparing plans. The issue was whether the defendant had a

court cases that imposed liability against architects in favor of other parties, such as contractors.[85] In a Texas case,[86] the court held against the contractor on the basis that the architect was the agent of the owner.

One court held that, as the owner's agent, the architect can direct the contractor to make changes in the schedule and scope of the work, authorize additional costs, waive bidding requirements, and waive the requirements of written change orders.[87] This is, of course, not recommended and is contrary to other provisions of the agreement.

Other courts have held that even under prior versions of this agreement that gave the architect general supervisory authority, the architect was not the owner's agent

---

duty to warn the owner of a potential hazard related to the product in question. The court found that the architect was aware of the risk of mold growth associated with the product. Therefore, by operation of law, the owner was also charged with this knowledge. Therefore, there was no duty to warn the owner.

[85] Mayor & City Council v. Clark-Dietz, 550 F. Supp. 610 (N.D. Miss. 1982); Owen v. Dodd, 431 F. Supp. 1239 (N.D. Miss. 1977). Shoffner Indus. v. W.B. Lloyd Constr. Co., 42 N.C. 259, 257 S.E.2d 50 (Ct. App. 1957), involved economic damages to the contractor as a result of the collapse of roof trusses. Under the doctrine of Moorman Mfg. Co. v. National Tank Co., 91 Ill. 2d 69, 435 N.E.2d 443 (1982), such damages are not recoverable by the contractor against the architect unless there was a contractual relationship. *See* John Martin Co. v. Morse/Diesel, Inc., 819 S.W.2d 428 (Tenn. 1991) (survey of recent decisions regarding application of the *Moorman* or economic loss doctrine).

In Tri-City Constr. Co. v. A.C. Kirkwood & Assocs., 738 S.W.2d 925 (Mo. Ct. App. 1987), the engineer was the agent of the owner. A release agreement between the contractor and owner applied to the engineer and barred the contractor's action against the engineer.

[86] Bernard Johnson, Inc. v. Continental Constructors, Inc., 630 S.W.2d 365 (Tex. App. 1982). In another Texas case, Alamo Community Coll. Dist. v. Browning Constr. Co., 131 S.W.3d 146 (Tex. App. 2004), the court held that the architect was the owner's agent, based on various authorized duties of the architect identified in the contracts.

[87] Love v. Double "AA" Constructors, Inc., 117 Ariz. 41,570 P.2d 812 (Ct. App. 1977). This was a unique case in which the owner was unavailable during construction. The architect authorized changes based on an extremely tight time schedule.

In Absher Constr. Co. v. Kent Sch. Dist. No. 415, 77 Wash. App. 137, 890 P.2d 1071 (1995), the phrase stating the architect would be the owner's representative was deleted. The contract also stated that notice to the architect would not be deemed notice to the owner. Thus, when the contractor discussed problems with the engineer and received guidance, there was no waiver of the notice requirements and the contractor was not entitled to "extras" because of his failure to timely notify the owner.

The case of Pete Wing Contracting, Inc. v. Port Conneaut Investors Ltd. P'ship, 1995 Ohio App. LEXIS 4341 (Sept. 29, 1995), illustrates what can happen when the owner is too cheap to hire the architect for the normal construction phase services. The architect issued a certificate of substantial completion to the contractor, but the owner refused to accept it or release the retainage. The problem was that the original bid plans called for the ceiling height in the loft area to be seven feet eight inches. When the construction plans were issued, the height was apparently given as 78" instead of 7'8". The architect testified that it was understandable that the contractor interpreted the drawings in that manner, especially because the owner reduced the architect's scope of services during construction, leaving the contractor to deal directly with the owner instead of the architect. Thus, the ceiling height problem was not the contractor's fault, and the contractor attained substantial completion of the project.

for all purposes.[88] Thus, the architect had only the authority given to it in the owner-architect and owner-contractor contracts, which did not include authority to make alterations in the plans and specifications.[89]

In a case involving an architect who wrote letters to the contractor's surety stating that the contractor was negligent in performing the work, the architect was not liable for defamation.[90] As the owner's agent, he was privileged to make such statements if they were not made in bad faith and were within his duty to the owner. In a Connecticut case,[91] the architect was sued by the general contractor pursuant to an assignment by the owner. The contractor claimed that the architect tortiously interfered with the owner-contractor agreement by preparing a certificate (per ¶ 14.2.2 of A201-1997) that the contractor was in default, even though substantial completion had been achieved, as evidenced by the fact that the owner had moved in. The court held that this was an issue for the trial court to determine, with a finding of bad faith, malice, wantonness, or acting beyond the scope of authority—a requirement before liability is imposed on the architect. The court dismissed a negligence count on the basis that the architect cannot owe a duty to the contractor without breaching its agency agreement with the owner. *See also* ¶ 2.6.2.2, which states that no duty exists by the architect toward the contractor.

The sentence concerning "construction means, methods" is one of the most important in the agreement for the protection of the architect. The contractor is solely responsible for construction means and methods and for safety.

In a Wisconsin case, a roof collapsed during construction because the contractor did not follow proper procedures for installation of bridging.[92] The contractor

---

[88] *E.g.*, Kirk Reid Co. v. Fine, 205 Va. 778, 139 S.E.2d 829 (1965).

In Blue Cross & Blue Shield v. W.R. Grace & Co., 781 F. Supp. 420 (D.S.C. 1991), the court held that an architect was both an agent of the owner and an independent contractor, depending on the function being performed. As to matters concerning preparation of the plans and specifications, an architect is an independent contractor. This is because the owner has control over the final product that the architect designs but not the means by which the design work is performed. As to matters regarding the architect's "supervision" over the construction of the building, the architect is the owner's agent. This would apply generally to what A201 refers to as "administration of the contract." In this case, the architect's knowledge of the selection of building materials could not be imputed to the owner, because the architect was not an agent of the owner for that item.

In Caribbean Lumber Co. v. Anderson, 422 S.E.2d 267, 269 (Ga. Ct. App. 1992), the court stated that "it is axiomatic that the owners are only bound by the acts of the engineer which are within the scope of his actual or apparent authority" (interpreting non-AIA contract).

[89] Norair Eng'g v. St. Joseph's Hosp., 147 Ga. App. 595, 249 S.E.2d 642 (1978) (*citing* Mallard, Stacy & Co. v. Moody & Brewster, 105 Ga. 104, 31 S.E. 45 (1898); Cannon v. Hunt, 113 Ga. 501, 38 S.E. 983 (1901)); Coin v. Board of Educ., 298 Ky. 645, 183 S.W.2d 819 (1944) (contractor not entitled to additional money for work performed after a flood, although architect gave contractor assurances that he would be paid; court also found that this was not an emergency); Kirk Reid Co. v. Fine, 205 Va. 778, 139 S.E.2d 829 (1965).

[90] Alfred A. Altimont v. Chatelain, Samperton, 374 A.2d 284 (D.C. 1977).

[91] United Stone Am., Inc. v. Frazier, Lamson, Budlong, P.C., 2002 WL 241370 (Conn. Super., Jan. 30, 2002).

[92] Vonasek v. Hirsch & Stevens, Inc., 65 Wis. 2d 1, 221 N.W.2d 815 (1974) (construing AIA Document B131 (1963)).

argued that the architect had a duty to warn the contractor of the possible construction hazards of the contractor's method of erection. The court held that under the AIA contract, the contractor was solely responsible for construction methods and, unless the construction involved a unique or novel design, the architect had no duty to warn the contractor of such hazards.

In cases involving injuries to workers, this language has been held to exculpate the architect from injuries to such third parties.[93] An Illinois case[94] involved a former mental patient who visited the construction site and made his way to the sixth floor, from which he either fell or jumped to his death. His wife sued the hospital, the architect, and several contractors under the Illinois Structural Work Act.[95] The court found that the architect had no duty to this person, because the architect's duties were defined by the contract. In cases involving damage to property, this clause has relieved the architect from liability for a contractor's negligence.[96] It has also

---

[93] Jaroszewicz v. Facilities Dev. Corp., 115 A.D.2d 159, 495 N.Y.S.2d 498 (1985). Plaintiff filed suit against architect under common law negligence theory for death of maintenance worker after completion of construction. Although the architect had contracted to provide "supervision and inspection" of the project and had employed a clerk of the works, there was no liability to third parties absent proof of active malfeasance that contributed to the accident. *See* Wheeler & Lewis v. Slifer, 195 Colo. 291, 577 P.2d 1092 (1978) (no duty on part of architect toward injured worker, even though architect had right to stop the work); Yow v. Hussey, Gay, Bell & Deyoung Int'l, Inc., 412 S.E.2d 565 (Ga. Ct. App. 1991); Diomar v. Landmark Assocs., 81 Ill. App. 3d 1135, 401 N.E.2d 1287 (1980); Fruzyna v. Walter C. Carlson Assoc., 78 Ill. App. 3d 1050, 398 N.E.2d 60 (1979); Burns v. Black & Veach Architects, Inc., 854 S.W.2d 450 (Mo. 1993); Case v. Midwest Mech. Contractors, 876 S.W.2d 51 (Mo. Ct. App. 1994) (architect not liable to injured worker even though architect had duty to review safety program of contractors); Dillard v. Shaughnessy, Fickel & Scott Architects, 864 S.W.2d 368 (Mo. Ct. App. 1993); Fox v. Jenny Eng'g Corp., 122 A.D.2d 532, 505 N.Y.S.2d 270 (1986); Kerr v. Rochester Gas & Elec. Corp., 113 A.D.2d 412, 496 N.Y.S.2d 880 (1985) (engineer hired to supervise work not liable to injured worker and not agent of the owner); Romero v. Parkhill, Smith & Cooper, Inc., 881 S.W.2d 522 (Tex. App. 1994); Garcia v. Federics, 2007 WL 2367672 (Mass. Super., July 9, 2007).

[94] Kelly v. Northwest Community Hospital, 66 Ill. App. 3d 679, 384 N.E.2d 102 (1978).

[95] Ill. Rev. Stat. ch. 48, para. 60 (repealed in Feb. 1995).

[96] Jewish Bd. of Guardians v. Grumman Allied Indus., 96 A.D.2d 465, 464 N.Y.S.2d 778 (1983) (general contractor had duty to protect prefab units that were damaged by a severe rainstorm on-site during construction; architect had no duty to supervise and no responsibility for this damage). *See also* Hernandez v. Racanelli Constr. Co., Inc., 33 A.D.3d 536, 823 N.Y.S.2d 377 (N.Y.A.D. 1 Dep't 2006), where the architect was completely relieved of responsibility for construction defects and safety precautions under its contract with the owner, thus avoiding responsibility to the demolition contractor where a plywood fence collapsed during demolition, injuring passers-by. *See also* Moundsview Indep. Sch. Dist. No. 621 v. Buetow & Assocs., 253 N.W.2d 836, 839 (Minn. 1977). An architect's drawings and specifications required that wooden plates were to be fastened to the concrete walls of the building with washers and nuts attached to 1/2-inch studs. During a severe windstorm, a portion of the roof blew off and it was discovered that the washers and nuts were missing. The architect had made some 90 visits to the construction site but had not observed the omission of the washers and nuts. The court stated:

> [I]t is apparent that by the plain language of the contract an architect is exculpated from any liability occasioned by the acts or omissions of a contractor. The language of the contract is

been cited in finding that the contractor was not a third-party beneficiary of the owner-architect agreement.[97]

The architect has no liability to a contractor's surety when the contractor fails to perform.[98] The architect has no duty to third parties when the contractor deviates from the drawings and the architect has no contractual duty to supervise the construction.[99]

---

unambiguous. The failure of a contractor to follow the plans and specifications caused the mishap. By virtue of the aforementioned contractual provisions, Buetow is absolved from any liability, as a matter of law, for a contractor's failure to fasten the roof to the building with washers and nuts.

Moundsview was not followed in the case of Hunt v. Ellisor & Tanner, Inc., 739 S.W.2d 933 (Tex. App. 1987). There, the court found the exculpatory language in the AIA contract to mean only that the architect was not the insurer or guarantor of the general contractor's obligation to carry out the work in accordance with the contract documents. The architect breached his duty to observe the work and to report the contractor's deficiencies to the owner. The architect was held 5% liable for the owner's damages. *Accord* Diocese of Rochester v. R-Monde Contractors, Inc., 148 Misc. 2d 926, 562 N.Y.S.2d 593 (Sup. Ct. 1989).

The court in Gables CVF, Inc. v. Bahr, Vermeer & Haecker Architect, Ltd., 244 Neb. 346, 506 N.W.2d 706 (1993), followed *Hunt* in determining that an architect was not entitled to summary judgment. One of the partners of the owner was also the general contractor. All of the provisions in the construction phase of the 1977 version of B 141 except this one were stricken. The court found an ambiguity existed as to what the architect's duties were in observing the construction. The architect was to be paid $125 per trip for such observations, and the owner alleged that the architect failed to notify the owner of defects in the work. The court stated that the language of this provision provides that the architect is not the insurer or guarantor of work completed by the contractor. However, such language does not absolve the architect from liability for a breach of the architect's duty, if one exists, to inform the owner of deviations from the building plans when the architect has agreed to make periodic observations.

Moundsview was followed in Shepherd Components, Inc. v. Brice Petrides-Donohue & Assocs., 473 N.W.2d 612 (Iowa 1991) ("[A] general rule applicable to this case is that an engineer does not, by reason of its duty to inspect the construction site, assume responsibility for either day-to-day construction methods utilized by the contractor or the contractor's negligence.").

In Weill Constr. Co. v. Thibodeaux, 491 So. 2d 166 (La. 1986), a skating rink was damaged by water that entered the foundation after construction. The architect was found to have a limited duty that did not include construction supervision, based on a contractual provision modeled after the standard AIA clause.

In Board of Educ. v. Sargent, Webster, 146 A.D.2d 190, 539 N.Y.S.2d 814 (1989), the architect was liable despite the exculpatory clause because he knew of the defective work.

[97] Sheetz, Aiken & Aiken v. Spann, Inc., 512 So. 2d 99 (Ala. 1987); Linde Enters., Inc. v. Hazelton City Auth., 602 A.2d 897 (Pa. Super. Ct. 1992).

[98] J & J Elec., Inc. v. Gilbert H. Moen Co., 9 Wash. App. 954, 516 P.2d 217 (1974). *But see* Westerhold v. Carroll, 419 S.W.2d 73 (Mo. 1967) (architect owes a duty to surety to correctly certify applications for payment). Following *Westerhold*, the court in Aetna Ins. Co. v. Hellmuth, Obata & Kassabaum, 392 F.2d 472 (8th Cir. 1968), allowed a surety to maintain an action against an architect for negligent supervision when the architect is obligated by the contract to supervise the construction. The AIA contracts do not provide for such supervision.

[99] Goette v. Press Bar & Cafe, Inc., 413 N.W.2d 854 (Minn. Ct. App. 1987).

In Rian v. Imperial Mun. Serv. Group, Inc., 768 P.2d 1260 (Colo. Ct. App. 1988), the court held that the architect had no supervisory duties (apparently the contract was not an AIA contract). An

Other courts have stated that the right to stop the work under a general supervision or inspection clause is not the same as a duty when such a duty is not contemplated by the contract or the parties.[100]

Even though the architect is not responsible for the contractor's failure to perform the work according to the Contract Documents, if the architect fails to guard the owner against defects and deficiencies, it may be liable to the owner for damages,[101] particularly if the architect knows that the construction is not

---

injured worker could not rely on an architect's supervision to protect him. The court, however, did not permit summary judgment on the issue of whether the architect breached a duty by failing to specify the federal safety regulations and construction practices that were to be complied with by the contractors. Note that, in this case, the architect was permitted to introduce an affidavit of a licensed architect in the motion for summary judgment to the effect that architects do not customarily supervise construction.

In Atherton Condominium Apartment-Owners Ass'n v. Blume Dev. Co., 115 Wash. 2d 506, 799 P.2d 250 (1990), the architect's involvement ended when building permits were obtained. When the contractor deviated from the drawings, the architect had no liability for the defects. In Baumeister v. Automated Prods., Inc., 673 N.W.2d 410 (Wis. App. 2003), aff'd, 690 N.W.2d 1 (Wis. 2004), an injured worker claimed that the architect negligently failed to "design" or "approve" safe temporary truss bracing for use during construction. The worker also argued that the architect's superior knowledge regarding the temporary bracing needed for safe installation created a duty to provide temporary truss bracing instructions. The court held in favor of the architect, finding no duty.

[100] Reber v. Chandler High Sch. Dist. #202, 13 Ariz. App. 133, 474 P.2d 852 (1970).

In Wheeler & Lewis v. Slifer, 195 Colo. 291, 577 P.2d 1092 (1978), an architect designed a high school pursuant to a contract. A worker was injured when a section of the roof collapsed because of insufficient bracing and shoring of the roof. The worker sued the architect, contending that a contractual provision permitting the architect to stop the work imposed a duty on the architect "to see that reasonable precautions were taken in protecting the workmen on the site from unsafe conditions." The appellate court held for the worker and the Colorado Supreme Court reversed, finding that there was no duty to stop the work for the safety of the worker. The contract contemplated that the contractor would be in charge of safety. The power to stop the work was to insure that the work conformed to the contract documents, not to insure safety. The court specifically disagreed with Miller v. DeWitt, 37 Ill. 2d 273, 226 N.E.2d 630 (1967). See Walker v. Wittenberg, Delony & Davidson, Inc., 241 Ark. 525, 412 S.W.2d 621 (1966).

In Emberton v. State Farm Mut. Auto. Ins. Co., 44 Ill. App. 3d 839, 358 N.E.2d 1254 (1976), rev'd, 71 Ill. 2d 111, 373 N.E.2d 1348 (1978), further proceedings, 85 Ill. App. 3d 247, 406 N.E.2d 219 (1980), the court found the architect in charge of the work for Structural Work Act purposes. This case involved an earlier version of the AIA documents in which he had authority to stop the work.

[101] Campbell County Bd. of Educ. v. Brownlee-Kesterson, 677 S.W.2d 457 (Tenn. Ct. App. 1984).

In Roland A. Wilson v. Forty-O-Four Grand Corp., 246 N.W.2d 922, 924-25 (Iowa 1976), the architect discovered before final completion that the windows of a building leaked. It was agreed that the contractor was to caulk all the windows. The architect then made a final inspection but did not test the windows, and it notified the owner that the construction work passed the final inspection. The owner paid the contractor and later discovered that the windows still leaked. The owner refused to make a final payment to the architect, and the architect sued to collect his fees. The court found that the architect should not have approved final payment to the contractor without a water test on the windows to see if the leaking problem had been corrected by caulking. The court stated: "This contract imposed two principal duties on plaintiff. One was to prepare plans and specifications. The other was to see that the plans and specifications were carried out by the contractor to achieve the result planned for."

In Westerhold v. Carroll, 419 S.W.2d 73 (Mo. 1967), the architect was contractually obligated to inspect and supervise performance of the work. The architect improperly certified work, resulting in overpayment to the contractor. The contractor defaulted, and the surety was required to complete the project with insufficient funds remaining. The architect argued that his negligence was not the proximate cause of the loss, but the default of the contractor was an intervening cause. The court stated that the purpose of the inspection and certification was to protect the owner and surety against payments to the contractor for work not performed and materials not delivered in case of default.

In Lee County v. Southern Water Contractors, Inc., 298 So. 2d 518 (Fla. Dist. Ct. App. 1974), the court found that an engineer had improperly supervised and inspected the laying of a pipeline. The pipeline was later damaged because it was covered by only two feet of backfill instead of six. The owner claimed against the supervising engineer for the resulting damages. Although the engineer is not a guarantor of the contractor's work, if he fails to use reasonable care to prevent material deviations from the plans and specifications and to prevent substandard workmanship, he is liable to the owner for the defects that could have been eliminated if he had properly performed his work.

In Shepard v. City of Palatka, 414 So. 2d 1077, 1078-79 (Fla. Dist. Ct. App. 1981), the contractor used the wrong type of wallboard and the veneer plaster soon started to peel. The appellate court held for the architect on the basis of the AIA provision. It was clear that the architect had made periodic inspections of the site:

> In a proper factual situation, where it is demonstrated that the architect ignored his contractual duty to make periodic visits to the site, liability could possibly lie regardless of such exonerating language. Such was not the case here. . . . While he admitted that he had not discovered the misuse of the wallboard, the contract clearly protected him because it imposed no duty upon him to discover the omission of the contractor and clearly absolved him of liability if there were, in fact, such omissions.

See also Skidmore, Owings & Merrill v. Connecticut Gen. Life Ins. Co., 25 Conn. Supp. 76, 197 A.2d 83 (1963); Willner v. Woodward, 107 Ga. App. 876, 132 S.E.2d 132 (1959); First Nat'l Bank v. Cann, 503 F. Supp. 419 (N.D. Ohio 1980).

In Gables CVF, Inc. v. Bahr, Vermeer & Haecker Architect, Ltd., 244 Neb. 346, 506 N.W.2d 706, 709 (1993), the architect agreed to provide a set of "builders plans," an abbreviated set of drawings and specifications, that cost the owner less than a full set of documents. The architect also agreed to be paid $125 per trip for "observation of construction" of the project. At substantial completion, the architect certified that the project was "generally in accordance with the intent of the drawings and specifications." However, there were deviations, including different materials used in the decks and balconies of the building, lack of a required waterproof membrane, and studs spaced at 24 rather than 16 inches on center. The court found that the architect was not entitled to summary judgment and might be liable to the owner for failing to detect these problems, despite the architect's argument that it should not be liable for damages resulting from unreported deviations from the plans.

In this case, one of the owners was also the general contractor. The architect argued that, because the general contractor has notice of any defects and deviations from the plans and specifications, and the other partners of the general contractor had constructive knowledge of these defects, the owner should not now complain that it was damaged by the architect's failure to observe the defects. The court rejected this argument, saying that the general contractor does not necessarily have notice of defects created by subcontractors.

However, in Great Northern Ins. Co. v. RLJ Plumbing & Heating, Inc., 2006 WL 1526066 (D.Conn., May 25, 2006), the court held for the architect. The contract between the owner and architect was a B155 with similar provisions. The drawings for the new home established the

according to the contract documents and fails to advise the owner of the deficiency.[102]

---

location of the bathroom and plumbing fixtures in the second floor bathroom but did not detail how the piping to the fixtures would be run or make any provisions for freeze protection. The architect never inspected this portion of the work. The pipes later froze, causing extensive damage. According to the court, "because it was not [the architect's] responsibility under the contract to inspect the 'construction means, methods, techniques, sequences and procedures' used in the plumbing of the second floor bathroom, it was not in a position to prevent the harm that allegedly occurred in the instant case." In reviewing all of the contracts, the court added that "in addition, this responsibility was accounted for within the contract scheme, and it fell on the general contractor . . ."

[102] In Brown v. McBro Planning & Dev. Co., 660 F. Supp. 1333 (D.V.I. 1987), the court held that the owner's acceptance of the project did not shift liability when a hospital technician injured his knee in a fall on the hospital floor. The owner's acceptance of full responsibility and control of the hospital was not a superseding cause of the injury. The architect was liable because he was aware of the defect early on and failed to order the contractor to correct the defect and continued to certify payments.

In Dan Cowling & Assocs. v. Board of Educ., 273 Ark. 214, 618 S.W.2d 158 (1981), the court held that the architect breached his duty to the owner by approving a wall in an obviously defective condition.

In City of Houma v. Municipal & Indus. Pipe Serv., Inc., 884 F.2d 886 (5th Cir. 1989), the engineer was found liable to the owner for failing to detect incomplete work. The engineer had full-time personnel on the project, although they were poorly trained and "employed no reasonable, analytical method calculated to spot performance deficiencies." The engineer was not, however, liable for the contractor's criminal fraud.

In Board of Educ. v. Sargent, Webster, 146 A.D.2d 190, 539 N.Y.S.2d 814, 818 (1989), the architect was liable because he knew that the roofing system was not installed per the specifications. The court stated:

When, as a result of periodic inspection, an architect discovers defects in the progress of the work which the owner, if notified, could have taken steps to ameliorate, the imposition of liability upon the architect for failure to notify would be based on a breach of his own contractual duty, and not as a guarantor of the contractor's performance.

In U.R.S. Co. v. Gulfport-Biloxi Airport Auth., 544 So. 2d 824 (Miss. 1989), the architect was liable when it failed to notify the owner that the roof was defective. The architect had employed a full-time field representative, but that representative was not competent as an inspector of the roof work. Perhaps most damaging was the fact that the improper roofing work was pointed out to the field representative and to the architect's office on a number of occasions. The architect certified the work anyway. The architect was also found liable to the surety because of the improper certifications for payment.

The court in Watson, Watson, Rutland v. Board of Educ., 559 So. 2d 168, 174 (Ala. 1990), stated:

We conclude that, although the Architect had a duty under the contract to inspect, exhaustive, continuous on-site inspections were not required. We also hold, however that an architect has a legal duty, under such an agreement, to notify the owner of a known defect. Furthermore, an architect cannot close his eyes on the construction site and refuse to engage in any inspection procedure whatsoever and then disclaim liability for construction defects that even the most perfunctory monitoring could have prevented.

In City of Charlotte v. Skidmore, Owings & Merrill., 407 S.E.2d 571 (N.C. 1991), the architect was liable to the owner despite the contractor's deviation from the plans and specifications. This case is of some concern to architects, as most people in the construction industry assume

3.6.1.3 *Subject to Section 4.3, the Architect's responsibility to provide Construction Phase services commences with the award of the Contract for Construction and terminates on the date the Architect issues the final Certificate for Payment.* **(2.6.1.2; modifications; deleted reference to entitlement to additional fees after 60 days after Substantial Completion)**

This provision clarifies the end of the architect's basic services, which may have consequences beyond payment.[103] The date referred to is the date when the architect issues the final certificate for payment, and not necessarily the date the owner makes that payment. The 1997 version of B141 had a provision that if the work is not completed 60 days after the date of substantial completion, the architect would be entitled to additional payments. Although it may appear from this paragraph that this has been eliminated, it has not. Note that this paragraph provides that the architect will continue services until completion of the project, but ¶ 4.2.2.6 still states that the architect will be paid for such services as Additional Services after 60 days after substantial completion.

§ 2.14   Evaluations of the Work: ¶ 3.6.2

3.6.2 EVALUATIONS OF THE WORK

3.6.2.1 *The Architect shall visit the site at intervals appropriate to the stage of construction or as otherwise required in Section 4.3.3, to become generally familiar with the progress and quality of the portion of the Work completed, and to determine in general if the Work observed is being performed in a manner indicating that the*

---

automatically that any deviation from the plans and specs insulates the architect from liability. This is not necessarily the case. Here, the plans and specifications were defective and the contractor deviated from the plans. The court found the architect liable for most of the damages.

The fact that the contractor deviated did not shift the blame away from the architect because the contractor was able to demonstrate that, even if the plans and specifications had been followed exactly, the damage would still have resulted. A contractor can avoid liability if it follows the plans and specifications exactly. However, a contractor's failure to follow them does not protect the architect if they are improper.

Under this provision, in Welch v. Grant Dev. Co., 120 Misc. 2d 493, 466 N.Y.S.2d 112 (1983), although the architect did not owe a duty to the worker, he had a duty to report to the owner any work not being done in a workmanlike manner, including unprotected openings in floors, if the architect had knowledge of such problems.

[103] In Butler v. Mitchell-Hugeback, Inc., 895 S.W.2d 15 (Mo. 1995), the court held that the waiver of subrogation (currently found in ¶ 8.1.2) is effective so long as insurance was required to be maintained.

Even though the contractor sent the owner an invoice stating that the work was 100% complete, there was no evidence that a final certificate of payment had been issued. Further, under ¶ 2.6.1 of B141 (1987 version), the architect's services are not concluded until final payment to the contractor or 60 days after substantial completion, whichever occurs first. The building collapsed suddenly, shortly after completion of construction but within 60 days of substantial completion. The court held that insurance was required to be maintained during this time, so that the waiver of subrogation was effective to protect the architect.

Work, when fully completed, will be in accordance with the Contract Documents. However, the Architect shall not be required to make exhaustive or continuous on-site inspections to check the quality or quantity of the Work. On the basis of the site visits, the Architect shall keep the Owner reasonably informed about the progress and quality of the portion of the Work completed, and report to the Owner (1) known deviations from the Contract Documents and from the most recent construction schedule submitted by the Contractor, and (2) defects and deficiencies observed in the Work. **(2.6.2.1, 2.6.2.2; substantial modifications)**

The architect has a duty to the owner during the construction phase to report any serious problem with the design before construction is completed if the architect reasonably should know of the problem.[104] The first part of this paragraph has been read as requiring only general supervision as opposed to the more detailed "clerk-of-the-works" inspection services.[105] This is a factor in determining whether the architect "is in charge of the work" for purposes of determining liability to injured workers under statutes such as the Illinois Structural Work Act.[106] The contract

---

[104] Comptroller v. King, 217 Va. 751, 232 S.E.2d 895 (1977). The architect should have discovered that the plans were faulty in that they did not provide for proper drainage in a project with limestone and flagstone exterior areas. After completion, these areas deteriorated because of standing water. Although the statute of limitations had run for improperly preparing the drawings, it had not run for negligence in observation of the construction. The owner can reasonably expect the architect to disclose defects observed during construction, or that reasonably should be observed, whether from errors in design or deviations from plans and specifications.

In Welch v. Grant Dev. Co., 120 Misc. 2d 493, 466 N.Y.S.2d 112 (1983), although the architect did not owe a duty to the worker, he had a duty to report to the owner any work not being done in a workmanlike manner, including unprotected openings in floors, if the architect had knowledge of such problems.

[105] Moundsview Indep. Sch. Dist. No. 621 v. Buetow & Assocs., 253 N.W.2d 836 (Minn. 1977). In Central Sch. Dist. No. 2 v. Flintkote Co., 56 A.D.2d 642, 391 N.Y.S.2d 887, 888 (1977), the owner employed a clerk of the works. Under this provision, the architect could not rely solely on the assumption that the clerk of the works was properly inspecting the work of the roofing subcontractor:

If an architect can escape the obligation and supervisory duties he contracted to perform merely by accepting, at face value and without verification, the approval of the "Clerk of the Works" as to the progress of the work, the owner would be deprived of the professional judgment which he had the right to expect. The owner's retainer of a "Clerk of the Works" for full-time, on-site services, constituted a protection that is an addition to and not a substitute for the contractual and professional obligations of the architect.

[106] Ill. Rev. Stat. ch. 48, para. 60-69 (repealed in 1995). Under the Illinois Act, one element required to prove liability was that the defendant have "charge" of the work. This is not limited to direct supervision, Larson v. Commonwealth Edison, 33 Ill.2d 316, 211 N.E.2d 247 (1965), but also includes any contractual right or obligation to supervise the work, stop the work, or otherwise control the work. More than one party can be "in charge of the work in Illinois. Fruzyna v. Walter C. Carlson Assoc., 78 Ill. App. 3d 1050, 398 N.E.2d 60 (1979); Diomar v. Landmark Assocs., 81 Ill. App. 3d 1135, 401 N.E.2d 1287 (1980).

In Jaroszewicz v. Facilities Dev. Corp., 115 A.D.2d 159, 495 N.Y.S.2d 498 (1985), the architect was responsible to provide "supervision and inspection" and employed a clerk of the works.

must be examined in answering the question of whether an architect has breached a duty to supervise.[107]

Note that under ¶ 12.3.1 of AIA Document A201, only the owner can accept nonconforming work. The architect may want to delete the language "and shall endeavor to guard the owner against defects and deficiencies in the Work." This

---

There was no liability by the architect when a worker was electrocuted because there was no active malfeasance or right to control the manner in which the construction was performed.

In Hausam v. Victor Gruen & Assocs., 86 Ill. App. 3d 1145, 408 N.E.2d 1051 (1980), a non-AIA contract was used but the language was similar. The architect was not in charge of the work. *See also* Pugh v. Butler Tel. Co., 512 So. 2d 1317 (Ala. 1987) (engineer had no duty to injured worker when contract was for benefit of owner and was to ensure compliance with the plans and specifications, and had nothing to do with safety of workers); Young v. Eastern Eng'g & Elevator, 381 Pa. Super. 428, 554 A.2d 77, 81 (1989) (architect not liable when worker fell through hole in drywall surrounding elevator shaft: "an architect is not under a duty to notify workers or employees of the contractor or subcontractors of hazardous conditions on the construction site."). Unless an architect has undertaken by conduct or contract to supervise a construction project, he is under no duty to notify or warn workers or employees of the contractor or subcontractor of hazardous conditions on the construction site. Jones v. James Reeves Contractors, Inc., 701 So. 2d 774 (Miss. 1997), *following* Young. Hobson v. Waggoner Eng'g, Inc., 2003 WL 21789396 (Miss. App., Aug. 5, 2003).

[107] Moundsview Indep. Sch. Dist. No. 621 v. Buetow & Assocs., 253 N.W.2d 836 (Minn. 1977). In Slifer v. Wheeler & Lewis, 39 Colo. App. 269, 567 P.2d 388, 391 (1977), the architect was liable for the injuries of workers because his contract required that he provide supervision:

> The Architects shall supervise the construction of the work in such manner as to assure the District performance of all contracts in accordance with the terms thereof; and the Architect shall exercise due diligence so that the construction shall be strictly in accordance with the final approved plans and specifications or any authorized changes thereto, of good workmanship and of materials of the kinds specified in each instance. The Architects shall personally devote whatever time is necessary adequately to supervise, and employ at their own expense, whatever help may be necessary to fully supervise the construction of the work to the entire satisfaction of the District.

In Great Northern Ins. Co. v. RLJ Plumbing & Heating, Inc., 2006 WL 1526066 (D. Conn. May 25, 2006), frozen water pipes burst in a newly constructed house. In finding in favor of the architect, the court reviewed these contract provisions and stated that the architect "had a duty to inspect the construction to ensure that it conformed to the contract documents but was not charged with inspecting the workmanship of the general contractor or its subcontractors to ensure that the work was performed according to specific standards and municipal codes. Moreover, responsibility for the quality of the workmanship clearly fell on the general contractor. There was, therefore, no duty on [the architect] to inspect the insulation of the plumbing in the second floor bathroom. . . ."

In Goette v. Press Bar & Cafe, Inc., 413 N.W.2d 854 (Minn. Ct. App. 1987), the architect was not liable for a worker's injuries when the architect did only conceptual drawings and had no contractual duty to supervise and the contractor deviated from the drawings.

In South Burlington Sch. Dist. v. Calcagni-Frazier-Zajchowski Architects, Inc., 138 Vt. 33, 410 A.2d 1359 (1980), this provision was examined by the court. After completion of the project, the roof leaked and the school district brought suit against the architect and others. The court held that the architect did not have a duty to supervise the construction of the project.

The duty to inspect the work does not equate to a duty to protect workers on the job. Burns v. Black & Veach Architects, Inc., 854 S.W.2d 450 (Mo. 1993).

may help to remove any ambiguity that a court may use to establish a higher duty by the architect to the owner than is intended.

Some architects insert language in this provision to the effect that they will visit the site at critical periods, such as during concrete pours. Such language is generally not a good idea, because a problem may occur at such a time that the architect does not observe. Worse yet, the architect may not actually be present at such a time, and a construction defect may later be found. For instance, would the architect be present during the entire time that concrete is being poured, as well as immediately before and after the pour? The owner may well accuse the architect of a breach of duty to be there at the specified time if something goes wrong. The number of site visits can be inserted in ¶ 4.2.3.2.

It is also a useful practice to prepare reports of each site visit, indicating the date and time of the visit, who was observed to be on the project, what work was being performed at the time, and any comments as to the quality of completed work. Comments should not be made as to the contractor's methods of construction or his equipment. (For example, the architect should not be concerned with the type of ladders or scaffolding used by the contractor, because these are not incorporated into the work.) These reports should be forwarded to the owner and contractor.

In agreements in which the architect undertakes to supervise the construction, the architect may be liable if the construction is improperly performed.[108] Other courts have found that an architect had breached an implied warranty to supervise the construction.[109] Also, the architect might voluntarily undertake the duty to be responsible for safety. Thus, if the architect attends safety meetings and instructs the contractors in safety matters, the architect may become responsible for safety.[110] This, of course, would be contrary to the normal practice and counter

---

[108] Corbetta Constr. Co. v. Lake County Pub. Bldg. Comm'n, 64 Ill. App. 3d 313, 381 N.E.2d 758 (1978). This case also stated that an owner may have a cause of action against both an architect and a contractor if both faulty construction and faulty design appear in the same project. The contractor cannot interpose the defense of improper design if he does not follow the drawings or specifications.

[109] E.g., Tamarac Dev. v. Delamater, Freund, 234 Kan. 618, 675 P.2d 361, 365 (1984). A developer alleged that the architect orally agreed to supervise the grading construction and to check the grades of a trailer park on completion. The grades were later found to be improper, causing drainage problems. A written contract also existed, but no breach of that was alleged. The court held that an action in express contract, implied warranty, or negligence could be maintained. The court stated that although the work of some professionals such as lawyers or doctors is not an exact science, that of an architect or engineer is, giving rise to an implied warranty. "A person who contracts with an architect or engineer for a building of a certain size and elevation has a right to expect an exact result." Although this is true in the context of supervising grading, it is not true of many architectural or engineering projects.

[110] In Teitge v. Remy Constr. Co., 526 N.E.2d 1008 (Ind. Ct. App. 1988), the architect did not assume responsibility for safety, even though on two occasions the architect had instructed workers "to observe certain obvious safety practices." Because not every court would be so enlightened, architects should avoid even discussing safety except in emergencies. See also Case v. Midwest Mech. Contractors, 876 S.W.2d 51, 53 (Mo. Ct. App. 1994) (architect not liable to injured worker even though architect had duty to review safety program of contractors; "It is a far cry from being responsible to review the safety programs of others to specifically assuming responsibility for safety precautions to be taken.").

to the clear intent of the parties to the contract. Because the architect is normally not on the job on a full-time basis, it is in no position to deal with safety issues. That responsibility can be properly carried out only by the contractor.

One court has stated that "supervision" connotes the same standard of duty as "inspection."[111] In that case, the specifications prepared by the architect for a factory required the architect to oversee the placing of backfill against the foundation walls by inspecting and approving a number of items, including debris removal, reinforcing splices, all areas to receive concrete, and pipe lines. A soils report had indicated a layer of expansive clay from five to seven feet below grade. The engineer had recommended that surface grading be kept to a minimum; that piers be sunk to at least 20 feet below grade; that voids filled with vermiculite be placed below the grade beams to allow for swelling; that soil under the floor slab be of a select clay; and that ground levels on the east side of the building provide for rapid drainage.

Soon after the building was occupied, severe cracking of the slab started, the roof leaked, walls cracked, and other problems developed, to the point that the building became unsafe. It was found that the water table had risen and the clay had expanded, causing the cracking. Much testimony was offered as to the reason for the water. The architect contended that the owner allowed chemicals to corrode the plastic piping under the slab, causing leaking that saturated the soil. The owner offered evidence that the recommendations of the soils engineer were not followed. The architect was held liable because he did not inspect the construction properly, and the contractor was found liable for not following the plans and specifications. The court stated "the requirements that the (architect) be present at certain stages of construction and to certify progress payments and to make a final inspection and certification means that the (architect) will use reasonable care to see that the Contractor constructs the building in substantial compliance with the plans and specifications."[112] The key here was probably the fact that the specifications were

---

[111] Dickerson Constr. Co. v. Process Eng'g, 341 So. 2d 646 (Miss. 1977).

[112] *Id.* at 652. *Accord*, Skidmore, Owings & Merrill v. Connecticut Gen. Life Ins. Co., 25 Conn. Supp. 76,197 A.2d 83 (1963); Lee County v. Southern Water Contractors, Inc., 298 So. 2d 518 (Fla. Dist. Ct. App. 1974); Willner v. Woodward, 107 Ga. App. 876, 132 S.E.2d 132 (1959); Roland A. Wilson v. Forty-O-Four Grand Corp., 246 N.W.2d 922 (Iowa 1976); Westerhold v. Carroll, 419 S.W.2d 73 (Mo. 1967).

In Gables CVF, Inc. v. Bahr, Vermeer & Haecker Architect, Ltd., 244 Neb. 346, 506 N.W.2d 706, 709-10 (1993), the architect agreed to provide a set of "builders plans," an abbreviated set of drawings and specifications that cost the owner less than a full set of documents. The architect also agreed to be paid $125 per trip for "observation of construction" of the project. At substantial completion, the architect certified that the project was "generally in accordance with the intent of the drawings and specifications." However, there were deviations, including different materials used in the decks and balconies of the building, lack of a required waterproof membrane, and studs spaced at 24 rather than 16 inches on center. The court found that the architect was not entitled to summary judgment and might be liable to the owner for failing to detect these problems, despite the architect's argument that it should not be liable for damages resulting from unreported deviations from the plans, based in part on this contract provision. The court stated that this provision provides that the architect is not the insurer or guarantor of work completed by

quite specific about the approval required by the architect at certain points of the construction of the foundation.

One court has construed the general duty of the architect under an earlier version of the contract in this way:

> As we view the matter, the primary object of this provision was to impose the duty or obligation on the architects to insure to the owner that before final acceptance of the work the building would be completed in accordance with the plans and specifications; and to insure this result the architects were to make "frequent visits to the work site" during the progress of the work. Under the contract they as architects had no duty to supervise the contractor's method of performing his contract unless such power was provided for in the specifications. Their duty to the owner was to see that before final acceptance of the work the plans and specifications had been complied with, that proper materials had been used, and generally that the owner secured the building it had contracted for.[113]

A New York court held that when an engineer had contractual authority to stop work if the contractor failed to correct conditions unsafe for workers, the engineer had no duty to the workers. An injured worker was not permitted to recover from the engineer on a negligence theory.[114]

An architect or engineer should not accept authority to stop the work without being aware of the high risk this entails. In a case in which the architect was also the construction manager, this language did not insulate the architect from liability to workers injured in a fall.[115] The court found that this exculpatory language was not

---

the contractor. However, such language does not absolve the architect from liability for a breach of the architect's contractual duty, if one exists, to inform the owner of deviations from the building plans when the architect has agreed to make periodic observations.

[113] Day v. National U. S. Radiator Corp., 241 La. 288, 128 So. 2d 660, 666 (1961).

[114] Fox v. Jenny Eng'g Corp., 122 A.D.2d 532, 505 N.Y.S.2d 270 (1986). The engineer had no authority to supervise or control the injured worker or to direct construction procedures or safety measures. No active negligence was alleged. In another New York case, Jaroszewicz v. Facilities Dev. Corp., 115 A.D.2d 159, 495 N.Y.S.2d 498 (1985), this same reasoning was used in finding for an architect when a worker was electrocuted while working on a parking lot lighting system. A short circuit in a line was probably caused by postinstallation excavation work. There was no allegation of a design defect by the architect. The court found that the architect had no duty or right to control the manner in which the work was performed and no proof that the architect engaged in any active malfeasance that contributed to the accident. New York also will not impose liability on an engineer who is engaged to assure compliance with construction plans and specifications for an injury sustained by a worker, unless the engineer commits an affirmative act or negligence, or unless such liability is imposed by a clear contractual provision. Brooks v. A. Gatty Serv. Co., 27 A.D.2d 553, 511 N.Y.S.2d 642 (1987); Davis v. Lenox Sch., 151 A.D.2d 230, 541 N.Y.S.2d 814, 815 (1989) ("In the absence of any contractual right to supervise and control the construction work, as well as site safety, Petrarca, as the architect, cannot be held liable in negligence for plaintiff's injuries."); Welch v. Grant Dev. Co., 120 Misc. 2d 493, 466 N.Y.S.2d 112 (1983). See Swarthout v. Beard, 33 Mich. App. 395, 190 N.W.2d 373 (1971) (architect had right to stop work and was held liable for death of a worker when architect had knowledge of improper shoring).

[115] Wenzel v. Boyles Galvanizing Co., 920 F.2d 778, 779, 780 (11th Cir. 1991). The architect/ construction manager contracted to "provide, implement and administer a site safety and health

sufficient to disclaim liability. The court found that, under Florida law, "an architect's duty to provide for the safety of workmen employed in the construction of its designs may arise from the contractual assumption of such responsibility." [Citation omitted] This assumption of responsibility need not be explicit.[116] Here, the fact that the architect/construction manager contracted for the responsibility to oversee the safety of the project rendered the exculpatory language largely meaningless. Once a party becomes involved in job site safety issues, attempts to insulate that party from liability for workers' injuries becomes difficult or impossible.

In a federal case,[117] OSHA attempted to impose a fine on a structural engineer following an accident at a construction site. Based in part on this provision, the

---

program consisting of . . . [p]rovid[ing] daily surveillance of contractor work areas for compliance with safety program." The architect/CM also had authority to issue notices of safety violations, insist that a particular employee be removed from the job, and stop the work of contractors. In an effort to utilize the AIA language to insulate himself from liability, the contract provided that the Construction Manager shall have no responsibility or right to exercise any actual or direct control over employees of the Contractors. The obligations assumed by the Construction Manager hereunder run to and are for the sole benefit of the Commission. . . . [T]he furnishing of such services shall not make the Construction Manager responsible for construction means, methods, techniques, work sequences or procedures.

[116] *Id.* at 781.

[117] Reich v. Simpson, Gumpertz & Heger, Inc., 3 F.3d 1 (1st Cir. 1993). In Peck v. Harrocks Eng'rs, Inc., 106 F.3d 949 (10th Cir. 1997), citing this language, the court found no duty by the engineer towards an injured worker where there was an OSHA violation of trenching regulations. Architects and engineers are subject to OSHA regulations if they undertake safety duties at a construction project. On the other hand, OSHA does not apply to design professionals if they do not have a duty to oversee safety. CH2M Hill, Inc. v. Alexis Herman, Secretary of Labor and Occupational Safety and Health Review Comm'n, 192 F.3d 711 (7th Cir. 1999). In 1977, the Milwaukee Metropolitan Sewerage District ("MMSD") undertook a $2.2 billion pollution abatement program that included construction of some 80 miles of sewer tunnels. CH2M Hill was the lead engineering consultant firm and Healy was the contractor for one of the tunnels, CT-7. In November 1988, the tunnel boring machine's methane monitor detected a high concentration of methane. The tunnel was evacuated, but the electrical power was not turned off. Contrary to Healy's evacuation plan, three supervisors reentered the tunnel only 17 minutes after the methane was detected. Apparently, one of the three attempted to operate the grout pump, which was not explosion-proof. The gas exploded, killing all three.

The engineer was cited for willful violation of OSHA standards. At the initial trial before an administrative law judge, the engineer was found not guilty because it was not engaged in construction work, but was engaged in "typical engineering services." This finding was reversed by a review panel. Ultimately, the case was brought before the Seventh Circuit, which reinstated the original finding that the engineer was not liable under the OSHA regulations. The court did not hold that an engineer could never be liable under OSHA, but rather, that each case must be examined to determine the scope of the engineer's duty on each particular construction project. The court found that OSHA was intended to apply to those employers who were best suited to alleviate hazards at the construction site.

With this intent in mind, the Seventh Circuit examined the contractual duties of the engineer, as well as how the project was actually run, to determine if this engineer was in a position to "alleviate hazards at the construction site." The starting point was the contract. The court stated that "contracts represent an agreed upon bargain in which the parties allocate responsibilities

court found that the construction site was not a "place of employment" which the engineer had a duty under OSHA to protect.

If the design professional allows the contractor to deviate from the drawings and specifications and that deviation results in injury to third parties, ¶ 2.6.2.1 will not protect the designer.[118]

2.6.2.2 The Architect has the authority to reject Work that does not conform to the Contract Documents. Whenever the Architect considers it necessary or advisable, the Architect shall have the authority to require inspection or testing of the Work in accordance with the provisions of the Contract Documents, whether or not such Work is fabricated, installed or completed. However, neither this authority of the Architect nor a decision made in good faith either to exercise or not to exercise such authority shall give rise to a duty or responsibility of the Architect to the Contractor, Subcontractors, material and equipment suppliers, their agents or employees or other persons or entities performing portions of the Work. **(2.6.2.5; minor modification)**

---

based on a variety of factors. To ignore the manner in which the parties distributed the burdens and benefits is contrary to our notion of contract law."

The engineer's contract forbade it from supervising and directing Healy or any other contractor. The engineer was to "administer the contracts in accordance with the requirements of the Contract Documents and act as the DISTRICT's representative during construction." The contract also stated that "the construction contractors shall remain responsible for construction means, methods, techniques, procedures and safety precautions on the construction portions of the PROGRAM." These provisions are similar to the language found here.

Additional language in the engineer's contract referred to site visits and observations by the engineer during construction, which included this language: the visits "shall not relieve a construction contractor of its obligation to conduct comprehensive inspections of the work sufficient to ensure conformance with the intent of the construction contract documents, and shall not relieve a construction contractor of its responsibility for means, methods, techniques, sequences and procedures necessary for coordinating and completing all portions of the work under the construction contract and for all safety precautions incidental thereto."

The court also examined whether or not the engineer actually exercised substantial control or supervision so as to place it in charge of safety and concluded that there was no evidence of such control or supervision. While the engineers did enter the tunnel, and were further involved in issuing contract modifications with regard to discovery of methane gas in another tunnel, the owner retained the right to approve any modification dealing with safety. The engineer could not act on its own. The court then went on to state:

In addition, the fact that Healy turned to CH2M Hill for advice does not indicate that CH2M Hill was acting as the de facto director of safety. Before it issued the clarification, CH2M Hill had to check with MMSD. From the record it is clear that CH 2M Hill functioned as an intermediary between Healy and MMSD. MMSD directed the modifications; it ordered CH2M Hill to conduct an inquiry, reviewed CH 2M Hill's suggestions and granted the final approval to the draft language, as well as the clarification of it requested by Healy. With this fuller picture, it would be disingenuous to say that the record supports the Commission or the Secretary's conclusion that CH2M Hill played a "central role" as the "nerve center" for developing and implementing safety practices for tunnel CT-7.

[118] *See* Campbell v. The Daimler Group, Inc., 115 Ohio App. 3d 783, 686 N.E.2d 337 (1996) (engineer who approved substitution of expansion anchor bolts in place of the originally specified imbedded anchor bolts was not protected by this provision).

The architect can order tests to be performed without being held liable to the contractor for interference with the owner-contractor relationship, unless it is done in bad faith.[119] If these tests will hold up the project, the architect should keep the owner closely informed. If these tests delay the contractor, the contractor would be entitled to a change order if the tests showed that the work was in accordance with the contract documents, but not if there were any deficiencies in the work.

The last sentence of this paragraph was added in the 1987 version to prevent suits against the architect by third parties, such as contractors, subcontractors, and others.[120] The scope of the architect's duty to the contractor is defined by the architect's contract with the owner.[121] Thus, it appears that the courts would honor language in the agreement limiting the architect's duty to the contractor. To be effective, however, the owner-contractor agreement must similarly limit the owner's liability toward the contractor for the same duties. Otherwise, the contractor would sue the owner who would, in turn, sue the architect. Generally, the architect does not owe any contractual duties to the contractor (except as noted above).[122] This language also protects the architect from the various contractors if the architect chooses not to perform any particular testing. The architect should not agree to be bound by the terms of any other contract, such as the owner-contractor agreement.

Caution should be exercised when rejecting work not to give the contractor any direction regarding methods of construction. The 1967 version of AIA Document B131 included the architect's authority to stop the work. This was a basis upon

---

[119] Certified Mech. Contractors v. Wight & Co., 162 Ill. App. 3d 391, 515 N.E.2d 1047, 1053 (1987) (*citing* Santucci Constr. Co. v. Baxter & Woodman, Inc., 151 Ill. App. 3d 547, 104 Ill. Dec. 474, 502 N.E.2d 1134 (1986)); Waldinger Corp. v. CRS Group Eng'rs, Inc., 775 F.2d 781 (7th Cir. 1985); George A. Fuller Co. v. Chicago College of Osteopathic Med., 719 F.2d 1326 (7th Cir. 1983)) ("It has been established that an architect or engineer has a conditional privilege to interfere with the construction contract of its principal."). In United Stone Am., Inc. v. Frazier, Lamson, Budlong, P.C., 2002 WL 241370 (Conn. Super., Jan. 30, 2002), the architect was sued by the general contractor pursuant to an assignment by the owner. The contractor claimed that the architect tortiously interfered with the owner-contractor agreement by preparing a certificate (per ¶ 14.2.2 of A201) that the contractor was in default, even though substantial completion had been achieved, as evidenced by the fact that the owner had moved in. The court held that this was an issue for the trial court to determine, with a finding of bad faith, malice, wantonness, or acting beyond the scope of authority a requirement before liability is imposed on the architect. The court dismissed a negligence count on the basis that the architect cannot owe a duty to the contractor without breaching its agency agreement with the owner. *See also* ¶ 3.6.2.2, which states that the architect has no duty toward the contractor.

[120] A similar provision absolved an engineer of liability to a contractor in Farrell Constr. Co. v. Jefferson Parish, 693 F. Supp. 490 (E.D. La. 1988).

[121] Bates & Rogers Constr. v. North Shore, 92 Ill. App. 3d 90, 414 N.E.2d 1274, (1980); Mississippi Meadows, Inc. v. Hodson, 13 Ill. App. 3d 26, 299 N.E.2d 359 (1973); Frischhertz Elec. Co. v. Housing Auth., 534 So. 2d 1310 (La. Ct. App. 1988); Magnolia Constr. v. Mississippi Gulf S. Eng'g, 518 So. 2d 1194 (Miss. 1988); Davidson & Jones, Inc. v. County of New Hanover, 41 N.C. App. 661, 255 S.E.2d 580 (1979).

[122] Chicago College of Osteopathic Med. v. George A. Fuller Co., 719 F.2d 1335 (7th Cir. 1983); Linde Enters., Inc. v. Hazelton City Auth., 602 A.2d 897 (Pa. Super. Ct. 1992).

which many courts found the architect liable for injuries sustained by workers.[123] Under more recent versions, the architect can only reject work that does not conform to the contract documents. It has no power to stop the work, and it has no authority as to the method of construction, only as to the finished product.

The architect can reject work at any time if it appears that the final product will not conform to the contract documents. It is to everyone's advantage for this to occur as early as possible, so that the contractor can correct the work. Paragraph 13.5 of AIA Document A201 covers tests and inspections referred to here.

3.6.2.3 The Architect shall interpret and decide matters concerning performance under, and requirements of, the Contract Documents on written request of either the Owner or Contractor. The Architect's response to such requests shall be made in writing within any time limits agreed upon or otherwise with reasonable promptness. **(2.6.1.7; minor modification)**

The architect is made the interpreter of the contract and, at the request of either the owner or contractor, is authorized to make all determinations regarding acceptability of work and material provided under the contract.[124] Some owners may

---

[123] In Emberton v. State Farm Mut. Auto. Ins. Co., 44 Ill. App. 3d. 839, 358 N.E.2d 1254 (1976), *rev'd*, 71 Ill. 2d 111, 373 N.E.2d 1348 (1978), *further proceedings*, 85 Ill. App. 3d 247, 406 N.E.2d 219 (1980), which involved an earlier version of the AIA documents in which architect had authority to stop the work, the court found the architect in charge of the work for Structural Work Act purposes. This case was later analyzed in Fruzyna v. Walter C. Carlson Assocs., 78 Ill. App. 3d 1050, 398 N.E.2d 60 (1979), and compared against the newest AIA provisions, which give the architect no such authority.

[124] Helmer-Cronin Constr., Inc. v. Central Sch. Dist. No. 1, 51 A.D.2d 1085, 381 N.Y.S.2d 347 (1976). However, in this case, the parties intended to withhold from the architect the absolute power to determine the suitability of on-site fill where the specifications required certain compaction criteria.

The architect's authority extends to all items that are part of the contract, so long as they are clearly expressed. This includes alternatives that are part of the contract. Bert C. Young & Sons Corp. v. Association of Franciscan Sisters of the Sacred Heart, 47 Ill. App. 3d 336, 361 N.E.2d 1162 (1977).

In interpreting a similar clause, a North Carolina court, in Bolton Corp. v. T.A. Loving Co., 94 N.C. App. 392, 380 S.E.2d 796 (1989), held that the architect's determination of the contractor's performance is prima facie correct, and the burden is on the other parties to show fraud or mistake. In other words, if the architect makes a determination of whether the contractor's performance is defective, that determination will stand unless the other party can show clearly that the architect is wrong.

In Mayfair Constr. v. Waveland Assocs., 249 Ill. App. 3d 188, 619 N.E.2d 144 (1993), the contract between the owner and architect specifically excluded this provision. The General Conditions, on the other hand, contained this provision. The architect pointed out to the owner that there were significant conflicts between the construction contract and the architect's contract, but these did not get resolved. When a dispute arose between the owner and contractor, the contractor attempted to submit the dispute to the architect. The owner threatened the architect with a lawsuit if the architect attempted to resolve the dispute, and the architect then refused to be involved in the dispute. The dispute was then submitted to the trial court, which found that the owner breached its agreement with the contractor by refusing to allow the architect to initially resolve the dispute. The contract between the owner and contractor required that the architect be

want to delete or modify this provision because they believe that the architect, hired by the owner, should be strictly accountable to the owner. This ignores, however, the frequently effective arbitration role that the architect plays. If this role is reduced, the architect would no longer enjoy an arbitrator's immunity and all controversies would bypass the architect. This usually results in delays and increased costs to resolve disputes. Owners should carefully review this issue before removing this protection.

3.6.2.4 Interpretations and decisions of the Architect shall be consistent with the intent of and reasonably inferable from the Contract Documents and shall be in writing or in the form of drawings. When making such interpretations and decisions, the Architect shall endeavor to secure faithful performance by both Owner and Contractor, shall not show partiality to either, and shall not be liable for results of interpretations or decisions rendered in good faith. The Architect's decisions on matters relating to aesthetic effect shall be final if consistent with the intent expressed in the Contract Documents. **(2.6.1.8, 2.6.1.9; minor modifications)**

It is not improper that the representative of one of the contracting parties is the one designated as the mediator whose decision is to be recognized as final and binding.[125] One court held that this provision did not bar a claim by a

---

the initial judge of performance. It made no difference to the contractor that the owner-architect agreement conflicted with these provisions. In Ostrow Elec. Co. v. J.L. Marshall & Sons, Inc., 798 N.E.2d 310 (Mass. App. 2003), the court stated that contract terms are not changed when the architect decides in the course of a job what a specification means or whether one specification trumps another. The subcontractor had argued that the architect's decision had no rational basis and was, therefore, arbitrary and capricious. The court held that the architect's resolution did not constitute bad faith. The subcontractor's claim against the architect was dismissed.

In Fontaine Bros. v. City of Springfield, 35 Mass. App. Ct. 155, 617 N.E.2d 1002 (1993), the arbitration provisions had been deleted from the owner-contractor agreement. This left the architect as the sole arbiter of disputes, and his decision was final and binding. The architect in that case had ruled that the contractor was not entitled to extra compensation for certain work. The contractor brought a court action, and this court held that the owner was entitled to a directed verdict based on the architect's decision.

In Absher Constr. Co. v. Kent Sch. Dist. No. 415, 77 Wash. App. 137, 890 P.2d 1071 (1995), the architect had no authority to waive notice because he could not change the contract requirements. The architect was not authorized in the contract to receive notice on behalf of the owner and his role as authorized representative was stricken from the contract. Thus, when the contractor discussed problems with the engineer and received guidance, there was no waiver of the notice requirements, and the contractor was not entitled to extras because of his failure to timely notify the owner.

In A. Prete & Son Constr. v. Town of Madison, No. CV 91-03103073-5, 1994 Conn. Super. LEXIS 2532 (Oct. 4, 1994), the owner failed to obtain the architect's disapproval of a gymnasium floor before rejecting the floor, a provision cited by the court in holding for the contractor. The owner, the court said, should have followed the procedures in the AIA documents, which required the architect to make a finding that the contractor failed to comply with the contract documents.

[125] Joseph F. Trionfo & Sons v. Ernest B. LaRosa, 38 Md. App. 598, 381 A.2d 727 (1978). In Fru-Con Constr. Co. v. Southwestern Redevelopment Corp., 908 S.W.2d 741 (Mo. Ct. App. 1995), the contract was amended to allow arbitration only for claims that were less than $200,000. The contractor submitted a claim consisting of numerous claims, and the architect determined

contractor.[126] The architect had ruled that the brickwork was artistically deficient and withheld final payment from the contractor. Although the artistic decision was final, the decision to withhold payment was subject to review.

3.6.2.5 Unless the Owner and the Contractor designate another person to serve as an Initial Decision Maker, as that term is defined in AIA Document A201-2007, the Architect shall render initial decisions on Claims between the owner and Contractor as provided in the Contract Documents. **(new as to Initial Decision Maker, 2.6.1.9)**

The owner and contractor may elect a different person to serve as the Initial Decision Maker. If no selection is made, the architect will serve this function.

§ 2.15   Certificates for Payment to Contractor: ¶ 3.6.3

3.6.3 CERTIFICATES FOR PAYMENT TO CONTRACTOR

3.6.3.1 The Architect shall review and certify the amounts due the Contractor and shall issue certificates in such amounts. The Architect's certification for payment

---

that only one of these claims exceeded the limit. The contractor filed suit, and the owner sought to arbitrate, asking the court to stay the lawsuit pending the arbitration. The trial court declined to stay the proceeding. The appellate court reversed, saying that the architect had the power to make the initial determination under the contract. Because the architect determined that all the claims except one were subject to arbitration, the court proceeding should have been stayed. The contractor also argued that it should not have to entrust its "ticket to the courthouse" to the architect. The court rejected this contention, because the contractor had agreed to the architect's role up front.

In Britt Constr., Inc. v. Magazzine Clean LLC, 2006 WL 240619 (Va. Cir. Ct. Feb. 1, 2006), the contractor asserted a claim for tortious interference with contract against the architect, based on this provision. The court found that the language "not to show partiality to either [the owner or contractor]" when making interpretations and initial decisions meant that the architect's claim of privilege could not stand.

[126] NSC Contractors v. Borders, 317 Md. 394, 564 A.2d 408 (1989). In Aetna Cas. & Sur. Co. v. Leo A. Daly Co., 870 F. Supp. 925 (S.D. Iowa 1994), the court found that the architect could limit his liability contractually so long as third parties were not injured. The court held that the architect had no duty to evaluate the performance of a proposed design substitution, particularly if the change did not involve artistic effect.

In May Constr. Co. v. Benton Sch. Dist. No. 8, 320 Ark. 147, 895 S.W.2d 521 (1995), the court denied arbitration on the basis of this provision. The contractor had requested the use of a substitute for certain curing materials to be used as a sealer on the concrete floors, stating that the proposed substitute met or exceeded the quality levels of the original materials. After the application of the substitute, the finish on the concrete floor began experiencing gross and unsightly scuff marks from student traffic, to the extent that the floors' appearance became totally unacceptable; the sealer was magnifying all traffic in an unsightly manner. Further, the floors became sticky in some areas. The owner sued and the contractor sought arbitration. The owner objected, stating that the claim related only to the aesthetic appearance of the floors; the court agreed. The contractor's argument, that the claim was really one for breach of contract, was rejected.

See also Meco Sys., Inc. v. Dancing Bear Entm't, Inc., 948 S.W.2d 185 (Mo. Ct. App. 1997).

shall constitute a representation to the Owner, based on the Architect's evaluation of the Work as provided in Section 3.6.2 and on the data comprising the Contractor's Application for Payment, that, to the best of the Architect's knowledge, information and belief, the Work has progressed to the point indicated and that the quality of the Work is in accordance with the Contract Documents. The foregoing representations are subject (1) to an evaluation of the Work for conformance with the Contract Documents upon Substantial Completion, (2) to results of subsequent tests and inspections, (3) to correction of minor deviations from the Contract Documents prior to completion, and (4) to specific qualifications expressed by the Architect. **(2.6.3.1; modified)**

3.6.3.2 The issuance of a Certificate for Payment shall not be a representation that the Architect has (1) made exhaustive or continuous on-site inspections to check the quality or quantity of the Work, (2) reviewed construction means, methods, techniques, sequences or procedures, (3) reviewed copies of requisitions received from Subcontractors and material suppliers and other data requested by the Owner to substantiate the Contractor's right to payment, or (4) ascertained how or for what purpose the Contractor has used money previously paid on account of the Contract Sum. **(2.6.3.2; no change)**

The architect has a duty to exercise reasonable care to see that the work is done in a proper manner with proper materials, and to certify payment to the contractor only upon verification that the contractor has performed the work as stated in the application for payment.[127]

---

[127] Skidmore, Owings & Merrill v. Connecticut Gen. Life Ins. Co., 25 Conn. Supp. 76, 197 A.2d 83 (1963); Lee County v. Southern Water Contractors, Inc., 298 So. 2d 518 (Fla. Dist. Ct. App. 1974); Willner v. Woodward, 107 Ga. App. 876, 132 S.E.2d 132 (1959); Eventide Lutheran Home v. Smithson Elec., 445 N.W.2d 789 (Iowa 1989); Roland A. Wilson v. Forty-O-Four Grand Corp., 246 N.W.2d 922 (Iowa 1976); City of Mound Bayou v. Roy Collins Constr. Co., 499 So. 2d 1354 (Miss. 1986) (architect has duty to protect owner, but he must act in good faith); Westerhold v. Carroll, 419 S.W.2d 73 (Mo. 1967). Mound Bayou was cited in Lyndon Property Ins. Co. v. Duke Levy and Assocs., LLC, 475 F.3d 268 (5th Cir. 2007), where the surety for a county sewer project sued the engineer for the amount the surety paid to fix and test work that had previously been approved by the engineer. The court stated that an engineer may be held liable at least in some egregious cases for failure to inspect and for improper recommendations of payment if the engineer failed to meet the standard of employing ordinary professional skills and diligence, even in the absence of contractual provisions specifically guaranteeing the contractor's work.

In one case, Palmer v. Brown, 127 Cal. App. 2d 44, 273 P.2d 306 (1954), an architect was held liable for the contractor's failure to pay the subs when the person who signed the certificate for payment as an "architect" was not licensed until the following day.

In St. John Pub. Sch. Dist. v. Engineers-Architects, 414 N.W.2d 285 (N.D. 1987), the architect was not liable to an owner when the architect was hired directly by the contractor and did not sign the certificate for payment.

In First Nat'l Bank v. Cann, 503 F. Supp. 419 (N.D. Ohio 1980), although this paragraph means that the architect warrants only that to the best of his knowledge the quality of the work is consistent with the plans, the architect was liable to the owner. The court determined that "even the most cursory discharge of the supervisory duties would at least have given notice" of some of the defects in the construction.

These certificates may be withdrawn at a later date if the work does not conform to the contract documents. The architect should use the AIA form of certificate, G702, which contains the same qualifying language as this paragraph, that "to the best of the Architect's knowledge, information and belief," the work has progressed to the point indicated. Some owners want unqualified language in a certificate. This amounts to a guarantee by the architect of the work and is a source of potential liability. Note that some professional liability insurance policies exclude coverage for guarantees. Paragraph 10.4 requires the owner to submit proposed language of any certificates that the owner requires at least 14 days before the certificate is needed. This allows the architect to seek legal advice if a non-AIA certificate is requested.

The architect is required to issue certificates for payment. These certificates relate only to the work performed and not to how any money was spent.[128] It is up

---

In U.R.S. Co. v. Gulfport-Biloxi Airport Auth., 544 So. 2d 824 (Miss. 1989), the architect was liable when it failed to notify the owner that the roof was defective. The architect had employed a full-time field representative, but that representative was not competent as an inspector of the roof work. Perhaps most damaging was the fact that the improper roofing work was pointed out to the field representative and to the architect's office on a number of occasions. The architect certified the work anyway. The architect was also found liable to the surety because of the improper certifications for payment.

In Alberti, LaRochelle & Hodson Eng'g Corp. v. FDIC, 844 F. Supp. 832, 843 (D. Me. 1994), the engineering and construction management firm was liable to the lender for negligence and negligent misrepresentation regarding the developer's budget prior to the loan and in issuing certificates for payments. The court was convinced that the engineer knew, or should have known, that several items were missing from the budget presented to the lender. The court stated:

> [W]hen [the engineers] signed the [certificates for payment], they were giving the Bank assurances about the continued accuracy of the construction budget when they knew the budget was substantially less than the amount of money needed to complete Phase I of the project. These assurances about the accuracy of the budget led the Bank to wrongfully disburse loan proceeds to the Developer. The testimony in this case underscores the responsibilities of construction and design professionals] when giving information to lenders. After careful review of all the evidence in this case, the Court concludes that under these circumstances, [the engineer] owed a duty of care to the Bank.

Most courts hold that negligent misrepresentation is an exception to the economic loss doctrine, allowing a party that has no direct contract with the architect to sue, even if the damages are solely economic. Resurgence Prop., Inc. v. W.E. O'Neil Constr. Co., No. 92C6618, 1995 U.S. Dist. LEXIS 4939 (N.D. Ill. Apr. 14, 1995).

[128] In a Missouri case, Fabe v. WVP Corp., 760 S.W.2d 490, 492 (Mo. Ct. App. 1988), the court stated:

> When contractor presented a payment request, the duties of the architect consisted of determining that the work not only had progressed to the point indicated by the contractor but also had been performed in conformity with contract specifications. Once the architect authorized payment to the contractor, the contract expressly relieved the architect of any responsibility to ascertain how the contractor used the moneys.

In Gary Boren v. Thompson & Assocs., 2000 OK 3, 999 P.2d 438 (2000), the school board hired the architect for a school library project that required both performance and payment bonds by the contractor. After approving several payments to the contractor without any bonds, the

to the owner and the owner's agents to make sure that the money paid to the contractor is used properly.

If the contractor neglects to pay its subcontractors, the owner may find that these subcontractors have filed mechanic's liens and the owner could be forced to pay twice. A *mechanic's lien* is a filing by a party that has performed work on a project and not been paid in full for that work. In order to be effective, the lien must be perfected. This means that it must be filed in a timely manner with the proper authority, and the document itself must contain the proper information required by that state's statutes. Once the lien has been properly filed, the lien is of record against the property, and any party who purchases the property would seek to have the lien taken care of before the purchase is finalized. This lien right is valuable because the property itself has value, unlike the owner who may have no other assets.

One problem that occurs is that lien rights usually extend to sub-subcontractors and suppliers. It may be very difficult for the owner or architect to know just who is furnishing labor or materials for a project. The architect cannot protect the owner against mechanic's lien filings. Many lenders require that a title company examine each request for payment, conduct a title search, and examine the waivers of lien that normally accompany the request. Of course, without knowing exactly who furnished labor or materials, it is unlikely that the owner

---

architect sent a letter to the contractor reminding him that bonds were required. Thereafter, the architect received a copy of the performance bond, but not the payment bond. Later, a lawyer contacted the architect on behalf of a subcontractor, informing the architect of a problem with payments. At that time, the architect realized that he did not have a copy of a payment bond. Nevertheless, the architect continued certifying payments. After the contractor failed to make payment to the subcontractors, the subcontractors filed suit against the architect for negligence in certifying payments to the contractor in the absence of the statutorily required payment bond. The architect moved to dismiss on three bases: (1) no duty to the subcontractors, (2) no privity of contract, and (3) the subcontractors had constructive knowledge of the contractor's duty to procure the bonds, making the proximate cause of the loss their own negligence. The court denied the motion. At trial, the subcontractors obtained an award against the architect. Finding that the purpose of bonding statutes would not be served by protecting a private, for-profit company (the architect) engaged in the business of designing and overseeing public construction projects from potential liability. The damage arose from the architect's negligence in failing to ascertain that there was no payment bond and in making unauthorized payment to the contractor after he discovered that no payment bond existed. Further, once a public entity has contracted with a private party to oversee a construction project, subcontractors should be able to assume that the private party responsible for certifying payments has verified the existence of the bonds. The court stated that the architect ought to labor under a duty to subcontractors to refrain from paying the contractor without the required bond being secured.

In Broadmoor/Roy Anderson Corp., v. XL Reinsurance Am., Inc., 2006 WL 2873473 (USDC, W.D. La., Oct. 4, 2006), the architect certified payment after an arbitration award and after the architect was no longer being paid. Apparently, the surety argued that such a late certification by an architect made the certification invalid and allowed the surety to escape its obligation to pay. The court rejected this argument. The architect had testified that, "When you're the architect, you're the architect forever." The delay in obtaining the certification was harmless and under-standable.

will obtain all waivers that should be furnished. Typically, the general contractor furnishes a sworn contractor's statement that lists each subcontractor and the amount due to each subcontractor. There should be a partial lien waiver for each of these subcontractors. However, each of these subcontractors may have sub-subcontractors or material suppliers. Each of these also should furnish lien waivers. This is often not done, either because it is too much trouble or because the subcontractors do not disclose these other parties. With proper waivers, the likelihood that a subcontractor would prevail against an owner is significantly reduced.

Unless the architect is very familiar with title procedures and obtains additional compensation, the architect should not give any opinion on the status of the contractor's payments or the propriety of the various lien waivers. These are best left to the owner's attorney and lender.

A contractor or subcontractor is not entitled to use the architect's certification as a defense against the owner or others when the architect has acted reasonably and in reliance on the contractor's statements in issuing a certificate for payment. Thus, if a contractor overstates the amount of work it has performed and the architect certifies that payment, the contractor may not later rely on the architect's certification as a defense if the owner later brings an action against the contractor for failing to perform the work.

An interesting case concerning certification of payouts is *Northwest Bank v. Garrison.*[129] The owner decided to build a custom home and hired a contractor. The contractor convinced the owner that he could save $60,000 by not hiring an architect. All references to the term "architect" were deleted from the construction contract and the term "lending institution" was substituted in its place (without the bank's approval). Apparently, the bank performed some normal lender's inspections during the first part of the construction. After problems with the construction developed, the bank refused to fund further draws. In the foreclosure action on the mortgage, the owner defended, arguing that the bank had undertaken a common law duty to perform inspections. As evidence, the owner submitted a G702 form that had been prepared by the owner and was signed by the bank in the space reserved for the architect. Because there was a dispute over this issue, summary judgment was not proper, and the case was remanded to the trial court for further evidence on the subject of whether the bank owed the owner a duty.

2.6.3.3 The Architect shall maintain a record of the Applications and Certificates for Payment. (**2.6.3.3; minor modifications**)

This provision places a duty on the architect to maintain a record of the payment applications by the contractor. Included with this information would be all the associated paperwork, such as reports of site visits, lien waivers, and so on.

---

[129] 874 S.W.2d 278 (Tex. App. 1994).

§ 2.16   Submittals: ¶ 3.6.4

3.6.4 SUBMITTALS

3.6.4.1 *The Architect shall review the Contractor's submittal schedule and shall not unreasonably delay or withhold approval. The Architect's action in reviewing submittals shall be taken in accordance with the approved submittal schedule or, in the absence of an approved submittal schedule, with reasonable promptness while allowing sufficient time in the Architect's professional judgment to permit adequate review.* (**2.6.4.1; substantial modifications**)

3.6.4.2 *In accordance with the Architect-approved submittal schedule, the Architect shall review and approve or take other appropriate action upon the Contractor's submittals such as Shop Drawings, Product Data and Samples, but only for the limited purpose of checking for conformance with information given and the design concept expressed in the Contract Documents. Review of such submittals is not for the purpose of determining the accuracy and completeness of other information such as dimensions, quantities, and installation or performance of equipment or systems, which are the Contractor's responsibility. The Architect's review shall not constitute approval of safety precautions or, unless otherwise specifically stated by the Architect, of any construction means, methods, techniques, sequences or procedures. The Architect's approval of a specific item shall not indicate approval of an assembly of which the item is a component.* (**2.6.4.1; substantial modifications**)

Shop drawings[130] are drawings, diagrams, schedules, and other data specially prepared for the work by the contractor, subcontractor, sub-subcontractor, manufacturer, supplier, or distributer that show some portion of the work. The shop drawings are not contract documents.[131] Normally, these are much more detailed than the architect's drawings and often provide a last check before materials are fabricated and substantial money is spent. However, not all materials or items incorporated into a project are shown on shop drawings. Some architects list those items for which shop drawings are required. It would be virtually impossible for most projects to be built without some shop drawings. The contractor takes a substantial risk if it does not submit most of these to the architect for review.

The procedure for shop drawing review is essentially the following. First, the supplier or subcontractor prepares the shop drawing based on the architect's contract documents and submits this to the contractor. Second, the contractor checks this against the contract documents and against other shop drawings that affect the same portion of the work. This is then forwarded to the architect,

---

[130] *See* Stubbs, *Shop Drawings: Minding Someone Else's Business*, Architecture 88 (Jan. 1987) Shop drawings are defined in AIA Document A201, U 3.12.1.

[131] AIA Document A201, ¶ 3.12.4. In Fauss Constr., Inc. v. City of Hooper, 197 Neb. 398, 249 N.W.2d 478 (1977), the drawings required solid core doors. The contractor submitted shop drawings that showed particle core doors. The fire marshal inspected the doors and rejected them. The contractor then demanded an extra for the change. The court rejected the claim. Even though the architect had approved the shop drawings (although that may have been an oversight), the architect had no authority to change the contract documents. No change order had been requested or issued, and this was not an oral modification of the written contract.

along with notations as to any corrections required (only if such corrections are minor), or the fact that no errors were found. If there are major corrections, the shop drawing should be returned by the contractor to the subcontractor or supplier. Some architects insist that only an approved shop drawing be forwarded by the contractor. The architect then checks this for conformance with the design intent and the contract documents, notes on each drawing the state of approval, as well as any comments or notations, and returns one or more copies to the contractor. The contractor then notes the approval status and returns one set to the supplier or subcontractor. Unless the drawing is marked "approved" or "approved as noted" or some variation of these words, the entire process is repeated with the revised shop drawings.

The architect's review of the shop drawings is "only for the limited purpose of checking for conformance with information given and the design concept expressed in the Contract Documents." Thus, the architect does not check for quantities, field dimensions, techniques of erection, and so on. The architect should refuse to approve shop drawings that contain statements from the contractor to the effect that approval of the shop drawing shifts responsibility for dimensions, quantities, or other such items to the architect. Also, the architect should refuse to review shop drawings that have not been reviewed by the contractor. Even if the architect approves shop drawings, the contractor is not relieved of responsibility for errors or omissions in those shop drawings.[132] Most architects have shop-drawing stamps such as the following:[133]

( ) Approved   Fabrication/installation may be undertaken.

( ) Approved as Corrected   Approval does not authorize changes to the
                                       Contract Sum or Contract Time.

( ) Revise and Resubmit   Fabrication and/or installation MAY NOT be
                                       undertaken. In resubmitting, limit corrections to
                                       items marked.

( ) Rejected

Review/approval neither extends nor alters any contractual obligations of the Architect or Contractor.

---

[132] AIA Document A201, ¶ 3.12.8. In Acmat Corp. v. Daniel O'Connell's Sons, 17 Mass. App. Ct. 44, 455 N.E.2d 652 (1983), *appeal denied*, 390 Mass. 1106, 459 N.E.2d 824 (1984), the contractor submitted a brochure for "Spraydon," an acoustical coating material that the subcontractor was proposing as a substitute for the originally specified material, "Cafco Soundshield." Pursuant to the contract documents, the contractor later submitted a sample of the product, which the architect rejected as not meeting certain requirements. In the resultant lawsuit, the architect's decision was upheld because the contract gave him broad rights to make final and conclusive decisions related to the work. Under the current AIA documents, the contractor would have the right to arbitration.

[133] This language is from AIA, B-7, Construction Contract Administration, in The Architect's Handbook of Professional Practice ch. 2.8, at 7 (1987), and uses language from AIA Document A201.

Note that some insurance carriers are insisting that the word "approved" be replaced with "reviewed." Given the additional contract language included in the shop-drawing stamp above and in the agreement, it is doubtful whether it makes any difference as to which word is used. "Approved," in the context of B101, has a limited meaning and should not be interpreted as an assumption of liability by the architect.

In a high-profile case,[134] the structural engineer who reviewed shop drawings for the walkways that collapsed at the Kansas City Hyatt Hotel, killing over 100 people, asserted that the connections that failed were to be designed by the steel fabricator. The drawings failed to indicate this, and the court found the engineer responsible for the design of these connections. This case appears to make the design professional responsible for shop drawings to a greater degree than the AIA documents and custom provide. The court stated: "Custom, practice, or 'bottom line' necessity cannot alter that responsibility."[135]

> ***3.6.4.3** If the Contract Documents specifically require the Contractor to provide professional design services or certifications by a design professional related to systems, materials or equipment, the Architect shall specify the appropriate performance and design criteria that such services must satisfy. The Architect shall review Shop Drawings and other submittals related to the Work designed or certified by the design professional retained by the Contractor that bear such professional's seal and signature when submitted to the Architect. The Architect shall be entitled to rely upon the adequacy, accuracy and completeness of the services, certifications and approvals performed or provided by such design professionals.* **(2.6.4.3; minor modifications)**

If the contract documents require the contractor to submit certifications or provide design services, the contract documents must also specify the criteria by which such certifications or designs are to be judged.

The last sentence of this paragraph permits the architect to rely on submissions of engineers' certifications or designs required by the contract documents. These can be professional certifications of performance criteria of materials, systems, and equipment. An example of design services would include curtain wall designs, for which the contract documents would specify the loading, wind speed, water infiltration rates, and other criteria that the curtain wall must meet. The contractor would then hire a design firm to prepare a design that meets these criteria. If such a submittal is made, the architect need not check any of the calculations or other information submitted by the engineer.

> ***3.6.4.4** Subject to the provisions of Section 4.3, the Architect shall review and respond to requests for information about the Contract Documents. The Architect*

---

[134] Duncan v. Missouri Bd. for Architects, 744 S.W.2d 524 (Mo. Ct. App. 1988).

[135] *Id.* at 537.

shall set forth in the Contract Documents the requirements for requests for information. Requests for information shall include, at a minimum, a detailed written statement that indicates the specific Drawings or Specifications in need of clarification and the nature of the clarification requested. The Architect's response to such requests shall be made in writing within any time limits agreed upon, or otherwise with reasonable promptness. If appropriate, the Architect shall prepare and issue supplemental Drawings and Specifications in response to requests for information. **(2.6.1.5, 2.6.1.6; modified)**

This procedure is frequently referred to as a request for information (RFI). Some contractors abuse this process by sending in an excessive number of RFIs for the most trivial questions. If the question could be answered by an inspection of the drawings and specifications, then the request is improper, and if it is done too often, the contractor should be charged for the architect's time.

3.6.4.5 The Architect shall maintain a record of submittals and copies of submittals supplied by the Contractor in accordance with the requirements of the Contract Documents. **(2.6.4.2; no change)**

An accurate shop-drawing log is essential to minimize liability. This should identify each drawing or other item submitted, the date of the submittal, the party submitting it, the party preparing the submittal (the subcontractor or supplier), the action taken, the date it was returned to the contractor, and the reason for the action. Of course, a copy of each and every submittal must be retained for the architect's records. It would be wise to have a meeting before the start of construction with the contractor and major subs to review the procedure for shop drawing submittals, number of copies required, turn-around time expected, and the extent of information expected on the submittals.

§ 2.17   Changes in the Work: ¶ 3.6.5

3.6.5 CHANGES IN THE WORK

3.6.5.1 The Architect may authorize minor changes in the Work that are consistent with the intent of the Contract Documents and do not involve an adjustment in the Contract Sum or an extension of the Contract Time. Subject to the provisions of Section 4.2, the Architect shall prepare Change Orders and Construction Change Directives for the Owner's approval and execution in accordance with the Contract Documents. **(2.6.5.1; substantial modifications)**

A minor change in the work is a change that does not involve any change in the Contract Sum or Contract Time. While the architect is authorized to make such a change, the better procedure is to prepare a change order for such minor work and indicate no change in the cost or time, with the owner's signature on the change order form.

Under the construction change directive procedure, the contractor is authorized by the owner and architect to proceed with a change, even though the final cost of the change is not then known. The change order requires that the contractor also approve the change and that the cost be ascertained at the time of preparing the change order. This can result in serious delay in commencing the change. Thus, the construction change directive is utilized to fill the gap. The procedure for this is spelled out in AIA Document A201. Note that the architect can obtain additional compensation if supporting documentation and data are required, so long as the architect informs the owner that this additional compensation will be sought (**see** ¶ 4.1).

Courts have held that the architect can waive the requirement that the contractor submit supporting documentation in making a claim for delay.[136]

The architect should be thoroughly familiar with the change order procedure found in A201. Virtually every project has changes that affect the contract sum and time. If the request is incomplete or not otherwise "properly prepared," then the architect could return the request to the party who is asking for the change with directions to provide additional or more cohesive information. The architect should not have to prepare this documentation, only make a recommendation as to whether to approve or disapprove the request.

Most requests for changes in the Work come from the contractor. These requests must be "timely," per the claims provisions in A201 (**see** AIA Document A201, Article 15). There are a number of actions that the architect can take in response to a claim, and the architect should pay close attention to these options as well as the time limits associated with responding to claims.

This paragraph states that the architect will prepare change orders and construction change directives. Alternatively, the owner or contractor may prepare its own proposed change order for consideration by the other party. If signed by both sides, the change order becomes effective. If not, the owner may want to consider a construction change directive (¶ 7.3 of A201).

3.6.4.2 *The Architect shall maintain records relative to changes in the Work.* **(2.6.5.4; no change)**

The architect should keep a complete record of the project, including all requests for changes, change directives, supporting documents, memos, letters, and the like. In the event of later arbitration or litigation, this would be a source of documents to sort out the issues.

---

[136] In Bellim Constr. Co. v. Flagler County, 622 So. 2d 21 (Fla. Dist. Ct. App. 1993), the architect first approved a six-day extension of time requested by the contractor. When the owner withheld money from the contractor, the contractor sued. The architect then rescinded his earlier approval, citing lack of proof as to weather delays and triggering a liquidated damages provision. The court found this unreasonable and stated that the architect had waived the documentation required by ¶ 4.3.7.2 of A201.

§ 2.18   Project Completion: ¶ 3.6.6

3.6.6 PROJECT COMPLETION

3.6.6.1 The Architect shall conduct inspections to determine the date or dates of Substantial Completion and the date of final completion; issue Certificates of Substantial Completion; receive from the Contractor and forward to the Owner, for the Owner's review and records, written warranties and related documents required by the Contract Documents and assembled by the Contractor; and issue a final Certificate for Payment based upon a final inspection indicating the Work complies with the requirements of the Contract Documents. **(2.6.6.1; minor modifications)**

These two "inspections" are the only ones referred to in the agreement. An inspection connotes that the architect has done a more thorough review of the work than mere observation. When the architect visits the site at other times, an inspection is not being conducted. Care should be taken in using this terminology in letters and reports of the architect. In one case, the "inspecting architect" was held liable to the owner for failing to notify the owner of deviations (including the use of smaller wires and the use of 100-amp instead of 125-amp breakers) from the electrical drawings by the contractor.[137]

The architect has no authority to accept work that is incomplete or deviates from the contract documents.[138] It cannot change the specifications or interpolate something into the contract that is not justified by any fair interpretation of its terms.[139] The architect is not liable to the contractor for failing to issue a certificate of substantial completion,[140] absent bad faith. At least one court has held that the

---

[137] South Union v. George Parker & Assocs., 29 Ohio App. 3d 197, 504 N.E.2d 1131 (1985).

[138] Hurley v. Kiona-Benton Sch. Dist. No. 27, 124 Wash. 537, 215 P. 21 (1923). All parties knew that construction was not completed. Issuance of the final certificate by the architect constituted fraud and was not binding on the owner.

In Whitfield Constr. Co. v. Commercial Dev. Corp., 392 F. Supp. 982 (D.V.I. 1975), the court stated that the architect has no authority to bind the owner to alterations in the plans and specifications except as provided in the contract documents.

Even if the architect issues a certificate of substantial completion, the contractor is not relieved from liability if the building is not constructed according to the plans and specifications. This occurred in May v. Ralph L. Dickerson Constr. Corp., 560 So. 2d 729 (Miss. 1990), wherein the court held that the fact that the certificate had been issued did not mean that the building had been properly built, or even that the architect was acting as arbiter between the owner and contractor for the purpose of making such a determination.

[139] Northwestern Marble & Tile Co. v. Megrath, 72 Wash. 441, 130 P. 484 (1913).

[140] Blecick v. Sch. Dist. No. 18, 2 Ariz. App. 115, 406 P.2d 750 (1965). This issue was decided on the basis of the architect's role as arbitrator.

One court held that an architect's determination of the date of substantial completion is not necessarily final. In Allen v. A&W Contractors, Inc., 433 So. 2d 839, 841 (La. Ct. App. 1983), the court found that an arbitrator had the power to fix a different date of substantial completion than the architect had determined. The fact that the contract provides that the architect will establish

architect has exclusive authority to determine the date of substantial completion under this language.[141]

A New York architect issued a final certificate for payment one week after notifying the general contractor that roof repairs were required because of roof leakage. After several years, the owner sought damages from the architect and demanded arbitration with the architect based on the roof problem. The architect sought to stay the proceeding because the three-year statute of limitation had run. The court found in favor of the architect, denying the owner's argument that a longer statute should apply because issuance of the certificate amounted to fraud. There was no attempt to conceal the facts, and the leakage was not caused by the issuance of the certificate.[142]

This provision can be modified. In a Tennessee case, the owner sued the architect and contractor after two defective balconies at the apartment complex caused serious injuries, some four years after the initial construction.[143] The issue was whether the trial court properly applied the four-year statute of limitations. The AIA contracts were modified by certain HUD amendments. The architect issued his certificate of substantial completion on June 11, 1986. However, the owner and contractor agreed that the date of substantial completion would be the date the HUD representative signed the final HUD Representative's Trip Report. That occurred on August 21, 1986. The suit filed on August 20, 1990, was therefore timely filed.

3.6.6.2 The Architect's inspections shall be conducted with the Owner to check conformance of the Work with the requirements of the Contract Documents and to verify the accuracy and completeness of the list submitted by the Contractor of Work to be completed or corrected. (**2.6.6.2; minor modification**)

The architect is required to perform these inspections with the owner or its representative. The architect should, at the time of the inspection for substantial completion, have the contractor's punchlist for review.

3.6.6.3 When the Work is found to be substantially complete, the Architect shall inform the Owner about the balance of the Contract Sum remaining to be paid the Contractor, including the amount to be retained from the Contract sum, if any, for final completion or correction of the Work. (**2.6.6.3; minor modifications**)

The architect bases this balance on the contractor's applications for payment that were previously received, as well as the extent of the punchlist and an

---

the date of substantial completion is "not sacrosanct if the facts show substantial completion at a date earlier than that certified by the owner's architect."

[141] American Prods. Co. v. Reynolds & Stone, 1998 Tex. App. LEXIS 7387 (Nov. 30, 1998).

[142] Naetzker v. Brocton Cent. Sch. Dist., 50 A.D.2d 142, 376 N.Y.S.2d 300 (1975).

[143] Brookridge Apartments, Ltd. v. Universal Constructors, Inc., 844 S.W.2d 637 (Tenn. Ct. App. 1992).

evaluation of the cost to complete that punchlist. Hopefully, the amount of the contract sum, including any retainage remaining, is greater than the cost of the work still to be completed.

3.6.6.4 The Architect shall forward to the Owner the following information received from the Contractor: (1) consent of surety or sureties, if any, to reduction in or partial release of retainage or the making of final payment; (2) affidavits, receipts, releases and waivers of liens or bonds indemnifying the Owner against liens; and (3) any other documentation required of the Contractor under the Contract Documents. **(2.6.6.4; added (3); minor modifications)**

If there is a surety, the architect should obtain written permission to release retainage and/or make final payment. The second item should be coordinated with the owner's title company to ensure that the proper waivers, as well as any other documents listed, have been received.

3.6.6.5 Upon request of the Owner, and prior to the expiration of one year from the date of Substantial Completion, the Architect shall without additional compensation, conduct a meeting with the Owner to review the facility operations and performance. **(2.7.2; minor modifications)**

There is a need for architects to assist owners in the operations of their facility in the first year after construction and to determine if there are any latent defects in the construction work that would require the contractor's correction. This provision allows the owner to request that the architect meet with the owner prior to the one-year corrections period (**see** ¶ 12.2.2 of AIA Document A201) in order to review the operations of the facility and the performance of the building's systems. This means that the architect should be evaluating the building for conformance with the design within the one-year corrections period so that, if necessary, the contractor can be called back to make corrections to the work.

§ 2.19   Article 4: Additional Services

ARTICLE 4 ADDITIONAL SERVICES

4.1 Additional Services listed below are not included in Basic Services but may be required for the Project. The Architect shall provide the listed Additional Services only if specifically designated in the table below as the Architect's responsibility, and the Owner shall compensate the Architect as provided in Section 11.2.

(Designate the Additional Services the Architect shall provide in the second column of the table below. In the third column indicate whether the service description is located in Section 4.2 or in an attached exhibit. If in an exhibit, identify the exhibit.) **(2.8.3; substantial revisions)**

Additional Services	Responsibility (Architect, Owner, or Not Provided)	Location of Service Description (Section 4.2 below or in an exhibit attached to this document and identified below)
§ 4.1.1 Programming		
§ 4.1.2 Multiple preliminary designs		
§ 4.1.3 Measured drawings		
§ 4.1.4 Existing facilities surveys		
§ 4.1.5 Site Evaluation and Planning (B203™–2007)		
§ 4.1.6 Building information modeling		
§ 4.1.7 Civil engineering		
§ 4.1.8 Landscape design		
§ 4.1.9 Architectural Interior Design (B252™–2007)		
§ 4.1.10 Value Analysis (B204™–2007)		
§ 4.1.11 Detailed cost estimating		
§ 4.1.12 On-site project representation		
§ 4.1.13 Conformed construction documents		
§ 4.1.14 As-designed record drawings		
§ 4.1.15 As-constructed record drawings		
§ 4.1.16 Post occupancy evaluation		
§ 4.1.17 Facility Support Services (B210™–2007)		
§ 4.1.18 Tenant-related services		
§ 4.1.19 Coordination of Owner's consultants		
§ 4.1.20 Telecommunications/data design		
§ 4.1.21 Security Evaluation and Planning (B206™–2007)		
§ 4.1.22 Commissioning (B211™–2007)		

Additional Services	Responsibility (Architect, Owner, or Not Provided)	Location of Service Description (Section 4.2 below or in an exhibit attached to this document and identified below)
§ 4.1.23 Extensive environmentally responsible design		
§ 4.1.24 LEED® Certification (B214™–2007)		
§ 4.1.25 Fast-track design services		
§ 4.1.26 Historic Preservation (B205™–2007)		
§ 4.1.27 Furniture, Finishings, and Equipment Design (B253™–2007)		

This table lists various possible services that may be required for a project. The second column allows the parties to designate whether each of the listed services is to be the responsibility of the owner, the architect, or not provided at all. The third column lists where in this, or an attached document, a more detailed description of each such service is to be found. Where the architect is to provide any of these services, the description can be inserted at Section 4.2, below.

§ 2.20   Description of Additional Services: ¶ 4.2

4.2 DESCRIPTION OF ADDITIONAL SERVICES

Insert a description of each Additional Service designated in Section 4.1 as the Architect's responsibility, if not further described in an exhibit attached to this document. (**new**)

For any services listed in the table as being performed by the architect, a description of the services is required. This can be set forth here on in an attached exhibit.

4.3 Additional Services may be provided after execution of this agreement, without invalidating the Agreement. Except for services required due to the fault of the Architect, any Additional Services provided in accordance with this Section 4.3 shall entitle the Architect to compensation pursuant to section 11.3 and an appropriate adjustment in the Architect's schedule. (**1.3.3.1; extensive modifications**)

This section covers situations where the architect will be entitled to additional fees if any of the following conditions occur. The only exception is where the architect is at fault for the additional services.

> **4.3.1** *Upon recognizing the need to perform the following Additional Services, the Architect shall notify the Owner with reasonable promptness and explain the facts and circumstances giving rise to the need. The Architect shall not proceed to provide the following services until the Architect receives the Owner's written authorization:* **(1.3.3.2, 2.8.2; substantial revisions)**

Before the architect proceeds under the following conditions, the architect must give notice to the owner that these services are required. The architect should not start work on any of these additional services until written approval is received from the owner.

> **.1** *services necessitated by a change in the initial information, previous instructions or approvals given by the Owner, or a material change in the Project including, but not limited to, size, quality, complexity, the owner's schedule or budget for Cost of the Work, or procurement or delivery method;* **(1.3.3.2.1, 1.3.3.2.4; modifications)**

With this provision, the architect is entitled to additional fees if the owner revises the plans from those previously approved or there is a change in the initial information provided on Exhibit A to the agreement. For this reason, it is important to obtain written approvals of drawings for each stage of the project to avoid any question of whether such additional fees are warranted. Also, changes to the delivery method or other major changes to the project qualify. This provision could encompass many types of changes. In general, if a change is made that differs significantly from the understanding of the parties at the time the owner approved the design development documents, it is likely that this paragraph would apply. Also, if the owner wants to use general conditions other than AIA Document A201, this provision applies.

This provision, under a prior version of this document, was examined by a New York court.[144] The owner had hired the architect to design the exterior renovation of a building. One year later, the owner decided to remodel the lobby of the building. It requested proposals from the architect and other architects for this work. Before the owner selected a firm, the architect did some work on the lobby and billed the owner on the same invoices as for the exterior work. The owner selected another design firm for the lobby work. When a dispute developed over fees, the architect filed a demand for arbitration. The owner sought a stay of arbitration for that part of the claim related to the lobby work on the basis that there was no agreement to arbitrate that work. The court held that the lobby work was not

---

[144] Degi Deutsche Gesellschaft Fuer Immobilienfonds MBH v. Jaffey, No. 95 Civ. 6813 (MBM), 1995 U.S. Dist. LEXIS 16737 (S.D.N.Y. Nov. 3, 1995).

part of the same contract but was an entirely new project, not part of the first project.

The architect may want to formalize the approval procedure with the use of a form such as the following:[145]

NOTICE OF APPROVAL AND AUTHORIZATION TO PROCEED.

The undersigned, as agent for the Owner, hereby approves the following documents: (list drawing numbers and dates, and specifications, if any). The Architect is hereby authorized and directed to proceed to the _____ phase of the Project. Any subsequent changes that deviate from this approval shall be subject to additional compensation to the Architect per Article 4 of the Agreement.

Owner: _____

Date of approval: _____

> .2 *services necessitated by the Owner's request for extensive environmentally responsible design alternatives, such as unique system designs, in-depth material research, energy modeling, or LEED® certification;* (**new**)

Under this provision, the architect will be entitled to additional fees if the owner requires any of these services. LEED is an acronym for Leadership in Energy and Environmental Design. There is a rating system that is the nationally accepted benchmark for the design, construction, and operation of high performance green buildings.[146]

> .3 *changing or editing previously prepared Instruments of Service necessitated by the enactment or revisions of codes, laws or regulations or official interpretations;* (**new**)

---

[145] This provision was examined by the court in Degi Deutsche MBH v. Jaffey, No. 95 Civ. 6813 (MBM), 1995 U.S. Dist. LEXIS 16737 (S.D.N.Y. Nov. 3, 1995). The owner had hired the architect to design the exterior renovation of a building. One year later, the owner decided to remodel the lobby of the building. It requested proposals from the architect and other architects for this work. Before the owner selected a firm, the architect did some work on the lobby and billed the owner on the same invoices as for the exterior work. The owner selected another design firm for the lobby work. When a dispute developed over fees, the architect filed a demand for arbitration. The owner sought a stay of arbitration for that part of the claim related to the lobby work on the basis that there was no agreement to arbitrate that work. The court held that the lobby work was not part of the same contract but was an entirely new project, not part of the first project. The court found that the lobby work was not covered by ¶ 3.4.13 (1987 version) or ¶ 3.4.20 because there was no writing. Subparagraph 3.3.2 was not applicable because of lack of notice.

[146] Overseen by the U.S. Green Building Counsel, a non-profit composed of leaders from every sector of the building industry.

If any revisions to any laws, codes, etc., are enacted after drawings or specifications are prepared and such revisions require a change in the plans or specifications, the architect will be entitled to additional fees. This is not necessarily limited to the construction documents phase. For instance, if a building or zoning code is changed, requiring a redesign of schematic drawings, the architect will be entitled to additional compensation. Also, if a code official makes an interpretation that is different from that of the architect or its consultants, requiring changes to the plans or specifications, that is an additional service, unless it is shown that the architect or consultant was in error. It is not uncommon for building officials to have widely varying interpretations of codes, so that even if the architect obtains an early opinion from one plan examiner, that does not mean that a different, or even the same, plan examiner will not take a contrary position at the time the plans are reviewed for permit.

> *.4* *services necessitated by decisions of the owner not rendered in a timely manner or any other failure of performance on the part of the Owner or the Owner's consultants or contractors;* (**1.3.3.2.3, 1.3.3.2.5; modifications**)

If an owner fails to make timely decisions, the project may be delayed and additional changes may be required. The architect is entitled to additional fees when this happens.

One issue here is whether this failure of performance by the owner or a contractor needs to be material, or whether any failure will trigger this provision. This paragraph applies when a contractor fails to live up to the requirement of the contract documents. In some cases this could mean an extensive number of punchlist[147] items or the failure of a major system. It certainly includes time spent by the architect if the owner sues the contractor (**see** ¶ 4.3.1.8) or if special testing is done because a component of the work is suspected of being deficient.

In a Texas case, the court found that an architect's field representative had no apparent authority to bind the owner to pay a subcontractor to finish work after the general contractor defaulted.[148] The clause in the agreement that provided that the

---

[147] A "punchlist" is the list that the architect, owner, or contractor prepares about the time of substantial completion of the project. This is the complete list of all items required of the contractor for final completion. If the contractor has done a reasonably good job, the punchlist should be relatively short. Note that AIA Document A201 requires the contractor to prepare its own punchlist when it requests a certificate of substantial completion. The architect then inspects the work and reviews the contractor's punchlist, amending it as necessary. In Newman Marchive P'ship, Inc. v. City of Shreveport, 944 So. 2d 703 (La. App. 2d Cir. 2006), the architect performed services more than 60 days after substantial completion and charged for these additional services. The owner argued that the architect failed to provide notice of these services. According to the court, however, the evidence showed that the owner not only was aware that the architect was performing these services, but that the owner was working with the architect in getting the contractor to perform the punchlist work. Because the owner had notice and actively sought the architect's assistance in completing the punchlist items, the architect was entitled to additional fees.

[148] Marble Falls Hous. Auth. v. McKinley, 474 S.W.2d 292 (Tex. 1971).

architect would be paid for services rendered for arranging for work to proceed upon the default of a contractor did not give this employee of the architect the required authority to bind the owner. There was no evidence that the subcontractor knew of the existence of this clause or that the owner knew that this person was speaking for the owner. The field representative was clearly overstepping his bounds by telling the subcontractor to continue with the work and that payment would be made.

When a contractor defaults, the architect must obtain approval from the owner to issue any orders to contractors to proceed with the work. The better course is for the owner to direct each contractor in writing and to take over the contracts. AIA Document A201 provides for a collateral assignment of all the subcontracts, so that in case of default by the general, the contracts are assigned to the owner. The architect should also confirm in writing that additional fees are due under this part of the agreement because of such a default. The architect must be careful to keep adequate records to justify additional fees under these provisions. In a Kansas case,[149] the contractor defaulted, causing the architect additional work. The architect submitted bills per a prior version of this provision, but the trial court denied his claim. The appellate court upheld the judgment, finding that the architect had failed to properly prove his entitlement to money. The court found that it was impossible, based on the architect's records, to differentiate between his performance of basic services and contingent additional services. The architect did not keep daily time records of tasks performed, nor did he keep detailed records of time spent on contingent additional services. Had he kept proper records, it is likely that the architect could have recovered a substantial fee due to the contractor's default.

> .5   *preparing digital data for transmission to the owner's consultants and contractors, or to other owner authorized recipients;* **(new)**

This new section is intended to cover the issues related to the transfer of computer files such as CAD drawings to the contractor and subcontractors. Among the issues that concern the architect in allowing such transfers are prevention of use of the drawings on other projects, possible errors created by the transmission itself, use of the wrong version of a drawing, and the continuation of mistakes in the architect's drawings onto the shop drawings. For instance, subcontractors normally prepare shop drawings from the drawings created by the architect and its consultants. In the preparation of these shop drawings, there is an opportunity to catch and correct errors because the subcontractor is preparing new drawings. If, on the other hand, the subcontractor is using the architect's electronic files to create the shop drawings, there is a much greater likelihood that the architect's mistakes will not be caught in this process.

Note that as of 2007, the AIA has developed two new documents related to digital data. The first is AIA Document E201-2007, Digital Data Protocol Exhibit. This document is intended to cover procedures regarding the exchange of digital data. A second document, C107-2007, Digital Data Licensing Agreement, is for

---

[149] Belot v. U.S.D. No. 497, 4 P.3d 626 (Kan. Ct. App. 2000).

use when the parties do not already have an AIA agreement in place, such as when the architect sends digital documents to a subcontractor.

> **.6** *preparation of design and documentation for alternate bid or proposal requests proposed by the Owner;* (**new**)

If the architect needs to prepare additional or different drawings or specifications for alternate bids, or other situations other than the normal single general contractor bid situation, the architect will be entitled to additional fees.

> **.7** *preparation for, and attendance at, a public presentation, meeting or hearing;* (**1.3.3.2.6; minor modifications**)

The owner may want to include attendance at public hearings, such as zoning hearings, as part of basic services. Because the time required for such hearings may be difficult to determine, the architect would want to be reimbursed at an hourly rate for this service.[150]

> **.8** *preparation for, and attendance at a dispute resolution proceeding or legal proceeding, except where the Architect is a party thereto;* (**1.3.3.2.6; minor modifications**)

Unless the architect is a plaintiff or defendant, the architect is entitled to additional fees for services related to arbitration, litigation, or litigation.

> **.9** *evaluation of the qualifications of bidders or persons providing proposals;* (**new**)

Unless the architect is asked, as an additional service, to evaluate prospective bidders, this would be the owner's responsibility.

> **.10** *consultation concerning replacement of Work resulting from fire or other cause during construction; or,* (**2.8.2.4; modifications**)

These damages are not limited to fire. They can be damages due to any cause, such as wind, lightning, or flood. These costs may be covered by the owner's insurance.

> **.11** *assistance to the Initial Decision Maker, if other than the Architect.* (**new**)

If the owner and contractor have designated anyone other than this architect as the Initial Decision Maker per Section 15.2 of A201, the architect will be entitled to additional fees for rendering assistance to this person.

---

[150] In Robert M. Swerdroe, Architect/Planners v. First Am. Inv. Corp., 565 So. 2d 349 (Fla. Dist. Ct. App. 1990), the architect sought to foreclose a lien for his services as an expert witness under this provision, as well as for interior design services under ¶ 3.4.13 and services under ¶ 3.3.6 (1987 version of B 141). The court held that these were not, by themselves, lienable.

4.3.2 To avoid delay in the Construction Phase, the Architect shall provide the following Additional Services, notify the Owner with reasonable promptness, and explain the facts and circumstances giving rise to the need. If the Owner subsequently determines that all or parts of those services are not required, the Owner shall give prompt written notice to the Architect, and the Owner shall have no further obligation to compensate the Architect for those services: **(2.8.2; modifications)**

For these additional services, the architect must give notice of these services with "reasonable promptness." The owner can then decide that the services are not required and notify the architect. The architect will be paid for such services until receipt of the owner's notice to stop. After such notification, the architect will not receive further compensation, even if the architect does additional work on such services.

> *.1   Reviewing a Contractor's submittal out of sequence from the submittal schedule agreed to by the Architect;* **(2.8.2.1; minor modifications)**

The contractor is required to prepare a submission schedule at the outset of the project. ¶ 3.10.2 of A201. If the actual submissions are not in accordance with that schedule, the architect may be entitled to additional compensation.

> *.2   Responding to the Contractor's requests for information that are not prepared in accordance with the Contract Documents or where such information is available to the Contractor from a careful study and comparison of the Contract Documents, field conditions, other Owner-provided information, Contractor-prepared coordination drawings, or prior Project correspondence or documentation;* **(2.8.2.2; modifications)**

Sometimes contractors fail to carefully review the drawings and specifications or compare them to the site conditions. Such a contractor may then ask for information that is already indicated on the contract documents. The architect is entitled to an additional fee for such requests.

> *.3   Preparing Change Orders and Construction Change Directives that require evaluation of Contractor's proposals and supporting data, or the preparation or revisions of Instruments of Service;* **(2.8.2.3; modifications)**

The preparation of the actual change order or construction change directive form is not an additional service, but any other time spent on such a change would be covered under this clause. If the contractor submits proposed substitutions, the architect should inform the owner in writing that it intends to make these evaluations. If the owner then tells the architect not to expend any billable time on these evaluations, the architect will be in a good position if the owner later makes a claim against the architect for any reason related to this item, assuming the owner accepts the contractor's proposed substitution.

> *.4   Evaluating an extensive number of Claims as the Initial Decision Maker;*
> **(2.8.2.5; add Initial Decision Maker)**

The term *extensive* is not defined here. It may be that the contractor lumps all claims into a single large claim. The architect may consider striking the words "an extensive number of" so that it is paid for evaluating all claims. Alternatively, an allocation of hours for such evaluations may be inserted in the agreement as part of basic services. If the architect is not the Initial Decision Maker, this will not apply.

> *.5   Evaluating substitutions proposed by the Owner or Contractor and making subsequent revisions to Instruments of Service resulting therefrom; or* **(2.8.2.6; modifications)**

Contractors often propose substitutions that are not listed in the contract documents. Often this is for the convenience of the contractor, or because the contractor can obtain the substituted materials for less money. If such substitutions are accepted and the architect is required to revise the drawings or specifications, the architect will be entitled to additional fees, as well as for the architect's time spent in evaluating the proposals.

> *.6   To the extent the Architect's Basic Services are affected, providing Construction Phase Services 60 days after (1) the date of Substantial Completion of the Work or (2) the anticipated date of Substantial Completion identified in Initial Information, whichever is earlier.* **(2.8.2.8; modifications)**

This marks the end of the architect's basic services. It would be rare for the architect to actually end its services at this point, however. The owner should be kept aware of this time frame and when additional costs will start. The owner's written approval of such costs is not required for the architect to collect its fees, but the owner has the option of notifying the architect to cease such additional services. If this occurs, the architect would cease all further work on the project.

> **4.3.3** *The Architect shall provide Construction Phase Services exceeding the limits set forth below as Additional Services. When the limits below are reached, the Architect shall notify the Owner:* **(2.8.1; modifications)**

Here, the parties should anticipate the time required for each of the listed services. Time spent beyond these limits will entitle the architect to additional fees, but only after the architect has notified the owner that the relevant limit has been reached.

> *.1   (_____) reviews of each shop drawing, product data item, sample and similar submittal of the Contractor* **(2.8.1.1; delete "up to")**

Ideally, the architect should only have to review each shop drawing and other submittal once. However, that is probably unrealistic. On the other hand, it is not realistic for the architect to have to repeatedly review the same shop drawing because of an incompetent contractor. The architect would want to insert "one" in this space, while the owner may want at least "two."

> .2   (_____) *visits to the site by the Architect over the duration of the* *Project during construction* (**2.8.1.2; delete "up to"**)

The number of site visits depends on the scope of the project. Obviously, the greater the number of site visits, the more the owner is protected, but the higher the cost. If this number is exceeded, the architect will be entitled to additional compensation.

> .3   (_____) *inspections for any portion of the Work to determine* *whether such portion of the Work is substantially complete in accordance* *with the requirements of the Contract Documents* (**2.8.1.3; delete "up to"**)

This allows for the possibility that several inspections will be required in order to certify substantial completion. This might occur when the contractor determines that the project is substantially complete but the architect's inspection reveals serious deficiencies. This problem would be minimized with a competent contractor. The architect would want to insert "one" while the owner may want a higher number. Additional inspections require additional payments to the architect.

> .4   (_____) *inspections for any portion of the Work to determine final* *completion* (**2.8.1.4; delete "up to"**)

As with the prior paragraph, the competence of the contractor will largely determine the number of these inspections that are actually required.

> **4.3.4** *If the services covered by this Agreement have not been completed* *within _____ (_____) months of the date of this Agreement, through* *no fault of the Architect, extension of the Architect's services beyond that time shall* *be compensated as Additional Services.* (**1.5.9; revise reference to Additional** **Services**)

A reasonable time to complete the project must be inserted here.[151] If delays that are not the fault of the architect are encountered, the architect should inform the owner at the end of the initial time period that the architect's compensation rates are being adjusted with a new schedule of rates. The owner's signed agreement to these changes should be obtained at this point.

---

[151] In Quinn Assocs. v. Town of E. Hampton, No. CV 92-0 54 46 09, 1995 Conn. Super. Ct. LEXIS 1378 (May 5, 1995), the owner argued that there was a mutual mistake and no meeting of the minds regarding this provision. The original contract required the services to be completed within 24 months. Later, the architect submitted a replacement page showing this period to be 36 months. The court found that there had been an agreement as to the extension.

Not only can the hourly rates be adjusted in this situation, but the architect would also be permitted compensation for time expended after this date. Thus, if the architect is delayed and the delay is not caused by the architect's fault, the architect will be entitled to additional compensation, depending on what is inserted in ¶ 11.3.[152]

At least one court held that an owner can enforce a no damages for delay clause against an engineer.[153] If an owner inserts such a clause, the architect and the architect's consultants would not be entitled to any additional fees if the project is delayed for any reason. Of course, such a clause would conflict with other provisions of this document, so significant editing would be required to make this effective.

§ 2.21   Article 5: Owner's Responsibilities

ARTICLE 5 OWNER'S RESPONSIBILITIES

5.1** Unless otherwise provided for under this Agreement, the Owner shall provide information in a timely manner regarding requirements for and limitations on the Project, including a written program which shall set forth the Owner's objectives, schedule, constraints and criteria, including space requirements and relationships, flexibility, expandability, special equipment, systems and site requirements. Within 15 days after receipt of a written request from the Architect, the Owner shall furnish the requested information as necessary and relevant for the Architect to evaluate, give notice of or enforce lien rights. **(1.2.2.1; modifications)

This provision requires the owner to furnish the architect with any information about the project, the site, budget, program, entities working on the project, and so on. This must be done whether or not the architect requests such information. If the owner has information that would be important to the architect and fails to furnish this information to the architect in a timely manner, the owner cannot later complain that the architect failed to properly perform under this agreement. The owner

---

[152] In SKM Partners v. Associated Dry Goods Corp., 577 N.Y.S.2d 23 (App. Div. 1991), the parties contemplated a six-month project. It actually took 26 months. It was agreed that normally an architect would have taken 1,400 hours to perform this work but the architect actually expended 5,098 hours. The court held that the architect was entitled to additional fees for this time because the owner actively interfered with the architect's work. This case also underscores the need to keep accurate time records, even when the architect is working on other than an hourly basis.

[153] In State Highway Admin. v. Greiner Eng'g Sciences, Inc., 83 Md. App. 621, 577 A.2d 363 (1990), the contract between the owner and engineer read:

The Consultant agrees to prosecute the work continuously and diligently and no charges or claims for damages shall be made by him for any delays or hindrances, from any cause whatsoever during the progress of any portion of the services specified in this Agreement. Such delays or hindrances, if any, may be compensated for by an extension of time for such reasonable period as the Department may decide. Time extensions will be granted only for excusable delays such as delays beyond the control and without the fault or negligence of the consultant.

The court held that enforcement of this clause was not unconscionable.

is also required to furnish to the architect, upon request, information sufficient for the architect to file and enforce lien rights. This would include a legal description, the actual identity of the owner of the land, and other information required by the jurisdiction to perfect a lien. If the owner fails to do so, it would be a breach of the contract, entitling the architect to stop work on the project.

5.2 The Owner shall establish and periodically update the Owner's budget for the Project, including (1) the budget for the Cost of the work as defined in Section 6.1; (2) the Owner's other costs; and, (3) reasonable contingencies related to all of these costs. If the Owner significantly increases or decreases the Owner's budget for the Cost of the work, the Owner shall notify the Architect. The Owner and the Architect shall thereafter agree to a corresponding change in the Project's scope and quality. **(1.2.2.2; revisions)**

It is the owner's duty to initially establish a budget. Thereafter, the owner is under a continuing obligation to update the project budget and cannot "significantly" increase or decrease the budget for the part of the work that affects the architect without the architect's agreement. This means that if the budget is changed by more than a minor amount, the architect may have to change the design and/or drawings and specifications, resulting in additional compensation for the architect. The failure by the owner to perform these updates, if required, may constitute a breach of the agreement if the architect is affected thereby.

5.3 The Owner shall identify a representative authorized to act on the Owner's behalf with respect to the Project. The Owner shall render decisions and approve the Architect's submittals in a timely manner in order to avoid unreasonable delay in the orderly and sequential progress of the Architect's services. **(1.2.2.3; revisions)**

This is particularly important when dealing with large organizations and governmental entities. In the case of public works, the architect should ask for copies of appropriate resolutions clarifying that the owner's representative is authorized to make binding decisions. The owner's representative can be a key person in determining whether a project is successful.[154] The Initial Information refers to Exhibit A, but the information can also be shown in Section 1.1.

5.4 The Owner shall furnish surveys to describe physical characteristics, legal limitations and utility locations for the site of the Project, and a written legal description of the site. The surveys and legal information shall include, as applicable, grades and lines of streets, alleys, pavements and adjoining property and structures; designated wetlands; adjacent drainage; rights-of-way, restrictions, easements, encroachments, zoning, deed restrictions, boundaries and contours of the site; locations, dimensions and necessary data with respect to existing buildings, other improvements and trees; and information concerning available utility services and lines, both public and private, above and below grade, including inverts and depths. All the information

---

[154] In Whitfield Constr. Co. v. Commercial Dev. Corp., 392 F. Supp. 982 (D.V.I. 1975), the project representative had an unclear concept of his duties. This may have contributed to this strange case.

on the survey shall be referenced to a Project benchmark. **(2.2.1.2; added "wetlands")**

The owner must furnish a proper survey for the architect's use. In an Illinois case,[155] a developer of a high-rise condominium furnished the architect a survey of the site that showed a dashed line and the notation "10 ft. building line." The architect followed the building and zoning code and located the building 6 feet 7 inches from the property line. The zoning permitted the building to be closer to the property line than the recorded building line, which was more restrictive. This building line was apparently a private restriction and not part of any code. After the caissons and some foundation walls were poured, the discrepancy was found. The cost of correction was estimated at some $1 million. The developer sued the architect, and the jury returned a verdict for the architect. On review, the appellate court found that because there was a substantial question of whether the developer fulfilled his contractual obligations to provide the architect with a plat of survey that showed restrictions and that was complete and accurate, the jury's verdict would be sustained. Placing the structure over the 10-foot building line was not a breach of duty as a matter of law. The fact that the developer had information that

---

[155] Cadral Corp. v. Solomon, Cordwell, Buenz, 147 Ill. App. 3d 466, 497 N.E.2d 1285 (1986). In Hempel v. Bragg, 856 S.W.2d 293 (Ark. 1993), this provision protected the architect, when the owner had given the architect a copy of the recorded plat of subdivision based on an earlier erroneous survey made for the subdivision developer. The architect relied on the plat and statements of the owner as establishing the boundaries of the lot. In Taylor v. DeLosso, 725 A.2d 51 (N.J. Super. Ct. App. Div. 1999), the owner hired a surveyor to prepare a survey of her property for a variance for a building she was planning. This survey showed a 30-inch-diameter maple tree on the site. The variance was rejected by the local municipality. Thereafter, the owner hired an architect to prepare a full site plan. The architect relied on the earlier survey for the location of the tree. Based on the new site plan, the variance was eventually granted. After construction started, it was found that the maple tree was located in the proposed driveway area instead of as shown on the architect's site plan, a difference of more than 11 feet. Construction was halted, and the owner was forced by the municipality to submit a new site plan. She then hired a new architect to start from scratch. The contract between the defendant architect and owner had contained language substantially similar to the AIA language found here. The court accepted testimony that the AIA language defines the standard of care relative to surveys. The court found that the "reliance" language in the contract gave the architect an unqualified right to rely on the survey furnished by the owner and that the architect had no duty to make a site inspection to check the accuracy of the physical features found on that survey.

In QB, LLC v. A/R Architects, LLP, 19 A.D.3d 675, 797 N.Y.S.2d 552 (2005), the owner retained the architect to provide services in connection with the design and construction of a building on a vacant lot. It was to be constructed in compliance with existing zoning regulations so as not to require any variance. The architect obtained information from one of its contacts that the maximum height requirement for this zoning classification was 60 feet or four stories, whichever was less. Nevertheless, the architect prepared preliminary plans for a six-story building. Subsequently, the architect submitted an AIA agreement to the owner containing these provisions. Following several years of work on the final plans, the error was discovered. The architect defended the suit based on these provisions. The court rejected this defense at the summary judgment stage.

clarified this building line and did not give it to the architect undoubtedly helped the defendant.

If information is furnished, either by the owner or by consultants hired by the architect directly, the failure to follow the recommendations contained in such reports can constitute negligence,[156] or a breach of contract.

The owner is responsible for obtaining zoning and related approvals. Frequently, the architect is asked to assist in these efforts. This is an additional service and not a basic service. Because there is no assurance that zoning variances and similar approvals will be granted, the architect should never agree to secure necessary zoning or related approvals, but only to "assist."

> **5.5** *The Owner shall furnish services of geotechnical engineers which may include but are not limited to test borings, test pits, determinations of soil bearing values, percolation tests, evaluations of hazardous materials, seismic evaluation, ground corrosion tests and resistivity tests, including necessary operations for anticipating subsoil conditions, with written reports and appropriate recommendations.* **(2.2.1.3; minor modifications)**

Depending on the nature of the project, the services specified in this paragraph may be needed for the project. Most projects will not need all of these services, but they may need one or more, or different special services. These are usually furnished by a specialty engineer hired by the owner. If the owner wants the architect to subcontract for these services, this paragraph should be amended, although this substantially increases the architect's liability exposure and is not a good idea.

> **5.6** *The owner shall coordinate the services of its own consultants with those services provided by the Architect. Upon the Architect's request, the Owner shall furnish copies of the scope of services in the contracts between the Owner and the Owner's consultants. The Owner shall furnish the services of consultants other than those designated in this Agreement, or authorize the Architect to furnish them as an Additional Service, when the Architect requests such services and demonstrates that they are reasonably required by the scope of the Project. The Owner shall require that its consultants maintain professional liability insurance as appropriate to the services provided.* **(1.2.2.4; substantial modifications)**

Note that the architect is providing certain engineering services under the Project Team provision, ¶ A.2. The owner will provide consultants listed in ¶ A.2.3. Beyond that, the architect may later identify additional services that are required for the project, including those of geotechnical engineers, acoustic engineers, lighting consultants, and others. The owner can either furnish those services or ask the architect to hire consultants at an additional cost to the owner. If the

---

[156] Reiman Constr. Co. v. Jerry Miller, 709 P.2d 1271 (Wyo. 1985) (soils report recommendations were not followed, resulting in cracking of concrete and other structural damage to completed building).

owner hires consultants, the owner must require "appropriate" insurance of these consultants.

5.7 The Owner shall furnish tests, inspections and reports required by law or the Contract Documents, such as structural, mechanical, and chemical tests, tests for air and water pollution, and tests for hazardous materials. **(1.2.2.5; minor revisions)**

The architect is not responsible for any hazardous materials (**see** ¶ 10.6). If the architect suspects asbestos or other such hazardous materials, it should request, in writing, that the owner hire a specialist to test for such materials.

5.8 The Owner shall furnish all legal, insurance and accounting services, including auditing services, that may be reasonably necessary at any time for the Project to meet the Owner's needs and interests. **(1.2.2.6; no change)**

The reference to insurance services in this clause means that the owner should furnish to the architect the insurance requirements for the project to be incorporated into the contract documents, as well as reviews of certificates of insurance by the various contractors. Any other insurance matters should also be handled by the owner's insurance representative. The term *legal services* means all the normal owner's legal services, plus review of the contract documents by the owner's attorney for compliance with the owner's legal requirements, and any other miscellaneous legal services required throughout the project. The term *accounting services* relates to any reviews of payment applications, accounting work related to changes, and any other matters that fall into the accounting category.

5.9 The Owner shall provide prompt written notice to the Architect if the Owner becomes aware of any fault or defect in the Project, including errors, omissions or inconsistencies in the Architect's Instruments of Service. **(1.2.2.7; minor modification)**

The owner must inform the architect immediately if the owner becomes aware that the contract documents are in error or do not comply with the program. The owner cannot stand by and do nothing. If the owner does not give such notification, the owner cannot later claim damages that could have been avoided had the notification been given.

5.10 Except as otherwise provided in this Agreement, or when direct communications have been specially authorized, the Owner shall endeavor to communicate with the Contractor and the Architect's consultants through the Architect about matters arising out of or relating to the Contract Documents. The Owner shall promptly notify the Architect of any direct communications that may affect the Architect's services. **(2.6.2.4; minor modifications)**

This paragraph provides for orderly communications on the project. Unless the architect is kept informed of such communications, the probability of errors and increased costs is heightened. An owner who ignores this provision does so at its own peril and expense.

5.11 Before executing the Contract for Construction, the Owner shall coordinate the Architect's duties and responsibilities set forth in the Contract for Construction with the Architect's services set forth in this Agreement. The Owner shall provide the Architect a copy of the executed agreement between the Owner and Contractor, including the General Conditions of the Contract for Construction. **(new)**

It is important to the owner to coordinate the services required of the architect in this agreement with those indicated in the owner-contractor agreement. If standard AIA documents are used throughout, this should not be an issue. It becomes more difficult if substantial modifications are made, or if custom documents are used. If the owner fails to provide the architect with copies of the referenced documents, it will be difficult for the architect to perform its construction phase duties.

5.12 The Owner shall provide the Architect access to the Project site prior to commencement of the Work and shall obligate the Contractor to provide the Architect access to the Work wherever it is in preparation or progress. **(2.6.2.3; modifications)**

This includes any portion of the Work that is performed off-site.

§ 2.22 Article 6: Cost of the Work

ARTICLE 6 COST OF THE WORK

6.1 For purposes of this Agreement, the Cost of the Work shall be the total cost to the Owner to construct all elements of the Project designed or specified by the Architect and shall include contractors' general conditions costs, overhead and profit. The Cost of the Work does not include the compensation of the Architect, the costs of the land, rights-of-way, financing, contingencies for changes in the work or other costs that are the responsibility of the Owner. **(1.3.1.1, 1.3.1.2, 1.3.1.3; substantial modifications)**

This is the estimated cost as provided by the architect before bidding. After bids are received or a contract is negotiated with one or more contractors, it is the total of the contract costs, but not including contingencies. Once construction has been completed, it is the actual cost of the work as designed by the architect, whether or not the owner actually pays all of these costs.

In a Nebraska case, the bids came in too high.[157] The architect refused to change the plans until the owner paid his prior bills, and the owner blamed the architect for the high cost. The owner took the architect's plans to a draftsman to be redone, after which the architect sued to recover his fee. The court interpreted this paragraph to

---

[157] Getzschman v. Miller Chem. Co., 232 Neb. 885, 443 N.W.2d 260 (1989).

mean that the construction cost upon which the architect's fee was based is the lowest bid for the work designed by the architect, not as changed by someone else. In a New York case,[158] the court found that the term "cost of construction" is a term of art that has an accepted meaning among architects and the AIA. Even though the contract in that case was not an AIA contract, this term had the same meaning as defined in AIA contracts since the parties failed to specify a different meaning in their contract. The current AIA documents use the term *Cost of the Work*, which is defined in this paragraph. It essentially has the same meaning as "cost of construction." The term *Work* is defined in A201 (**see** ¶ 10.2 herein, referencing A201 for definitions of terms).

This provision should be discussed with any owner who is furnishing its own labor or equipment and when the contract is a percentage contract. Prior to 2007, the definition of Cost of the Work included "the cost at current market rates of labor and materials furnished by the Owner and equipment designed, specified, selected or specially provided for by the Architect, including the costs of management or supervision of construction or installation provided by a separate construction manager or contractor, plus a reasonable allowance for their overhead and profit." Thus if the owner obtained labor at a reduced rate or free (for instance, if the owner were a church), the definition of Cost of the Work included the fair market value of the free labor or free materials. This meant the architect would be fairly compensated at the same rates as if the owner was paying for all of the labor and materials. Under the 2007 version of A201, this has been changed to be only the actual cost to the owner of labor and materials. In the situation where the owner obtains free labor or materials, the cost to the owner would be less than the fair market value, and the architect would be paid a fee based on a percentage of this reduced rate. If the owner is a nonprofit, or if the owner will be performing work designed by the architect using the owner's own forces, where the true cost to the owner would be difficult to determine, then this language should be modified if the architect is to be paid based on a percentage of the Cost of the Work. In such a case, the architect may want to be compensated at a fixed fee.

The prior version of this section also allowed the architect to include a "reasonable allowance for contingencies." This allowance would be deleted once the actual contract for construction was signed. The contingency language is important to cover unforeseen conditions that occur on practically every project. Owners should realize that it is impossible to foresee every conceivable contingency and that a reasonable cost allowance must be provided. In the current version, this allowance is moved to ¶ 6.3.

Under ¶ 6.1, the construction cost does not include any engineering or other consultants' costs, the cost of soil borings, or similar services. It also would not include the costs of interior designers or other services paid for directly by the owner and for which the owner is responsible. The architect should be sure to

---

[158] Harza Northeast, Inc. v. Lehrer McGovern Bovis, Inc., 680 N.Y.S.2d 379 (App. Div. 1998).

include costs of coordinating the work of consultants retained by the owner, along with a reasonable markup, in its fees.

> *6.2 The Owner's budget for the Cost of the Work is provided in Initial Information, and may be adjusted throughout the Project as required under Sections 5.2, 6.4 and 6.5. Evaluations of the Owner's budget for the Cost of the Work, the preliminary estimate of the Cost of the Work and updated estimates of the Cost of the Work prepared by the Architect, represent the Architect's judgment as a design professional. It is recognized, however, that neither the Architect nor the Owner has control over the cost of labor, materials or equipment; the Contractor's methods of determining bid prices; or competitive bidding, market or negotiating conditions. Accordingly, the Architect cannot and does not warrant or represent that bids or negotiated prices will not vary from the Owner's budget for the Cost of the Work or from any estimate of the Cost of the Work or evaluation prepared or agreed to by the Architect.* **(2.1.7.2; minor modifications; add first sentence)**

If the bids come in too high, the owner has several remedies, including having the architect redraw the project at no additional cost (**see ¶** 6.6). The architect is not, however, liable to the owner for consequential damages (**see ¶** 8.1.3), such as interest costs, real estate taxes, or other costs associated with ownership of the property, even if the project is abandoned because the bids came in too high.

> *6.3 In preparing estimates of the Cost of the Work, the Architect shall be permitted to include contingencies for design, bidding and price escalation; to determine what materials, equipment, component systems and types of construction are to be included in the Contract Documents; to make reasonable adjustments in the program and scope of the Project; and to include in the Contract Documents alternate bids as may be necessary to adjust the estimated Cost of the Work to meet the Owner's budget for the Cost of the Work. The Architect's estimate of the Cost of the Work shall be based on current area, volume or similar conceptual estimating techniques. If the Owner requests detailed cost estimating services, the Architect shall provide such services as an Additional Service under Article 4.* **(2.1.7.3; add last two sentences; minor modifications)**

The architect should include contingencies in the cost estimates. The architect is also expected to exercise judgment in making assumptions as to materials, equipment, and other elements of the project for cost estimating purposes. One way of keeping costs down is to have alternates in the bids. For instance, the building could be priced with a masonry or alternative exterior, or bronze or aluminum cladding. If the owner increases the contract sum by a change order after the construction contract is signed, the budget is automatically adjusted.

Under the 1976 edition, this estimate was termed a Statement of Probable Cost.[159] In reality, it was simply an estimate, and frequently it was a pure

---

[159] In Brown v. Cox, 459 S.W.2d 471 (Tex. 1970), the court held that this estimate did not require a detailed cost estimate because other provisions of the agreement provide for additional fees to the architect for such detailed estimates. Further, the estimate was not required to be in unit costs but could merely be a lump sum estimate.

guess. Architects generally depend on prior similar projects, reference books, or input from contractors to come up with unit costs. Owners usually do not realize the extent to which these estimates are inexact. Not only is the architect generally unskilled in this area, but construction costs are subject to wide fluctuation. This can lead to serious conflict between the owner and architect. The language in this and the next subparagraph tells the owner that the architect's estimate is not a guarantee but is simply an educated guess at the cost.

The reference to "detailed cost estimating services" in the last sentence generally requires the architect to hire an outside cost consultant. This would be an additional service.

6.4 If the Bidding or Negotiation Phase has not commenced within 90 days after the Architect submits the Construction Documents to the Owner, through no fault of the Architect, the Owner's budget for the Cost of the Work shall be adjusted to reflect changes in the general level of prices in the applicable construction market. **(2.1.7.4; modifications)**

This covers the situation where there is a substantial delay between delivery of the Construction Documents to the owner and the bidding or negotiation. If more than 90 days have passed, it is reasonable to anticipate that the budget will need to be increased. If the owner is unwilling to increase the budget, the architect should not be held to the restriction of paragraphs 6.5 or 6.7.

6.5 If at any time the Architect's estimate of the Cost of the Work exceeds the Owner's budget for the Cost of the Work, the Architect shall make appropriate recommendations to the Owner to adjust the Project's size, quality or budget for the Cost of the Work, and the Owner shall cooperate with the Architect in making such adjustments. **(2.1.7.1, minor modifications)**

If the anticipated cost of the project exceeds the updated budget, the project needs to be brought back into line. The following section gives various options for accomplishing this if this occurs after the completion of the construction documents, including increasing the budget. This situation normally occurs if the bids come in too high. If this occurs prior to this stage of the project, the architect will work with the owner to modify the design and the documents to bring the cost back down to the budget amount.

6.6 If the Owner's budget for the Cost of the Work at the conclusion of the Construction Documents Phase is exceeded by the lowest bona fide bid or negotiated proposal, the Owner shall: **(2.1.7.5; no change)**

   .1   give written approval of an increase in the budget for the Cost of the Work; **(2.1.7.5.1; no change)**

In Gunter Hotel Inc. v. Buck, 775 S.W.2d 689 (Tex. App. 1989), one issue revolved around whether a construction budget had been established. The architect argued that this budget included only those items over which the architect had control, but the owner argued that this budget included all costs. A jury found in favor of the architect.

.2   *authorize rebidding or renegotiating of the Project within a reasonable time;* (**2.1.7.5.2; no change**)

.3   *terminate in accordance with Section 9.5;* (**2.1.7.5.3; change reference**)

.4   *in consultation with the Architect, revise the Project program, scope, or quality as required to reduce the Cost of the Work; or* (**2.1.7.5.4; minor modifications***)

.5   *implement any other mutually acceptable alternative.* (**new**)

If the low bid comes in over the latest revised budget, the owner has several choices in how to proceed. These choices range from simply increasing the budget and, presumably, accepting the low bid, to rebidding or abandoning the project. The owner can also ask the architect to revise the Construction Documents to bring the price down, in which case the following provision will apply.

6.7** If the Owner chooses to proceed under Section 6.6.4, the Architect, without additional compensation, shall modify the Construction Documents as necessary to comply with the Owner's budget for the Cost of the Work at the conclusion of the Construction Documents Phase Services, or the budget as adjusted under Section 6.6.1. The Architect's modification of the Construction Documents shall be the limit of the Architect's responsibility under this Article 6.* (**2.1.7.6; modifications)

This provision states that if the budget is exceeded by the lowest bid and the owner wishes to proceed and modify the contract documents, the architect shall have authority to modify the contract documents as necessary to bring the project within cost. The owner must be reasonable in cooperating with the architect in making such changes.

The agreement states that the architect is not a guarantor of the construction cost (¶ 6.2). There was no cost limitation under B141-1987 or B151-1997,[160] but the current version of B101 (as well as B141-1997) limits the cost to the budget. A problem that often arises is that bids come in substantially over the architect's estimate and the owner then does not want to pay the architect its fee. Unless negligence is shown, the architect is usually entitled to its fee,[161] at least under

---

[160] In Getzschman v. Miller Chem. Co., 232 Neb. 885, 443 N.W.2d 260, 270 (1989), in which the owner agreed to provide estimating services, the court stated: "However, when an architect has no express contractual obligation to design a structure within a specified budget or to estimate the construction cost of a proposed project, construction at a cost greater than anticipated by or acceptable to the owner is no defense to an architect's action to recover a fee." *Getzschman* was cited in Kahn v. Terry, 628 So. 2d 390 (Ala. 1993). There, under a prior version of B141, the parties disagreed as to whether there had been a fixed limit of construction cost. The court held that this was a matter for the jury to determine and upheld the jury's finding in favor of the architect.

In Anderzhon Architects, Inc. v. 57 Oxbow II P'ship, 250 Neb. 768, 553 N.W.2d 157 (1996), the court held that if the standard AIA no-fixed-limit-language is used, the architect is entitled to its fee. The fact that the construction cost exceeded the anticipated cost by more than 35% is irrelevant.

[161] Pipe Welding Supply v. Haskell, Connor & Frost, 96 A.D.2d 29, 469 N.Y.S.2d 221 (1983), *aff'd*, 61 N.Y.2d 884, 462 N.E.2d 1190, 474 N.Y.S.2d 472 (1984) (bids came in 33 to 45% above estimate, and architect was awarded his fees). *See also* Moossy v. Huckabay Hosp., Inc., 283 So. 2d 699 (La. 1973); Jetty, Inc. v. Hall-McGuff Architects, 595 S.W.2d 918 (Tex. App. 1980).

the prior version of these documents. When the estimate "substantially" exceeds the estimated construction cost, architects have often been denied their fees.[162] This appears to be contrary to the reality of the construction industry. Architects often give estimates to owners before the construction drawings are completed. When contractors bid on final sets of drawings, the bids often range significantly. How, then, can the architect hope to give a reasonable estimate? And how can the courts hold the architect to such an estimate?

When, as in the 2007 edition of B101, a limit to the construction cost is set, the architect might still be entitled to a fee if the bids come in over the limit by less than a "substantial" amount.[163] There is no definition for what a substantial amount is.

In Anderson-Nichols & Co. v. Page, 309 N.H. 445, 309 A.2d 148 (1973), the architect was hired by the contractor for a turnkey project. The contractor submitted an improper bid for the work, lost the project, and refused to pay the architect's fee. The court found that there was no contractual provision requiring the architect to assist in preparing the bid. In such a situation, the standard AIA documents must be modified.

In Brenner v. DeBruin, 589 N.Y.S.2d 56, 57 (App. Div. 1992), the owner hired an engineer to design a single-family house and advised the engineer that he had a budget of $100,000. The contract stated that "all work is predicated upon [the plaintiffs] intention to maintain a budget for the cost of construction in the area of $100,000." When the lowest bid came in at $160,000, the owner sued for breach of contract and fraud. The court found for the engineer because the engineer never represented that he was an expert with respect to construction costs or that the cost of constructing the residence as proposed and approved by the owner would not exceed $100,000. The court found that several changes had been made to the design subsequent to the agreement, that the owner had approved the changes, and that the parties had never discussed the potential costs of these changes.

In Housing Vt. v. Goldsmith & Morris, 685 A.2d 1086 (Vt. 1996), the final grading plan prepared by the architect failed to comply with the standard of care required of architects for such plans. The architect has prepared a site plan that contained grading information used for the grades at a ravine. When these steep slopes proved to be unstable, they had to be redesigned and stabilized at a substantial additional cost. The architect argued that it was not responsible for the grading plan, but the court held otherwise. This was based in part on the fact that the architect had sealed the site plans with the grading information.

The architect also argued that prior language from this document (now ¶ 6.2 and 6.3) shielded it from liability for cost overruns caused by its malpractice. The court held that these provisions are "broadly worded disclaimers . . . [which] contain no reference to negligence or wrongful conduct of any kind, and thus cannot insulate defendant from liability for malpractice."

[162] Durand Assocs., Inc. v. Guardian Inv. Co., 186 Neb. 349, 183 N.W.2d 246 (1971) (bid was $652,728, estimate was $420,000; court found that it would be against public policy to absolve architect from any liability whatsoever from his estimates). *See also* Stanley Consultants v. H. Kalicak Constr. Co., 383 F. Supp. 315 (E.D. Mo. 1974); Kostohryz v. McGuire, 298 Minn. 513, 212 N.W.2d 850 (1973); Williams & Assocs. v. Ramsey Prod. Corp., 19 N.C. App. 1, 198 S.E.2d 67 (1973); Peteet v. Fogarty, 297 S.C. 226, 375 S.E.2d 527 (Ct. App. 1988) (owner verbally set $150,000 maximum cost, bid was $307,000; owner entitled to return of architectural fees); Capitol Hotel Co. v. Rittenberry, 41 S.W.2d 697 (Tex. Civ. App. 1931).

[163] *See* Annotation, *Building Contract Cost Limitations*, 20 A.L.R.3d 778 (1968). In Clark v. Madeira, 252 Ark. 157, 477 S.W.2d 817, 818 (1972), an architect had agreed to a construction cost of $23,000. The final cost was $43,000 and the architect sued to recover his fee based on 6% of the larger construction cost. The court stated that one cannot profit from one's own wrong and

It can be argued that the language in the 2007 version of B101 indicates an absolute limit which is exceeded by even a trivial amount. The architect may want to change this provision to permit greater latitude in the cost.

In some jurisdictions, such a limit is absolute and no fee is due if the amount is exceeded, even if the additional fee was a result of additional services.[164] This view reflects the intent of a fixed-limit provision. Of course, the agreement permits the architect to modify the drawings to reduce the scope of the project, and the owner must cooperate with the architect.[165] One court, however,[166] held that the owner was under no such requirement. There, the architect had agreed to design a prestressed concrete building of a certain size for $250,000. The low bid came in at $317,000. Because the parties contracted only for a specific type of building at a given price, there was no obligation on the owner to change the size or type of building.

If the owner makes changes that increase the cost, there may not be any cost limitation.[167] Some courts have permitted an owner to prove that a fixed limit was

---

refused to allow the additional fee: "An architect whose cost estimate is culpably below the actual cost of the job is not entitled to a commission upon the excess."

In Griswold & Rauma, Architects, Inc. v. Aesculapius Corp., 301 Minn. 121, 221 N.W.2d 556, 560 (1974), the court stated:

> One very significant factor is whether the agreed maximum cost figure was expressed in terms of an approximation or estimate rather than a guarantee. Where the figure was merely an approximation or estimate and not a guarantee, courts generally permit the architect to recover compensation provided the actual or probable cost of construction does not substantially exceed the agreed figure.

[164] Harris County v. Howard, 494 S.W.2d 250 (Tex. 1973). The owner had included a clause specifying that it had an absolute limitation on the amount that would be paid to the architect. The architect was requested to perform additional services by the owner's representatives, but no written modification of the contract was made. The fact that the owner was a governmental body undoubtedly was a factor in the decision, but the court based its finding on the clause limiting the architect's fee, finding that clause unambiguous and finding the contract contemplated additional services. In a private contract situation, the architect might have argued that the additional services were a subsequent modification of the contract.

[165] See Griswold & Rauma, Architects, Inc. v. Aesculapius Corp., 301 Minn. 121, 221 N.W.2d 556, 561 (1974) ("While it might be against public policy to allow an architect, under such a provision, to reduce substantially the area of a proposed project, it seems that, barring a specifically guaranteed area, a reasonable reduction in the size of the project should be allowed when necessary to meet the construction cost.").

[166] Stevens v. Fanning, 59 Ill. App. 2d 285, 207 N.E.2d 136 (1965).

[167] Cobb v. Thomas, 565 S.W.2d 281 (Tex. 1978). The architect estimated the cost of a house at $500,000. Due to extras ordered by the owner, a six-month delay in construction, and the fact that the owners entered into a cost-plus contract, the architect was not guilty of negligence when the cost finally came in at $660,000. There was no evidence that the architect promised that the cost would not exceed the stipulated sum, merely that he gave the owners the "probable" cost of the construction. There was no evidence that the architect's statement that the house could be built for $500,000 was false at the time it was made, and therefore, there was no fraud by the architect. See also Schwender v. Schrafft, 246 Mass. 543, 141 N.E. 511 (1923).

In Moore v. Bolton, 480 S.W.2d 805 (Tex. 1972), an architect sent a letter confirming the oral understanding of the parties that he would design a house for the owners for a fee of 10% of the

agreed to even though it was not stated in the agreement.[168]

Under this document, the architect is required to furnish estimates to the owner during the Schematic Design Phase (¶ 3.2.6) and thereafter as the design progresses through the end of the construction documents phase. Construction costs that are estimated to be too high for the owner's budget may be a proper cause for termination of an architect's services, even if no fixed limit had been agreed to between the owner and architect.[169] When an owner terminates the architect for this reason, the architect is entitled to all amounts properly due under the contract up to the time of termination. However, if the contract called for the owner's approval of a "preliminary study plan" and a fee for the workup to such a plan,

---

construction cost. The agreement was never signed by the owners, although they told the architect to proceed and worked with him throughout. The owners testified that there had been an agreement that the cost of the house would be $75,000. The owners, however, wanted a number of changes, and the architect advised them in writing that the costs were increasing. The final bids came in at $108,000 and the owners abandoned the project. The court held that the architect was entitled to his fee based on the higher amount, less 20%, which represented the cost of construction administration.

In Willner & Millkey v. Shure, 124 Ga. App. 268, 183 S.E.2d 479 (1971), an architect contracted to design a house costing $60,000. The owner made a number of changes. When the architect advised that these changes would increase the cost, the owner told him not to worry, that they would work it out. The lowest bid came in at $73,000 and the owner abandoned the project. The fee was based on a percentage of the construction cost, and the architect was permitted to proceed with the action. An interesting footnote is that the work was done in 1962. The architect testified that, because of an oversight, he did not bill the owner until late in 1967. A statute of limitations argument was raised, the owner contending that the contract was abandoned and the action could only be brought on quantum meruit. The court held, however, that it was based on the longer contract statute of limitations, although quantum meruit was the rule of evidence to be used to determine the value of the services under the contract.

In Circus Studio, Ltd. v. Tufa, 145 Vt. 219, 485 A.2d 1261 (1984), the architect drafted a letter to the homeowner as the basis for the fee. The letter was never signed by the owner, who did not use the architect after the initial design, claiming that the estimate of construction was too high. The court held that the architect was entitled to the reasonable value of his work because the owner accepted the architect's work product and utilized it for his own use.

In Brenner v. DeBruin, 589 N.Y.S.2d 56 (App. Div. 2d Dep't 1992), the written agreement between the owner and engineer contained a provision that "all work is predicated upon [the plaintiffs] intention to maintain a budget for the cost of construction in the area of $100,000." The lowest bid came in at $160,000. The court held in favor of the engineer, finding that several changes had been made to the design of the residence subsequent to the parties' agreement, that the owner approved those changes, and that the parties never discussed the potential costs of the changes. There was no breach of the contract because the engineer never represented that the cost would not exceed the original budget cost.

[168] Bair v. School Dist. No. 141, 94 Kan. 144, 146 P. 347 (1915); Bleck v. Stepanich, 64 Ill. App. 3d 436, 381 N.E.2d 363 (1978); Williams & Assocs. v. Ramsey Prod. Corp., 19 N.C. App. 1, 198 S.E.2d 67 (1973); Fishel & Taylor v. Grifton United Methodist Church, 9 N.C. App. 224, 175 S.E.2d 785 (1970). *But see* Kurz v. Quincy Post No. 37, 5 Ill. App. 3d 412, 283 N.E.2d 8 (1972) (court found no agreement as to cost when architect's several revisions resulted in increased costs, all apparently with owner's knowledge).

[169] Shanahan v. Universal Tavern Corp., 585 P.2d 1314 (Mont. 1978).

the architect would only be entitled to that initial fee if the owner decides to terminate the services because the estimated construction cost is too high.[170]

One owner has argued that the "construction cost" is the amount that the owner actually spent toward the cost of construction, which was nothing in that case because the owner abandoned the project.[171] The court rejected this interpretation, saying that because the architect had almost fully performed his part of the contract, it was reasonable to infer that the parties intended "construction cost" to mean the amount that construction of the project would cost as determined by the bid and the owner was not at liberty to escape liability by changing his intention to build.

In an Arkansas case, an owner sued the architect several years after the building was finished.[172] The owner attempted to sell the building and found that the area of the building was less than he thought. He sued the architect, seeking damages for the architect's failure to accurately estimate the total number of square feet in the building. The court held for the architect, finding that the difference in area had no effect on the market value of the building.

§ 2.23   Article 7: Instruments of Service

ARTICLE 7 INSTRUMENTS OF SERVICE

7.1 The Architect and the Owner warrant that in transmitting Instruments of Service, or any other information, the transmitting party is the copyright owner of such information or has permission from the copyright owner to transmit such information for its use on the Project. If the Owner and Architect intend to transmit Instruments of Service or any other information or documentation in digital form, they shall endeavor to establish necessary protocols governing such transmissions. (**new**)

Many architects provide drawings and specifications on computer disks or other electronic format (e-mail, Internet, etc.) to the owner. The owner also sometimes provides the architect with such information. An example of this might occur when a large corporation that has had a number of architects working on its facility furnishes a new architect with electronic data for background and other information. The parties should carefully consider the use that might be made of this information and carefully allocate risks associated with these uses, as well as provide for fees to be paid in case of reuse of this information.

---

[170] *Id.*

[171] Michalowski v. Richter Spring Corp., 112 Ill. App. 2d 451, 251 N.E.2d 299 (1969). *See also* Willner & Millkey v. Shure, 124 Ga. App. 268, 183 S.E.2d 479 (1971); Orput-Orput Assocs., Inc. v. McCarthy, 12 Ill. App. 3d 88, 298 N.E.2d 225 (1973).

In Stark v. Ralph F. Roussey & Assocs., 25 Ill. App. 3d 659, 323 N.E.2d 826 (1975), the court held that a contract provision for a fee of 6% of the construction cost should form the basis of the architect's recovery, rejecting the trial court's quantum meruit basis for recovery.

[172] Winrock Homes, Inc. v. Dean, 14 Ark. App. 26, 684 S.W.2d 286 (1985).

This paragraph contains a warranty by both the owner and the architect that, if either of them transmit documents or information, the party sending such documents or information either is the owner of the copyright (usually the author), or has permission from the actual owner to transmit such documents or information to be used for the project.

Among the issues that concern the architect in allowing such transfers are prevention of use of the drawings on other projects, possible errors created by the transmission itself, use of the wrong version of a drawing, and the continuation of mistakes in the architect's drawings onto the shop drawings. For instance, subcontractors normally prepare shop drawings from the drawings created by the architect and its consultants. In the preparation of these shop drawings, there is an opportunity to catch and correct errors because the subcontractor is preparing new drawings. If, on the other hand, the subcontractor is using the architect's electronic files to create the shop drawings, there is a much greater likelihood that the architect's mistakes will not be caught in this process.

Note that as of 2007, the AIA has developed two new documents related to digital data. The first is AIA Document E201-2007, Digital Data Protocol Exhibit. This document is intended to cover procedures regarding the exchange of digital data. A second document, C107-2007, Digital Data Licensing Agreement, is for use when the parties do not already have an AIA agreement in place, such as when the architect sends digital documents to a subcontractor.

The term *digital form* generally refers to the numerous programs available to the architect for CAD systems, although there are other electronic formats, such as the ".pdf" format developed by Adobe. It is the CAD formats that are of greatest concern, since the CAD format can easily be changed or used for other projects. If the owner wants electronic documents, the owner and architect must coordinate that they have the same systems that can read each other's documents.

7.2 *The Architect and the Architect's consultants shall be deemed the authors and owners of their respective Instruments of Service, including the Drawings and Specifications, and shall retain all common law, statutory and other reserved rights, including copyrights. Submission or distribution of Instruments of Service to meet official regulatory requirements or for similar purposes in connection with the Project is not to be construed as publication in derogation of the reserved rights of the Architect and the Architect's consultants.* **(1.3.2.1; 1.3.2.3; add reference to consultants and electronic form)**

There is a difference between providing a service and selling drawings. According to B101, the architect is providing a service and not merely selling a set of drawings. In one case, a firm provided a set of drawings for an owner who was his own general contractor.[173] When the foundation settled because of inadequate design, the architect was held liable on a theory of implied warranty of fitness for purpose. He was held to warrant that the drawings were fit for the purpose for which they were sold, and by implication, that the foundation would be adequate to

---

[173] O'Dell v. Custom Builders Corp., 560 S.W.2d 862 (Mo. 1978).

support the house if the design were followed. Because the architect had sold plans, and the plans were inadequate, he was liable on this warranty theory. As a footnote, the case suggests that proper supervision of the construction might have avoided the problem. The person who poured the foundation testified that he did not know that the fill was new. If he had, he would not have proceeded. Possibly, if the architect had been present during construction, he might also have noticed the problem. Merely selling plans without any input during the construction process is an invitation to a lawsuit.

The 1977 contract language gave the architect both ownership of the drawings and ownership of the copyright. These are two distinct terms. One person can own the copyright and another own the drawings. Without the language giving the architect ownership of the drawings, the owner who employs the architect would become the owner of the drawings and specifications.[174] Because the architect is providing a service and not a product (the drawings are not an end product, the structure is; the drawings are merely a means to construct the building), it is not "selling" the drawings or specifications.

Beginning with the 1987 version, ownership of the documents is no longer important, because use of the documents is restricted. The service aspect of the documents is emphasized, which is similar to an accountant's work papers or a hospital chart. The owner is only entitled to the building and not the drawings or specifications. Indeed, the architect should closely question the owner's motives if the owner insists on ownership of the drawings or the copyright. If the owner's intention is to use these drawings on another building without the architect's input, there is a real possibility for trouble in the construction. Every architect, but not every owner, knows that no two sites or situations are identical. Serious damage or injury can result if an unscrupulous owner tries to save a fee and reuse drawings. The architect is also open to some liability if a third party is injured. The architect might find itself sued because its drawings were used. Its defense, of course, might be perfectly good, but the initial legal costs can still be high.

Copyrights are governed almost exclusively by the Copyright Act of 1976.[175] The Act was amended by the Architectural Works Copyright Protection Act ("AWCPA"),[176]

---

[174] Tumey v. Little, 18 Misc. 2d 462, 186 N.Y.S.2d 94 (Sup. Ct. 1959).

[175] 17 U.S.C. § 101.

[176] One of the first published opinions based on the AWCPA is Value Group, Inc. v. Mendham Lake Estates, L.P., 800 F. Supp. 1228, 1233 (D.N.J. 1992). In that case, the architect designed unique single-family homes for a housing development. In 1992 the architect was contacted by the developer of a competing development, seeking permission to use one of the plans. The architect denied the request, following up with a letter reiterating that the architect owned the plans. The developer proceeded with the house, claiming that it had derived a new design through modification and a total redrawing of the original plans. The architect subsequently learned that the plans filed with the building authority were virtually identical to his plans. The architect and his developer then brought a copyright infringement action, also seeking a temporary restraining order.

effective as of November 30, 1990.[177] The Act applies to "any architectural work

---

The court agreed that there was an infringement and issued the restraining order. The opinion gives a good overview of copyright law as applied to architectural works. Among the evidence that proved infringement was the fact that the defendants included a photocopy of the first architect's brochure in their sales contract with the infringing buyer. In determining substantial similarity, the court used this test: sufficient similarity is found where the work is recognizable by the ordinary observer as having been taken from the copyrighted source. . . . When viewing Value Group's and Mendham Lake's plans side by side, there is no question that the plans are "substantially similar." Since the standard for originality and therefore copyrightability is so low, copying the floor plan alone infringes the copyright. Additionally, many of the details on each set of plans are exactly the same. For example, the placement, design, and size of many windows are identical.

In Rottlund Co., Inc. v. Pinnacle Corp., 452 F.3d 726 (8th Cir. 2006), a new trial was ordered because the trial judge had improperly allowed the expert testimony of an architectural expert. With reference to the issue of substantial similarity, the expert gave an opinion that there was no direct evidence of copying the architectural plans and described the differences between the plans. The appeals court found that expert testimony was not appropriate. "Expert opinion and analytical dissection are inadmissible to prove similarity of expression."

[177] The Copyright Office, Library of Congress, issued final regulations concerning architectural works, effective as of October 1, 1992. Some of the important provisions are:

(2) The term building means humanly habitable structures that are intended to be both permanent and stationary, such as houses and office buildings, and other permanent and stationary structures designed for human occupancy, including but not limited to churches, museums, gazebos, and garden pavilions.

(c) Registration

(1) Original design. In general, an original design of a building embodied in any tangible medium of expression, including a building, architectural plans, or drawings, may be registered as an architectural work.

(2) Registration limited to single architectural work. For published and unpublished architectural works, a single application may cover only a single architectural work. A group of architectural works may not be registered on a single application form. For works such as tract housing, a single work is one house model, with all accompanying floor plan options, elevations, and styles that are applicable to that particular model.

(3) Application form. Registration should be sought on Form VA. Line one of the form should give the title of the building. The date of construction of the building, if any, should also be designated. If the building has not yet been constructed, the notation "not yet constructed" should be given following the title.

(4) Separate registration for plans. Where dual copyright claims exist in technical drawings and the architectural work depicted in the drawings, any claims with respect to the technical drawings and architectural work must be registered separately.

(5) Publication. Publication of an architectural work occurs when underlying plans or drawings of the building or other copies of the building design are distributed or made available to the general public by sale or other transfer of ownership, or by rental, lease, or lending. Construction of a building does not itself constitute publication for purposes of registration, unless multiple copies are constructed.

(d) Works excluded. The following structures, features, or works cannot be registered:

(1) Structures other than buildings. Structures other than buildings, such as bridges, cloverleafs, dams, walkways, tents, recreational vehicles, mobile homes, and boats.

(2) Standard features. Standard configurations of spaces, and individual standard features, such as windows, doors, and other staple building components.

that, on the date of the enactment of this Act, is unconstructed and embodied in unpublished plans or drawings." Under this Act, an *architectural work* is

> the design of a building as embodied in any tangible medium of expression, including a building, architectural plans, or drawings. The work includes the overall form as well as the arrangement and composition of spaces and elements in the design, but it does not include individual standard features.

In other words, architects are now protected from someone's copying the building itself,[178]

---

(3) Pre-December 1, 1990 building designs. The designs of buildings where the plans or drawings of the building were published before December 1, 1990, or the buildings were constructed or otherwise published before December 1, 1990.

[178] In Guillot-Vogt Assocs. v. Holly & Smith, 848 F. Supp. 682 (E.D. La. 1994), the architect on a state project hired a firm to design the electrical and mechanical portions of the project. The architect then advised the state that, because of failing health, he wanted his contract terminated. The architect had been paid in full for his work, but he had failed to fully pay his subcontractor for the engineering work. When the replacement architect wanted the subcontractor to continue its work, the subcontractor refused because both the state and the second architect refused to pay the balance owed for the work previously done. The subcontractor then filed for copyright registration for its drawings and filed suit for copyright infringement.

The court reviewed the 1990 Act and stated: "The Court interprets the 1990 Amendment primarily as providing previously lacking copyright protection to physical architectural works, not drawings of such works. Indeed, drawings, whether of architectural works, or otherwise, have long enjoyed copyright protection."

The state also argued, based on 17 U.S.C. § 120(b), that copyright protection was lacking because the project involved an alteration to an existing structure. The court held that this section is not intended to nullify copyright protection in drawings. The drawings in question are not a building, which is the subject of that section of the Act.

Finally, the state argued that the drawings were not copyrightable because they were utilitarian or mechanical in nature. This was rejected on the basis that there was no showing that the drawings in question were the only meaningful way to depict such articles. *See* Javelin Inv., LLC v. McGinnis, 2007 WL 781190 (S.D.Tex., Jan. 23, 2007), distinguishing *Guillot-Vogt* and interpreting Section 120(b) and finding that this Section allows an owner to use an architect's plans to complete a house following termination of the architect. This was a case, however, with bad facts creating bad law and should not be relied on in permitting an owner to make copies of an architect's plans without a license.

In Hunt v. Pasternak, 179 F.3d 683 (9th Cir. 1999), the Ninth Circuit overruled the trial court, finding that the AWCPA allowed a valid copyright in an architectural work even when the work has not been constructed. The owner had hired an architect to design a pizza restaurant. Dissatisfied with the plans, the owner then hired the plaintiff architect, who prepared a new set of plans. Again dissatisfied, the owner hired a third architect, who prepared plans presumably based on those prepared by the second architect. The second architect obtained a copyright registration for the architectural work embodied in his plans and filed suit for copyright infringement against the owner and third architect. It appears that there was no separate registration of the drawings as technical drawings, only as an architectural work. This appeared to confuse the trial court, which dismissed the action, believing that it was based on alleged infringement of the plans, not the architectural work. The Ninth Circuit, however, read the complaint as alleging infringement of the design of the architectural work, not of a separate copyright in the plans or drawings themselves.

In National Med. Care, Inc. v. Espiritu, 284 F. Supp. 2d 424 (S.D. W. Va. 2003), a dialysis service provider sued various parties for infringement of standard details, relating to things such as cabinets. The drawings were registered as technical drawings. The court held that an as-built structure or feature cannot be an infringing copy of a technical drawing. Copyright protection extends to as-built structures only when the copyright is registered under the AWCPA.

The *Espiritu* case was discussed in Oravec v. Sunny Isles Luxury Ventures L.C., 469 F. Supp. 2d 1148 (S.D. Fla., July 24, 2006). The plaintiff was an architect in Czechoslovakia who came to the United States in 1983 where he was never licensed as an architect. Starting in 1996, he created several building designs which he registered with the Copyright Office over several years. In 2000, the defendants completed a project that the plaintiff believed infringed on his copyrights. One of the issues in this case was whether the completed buildings could infringe on one of the copyrights, specifically a copyright that was not for an architectural work. The court followed *Espiritu* in finding that there could be no infringement of this particular copyright, finding that an as-built structure does not infringe upon the architectural drawings and plans that depict the design. Apparently anticipating this result, the plaintiff asked for leave to amend his complaint based on a copyright registration that he had just filed, as an architectural work. The court rejected this as being too late. This case also provides a good analysis of several concepts central to architectural copyright issues.

See Rottlund Co., Inc. v. Pinnacle Corp., 2004 WL 1879983 (D. Minn., Aug. 20, 2004), *rev'd*, 452 F. 3d 726 (8th Cir. 2006).

In T-Peg, Inc., v. Isbitski, 2005 WL 768594 (D.N.H., April 6, 2005), the plaintiff sued a timberframe contractor for infringement of its architectural work. The interesting point is that the alleged infringement was for the design and construction of a structural frame for a building, namely the timberframe. The defendant did not design or build the house that was found to be substantially similar to plaintiff's work; it only designed and built the timberframe. This timberframe was capable of supporting many different arrangements of windows, doors and other elements.

While VTW's timberframe allows for rooms, doors, and windows to be placed as depicted in the second preliminary plans, nothing in VTW's timberframe, either as designed in the shop drawings or as built, *requires* those elements to be so placed. . . . In short, VTW's timerberframe does not reflect any particular "overall form" or "arrangement and composition of spaces and elements" because one of its attractive features is its flexibility—that frame can accommodate multiple building designs including, but certainly not limited to, the design embodied in the second preliminary plans.

The court concluded that plaintiff did not demonstrate the necessary substantial similarity.

T-Peg was reversed on appeal. 459 F.3d 97 (1st Cir. 2006). The court did an extensive analysis of the Architectural Works Protection Act. Under prior law, only the plans were protected as a pictorial, graphic, or sculptural work. However, there was a requirement that the protectable elements of such a work be separated from the "utilitarian aspects" of the work under 17 USC § 102(a)(5). By contrast, however, there is no such separability test for architectural works. The copyright in question here was one as an architectural work. There was no separate copyright as technical drawings. Whether this court totally grasped this distinction is open to question. The court did an analysis of substantial similarity between the competing works and held that there was an issue of material fact that would preclude summary judgment. The lower court had also emphasized that the defendant here had designed only a frame, while Timberpeg's plans did not contain a complete frame design. This was of no consequence as the question here was whether a reasonable jury could conclude that the infringing frame as drawn and built is substantially similar to Timberpeg's architectural work, which includes "the overall form as well as the arrangement and composition of spaces and elements in the design." The appellate court also rejected an argument by the defendant that the defendant's work was not a "building" for purposes of an architectural work. The statute does not require that the infringing work meet the definition of an architectural work.

not only the drawings.[179] The term *individual standard features* presumably refers to details that are commonly used, such as typical flashing details, standard doors and windows, toilets, and the like. The copyright protection does include the arrangement of such standard details into a new and original building shape.[180] The AWPCA applies to "buildings," meaning habitable structures.[181] Non-habitable structures, such as bridges, are not covered by the AWCPA. Buildings

---

In Cornerstone Home Builders, Inc. v. Lemanski, 2005 WL 1863387 (M.D. Fla. Aug. 5, 2005), the builder and homeowner entered into a contract to construct a house using one of the builder's copyrighted plans. When a dispute arose during construction, the homeowner terminated the builder and hired another contractor to complete the house. The builder filed suits in state court for breach of contract and in federal court seeking a declaratory judgment regarding its right to protect copyright interests, then sought an injunction to prevent the owner from completing the structure. The court declined to issue the injunction because there were unresolved factual issues. The homeowner denied copying the plans because the work being completed did not use the copyrighted plans. Rather, according to the homeowner, the work involved the installation of finish treatment according to local code requirements.

[179] In Shalom Baranes Assocs., P.C. v. 900 F St. Corp., 940 F. Supp. 1 (D.C. 1996), the parties had a letter memorandum that referred to the standard AIA document by stating that "this proposal letter temporarily substitutes for the AIA Agreement." After having prepared a feasibility study and some architectural drawings and not being fully paid, the architect registered its copyright in both and sued. The owner hired a new architect to prepare drawings. The court held that the feasibility study did not merit copyright protection and that claim fell. Apparently, there was not proof offered of copyright infringement in the drawings. The court did hold for the architect on its contract claim.

[180] In Lindal Cedar Homes, Inc. v. Ireland, 2004 WL 2066742 (D. Or., Sept. 14, 2004), the court rejected the defendant's argument that the work in question was not entitled to copyright protection because it is merely a compilation of standard features in the public domain.

Defendant focuses on the separate unprotectable elements of the Parkside design, e.g., the fact that the home has bedrooms and other standard features, and leaps to the conclusion that the design is not protected. However, the originality of the design is in the selection of its elements and in the coordination and arrangement of those elements into a design. Although the underlying unoriginal component parts of a creation are not subject to protection, a creator's selection and arrangement of these parts, where independently made, is original and therefore copyrightable.

In Jeffrey A. Grunsenmeyer & Assocs., Inc. v. Davidson, Smith & Certo Architects, 2007 WL 62620 (6th Cir. Jan. 8, 2007), the plaintiff sued for copyright infringement, claiming that the defendant used the plaintiff's copyrighted "existing site and building conditions" without permission. These drawings, prepared by the plaintiff for the owner pursuant to a contract for master planning, depicted existing conditions at the site. The defendant contended that the drawings were a compilation of facts and did not possess the requisite degree of creativity necessary for copyright protection. The court agreed, finding that the plaintiff's drawings set forth the existing physical characteristics of the facilities and, as such, set forth facts. These were not sufficiently unique so as to render the drawings protected. As a follow-up, the court also found that, even if the drawings were unique and entitled to copyright protection, the defendant would still be entitled to summary judgment because the owner had bargained for drawings to be used for a master plan that would be subject to later implementation. The defendant was working on the implementation and would be entitled to use the drawings. Basically, the defendant enjoyed a license to use the drawings.

[181] The Copyright Office cited the definition of *building* taken from 37 C.F.R. § 202.11(b)(2) in Viad Corp. v. Stak Design, Inc., 2005 WL 894853 (E.D. Tex., Apr. 14, 2005), as "humanly habitable structures that are intended to be both permanent and stationary, such as houses and office buildings, and other permanent and stationary structures designed for human occupancy,

existing prior to the date of the Act, 1990, are not covered by the AWCPA, and photographing the building is not prohibited. Also, the owners of such buildings may alter or demolish them without consent of the architect.[182] Note that it is the expression of an idea that is subject to copyright, not the idea itself.[183]

Unless the architect specifically transfers the copyright to another party, it keeps the copyright to the drawings and specifications.[184] To protect itself, the architect

---

including but not limited to churches, museums, gazebos, and garden pavilions." Viad had sought to register 17 kiosks as architectural works, but the application was rejected by the Copyright Office. Viad argued to the court that the Copyright Office's definition was improperly narrow. The court agreed that the term "building" is ambiguous, but declined to hold that a kiosk qualified as a "building." The Copyright Office's definition of a "building" was entitled to some deference and was reasonable in this case.

A parking garage designed for a university is a "building" within this definition. Moser Pilon Nelson Architects, LLC v. HNTB Corp., 2006 WL 2331013 (D.Conn., Aug. 8, 2006). The infringing work does not need to fit the definition of a building, only the infringed-upon work. T-Peg, Inc., v. Vermont Timber Works, Inc., 459 F.3d 97 (1st Cir. 2006).

[182] Javelin Inv., LLC v. McGinnis, 2007 WL 781190 (S.D. Tex., Jan. 23, 2007).

[183] In Eales v. Envtl. Lifestyles, Inc., 958 F.2d 876 (9th Cir. 1992), an architect had prepared plans for several homes in Arizona for a developer. The developer originally hired a California architect to prepare plans, but as the construction began, the developer became concerned that the plans were not suitable for Arizona. The developer then consulted with a local architect, who reported that the original plans should be changed. At that time, the foundations for the homes had been completed. The new architect did not use the old plans but had to design the homes based on the existing foundation locations and other constraints. Apparently, an AIA owner-architect contract was used, because the architect retained ownership of the plans. New permits were issued and construction resumed. However, financial problems again stopped construction, and the developer went out of business.

One of the lots had been sold to a buyer who desired that a certain model be built on that site. The buyer contacted one of the people who had worked for the developer to construct the home because the developer was no longer in business. Nobody contacted the architect to ask permission to use the plans.

The architect found out that her plans were being used to construct the home, and she sent a notice to the contractor, stating that her plans were being used without permission and demanding her usual fee of $4.00 per square foot. No response was made, so she copyrighted her plans and then filed suit in federal court, claiming copyright infringement.

The court awarded the architect not only her normal fee but also an amount equivalent to the contractor's profit on the home, for a total of over $57,000. The defendants had argued that the plans fell under the "useful article" exception to the copyright code, but that argument was summarily rejected. The next argument was that the plans did not substantially differ from the earlier drawings prepared by the California architect. Because the second set of plans was created without consulting the prior drawings and the plans were, in fact, different, this argument was likewise rejected. The court found that the plans were entitled to copyright protection and that they constituted an expression of an idea. The infringer in this case copied the plans and not the idea.

See Hunt v. Pasternak, 179 F.3d 683 (9th Cir. 1999), *further proceedings*, 192 F.3d 877 (9th Cir. 1999), in which the Ninth Circuit clarified its decision in the *Eales* case relating to whether an architectural work can be embodied in plans or drawings as well as a building.

[184] In McCormick v. Amir Constr., Inc., 2006 WL 784770 (C.D. Cal., Jan 12, 2006), the issue was whether the architect had transferred the exclusive rights to the drawings to the owner. The architect, WHA, had entered into a contract with the owner for the design of a residence. The contract included language substantially similar to the AIA copyright provisions. The architect

must comply with the provisions of the Act.[185] Although it is no longer required, architects should comply with the notice provision that all publicly distributed copies[186] of the drawings or specifications should contain three elements: first, the word "copyright," the abbreviation "Copr," or the symbol "c" in a circle: "©"; second, the year of the first publication of the work; and third, the name or a recognizable abbreviation of the copyright owner for identification. For the year of publication, the architect could use the year in which the drawings are issued for permit or, if they are never issued for permit, the date of the final set of drawings. This is because, for architectural plans, there is usually no actual "publication." If the architect complies with this, it will have an easier time prevailing in a copyright infringement action.[187]

In a Nebraska case, an architect prepared a set of drawings for the owner for the construction of a 22-unit apartment complex, based on an oral contract.[188]

---

later filed suit against third parties for copyright infringement and won substantial damages. The owner filed a separate copyright infringement action against the same third parties and the architect, alleging that the owner enjoyed an exclusive license. The court was not persuaded that the architect gave the owner an exclusive license, based on the language that "the Architect shall be deemed the author of these documents and shall retain all the copyrights thereto." Since only the copyright owner or the owner of exclusive rights under the copyright has standing to bring an infringement action, and this owner did not have either, the court dismissed the case. Note that the 1997 version of B141 specifies that the license granted to the owner is a nonexclusive license.

[185] For an interesting case decided before the 1976 Act, *see* Nucor Corp. v. Tennessee Forging Steel Serv., Inc., 476 F.2d 386 (8th Cir. 1973). There, a senior officer of a joist manufacturer took a set of plans for a joist plant from his company. After he became the president of another joist manufacturing company, he had them slightly modified, and a new plant was constructed based on these drawings. When the first company learned of this, it initiated the suit to enjoin the use of the plans, based on infringement of common law copyright and improper disclosure of confidential information. Although the plant was finished, recovery was allowed against the defendant.

[186] In the case of architectural drawings, publicly distributed means drawings that are distributed outside the architect's own office. It is recommended that the notice appear on any drawing that goes out of the architect's office, whether to the client, the municipality, an engineer, or to anyone else. This is not required by law, but is a practical consideration in helping to protect the architect's intellectual property, and it is very easy to do.

[187] *See* Bryce & Palazzola Architects & Assocs., v. A.M.E. Group, 865 F. Supp. 401 (E.D. Mich. 1994), for a discussion of the innocent infringer defense where there was no notice.

[188] Aitken, Hazen, Hoffman v. Empire Constr. Co., 542 F. Supp. 252 (D. Neb. 1982). The court discussed a number of issues relating to architects. The court held that the drawings were not "works made for hire" because the architect was an "independent contractor" and not an employee of the owner. The owner argued that it had joint ownership of the drawings because it contributed ideas, directed certain changes to be made, and exercised approval of the plans. The court found that such involvement by a client in the preparation of architectural plans is normally to be expected and does not ordinarily render the client the author of the work. In determining whether there is joint authorship, the courts must determine the intention at the time the plans are prepared. Finding no such intention in this case, the court held that there was no joint authorship of the plans. The owner also argued that it was an "innocent infringer." Under this doctrine, if a party innocently infringes a copyright of plans from which the copyright notice is omitted, the party is not liable for damages to the copyright owner (although the court may allow recovery of

The owner paid the architect for his services based on an hourly rate. None of the documents contained any copyright notice. The following year, the owner decided to build another, identical apartment complex next to the first one. The first drawings were taken to another architect, who placed his seal on them, and the new project was built. When the first architect learned of this, he sent the owner a bill for the second project, but the bill was rejected. The architect then placed a copyright notice on the originals and registered the copyright with the United States Copyright Office. He then notified the parties of the copyright infringement and filed suit in federal court. The court found in favor of the architect.

Drawings or specifications need not be registered unless an action for infringement is planned, in which case registration is mandated to obtain certain remedies. Such registration can be made at any time on forms furnished by the Register of Copyrights. Note, however, in the event of a possible infringement that there are filing time limits. Also, registration, if done timely, will permit an award of statutory damages (so that the claimant does not need to prove actual damages) and attorneys' fees.[189]

Until passage of the Architectural Works Copyright Protection Act, copyright protection only prevented the copying of the drawings or specifications. It did not prevent another person from using the ideas shown in them. Also, if another person examined the building, he could build another structure just like it as long as he did not copy the drawings or specifications. For new buildings subject to this Act, that is no longer true. Note that it is the expression of ideas that is protected. Ideas themselves are not accorded any copyright protection.[190] A building represents the

---

the infringer's profits). In this case, the owner was not innocent, because it was fully aware of all the circumstances of the infringement. The owner also argued that it was the owner of the plans, because there was no written contract between the owner and architect. The architect was allowed to introduce evidence of the custom in the industry, including AIA Document B141, by which the architect remains the owner of the plans. The court also rejected the owner's contention that it had paid for the plans, relying on the architect's invoices which stated "for architectural and engineering services rendered." The architect was paid for his services and not the plans.

Another case where joint ownership was argued is Gordon v. Lee, 2007 WL 1450403 (N.D. Ga., May 14, 2007). In that case two architects and another person formed a partnership to design and develop properties. The firm later split up and one of the former partners registered copyrights in various plans and filed suit against the partner who left. The court found that there was joint ownership by the two architects.

[189] In Holm v. Pollack, 2000 U.S. Dist. LEXIS 16007 (E.D. Pa. Nov. 1, 2000), the architect sued his homeowner client for copyright infringement after the client failed to make payment. Because the architect did not register the plans until more than four months after the plans were first used, he was not entitled to statutory damages and attorneys' fees. See 17 U.S.C. § 412.

[190] In Eli Attia v. Society of New York Hospital, 201 F.3d 50 (2d Cir. 1999), an architect was hired to prepare a concept for a hospital expansion over a highway. He prepared a series of concept drawings for which he was paid. Several years later, the hospital hired another architect to utilize the concept for the actual hospital construction. A copyright infringement action followed. The court examined the distinction between unprotected ideas and protected expressions of ideas in finding no infringement. The drawings prepared by the first architect contained creative ideas, but they were highly preliminary and generalized. Although there were similarities, these similarities did not go beyond the concepts and ideas contained in the first architect's drawings. Another

expression of the architect's ideas, just like the drawings. Making a copy of the building is thus a violation just like making a copy of the drawings, at least for buildings constructed since the AWCPA was enacted.

In a Minnesota case, a developer and architectural firm brought a declaratory judgment action against a competitor, seeking a determination that their plans did not infringe on the competitor's drawings.[191] This instructive case involves not only copyrights but also the Lanham Act, unfair competition, and other matters. Portions of the court's opinion are reproduced here:

> Copyright protection automatically arises from the moment a work is created. In order to establish a claim of copyright infringement, Everest must prove ownership of a valid copyright and copying, or infringement, of protected portions of its copyrighted work by the plaintiffs. Architectural plans and drawings are protected by copyright law as "pictorial, graphic, and sculptural works" and "architectural works." 17 U.S.C. § 102(a)(5) & (8). An "architectural work" is the design of a building as embodied in any tangible medium of expression, including a building, architectural plans, or drawings. Id. § 101. The work includes the overall form as well as the arrangement and composition of spaces and elements in the design, but does not include individual standard features.
>
> A comparison of the objective details of the Everest plans and the two sets of CSM plans shows there is a similarity of ideas between them. Both Everest and CSM proposed and designed a rectangular office showroom for the Roseville site. Infringement of expression requires the court to evaluate the response of the ordinary, reasonable person to the two forms of expression. The essential inquiry is whether the total concept and feel of the works in question are substantially similar.
>
> When viewing the Everest plans and the first set of CSM plans side by side, it is clear that the plans are substantially similar. The building design and the site plan are remarkably similar. Both buildings include unusual features such as sawtooth loading doors and a parapet wall. The ordinary observer would also conclude that the length of the buildings, the use of brick on frontage and back of the building and the floor elevations are very similar. The court notes that the placement of cars and trees as well as the lettering and numbering in some of the CSM drawings is identical to the Everest plans. The placement of the building on the site, the landscaping, the parking areas and vehicular circulation are also virtually identical. . . . The similarities between the Everest and CSM plans are substantial and cannot be considered inconsequential.

The court also rejected the argument that there was no infringement of copyright because there was a "fair use" of the plans. The court found that the plans were not

---

allegation was that some plans were traced from the plaintiff's drawings. The court held that because the traced drawings related to the existing structure, this constituted simple statements of existing facts, such as the location of existing walls. Facts, like ideas, are in the public domain and not subject to copyright. The court stated that the copying of a line that has no expressive content but has as its sole purpose to identify the position of an existing wall takes only fact and nothing of expression and does not infringe copyright.

[191] CSM Investors, Inc. v. Everest Dev., Ltd., 840 F. Supp. 1304, 1309–12 (D. Minn. 1994).

used for nonprofit or educational purposes but rather for commercial gain. The fact that the original drawings had not been published also led to the conclusion that there was no fair use.

Even though a builder presents ideas and concepts to the architect who prepares the drawings, and even though the builder may include sketches and maintain approval authority, such participation is insufficient to establish a claim of co-authorship, and the builder would not be entitled to a copyright.[192]

In a West Virginia case,[193] the court reviewed a case involving copyright infringement of a single-family home design:

> Architectural structures and plans are subject to copyright protection under 17 U.S.C. § 102(a)(5) and (8) where the author has independently created the works and the works reflect creativity, regardless of how simple the design. [Citation omitted] If a house design is sufficiently original, copyright protection is not precluded because the design is also utilitarian. Architectural works are by definition at least partly designed to serve a "useful function."[Citation omitted]
>
> The level of originality required for copyright protection is not high. [Citation omitted] Rather, a work need only exhibit a "minimal degree of creativity." . . .
>
> Although the underlying unoriginal component parts of a creation are not subject to protection, a creator's choice of selection and arrangement of these parts, where

---

[192] Fred Riley Home Bldg. Corp. v. Cosgrove, 864 F. Supp. 1034 (D. Kan. 1994). *See* Rottlund Co., Inc. v. Pinnacle Corp., 2004 WL 1879983 (D. Minn., Aug. 20, 2004), *rev'd*, 452 F. 3d 726 (8th Cir. 2006); M.G.B. Homes, Inc. v. Ameron Homes, Inc., 903 F.2d 1486 (11th Cir. 1990).

[193] Richmond Homes Mgmt., Inc. v. Raintree, Inc, 862 F. Supp. 1517 (W.D. Va. 1994). The case of Van Brouck & Assocs., Inc. v. Darmik, Inc., 329 F. Supp. 2d 924 (E.D. Mich., Aug. 4, 2004), involving the infringement of plans for a single-family home, is instructive with regards to damages available from an infringer. The original architect had designed a home in 1997, for a fee of $15,000. In 1999, the defendants attempted to purchase these plans from the plaintiff, who refused. At that time, plaintiff charged $4.00 per square foot for plans. The court found that this was an accurate measure of the value of the plans, namely, 6,238 square feet times $4 equals $24,952. In addition, the architect was entitled to the infringer's profits. Evidence was presented as to the value of the infringing residence. The court found that this value, including the land, was $711,900. The burden then shifted to the defendants to prove the expenses. The court carefully examined all the evidence and found that the total amount of expenses proved by the defendants was $555,351.54, leaving a difference of $156,548.46, for a total award of $181,500.46!

In Tiseo Architects, Inc. v. SSOE, Inc., 431 F. Supp. 2d 735 (E.D. Mich. 2006), the architect hired to prepare the initial design for a development project brought suit against the designer of a grocery store that was part of the development. The designer sought to dismiss, based on several arguments. First, the designer claimed that the architect did not possess a valid copyright, relying on the premise that this design was a derivative of a prior work that the architect had done for another development. The designer asserted that because on this copyright application the architect had failed to indicate the existing work and identify what new material was added, the registration was allegedly unenforceable. The court accepted the architect's argument that the court lacked jurisdiction to cancel the architect's registration, finding that, under the Act and Regulations (37 C.F.R. § 201, setting forth procedures for cancellation of a copyright certificate), only the Copyright Office, not a court, can cancel a copyright certificate. The court rejected the designer's argument that there was insufficient originality shown in the design, based on the architect's demonstration of differences between this design and the prior one. *See also* David & Goliath Builders, Inc. v. Elliott Constr., Inc., 2006 WL 1515618 (W.D. Wis., May 25, 2006).

independently made, may be original and therefore copyrightable. [Citation omitted] Moreover, just as compilations of facts may be copyrighted, [citation omitted] original creations that are derived from preexisting works are also subject to copyright protection as "derivative works." 17 U.S.C. § 101. The court looks to the new material added to the underlying elements in the derivative work to resolve the question of originality. [Citation omitted] In line with the aforementioned principles, a derivative work will be deemed sufficiently original if it has a "'faint trace of originality' and if it provides a 'distinguishable variation' " from the original material.

In a Georgia case, with apparently no contract between the owner and architect, the architect had prepared a set of drawings for the owner.[194] Later, the architect learned that the owner was proceeding with the construction without paying the architect for the plans. The architect sued to recover his fees. The owner apparently claimed that the drawings used for the construction were not the same, although the trial court found that by overlaying the two sets of drawings, the plans were identical. The architect prevailed. This situation is similar to those in which an owner comes to the architect with a set of "stock" plans, or plans prepared by another architect. If the other architect can be identified, the new architect should insist on obtaining a written release or permission from the prior architect. Otherwise, the owner should provide a written representation that the owner has the authority to use and copy the drawings, and that the owner will indemnify the architect from any action that may be brought by any party related to the use of such drawings.

A related issue is patent protection. There are three requirements before a patent can be issued: the object must be novel, utilitarian, and nonobvious.[195] It is almost impossible to obtain a patent for a building in its entirety because of the requirements of novelty and nonobviousness. Components can, however, be patented under the proper circumstances.[196] If an architect wants to obtain a patent on a component of a building, it should obtain the services of a competent patent attorney, because the requirements are strict and the patent could be lost.

Computers and software have been the subject of litigation. In *NIKA Corp. v. City of Kansas City*,[197] an architectural firm had developed a system to prepare specifications and cost estimates for the rehabilitation of old houses. The firm entered into a contract with the city to provide the system for an urban renewal project. At the end of the contract term, the city refused to return the components, arguing that the contract provided that the ownership had shifted to the city. Although the firm was able to recover some costs at trial, it was unable to force the return of the system. Based on this case, a design firm should include specific language in its contract that provides for retention of ownership of all hardware and software, as well as all manuals, data, and other information. Paragraph 1.3.2.1 of

---

[194] G.E.C. Corp. v. Levy, 126 Ga. App. 604, 191 S.E.2d 461 (1972).

[195] 35 U.S.C. §§ 101-103.

[196] Blumcraft of Pittsburgh v. United States, 372 F.2d 1014 (Ct. Cl. 1967) (handrail found to be patentable and action for infringement upheld).

[197] 582 F. Supp 343 (W.D. Mo. 1984).

B141-1997 did include software under "instruments of service." This was eliminated in the 2007 versions of the AIA owner/architect agreements. The definition found in ¶ 7.2, however, is broad enough to include software if the architect or its consultants authored such software.

> *7.3* *Upon execution of this Agreement, the Architect grants to the Owner a nonexclusive license to use the Architect's Instruments of Service solely and exclusively for purposes of constructing, using, maintaining, altering and adding to the Project, provided that the Owner substantially performs its obligations, including prompt payment of all sums when due, under this Agreement. The Architect shall obtain similar nonexclusive licenses from the Architect's consultants consistent with this Agreement. The license granted under this section permits the Owner to authorize the Contractor, Subcontractors, Sub-subcontractors, and material or equipment suppliers, as well as the Owner's consultants and separate contractors, to reproduce applicable portions of the Instruments of Service solely and exclusively for use in performing services or construction for the Project. If the Architect rightfully terminates this Agreement for cause as provided in Section 9.4, the license granted in this Section 7.3 shall terminate.* **(1.3.2.2; 1.3.2.3; substantial modifications)**

If the architect terminates this agreement "rightfully," then the license granted to the owner and other parties working on the project is terminated. The question of whether the architect's termination was rightful or not is usually a fact question that would be determined by a court or arbitrator. If the owner terminates the architect, the architect should promptly issue its own termination notice in order to trigger this provision and revoke all licenses. Immediately after most terminations, it is unclear who would ultimately prevail on the issue of who was "right" and the owner proceeds to use the architect's Instruments of Service with some peril.

Such a termination often occurs late in the project, when there is great pressure to start construction. Should the owner take the architect to court and obtain a judgment from the court that the architect is in breach of the contract before the owner uses the architect's documents? A decision in the owner's favor also validates the owner's license to use the documents. That would, of course, greatly slow the project and potentially cause the owner extensive damages. The owner may try to use the architect's documents to complete the project by others. The risk to the owner is that a court may very well enjoin such use and award the architect damages above and beyond contract damages, such as attorneys' fees in a copyright infringement suit. Note the waiver of consequential damages found in ¶ 8.1.3 that may impact this analysis.

There are two types of licenses: exclusive and nonexclusive. When an architect grants an exclusive license to someone, the transferee has the sole right to use the drawings within the definition of the license agreement. The exclusivity may be for a geographic area or a time frame, or both. A nonexclusive license, on the other hand, means only that the transferee has a right to use the drawings, but others may also have the same rights.[198]

---

[198] In Home Design Serv., Inc., v. Park Square Enter., Inc., 2005 WL 1027370 (M.D. Fla., May 2, 2005), the plaintiff was an architectural firm that designs and sells architectural plans for

In a California case,[199] the issue was whether the architect had transferred the exclusive rights to the drawings to the owner. The architect had entered into a contract with the owner for the design of a residence. The contract included language substantially similar to the AIA copyright provisions. The architect later filed suit against third parties for copyright infringement and won substantial damages. The owner filed a separate copyright infringement action against the same third parties and the architect, alleging that the owner enjoyed an exclusive license. The owner argued that the language "the drawings, specifications and documents prepared by the Architect for this Project are instruments of the Architect's service for use solely with respect to this project . . ." meant that the architect was precluded from using the copyrighted drawings for any other project, while the provision that "the Owner shall be permitted to retain copies, including reproducible copies of Architect's drawings, specifications, and documents for information and reference in connection with Owner's use and occupancy of the Project" provided for the owner's use of the drawings. The owner argued that these provisions, read together, gave the owner an exclusive license to use the copyrighted drawings. The court rejected this theory and found that the owner did not have an exclusive license. Since only the copyright owner or the owner of exclusive rights under the copyright has standing to bring an infringement action, and this owner did not have either, the court dismissed the case. Note that the current version of B101 specifies that the license granted to the owner is a nonexclusive license.

Where the licensee exceeds the scope of the license granted by the copyright holder, the licensee is liable for infringement.[200]

---

residential homes. It entered into an agreement whereby a builder obtained the exclusive right to build homes using plaintiff's plans: "We will not sell the aforementioned plans to any builder within a fifty mile radius of the intersection of SR-50 and I-4, an area of approximately 7850 square miles. This policy will cover all past, present and future designs from this date forward." When plaintiff discovered another builder constructing an infringing home, it sued that builder, who filed a motion for summary judgment, contending that plaintiff relinquished the right to sue for copyright infringement because of the transfer of rights quoted above. The court rejected this argument, holding that "the memorandum should, at a minimum, make mention of the copyright which is being transferred. . . . Because [plaintiff's] letter makes no such reference and otherwise fails to evince an intent by Home Design to transfer copyright in the . . . design, Defendants' argument that Home Design lacks standing to bring its . . . infringement claim fails."

[199] McCormick v. Amir Constr., Inc., 2006 WL 784770 (C.D. Cal., Jan. 12, 2006)

[200] In LGS Architects, Inc. v. Concordia Homes of Nevada, 434 F.3d 1150 (9th Cir. 2006), the architect entered into a licensing agreement with Concordia to use two of the architect's plans to construction Arbor Glen I, a master-planned community of 80 houses. The contract contained a provision for a "reuse fee." Later, Concordia decided to use the architect's plans to construct houses in the adjacent Arbor Glen II project. The architect requested a new agreement, but Concordia refused, apparently believing that the original contract covered the reuse provision. The architect sued, and the issue was whether the owner exceeded the scope of its license. The court found that the license was limited to the Arbor Glen I project, and that the scope of the license had been exceeded. Concordia's argument that the architect breached the covenant of good faith and fair dealing by refusing to authorize reuse was also rejected.

7.3.1 In the event the Owner uses the Instruments of Service without retaining the author of the Instruments of Service, the Owner releases the Architect and Architect's consultant(s) from all claims and causes of action arising from such uses. The Owner, to the extent permitted by law, further agrees to indemnify and hold harmless the Architect and its consultants from all costs and expenses, including the cost of defense, related to claims and causes of action asserted by any third person or entity to the extent such costs and expenses arise from the Owner's use of the Instruments of Service under this Section 7.3.1. The terms of this Section 7.3.1 shall not apply if the Owner rightfully terminates this Agreement for cause under Section 9.4. (**1.3.2.2; revisions**)

The analysis of the last sentence is the inverse of the analysis under paragraph 7.3, above. If the owner, as opposed to the architect, "rightfully" terminates, then the license to use the documents continues, and there is no indemnification of the architect under this paragraph. This paragraph is not intended to waive any rights to bring an infringement action by the architect against the owner. It is an additional right or remedy by the architect against the owner.

7.4 Except for the licenses granted in this Article 7, no other license or right shall be deemed granted or implied under this Agreement. The Owner shall not assign, delegate, sublicense, pledge or otherwise transfer any license granted herein to another party without the prior written agreement of the Architect. Any unauthorized use of the Instruments of Service shall be at the Owner's sole risk and without liability to the Architect and the Architect's consultants. (**1.3.2.2; substantial modifications**)

Under one provision of the Copyright Act, 17 U.S.C. § 204, any transfer of interest in a copyright must be in writing,[201] with the only exception being a nonexclusive license. In nonstandard contracts, such as letter form agreements, this may be a problem, however. It could very well be argued that the AIA language expresses the custom and usage among architects that no license is implied by the rendering of architectural services. It may well be impossible to create a nonexclusive license for a work of architecture, because generally each work of architecture is for a unique site. Although an exception may exist for design concepts, when it comes to working drawings that are done for a specific site and specific building and zoning codes, the design is not meant to be constructed multiple times on the same site. Any license in such a case must be exclusive. (**See** the discussion of license issues under ¶ 7.3.)

In a case interpreting the 1974 version of this provision, a court held that the owner was able to use the architect's drawings to build the same project, even though the architect was terminated.[202] The architect in that case was paid for his services to the point of termination, plus termination fees.

---

[201] *See* Arthur Rutenberg Homes, Inc. v. Drew Homes, Inc., 29 F.3d 1529 (11th Cir. 1994) (discussing transfer of ownership of copyrights in drawings).

[202] Reeves v. Hill Aero, Inc., 231 Neb. 345, 436 N.W.2d 494 (1989).

The *works for hire* provision of the Copyright Act has been the subject of several cases.[203] Owners sometimes want to substitute a work-for-hire clause in place of

---

In a curious decision that should be limited to the unique facts of that case, a federal court found that the architect's plans were "tangible property" in the context of a declaratory judgment action by an insurer. State Farm Fire & Cas. Co. v. White, 777 F. Supp. 952 (N.D. Ga. 1991).

In Michael Eiben v. A. Epstein & Sons Int'l, Inc., 57 F. Supp. 2d 607 (N.D. Ill. 1999), Eiben was hired in 1983 by a county to develop plans for a building. A pre-1997 AIA form was used. Ten years later, the county hired other architects to work on an adjacent project. One portion of this new project involved the renovation of a wing of the project originally designed by Eiben. To do this work, the new architects obtained Eiben's original drawings from the county. Two of these drawings were included in the bid documents and were labeled "Available Information." Also included in the bid documents were plans that were allegedly closely based on Eiben's original drawings. Eiben filed a copyright action against the new architects and others. In addressing several motions, the court analyzed the AIA language to determine the extent of the permission granted to the county to use the Eiben drawings. The court found no ambiguity in the AIA language. The court found that the new use of the drawings did not constitute "completion of the Project by others," nor did it constitute an "addition" to Eiben's project. It could also not be considered an "other project." The court found that the new use was activity "in connection with the County's use and occupancy of the Project."

In Womack + Hampton Architects, L.L.C. v. Metric Holdings Ltd. P'ship, 102 Fed. Appx. 374, 2004 Copr. L. Dec. ¶ 28,830, 71 U.S.P.Q.2d 1209 (5th Cir. (Tex.) 2004), the parties had a contract containing modified AIA language as follows:

A. The Drawings, Specifications, and other documents prepared by the Architect for this project are the instruments of the Architect's service for use solely with respect to this project and the Architect shall be deemed the author of these documents and shall retain all common law, statutory, and other reserved right [sic], including copyright.

. . .

B. The Owner agrees not to use, copy or cause to have copied, the drawings and specifications prepared for this project on subsequent phases or other sites without proper compensation to the Architect, which shall be based upon a mutually agreed upon of [sic] $150.00 per unit (base architectural fee), plus engineering services, plus contingent hourly charges and expenses for plan modifications necessary to adapt these plans and specifications to other sites.

After the original architect (plaintiff) completed its work, the owner hired other architects to use plaintiff's drawings for other projects. Plaintiff sent the owner a letter demanding payment for the reuse, but the owner failed to pay. Plaintiff sued for copyright infringement. The defendants contended that the contract permitted reuse, and the lack of payment was only a breach of contract, not copyright infringement. The court agreed with the defendants, finding that there was a license that permitted reuse. The plaintiff argued that payment of the license fee was a condition precedent, but the court also rejected this argument. The final argument—that the license did not extend to the subsequent architects—was also rejected on the basis that the contract anticipated that third parties would be involved in the reuse.

[203] In Kunycia v. Melville Realty Co., 755 F. Supp. 566 (S.D.N.Y. 1990), the architect had worked for an industrial design firm that had been hired by a chain of toy stores. A number of plans were prepared. The architect then left to start his own business and the toy store chain became his client. There was no agreement that addressed copyrights. Several years later, the architect suspected that the store was using his designs without permission and he copyrighted his plans. He then brought an action for copyright infringement against the store. The court found that the architect was an independent contractor; therefore, the work was not made for hire and proper ownership rested with the architect. The court also stated that "an architect owns his drawings, unless

the AIA language in order to obtain the copyrights to the architect's drawings. Because the work that design professionals create does not fit within the categories of 17 U.S.C. § 101,[204] such a strategy does not work. Instead, if the owner wants to

---

expressly agreed otherwise by the parties." The court also found that failure to add the copyright notice to the drawings given to contractors was a limited publication. Thus, the architect was allowed to try to recover under the copyright act for infringement by the owner. The *Kunycia* case was discussed in John G. Danielson, Inc. v. Winchester-Conant Prop., Inc., 186 F. Supp. 2d 1 (D. Mass. 2002). *See also* M.G.B. Homes, Inc. v. Ameron Homes, Inc., 903 F.2d 1486 (11th Cir. 1990).

In Harris Custom Builders, Inc. v. Hoffmeyer, 834 F. Supp. 256 (N.D. Ill. 1993), the defendant argued that a builder who claimed a copyright was guilty of a criminal conspiracy because the copyright application listed the builder as the author of a drawing as a "work made for hire." However, the Supreme Court, in Community for Creative Non-Violence v. Reid, 490 U.S. 730, 109 S. Ct. 2166 (1989), had held that the work-made-for-hire provision generally applies only to works produced by employees of an entity, not to those created by independent contractors. The *Hoffmeyer* court found that this was not significant because the *Reid* decision was issued after the copyright application was filed. The court also held that the builder had no duty to report unpublished works on the copyright application form as "preexisting works."

In Bonner v. Dawson, 2003 WL 22432941 (W.D. Va., Oct. 14, 2003), the court rejected the work for hire defense because the architect was not employed by the owner, but as an independent contractor, nor was there anything in the contract to indicate copyright ownership. Bonner was awarded $10,707 by the jury. The jury found that Bonner was not entitled to infringer's profits. In an appeal of that award, Bonner v. Dawson, 404 F.3d 290 (4th Cir. 2005), Bonner asked the court to overturn the jury's award denying him infringer's profits. The appellate court held that Bonner had met his initial burden of establishing the required causal connection between the infringement and the profit stream, but that the jury's award would not be disturbed.

In The Rottlund Co., Inc. v. Pinnacle Corp., 2004 WL 1879983 (D. Minn., Aug. 20, 2004), the plans in question were drawn by an architect working for AFI. AFI orally agreed that AFI would assign the plans to Rottlund. Some years later, this agreement was reduced to writing. Later, Rottlund registered the copyrights and checked the box marked "work for hire." At trial, the defendant argued that Rottlund did not have a valid copyright because the date of creation predated the date of the written assignment. The court rejected this argument on the basis that an oral assignment of a copyright interest could be confirmed by a later writing to satisfy § 204 of the Copyright Act. The opinion examines the test for fraud on the Copyright Office and goes through the elements required for proof of infringement.

In Trek Leasing, Inc. v. United States, 62 Fed. Cl. 673 (2004), the United States government entered into a lease with the plaintiff for the construction of a post office. The plaintiff hired an architect to prepare the plans. The architect and plaintiff executed an AIA B141 agreement. When the plaintiff learned that the U.S. government had built another post office using the plaintiff's copyrighted plans, it brought this action. The government moved to dismiss, alleging that the plaintiff did not own the copyright and therefore did not have standing to sue. The government argued that the plain language of the AIA contract gave the architect ownership of the copyright, while the plaintiff argued that the contract was a mere formality that was executed only to obtain financing for the post office project, and that the architect was actually an employee of plaintiff, triggering the work for hire doctrine. Since both parties to the contract asserted the contract's invalidity, the court agreed with the plaintiff.

[204] That definition is as follows:

A "work made for hire" is—
(1) a work prepared by an employee within the scope of his or her employment; or
(2) a work specially ordered or commissioned for use as a contribution to a collective work, as a part of a motion picture or other audiovisual work, as a translation, as a supplementary

own the copyright, an outright assignment is the correct way to obtain the copyright.

In an important decision that strips the architect of copyright protection in the absence of good contract language such as the language in the 1997, and subsequent, AIA documents, the court in *I.A.E. v. Shaver*[205] held that the architect

---

work, as a compilation, as an instructional text, as a test, as answer material for a test, or as an atlas, if the parties expressly agree in a written instrument signed by them that the work shall be considered a work made for hire. For the purpose of the foregoing sentence, a "supplementary work" is a work prepared for publication as a secondary adjunct to a work by another author for the purpose of introducing, concluding, illustrating, explaining, revising, commenting upon, or assisting in the use of the other work, such as forewords, afterwords, pictorial illustrations, maps, charts, tables, editorial notes, musical arrangements, answer material for tests, bibliographies, appendixes, and indexes, and an "instructional text" is a literary, pictorial, or graphic work prepared for publication and with the purpose of use in systematic instructional activities.

[205] 74 F.3d 768, 37 U.S.P.Q.2d (BNA) 1436, Copyright L. Rep. (CCH) ¶ 27,479 (7th Cir. Jan. 17, 1996).

In Johnson v. Jones, 921 F. Supp. 1573 (E.D. Mich. 1996), the court rejected a nonexclusive license argument and awarded some $107,125 in damages to the first architect against the second architect and the contractor. The owner had hired the first architect to design a home for her. Although the architect submitted a standard AIA contract, it was never signed. The architect presented several drawings to the owner, all containing a copyright notice. Later, the owner decided to hire another architect, who traced the first architect's drawings because of the owner's tight time schedule. The interesting thing about this case is that the second architect was concerned over the copyright issue but was assured by the owner's attorney that they had a right to use the drawings. What was incredible was that the court found the owner not liable for copyright infringement because she had done nothing more than give the drawings to the second architect. The court found it unreasonable for the second architect and contractor not to seek the advice of their own attorneys and, instead, rely on the owner's attorney. For the court not to have held the owner liable seems totally unfair, particularly when it was the owner who most benefitted from these misdeeds.

The *Shaver* case was distinguished in Saxelbye Architects, Inc. v. First Citizens Bank & Trust Co., 129 F.3d 117, 44 U.S.P.Q.2d 1634 (4th Cir. 1997), wherein it was found that the implied-license exception to the requirement of a writing is a limited one. The case was remanded to the circuit court for a determination of factual issues regarding the scope of the architect's contract.

In Foad Consulting Group, Inc. v. Musil Govan Azzalino, 270 F.3d 821 (9th Cir. 2001), an engineering firm was hired in 1995 to create a plot plan for a shopping center project. After the plans were completed and city approval was obtained, the owner transferred its interests in the project to a new developer who hired another firm to provide architectural and engineering services. In doing those services, the firm copied much of Foad's plot plan. Foad then filed a copyright infringement lawsuit against various defendants. Summary judgment in favor of the defendants was upheld on appeal.

The court held that the contract (this was apparently not an AIA form contract) granted to the owner "an implied license to use the revised plot plan to complete development of the project, to hire another firm to create derivative works using the revised plot plan for the purpose of completing the project, and to publish the resulting work." Foad was paid a fee of $175,000 to prepare various drawings and to process them with the city. Given this amount of money, the court found that:

it would be surprising if the parties had intended for [the owner] to seek Foad's permission before using the plans to build the project. Had that been the parties' intention, one would expect to see the requirement spelled out explicitly in the agreement. But nowhere does the contract state that after the city had granted its approval, [the owner] would need to obtain Foad's permission before commencing work. We conclude that the contract gives [the owner] an implied license to use the plot plan to build the project.

The court also stated that:

[i]f accepted, Foad's claim that although it was hired to create documents for the project, [the owner] had no right to use the documents to build the project, would allow architectural or engineering firms to hold entire projects hostage, forcing the owner either to pay the firm off, continue to employ it, or forego the value of all work completed so far and start from scratch. . . . One would expect project owners to think long and hard before placing their fortunes so entirely in the hands of a single firm. And one would expect that a firm that intended to exercise such ongoing control over a project would clearly specify this in a contract.

The engineer also argued that a legend placed on the drawings supported its position:

All ideas, designs, arrangements and plans indicated or represented by this drawing are owned by, and the property of Foad Consulting Group, Inc. and were created, evolved and developed for use on, and in connection with the specified project. None of such ideas, designs, arrangements or plans shall be used without written permission of Foad Consulting Group, Inc.

The court held that this legend would only apply, if at all, to projects other than the specified project. Because it was created after the contract, the legend would not affect the owner. Finally, the court noted that ideas cannot be copyrighted.

In another case involving site development plans, the court appears to accept that plans that are part of a municipality's annexation agreement (which apparently requires that the plans be followed) cannot be copied by a subsequent developer who takes over the project after the first developer fails to close on the purchase of the property. Ocean Atlantic Woodland Corp. v. DRH Cambridge Homes, Inc., 262 F. Supp. 2d 923 (N.D. Ill. 2003). The initial developer lost the right to develop the property and obtained the copyrights to the plans. It then sued the subsequent developer who was developing the land in accordance with the copyrighted plans that had been adopted in the annexation plan and incorporated by the municipality.

In John G. Danielson, Inc. v. Winchester-Conant Prop., Inc., 322 F.3d 26 (1st Cir. 2003), a developer hired an architect to prepare site plans for a project. These were incorporated into a restrictive covenant between the owner and the local municipality. Soon after construction began, the developer encountered financial difficulties and abandoned the project, leaving the architect unpaid for its work. Several years later, a new developer acquired the land and attempted to change the covenant. After failing in several attempts to alter the covenant, the new developer decided to develop the site according to the previously approved covenant. The first architect learned of the construction and filed suit for copyright infringement.

This case raised several issues. The defense of "publication" was rejected (note that the applicable time period was after the 1976 Act and before the 1990 Act) on the basis that there was limited publication. The second defense, also rejected, was that the drawings, by virtue of their inclusion in the restrictive covenants approved by the municipality, had become "laws," which are in the public domain and uncopyrightable. The court held that this restrictive covenant, even though entered into with a municipality, is essentially a private agreement and not a generally applicable law. Therefore, the drawings did not pass into the public domain. The next argument—that an implied license was granted by the architect—also failed because the architect had executed an AIA agreement.

The court also reviewed the "merger doctrine." Under this concept, the restrictive covenant meant that there was only one way to build on the land, and the covenant drawings merge with that idea. Since ideas are not protected by copyright, the merger doctrine would mean the copyright claim would fail. The court rejected this argument, finding that, while the restrictive covenant

had given an implied license to use his drawings. The architect was hired by an engineer to design an air cargo building for an airport. Believing that this was a contract for the schematic design phase of the project and that he would be retained to perform the remaining architectural services, the architect agreed to accept $10,000 for this first phase. The letter form of contract made no mention of copyright or license. After he prepared several schemes (with copyright notices) and one was chosen by the owner, the architect was terminated from the project by the engineer, who had contracted with another architect to do the remaining work. The first architect was paid only $5,000 of the fee. When he learned of the second architect's use of his drawings, he registered his drawings with the Copyright Office and notified all parties that his drawing could not be used. The engineer then brought a declaratory judgment action against the first architect, contending that he had a right to use the drawings.

The court agreed with the engineer, finding that the architect knew that his drawings would be used for the construction of the building. The fact that he was not paid the contract sum made no difference in a copyright action. The architect's attempted revocation of any implied license was ineffective. The court held that such an implied license could not be revoked. The architect also argued that the license was an exclusive implied license that required a writing in order to

---

made one method of developing the site legally cheaper and easier than others, it did not transform the covenant drawings into the only physically possible means to express ideas for such development.

In Nelson-Salabes, Inc. v. Morningside Dev., LLC, 284 F.3d 505 (4th Cir. 2002), the original architect who had been hired by a developer to provide planning assistance for a proposed assisted-living facility brought a copyright infringement action against the successor developer. The first architect had prepared some drawings and was then not retained for the final project, but the drawings were nonetheless used. The proposed AIA contract was never executed. The court distinguished *Shaver* and *Foad* on the basis that in this case, the architect made it clear that its drawings were not to be used to complete the project by others. The court stated:

> Our analysis of these decisions thus suggests that the existence of an implied nonexclusive license in a particular situation turns on at least three factors: (1) whether the parties were engaged in a short-term discrete transaction as opposed to an ongoing relationship; (2) whether the creator utilized written contracts, such as the standard AIA contract, providing that copyrighted materials could only be used with the creator's future involvement or express permission; and (3) whether the creator's conduct during the creation or delivery of the copyrighted material indicated that use of the material without the creator's involvement or consent was permissible.

In John G. Danielson, Inc. v. Winchester-Conant Prop., Inc., 186 F. Supp. 2d 1 (D. Mass. 2002), the court found no implied license because the architect had submitted a proposed AIA contract.

The *Johnson* case was cited in Tiseo Architects, Inc. v. SSOE, Inc., 431 F. Supp. 2d 735 (E.D. Mich. 2006) and in Jeffrey A. Grunsenmeyer & Assocs. v. Davidson, Smith & Certo Architects, 2/2 Fed. Appx. 510 2007 WL 62620 (6th Cir., Jan. 8, 2007). There, the plaintiff architect was hired to prepare a master plan. As part of this work, the architect prepared a set of drawings showing the existing conditions. The owner then hired another architect and gave this architect the plaintiff's drawings to use for reference. The plaintiff sued this second architect for copyright infringement. The court found in favor of the defendant, finding that the contract permitted the use of the plans.

be effective. The court rejected this argument on the grounds that the architect could have licensed this particular design to any number of others.

The result of this case is that design professionals should not consider that they have any action under the Copyright Act against anyone other than a total stranger unless specific language is used to restrict such a license, such as is found in AIA documents. In the *Shaver* case, the court held that the architect had given an implied license to everyone who had anything to do with the project, including the second architect, of whom Shaver was not even aware. The mere act of delivering drawings constituted the grant of an implied license.

Approval of a site plan by a governmental agency does not invalidate a copyright. In one case, an architect prepared a subdivision plat that was approved by the municipality.[206] The architect subsequently registered the copyright and filed suit for infringement against the successor owner of the subdivision. The owner argued that once the municipality approved the plat, it was not copyrightable. The court found the copyright valid.

§ 2.24   Article 8: Claims and Disputes

ARTICLE 8 CLAIMS AND DISPUTES

8.1 GENERAL

8.1.1 The Owner and Architect shall commence all claims and causes of action, whether in contract, tort, or otherwise, against the other arising out of or related to this Agreement in accordance with the requirements of the method of binding dispute resolution selected in this Agreement within the period specified by applicable law, but in any case not more than 10 years after the date of Substantial Completion of the Work. The Owner and Architect waive all claims and causes of action not commenced in accordance with this Section 8.1.1. (**1.3.7.3; completely rewritten; established ten year outside limit on all claims**)

The prior version of this paragraph (in the 1997 documents) set the start of the applicable statute of limitations (the *accrual* date) at the date of substantial completion. This paragraph states that the parties must look to applicable state law for the period in which any lawsuit, arbitration, claim, or cause of action must be commenced. It sets an outside time limit for this at ten years after the date of substantial completion.[207] In determining the proper time limit within a suit or arbitration must be filed, several considerations must be given.

---

[206] Del Madera Prop. v. Rhodes & Gardner, Inc., 637 F. Supp. 262 (N.D. Cal. 1985). *See also* John G. Danielson, Inc. v. Winchester-Conant Prop., Inc., 322 F.3d 26 (1st Cir. 2003).

[207] In Louisville/Jefferson County Metro Gov't v. HNTB Corp., 2007 WL 1100743 (W.D.Ky., April 11, 2007), it was argued that there had not been substantial completion so as to trigger the running of the statute of limitations. The project was for a baseball field, and the plaintiff argued that the project had failed to meet the standards of the governing body for baseball. The court

First, there is the applicable statute of limitations itself. Unless the contract says otherwise, this will be found in the laws of the state where the project is located. Most states have specific statutes of limitations for construction and design, which are often different time limits than simple breach of contract claims. Typically, construction statutes of limitation are in the range of two to four years.

Second, is the concept of "discovery," which is a judicially-created exception to the statute of limitations. This means that the statute of limitations does not even start to run until the problem is "discovered," meaning either that the party knew of the problem or should have known of the problem. Latent defects in construction usually mean that this consideration comes into play, since they are the types of defects that an owner may not learn about until years after completion of construction. Whether a party should have known of a problem can be a difficult question and often results in litigation over this issue. Once the owner learns of a problem, the statute of limitations starts to run.

Finally, the statute of repose comes into play. Statutes of repose were passed by the states as a way to limit the discovery exception. Without a statute of repose, a plaintiff might have an unlimited amount of time in which to sue if the problem is not discovered. A statute of repose places an outside limit in which a plaintiff can discover a defect and in which the statute of limitations starts. A typical statute of repose limit is ten years. If, within this time period, a plaintiff learns of a problem, the statute of limitations starts to run. If the problem is discovered after this repose time, the claim is cut off.[208] Notice that a plaintiff will have the benefit of the full statute of limitations period following discovery of the problem, but only if the problem is discovered within the statute of repose period.

For example, if an owner discovers a roof leak nine years and eleven months after substantial completion, he will have a further four years to sue the contractor and architect, assuming the typical time limits. If he discovers the problem two months later—ten years and one month after substantial completion—the statute of repose would cut off the claim.

Under this document, there is an absolute ten year cut-off of claims. If the owner discovers a roof leak nine years and eleven months after substantial completion, he would have one month within which to file suit against the architect and contractor. If he discovered the leak two months later, the suit would be absolutely barred. However, state law might shorten that time if the plaintiff discovers the problem at

---

rejected this argument on the basis that baseball games were being played at the facility without apparent problems, so that it was sufficiently complete for the owner to occupy and use the facility. This was under an earlier version of A201, which actually set the start of the statute of limitations upon the date of substantial completion.

[208] For a case that examines the history of statutes of repose and limitations as applied to architects and builders, *see* Horton v. Goldminer's Daughter, 785 P.2d 1987 (Utah 1989). In that case, the Utah Supreme Court struck down a statute of repose. In Ohio, the court in Sedar v. Knowlton Constr. Co., 49 Ohio St. 3d 193, 551 N.E.2d 938 (1990), upheld a statute of repose.

In Ball v. Harnischfeger Corp., 877 P.2d 45 (Okla. 1994), the court reviewed statutes of repose in various jurisdictions and found that the activity, not the status, of a party controlled in determining whether a construction statute of repose applied to the manufacturer of a crane.

an earlier date. For example, if the owner discovers a roof leak one year after substantial completion and the state's construction statute of limitations is four years, the claim is barred five years after substantial completion. Claims made after this ten year period are waived.

The parties can fix limitations periods that differ from those in the statutes.[209] In one case, a court stated that when the contract requires the architect to conduct inspections to determine completion dates and issue a final certificate, the issuance of that certificate represents a significant contractual right of the owner and a concomitant obligation of the architect.[210] The architect's issuance of the certificate marks the completion of its performance and the point when the statute of limitations starts to run for a breach of its contractual undertaking. However, that same result does not obtain when the owner controls the issuance of the final certificate. The final certificate in such circumstances may indicate the owner's acceptance of the work for purposes of contractual guarantees or equitable price adjustments, but it does not represent the completion of the contractual obligations of the architect or general contractor for purposes of triggering the statute of limitations. A New York case held that normally an action against an architect accrues on the date the final certificate of completion is issued by the architect, although in that case the architect had agreed to inspect the completed building annually for three years and to advise the owner of any possible defects.[211] When the roof leaked, the owner sued the architect. Based on the continuous treatment

---

[209] *See, e.g.*, People ex rel. Skinner v. FMG, Inc., 166 Ill. App. 3d 802, 520 N.E.2d 1024 (1988) (agreement whereby roofing manufacturer agreed to a ten-year period).

[210] State v. Lundin, 60 N.Y.2d 987, 459 N.E.2d 486, 471 N.Y.S.2d 261 (1983). *See also* Rose v. Fox Pool Corp., 335 Md. 351, 643 A.2d 906 (1994).

[211] Board of Educ. v. Thompson Constr. Corp., 111 A.D.2d 497, 488 N.Y.S.2d 880 (1985). In Irwin G. Cantor, P.C. v. Swanke Hayden Connell & Partners, 186 A.D.2d 71, 588 N.Y.S.2d 19 (1992), this provision was applied to an arbitration case. The owner obtained a large arbitration award against the architect for the Trump Tower. The architect then sought arbitration against the engineer. The engineer argued that the six-year statute of limitations had run and arbitration could not be initiated. The court held, however, that one of the claims, for indemnification, did not accrue until payment by the architect. Because that claim was not time-barred, all matters should be presented to the arbitrator.

In Armenia v. Carini, 174 A.D.2d 1040, 572 N.Y.S.2d 205 (1991), the court stated that "it is well settled that an owner's claim against an architect arising out of alleged defective construction of a building, however denominated, accrues for purposes of the Statute of Limitations upon completion of the construction."

In In re Oriskany Cent. Sch. Dist. v. Booth Architects, 615 N.Y.S.2d 160 (App. Div. 1994), *aff'd*, 85 N.Y.2d 995, 654 N.E.2d 1208, 630 N.Y.S.2d 960 (1995), the court agreed with the architect that the owner had brought an action too late under this provision. Even though the architect had not issued a certificate of substantial completion, the statute started to run when the owner occupied the building. In Northridge Homes, Inc. v. John W. French & Assocs., Inc., 1999 Mass. Super. LEXIS 435 (Nov. 15, 1999), the owner sued the architect after severe leaks were found in the roofs of the townhouses designed by the defendant. The architect provided construction administration for the project and issued a Certificate of Substantial Completion effective February 17, 1993. Massachusetts has a three-year statute of limitations. Suit was filed on April 30, 1997.

Starting in 1994, and continuing into 1995 and 1996, the architect met with the owner for the purpose of resolving the roof leak problems. After the lawsuit was filed, the architect moved for summary judgment based on the AIA language that requires that suit be filed within the applicable three-year period following the date of substantial completion. The court found that, for actions that occurred before the date of substantial completion, the time had expired. The court permitted the lawsuit to proceed to give the owner a chance to demonstrate that the architect misrepresented facts, thereby misleading the owner into waiting to file suit. This is referred to as the doctrine of equitable estoppel, and places a substantial burden of proof on the owner. If the owner can prove that the architect misrepresented facts with the intent to mislead the owner, then the limitations period is extended.

One argument of the owner was that the Certificate of Substantial Completion was never signed by the owner. Therefore, it should be disregarded. The court rejected this contention, based on the language found in AIA Document A201 (now §§ 9.8.4 and 9.8.5). The court held that, based on these provisions, by failing to sign the Certificate, the owner does not negate the date of Substantial Completion. This language places the responsibility for determining the date of Substantial Completion entirely on the architect, and does not require the owner's signature to set that date. At most, a failure to sign reflects the owner's refusal to accept the responsibilities assigned under the Certificate.

In another case, College of Notre Dame of Maryland v. Morabito Consultants, 132 Md. App. 158, 752 A.2d 265 (2000), an owner sued the structural engineer hired by the architect for breach of contract and negligence. The engineer had been hired to inspect a building to be renovated in 1991. In 1997, significant movement in the building was detected and a subsequent engineer concluded that the first engineer had failed to properly calculate the loads. In December 1998, the owner filed suit and the engineer moved to dismiss on the grounds that its contract contained the limitations provisions found above. In fact, both the owner-architect agreement and the architect-consultant agreement contained this language.

The owner argued that the accrual clause is like an exculpatory clause and should be read more stringently. The court held that this is not an exculpatory clause because it does not relieve any party from liability. Instead, it alters the time for accrual of a cause of action from what the law would otherwise impose. Such contractual modifications are generally not disfavored in law. Here, the Maryland three-year statute of limitations barred the action.

In Harbor Court Associates v. Leo A. Daly Co., 179 F.3d 147 (4th Cir. 1999), construction on a project—a combination office tower, hotel, and garage located in Baltimore—started in mid-1984, and a final Certificate of Completion was issued on September 11, 1987. Other than some minor chipping and cracking of the outer brick veneer, no problems appeared until April 1996, when a 15-square-foot section of brick suddenly and without warning exploded off the face of the structure. The consulting engineers concluded that the structure suffered from fundamental and latent defects in design and construction.

In the resulting lawsuit by the owner, the defendant architect moved for summary judgment based on the AIA language and the fact that Maryland had a three-year statute of limitations. The plaintiff argued that the "discovery rule" should apply whereby the statute of limitations does not start to run until a defect is discovered. The federal court, however, held that the parties can contractually set their own rule that departs from the discovery rule, and that the AIA clause did just that. If the discovery rule had applied, the statute of limitations would not even start to run until the owner either knew or should have known that there was a problem. In this case, that would probably have meant that the three-year period would have started in April of 1996, thus permitting the architect to be sued. Instead, the court found that the three-year period started on the date of substantial completion and ended three years later, in September of 1990.

In Aldrich v. ADD Inc., 437 Mass. 213, 770 N.E.2d 447 (2002), the trustees of a condominium trust sued the architect after completion of the project. Water was entering the structure, causing significant damage. The architect argued that the plaintiffs were the successor to the developer under the architect's contract. Under this provision, the applicable statute of limitation would bar

doctrine,[212] also called the continuing relationship doctrine, the statute of limitations did not start to run until such time as the confidential professional relationship existed between the owner and architect.

In another case, the statute of limitations for damages caused by defective design was deemed to start running on the date the owner gave final approval of design drawings.[213] However, in that same case, there was evidence that

---

the action. In reversing the trial court, the appellate court held that the trust was not a successor (*see* ¶ 1.3.7.9) to the contract and the action was not time-barred.

In Gustine Uniontown Assocs., Ltd. v. Anthony Crane Rental, Inc., 842 A.2d 334 (Pa. 2004), this provision set the start of the limitations period. The case revolved around what the length of the limitations period should be. Note that the AIA language merely establishes the start of the applicable statute of limitations and extinguishes any "discovery period," but does not set an actual limitations period. On remand, the court again held that this language precludes application of the discovery rule.

In Newman Memorial Hospital v. Walton Construction Company, 149 P.3d 525 (Kan. App. 2007), a county hospital sued its architect claiming breach of contract and breach of implied warranty. Substantial completion occurred on February 18, 1997. By July 17, 1997, the hospital had knowledge of water leaking through or around windows and of resultant damage. On July 31, 2002, this action was brought. The applicable statute of limitations was five years. The architect moved for summary judgment based on this provision. The hospital argued an estoppel theory, alleging that the architect induced delay and inaction by the hospital. The trial court ruled for the hospital and there was a substantial award in favor of the hospital at trial. The appellate court reversed, finding that the hospital filed the action beyond the statute of limitations and that the hospital had failed to specify any actions by the architect that the hospital relied on in delaying the filing of the suit.

[212] *See* Northern Mont. Hosp. v. Knight, 811 P.2d 1276 (Mont. 1991); Borgia v. City of N.Y., 12 N.Y.2d 151, 187 N.E.2d 777, 237 N.Y.S.2d 319 (1962). In Senior Hous. v. Nakawatase, 192 Ill. App. 3d 766, 549 N.E.2d 604 (1989), the architect was estopped from asserting the statute of limitations defense when he actively engaged in efforts to stop building leaks after construction was completed. Apparently, the cause of the leaks was improper detailing by the architect. *See also* Greater Johnstown v. Cataldo & Waters, 159 A.D.2d 784, 551 N.Y.S.2d 1003 (1990) (architect sought to identify cause of condensation over a period of years; statute of limitation tolled).

In Russo Farms, Inc. v. Vineland Bd. of Educ., 675 A.2d 1077 (N.J. 1996), farmers sued for damages caused by improper siting and construction of a school across from their property and by an inadequate drainage system. Among the defendants were the architect and contractor, who were sued in tort. They defended based on the running of the statute of limitations. The plaintiffs alleged a continuing tort theory. In rejecting this argument, the court held that the architect and contractor had no control over the school property after the end of construction. It is only when the new injury results from a new breach of duty that a new cause of action accrues. "For there to be a continuing tort there must be a continuing duty. . . . That they never corrected the problem does not render the tort continuing." Their "mere failure to right a wrong and make plaintiff whole cannot be a continuing wrong which tolls the statute of limitations, for that it is the purpose of any lawsuit and the exception would obliterate the rule." Thus, the new damages are simply continuing damages, not a new tort. As a result, the only tort alleged is negligence, and that tort accrued in 1980-81 when the injury first occurred.

[213] Comptroller v. King, 217 Va. 751, 232 S.E.2d 895 (1977). Architects designed an alumni building containing significant amounts of limestone and flagstone masonry. Shortly after construction was completed, deterioration of the stonework was noticed. Testimony was given that the design was faulty in that the exterior stonework did not have adequate drainage, permitting water to

some of the damages were caused by the architect's negligent failure to perform his duties of supervision during construction. For that negligence, the statute did not start to run until the installation of that component was completed.

> *8.1.2 To the extent damages are covered by property insurance, the Owner and Architect waive all rights against each other and against the contractors, consultants, agents and employees of the other for damages, except such rights as they may have to the proceeds of such insurance as set forth in AIA Document A201-2007, General Conditions of the Contract for Construction. The Owner or the Architect, as appropriate, shall require of the contractors, consultants, agents and employees of any of them similar waivers in favor of the other parties enumerated herein.* **(1.3.7.4; deleted limitation of time to "during construction"; changed reference)**

This paragraph should be discussed with insurance representatives on a job-by-job basis.[214] See the discussion of this provision under ¶ 11.3.7 of AIA Document

---

accumulate continuously, resulting in deterioration and discoloration. Also, the slope of the paving areas directed water toward a limestone wall without any provision for removing the water or waterproofing the stone, and although expansion joints were called for in the specifications, none were shown on the drawings. The court found that, although the statute of limitations had run and no action could be brought against the architect for improper design, there was a continuing duty to discover faulty design during the construction phase. The architect should have found these problems during construction and requested changes. The supervisory obligation of the architect could not be satisfied with rigid adherence to the plans when inspection of the project revealed obvious errors.

In Naetzker v. Brocton Cent. Sch. Dist., 50 A.D.2d 142, 376 N.Y.S.2d 300, 307 (1975), the court stated that: the rule in cases where the gravamen of the action is professional malpractice is and always has been that the cause of action accrues upon the performance of the work by professionals. [Citations omitted] Where the action seeks damages for architectural malpractice, the cause of action accrues upon completion of the building. An exception to the general rule is the "continuous treatment" doctrine. Where applicable, the doctrine suspends accrual of the malpractice cause of action until tresatment ends or the professional relationship is terminated. [Citations omitted].

[214] Jewish Bd. of Guardians v. Grumman Allied Indus., 96 A.D.2d 465, 464 N.Y.S.2d 778 (1983) (general contractor had duty to protect prefab units, which were damaged by a severe rainstorm on-site during construction; waiver of rights presents a question of law, which may be dispositive of rights of parties, except the architect, who had no responsibility for damage). *See also* Hernandez v. Racanelli Construction Company, Inc., 33 A.D.3d 536, 823 N.Y.S.2d 377 (N.Y.A.D. 1 Dep't 2006), where the architect was completely relieved of responsibility for construction defects and safety precautions under its contract with the owner, thus avoiding responsibility to the demolition contractor where a plywood fence collapsed during demolition, injuring passers-by.

In Blue Cross of S. W. Va. v. McDevitt & St. Co., 234 Va. 191, 360 S.E.2d 825 (1987), the owner had occupied a building that was substantially complete. Water pipes burst, causing damage. The owner's insurer sued the architect and contractor, alleging that this waiver provision was not effective because the loss occurred after substantial completion and occupancy. The court held that this provision was effective until final payment.

In Travelers Indem. Co. v. Losco Group, Inc., 136 F. Supp. 2d 253 (S.D.N.Y. 2001), the owner's insurance carrier brought a subrogation action against the architect and others arising out of a construction accident. The issue was whether a count for gross negligence should be dismissed based on the waiver of subrogation language. The court held that when the losses are

A201. In a North Carolina case in which the parties had included a provision that the architect be required to obtain professional liability insurance, the court found an ambiguity between this paragraph and the professional liability insurance requirement.[215] In view of this case, an architect should modify any proposed insurance clause to provide that ¶ 8.1.2 is not meant to be waived. Note that if the parties intend to use other general conditions, this provision should be modified, or the appropriate language from AIA Document A201 should be added to those other general conditions.

> *8.1.3 The Architect and Owner waive consequential damages for claims, disputes or other matters in question arising out of or relating to this Agreement. This mutual waiver is applicable, without limitation, to all consequential damages due to either party's termination of this Agreement, except as specifically provided in Section 9.7.*
> **(1.3.6; minor modifications)**

This paragraph is a waiver of certain damages known as consequential damages. If either the owner or architect is damaged, they waive any recovery for any consequential damages. In A201, a list specified in ¶¶ 15.1.6.1 and 15.1.6.2 provides some examples of consequential damages. Note that this list is not exclusive, so that if other damages are incurred that are deemed by a court to be consequential

---

the result of gross negligence, a waiver clause (such as the waiver of subrogation) will not preclude recovery in tort or contract. Here, the insurance carrier had set forth sufficient allegations that, if proved, would constitute gross negligence.

In Town of Fairfield v. Commercial Roofing, 2003 WL 21404027 (Conn. Super., June 5, 2003), the owner sued several defendants, including the architect, for damages relating to a school re-roofing project that led to extensive water damage. The architect moved for summary judgment, claiming that the waiver of subrogation provision in the A201 was applicable. The issue was to what extent the waiver applied. The claim presented by the owner to the insurance carrier was for $5 million. These parties then settled for $1.2 million. The difference resulted from the fact that the overall damages included closing the school, transporting the children to other schools, and other costs, including remediation of mold, which was subject to an exclusion in the insurance policy. The architect's argument, that ¶ 11.4.1 of A201 required the owner to purchase an "all risk" policy meant that all risks should have been covered, was rejected by the court: "However, despite its title, an 'all risk' policy does not insure against every conceivable risk. In general, an 'all risk' insurance policy insures against 'all fortuitous losses, unless specifically excluded, not resulting from the wilful misconduct or fraud of the insured.' " Because the architect failed to present any evidence that the owner's claims in the case were covered by the policy, summary judgment was, therefore, denied as to costs beyond the $1.2 million.

[215] St. Paul Fire & Marine v. Freeman-White, 322 N.C. 109, 366 S.E.2d 480 (1988). A hospital under construction collapsed. The owner's insurance carrier filed suit against the architect, alleging negligence. The trial court had dismissed the case on the basis of the AIA language, finding a waiver. On appeal, the dismissal was reversed, finding that there was an ambiguity between the standard waiver language and the additional language which required the architect to obtain professional liability insurance. The court apparently did not observe that the primary purpose of requiring an architect to carry such liability insurance in addition to the waiver of this paragraph is to protect the owner from third-party claims, as well as to cover the owner in a variety of situations in which the waiver simply does not apply. The dissent in this case was clearly the more logical position, as it did acknowledge these issues.

damages,[216] those will also be waived. This is a way of limiting damages, and courts have generally upheld such provisions.[217] The provision that the waiver is applicable to termination is not a limitation, because the opening language of "arising out of or relating to" is considered to be very broadly inclusive (**see** discussion as to broad form arbitration clauses at ¶ 8.3.1). If the contract

---

In Butler v. Mitchell-Hugeback, Inc., 895 S.W.2d 15 (Mo. 1995), the court held that a waiver of subrogation is effective so long as insurance was required to be maintained. In this case, even though the contractor sent the owner an invoice stating that the work was 100 percent completed, there was no evidence that a final certificate of payment had been issued. Further, under ¶ 2.6.1 of B141, the architect's services are not concluded until final payment is made to the contractor or 60 days after substantial completion, whichever occurs first. The building collapsed suddenly shortly after completion of construction, but within 60 days of substantial completion. The court held that insurance was required to be maintained during this time, so that the waiver of subrogation was effective to protect the architect.

[216] Section 2-715 of the Uniform Commercial Code defines *consequential damages* as:

(a) any loss resulting from general or particular requirements and needs of which the seller at the time of contracting had reason to know and which could not reasonably be prevented by cover or otherwise; and

(b) injury to person or property proximately resulting from any breach of warranty."

Thus, if there is an injury to the owner's property caused by the breach of any warranty by the contractor, this provision may waive recovery for such damages.

[217] *See, e.g.*, Adams Lab., Inc. v. Jacobs Eng'g Co., 486 F. Supp. 383 (N.D. Ill. 1980). Note that parties sometimes attempt to circumvent this limitation by asking for recovery in tort. In Lincoln Pulp & Paper Co. v. Dravo Corp., 436 F. Supp. 262 (D. Me. 1977), the owner was permitted tort recovery against the engineer despite such language. Travelers Cas. & Sur. Co. v. Dormitory Auth. of the State of N.Y., 2005 WL 1177715 (S.D.N.Y., May 19, 2005).

In City of Milford v. Coppola Constr., 2004 WL 3090680 (Conn. Super., Dec. 1, 2004), the city hired the contractor to lift several houses. The contractor brought specialized equipment to the site to perform the work, but the city needed to correct the plans to obtain the required permits, resulting in a delay of some seven months during which the equipment lay idle. The contractor's claims were arbitrated, with the arbitrator awarding the contractor damages. The city then brought an action to modify and correct the award on the ground that idle equipment and materials were not matters submitted to the arbitrator. The court examined the AIA contract, including the waiver of consequential damages provision, and held that the arbitrator's award should stand. These provisions "indicate that not all types of damages are amendable to arbitration." Since the parties waived consequential damages, the arbitrator could not award such damages. The issue for the court was "whether the idle equipment and materials were consequential damages, and therefore not arbitrable, or whether they were other than consequential damages, such as liquidated direct damages, and therefore properly arbitrable." The court found that the loss from this contractor's idle equipment in this case arose in the usual course of the performance of the contract, was foreseeable, and was the direct result of the city's breach.

In Pitts v. Watkins, 905 So. 2d 553 (Miss. 2005), involving a home inspection agreement, the Mississippi Supreme Court held the limitation of liability provision and arbitration provision unconscionable. The agreement required arbitration of any claim by the homeowner, but the inspector could bring a court action. The inspector's liability was capped at his fee—$265. That provision stated: "There will be no recovery for consequential damages." "The limitation of liability clause, when paired with the arbitration clause, effectively denies the plaintiff of an adequate remedy and is further evidence of substantive unconscionability."

documents contain a liquidated damages clause, this provision does not waive such damages.

There is an apparent conflict between this provision and that found at ¶ 9.7, which defines termination expenses due the architect in the event the owner terminates the contract for convenience. Here, the architect waives any claims for consequential damages, which presumably includes lost profits and similar costs. However, in ¶ 9.7, the architect is entitled to lost profits on the services not performed at the time of termination not the architect's fault. This apparent conflict can be reconciled by finding that ¶ 9.7 is an exception to the general waiver of consequential damages, but only if the architect is not at fault. This may well cause owners to trump up some "cause" in order to avoid paying the architect the proper lost profits.

§ 2.25   Mediation: ¶ 8.2

8.2 MEDIATION

8.2.1 Any claim, dispute or other matter in question arising out of or related to this Agreement shall be subject to mediation as a condition precedent to binding dispute resolution. If such matter relates to or is the subject of a lien arising out of the Architect's services, the Architect may proceed in accordance with applicable law to comply with the lien notice or filing deadlines prior to resolution of the matter by mediation or by binding dispute resolution. (**1.3.4.1; modified to add binding dispute resolution process**)

Although essentially a voluntary procedure, mediation is now a precondition to arbitration or litigation. If one party refuses to participate in the mediation, the other party could presumably ask the arbitrator or court for a summary finding in its favor based on a breach of this provision. As of 2007, arbitration is no longer the default dispute resolution process if mediation fails. Section 8.2.3 gives the parties the option to choose arbitration, litigation, or other process, with litigation the default if no choice is made. However, mediation is still required, no matter what choice is made in Section 8.2.3.

8.2.2 The Owner and Architect shall endeavor to resolve claims, disputes and other matters in question between them by mediation which, unless the parties mutually agree otherwise, shall be administered by the American Arbitration Association in accordance with its Construction Industry Mediation Procedures in effect on the date of the Agreement. A request for mediation shall be made in writing, delivered to the other party to the Agreement, and filed with the person or entity administering the mediation. The request may be made concurrently with the filing of a complaint or other appropriate demand for binding dispute resolution but, in such event, mediation shall proceed in advance of binding dispute resolution proceedings, which shall be stayed pending mediation for a period of 60 days from the date of filing, unless stayed for a longer period by agreement of the parties or court order. If an arbitration proceeding is stayed pursuant to this section, the parties may nonetheless proceed to the selection of the arbitrator(s) and agree upon a schedule for later proceedings. (**1.3.4.2; substantial modifications**)

Mediation must be held prior to arbitration or litigation. If no selection is made by the parties as to a dispute resolution procedure in ¶ 8.2.3, then litigation will be the default dispute resolution procedure and mediation will be a precondition to litigation. Note the 60-day stay provision, which provides for a stay of either litigation or arbitration for a period of at least 60 days in order to permit the mediation to proceed.[218]

8.2.3 The parties shall share the mediator's fee and any filing fees equally. The mediation shall be held in the place where the Project is located, unless another location is mutually agreed upon. Agreements reached in mediation shall be enforceable as settlement agreements in any court having jurisdiction thereof. (**1.3.4.3; no change**)

If the parties reach a settlement (the goal of mediation), that settlement may be enforced in a court without the need for arbitration.

8.2.4 If the parties do not resolve a dispute through mediation pursuant to this Section 8.2, the method of binding dispute resolution shall be the following:

(Check the appropriate box. If the Owner and Architect do not select a method of binding dispute resolution below, or do not subsequently agree in writing to a binding dispute resolution method other than litigation, the dispute will be resolved in a court of competent jurisdiction.) (**new**)

☐ *Arbitration pursuant to Section 8.3 of this Agreement*

or

☐ *Litigation in a court of competent jurisdiction*

or

☐ *Other (Specify)* (**new**)

For the first time, the 2007 AIA documents have eliminated arbitration as the default mechanism for resolution of disputes, following mediation. The parties should check one of the three boxes to select arbitration, litigation, or some other dispute resolution method. If all boxes are left blank, or if more than one is checked or it is otherwise unclear as to which method is selected, then litigation is the final process and there will not be arbitration. Of course, mediation is still a condition precedent to any of these processes (**see** Section 8.2). Users of the AIA

---

[218] According to the court in LBL Skysystems (USA), Inc. v. APG-America, Inc., 2005 WL 2140240 (E.D. Pa., Aug. 31, 2005), the 60-day provision means that a demand for mediation stays any legal or equitable proceedings for 60 days. In that case, when one party filed a request for mediation, the other party took the position that mediation was not a reasonable alternative at the time. When litigation was initiated, there was no request for a stay of the legal action pending mediation. This constituted a waiver of the mediation requirement.

Electronic Documents should note that arbitration may be automatically checked unless the user manually changes this selection.

§ 2.26   Arbitration: ¶ 8.3

8.3 ARBITRATION

Arbitration is a method of resolving disputes whereby an impartial arbitrator issues an award that is enforceable by the judicial system. All parties to arbitration must agree, at some point, to engage in arbitration. This agreement can be made at the beginning of a relationship, such as when the parties execute a standard AIA contract that contains an arbitration clause. It can also be made after a dispute has arisen, in which case the parties submit the dispute to arbitration.

Arbitrators are not required to be attorneys, although many are, and many cases are not decided on legal principles. Arbitrators should be persons who have knowledge of the construction industry and construction cases. Arbitrators should not be biased or have an interest in the outcome of the case.[219] Arbitrators must give each side an opportunity to present its case, after which the award is made, presumably on the basis of the evidence presented. One of the advantages of arbitration is that the rules of evidence, which can prolong a trial, are not necessarily adhered to. This also means that the parties do not need an attorney to represent their interests, although most parties choose to retain an attorney.

If a party who has signed an arbitration agreement subsequently refuses to participate in the arbitration proceeding, the courts can compel arbitration.[220] However, if

---

[219] In Carteret County v. United Contractors of Kinston, Inc., 462 S.E.2d 816 (N.C. Ct. App. 1995), the court found that an arbitration panel of three contractors was not fundamentally unfair or biased in favor of the contractor and against the owner. The court found that the only link between the arbitrators and the contractor is that they have the same occupation. It likened this to an argument that three judges could not impartially decide a matter involving an attorney because they are members of the same profession.

In Wilharm v. M.J. Constr. Co., 1997 Ohio App. LEXIS 591 (Feb. 20, 1997), the question was whether the court or the arbitrator should hear a claim for rescission of the contract based on frustration of purpose. The court held that this was a question for the arbitrator to determine.

[220] *But see* Palmer Steel Structures v. Westech, Inc., 178 Mont. 347, 584 P.2d 152 (1978) (contrary to prevailing rule, court relied on Montana statute to hold that any party's consent to arbitration may be withdrawn up to the time an arbitration award is entered in the district court). In Omega Constr. Co. v. Altman, 147 Mich. App. 649, 382 N.W.2d 839 (1985), the project manual stated that AIA Document A201 was incorporated. However, the owner-contractor agreement did not contain an arbitration clause. The court held that the contractor could not compel arbitration because the project manual had information for a limited purpose that did not include a contractual arbitration clause.

In Joseph Francese Inc. v. Enlarged City Sch. Dist. of Troy, 263 A.D.2d 582, 693 N.Y.S.2d 280 (App. Div. 1999), the primary arbitration provision in the owner/architect agreement was stricken, although there remained a reference to arbitration in the claims section of the contract. When a dispute arose, the contractor demanded arbitration. Defendant then moved to stay the arbitration, and the trial court granted the stay. The appellate court confirmed. The contractor then

both parties proceed in court, they may have waived arbitration.[221] Sometimes, both parties proceed to trial on a construction issue and the losing party then wants to compel arbitration. In that case, courts have held that there was a waiver of arbitration.[222]

---

filed the instant action, and the defendant moved for summary judgment on the basis that the action was filed after the one-year statute of limitations had expired, thereby barring the action. Both the trial court and the appellate court agreed. The filing of the arbitration demand did not toll the statute of limitations because there was no good-faith basis for the contractor to believe that arbitration was an available forum. The undeleted references to arbitration were rendered meaningless by virtue of the fact that the entire arbitration section had been deleted. *See also* Chianelli Constr. Co., Inc. v. Town of Durham, 1999 Conn. Super. LEXIS 1707 (July 2, 1999).

In Summit Contractors, Inc. v. Legacy Corner LLC, 147 Fed. Appx. 798 (2005), the owner and general contractor had two agreements: First, a standard AIA "cost-plus" agreement that incorporated A201; second, an "Identity of Interests Agreement" in which the contractor agreed to accept a fixed fee in lieu of a percentage profit. Both contracts had an integration clause, and the Identity Agreement did not contain an arbitration provision. Following a dispute over payments, the contractor filed suit based on the Identity Agreement. The owner initiated mediation and arbitration with AAA, and filed a motion with the court to stay and compel arbitration. The court denied the motion, stating that merger clauses are strong evidence that the parties did not intend to include terms not expressly incorporated into the document containing the clause. The court also found even more telling the fact that the AIA agreement failed to incorporate or even mention the Identity Agreement. Thus, the court found that the contractor had not agreed to arbitrate disputes under the Identity Agreement.

In Twin Oaks at Southwood, LLC v. Summit Constructors, Inc., 941 So. 2d 1263 (Fla.App. 1 Dist., 2006), the parties entered into a "Project Completion Agreement" several months after the completion date contemplated by the contract. This document contained an integration clause, but also stated that the "terms and conditions of the Contract shall remain in full force and effect." The contractor filed suit and the owner filed counterclaims. The contractor then filed a motion to dismiss, asserting that the counterclaims arose under the contract, which required arbitration, but the complaint was filed pursuant to the PCA which did not require arbitration. The owner then asserted that arbitration had been waived by both parties. The appellate court held that there was no waiver of arbitration, that the PCA met the AIA definition of a modification and was therefore subject to arbitration.

In F.L. Crane & Sons, Inc., v. Malouf Construction Corporation, 953 So. 2d 366 (Ala., 2006), the contract between the contractor and subcontractor incorporated the A201, which required mediation and arbitration. The subcontract itself, however, stated that "no such ADR procedures shall be mandatory unless both contractor and subcontractor enter into a separate agreement to submit their disputes to ADR." The court held that the mandatory provisions in the A201 governed only the owner and general contractor and did not abolish the different dispute-resolution procedure found in the subcontract.

[221] Healy v. Silverhill Construction Co., 2007 WL 2769799 (De.Com.Pl., Sept. 19, 2007) (arbitration waived where defendant waited until the day of trial to raise arbitration. The parties are presumed to know what is in their contract).

[222] Batter Bldg. Materials Co. v. Kirschner, 142 Conn. 1, 110 A.2d 464 (1954). *See also* Thomas Wells & Assocs. v. Cardinal Properties, 192 Colo. 197, 557 P.2d 396 (1976), in which an architect sought to collect fees. The defendant moved to dismiss on the basis that the architect had failed to arbitrate. The trial court should have heard the architect's evidence that arbitration had been waived by the parties.

In Board of Educ. Taos Mun. Sch. v. The Architects, Taos, 103 N.M. 462, 709 P.2d 184 (1985), the architect waived arbitration by availing itself of discovery procedures, even though the right to arbitrate was raised as an affirmative defense. In Sanford Constr. Co. v. Rosenblatt, 25 Ohio

Misc. 99, 266 N.E.2d 267 (1970), the owner told the contractor: "If you want to collect, sue us!" The court took this as a waiver of the arbitration clause.

An owner waived arbitration by participating in litigation, although the owner had affirmatively pled the arbitration provision. North W. Mich. Constr., Inc. v. Stroud, 185 Mich. App. 649, 462 N.W.2d 804 (1990). The court stated that had the owner filed a motion to compel arbitration at the beginning of the proceeding, a waiver would not have resulted.

In I.D.C., Inc. v. McCain-Winkler Partnership, 396 So. 2d 590 (La. Ct. App. 1981), a case in which the general contractor brought an action against the owner and architect, waiting to demand arbitration until the day of trial was too late. In addition, the architect was not entitled to arbitration against the general contractor, because there was no contract between those parties.

In Triangle Air Conditioning, Inc. v. Caswell County Bd. of Educ., 291 S.E.2d 808 (N.C. Ct. App. 1982), the court held that, under the 1970 version of the AIA document language, because neither party demanded arbitration, the parties were not bound to arbitrate. *See also* Multi-Service Contractors, Inc. v. Town of Vernon, 435 A.2d 983, 985 (Conn. 1980) ("The arbitration clause in this case does not require, either by express language or by necessary implication, arbitration as a condition precedent to court action.").

In Paul Mullins Constr. Co. v. Alspaugh, 628 P.2d 113, 113 (Colo. Ct. App. 1980), the owner filed a demand for arbitration against the contractor after a dispute developed. The contractor filed a response that reserved the right to arbitrate "only as a condition precedent to a possible court action." The owner then filed suit, alleging a wrongful attempt by the contractor to avoid finality of submission to arbitration. The appellate court found that the trial court should have entered an order staying the civil action pending arbitration, because the owner did not waive its right to arbitration by initiating litigation when the principal assertion was to compel arbitration. The court also found that submission of a dispute to arbitration was a condition precedent to the pursuit of legal action.

In McCarney v. Nearing, Staats, Prelogar & Jones, 866 S.W.2d 881, 890 (Mo. Ct. App. 1993), the court stated that a finding that arbitration has been waived is not favored; there is a presumption against waiver. The court used a three-factor test to determine if there was a waiver: "The test requires that the party seeking to establish waiver bear the burden of demonstrating that the alleged waiving party: (1) had knowledge of the existing right to arbitrate; (2) acted inconsistently with that existing right; and (3) prejudiced the party opposing arbitration by such inconsistent acts." This test seems consistent with other decisions concerning waiver.

Ohio applies a two part test: (1) knowledge of an existing right to arbitrate and (2) acting inconsistently with that right, given the totality of the circumstances. Bedford City Sch. Dist. v. Trane Co., 1997 Ohio App. LEXIS 1107 (Mar. 20, 1997). In MGM Landscaping Contractors, Inc. v. Robert Berry, 2000 Ohio App. LEXIS 1117 (March 22, 2000), a contractor brought a mechanic's lien foreclosure action against the owner. The owner filed a counterclaim against the contractor and various other claims against third parties. An AIA form contract containing an arbitration clause was bound into a "spec book." Following some court proceedings, the contractor moved to stay the proceedings and compel arbitration. One issue was whether the contractor had waived arbitration. The contractor argued that it had not waived arbitration because it had not seen the arbitration clause in the contract. The appellate court held that there was a waiver because the contractor was deemed to know the terms of the contract from its inception. This is so whether or not the contractor had actually seen the contract, since the contractor must make a reasonable effort to know the contents of that contract.

In James J. Gory Mech. Contracting, Inc. v. Philadelphia Housing Authority, 2001 WL 1736483 (Pa. Comm. Pl., July 11, 2001), arbitration was waived because the defendant waited until two months before trial to bring up the issue.

In East 26th Street & Park Realty v. Shaw Indus., Inc., 6 Misc. 3d 1036A (N.Y. Sup. 2004), the contractor waived arbitration by answering the complaint, filing a counterclaim, and participating in extensive discovery proceedings.

In T.A. Tyre Contractor, Inc. v. Dean, 2005 WL 1953036 (Del. Super., June 14, 2005), a dispute arose, the contractor submitted a number of claims to the architect and subsequently

If the opposing party has not been prejudiced, arbitration is not waived even after the filing of a lawsuit.[223]

---

initiated litigation. The owners did not raise the architect's decision in their pleadings and filed a counterclaim. Some time later, the owners moved to dismiss based on the arbitration provision and that the decision of the architect was binding (A201, ¶ 4.4.6). The court denied the motion based on waiver. *See also* Jones-Williams Constr. Co., Inc. v. Town & Country Prop., LLC, 923 So. 2d 321 (Ala. Ct. App. 2005), finding waiver.

In Wright v. Jerry Fulks & Co., Inc., 2005 WL 2840335 (Wash. App. Div. 1, Oct. 31, 2005), on the first day of trial, after the trial court asked whether the claims between the owner and contractor were subject to arbitration, the contractor asked the court to compel arbitration. The court did so and certified the order for immediate appeal. The appellate court reversed, finding waiver. The court found that the delay in demanding arbitration by the contractor caused the owner significant and unnecessary additional expense, resulting in prejudice to the owner.

In Aberdeen Golf & Country Club v. Bliss Constr., Inc., 932 So. 2d 235 (Fla. App. Dist. 4 Sept. 2005), the appellate court affirmed a trial court's finding of waiver of the arbitration provision. When mold was discovered in the building at issue, the contractor gave notice to the architect and made a claim for extras. The architect confirmed the existence of the mold and agreed that it had caused delays that could affect the price. Rather than initiate mediation, the owner refused to pay the contractor's draw request and terminated the contract. Several weeks later, the owner wrote to the contractor asking for arbitration, but, apparently, never initiated arbitration with the AAA. The court analyzed the ADR provisions in the AIA agreement, finding that arbitration was not set forth as a precondition to litigation and that the ADR system was meant to function in place of the courts while progress was being made on the contract. "If either party terminated, all bets would be off and either could have its day in court." The ADR provision failed of its essential purpose.

In Silver Dollar City v. Kitsmiller Constr., 874 S.W.2d 526 (Mo. Ct. App. 1994), following *McCarney* in a case interpreting the Federal Arbitration Act, the court held that there was no waiver. The owner claimed that the contractor had failed to present its claim to the architect, thereby waiving arbitration. The evidence showed that the owner was also the "architect" under the contract and that some of the claims had been presented to the owner. Thus, there was no waiver.

In a follow-up case, Silver Dollar City v. Kitsmiller Constr., 931 S.W.2d 909 (Mo. Ct. App. 1996), the court compelled arbitration. The owner thought it was contracting with a joint venture consisting of two parties. The first page of the contract named both entities the "Contractor" but the signature page listed only one. There was thus an apparent ambiguity and the contract could have been voided by the owner at that time, particularly because the contractors knew of the mistake. However, by failing to take prompt action to rescind or correct the contract, the court found that the owner had ratified the contract.

In Plan Pac., Inc. v. Andelson, 6 Cal. 4th 307, 962 P.2d 158, 24 Cal. Rptr. 2d 597 (1994), the Supreme Court of California thoroughly analyzed the concept of waiver in the context of construction arbitration. Although in that case the parties had a contract that required arbitration to be initiated by a specified date, and the issue was whether failure to file by that date could be excused, the analysis is instructive in any situation in which waiver becomes an issue.

In Robert J. Denley Co., Inc., v. Neal Smith Const. Co., Inc., 2007 WL 1153121 (Tenn.Ct.App., April 19, 2007), the court compelled arbitration where A101 incorporated A201. The bonding company had standing to demand arbitration and there was no waiver by filing an answer.

[223] D.M. Ward Constr. Co. v. Electric Corp., 15 Kan. App. 2d 114, 803 P.2d 592 (1991); District Moving & Storage v. Gardinier & Gardinier, 63 Md. App. 96,492 A.2d. 319 (1985); Adams v. Nelson, 313 N.C. 442, 329 S.E.2d 322 (1985). In Valley Constr. Co. v. Perry Host Management Co., 796 S.W.2d 365 (Ky. 1990), there was no waiver, even though the contractor participated in a separate foreclosure action and a federal court action. In Clinton Nat'l Bank v. Kirk Gross Co., 559 N.W.2d 282 (Iowa 1997), perfecting a lien did not constitute waiver. For instances in which litigation proceeded too far, see Red Sky Homeowners Ass'n v. Heritage Co., 701 P.2d 603

There is some authority that arbitration, once waived, cannot be revived over the objections of the opponent.[224] If a party participates in an arbitration proceeding,

---

(Colo. Ct. App. 1984); Servomation Corp. v. Hickory Constr. Co., 74 N.C. App. 603, 328 S.E.2d 842 (1985). In People ex rel. Delisi Constr. Co. v. Board of Educ., 26 Ill. App. 3d 893, 326 N.E.2d 55 (1975), the contract incorporated the AAA rules (as do the AIA contracts). Those rules state that no judicial proceedings should be deemed a waiver of the right to arbitrate. Arbitration was allowed.

The court in Wasserstein v. Kovatch, 261 N.J. Super. 277, 618 A.2d 886, 893 (1993), held that "election of remedies is not irrevocable unless and until either a court proceeding goes to judgment or an arbitration proceeding consummates in an award."

There was no waiver in Commercial Union Ins. Co. v. Gilbane Bldg. Co., 992 F.2d 386 (1st Cir. 1993), in which there was no showing of prejudice. *See also* B&S Equip. Co. v. Woodward, 620 So. 2d 347 (La. Ct. App. 1993) (arbitration requested eight months after court petition filed); Parmer v. Pre-Fab Constr., Inc., 1997 Tex. App. LEXIS 86 (Jan. 9, 1997); D. Wilson Constr. Co. v. McAllen Indep. Sch. Dist., 848 S.W.2d 226 (Tex. Ct. App. 1992) (arbitration requested 13 months after dispute arose). In Lee v. Yes of Russellville, Inc., 784 So. 2d 1022 (Ala. 2000), there was no waiver after filing a complaint, moving for a stay to compel arbitration, and filing an amended complaint. *See also* Zapor Architects Group, Inc., v. Riley, 2004 WL 1376236 (Ohio App., June 16, 2004) ("A party can hardly be deemed to have waived arbitration by raising the matter in its initial pleading with the court."); Patten Grading & Paving, Inc. v. Skanska USA Building, Inc., 380 F.3d 200 (4th Cir. 2004) (four-month delay was not prejudicial).

In Contract Constr., Inc. v. Power Tech. Ctr. Ltd. P'ship, 100 Md. App. 173, 640 A.2d 251 (1994), the court found that there was no waiver. A demand for arbitration is not a claim. The other side argued that by failing to submit its claim to the architect prior to demanding arbitration, there was a waiver. This case involved a construction accident. The decedent's estate filed a complaint against several parties. The general contractor was dismissed because it was the decedent's employer and thus immune under the worker's compensation act. The owner then filed a third-party complaint against the general for indemnification and breach of contract, whereupon the general demanded arbitration. The court held that arbitration was proper.

In City of Centralia v. Natkin & Co., 630 N.E.2d 458, 461 (Ill. App. Ct. 1994), a dispute developed between the owner and the contractor. For three years, before the owner filed suit, the parties attempted to work out a settlement. When the contractor moved to stay the litigation pending arbitration, the owner argued that it had been prejudiced by the delay. The appellate court ordered arbitration:

> Although it is well-settled law that a contractual right to arbitrate can be waived like any other contract right, waiver will only be deemed to have occurred when a party's conduct has been inconsistent with the arbitration clause so as to indicate that he has abandoned his right to avail himself of such right. . . . In Illinois, waiver has been found when: (1) a party has instituted legal proceedings and participated in a trial on the merits . . . (2) a party has filed an answer without asserting his right to arbitrate . . . and (3) a party that has sought arbitration files a motion for summary judgment on an arbitrable issue. . . . Hence, we refuse to punish Natkin merely because it attempted to resolve a dispute between itself and the City before filing a demand for arbitration. Therefore, we find that Natkin did not act inconsistently with its right to arbitration, and that a three-year delay is not an inordinate period of time to wait before resorting to arbitration.

In Scott Addison Constr. v. Lauderdale County Sch. Sys., 789 So. 2d 771 (Miss. 2001), there was a waiver by the contractor. The bid documents included an unmodified A201. However, following the bid opening, the school deleted the arbitration provisions from the contract. The court found that the contractor had waived any objection to the deletion by virtue of his subsequent conduct.

[224] Standard Co. v. Elliott Constr. Co., 359 So. 2d 224 (La. Ct. App. 1978) (*construing* 7.10.1 of AIA Documents A201-1970 and A401-1972). *See* Gutor Int'l A.G. v. Raymond Packer Co., 493 F.2d 938 (1st Cir. 1974) ("Submission of part of an arbitrable matter to a court waives the submittor's

that party cannot later object to it.[225] Even if a party has not signed the contract, it may still be bound by the arbitration provision,[226] particularly if the arbitration

---

right upon arbitration of the remainder."); Alspaugh v. District Court, 190 Colo. 282, 545 P.2d 1362 (1976); Denihan v. Denihan, 34 N.Y.2d 307, 313 N.E.2d 759, 357 N.Y.S.2d 454 (1974).

[225] Ruffin Woody & Assocs. v. Person County, 92 N.C. App. 129, 374 S.E.2d 165 (1988) (plaintiff had not waived its objection because the objection was filed before commencement of arbitration).

In Degi Deutsche Gesellschaft Fuer Immobilienfonds MBH v. Jaffey, No. 95 Civ. 6813 (MBM), 1995 U.S. Dist. LEXIS 16737 (S.D.N.Y. Nov. 3, 1995), the architect filed a demand for arbitration, and the owner answered the demand and later sought a stay of arbitration in court. The court held that the owner had not waived his rights to seek a stay. The answer had explicitly disavowed submission to the arbitration.

[226] Hurley v. Fox, 559 So. 2d 887 (La. 1990). In Garaman, Inc. v. Danny Williams, A.I.A., 912 P.2d 1121 (Wyo. 1996), the architect prepared a contract that was never executed. The architect performed services under the first phase and part of the second. When the owner failed to pay, the architect sued. The owner moved to dismiss on the basis that the contract required arbitration. The architect contested the motion on the basis that the contract was never executed, and the owner moved to dismiss the motion. At trial, however, the architect admitted that he had prepared the written contract, which the owner did not execute, and that that contract governed the parties' relationship. The trial court refused to dismiss the case and order arbitration. The architect argued that the owner had waived the arbitration provision. The Wyoming Supreme Court agreed, saying that the owner could have demanded an immediate determination of this issue when the architect initially denied that an agreement to arbitrate existed. By choosing not to pursue this matter at that time, the owner waived the right to arbitrate.

In Liberty Management v. Fifth Avenue & Sixty-Sixth St. Corp., 208 A.D.2d 73, 620 N.Y.S.2d 827 (1995), the contractor never signed the AIA contract that incorporated the arbitration clause by reference. The court found that the arbitration agreement was still valid, holding that the state arbitration statute merely required a written agreement to arbitrate. The statute does not require that the writing be signed so long as there is other proof that the parties actually agreed on it. Further, change orders are also subject to arbitration.

However, the court refused to order arbitration in Wooster Assocs. v. Walter Jones Constr. Co., No. C.A. No. 2900, 1994 Ohio App. LEXIS 5911 (Dec. 21, 1994). The contractor claimed a right to arbitration on the basis that one of the exhibits to his contract contained an index to drawings, which in turn contained outline specifications, which in turn contained a provision incorporating AIA Document A201, which further contained the arbitration provision. The court found this too far removed.

In Brooks & Co. General Contractors, Inc. v. Randy Robinson Contracting, Inc., 513 S.E.2d 858 (Va. 1999), the general contractor obtained a bid from a subcontractor. There was a delay in the start of the project, and the general contractor subsequently orally confirmed the bid with the subcontractor. The general contractor then sent to the subcontractor a standard AIA subcontractor agreement that contained an arbitration agreement. The bid documents had not contained an arbitration agreement. The written agreement was never executed. The subcontractor started work at the site and three weeks later used a front-end loader to demolish the general contractor's job-site trailer, whereupon it left the site, never to return. The general contractor then demanded arbitration with the subcontractor and the subcontractor denied that there was an agreement to arbitrate. The general contractor argued that the subcontractor had agreed to the terms of the AIA form contract by performance. The court found that based on these facts, there was only an oral agreement, but no meeting of the minds as to the subsequent written AIA contract that included the arbitration provision. The court refused to order arbitration.

In Todd Habermann Constr. v. David Epstein, 70 F. Supp. 2d 1170 (D. Colo. 1999), the owner (who was an attorney and state Court of Claims Judge) hired an architect to prepare plans and specifications for a summer house. A201 was included in the bidding documents and specifications. The contractor submitted its proposal, which was verbally accepted through the architect; however, a written contract was never signed. After litigation was initiated, the contractor moved to compel arbitration. The owner objected on the basis that there was no signed agreement to arbitrate. The court found the arbitration provision binding on the owner based on the course of conduct between the parties, which established an agreement on the AIA contract, including the arbitration provision. When the contractor's bid was received, the owner directed the architect to accept the bid offer, thereby forming a contract. The owner also argued that, under the Federal Arbitration Act, which requires an arbitration agreement to be in writing, this arbitration agreement could not be enforced since it was not signed. The court found that there was no requirement that such an agreement actually needs to be signed so long as it was in writing.

In Marinos v. Building Rehabilitations, 2000 Conn. Super. LEXIS 1601 (June 20, 2000), the proposed AIA contract was never signed. One month before the architect was terminated, she had sent a letter to the owner as a reminder that they were working without a contract. Based on these facts, the court found that the arbitration provision of the draft contract was not enforceable.

In Watkins Engineers & Constructors, Inc. v. Deutz AG, 2001 WL 1545738 (N.D. Tex., Dec. 3, 2001), the court used the principles of equitable estoppel to order arbitration where a guarantee did not contain an arbitration agreement, but did refer to general "obligations" of another document that contained an arbitration agreement.

The doctrine of equitable estoppel was invoked by a general contractor in R.J. Griffin & Co. v. Beach Club II Homeowners Ass'n, 384 F.3d 157 (4th Cir. 2004), to force arbitration of claims brought by a homeowners' association. The court examined the association's claims to determine whether it was seeking a direct benefit from the provisions of the general contract which it did not sign. If it was, then arbitration would be appropriate. In this case, the court found against arbitration because the claims did not hinge on any rights it might have under the general contract.

In In re Kellogg Brown & Root, Inc., 166 S.W.3d 732 (Tex. 2005), the court identified six theories from federal courts, "arising out of common principles of contract and agency law, that may bind nonsignatories to arbitration agreements: (1) incorporation by reference; (2) assumption; (3) agency; (4) alter ego; (5) equitable estoppel, and (6) third party beneficiary." In this case, the theory of "direct benefits estoppel" (by which a nonsignatory plaintiff seeking the benefits of a contract is estopped from simultaneously attempting to avoid the contract's burdens, such as the obligation to arbitrate disputes) did not apply.

In Habitat Architectural Group, P.A. v. Capital Lodging Corp., 28 Fed. Appx. 242, 2002 WL 86682 (4th Cir. (N.C.), Jan. 23, 2002), an architect entered into an agreement for architectural services to design Comfort Suite hotels. The architect drafted a letter agreement that incorporated the AIA "Standard Form of Agreement between Owner and Architect." The letter agreement was sent to George Justice for signature. Justice signed his name in the "By" line and in the "Title" line scrawled the abbreviation "Pres." At that time, Justice was president of two companies: Capitol Lodging Corporation and Capitol Lodging Development Corporation.

After a dispute over the architectural services arose, Capitol Lodging Corporation filed a demand for arbitration. The architect at first agreed to arbitrate, but then voiced concerns that Capital Lodging Corporation was not a proper party to the contract. In response to those concerns, the owner amended the Demand for Arbitration to add Capitol Lodging Development Corporation and Carolina Partners I as claimants. Almost a year later, the architect filed an action in state court to stay the arbitration proceedings. It argued that none of the claimants was a clear party to the arbitration agreement, and therefore, none of those parties had the authority to force the architect into binding arbitration.

Both the trial and the appellate courts found that the architect was required to arbitrate. The court first noted that, had the architect been concerned about the owner's identity, it could have made inquiries prior to the dispute. Instead, this interest did not surface until the arbitration. The architect conceded that it agreed to arbitrate when it signed the letter agreement. It admitted that it agreed to arbitrate with someone, but the architect never provided a clear answer as to who that someone was. Instead, the architect argued that it could not be compelled to arbitrate because it did not know

clause is incorporated by an agreement that the party did sign.[227] One question that frequently arises is whether it is the arbitrator or the court that initially determines

---

with whom it agreed to arbitrate. The court did not accept this argument. The court further held that, under local law (followed in most other jurisdictions), a party does not have to be positively identified for a contract to exist or for that previously unknown party to enforce that contract. Thus, because the architect failed to follow up and clarify any possible ambiguity in the agreement that it itself drafted, the architect could not later complain about the identity of the owner.

In Brothers Bldg. Co. of Nantucket, Inc. v. Yankow, 56 Mass. App. 688, 779 N.E.2d 991 (2002), Brothers Nantucket executed an AIA agreement with the Yankows whereby Brothers would act as contractor. After a dispute arose, Brothers Nantucket filed a demand for arbitration and the Yankows counterclaimed against a related entity, Brothers Building Co. The subsequent award was entered against Brothers Nantucket and Brothers Vermont, another related entity. The court vacated the award, finding that the arbitrator had exceeded his authority in relying on a form of "alter ego" or "disregard of corporate identity" theory in holding Brothers Vermont liable.

In Guarantee Trust Life Ins. Co. v. America United Life Ins. Co., 2003 U.S. Dist. LEXIS 22777 (N.D. Ill., Dec. 17, 2003), the court upheld arbitration based on a "slip" agreement that contained the words "arbitration clause." The court held that this two-word clause was sufficient to constitute a valid agreement to arbitrate. Based on the strong federal and state policies in favor of arbitration, the party seeking to avoid arbitration had to demonstrate a genuine issue of material fact as to whether an agreement to arbitrate exists. This is a high standard comparable to the standard for summary judgment.

In Smay v. E.R. Stuebner, Inc., 864 A.2d 1266 (Pa. Super. 2004), an employee of the general contractor was injured and filed a Workers' Compensation claim against the contractor and a lawsuit against the owner and architect. The owner and architect brought an action against the contractor based on the indemnification provision of A201-1987. The contractor sought to arbitrate these claims and the owner and architect opposed the request. The appellate court reversed the trial court and granted arbitration, finding that the standard AIA nonconsolidation language had been replaced by this provision that allowed the owner to consent to the architect's participation: "No arbitration arising out of or relating to the Contract Documents shall include by consolidation or joinder or in any other manner, the Architect, the Architect's employees or consultants, except by written consent of the Owner." The court failed to explain why the architect should be bound by this provision when the architect opposed arbitration and was not a party to the agreement between the owner and contractor.

In God's Battalion of Prayer Pentecostal Church, Inc. v. Miele Assocs., LLP, 6 N.Y.3d 371, 812 N.Y.S.2d 435 (2006), the court held that an arbitration agreement in a written contract is enforceable, even if the agreement is not signed, when it is evident that the parties intended to be bound by the contract. The architect had forwarded an AIA contract to the church, which had retained it unsigned. When the church sued the architect, the architect demanded arbitration. The court found the church's argument that the contract was not signed to be unpersuasive, particularly because the church's pleadings alleged that the architect had breached that very agreement.

In Framan Mech., Inc. v. Lakeland Reg'l High Sch., 2005 WL 2877923 (N.J. Super. A.D., Nov. 3, 2005), the contractor was awarded the contract after its protest resulted in the disqualification of the lowest bidder. A court barred the school from rebidding and required that the contractor be hired. As a result of the delay involved in the protest and litigation, when the school forwarded the contract to the contractor, the contractor protested the school's failure to adjust the completion schedule and refused to sign the contract. When this dispute arose, the contractor argued that there was no arbitration agreement because it had not signed the contract. The appellate court held that, because a court order required the school to issue the contract, which included an arbitration provision, and because the contract had been substantially performed, the arbitration provision applied.

[227] In Rutter v. McLaughlin, 101 Idaho 292, 612 P.2d 135, 136 (1980), the general conditions were not specifically incorporated into the contract. When a dispute arose, the owners sought

arbitration under the provisions of A201. The contractor argued that A201 was not incorporated and, because only that document contained an arbitration agreement, it was not bound to arbitrate. The court stated that although the General Conditions were not explicitly incorporated by reference into the parties' written contract by enumeration in Article 16 of the agreement, provisions of the General Conditions were referred to in several important articles of the written agreement as if the General Conditions were part of the parties' agreement. The contract must be read as a whole to ascertain the intent of the parties.

In Giller v. Cafeteria of South Beach Ltd., LLP, 967 So. 2d 240 (Fla. Dist., Ct. App. 2007), the owner sued a principal of the architectural firm with which it had a contract for professional malpractice. There was no contract between the owner and the individual. The architect sought to compel arbitration and the appellate court agreed that the arbitration provision of A201 applied. The court reviewed the following provision from A201: "The Architect is the person lawfully licensed to practice architecture or an entity lawfully practicing architecture identified as such in the Agreement and is referred to throughout the Contract Documents as if singular in number. The term "Architect" means the Architect or the Architect's authorized representative." This applied because of the language of ¶ 10.2 references definitions of A201, and that definition included the individual architect.

In Cleveland Jet Ctr., Inc. v. Structural Sales Corp., 1995 Ohio App. LEXIS 4113 (Sept. 22, 1995), the court found that A201 was not incorporated since the parties elected to use the 1978 edition of A111 instead of the newer version that expressly incorporated A201.

In DJM Constr., Inc. v. Rust Eng'g Co., 1996 U.S. Dist. LEXIS 5455 (N.D. Ill. Apr. 24, 1996), the court denied a subcontractor's demand for arbitration. An attachment to the contract provided that A401 and A201 were to be part of the contract. However, the contract also incorporated the owner-general contract, which required litigation of all disputes. The court found that this was a more specific clause that took precedence over the standard general arbitration language in the AIA documents.

In RTKL Assocs., Inc. v. Baltimore County, 147 Md. App. 647, 810 A.2d 512 (2002), the architect's proposal read as follows: "RTKL fees are based upon the Detailed Scope of Services and the Standard Form of Agreement between Client and Architect 1987 Edition, AIA Document B-141. All Client-generated contracts will require review and acceptance by RTKL's legal counsel before any work may proceed." This proposal was incorporated into the non-AIA contract drafted by the owner and signed by the parties. The disputes provision in that contract was stricken. When a dispute arose, the architect wanted to compel arbitration, arguing that, by striking the disputes clause in the owner's form, the disputes provision in B141 should apply. The court rejected this and held that there was no valid arbitration clause.

In Dunn Indus. Group, Inc. v. City of Sugar Creek, 112 S.W.3d 421 (Mo. 2003), during a project a change order was executed that stated "Lafarge and DIG agree to first attempt to resolve the items marked on the PCO List by negotiation; however, either party, at any time, may resort to their respective contract remedies or remedies as provided by law." After a dispute arose, the contractor argued that this provision waived arbitration. The court held that when an arbitration clause is broad (like the AIA language) and contains no express provision excluding a particular grievance from arbitration, only the most forceful evidence of a purpose to exclude the claim from arbitration can prevail. The court ordered arbitration. The court also held that the parent company of the contractor was not bound to arbitrate under its guarantee of performance.

In TC Arrowpoint, L.P. v. Choate Constr. Co., 2006 WL 91767 (W.D.N.C., Jan. 13, 2006), the assignee of the original owner brought an arbitration action against the contractor. The contractor argued that the assignee was not a signatory to the agreement and could not enforce the arbitration agreement. The court disagreed, citing ¶ 13.2.1 of A201, which allows for assignments. The court also stated that "Concerning whether an otherwise valid and applicable arbitration clause may be enforced by a nonsignatory to the contract containing the arbitration provision, it is well settled that 'the obligation and entitlement to arbitrate does not attach only to one who has personally signed the written arbitration provision.' . . . Rather, 'well-established common

whether the issue is subject to arbitration.[228] Most jurisdictions restrict the court to a determination of whether or not there is an agreement to arbitrate between the parties. If so, all further matters are for the arbitrator to initially determine, but subject to court review after the arbitration is concluded, or possibly after the entry of an interim award.

Arbitration clauses have been upheld even when the underlying contract was inoperative. In one case, a general contractor declared his contract with a subcontractor null and void.[229] When the subcontractor sued, the general sought to compel arbitration based on the arbitration clause in the contract. This right was upheld by the court, which said that the arbitration clause was separate and independent from the rest of the contract. Even a waiver by the contractor of the contract does not constitute a waiver of the right to compel arbitration.[230] In another case involving a contract between a county and an architect, the contract was found to be void, even though it had been signed by the chairman of the board of supervisors and another board member.[231] The court stated that the entire board had to approve the contract. Without that, the contract was void, as was the arbitration clause.

---

law principles dictate that in an appropriate case a nonsignatory can enforce, or be bound by, an arbitration provision within a contract executed by other parties.' "

[228] *See, e.g.*, Robbinsdale Pub. Sch., Indep. Sch. Dist. No. 281 v. Haymaker Constr., 1997 Minn. App. LEXIS 1155 (Oct. 14, 1997); Congress Constr. Co., Inc. v. Geer Woods, Inc., 2005 WL 3657933 (D. Conn., Dec. 29, 2005). In O'Keefe Architects, Inc., v. CED Constr. Partners, Ltd., 944 So. 2d 181, (Fla., 2006), the issue was whether the demand for arbitration was filed within the statute of limitations and whether the arbitrator or the court decided this issue. This was a question for the arbitrator to determine.

[229] United States Insulation Inc. v. Hilro Constr. Co., 146 Ariz. 250, 705 P.2d 490 (1985).

In Lee v. Yes of Russellville, Inc., 784 So. 2d 1022 (Ala. 2000), the owner entered into a contract with a contractor (American Quality Service) to build a building, incorporating A201 into the contract. Hemingway signed the contract as agent for American Quality. When a dispute arose, the contractor sued the owner for unpaid money. The owner moved to dismiss, based on the fact that this contractor was unlicensed. Lee then moved to amend the complaint to show her as the plaintiff, doing business as American Quality, with Hemingway as her agent. This motion was granted, and Lee moved to stay the proceedings pending arbitration. This was opposed by the owner on the basis that they did not have a contract with Lee. The appellate court found that the contract was ambiguous as to whether Hemingway was an agent. Therefore, it ordered a trial on that issue. If the jury were to find that an agency did not exist, then the contract would be void because the contractor was not licensed. If an agency relationship did exist, then the dispute must proceed to arbitration because Lee would be a valid party to the contract.

In McCarl's, Inc. v. Beaver Falls Mun. Auth., 2004 WL 793202 (Pa. Cmwlth., Apr. 15, 2004), a dispute arose, the parties entered into a letter settlement agreement, and work continued. Subsequently, the owner refused to make further payments to the contractor and the contractor filed suit. The owner demanded arbitration. The contractor argued that the letter agreement superseded and was substituted for the original agreement that contained the arbitration clause. The court held for arbitration. The original contract stated that all modifications issued subsequent to the contract were to be considered part of the contract. In order to cancel the arbitration provisions of the original contract, the letter agreement must expressly cancel or otherwise nullify the arbitration provisions.

[230] Batter Bldg. Materials Co. v. Kirschner, 142 Conn. 1, 110 A.2d 464 (1954).

[231] County of Stephenson v. Bradley & Bradley, Inc., 2 Ill. App. 3d 421, 275 N.E.2d 675 (1971). In Judelson v. Christopher O'Connor, Inc., No. 95 CV 0371181, 1995 Conn. Super. LEXIS 1375

Courts will sometimes find an arbitration provision to be unconscionable.[232] A general partner of a partnership that entered into a B141 agreement was held subject to the arbitration provisions.[233] Another court has held that an undisclosed principal to a contract may prosecute an arbitration in his own name.[234] In that case, a contractor and an individual entered into a construction agreement. Legal title to the property, however, was held by another party. When a dispute developed, that other party sought to enforce the arbitration agreement, and the court permitted it to do so.

When public bodies are involved, arbitration may not be possible. One contractor brought an action to enforce an arbitration award against the Illinois Capital Development Board.[235] A dispute had arisen over the construction and the board refused to release the retainage on the project. The contractor demanded arbitration and the board refused to participate. The arbitrator found for the contractor, who then filed a complaint in court to enforce the award. Under Illinois law, the Illinois Court of Claims has exclusive jurisdiction to hear and determine all claims against the state founded upon any contracts entered into with the state. Because the board is a state agency, any suit against it is a suit against the state. Therefore, the court lacked jurisdiction to enforce the arbitration award. The contractor argued that the board had waived its right to have such suits decided by the Court of Claims. The court stated that such waiver can be made only by the legislature, not by any officer or agency.[236] In a Louisiana case, a public body authorized the signing of a contract

---

(May 2, 1995), a state statute required the contractor to be licensed as a home improvement contractor. Because the contractor was not properly licensed, the arbitration provision was not enforceable by the contractor.

[232] In Pitts v. Watkins, 905 So. 2d 553 (Miss. 2005), involving a home inspection agreement, the Mississippi Supreme Court held the limitation of liability provision and arbitration provision unconscionable. The agreement required arbitration of any claim by the homeowner, but the inspector could bring a court action. The inspector's liability was capped at his fee—$265. That provision stated that "[t]here will be no recovery for consequential damages." "The limitation of liability clause, when paired with the arbitration clause, effectively denies the plaintiff of an adequate remedy and is further evidence of substantive unconscionability."

In Ragucci v. Prof'l Constr. Serv., 25 A.D.3d 43, 803 N.Y.S.2d 139 (2005), the court found that the standard AIA arbitration provision was unenforceable because the parties' contract fell within the prohibition against mandatory arbitration clauses in contracts for the sale or purchase of consumer goods.

[233] In Keller Constr. Co. v. Kashani, 220 Cal. App. 3d 222, 269 Cal. Rptr. 259 (1990), the architect entered into a contract with a limited partnership. Kashani, the sole general partner, signed the agreement on behalf of the partnership. A dispute developed and the partnership filed bankruptcy. The architect sought arbitration against Kashani. The court held that Kashani was bound by the arbitration agreement.

[234] American Builder's Ass'n v. Au-Yang, 226 Cal. App. 3d 170, 276 Cal. Rptr. 262 (1990).

[235] J.L. Simmons Co. v. Capital Dev. Bd., 98 Ill. App. 3d 445, 424 N.E.2d 71 (1981).

[236] See also Canon Sch. Dist. No. 50 v. W.E.S. Constr. Co., 177 Ariz. 526, 869 P.2d 500 (1994) (school district sought to have the arbitration clause declared invalid on the basis that a state statute provided an exclusive remedy for dispute resolution; court, after extensive discussion of statutory construction, held arbitration provision applied).

for construction but did not authorize the signing of a specific contract.[237] The arbitration clause was held inapplicable because it was not specifically authorized.

In Foster Wheeler Passaic, Inc. v. County of Passaic, 266 N.J. Super. 429, 630 A.2d 280 (App. Div. 1993), the contract was not an AIA contract. There was, however, an arbitration provision. There was also a provision that allowed the owner to terminate the contract. Prior to actual construction but after the contractor had performed some work, including obtaining certain permits, the owner properly terminated the contract. The issue was whether the arbitration clause survived the termination. The contractor argued that it did and the trial court agreed. The appellate court reversed, finding that the specific language allowing termination did not refer to arbitration as a vehicle for resolution of any claims in the event of termination.

In an interesting case, City of Chamberlain v. R.E. Lien, Inc., 521 N.W.2d 130 (S.D. 1994), a state statute mandated that the arbitration provisions of the AIA standard form be automatically incorporated into every construction contract. In this case, the contractor sought arbitration against a municipality when a dispute arose. The municipality sought to stay the arbitration on the basis that the law was an unconstitutional delegation of its authority to contract. The court agreed, finding the law unconstitutional and leaving the contractor with litigation as its only remedy. *See also* Carteret County v. United Contractors of Kinston, Inc., 462 S.E.2d 816 (N.C. Ct. App. 1995) (county bound to arbitrate).

In C & L Enters. v. Citizen Band Potawatomi Indian Tribe, 121 S. Ct. 1589, 149 L. Ed. 2d 623 (2001), a federally recognized Indian Tribe entered into a contract with a contractor to install a roof on a Tribe-owned commercial structure located outside the Tribe's reservation. When a dispute arose, the contractor demanded arbitration and the Tribe refused to participate on the grounds of sovereign immunity. When the contractor sought to confirm the award, the case went all the way to the Supreme Court, which held that, because of the arbitration agreement and choice-of-law provision in the standard AIA agreement, the parties waived sovereign immunity.

Architects and contractors dealing with sovereign entities such as Indian Tribes and governments must be especially careful that they have an actual remedy in the event of a breach.

In M. O'Connor Contracting, Inc. v. City of Brockton, 809 N.E.2d 1062 (Mass. App. Ct. 2004), the arbitrator awarded a contractor double damages, under the unfair trade practices statute, for breach of contract. The court held that an award of double damages was against public policy on the basis of sovereign immunity and directed that the award be modified to only the contractor's actual damages.

[237] Landis Constr. Co. v. Health Educ. Auth., 359 So. 2d 1045 (La. Ct. App. 1978) (court interpreted state statute as not allowing apparent authority to bind the parties to arbitration). In a subsequent case, Standard Co. v. Elliott Constr. Co., 363 So. 2d 671 (La. 1978), the Supreme Court of Louisiana held that the state ratified a contract containing an arbitration clause.

In Board of County Comm'rs v. L. Robert Kimball & Assocs., 860 F.2d 683 (6th Cir. 1988), the court held that the county's defense that the contract was ultra vires was subject to arbitration. In W.M. Schlosser Co. v. School Bd. of Fairfax County, 980 F.2d 253 (4th Cir. 1992), under Virginia law, the school board did not have authority to agree to arbitrate construction contract disputes. The *Schlosser* decision was followed in Chattanooga Area Reg'l Transp. Auth. v. T.U. Parks Constr. Co., 1999 Tenn. App. LEXIS 58 (Jan. 28, 1999). In D. Wilson Constr. Co. v. McAllen Indep. Sch. Dist., 848 S.W.2d 226 (Tex. Ct. App. 1992), the school board sought to block arbitration on the grounds that it never discussed arbitration in the contract negotiations and never saw the General Conditions that contained the arbitration clause. This argument was quickly rejected.

In Kaplan v. Tabb Assocs., 657 N.E.2d 1065 (Ill. App. Ct. 1995), the owner entered into a contract for architectural services with a corporation. The sole shareholder of the corporation was properly licensed, but the corporation itself failed to comply with Illinois's Architecture Act by failing to be properly licensed. When a dispute over payment arose, the architectural corporation demanded arbitration, and the owner sought an injunction against the arbitration on the grounds

In a Louisiana case,[238] a hospital filed suit against the architect for negligence and breach of contract regarding an addition for medical records storage. The architect sought a stay pending arbitration pursuant to AIA B141-1977. The agreement, however was for a project that had been completed several years earlier. The appellate court rejected the architect's argument that the arbitration provision applied to all future work.

One deterrent to arbitration may be the cost of the initial filing fee, which is paid by the party filing the claim.[239] Under the American Arbitration Association rules as of September 15, 2005, the following fee schedule is in effect[240] (subject to change):

Amount of Claim	Total Fees
Up to $10,000	$950
Above $10,000 to $75,000	$1,250
Above $75,000 to $150,000	$2,550
Above $150,000 to $300,000	$4,000
Above $300,000 to $500,000	$6,000
Above $500,000 to $1,000,000	$8,500
Above $1,000,000 to $5,000,000	$11,250
Undetermined	$4,500

Cases are divided into three tracks: fast track (expedited) for cases up to $75,000; regular track for cases up to $500,000; and large, complex track for cases over $500,000.

Fast-track cases should be completed within 60 days. Arbitrators have more powers and discovery is expanded. Equitable relief is now fully available. A copy

---

that the contract was void because the architect was not properly licensed. The appellate court agreed with the owner, because the Act carries a criminal penalty for noncompliance. Because the contract was void, there was no agreement to arbitrate.

[238] Town of Homer, Inc. v. General Design, Inc., 960 So. 2d 310 (La. Ct. App. 2007).

[239] In Francis v. Westlan Constr., Inc., 2004 WL 1202987 (Cal. App. 2d Dist., June 2, 2004), a homeowner sued the architect and contractor for damages relating to a residential remodeling project. The defendants moved to stay litigation and compel arbitration. The court granted the motion. The owner then sent the defendants a letter stating that he preferred litigation and demanded that the defendants initiate the arbitration and pay the initial filing fee of $8,000. The defendants wrote back stating that they had no intention of initiating arbitration and that it was the owner's obligation to initiate arbitration proceedings and pay the initial filing fee. The trial court then dismissed the matter. On appeal, the appellate court reversed, holding that the owners may be deprived of the ability to resolve their claim. "Since respondents selected the provider, it is only fair that they bear the burden of paying the unusually high fees. If they do not want to pay the fees, they have the option of waiving arbitration. Or, with appellant's agreement, they may select another provider, whose fees compare favorably with the civil filing fees of the Superior Court." This case is contrary to the weight of modern authority.

[240] Note that there may be additional fees imposed for things such as room rental, etc.

of these new rules can be obtained from the AAA. Note that arbitrators now have expanded powers, which include ordering the parties to conduct discovery. For cases under $50,000, the new fast track may prove to be a very quick and inexpensive procedure.

These fees are much greater than court filing fees. In addition, if the arbitration proceeds beyond one day, the arbitrators are usually paid a fee for their time. In a lengthy proceeding, this could be substantial. The major cost saving of arbitration is the lack of discovery.

Discovery is a pretrial process whereby the parties attempt to obtain evidence and facts relevant to the case. Three main aspects of discovery are depositions, interrogatories, and requests for production of documents. Depositions involve the interrogation of a witness with attorneys for all parties present. The questions and answers are recorded by a court reporter, and these may be used during the trial. Interrogatories are written questions submitted by one party to another. A party can also request the production of relevant documents.

Discovery procedures are done under the supervision of the courts, except in arbitration cases. They can be very time-consuming and expensive, and disputes over discovery can result in numerous motions and court hearings. The discovery process can often take years to determine whether the case is economically worthwhile to pursue. An arbitration agreement could provide for discovery,[241] although

---

[241] There is some disagreement over whether the AAA rules allow for depositions. As of September 1, 2007, Rule 22 reads as follows:

R-22. Exchange of Information

(a) At the request of any party or at the discretion of the arbitrator, consistent with the expedited nature of arbitration, the arbitrator may direct
  (i) the production of documents and other information, and
  (ii) the identification of any witnesses to be called.

(b) At least five business days prior to the hearing, the parties shall exchange copies of all exhibits they intend to submit at the hearing.

(c) The arbitrator is authorized to resolve any disputes concerning the exchange of information.

(d) There shall be no other discovery, except as indicated herein or as ordered by the arbitrator in extraordinary cases when the demands of justice require it.

If the arbitrator rules that extensive discovery should take place, the party opposing it does so at the risk of an adverse ruling. The better practice is to write discovery provisions into the contract. In recent years, discovery in arbitrations has become rampant, with arbitrators and the courts doing little to curb this expensive practice. When parties to an arbitration try to obtain discovery from third parties, this becomes particularly problematic. In Hay Group, Inc. v. E.B.S. Acquisition Corp., 360 F.3d 404 (3d Cir. (Pa.) 2004), PriceWaterhouseCoopers and E.B.S., nonparties to an arbitration, sought to avoid compliance with an arbitration panel's subpoena requiring them to turn over documents prior to the panel's hearing. The trial court ordered these third parties to comply with the subpoenas, but the Third Circuit reversed. The court held that § 7 of the Federal Arbitration Act allowed for the arbitrator to compel a nonparty to bring documents to an actual arbitration hearing, but does not permit the production of documents for discovery. "Thus, Section 7's language unambiguously restricts an arbitrator's subpoena power to situations in which the nonparty has been called to appear in the physical presence of the arbitrator and to hand over the documents at that time." *Accord* Odfjell ASA v. Celanese AG, 328 F. Supp. 2d 505

that would normally be counterproductive to the aims of arbitration. Arbitrators can have widely varying views on discovery and the extent of the parties' entitlement to discovery in arbitrations. The effects of this should be given consideration in determining whether to agree to arbitration and to what extent to permit or limit discovery in the agreement.

Arbitration can be compelled by either party to the agreement and encompasses all claims arising out of the contract. A New York court held that a contract containing an arbitration clause was assignable by an owner to a tenant, permitting the tenant to pursue a claim inherited from the prior owner.[242]

The arbitration clause could provide for particular locations for one arbitration. One court, however, found that the arbitration clause was voidable when the location of the arbitration was in another state.[243]

In a Florida case involving a suit against an architect that alleged misrepresentation, breach of fiduciary duty,[244] and negligence, the court ordered arbitration of

---

(S.D.N.Y. 2004) ("Arbitration, which began as a quick and cheap alternative to litigation, is increasingly becoming slower and more expensive than the system it was designed to displace, and permitting pre-hearing discovery of nonparties would only make it more so.").

[242] John W. Cowper Co. v. Clintstone Prop., 120 A.D.2d 976, 503 N.Y.S.2d 205 (1986).

[243] Damora v. Stresscon Int'l, Inc., 324 So. 2d 80 (Fla. 1976). One way to look at this case is that the state court was being asked to enforce the arbitration clause. That clause provided for arbitration to take place in another state. Therefore, the Florida Arbitration Code would not apply and the state court could not be used to enforce the arbitration provision. There was also an apparent failure to specify that Florida law or that the Florida arbitration procedure would apply. Proper drafting might have resolved the issue.

In Johnson County v. R.N. Rouse & Co., 331 N.C. 88, 414 S.E.2d 30 (1992), the contract between the owner and contractor contained the usual arbitration clause. In addition, it contained the following provision in the supplementary general conditions:

By executing a contract for the Project the Contractor agrees to submit itself to the jurisdiction of the courts of the State of North Carolina for all matters arising or to arise hereunder, including but not limited to performance of said contract and payment of all licenses and taxes of whatever nature applicable thereto.

When the contractor sought arbitration of a dispute, the owner convinced the trial and appellate courts that this provision required the contractor to go to court with any dispute and that the arbitration provision was no longer operative. The North Carolina Supreme Court reversed, finding that this provision was a consent-to jurisdiction clause that did not conflict with the arbitration provision. *See also* Carteret County v. United Contractors of Kinston, Inc., 462 S.E.2d 816 (N.C. Ct. App. 1995) (*following Johnson County*).

[244] *See* Strauss Veal Feeds v. Mead & Hunt, Inc., 538 N.E.2d 299, 303 (Ind. Ct. App. 1989) ("an architect does not owe a fiduciary duty to its employer; rather, the architect's duties to its employer depend upon the agreement it has entered into with that employer."). In Teal Constr. v. Darren Casey Int'l, 46 S.W.2d 417 (Tex. App.—Austin, 2001), the owner's claim of fraudulent inducement of the construction contract did not render the arbitration clause invalid. Issues of payment and performance under the contract were the basis of the owner's claim, and these were arbitrable. In Carlson v. SALA Architects, Inc., 732 N.W.2d 324 (Minn. App., 2007), the stated that "the relationship of architect and client is not a fiduciary one. However, whether a fiduciary relationship exists is a fact question." AIA documents do not impose such a fiduciary relationship. In E-Med, Inc., v. Mainstreet Architects & Planners, 2007 WL 1536803 (Cal. App. 2 Dist., June 19, 2007), the court assumed the existence of a fiduciary duty. Apparently, no party argued otherwise.

all these counts under a standard arbitration clause.[245] Another interesting case
involved an arbitration proceeding between the architect and owner.[246] During the

---

In Garten v. Kurth, 265 F.3d 136 (2d Cir. 2001), after a dispute arose between the owners and
architect/builder, the architect moved to compel arbitration. The owners argued that arbitration
should not be upheld because the arbitration clause was part of the defendants' fraudulent scheme.
The architect/contractor had demanded substantial additional money to complete the project and
told the owners that they could not litigate and that he was very familiar with arbitration and it
would be very expensive for the owners. The court of appeals held that, without more, such threats
by a contractor that arbitration favors him do not suffice to show that the arbitration clause was
used by the contractor as a weapon in the broader fraud. Arbitration was thus required.

[245] Morton Z. Levine & Assocs. v. Van Deree, 334 So. 2d 287 (Fla. Dist. Ct. App., 1976). In Ronbeck
Constr. Co. v. Savanna Club Corp., 592 So. 2d 344 (Fla. 1992), the owner sought rescission
against the construction contractor, as well as breach of contract, damages for fraud, conversion,
civil theft, and conspiracy. The parties had used the 1970 AIA Document A107 which contained
the standard arbitration clause. Because all the claims arose out of that contract, the court ordered
arbitration. Note that even the rescission action was subject to arbitration, because the alleged
basis for the rescission did not include any allegation that the arbitration clause itself was secured
by fraud.

In Warwick Township Water & Sewer Auth. v. Boucher & James, Inc., 2004 WL 557597 (Pa.
Super., Mar. 23, 2004), the trial court refused to compel arbitration on the basis that the arbitration
provision applied only for disputes during construction, based in part on a provision similar to
¶ 4.4.1 of A201. Arbitration also applied to a count for negligence.

In BFN-Greeley, LLC v. Adair Group, Inc., 141 P.3d 937, (Colo. App., 2006), the court held
that arbitrators have authority to decide post-construction claims relating to time extensions and
change orders.

[246] Quinn Assocs., Inc. v. Borkowski, 41 Conn. Supp. 17, 548 A.2d 480, 482 (1988). The court
interpreted a standard AIA arbitration clause and found that the submission covered all claims in
question under the contract:

Thus, the award conforms to the submission and is within the authority of the arbitrator. . . . In this court's view, it makes no sense for the amount of the claim or relief sought
to constitute a restriction on the arbitrator. . . . But when the submission is general, as here,
and includes an agreement to decide by arbitration all disputes under the contract, the
arbitrator is free to award more or less than the amount claimed. The essence of the submission is that the arbitrator resolve all disputes. A statement of the amount claimed is a
guide to the arbitrator, but not a limitation on his power.

See also Village of Westville v. Loitz Bros. Constr. Co., 165 111. App. 3d 338, 519 N.E.2d 37
(1988) (case also held that right to arbitrate could be assigned despite nonassignability clause in
contract).

In Robert Lamb Hart Planners & Architects. v. Evergreen, Ltd., 787 F. Supp. 753 (S.D. Ohio
1992), the court followed *Village of Westville* in allowing an assignee to arbitrate claims in the place
of an original party to the contract, despite the nonconsolidation provision of 1.3.5.4. *See also*
Sisters of St. John v. Geraghty Constr., 67 N.Y.2d 997, 494 N.E.2d 102, 104, 502 N.Y.S.2d 997
(1986) ("disputes relating to extra work allegedly authorized and required for execution and
completion of respondent's renovation of the convent arise out of or relate to the 'Contract Documents' and thus fall generally within the arbitration clause."); County of Middlesex v. Gevyn
Constr. Corp., 450 F.2d 53 (1st Cir. 1971) (general contractor filed arbitration against municipality
for benefit of subcontractor; appellate court overturned lower court's stay of arbitration based on
broad form of arbitration clause); Ozdeger v. Altay, 66 Ill. App. 3d 629, 384 N.E.2d 82 (1978);
Roosevelt Univ. v. Mayfair Constr. Co., 28 Ill. App. 3d 1045, 331 N.E.2d 835 (1975).

In Hurley v. Fox, 587 So. 2d I (La. Ct. App. 1991), the court held that the architect's claim was
barred by the principle of *res judicata*, because of a prior arbitration. The architect argued that she

arbitration proceeding, the architect increased his demand beyond that in the original demand. The arbitrator awarded the higher amount. The owner then claimed that the arbitrator exceeded his powers. This argument was rejected by the court.

In many states, an arbitration award can be overturned only in very limited circumstances. One of these is fraud.[247] Sometimes the issue is whether the state arbitration act or the Federal Arbitration Act (FAA) applies to a matter.[248]

---

had never presented her claim to the arbitrator, but the court found that she had had the opportunity to present her claim, but failed to do so in a timely manner.

In Adena Corp. v. Sunset View, Ltd., 2001 Ohio App. LEXIS 1726 (April 10, 2001), *appeal denied*, 754 N.E.2d 259 (2001), the contractor filed suit against the owner. The owner moved for a stay and to compel arbitration. The arbitrator granted the owner a substantial award, which was confirmed by the trial court. The contractor then moved to vacate the award on the basis that the arbitrator had exceeded his authority. Apparently, the arbitrator had included attorneys' fees in the award to the owner. Because the arbitrator is authorized to award attorneys' fees in at least some incidences (here, the court cited the indemnification provision of AIA Document A201, ¶ 3.18.1 that permits attorneys' fees, although that was not the basis in this case), the arbitrator did not exceed his powers in rendering such an award.

Many arbitration cases involve attorneys' fees. These are often refused on the basis that there is not a specific reference to attorneys' fees in the contract. However, ¶ 3.18.1 may well provide a justification for an award of such fees, and this case would support such a contention.

[247] In a Kansas case, Prof'l Builders, Inc. v. Sedan Floral, Inc., 819 P.2d 1254 (Kan. Ct. App. 1991), the architect had a 50% ownership in the general contracting firm. After the owner lost in arbitration with the contractor, the owner sought to overturn the arbitration on the grounds of fraud. The court held that this was actually the issue before the arbitrator. The type of fraud necessary to overturn an arbitration award deals with fraud in the arbitration, not that which is totally outside the process of arbitration. The arbitration was upheld. *See also* Hercules & Co. v. Shama Rest. Corp., 613 A.2d 916 (D.C. 1992).

Another is the arbitrator's refusal to hear material evidence. In City of Bridgeport v. The Kasper Group, Inc., 899 A.2d 523, 278 Conn. 466 (2006), the plaintiff city invited architects to present proposals for the design of a new school. The defendant was awarded the contract. Subsequently, the scope of the project materially changed, and the city wanted to repeat the proposal and selection process. The defendant filed suit, which was stayed pending arbitration. During the arbitration process, the major shareholder of the design firm entered into a plea agreement, admitting to having engaged in a bribery and kickback scheme with the then-mayor to obtain various contracts. The plaintiff argued that this rendered the contract at issue void ab initio because it had been obtained by illegal means. The criminal trial started shortly after the last day of arbitration hearings. The plaintiff sought leave to submit additional evidence in the arbitration in the form of transcripts from the criminal trial. The arbitrator refused this evidence and ruled in favor of the design firm. On appeal, the court vacated the award, holding that the town was deprived of a full and fair arbitration hearing because the testimony from the criminal trial was highly incriminating and instrumental to the town's defense that the contract was void because it had been procured illegally.

In Positive Software Solutions v. New Century Mortgage, 476 F.3d 278 (5th Cir. 2007), the issue was whether an arbitrator's award should be vacated because the arbitrator failed to disclose a prior professional association with a member of one of the law firms that engaged him. The appellate court found that the past association was trivial and did not require vacatur of the award.

[248] In McCarney v. Nearing, Staats, Prelogar & Jones, 866 S.W.2d 881 (Mo. Ct. App. 1993), the state statute required that all contracts with an arbitration clause have a boldface warning above the signature lines that an arbitration clause was included. The parties used standard AIA contracts, which do not contain such warnings. When one party sought to avoid arbitration, the court held

that the FAA applied, because the construction involved interstate commerce. In this case, some of the participants were from states other than Missouri, and at least 29 suppliers of materials were in other states. Based on these facts, the court held that the contract involved interstate commerce, triggering the FAA.

In an important arbitration decision, the Third Circuit Court of Appeals held that "a generic choice-of-law clause [similar to AIA language], standing alone, is insufficient to support a finding that contracting parties intended to opt out of the FAA's default regime." Roadway Package Sys., Inc. v. Scott Kayser, 257 F.3d 207 (3d Cir. 2001). Although not a construction case, the holding would apply to standard AIA contracts. Of significance was the fact that the choice-of-law provision in that case stated that it "shall be governed by and construed in accordance with the laws of the Commonwealth of Pennsylvania." This is similar to ¶ 10.1 in B141. In this case, Kayser demanded arbitration under a contract. After the arbitrator awarded substantial damages, Roadway appealed and the court vacated the award on the grounds that the arbitrator exceeded the scope of his authority. The issue turned on whether the FAA or the arbitration rules of Pennsylvania applied. Kayser argued that the choice-of-law provision indicated that the parties wanted to opt out of the FAA. The Third Circuit disagreed and upheld the trial court's vacation of the award.

The court held that parties may agree that judicial review of an arbitrator's decision will be conducted according to standards borrowed from state law. Parties may opt out of the FAA standards and either create their own or use state law standards.

This is an important decision in light of the fact that in 2000 the National Conference of Commissioners on Uniform State Laws recommended the adoption of a Revised Uniform Arbitration Act. It is likely that most states will adopt this revision in the next few years. Readers should note that there are a number of significant differences between this revised act and the FAA. Drafters of contracts should decide which act should apply to their contract and take steps to specify in the contract that act. Parties should consider whether to include language in their agreement that would make an opt-out clear, if that is intended.

In Strain-Japan R-16 Sch. Dist. v. Landmark Sys., Inc., 51 S.W.3d 916 (Mo. Ct. App. 2001), the FAA applied where a major building component was shipped from another state. However, in F.A. Dobbs & Sons, Inc. v. Northcutt, 819 So. 2d 607 (Ala. 2001), the court held that there was insufficient evidence that interstate commerce was involved and refused to apply the FAA. The *Dobbs* court stated that a transaction must substantially affect interstate commerce in order for the FAA to apply, and that it is up to the party seeking the application of the FAA to present sufficient proof of such substantial effect on interstate commerce. Here, an unrebutted affidavit was apparently insufficient proof.

Blanton v. Stathos, 351 S.C. 534, 570 S.E.2d 565 (2002), deals with the South Carolina Arbitration Act, which requires that contracts with an arbitration provision contain a prominent notice on the first page of the contract, something the AIA forms do not. If this is not complied with, then the arbitration provision is not enforceable. The court held that the FAA applied.

In Dunn Indus. Group, Inc. v. City of Sugar Creek, 2002 WL 31548615 (Mo. App. Nov. 19, 2002), the court found that the FAA applied, rejecting a claim that the state's equitable mechanic's lien statutes place exclusive jurisdiction for the claims in the case in the state court and not in arbitration. The Supremacy Clause defeats such an argument.

In Huntsville Utilities v. Consolidated Constr. Co., 2003 WL 21205396 (Ala., May 23, 2003), the Supreme Court of Alabama continued that state's vehement stance against arbitration. After a dispute arose and suit was filed, various parties moved to stay the proceedings and compel arbitration. The trial court denied the motion and the state supreme court affirmed. The issue centered on whether the FAA applied or not. The court stated that the party moving for arbitration had the burden of proving the existence of a contract containing a written arbitration clause and relating to a transaction that substantially affects interstate commerce. The case of Sisters of the Visitation v. Cochran Plastering Co., 775 So. 2d 759 (Ala. 2000), was cited as precedent and identified a number of factors for a court to determine whether the FAA applied. The court said the movant did not meet that burden despite the fact that one party was incorporated in Delaware with its sole office in Alabama and some 499 invoices evidenced out-of-state material purchases.

As of the 2007 AIA documents, the FAA applies. (**See** ¶ 10.1 in this agreement). It should be noted that it would be difficult to find any construction project anywhere in the United States that did not use materials from many states. Courts use this argument to send a case to arbitration even when a state arbitration act dictates otherwise. There is a very strong trend toward arbitration of construction cases.

In 2000, the National Conference of Commissioners on Uniform State Laws approved the long-awaited Revised Uniform Arbitration Act (RUAA). Numerous states have adopted this new act as written or with some modifications to replace the old Uniform Arbitration Act (UAA). It may reasonably be anticipated that the majority of states will adopt some version of this within the next few years. Parties will need to carefully examine the actual law that affects their transaction and modify their agreement accordingly, or perhaps strike arbitration altogether. Here are some of the key points of the RUAA:

> The issue of who determines whether a particular dispute is subject to arbitration was not addressed by the UAA. Numerous cases have been filed over this issue.

Shortly after *Huntsville* was decided, the United States Supreme Court overturned *Sisters of the Visitation* in The Citizens Bank v. Alafabco, Inc. et al., 123 S. Ct. 2037 (2003), finding that the Alabama court interpreted the FAA too narrowly, and further finding that the five-part test in *Sisters of the Visitation* is improper where the FAA is involved. This case should sound the death knell for attempts to avoid the broad reach of the FAA in construction cases. There is virtually no construction project in the United States that would not fall within the reach of the FAA. Apparently the Alabama Supreme Court finally got the message in Huntsville Utils. v. Consolidated Constr. Co., 2003 WL 22064079 (Ala., Sept. 5, 2003).

In Mpact Constr. Group, LLC v. Superior Concrete Constructors, Inc., 785 N.E.2d 632 (Ind. App. 2003), *further proceedings*, 802 N.E.2d 901 (Ind. 2004), the court found that the FAA applied. "Even though a matter falls within the scope of the FAA, courts generally apply state law to the issue of whether the parties agreed to arbitrate their claims. . . . The FAA preempts state law, however, to the extent that they conflict." Here, the contract between the owner and general contractor incorporated A201, and arbitration was required between those parties. A second issue was whether certain subcontractors were subject to arbitration. The court held that they were not, apparently because the subcontract agreements were not standard AIA forms. The court, in a footnote, stated that "we note the possibility that this outcome was forecast by the General Instructions to AIA Form A 201 (General Conditions), which cautions the practitioner against the use of non-AIA forms to avoid inconsistency in language and intent."

In Foodbrands Supply Chain Serv., Inc. v. Terracon Inc., 2003 WL 23484633 (D. Kan., Dec. 8, 2003), the court held that the FAA applied because the contract involved interstate commerce. Under Kansas law, the matter would likely have been litigated in court because several parties were involved.

In an interesting case, Peterson Constr., Inc. v. Sungate Dev., L.L.C., 2003 WL 22480613 (Tex. App.—Corpus Christi, Oct. 30, 2003), the court held that the contract, which included the A201-1997, was governed by the FAA and not the Texas Arbitration Act. When the trial judge refused to compel arbitration, the contractor filed an appeal. The court held that such an order was not subject to interlocutory appeal, although an interlocutory appeal would have been available under the Texas Act. The dissent's analysis provides a good overview of this area of law.

In Paul Woolls v. Superior Court of Los Angeles County, 127 Cal. App. 4th 197, 25 Cal. Rptr. 3d 426 (2d Dist. 2005), the court held that the FAA did not apply to a single-family home project and that a statute requiring that the consumer be advised that he was giving up the right to have the dispute litigated in a court or jury trial rendered the arbitration invalid.

The RUAA attempts to codify these opinions—the court will determine whether an agreement to arbitrate exists or a matter is subject to arbitration, while the arbitrator determines whether a condition precedent has been met. An example of this would be whether the architect was to decide a claim prior to arbitration.

The UAA did not specify which, if any, provisions of that act could be waived by the parties. The RUAA now makes certain provisions non-waivable: the obligation of the arbitrator to disclose facts that would tend to indicate partiality; the right of a party to be represented by counsel; the obligation to give reasonable notice of the arbitration to the other side; certain rights of judicial review prior to the dispute; or the power of the courts to compel or stay arbitration. Also, one cannot interfere with the immunity of the arbitrator or with the powers of the court post-award.

Under the UAA, discovery was not available except by agreement of the parties. In particular, the arbitrator could not require depositions. This is changed under the RUAA. Evidence depositions may be taken at the direction of the arbitrator. Further, arbitrators are given the power to "permit such discovery as the arbitrator decides is appropriate under the circumstances." The arbitrator has the same powers as a state court judge to compel compliance with discovery orders, including impositions of sanctions. Incredibly, this law gives the arbitrator powers over nonparties as far as attendance at depositions and production of documents. With virtually no oversight, this will certainly lead to abuses and there is little that a third party will be able to do about it, except at significant cost.

The arbitrator is now empowered to award punitive damages! The arbitrator may award attorneys' fees if a court could award them. Most incredible of all, "an arbitrator may order such remedies as the arbitrator considers just and appropriate under the circumstances of the arbitration proceeding. The fact that such a remedy could not be or would not be granted by the court is not a ground for refusing to confirm an award. . . ." Presumably, the arbitrator could order one of the parties deported or jailed. The courts will likely be busy with this provision.

8.3.1 If the parties have selected arbitration as the method for binding dispute resolution in this Agreement, any claim, dispute or other matter in question arising out of or related to this Agreement subject to, but not resolved by mediation shall be subject to arbitration which, unless the parties mutually agree otherwise, shall be administered by the American Arbitration Association in accordance with its Construction Industry Arbitration Rules in effect on the date of this Agreement. A demand for arbitration shall be made in writing, delivered to the other party to this Agreement, and filed with the person or entity administering the arbitration. (1.3.5.3, **1.3.5.2; modified to reflect modifications to binding dispute resolution procedures; minor modifications)**

This "arising out of or relating to" language has been held to be a "broad form" of arbitration agreement as opposed to language such as "arising in connection with."[249] The use of such a broad form assures that virtually any dispute related to

---

[249] For cases involving an arbitration provision that states "all disputes arising in connection with this contract," *see* Harrison F. Blades, Inc. v. Jarman Mem'l Hosp. Bldg. F, 109 Ill. App. 2d 224, 248

the construction project will be subject to arbitration. An owner's argument that the inclusion of this arbitration agreement indicated a waiver by the architect of the right to a mechanic's lien was rejected by the court.[250] One court held that this broad arbitration provision allows arbitrators to award prejudgment interest even in the absence of a specific contractual provision to permit it.[251]

Some owners want to change the word "shall" to "may," which effectively deletes the entire arbitration process.[252]

N.E.2d 289 (1969); Silver Cross Hosp. v. S.N. Nielsen Co., 8 Ill. App. 3d 1000, 291 N.E.2d 247 (1972).

The broad arbitration provision applies to indemnification provisions in a construction contract. Contract Constr., Inc. v. Power Tech. Ctr. Ltd. P'ship, 100 Md. App. 173, 640 A.2d 251 (1994). *See also* McCarney v. Nearing, Staats, Prelogar & Jones, 866 S.W.2d 881 (Mo. Ct. App. 1993).

One court has held that the waiver of consequential damages provision in an AIA contract limits the damages that an arbitrator may award. In City of Milford v. Coppola Constr., 2004 WL 3090680 (Conn. Super., Dec. 1, 2004), the city hired the contractor to lift several houses. The contractor brought specialized equipment to the site to perform the work, but the city needed to correct the plans to obtain the required permits, resulting in a delay of some seven months while the equipment lay idle. The contractor's claims were arbitrated, with the arbitrator awarding the contractor damages. The city then brought an action to modify and correct the award on the ground that idle equipment and materials were not matters submitted to the arbitrator. The court examined the AIA contract, including the waiver of consequential damages provisions, and held that the arbitrator's award should stand. These provisions waiving consequential damages "indicate that not all types of damages are amendable to arbitration." Since the parties waived consequential damages, the arbitrator could not award such damages. The issue for the court was "whether the idle equipment and materials were consequential damages, and therefore not arbitrable, or whether they were other than consequential damages, such as liquidated direct damages, and therefore properly arbitrable." The court found that the loss from this contractor's idle equipment in this case arose in the usual course of the performance of this contract, was foreseeable, and was the direct result of the city's breach.

[250] Buckminster v. Acadia Village Resort, 565 A.2d 313 (Me. 1989).

[251] Gordon Sel-Way, Inc. v. Spence Bros., 438 Mich. 488, 475 N.W.2d 704 (1991).

[252] Of course, if the parties want to arbitrate they can do so by agreement after a dispute arises, no matter what this contract says. The issue normally is whether one party can force the other party to arbitrate. In Stevens/Leinweber/Sullens v. Holm Dev., 165 Ariz. 25, 795 P.2d 1308 (App. Ct. 1990), the contract stated: "If a claim or dispute arises . . . the Owner shall have the option of (i) submitting the dispute to arbitration in accordance with the Construction Industry Arbitration Rules of the American Arbitration Association then obtaining, or (ii) foregoing arbitration and filing a lawsuit." The court found that this arbitration agreement lacked mutuality and was void for lack of consideration.

In A. Dubreuil & Sons, Inc. v. Town of Lisbon, 215 Conn. 604, 577 A.2d 709 (1990), the parties changed "shall" to "may" in the standard AIA language. When one party moved to compel arbitration, the court found that this change meant that the parties had agreed to consensual rather than mandatory arbitration. The court refused to compel the arbitration.

In Ringwelski v. Pederson, 919 P.2d 957 (Colo. Ct. App. 1996), the contract contained the following arbitration provision:

Any disagreement arising out of this contract or from the breach thereof shall be submitted to arbitration, and judgment upon the award rendered may be entered in the court of the forum, state or federal, having jurisdiction. It is mutually agreed that the decision of the arbitrators shall be a condition precedent to any right of legal action that either party may have against the other. The arbitration shall be held under the Standard Form of Arbitration Procedure

If the article is deleted, the parties can always decide later to arbitrate, but it cannot be forced. If the architect wants to arbitrate possible future conflicts, it should not change this wording.

The language of the AIA agreement incorporates the Construction Industry Arbitration Rules of the American Arbitration Association. Thus, those rules are adopted by the parties to the agreement as if they were written into the agreement. These rules provide for such things as filing fees and how the arbitration is initiated.[253] Note that these rules are changed from time to time. Generally, the rules in

---

of the American Institute of Architects or under the Rules of the American Arbitration Association.

After a dispute developed, the parties went through arbitration and an award was made. Thereafter, the losing plaintiffs filed a court action with the same allegations as in the prior arbitration. The defendant moved to confirm the award and dismiss the court action. The trial court refused to do so, finding that the first part of the second paragraph of the arbitration clause indicated an intent that the arbitration be nonbinding. It also found persuasive the use of the word "may" in the first paragraph. The appellate court reversed, finding that such an interpretation would severely limit the applicability of the first sentence providing for entry of judgment on an award. The court went on to state that it construed this language to mean that the prevailing party has the option to obtain a judgment for the purpose of enforcement or collection if necessary.

In Cumberland Cas. & Sur. Co. v. Lamar Sch. Dist., 2004 WL 2294409 (Ark. App., Oct. 13, 2004), the supplementary conditions read as follows: "In subparagraph 4.5.1: Change the words 'shall be subject to arbitration upon the written demand of either party' to read: 'Shall be subject to arbitration if both parties agree to arbitrate.'" The general conditions form used was AIA Document A201-1997. Unfortunately, subparagraph 4.5.1 of the 1997 A201 does not contain this language. Apparently, the architect had mistakenly used the 1987 version of A201 for the modifications, thereby making the supplementary conditions ambiguous. The owner presented an affidavit of the architect stating that the architect had intended to modify the contract to eliminate arbitration. This was of no avail, however, and the court ordered arbitration.

In J. Caldarera & Co, Inc. v. State of Louisiana, 2006 WL 3813721 (La. App. 1 Cir., Dec. 28, 2006), the contract between the owner and contractor deleted the arbitration section. However, the agreement also provided that claims are to be initially reviewed by the architect, and the architect's final decision is "subject to arbitration." Because of this, the court found that the contract requires arbitration.

In State ex rel. Gayle Vincent v. Schneider, 194 S.W.3d 853 (Mo., 2006), the contract between home purchasers and seller contained an arbitration provision that provided that the arbitrator would be selected by the president of the local homebuilders association, who happened to be the owner of the seller. The issue was whether the arbitration provision was unconscionable. The court held that it was unconscionable because the clause required that an individual in a position of bias be the sole selector of an arbitrator, who must be unbiased.

[253] Rule 1 provides (as of September 1, 2007) that "These rules and any amendment of them shall apply in the form in effect at the time the administrative requirements are met for a demand for arbitration or submission agreement received by the AAA. The parties, by written agreement, may vary the procedures set forth in these rules." Congress Constr. Co., Inc. v. Geer Woods, Inc., 2005 WL 3657933 (D. Conn. Dec. 29, 2005). In Congress, the contractor argued that the rules in effect at the time of formation of the contract applied, and not the rules in effect at the time of the demand for arbitration. The court rejected this argument.

In a Florida case, Carpet Concepts v. Architectural Concepts, Inc., 559 So. 2d 303 (Fla. Dist. Ct. App. 1990), an arbitration was commenced and the respondent filed a counterclaim. Under Rule 8 (Rule 6 as of September 1, 2007) of the American Arbitration Association, the arbitrator could either allow or reject the counterclaim. In this case, the arbitrator refused to hear the counterclaim and dismissed it. An award was entered for the other party. The respondent then filed a new arbitration based on the counterclaim. When the other party moved to confirm the arbitration, the respondent asked the court to remand the matter to arbitration so that its claim could be heard. The appellate court held that the first arbitration must be confirmed and could be enforced despite the existence of the second arbitration. The current Rule R-6 (September 1, 2007), which replaced Rule R-8 reads as follows:

R-6. Changes of Claim

A party may at any time prior to the close of the hearing increase or decrease the amount of its claim or counterclaim. Any new or different claim or counterclaim, as opposed to an increase or decrease in the amount of a pending claim or counterclaim, shall be made in writing and filed with the AAA, and a copy shall be mailed to the other party, who shall have a period of ten calendar days from the date of such mailing within which to file an answer with the AAA.

After the arbitrator is appointed no new or different claim or counterclaim may be submitted to the arbitrator except with the arbitrator's consent.

In H.L. Fuller Constr. Co. v. Indus. Dev. Bd., 590 So. 2d 218 (Ala. 1991), on the other hand, the arbitrator, under Rule 8, allowed a counterclaim for fraud. The court found that the arbitrator did not abuse his discretion in allowing the claim.

In Marsh v. Loffler Hous. Corp., 102 Md. App. 116, 648 A.2d 1081 (1994), the court modified an arbitrator's award to include costs and attorneys' fees. No written request for attorneys' fees was filed, despite the apparent presentation of evidence. The award did not on its face award attorneys' fees, although the contractor's form contract, which included a provision for attorneys' fees and costs, was included by reference. The court found that this issue was not submitted for arbitration, per Rule 8, and the trial court could modify the award to include attorney fees and costs.

Rule 6 was cited in Commercial Renovations, Inc. v. Shoney's of Boutte, Inc., 797 So. 2d 183 (La. Ct. App. 2001), in which a contractor filed an arbitration demand against the owner. When the contractor petitioned the court to confirm the award in its favor, the owner defended on the grounds that it had not been properly served. The court held that the American Arbitration Association Rules that were incorporated permitted the service that was obtained and confirmed the award.

Rule 8(a) ("the arbitrator shall have the power to rule on his or her own jurisdiction") was reviewed in Diesselhorst v. Munsey Building, 2005 WL 327532 (D. Md., Feb. 9, 2005). The issue was whether the court or the arbitrator was to decide which claims were arbitrable. Contrary to the weight of modern authority, the court found that this Rule "does not confer authority on the arbitrator to decide which claims are arbitrable."

This provision (cited as Rule 9(a)) was cited in Congress Constr. Co., Inc. v. Geer Woods, Inc., 2005 WL 3657933 (D. Conn. Dec. 29, 2005). In Congress, the contractor argued that the rules in effect at the time of formation of the contract applied, and not the rules in effect at the time of the demand for arbitration. Apparently, Rule 9(a) did not go into effect until after the contract was executed. The court rejected this argument, based on Rule 1.

The current Rule R-12 (September 1, 2007) reads as follows:

R-12. Appointment from National Roster

If the parties have not appointed an arbitrator and have not provided any other method of appointment, the arbitrator shall be appointed in the following manner:

(a) Immediately after the filing of the submission or the answering statement or the expiration of the time within which the answering statement is to be filed, the AAA shall send simultaneously to each party to the dispute an identical list of 10 (unless the AAA decides that a different number is appropriate) names of persons chosen from the National Roster, unless the AAA decides that a different number is appropriate. The parties are encouraged to agree to an arbitrator from the submitted list and to advise the AAA of their agreement. Absent agreement of the parties, the arbitrator shall not have served as the mediator in the mediation phase of the instant proceeding.

(b)   If the parties are unable to agree upon an arbitrator, each party to the dispute shall have 15 calendar days from the transmittal date in which to strike names objected to, number the remaining names in order of preference, and return the list to the AAA. If a party does not return the list within the time specified, all persons named therein shall be deemed acceptable. From among the persons who have been approved on both lists, and in accordance with the designated order of mutual preference, the AAA shall invite the acceptance of an arbitrator to serve. If the parties fail to agree on any of the persons named, or if acceptable arbitrators are unable to act, or if for any other reason the appointment cannot be made from the submitted lists, the AAA shall have the power to make the appointment from among other members of the National Roster without the submission of additional lists.

(c)   Unless the parties agree otherwise when there are two or more claimants or two or more respondents, the AAA may appoint all the arbitrators.

This rule was interpreted in Sanders v. Maple Springs Baptist Church, 787 A.2d 120 (D.C. 2001), in which the AAA apparently did not follow its own rules by unilaterally appointing an arbitrator without first sending out a list of proposed candidates. After objection, a new arbitrator was appointed from a list and the award was upheld.

Under Rule 19, the American Arbitration Association "has an obligation to divulge to the parties any information given by an appointed neutral arbitrator which reveals any partiality, bias, or financial or personal interest, and that under Rule 12 any neutral arbitrator appointed shall be subject to disqualification for any of these reasons." Health Serv. Mgmt. Corp. v. Hughes, 975 F.2d 1253 (7th Cir. 1992). The court went on to state that the AAA has no discretion in determining whether to communicate these matters to the parties. Rule R-19 as of Oct. 15, 1997 read as follows:

R-19. Disclosure and Challenge Procedure

Any person appointed as neutral arbitrator shall disclose to the AAA any circumstance likely to affect impartiality, including any bias or any financial or personal interest in the result of the arbitration or any past or present relationship with the parties or their representatives. Upon receipt of such information from the arbitrator or another source, the AAA shall communicate the information to the parties and, if it deems it appropriate to do so, to the arbitrator and others. Upon objection of a party to the continued service of a neutral arbitrator, the AAA shall determine whether the arbitrator should be disqualified and shall inform the parties of its decision, which shall be conclusive.

See also Commercial Renovations, Inc. v. Shoney's of Boutte, Inc., 797 So. 2d 183 (La. Ct. App. 2001). Rule 19 has been replaced by Rule 17 which reads (as of September 1, 2007):

R-17. Disclosure

(a)   Any person appointed or to be appointed as an arbitrator shall disclose to the AAA any circumstance likely to give rise to justifiable doubt as to the arbitrator's impartiality or independence, including any bias or any financial or personal interest in the result of the arbitration or any past or present relationship with the parties or their representatives. Such obligation shall remain in effect throughout the arbitration.

(b)   Upon receipt of such information from the arbitrator or another source, the AAA shall communicate the information to the parties and, if it deems it appropriate to do so, to the arbitrator and others.

(c)   In order to encourage disclosure by arbitrators, disclosure of information pursuant to this Section R-17 is not to be construed as an indication that the arbitrator considers that the disclosed circumstances is likely to affect impartiality or independence.

Under Rule 30, arbitration may proceed in the absence of a party, under certain circumstances. In Hurley v. Fox, 559 So. 2d 887 (La. 1990), the owner refused to participate in the arbitration on the grounds that she had never signed the AIA agreement. The court found that the agreement applied and the arbitration award against the owner was enforceable. The current Rule R-30 (September 1, 2007) reads as follows:

R-30. Arbitration in the Absence of a Party or Representative

Unless the law provides to the contrary, the arbitration may proceed in the absence of any party or representative who, after due notice, fails to be present or fails to obtain a postponement. An award shall not be made solely on the default of a party. The arbitrator shall require the party who is present to submit such evidence as the arbitrator may require for the making of an award.

Under Rule 34, an arbitrator did not have authority to enter an interim order directing a party to provide security toward the payment of any award the arbitrator might eventually enter. Charles Constr. Co. v. Derderian, 412 Mass. 14, 586 N.E.2d 992 (1992). Rule 34 stated that "the arbitrator may issue such orders for interim relief as may be deemed necessary to safeguard the property that is the subject matter of the arbitration without prejudice to the rights of the parties or to the final determination of the dispute." Here, the court held, the dispute was over a breach of contract, and the owner and contractor claimed damages from each other. Thus, no property was the subject matter of the arbitration, only money. This Rule was replaced by Rule 35, which reads as follows (as of September 1, 2007):

R-35. Interim Measures

(a) The arbitrator may take whatever interim measures he or she deems necessary, including injunctive relief and measures for the protection or conservation of property and disposition of perishable goods.

(b) Such interim measures may be taken in the form of an interim award, and the arbitrator may require security for the costs of such measures.

(c) A request for interim measures addressed by a party to a judicial authority shall not be deemed incompatible with the agreement to arbitrate or a waiver of the right to arbitrate.

In BFN-Greeley, LLC v. Adair Group, Inc., 141 P.3d 937 (Colo. App. 2006), the court interpreted a prior version of this Rule and allowed an interim award by the arbitrator.

Rule 38 requires that a party learning that any provision of the AAA rules has not been complied with must immediately object in writing. Health Serv. Mgmt. Corp. v. Hughes, 975 F.2d 1253 (7th Cir. 1992). In that case, the court held that the owner waived the right to object to the arbitrators, on the basis of prior dealings with the architect, by failing to promptly object upon learning of this at the start of the arbitration.

Rule 38 reads as follows (as of September 1, 2007):

R-38. Waiver of Rules

Any party who proceeds with the arbitration after knowledge that any provision or requirement of these rules has not been complied with and who fails to state an objection in writing shall be deemed to have waived the right to object.

In Strain-Japan R-16 Sch. Dist. v. Landmark Sys., Inc., 51 S.W.3d 916 (Mo. Ct. App. 2001), Rule 46 was reviewed in the context of an arbitrator who awarded attorneys' fees incurred in the litigation prior to the arbitration. That Rule currently reads (Rule 44(a) as of September 1, 2007): "The arbitrator may grant any remedy or relief that the arbitrator deems just and equitable and within the scope of the agreement of the parties, including, but not limited to, equitable relief and specific performance of a contract." The court held that, because the agreement between the parties contained no express or written provision allowing for attorneys' fees from a prior litigation to be recovered, the arbitrator exceeded his powers by awarding such fees.

effect at the time of filing the demand for arbitration are the rules that will govern that particular arbitration. This paragraph states that the rules in effect on the date of the agreement apply. However, Rule 1 of the AAA, which is incorporated into this agreement if arbitration is selected, states that the rules in effect as of the date of filing the demand for arbitration will apply. Parties can only arbitrate by

---

This rule (Rule 45(a) at the relevant time), along with other rules, did not prohibit the arbitrators from assessing monetary sanctions for discovery violations. Superadio Ltd. P'ship v. Winstar Radio Prods. LLC, 446 Mass. 330, 884 N.E.2d 246 (2006). "Noteworthy in these rules is the absence of any language limiting the means by which an arbitrator or arbitration panel may resolve discovery disputes, or language restricting the application of the broad remedial relief of rule 45(a) to final awards (and precluding the grant of broad remedial relief to interim awards). The rules, construed together, and supported by the broad arbitration provision in the agreement and the absence of any limiting language prohibiting a monetary sanction for discovery violations, authorized the panel to resolve discovery dispute by imposing monetary sanctions. . . . To give arbitrators control over discovery and discovery disputes without the authority to impose monetary sanctions for discovery violations and noncompliance with appropriate discovery orders, would impede the arbitrators' ability to adjudicate claims effectively in the manner contemplated by the arbitration process."

Under Rule 47(c), parties consent to have judgment entered on the award by a court. In St. Lawrence Explosives Corp. v. Worthy Bros. Pipeline Corp., 916 F. Supp. 187 (N.D.N.Y. 1996), ¶ 1.3.5.4 was stricken from the form contract (in that case, an A401 subcontract). The issue was whether striking this language made the arbitration provision nonbinding. Based on the liberal federal policy in favor of arbitration, and based on this provision (which incorporates the rules of the American Arbitration Association, which in turn provide that judgment may be entered on the arbitration award), the court held that arbitration was binding. Rule 47(c) is now Rule R-49(c) (September 1, 2007), which reads as follows:

R-49. Applications to Court and Exclusion of Liability

(a)   No judicial proceeding by a party relating to the subject matter of the arbitration shall be deemed a waiver of the party's right to arbitrate.

(b)   Neither the AAA nor any arbitrator in a proceeding under these rules is a necessary or proper party in judicial proceedings relating to the arbitration.

(c)   Parties to these rules shall be deemed to have consented that judgment upon the arbitration award may be entered in any federal or state court having jurisdiction thereof.

(d)   Parties to an arbitration under these rules shall be deemed to have consented that neither the AAA nor any arbitrator shall be liable to any party in any action for damages or injunctive relief for any act or omission in connection with any arbitration under these rules.

This Rule was cited in C & L Enters. v. Citizen Band Potawatomi Indian Tribe, 121 S. Ct. 1589, 149 L. Ed. 2d 623 (2001), in which a federally recognized Indian Tribe entered into a contract with a contractor to install a roof on a Tribe-owned commercial structure located outside the Tribe's reservation. When a dispute arose, the contractor demanded arbitration and the Tribe refused to participate on the grounds of sovereign immunity. When the contractor sought to confirm the award, the case went all the way to the Supreme Court, which held that, because of the arbitration agreement and choice-of-law provision in the standard AIA agreement, as well as this rule, the parties waived sovereign immunity. *See also* Smith v. Hopland Band of Pomo Indians, 95 Cal. App. 4th 1, 115 Cal. Rptr. 2d 455 (2002).

Pursuant to Rule 49(a), arbitration was not waived when a party sought judicial resolution of arbitrable disputes. This rule was incorporated into the agreement per this provision. Leon Williams General Contractor, Inc. v. Hyatt, 2002 WL 192548 (Tenn. Ct. App., Feb. 7, 2002).

agreement. When, as in Section 8.3, there is a contract that provides for arbitration, there is agreement,[254] provided that arbitration was selected at ¶ 8.2.4.

8.3.1.1 A demand for arbitration shall be made no earlier than concurrently with the filing of a request for mediation, but in no event shall it be made after the date when the institution of legal or equitable proceedings based on the claim, dispute or other

---

[254] In Weitz Co. v. Shoreline Care Ltd. P'ship, 39 Conn. App. 641, 666 A.2d 835 (1995), the project consisted of two phases. The supplementary conditions had deleted the arbitration provision from the contract for phase one work, although phase two work was subject to arbitration. When a dispute arose, the contractor sued the owner, who sought arbitration. The court enjoined the owner from arbitration on the basis that the dispute pertained to phase one work, and there was no agreement to arbitrate that part of the work.

In Curtis G. Testerman Co. v. Buck, 340 Md. 569, 667 A.2d 649 (1995), the owner and contractor were involved in a dispute. The contractor sought arbitration and moved to dismiss the owner's complaint against him personally. The court refused to dismiss him personally. The contract had the name "Curtis G. Testerman, Inc." instead of the actual name of the corporation, "Curtis G. Testerman Company." The appellate court held that this was not fatal and that the owner understood that he was contracting with a corporation. The individual should have been dismissed and was not personally liable under the contract, and he could not be compelled to arbitrate. In Commercial Renovations, Inc. v. Shoney's of Boutte, Inc., 797 So. 2d 183 (La. Ct. App. 2001), a contractor filed an arbitration demand against the owner. When the contractor petitioned the court to confirm the award in its favor, the owner defended on the grounds that it had not been properly served. The court held that the American Arbitration Association Rules that were incorporated permitted the service that was obtained and confirmed the award.

In Beers Constr. Co. v. Pikeville United Methodist Hosp., 2005 Fed. Appx. 0325N, 2005 WL 977264 (6th Cir., Apr. 28, 2005), the parties changed the word "arbitration" to "legal proceedings." The owner argued that the architect's decision was final and binding. The court, however, held that the contract contemplated that legal proceedings could be undertaken by any party not satisfied by the architect's determination.

In Baywest Constr. Group, Inc. v. Premcar Co., Ltd., 2006 WL 242505 (Ohio App. 8th Dist., Feb. 2, 2006), the arbitration provision of the 1987 version of A201 had been modified as follows:

"All demands for arbitration and all answering statements thereto, which include any monetary claim, must contain a statement that the total sum or value in controversy as alleged by the party making such demand or answering statement is not more than $100,000 (exclusive of interest and arbitration fees and costs). The arbitrators will not have jurisdiction, power or authority to consider or make findings (except in denial of their own jurisdiction) concerning any controversy where the amount at issued [sic] is more than $100,000 (exclusive of interest and arbitration fees and costs) or to render a monetary award in response thereto against any party which totals more than $100,000 (exclusive of interest and arbitration fees and costs)."

A dispute arose, the claims exceeded $230,000. The trial court refused the defendant's motion to stay pending arbitration. On appeal, the court found that the amended language limited the jurisdiction over arbitrable claims to those where the total sum or value in controversy is $100,000 or less. The defendant argued that the $100,000 figure merely placed a cap on damages. The appellate court said that this interpretation was not supported by the plain language of the agreement. Responding to the argument that it was for the arbitrators to initially determine jurisdiction, the court stated that the agreement precludes arbitrators from exercising any authority unless the total sum of the demand is $100,000 or less. "To stay proceedings and require the parties to submit the matter to arbitration for a perfunctory order denying jurisdiction is improvident."

matter in question would be barred by the applicable statute of limitations. For statute of limitations purposes, receipt of a written demand for arbitration by the person or entity administering the arbitration shall constitute the institution of legal or equitable proceedings based on the claim, dispute or other matter in question.
(1.3.5.3; added last sentence; minor modifications)

The normal statutes of limitations apply here to arbitration as well as legal action.[255] This paragraph requires that a written demand for arbitration be actually

---

[255] In Belfatto & Pavarini v. Providence Rest Nursing Home, Inc., 582 N.Y.S.2d 744 (App. Div. 1992), the court found that certain causes of action based on breach of contract and architectural malpractice were barred by the statute of limitations. Those matters could not be arbitrated. However, the court allowed arbitration of an indemnification claim because the statute of limitations had not yet run on that claim.

In Irwin G. Cantor, P.C. v. Swanke Hayden Connell & Partners, 186 A.D.2d 71, 588 N.Y.S.2d 19, 20 (1992), the owner obtained a large arbitration award against the architect for the Trump Tower. The architect then sought arbitration against the engineer. The engineer argued that the six-year statute of limitations had run and arbitration could not be initiated. The court held, however, that one of the claims, for indemnification, did not accrue until payment by the architect. Because that claim was not time-barred, all matters should be presented to the arbitrator, according to the court: "Where several theories of liability are predicated upon the same facts and some are timely whereas others are not, the judiciary will not separate between the different types of claims but will allow them all to proceed to arbitration."

In Robbinsdale Pub. Sch., Indep. Sch. Dist. No. 281 v. Haymaker Constr., Inc., 1997 Minn. App. LEXIS 1155 (Oct. 14, 1997), the court held that the arbitrator and not the court should determine whether the claim for arbitration was timely filed.

In Indep. Sch. Dist. No. 775 v. Holm Brothers Plumbing, 660 N.W.2d 146 (Minn. App. 2003), the applicable statute of limitations was two years. On December 9, 1999, the owner's consultant informed the owner that certain components of the HVAC system had prematurely failed. On December 21, 2001, the owner served a demand for arbitration on the defendant. The trial court granted the defendant's motion for a stay, finding that the demand was untimely. The appellate court agreed. The owner argued that the AIA language required that the Construction Industry Arbitration Rules of the AAA were to be followed, and those rules state that a written demand for arbitration must contain a statement setting forth the nature of the dispute and the names and addresses of all other parties. Because not all of this information was known until February 14, 2000, the owner argued, the limitations period did not begin to run until that time. The appellate court held that the AAA Rules require the initiating party to abide by the limitations period established in the contract. The AIA contract, in turn, incorporated state law as the limitations period. Thus, the trial court had correctly found that the arbitration demand was not timely filed.

In The Hillier Group, Inc. v. Torcon, Inc., 932 So. 2d 499 (Fla. App. 2d Dist. May 31, 2006), the design-build contractor brought an action against its subcontractor architect. The architect moved to stay pending arbitration and the trial court denied the motion. The appellate court reversed, stating that the arbitration provisions in the contract do not fix a specific time limit for making a demand for arbitration. Instead, they only require that the demand be made "within a reasonable time." The trial court had held that arbitration had been waived by the defendant since the defendant had not filed a demand for arbitration. The defendant argued that, since it was not making any claims, it did not have to file anything. The appellate court agreed, stating that "we think that the adoption of the rule that a defending party waives the right to arbitration by failing to demand it prior to being sued would be unwise."

In O'Keefe Architects, Inc., v. CED Constr. Partners, Ltd., 944 So. 2d 181, (Fla., 2006), the issue was whether the demand for arbitration was filed within the statute of limitations and

received by the American Arbitration Association or other entity named in the agreement, otherwise the process has not actually been initiated. Parties getting close to the end of the statute of limitations must take care to get the demand to the proper party, since a delay of even one day past the running of the statute of limitations is grounds for dismissal of the claim. The parties could decide to shorten the statute of limitation by providing in this agreement for some shorter time limit in which to bring either arbitration or other legal action. A delay of 11 months has been held not to be an unreasonable delay under prior versions of this provision.[256]

> *8.3.2 The foregoing agreement to arbitrate and other agreements to arbitrate with an additional person or entity duly consented to by parties to this Agreement shall be specifically enforceable in accordance with applicable law in any court having jurisdiction thereof.* **(1.3.5.4-last sentence; no change)**
>
> *8.3.3 The award rendered by the arbitrator(s) shall be final, and judgment may be entered upon it in accordance with applicable law in any court having jurisdiction thereof.* **(1.3.5.5; no change)**

The award is normally converted into a standard judgment that is enforceable by the courts through normal means, including attaching the assets of the losing party. Arbitrators' awards are difficult to overturn. For example, a Louisiana case involved a cost overrun.[257] The architect designed a house for the owner. There was discussion that the house was to cost $60,000. The lowest bid was $110,270. The architect then prepared a second set of plans, with bids coming in at $79,240. The owner then told the architect that the project was postponed and the architect's fees would not be paid. The architect commenced arbitration and obtained an award based on the contractual percentage of the $110,270 bid. The owner appealed the award. The court held that unless a party can show that the arbitrator was somehow prejudiced or the award was procured by undue means (statutory grounds in the Louisiana Arbitration Act), the amount of the award could not be challenged.

---

whether the arbitrator or the court decided this issue. This was a question for the arbitrator to determine.

[256] Milton Schwartz & Assocs. v. Magness Corp., 368 F. Supp. 749 (D. Del. 1974). This case also stated that a party who threatens to counterclaim if sued is not prohibited from thereafter invoking the arbitration clause of the agreement.

In Bickerstaff v. Frazier, 232 So. 2d 190 (Fla. 1970), a delay of some four months constituted a waiver.

In Miller Bldg. Corp. v. Coastline Assocs. Ltd. P'ship, 411 S.E.2d 420 (N.C. 1992), the court held that the arbitration demand was timely when only two months had passed from the time the other party filed suit to collect unpaid fees.

In Des Moines Asphalt & Paving Co. v. Colcon Indus. Corp., 500 N.W.2d 70 (Iowa 1993), the owner timely filed for arbitration. Only two months elapsed between the time the owner was served with a foreclosure notice and the time it filed its motion to compel arbitration, and the trial court should have compelled the arbitration. The court further held that it was for the arbitrator to determine whether the owner had demanded arbitration within a reasonable time.

[257] Firmin v. Garber, 353 So. 2d 975 (La. 1977).

Even when the "apparent, or even the plain, meaning of the words of the contract has been disregarded," the arbitrator's interpretation of such a contract is impervious to judicial challenge.[258] Although a decision by a court can be

---

[258] Maross Constr. v. Central Reg'l Transp., 66 N.Y.2d 341, 488 N.E.2d 67, 497 N.Y.S.2d 321 (1985). However, in Tretina Printing, Inc. v. Fitzpatrick & Assocs., 262 N.J. Super. 45, 619 A.2d 1037 (1993), the court held that an arbitrator may not make an award that is wholly bereft of evidential support. The arbitrator had itemized his award, allowing the court to review the basis for each line item. The arbitrator had ignored the guaranteed maximum price in the contract, as well as other provisions. Most arbitrators will not give reasons or itemize awards, in order to avoid having the award overturned by a court.

In Forge Square Assocs. Ltd. P'ship v. Construction Serv., 43 Conn. Supp. 32, 638 A.2d 654, 655 (1993), following an arbitration award in favor of the contractor, the owner sought to vacate the award. Apparently, the architect had certified that grounds existed for termination of the contractor (see, e.g., ¶¶ 2.6.18, 2.6.19 of B141 (1987) and ¶ 14.2.2 of A201 (1987)). The owner argued that this constituted a manifest disregard for the law and so exceeded the arbitrators' powers that the award should be vacated. The court disagreed, stating the rule that "in order to prevail on this application to vacate, the plaintiff must demonstrate that the award reflects an egregious or patently irrational rejection of clearly controlling legal principles." Because the award did not give specific reasons, the court would have had to speculate on the specific reasons for the award.

In RaDec Constr., Inc. v. School Dist. No. 17, 248 Neb. 338, 535 N.W.2d 408 (1995), the court found that the architect's determination of the amount owed by a contractor for a change was patently erroneous and therefore legally equivalent to bad faith, so that the court could reverse the architect's determination.

In St. Lawrence Explosives Corp. v. Worthy Bros. Pipeline Corp., 916 F. Supp. 187 (N.D.N.Y. 1996), this provision was stricken from the form contract (in that case, an A401 subcontract). The issue was whether striking this language made the arbitration provision nonbinding. Based on the liberal federal policy in favor of arbitration, and based on § 6.1 (¶ 8.3.1 here) of the agreement (which incorporated the rules of the American Arbitration Association, which in turn provide that judgment may be entered on the arbitration award), the court held that arbitration was binding.

In Ringwelski v. Pederson, 919 P.2d 957 (Colo. Ct. App. 1996), the contract contained the following arbitration provision:

Any disagreement arising out of this contract or from the breach thereof shall be submitted to arbitration, and judgment upon the award rendered may be entered in the court of the forum, state or federal, having jurisdiction.

It is mutually agreed that the decision of the arbitrators shall be a condition precedent to any right of legal action that either party may have against the other. The arbitration shall be held under the Standard Form of Arbitration Procedure of the American Institute of Architects or under the Rules of the American Arbitration Association.

After a dispute developed, the parties went through arbitration, and an award was made. Thereafter, the losing plaintiffs filed a court action with the same allegations as in the prior arbitration. The defendant moved to confirm the award and dismiss the court action. The trial court refused to do so, finding that the first part of the second paragraph of the arbitration clause indicated an intent that the arbitration be nonbinding. It also found persuasive the use of the word "may" in the first paragraph. The appellate court reversed, finding that such an interpretation would severely limit the applicability of the first sentence's providing for entry of judgment on an award. The court went on to state that it construed this language to mean that the prevailing party has the option to obtain a judgment for the purpose of enforcement or collection if necessary.

In Hazelton Area Sch. Dist. v. Krasnoff, 672 A.2d 858 (Pa. Commw. Ct. 1996), a school district hired an architect under a fast-track schedule and for a flat fee. Following the election of a new

appealed for even relatively minor misinterpretations, an arbitrator can be grossly mistaken and the party against whom the decision is rendered is without recourse.

---

school board, the architect was terminated. He then submitted a bill for 15 different claims for additional services, totaling more than $464,000. The matter went to arbitration, where the architect submitted additional claims for interest and early termination of the contract. The evidence showed that the president of the school board had verbally agreed to the extras, with the knowledge of a majority of the board. The arbitration concluded with a sizable award in favor of the architect. The courts reversed the arbitrators on the ground that the contract, as well as state law, required written authorization for additional services. All 15 claims were rejected. The court stated that any modification of a contract that increased the indebtedness of a school district must comply with requirements for a formal vote and approval.

In Waterfront Marine Constr. Co. v. North End 49ers, 251 Va. 417, 468 S.E.2d 894 (1996), an award was entered in an arbitration over defective design and construction of a waterfront bulkhead. The award required additional work on the bulkhead, but because of another controversy, the work was never done. A storm then caused further damage, and the owner filed a second demand for arbitration, asking that the previous arbitration panel be reassembled. Following that arbitration, the contractor moved to vacate the award. The issue was whether the first arbitration was "final" so as to preclude a second arbitration. The court held that the language of this AIA provision reflects the parties' understanding that the arbitration process would end with the first arbitration award. The court rejected the owner's argument that the failure to comply with the first arbitration award constituted a breach of the construction contract, thus triggering the arbitration provision.

In Carteret County v. United Contractors of Kinston, Inc., 462 S.E.2d 816 (N.C. Ct. App. 1995), the court held that the arbitrators were not required to give a reason or clarify their award. In Orangefield Indep. Sch. Dist. v. Callahan & Assocs., 93 S.W.3d 124 (Tex. App.—Beaumont, 2001), the court held that the arbitrator may have made an evident mistake or violated the common law by failing to award the owner damages related to a driveway replacement. The case was remanded to the trial court for a determination of this issue. This case is clearly in the minority.

In Strain-Japan R-16 Sch. Dist. v. Landmark Sys., Inc., 51 S.W.3d 916 (Mo. Ct. App., 2001), the arbitration award was vacated because the arbitrator exceeded his authority. Prior to the arbitration, the parties litigated the issue of arbitrability. After that court ordered arbitration, Landmark amended its demand for arbitration to include a claim for attorneys' fees arising out of the litigation. The arbitrator awarded those legal fees to Landmark. The award was vacated with the appellate court holding that an arbitrator has no power to award attorneys' fees incurred prior to the arbitration under the "American Rule."

In Lauro v. Visnapuu, 351 S.C. 507, 570 S.E.2d 551 (Ct. App., 2002), the trial court vacated the arbitrator's award on the basis of a "manifest disregard for the law." The appellate court reversed. Although an award may be vacated for this reason, this non-statutory ground requires something more than a mere error of law, or failure on the part of the arbitrator to understand or apply the law. "An erroneous application of the law, however, does not constitute manifest disregard." In Medvalusa Health Programs, Inc. v. Memberworks, Inc., 273 Conn. 634, 872 A.2d 423 (2005), not a construction case, the arbitration panel ruled in favor of the plaintiff on all counts, but awarded no compensatory damages, finding that the plaintiff had failed to establish damages with reasonable certainty. However, the panel awarded $5 million in punitive damages. This award was upheld on appeal. The court found no well-defined public policy against the award of excessive punitive damages. It also found that the defendant's right to due process was not violated because an arbitration award does not constitute state action and is not converted into state action by the trial court's confirmation of that award; therefore, an arbitration panel's award of punitive damages does not implicate the Due Process Clause of the Fourteenth Amendment of the United States Constitution, regardless of how excessive that award may be.

In an unusual case, the contractor agreed to construct an addition to the defendant's house. When a dispute arose, the parties arbitrated.[259] The award provided that the contractor would obtain the approvals necessary from the local building department, including a certificate of occupancy. The award further provided that when the documents were received by the contractor, they would be turned over to the owner, who would then pay the contractor $101,000 as full and final payment. Neither party sought to confirm or vacate the arbitration award within the statutory time. The owner then thwarted the contractor's performance. The municipality refused to issue the occupancy certificate until an additional smoke detector was installed. The owner refused to allow the contractor to install the smoke detector and did not pay the money to the contractor. The contractor then brought a court action to either enforce the award or obtain a new arbitration. The owner defended on the grounds that the contractor had failed to confirm the award. The appellate court found that the statutory procedures are not the exclusive means for judicial enforcement of an arbitration award. It confirmed the award, also finding that the effect of the award was to render issues of fact and law precluded from further litigation.

In another case,[260] on the other hand, the court held that the arbitration act is the exclusive remedy to confirm an arbitration award. There, the architect and owner had a dispute that was arbitrated. The arbitrator awarded the architect $10,924. Nearly two years later, the architect sought to confirm the award. As a result of a motion to dismiss, it became apparent that the action was barred by the statute of limitations. The architect then abandoned relief under the arbitration act and sought recovery under common law. The appellate court held that, because the architect's exclusive remedy was under the arbitration act, there was no jurisdiction to hear her case under any other theory.

One issue that sometimes arises is whether the court confirming an arbitration award may award pre-judgment interest.[261] In the case of nonbinding arbitration, a court may decline to order a stay of a court proceeding.[262]

---

[259] Spearhead Constr. Corp. v. Bianco, 39 Conn. App. 122, 665 A.2d 86 (1995).

[260] Capron v. Buccini, 2001 Del. Super. LEXIS 69 (Feb. 28, 2001), aff'd, 782 A.2d 262 (Del. 2001)

[261] In Herrenknecht v. Best Road Boring, 2007 WL 1149122 (S.D.N.Y., April 16, 2007), the court granted prejudgment interest from the date of the award: "In light of the presumption in favor of awarding pre-judgment interest and the fact that the Agreement stated that the arbitration decision would be final and binding, an award of pre-judgment interest is warranted in this case." The court also awarded attorneys fees after finding that the defendant had offered no justification for refusing to participate in the arbitration, failed to satisfy the award, and failed to contest the award.

[262] In City of Bridgeport v. C.R. Klewin Northeast, LLC, 2005 WL 1869140 (Conn. Super., July 19, 2005), the arbitration provision stated that "the award rendered by the arbitrator or arbitrators shall not be final and additional legal remedies may be pursued." This provided for nonbinding arbitration. In that event, "the public policy undergirding our arbitration statutes is not advanced by requiring parties to engage in arbitration proceedings that are not final and that have no reasonable likelihood of resolving their dispute. That policy is in fact turned on its head by staying pending litigation and mandating arbitration in such situations. Rather than lessening time and expense and avoiding litigation, an order to participate in non-binding arbitration with no reasonable likelihood of success simply adds the time and expense of the arbitration to the costs of the litigation."

§ 2.27 Consolidation or Joinder: ¶ 8.3.4

8.3.4 CONSOLIDATION OR JOINDER

8.3.4.1 Either party, at its sole discretion, may consolidate an arbitration conducted under this Agreement with any other arbitration to which it is a party provided that (1) the arbitration agreement governing the other arbitration permits consolidation; (2) the arbitrations to be consolidated substantially involve common questions of law or fact; and (3) the arbitrations employ materially similar procedural rules and methods for selecting arbitrator(s).

8.3.4.2 Either party, at its sole discretion, may include by joinder persons or entities substantially involved in a common question of law or fact whose presence is required if complete relief is to be accorded in arbitration, provided that the party sought to be joined consents in writing to such joinder. Consent to arbitration involving an additional person or entity shall not constitute consent to arbitration of any claim, dispute or other matter in question not described in the written consent.

8.3.4.3 The Owner and Architect grant to any person or entity made a party to an arbitration conducted under this Section 8.3, whether by joinder or consolidation, the same rights of joinder and consolidation as the Owner and Architect under this Agreement. **(1.3.5.4; enabled joinder instead of prohibiting it)**

These joinder provisions are new for 2007. The owner can elect to join several parties with whom the owner has a contract into one single arbitration. Similarly, the architect can elect to join its consultants so long as the stated conditions are met. For instance, the owner can elect to have a single arbitration that includes the architect and contractor, as well as other consultants hired by the owner. Of course, the owner is not obligated to join these parties if the owner believes that there may be some advantage to doing so. Normally, it will be to the owner's strategic advantage to join any and all parties who may be responsible for the loss or damage incurred by the owner related to the dispute.

In addition, any party who is in the arbitration may, as a matter of right, bring in additional parties who consent to being joined in the arbitration. Such voluntary joinder is probably unlikely if there are already a number of parties in the arbitration, since most attorneys will not want to participate in such an arbitration voluntarily.

All prior versions of the AIA Owner-Architect agreements contained arbitration clauses that stated that no person who is not a party to the agreement may be joined in the arbitration, except by agreement. Thus, an owner may have a claim not only against the contractor for extras caused by delay or additional work but also against the architect, but because of the nonconsolidation language, both in the prior Owner-Architect agreement and in AIA Document A201(1997), ¶ 4.6.4, the owner could not pursue a single arbitration action against both the contractor and the architect unless all parties agree to it.[263] If the owner elects to pursue

---

[263] *But see* Episcopal Hous. Corp. v. Federal Ins. Co., 273 S.C. 181, 255 S.E.2d 451 (1979); County of Sullivan v. Nezelek, 42 N.Y.2d 123, 366 N.E.2d 72, 397 N.Y.S.2d 371 (1977).

The subcontractor cannot be named as an additional party to an arbitration between the architect and owner because the language in the agreement restricts the arbitration to matters arising out of "this agreement." Cumberland-Perry v. Bogar & Bink, 261 Pa. Super. 350, 396 A.2d 433 (1978).

In Garden Grove Community Church v. Pittsburgh-Des Moines Steel Co., 140 Cal. App. 3d 251, 191 Cal. Rptr. 15, 17 (1983), the owner's contract with the builder contained the following provision:

> All claims, disputes . . . arising out of, or relating to, this Contract or the breach thereof shall be settled by arbitration . . . provided, however, that Owner shall not be obligated to arbitrate any such claim, dispute, or other matter, if Owner, in order to fully protect its interests, desires in good faith to bring in or make a party to any such claim, dispute, or other matter, the Construction Manager, the Architect, or any other third party who has not agreed to participate in and be bound by the same arbitration proceeding.

The owner had standard AIA language in the contract with the architect. The court held that this escape clause relieved the owner from the duty to arbitrate if all the necessary parties were not joined in the arbitration proceeding.

In Maxum Founds., Inc. v. Salus Corp., 817 F.2d 1086, 1087 (4th Cir. 1987), the parties had a modified clause that read: "No arbitration shall include by consolidation, joinder or in any other manner, parties other than the Owner, the Contractor and any other persons substantially involved in a common question of fact or law, whose presence is required if complete relief is to be accorded in the arbitration." The court found that this provision permitted consolidation of arbitration. *See also* Del E. Webb Constr. v. Richardson Hosp. Auth., 823 F.2d 145 (5th Cir. 1987); Callahan & Assocs. v. The Honorable Patrick A. Clark, 1996 Tex. App. LEXIS 5527 (Dec. 12, 1996).

In Wasserstein v. Kovatch, 261 N.J. Super. 277, 618 A.2d 886, 891 (1993), the court ordered consolidation under AIA Document A107, 1987 edition. In a very questionable decision, the court allowed consolidation between the owner, general contractor, and various subcontractors. The owners opposed the motion by the other parties to arbitrate, but the court held that, by virtue of the general contractor's joining in the motion, all the various contractors had consented to arbitration. The court also relied on a provision in the General Conditions, similar but not identical to ¶ 5.3.1, that allowed the subcontractors "the benefit of all rights and remedies" available to the contractor by virtue of the contract. The better view is that this type of consolidation is simply not allowed under A201.

In Raffa Assocs. v. Boca Raton Resort & Club, 616 So. 2d 1096 (Fla. Dist. Ct. App. 1993), the contractor sought arbitration against the owner and the architect after an injured worker sued the owner and the architect for negligence. The owner and the architect then filed third-party complaints against the contractor. The contractor could force the owner to arbitrate its third-party complaint, but this provision barred arbitration against the architect.

In the case of Stallings & Sons, Inc. v. Sherlock, Smith & Adams, Inc., 670 So. 2d 861 (Ala. 1995), the owner-architect agreement was a non-AIA agreement that did not contain an arbitration agreement. The owner-contractor agreement, however, incorporated A201. The contractor sued the architect for suppressing information about the project and dereliction of duties. The trial court stayed that action pending arbitration of the matter. The contractor appealed on the basis that there was no arbitration agreement between it and the architect. The Supreme Court of Alabama agreed with the contractor, finding that not only was there no agreement to arbitrate, but also the arbitration provision of A201 (¶ 4.6.4) specifically prohibited joinder.

In Curtis G. Testerman Co. v. Buck, 340 Md. 569, 667 A.2d 649 (1995), the owner and contractor were involved in a dispute. The contractor sought arbitration and moved to dismiss the owner's complaint against him personally. The court refused to dismiss him personally. The contract had the name "Curtis G. Testerman, Inc." instead of the actual name of the corporation, "Curtis G. Testerman Company." The appellate court held that this was not fatal and that the owner understood that he was contracting with a corporation. The individual should have been

both actions, it would probably initiate an arbitration action and a court action under the 1997 document. It is then possible that the resolution of the first-decided action would affect the later action, based on the legal doctrine of res judicata or collateral estoppel. (These doctrines say, in effect, that when a case or issue has been resolved in court, it cannot be relitigated.) Likewise, an engineer could not be joined in an arbitration between the owner and architect under the 1997 documents.[264] The parties could waive that provision, just like any other provision of the contract.[265] Thus, when a case is consolidated without objection, one party cannot later complain that it was improperly consolidated contrary to the express

---

dismissed and was not personally liable under the contract, and he could not be compelled to arbitrate. The nonconsolidation language was one reason for not joining the individual.

In Callahan & Assocs. v. The Honorable Patrick A. Clark, 1996 Tex. App. LEXIS 5527 (Dec. 12, 1996), the appellate court upheld this clause and refused to allow consolidation.

[264] School Dist. v. Livingston-Rosenwinkel, 690 A.2d 1321 (Pa. 1997).

[265] In Ure v. Wangler Constr. Co., 232 Ill. App. 3d 492, 597 N.E.2d 759 (1992), the owner had entered into separate contracts with an architect and a contractor. Each contract had language similar to the AIA nonconsolidation provision. After the building was substantially built, disputes arose among the three parties, and the owner filed separate demands for arbitration with the AAA. The AAA assigned a single case number to the demands and treated them as a consolidated file. No objection to consolidation was raised by any party.

Over the course of several months, a number of administrative matters concerning the arbitration took place. Letters indicating the consolidation were sent to the parties, but no objection was made. Finally, at the first evidentiary hearing, the contractor objected to the consolidation for the first time, during opening statements. The arbitrator consulted with the tribunal administrator from the AAA and ruled that there had been a waiver of the nonconsolidation provision. The hearings continued, and an award was entered in favor of the owners. Wangler filed a petition to vacate the award, alleging that the arbitrator exceeded his authority by allowing the consolidation in the face of a contractual provision prohibiting consolidation.

The trial court confirmed the arbitration award and, on appeal, the appellate court found that there had been a waiver. "A contractual right with respect to arbitration can be waived as can any other contract right." The court also found that the arbitrator has the power to determine whether a waiver had occurred. "Where arbitrability is not contested, questions concerning timeliness and waiver in asserting contractual rights are for arbitrators to decide, rather than courts."

In Felman v. Fuote, 2002 WL 551007 (Cal. App. 2 Dist., April 15, 2002), Mark Fuote signed an AIA agreement as "Mark Fuote Architect AIA Inc." After disputes arose, the owner terminated Fuote's services and filed a demand for arbitration. The demand identified the respondent as "Mark Fuote, Architect AIA, Inc., and Arkineto Architects." The arbitrator issued an award in favor of the owner in the amount of $37,463.75. This award was captioned "Elliot Felman, M.D., Claimant vs. Mark Fuote, AIA, Respondent." The owner then filed a petition to confirm the award while the architect petitioned to set it aside. After a hearing, the trial court issued an order confirming the arbitration award and entering judgment in favor of the owner and against "the Respondent."

On the same day the architect filed his notice of appeal, he asked the trial court to correct the judgment to show that the respondent and the party against whom the judgment was entered was "Mark Fuote, Architect, AIA Inc." Ultimately, this request was denied. On appeal, the architect argued that the arbitrator exceeded his authority by entering an award against him as an individual and not against his professional corporation.

The appellate court declined to set aside the award or to alter the judgment because the architect failed to make a timely request to the arbitrator to correct the award. According to the appellate court, this was a factual issue that may have been disputed. Further, the actual name of the

provision of this paragraph. If a party does not want consolidation of its case, it must make an objection at the outset to any possible consolidation.

Sometimes an argument is made that a nonparty to the agreement is a third-party beneficiary of the contract and should be permitted to arbitrate.[266]

When there is a possibility of litigation following an arbitration proceeding, a court reporter at the arbitration would be a wise investment. This typically occurs when other parties are involved in the project but not in the arbitration proceeding, such as in a dispute among the owner, architect, and contractors. However, this type of record would not be available to someone who is not a party to the arbitration, because arbitrations are not public. Also, this record may not be conclusive as to what issues were actually decided.[267] Several courts have refused to overturn an arbitration award because there was no transcript of the hearings, and affidavits of what transpired were deemed insufficient.[268]

The finding of the arbitrator is usually extremely succinct, affording little insight other than the outcome. More recently, arbitrators have given more detailed awards, thereby providing more ammunition to a disappointed litigant to overturn the award.

At least one court has held that a surety is not entitled to enforce an arbitration agreement because it is, at most, only an incidental beneficiary to the contract.

In an Ohio case, the court allowed an assignee to arbitrate claims in the place of an original party to the contract, despite the nonconsolidation provision of an earlier version.[269] The prior version (non-consolidation) of this provision has

---

corporation as shown on its incorporation papers was "Mark Fuote, A.I.A. An Architectural Corporation." This was an issue that should have been addressed by the arbitrator.

[266] This was rejected in Duchess of Dixwell Avenue, Inc. v. The Neri Corp., 1999 Conn. Super. LEXIS 2077 (Aug. 4, 1999). There, a contractor was hired to repair a culvert carrying a stream beneath land that contained plaintiff, a restaurant. The contract did not list the restaurant as an "Owner." When a dispute arose and the contractor demanded arbitration, the restaurant sought to file a counterclaim in the arbitration for the contractor's delay and interference with its business during construction, claiming that it was a third-party beneficiary of the contract. Finding that the contractor did not even know the name of the restaurant corporation until the arbitration and that there was no evidence that the contractor intended any entity other than the named "Owner" to enforce the contract or benefit from the contractor's services, the court held that the restaurant was not a third-party beneficiary and could not participate in the arbitration.

[267] Paston & Coffman, v. Katzen, 610 So. 2d 512 (Fla. App. 4 Dist., 1992); Dolton v. Merrill Lynch, 935 A.2d. 295 (D.C. 2007). In *Paston*, each party contended the other was responsible for obtaining the transcript, so none was given to the court. As a result, the court accepted the findings of the arbitrators. In *Dolton*, apparently a transcript could have been provided, but was not. The court also accepted the findings of the arbitrators. These, and other, cases seem to imply that the outcome might be different if no court reporter was present, versus the situations here where there was a court reporter but a failure to furnish the transcripts. The better view is that a court could take a transcript into account, but only to determine whether one of the very limited grounds for reversal of an arbitration award is present.

[268] *Id.*

[269] Robert Lamb Hart Planners & Architects v. Evergreen, Ltd., 787 F. Supp. 753 (S.D. Ohio 1992).

been held to mean that the architect has no duty to arbitrate a dispute between a contractor and owner.[270]

§ 2.28 Article 9: Termination or Suspension

ARTICLE 9 TERMINATION OR SUSPENSION

9.1 If the Owner fails to make payments to the Architect in accordance with this Agreement, such failure shall be considered substantial nonperformance and cause for termination or, at the Architect's option, cause for suspension of performance of services under this Agreement. If the Architect elects to suspend services, the Architect shall give seven days' written notice to the Owner before suspending services. In the event of a suspension of services, the Architect shall have no liability to the Owner for delay or damage caused the Owner because of such suspension of services. Before resuming services, the Architect shall be paid all sums due prior to suspension and any expenses incurred in the interruption and resumption of the Architect's services. The Architect's fees for the remaining services and the time schedules shall be equitably adjusted. (**1.3.8.1; minor modifications**)

This paragraph allows the architect to terminate in the event the owner does not make payments according to the contract. It is important for the architect to request payments at regular intervals so that it can monitor the owner's financial status. This is probably the architect's best early warning system for problems with the project. If the architect terminates the contract under this provision and the architect is not in default, the architect will be entitled to the termination expenses found in ¶ 9.7.

If this paragraph is invoked, it may be wise to send a second written notice of the use of this paragraph to the owner so that any question of liability for delay can be resolved in favor of the architect. The architect may want to insert the word "any" in the first line, so that it reads: "If the Owner fails to make any payments." This should prevent the owner's argument that the architect has waived a prior technical failure of payment, such as when the owner is a few days late with payment.

9.2 If the Owner suspends the Project, the Architect shall be compensated for services performed prior to notice of such suspension. When the Project is resumed, the Architect shall be compensated for expenses incurred in the interruption and resumption of the Architect's services. The Architect's fees for the remaining services and the time schedules shall be equitably adjusted. (**1.3.8.2; delete reference to 30 days**)

Without a written notice, it may be difficult for the architect to make this paragraph very effective. The provisions of ¶ 9.1 may be much more helpful to

---

[270] Century Ready-Mix Co. v. Campbell County Sch. Dist., 816 P.2d 795 (Wyo. 1991). In that case, there was no mention of ¶ 2.6.1.7 of B141-1997 as being included in the contract. Presumably, the standard AIA contract would impose some duty on the architect to arbitrate such a dispute if one of the parties made a written request that it do so.

the architect in determining when to suspend operation on a project. Because the architect will be compensated only for services rendered before it receives the suspension notice, the architect should immediately suspend its services when it receives this notice. Of course, the architect should carefully document any additional costs incurred by a suspension. These costs typically include additional salary and overhead expenses and costs of laying off personnel. Upon resumption of the project, the architect should immediately inform the owner of the additional costs, as well as possible future increases in costs as a result of inflation and other reasons. If the project is never resumed or if the parties cannot agree on this reasonable compensation, the owner will be deemed to have terminated the architect according to ¶ 9.6 and will be responsible for termination expenses.

9.3 *If the Owner suspends the Project for more than 90 cumulative days for reasons other than the fault of the Architect, the Architect may terminate this Agreement by giving not less than seven days' written notice.* **(1.3.8.3; change consecutive to cumulative days; minor modification)**

The architect can terminate this contract if there is a suspension for any reason for more than 90 cumulative days, so long as the architect was not at fault for the suspension. The architect must give a seven-day notice, whereupon the owner could resume the project. However, in the event of such suspension, the architect will be entitled to expenses, per Section 9.2.

9.4 *Either party may terminate this Agreement upon not less than seven days' written notice should the other party fail substantially to perform in accordance with the terms of this Agreement through no fault of the party initiating the termination.* **(1.3.8.4; minor modifications)**

If either the owner or architect breaches the agreement in a "substantial" manner, the other party can send a written termination notice.[271] The agreement is then terminated seven days later, without further notice. Of course, litigation or arbitration frequently follows such a termination. A termination notice should be sent by registered or certified mail, return receipt requested, or by messenger. Although this is not required, it does provide proof of the termination. If the architect terminates, it is not required to continue services during the seven-day period.

9.5 *The Owner may terminate this Agreement upon not less than seven days' written notice to the Architect for the Owner's convenience and without cause.* **(1.3.8.5; minor modification)**

If the owner terminates under a termination for convenience clause, it must still pay the architect as stated in ¶ 9.6. A Texas court held that, under these provisions, the owner did not have an absolute right of termination but could only terminate under one of two conditions: if the architect failed to substantially perform or if the

---

[271] Herbert Shaffer Assocs. v. First Bank, 30 Ill. App. 3d 647, 332 N.E.2d 703 (1975).

project was permanently abandoned.[272] Otherwise, the owner breached the contract.

In the event the architect is terminated for the owner's convenience, the license granted by the architect to the owner to use the drawings and specifications under ¶ 7.2 is terminated and the owner must, within seven days of the termination, return the drawings and specifications to the architect. This means that the owner will be unable to complete the project with these drawings and specifications unless the architect is paid for its work. The owner must consider any termination to determine whether the project is actually being abandoned and whether the owner intends to rehire this architect in the event the project is revived. If the owner will be using a different architect later on, the owner should negotiate a license agreement for the use of the drawings and specifications.

> *9.6 In the event of termination not the fault of the Architect, the Architect shall be compensated for services performed prior to termination, together with Reimbursable Expenses then due and all Termination Expenses as defined in Section 9.7.* **(1.3.8.6; change reference)**

In one case in which the owner abandoned the project after the bids received were substantially over the estimated cost, the architect was able to recover fees based on his last estimate, not on the bids actually received.[273] This provision has been held by a Georgia court[274] to provide consideration for a modification of the agreement. One case held that an architect could not recover for services during the design development phase because he proceeded on the basis of unapproved schematic design drawings.[275] If the architect is terminated because it breached the contract, it is not entitled to termination expenses.

> *9.7 Termination Expenses are in addition to compensation for the Architect's services and include expenses directly attributable to termination for which the Architect is not otherwise compensated, plus an amount for the Architect's anticipated profit on the value of the services not performed by the Architect.* **(1.3.8.7; minor modifications)**

---

[272] Gunter Hotel v. Buck, 775 S.W.2d 689 (Tex. App. 1989). See the opinion and concurring opinion in Indian River Colony Club v. Schopke Constr. & Eng'g, Inc., 592 So. 2d 1185 (Fla. 1992), for an examination of this provision and ¶ 9.6 in the context of construction management.

[273] Moossy v. Huckabay Hosp., Inc., 283 So. 2d 699 (La. 1973).

[274] Reynolds v. Long, 115 Ga. App. 182, 154 S.E.2d 299 (1967). The architect argued that a subsequent modification of the contract would be without consideration. The court stated that this provision of the contract provided consideration, because the owner could abandon the project by paying the stated amount to the architect. An oral modification to include a maximum cost of the project would mean that the owner would not abandon the project and the architect would be paid to the end of construction. The reasoning of that decision has doubtful merit. *See also* Fishel & Taylor v. Grifton United Methodist Church, 9 N.C. App. 224, 175 S.E.2d 785 (1970).

[275] Standley v. Egbert, 267 A.2d 365 (D.C. 1970).

If the owner terminates the agreement, the architect can still recover its fees to the date of the termination, even if the project is abandoned.[276] An owner can terminate an architect because the cost of the proposed project, based on the architect's estimate of probable construction cost, is too high.[277]

In one case, a court found that an architect was not entitled to the portion of the fee attributable to construction supervision when the project was abandoned by the owner after the bidding phase.[278] The architect had testified that 25 percent of the total fee was for supervision. The award to the architect was reduced by this amount.

The architect is entitled to expenses that are directly attributable to a termination.[279] These expenses might include the cost to the architect of personnel that lack work after the termination; equipment and supplies; and other office expenses that cannot be directly attributed to the project but that nevertheless would not have been incurred but for the project. Another example is the cost of the architect's errors and omissions insurance. Many smaller firms do not carry it or carry small amounts of it. If a particular project comes in, they often obtain new or additional insurance. If the project is then terminated by the owner, the termination provision should at least partially reimburse the architect for this added expense.

Termination of the architect may expose the architect to increased liability, because the architect cannot complete the project and answer questions, solve problems, and so on. The same is true if the owner hires the architect for less than the normal full services.

There is an apparent conflict between this provision and that found at ¶ 8.1.3, which waives consequential damages. In ¶ 8.1.3, the architect waives any claims for consequential damages, which presumably includes lost profits and similar costs. Here, the architect is entitled to lost profits on the services not performed at the time of termination not the architect's fault. This apparent conflict can be reconciled by finding that ¶ 9.7 is an exception to the general waiver of consequential damages, but only if the architect is not at fault.

9.8 The Owner's rights to use the Architect's Instruments of Service in the event of a termination of this Agreement are set forth in Article 7 and Section 11.9. (**new**)

---

[276] Beller v. DeLara, 565 S.W.2d 319 (Tex. 1978) (architect sued to recover his fees to the date of termination, on the basis of express contract and quantum meruit, in the alternative; court upheld jury's finding for architect based on value of his services).

In Columbia Architectural Group v. Barker, 274 S.C. 639, 266 S.E.2d 428 (1980), the owner argued that the contract was indivisible and that the architect was not entitled to a fee because the project had been abandoned as not financially feasible. This was apparently not an AIA contract.

[277] Shanahan v. Universal Tavern Corp., 585 P.2d 1314 (Mont. 1978).

[278] Michalowski v. Richter Spring Corp., 112 Ill. App. 2d 451, 251 N.E.2d 299 (1969).

[279] The 1987 version of B141 had a method for determining termination expenses. In Indian River Colony Club v. Schopke Constr. & Eng'g, Inc., 592 So. 2d 1185 (Fla. 1992), the concurring opinion stated that this provision does not limit damages to the stated items, but merely lists those items for which compensation may be obtained. *See* Newman Marchive P'ship, Inc. v. City of Shreveport, 944 So. 2d 703 (La. App. 2 Cir., 2006), for an example of a calculation of termination expenses.

Section 11.9 refers to a license fee to use the architect's drawings and specifications. Article 7 is the copyright section of the agreement. Generally, if the architect "rightfully" terminates the contract, the license to use the drawings is terminated. The owner can pay the fee that the parties have presumably filled in at Section 11.9 and obtain a permanent license, but Section 11.9 applies only if the owner has terminated for convenience, or if the architect terminates per Section 9.3. If no amount has been filled in, the parties will have to negotiate this fee.

§ 2.29   Article 10: Miscellaneous Provisions

ARTICLE 10 MISCELLANEOUS PROVISIONS

10.1 This Agreement shall be governed by the law of the place where the Project is located, except that if the parties have selected arbitration as the method of binding dispute resolution, the Federal Arbitration Act shall govern Section 8.3. (**1.3.7.1; modifications**)

Frequently, the architect and owner are in different states and the project is in yet a third state. The contract can contain a clause that selects the forum for litigation in advance. The parties thus choose where a case is to be tried and whose law is to be used. The standard language only selects which laws are to be used but not where the case is to be heard. Thus, if the architect is in New York and the owner and project are in California, the owner can bring suit in California (although the laws of New York would apply), causing considerable inconvenience and pressure on the architect to settle.

A typical forum selection clause might be thus:

> The parties agree that any litigation arising out of this contract will be brought in the United States District Court for the Northern District of Illinois in Chicago. If jurisdiction is refused by such court for any reason, such case shall be brought in the Circuit Court of Cook County, Chicago, Illinois. Any such action shall be heard by the court without a jury.

Whether to waive a jury trial should be discussed with an attorney.

In a Florida case in which the parties contractually agreed that arbitration would be conducted in New York, the architect sought to compel arbitration.[280] The court

---

[280] Damora v. Stresscon Int'l, Inc., 324 So. 2d 80 (Fla. 1976).

    In Johnson County v. R.N. Rouse & Co., 331 N.C. 88, 414 S.E.2d 30 (1992), the court held the arbitration provision valid and not superseded by a consent to jurisdiction provision.

    In Roberts & Schaefer Co. v. Merit Contracting, Inc., 901 F. Supp. 1349, 1351 (N.D. Ill. 1995), the court considered the following forum selection clause: "This purchase order and all disputes between the parties hereto shall be governed by and construed according to the laws of Illinois whose Circuit Court shall have exclusive jurisdiction to determine all such issues." However, the general conditions that contained this forum selection clause were in a document that was not attached to the contract and was not signed. The court held that the clause was not incorporated

held that the arbitration clause was voidable by either party because it was a rejection of the Florida Arbitration Code. The architect sought to arbitrate in New York because the owners, residents of Florida, filed a Florida action against the architect. The outcome might have been different if the architect had originated the action in New York.

The forum should be a court with some expertise in construction litigation, so as to offer the fairest treatment for the parties. Several states have recently passed laws setting the venue for construction projects. In Illinois, for instance, Public Act 92-0657, effective July 16, 2002, states:

A provision contained in or executed in connection with a building and construction contract to be performed in Illinois that makes the contract subject to the

---

into the contract, so the lawsuit could be removed to the federal court. The obvious lesson here is to make sure that all provisions of the contract are actually attached to or otherwise properly incorporated into the correct contract. A problem that sometimes arises is that a party prepares a "standard" purchase order or rider that references particular sections of standard AIA documents but then neglects to change that language when new editions of the AIA documents come out.

See New Eng. Tel. & Tel. Co. v. Gourdeau Constr. Co., 419 Mass. 658, 647 N.E.2d 42 (1995), for a case that revolves around the issue of which state's statute of limitations would apply, citing the forum selection clause in the contract.

In New Concept Constr. Co., Inc. v. Kirbyville Consolidated Indep. Sch. Dist., 119 S.W.3d 468 (Tex. App. 2003), the contract incorporated the general conditions and included this provision:

9.1 All matters relating to the validity, performance, interpretation of [sic] construction of the contract documents or breach thereof shall be governed by and construed in accordance with the laws of the state of Texas. The Contractor shall not institute any action of [sic] proceeding in any way relating to this agreement against the Owner except in a court of competent jurisdiction in the County in which the work was performed.

The court went to great pains to harmonize this provision with the arbitration clauses in A201: "We interpret the forum selection clause and the arbitration provision together to mean that the contractor must file any court proceeding not precluded by the arbitration provision in the County in which the work was performed—for example, an action to enforce arbitration and render judgment on an arbitration award, or an action where arbitration has been waived." The dissent provided a more cogent argument against arbitration against the "misguided slavish adherence to the 'arbitrate at every opportunity' mentality."

In F.L. Crane & Sons, Inc. v. Malouf Constr. Corp., 953 So. 2d 366 (Ala., 2006), the contract contained the following choice-of-law provision:

The Subcontract is deemed to have been made in the State of Mississippi regardless of the place of actual negotiation, execution, and performance, and the parties hereby agree that the laws of the State of Mississippi shall govern the interpretation, effect, enforcement, and all other aspects of both this Subcontract and the rights and obligations of Contractor and Subcontractor hereunder. This consensual choice of law shall apply irrespective of principles of conflict of laws, envoi, forum non conveniens, or other considerations.

There was also a forum selection clause:

Any dispute between the parties shall be resolved by litigation in a court of competent jurisdiction located either in Madison County, Mississippi, if a state court action, or in the Southern District of Mississippi, if a federal court action. . . . By execution of this Subcontract, Subcontractor irrevocably waives any objection, now or in the future, to such venue and agrees not to assert in any such court that any action or proceeding has been brought in an inconvenient or inappropriate forum.

laws of another state or that requires any litigation, arbitration, or dispute resolution to take place in another state is against public policy. Such a provision is void and unenforceable.

While this is consistent with the A201 venue provision in ¶ 13.1.1, it may not be consistent in relation to this venue provision. Note that the AIA clauses are not "forum selection" clauses, but merely define which state's laws apply. The Illinois act cited above actually requires both that the laws of Illinois apply and that hearings must be held in that state if the project is located there. Laws such as this will likely be challenged in the future as an impingement on the parties' ability to contract and the Commerce Clause.

10.2 *Terms in this Agreement shall have the same meaning as those in AIA Document A201-2007, General Conditions of the Contract for Construction.* **(1.3.7.2; modified reference)**

This provision is effective even if another general conditions form is used, unless specific terms are different in these other general conditions. In that event, there will be an ambiguity. If, at the time this agreement is prepared, the owner and architect know that other general conditions will be used, then this provision should be modified to reflect those other general conditions. If nothing is said in this agreement, it must be assumed that the parties did not contemplate other general conditions or, if they did, that they still intended the A201 General Conditions to apply to this particular agreement. It is not clear whether the word "terms" is limited to defined terms in A201. A201 has defined a number of terms that are then capitalized, such as "Work," "Contract Documents," and so on. Other terms are not capitalized but have some definition, such as the term "consequential damages" found at ¶ 15.1.6 of A201. While there is not a clear definition of that term, there is a list of examples of consequential damages. Although B101 also refers to consequential damages at ¶ 8.1.3, there is no corresponding list. Does the language in this paragraph help to define this term? Arguably it does. Paragraph 1.3.1 of A201 refers to capitalized terms in the General Conditions as being specifically defined. When one of those capitalized terms is used in B101, the definition found in A201 would definitely apply.[281] When another term is used, the language of A201 would at least be persuasive.

---

[281] An interesting case involving this provision is Giller v. Cafeteria of S. Beach Ltd., LLP, 967 So. 2d 240 (Fla. Dist. Ct. App. 2007). There, the owner sued a principal of the architectural firm with which it had a contract for professional malpractice. There was no contract between the owner and the individual. The architect sought to compel arbitration and the appellate court agreed that the arbitration provision of A201 applied. The court reviewed the following provision from A201: "The Architect is the person lawfully licensed to practice architecture or an entity lawfully practicing architecture identified as such in the Agreement and is referred to throughout the Contract Documents as if singular in number. The term "Architect" means the Architect or the Architect's authorized representative." This applied because of the language of this ¶ 10.2, which references definitions of A201, and that definition included the individual architect.

10.3 *The Owner and Architect, respectively, bind themselves, their agents, succes-*
sors, assigns and legal representatives to this Agreement. Neither the Owner nor the
Architect shall assign this Agreement without the written consent of the other, except
that the Owner may assign this Agreement to a lender providing financing for the
Project if the lender agrees to assume the Owner's rights and obligations under this
Agreement. (**1.3.7.9; modifications**)

The language of the first sentence of this provision does not confer personal responsibility on an agent of a disclosed principal in the absence of some other agreement to the contrary or other circumstances showing that the agent has expressly or impliedly incurred or intended to incur personal responsibility to the other party.[282]

An assignment occurs when one party to a contract transfers its rights and obligations under the contract to another party. It is a substitution of one party for another. For instance, the owner may want to assign its contract to a new owner, or the architect may want to assign the contract to another architect. This creates difficulties because architectural contracts are personal service contracts. The parties expect that they will deal with the same parties all the way through the contract. The owner hired a specific architect or architectural firm; the architect feels comfortable working with that particular owner. If the contract is assigned by the owner, the assignee can bring an action against the architect just as the original owner could.[283]

---

[282] Folgers Architects, Ltd. v. Pavel, 2001 Neb. App. LEXIS 9 (Jan. 16, 2001).

[283] Herlitz Constr. Co. v. Matherne, 476 So. 2d 1037 (La. 1985). An architect designed a hotel. During construction, a discrepancy between the architectural and mechanical drawings was discovered. It was decided to follow the mechanical drawings and revise the architectural and electrical drawings, resulting in revisions to room layouts and problems with television cables. The lender then called the loan because the work was not done in accordance with the approved drawings, resulting in additional financing costs to the owner of about $700,000. The owner's assignee sued the architect, claiming negligence. The court held for the architect on the basis that it was the owner's obligation to notify the lender whenever substantial changes were made to the plans. There was no evidence that if the owner had done so in a timely manner, the bank would not have approved the changes. The owner's contributory negligence was imputed to the assignee.

In Ford v. Robertson, 739 S.W.2d 3 (Tenn. Ct. App. 1987), the owner sold an apartment complex after substantial completion and after the architect was fully paid. The new owner sued the architect for breach of contract, negligence, and others. The architect defended on the basis of the nonassignability language of the 1987 version of B141. The court held that the agreement prohibited assigning an interest in the performance of the executory contract. The owner could assign his claim for damages for nonperformance, which was not prohibited by the agreement.

In Mears Park Holding Corp. v. Morse/Diesel, Inc., 427 N.W.2d 281 (Minn. Ct. App. 1988), the lender was assigned a development project when the developer defaulted. The lender brought suit against the architect and a subcontractor. The court held that the assignment was not valid because of this clause.

In Berschauer/Phillips Constr. Co. v. Seattle Sch. Dist. No. 1, 124 Wash. 2d 816, 881 P.2d 986, 994 (1994), the court found the assignment of the owner-architect contract effective despite the AIA language. "We therefore hold [that] a general assignment clause, one directed at performance of the contract, does not, after performance is completed, prohibit the assignment of a

This clause states that an assignment requires the approval of both parties to the contract.[284]

---

cause of action for breach of contract." The court found that the general assignment provision was meant to prohibit the exchange of contractual performance, but once performance was completed, the contract could be assigned. This means that an action for breach of the contract may be assigned, but not the performance of the contract itself.

In this case, the contractor could not sue the architect directly because of the economic loss rule. Apparently, the contractor suffered damages because the plans were inaccurate and incomplete. By obtaining an assignment of the contract from the owner, the contractor could bring an action for breach of contract against the architect.

[284] An Illinois case has held that the right to arbitrate could be assigned despite a nonassignability clause in the contract. Village of Westville v. Loitz Bros. Constr. Co., 165 Ill. App. 3d 338, 519 N.E.2d 37 (1988).

The court in Mississippi Bank v. Nickles & Wells Constr. Co., 421 So. 2d 1056 (Miss. 1982) held that a contractual provision prohibiting assignment was ineffectual if the assignment was for the purpose of creating a security interest under the Uniform Commercial Code. The Code specifically prohibits contractual restrictions on assignments under this circumstance.

In Elzinga & Volkers, Inc. v. LSSC Corp., 838 F. Supp. 1306, 1313 14 (N.D. Ind. 1993), additional proceedings, 852 F. Supp. 681 (N.D. Ind. 1994), the owner had assigned its rights to arbitrate. The contractor attempted to block the arbitration based on the nonassignment language in ¶ 1.3.7.9. The court found the assignment valid:

First the court notes that the meaning of a prohibition on assignment is a matter of interpretation for the court. . . . In addition, because the clause is a restriction on alienation, it must be strictly construed against the party urging the restriction. . . . Here the nonassignment provision appears internally inconsistent: in the first sentence it states that each party is bound to the assigns of the other while in the next sentence it states that neither party may assign the contract without the consent of the other. However it is important to note that consent is required only where "the contract as a whole" is assigned. Reading the two sentences together in order to give meaning to both leads to the interpretation that each party will be bound to the other's assignee for a partial assignment, even where consent is not given.

Next, case law of most states is in accord that a prohibition on assignment does not prohibit assignment of claims for money damages for nonperformance. . . . In the present situation, all the work had been performed and the money paid under the contract at the time of the assignment. The only right remaining under the construction contract was the right to sue for nonperformance or breach. Therefore this was the only right that was assigned to L & P and as such the assignment was not barred by the nonassignment clause.

In Wade M. Berry v. Absher Constr. Co., 1999 Wash. App. LEXIS 1853 (Oct. 29, 1999), a corporation entered into a contract with defendant. The corporation then merged into a partnership, with one partner assigning interest to plaintiff, the other partner. Defendant argued that the anti-assignment provision (similar to the language here) prohibits the assignment, thus the case should be dismissed. The court held that the AIA language did not prohibit successors and the plaintiff was a successor. The court found that, under the facts of the case where there was a family-owned corporation and an assignment between spouses, there was no assignment in violation of the contract.

In Folgers Architects, Ltd. v. Kerns, 9 Neb. App. 406, 612 N.W.2d 539 (2000), an assignment was proper when the only thing being assigned was a right to money damages for breach of contract under this provision.

In SME Indus., Inc. v. Thompson, Ventulett, Stainback & Assocs., Inc. 28 P.3d 669 (Utah 2001), the case was sent back to the trial court for a determination of whether the parties intended

An assignment by the owner may result in unexpected costs to the architect. Sometimes the effect of such an assignment is to assign the contract to a party that will not assume the original owner's obligation. This can occur when a lender asks for a collateral assignment of the agreement so that if the owner defaults under the loan agreement, the lender can take over the project. A *collateral assignment* becomes effective only if certain events occur, such as a default on a loan. Although unlikely, the architect must consider what will happen if the collateral assignee takes over the contract. The architect will then deal with that party as the new "owner," take direction from him, and be liable to him. In such cases, the assignment sometimes provides that the lender does not have to pay the architect. Such assignments should be closely reviewed by legal counsel.[285]

The owner may not want the architect to assign the contract, even on a contingent basis. The architect may want to use the contract as collateral in securing bank financing. It then assigns its rights under the contract to the bank, which may then contact the owner and demand that payments be forwarded directly to the bank.

In a Massachusetts case,[286] the trustees of a condominium trust sued the architect after completion of the project. Water was entering the structure, causing significant damage. The architect argued that the plaintiffs were the successor to the developer under the architect's contract. Under the provision of the AIA contract, the applicable statute of limitation starts to run as of the date of substantial completion (**see** ¶ 1.3.7.3 of B141-1997—now changed), which would bar the action. In reversing the trial court, the appellate court held that the trust was not a successor to the contract and the action was not time-barred.

10.4 If the Owner requests the Architect to execute certificates, the proposed language of such certificates shall be submitted to the Architect for review at least 14 days prior to the requested dates of execution. If the Owner requests the Architect to execute consents reasonably required to facilitate assignment to a lender, the Architect shall execute all such consents that are consistent with this Agreement, provided the proposed consent is submitted to the Architect for review at least 14 days prior to execution. The Architect shall not be required to execute certificates or consents that would require knowledge, services or responsibilities beyond the scope of this Agreement. **(1.3.7.8; modifications)**

The architect is frequently asked to give various certifications or consents for the owner's lenders, governmental bodies, and others. Often, these certifications differ from AIA documents in that the language tends to make the architect a guarantor. The architect should seek legal counsel when certificates other than standard AIA

---

the anti-assignment clause, which was substantially similar to the AIA language, to prohibit only the assignment of the performance of the contract, or whether it also prohibited the assignment of a cause of action seeking money damages for breach of the contract after the contract had been fully performed.

[285] **Ch. 3** contains a collateral assignment that is typically encountered, and an explanation of why such an assignment is not recommended.

[286] Aldrich v. ADD Inc., 437 Mass. 213, 770 N.E.2d 447 (2002).

documents are requested. The architect should never guarantee work other than its own, and even in that case it should not give an express guarantee. The architect can only give information to the best of its knowledge and belief. Note that some professional liability policies exclude claims arising out of express guarantees. The architect should also consider bringing its billing up to date at the time these certificates are requested, because it has considerable leverage at this time.

This language states that the architect does not have to execute any certifications that would extend its responsibility or liability beyond what it already is committed to. The time frame given permits the architect to carefully study the proposed certification to determine whether it does commit the architect to any further liability.

10.5 Nothing contained in this Agreement shall create a contractual relationship with or a cause of action in favor of a third party against either the Owner or Architect. (**1.3.7.5; no change**)

This paragraph is meant to prevent other parties, such as contractors and sub-contractors, from using the provisions of this agreement to claim that they are third-party beneficiaries of this agreement. A third-party beneficiary is someone who did not sign the contract but who is intended to obtain some direct benefit from the contract. In the event of a default by one party to the contract, a third-party beneficiary can bring suit against the defaulting party based on the contract. It is important for the architect that there be no third-party beneficiaries to any of the architect's contracts, in order to limit the architect's liability exposure. This provision merely affirms the position that many courts have previously taken.[287]

---

[287] Widett v. United States Fidelity & Guar. Co., 815 F.2d 885 (2d Cir. 1987) (*citing* Ultramares Corp. v. Touche, 255 N.Y. 170, 174 N.E. 441 (1941)) (in negligence action brought by subcontractor against architect, court held that professionals are not liable in tort or contract absent privity). *Accord* Peter Kiewit Sons' Co. v. Iowa S. Utils. Co., 355 F. Supp. 376 (S.D. Iowa 1973); Sheets, Aiken & Aiken v. Spann Inc., 512 So. 2d 99 (Ala. 1987); Reider Communities, Inc. v. Township of N. Brunswick, 227 N.J. Super. 214, 546 A.2d 563 (1988); James McKinney & Son, Inc. v. Lake Placid 1980 Olympic Games, Inc., 61 N.Y.2d 836, 462 N.E.2d 137, 473 N.Y.S.2d 960 (1984); Alvord & Swift v. Stewart M. Muller Constr. Co., 46 N.Y.2d 276, 385 N.E.2d 1238, 413 N.Y.S.2d 309 (1978); Underhill Constr. Corp. v. New York Tel. Co., 56 A.D.2d 760, 391 N.Y.S.2d 1000 (1977); Crow-Crimmins-Wolff & Munier v. County of Westchester, 90 A.D.2d 785, 455 N.Y.S.2d 390 (1982).

In Cullum Mech. Constr., Inc. v. South Carolina Baptist Hosp., 520 S.E.2d 809 (S.C. 1999), the mechanical subcontractor was not fully paid for its work and sued the architect, arguing that the architect owed it a professional duty distinct from any contractual duty. According to the subcontractor, this purported duty was breached by the architect's certification of several payment requests without obtaining written documentation that the subcontractor had been paid. The court rejected this argument, finding that this provision, as well as a similar provision in A201, did not support a legal duty on the part of the architect to assure payment to subcontractors.

This appellate decision was reversed in Cullum Mech. Constr., Inc. v. South Carolina Baptist Hosp., 344 S.C. 426, 544 S.E.2d 838 (2001). The court held that whether or not a design professional owes a duty depends on the facts and circumstances of each case, therefore, summary judgment was improper. Saying that "generally, an architect does not have a duty to assure payment to subcontractors," the court went on to say that "special conditions in these contract documents may have given rise to a special relationship with subcontractors, and therefore a duty of care."

Although the peculiar facts of this case undoubtedly influenced the reversal of this case, the appellate court made the correct finding. The AIA documents do not create any "special relationship" between the architect and any subcontractor. Only the actions or inactions by the architect may have given rise to such a relationship. Presumably, the court believed, the architect may have created a duty by leading the subcontractor astray. This is contrary to the intent of the AIA documents, which make it quite clear that there is no such duty. Only by pleading and showing actual intent to deceive a subcontractor by the architect should the subcontractor be able to make out a case of breach of duty against the architect.

The *Cullum* case was not followed in Gary Boren v. Thompson & Assocs., 2000 OK 3, 999 P.2d 438 (Okla. 2000). In *Gary Boren*, the school board hired the architect for a school library project that required both performance and payment bonds by the contractor. After approving several payments to the contractor without any bonds, the architect sent a letter to the contractor reminding him that bonds were required. Thereafter, the architect received a copy of the performance bond, but not the payment bond. Later, a lawyer contacted the architect on behalf of a subcontractor, informing the architect of a problem with payments. At that time, the architect realized that he did not have a copy of a payment bond. Nevertheless, the architect continued certifying payments. After the contractor failed to make payment to the subcontractors, the subcontractors filed suit against the architect for negligence in certifying payments to the contractor in the absence of the statutorily required payment bond. The architect moved to dismiss on three bases: (1) no duty to the subcontractors, (2) no privity of contract, and (3) the subcontractors had constructive knowledge of the contractor's duty to procure the bonds, making the proximate cause of the loss their own negligence. The court denied the motion. At trial, the subcontractors obtained an award against the architect. Finding that the purpose of bonding statutes would not be served by protecting a private, for-profit company (the architect) engaged in the business of designing and overseeing public construction projects from potential liability. The damage arose from the architect's negligence in failing to ascertain that there was no payment bond and in making unauthorized payment to the contractor after he discovered that no payment bond existed. Further, once a public entity has contracted with a private party to oversee a construction project, subcontractors should be able to assume that the private party responsible for certifying payments has verified the existence of the bonds. The court stated that the architect ought to labor under a duty to subcontractors to refrain from paying the contractor without the required bond being secured.

Construction workers are not third-party beneficiaries of the owner-architect agreement. Zukowksi v. Howard, Needles, Tammen & Bergendorr, Inc., 657 F. Supp. 926 (D. Colo. 1987) (injured construction workers sued architect for injuries suffered from collapsed viaduct; also holding that an architect is not liable for intentional infliction of mental suffering, absent an allegation of a pattern of conduct calculated to cause emotional distress).

In C.H. Leavell & Co. v. Glantz Contracting Corp., 322 F. Supp. 779 (E.D. La. 1971), contractor argued that he was a part of a tripartite contractual interrelationship that included the owner, architect, and contractor. The court stated that there was no contractual relationship between the architect and contractor, and that if the architect had contracted with the contractor, it would have constituted a breach of a duty the architect owed the owner, "to guard the Owner against defects and deficiencies in the work of contractors."

In Morse/Diesel, Inc. v. Trinity Indus., Inc., 859 F.2d 242 (2d Cir. 1988), *rev'g* 655 F. Supp. 346 (S.D.N.Y. 1987) (*citing* Widett v. United States Fidelity & Guar. Co., 815 F.2d 885 (2d Cir. 1987), (architects not liable either in tort or contract absent privity under New York law), the court held that damages are not available in a suit for negligent performance of a contractual duty when only economic injury is alleged.

In a New York case, the architect contracted with a party that was the developer but not the actual owner of the property.[288] The court allowed the owner to sue the architect, because the agreement had a rider attached that acknowledged the real owner's status. In such a case, the architect may want to add further language to make sure that the courts will apply this provision.

10.6 Unless otherwise required in this Agreement, the Architect shall have no responsibility for the discovery, presence, handling, removal or disposal of, or exposure of persons to hazardous materials or toxic substances in any form at the Project site. (**1.3.7.6; modifications**)

If the architect suspects the presence of hazardous or toxic substances, it should immediately inform the owner in writing and request that the owner hire specialists in this area to determine whether such substances are at the project and, if they are, to remove those substances. The architect's insurance generally does not cover any liability that results from hazardous or toxic substances. Presumably, this provision includes any material that governmental agencies such as the Environmental Protection Agency declare to be hazardous in the future.[289]

This provision applies only to the discovery of existing hazardous materials. If the architect designs or installs new hazardous materials as part of the work, then the architect may well be liable if the design or specification of such materials would constitute malpractice at the time of the design.

10.7 The Architect shall have the right to include photographic or artistic representations of the design of the Project among the Architect's promotional and

---

In John Day Co. v. Alvine & Assocs., 510 N.W.2d 462 (Neb. Ct. App. 1993), the owner was held not to be the beneficiary of the contract between the architect and the engineer. It was unclear as to whether an AIA contract was used in that case.

Architects may be liable to third parties under other theories. In Alberti, LaRochelle & Hodson Eng'g Corp. v. FDIC, 844 F. Supp. 832, 843 (D. Me. 1994), the engineer was liable to the lender when the construction budget submitted to the bank was grossly inadequate and the engineer attended a meeting with the bank but did not disclose the shortcoming. Later, the engineer certified several payments without disclosing the problem. The court found the engineer guilty of negligent misrepresentation. This provision will not protect against such a charge.

See also Resurgence Prop., Inc. v. W.E. O'Neil Constr. Co., No. 92C6618, 1995 U.S. Dist. LEXIS 4939 (N.D. Ill. Apr. 14, 1995) (architect not liable under third-party beneficiary theory).

[288] Sanbar Projects v. Gruzen P'ship, 148 A.D.2d 316, 538 N.Y.S.2d 532 (1989). In Travelers Cas. & Sur. Co. v. Dormitory Auth. of the State of N.Y., 2005 WL 1177715 (S.D.N.Y., May 19, 2005), the architect was alleged to have made numerous errors. The surety sued the architect and the court held that the surety brought the claim on behalf of the contractor, who had consented to the surety's pursuing the claim on its behalf. The claim was allowed to proceed.

[289] In Barnett v. City of Yonkers, 731 F. Supp. 594 (S.D.N.Y. 1990), an action was brought against a city following the death of a student allegedly as a result of exposure to friable asbestos. The city sued the architect for indemnity. The court held that the architect's performance must be examined under the standards of the architectural profession at the time of rendering of the services, which in this case was 1959. The court then found that an architect "could not reasonably have been expected to know of the deleterious effects of asbestos" in 1959; therefore, the architect was not liable.

professional materials. The Architect shall be given reasonable access to the completed Project to make such representations. However, the Architect's materials shall not include the Owner's confidential or proprietary information if the Owner has previously advised the Architect in writing of the specific information considered by the Owner to be confidential or proprietary. The Owner shall provide professional credit for the Architect in the Owner's promotional materials for the Project. (**1.3.7.7; no change**)

It would be prudent to advise the owner of each instance when the architect intends to use the owner's project in any publication or other publicity.

10.8 If the Architect or Owner receives information specifically designated by the other party as "confidential" or "business proprietary," the receiving party shall keep such information strictly confidential and shall not disclose it to any other person except to (1) its employees, (2) those who need to know the content of such information in order to perform services or construction solely and exclusively for the Project, or (3) its consultants and contractors whose contracts include similar restrictions on the use of Confidential Information. (**1.2.3.4; substantial modifications**)

If the owner or architect identifies certain information as confidential or "business proprietary," the architect and the architect's consultants (or owner and its consultants) are required to keep that information confidential. The exceptions include violation of law. This might occur if the owner asks the architect to deviate from building codes, or if one of the owner's consultants prepares designs contrary to building codes. If the owner is unwilling to bring such documents within the law, the architect can bring this to the proper authorities, even if the owner has identified this information as confidential. The same applies to situations in which there is a risk to the public safety.

If a dispute arises between the owner and architect, such as a suit for unpaid fees, the architect or owner have no obligation to maintain the confidentiality of any information that is relevant to that suit.

§ 2.30   Article 11: Compensation

ARTICLE 11 COMPENSATION

11.1 For the Architect's Basic Services as described under Article 3, the Owner shall compensate the Architect as follows:

(Insert amount of, or basis for, compensation.) (**1.5.1; change reference; minor modifications**)

This section describes the architect's compensation for the basic services. If the fee is to be a fixed fee, state, "A fixed fee of $X." If the fee is to be on a percentage, state, "Y percent (Y%) of the Cost of the Work." Many other variations are possible. In a Wisconsin case, the architect's contract had this provision: "The Owner agrees to pay the Architect for the above services, a fee of Five Hundred Dollars ($500.00) per apartment, for each apartment constructed."[290] The project was to consist of a number of condominium units. The initial bids came in too high, and after redesign and rebidding, the costs again came in too high. The owner then abandoned the project. The architect sought to collect a fee based on the work he had done. The appellate court held that the contract provision was defective, because the number of apartments to be built was indefinite. The parties had never agreed on an exact number of units to be constructed. Also, the parties had never agreed on an hourly basis for the architect's fee in the event the project was abandoned. The court did permit the architect to proceed on a quantum meruit basis.

> **11.2** *For Additional Services designated in Section 4.1, the Owner shall compensate the Architect as follows:*
>
> *(Insert amount of, or basis for, compensation. If necessary, list specific services to which particular methods of compensation apply.)* **(new)**

Section 4.1 included a table that designates services to be provided by the architect, the owner, or not at all. If any of the listed services are to be performed by the architect, the method of compensation would be listed here. Normally, this would be on an hourly basis, so that the architect could list a schedule of hourly rates for various classifications of employees.

> **11.3** *For Additional Services that may arise during the course of the Project, including those under Section 4.3, the Owner shall compensate the Architect as follows:*
>
> *(Insert amount of, or basis for, compensation.)* **(1.5.2; modifications)**

Here, again, a method of determining compensation should be inserted. The most common is as a multiple of direct personnel expense or as a flat hourly rate.

> **11.4** *Compensation for Additional Services of the Architect's consultants, when not included in Section 11.2 or 11.3 shall be the amount invoiced to the Architect plus _____ (_____), or as otherwise stated below:* **(1.5.3; modifications)**

Most architects mark up the services of their consultants to cover their overhead and increased liability exposure.

---

[290] Goebel v. National Exchangers, Inc., 88 Wis. 2d 596, 277 N.W.2d 755, 758 (1979).

11.5 Where compensation for Basic Services is based on a stipulated sum or percentage of Cost of the Work, the compensation for each phase of services shall be as follows:

Schematic Design Phase:	*percent (%)*
Design Development Phase:	*percent (%)*
Construction Documents Phase:	*percent (%)*
Bidding or Negotiation Phase:	*percent (%)*
Construction Phase:	*percent (%)*
Total Basic Compensation:	

These fees are fully earned at the conclusion of each respective phase. However, B101 requires that the owner pay monthly within each phase, so that the architect is paid for the progress within each phase. Some courts have held that contracts like these are divisible into the separate phases, so that an engineer is only entitled to the percentage fee for the phases actually completed.[291] If any particular service is not to be provided, that line should either be crossed out or the percentage indicated as "0%." There may also be other services that could be additional line items. Note that the services here are basic services, meaning the ones that the architect is required to perform for the base fee. Additional services are not part of this chart and are to be compensated per Section 11.2 and 11.3. The architect should carefully consider how these percentages are allocated, since if there is a dispute with the owner, the architect will likely be paid based on what is filled in here.

11.6 When compensation is based on a percentage of the Cost of the Work and any portions of the Project are deleted or otherwise not constructed, compensation for those portions of the Project shall be payable to the extent services are performed on those portions, in accordance with the schedule set forth in Section 11.5, based on (1) the lowest bona fide bid or negotiated proposal, or (2) if no such bid or proposal is received, the most recent estimate of the Cost of the Work for such portions of the Project. The Architect shall be entitled to compensation in accordance with this Agreement for all services performed whether or not the Construction Phase is commenced. **(new)**

This provision applies only if the architect's fee is based on a percentage of the construction cost.[292] The portions of the project that are deleted are billed up to the percentage of work completed. For instance, if the architect's fee on a $10 million project is six percent of the construction cost and the owner deleted 20 percent of

---

[291] Lake LBJ Mun. Util. Dist. v. Coulson, 771 S.W.2d 145 (Tex. App. 1988).

[292] A New York court interpreted this paragraph. Hess v. Zoological Soc'y, Inc., 134 A.D.2d 824, 521 N.Y.S.2d 903 (1987). The term *negotiated proposal* was held to mean a negotiated proposal between the owner and contractor, not between the owner and architect. In this case, the project never got as far as the bidding stage; the fee was based on the most recent estimate of construction cost.

In Holmquist v. Priesmeyer, 574 S.W.2d 173 (Tex. 1978), the court rejected the owner's contention that, because the house was not constructed and because the architect did not obtain a bid from a qualified builder that could serve as a basis for computing the construction cost of the project, the architect was not entitled to recover his fee. This argument was rejected.

the project and the architect has completed only the schematic phase (and ¶ 11.5 shows that the schematic accounts for 15 percent of the total), the calculation for the architect's services on the deleted work is as follows: $2 million (the deleted work) times six percent times 15 percent = $18,000. He is also entitled to his percentage fee for the remaining work. This often becomes a point of contention between the owner and architect where the project is downsized prior to bidding. Under this provision, the compensation is based on the architect's most recent estimate of construction costs. These estimates should have been prepared during Schematic Design (¶ 3.2) and during Design Development (¶ 3.3). Often, especially on smaller projects, there is no written record of any such estimate by the architect. Arguably, if this is the case, the applicable estimate would be the budget established between the owner and architect at the outset of the project.

11.7 The hourly billing rates for services of the Architect and the Architect's consultants, if any, are set forth below. The rates shall be adjusted in accordance with the Architect's and Architect's consultants' normal review practices.

(If applicable, attach an exhibit of hourly billing rates or insert them below.)

Employee or Category   Rate (1.5.6; modified)

Here, insert a table of hourly rates for the architect's personnel and those of the consultants. This provision allows the architect to increase hourly rates for personnel who are performing work on the project, usually on an annual basis.

§ 2.31   Compensation for Reimbursable Expenses: ¶ 11.8

11.8 COMPENSATION FOR REIMBURSABLE EXPENSES

11.8.1 Reimbursable Expenses are in addition to compensation for Basic and Additional Services and include expenses incurred by the Architect and the Architect's consultants directly related to the Project, as follows: **(1.3.9.2; modifications)**

These are primarily the architect's out-of-pocket[293] costs. Many architects mark up these expenses (**see** ¶ 11.8.2) to cover their overhead in administering them. Also, the architect, by advancing these costs for the client, acts as the client's

---

In Douglass v. First Nat'l Realty Corp., 437 F.2d 666 (D.C. 1970), the project was abandoned because financing was not obtained. Although the architect had not furnished a construction estimate, he could recover based on the owner's estimate used for the loan application.

In Getzschman v. Miller Chem. Co., 232 Neb. 885, 443 N.W.2d 260 (1989), the architect was entitled to the percentage based on the lowest bid for drawings prepared by the architect. Following a dispute, the owner took the plans to be redrawn by a draftsman and the actual construction cost was lower.

[293] *See* Tucker v. Whitehead, 155 Ga. App. 104, 270 S.E.2d 317 (1980) ("out-of-pocket expense" was ambiguous, and jury could determine what was included in such expenses).

banker and makes the client a loan of the funds. These costs are in addition to the architect's services and are in addition to the stated contract price.

> *.1 transportation and authorized out-of-town travel and subsistence;* **(1.3.9.2.1; modifications)**

Sometimes owners insist that local travel be included in the basic services. In such cases, the architect typically bills for its time on a portal-to-portal basis, that is, the architect's time starts when she leaves her office and ends when she returns. Of course, this only applies if she is paid on an hourly basis.

> *.2 long distance services, dedicated data and communication services, tele-conferences, Project Web sites, and extranets;* **(1.3.9.2.1; substantially expanded definition)**
>
> *.3 fees paid for securing approval of authorities having jurisdiction over the Project;* **(1.3.9.2.2; no change)**

This provision relates to permit fees, which are usually paid by the owner directly or by the contractor. The wise practice for the architect is to obtain the client's check in advance of obtaining the permit. If the permit fee is payable upon issuance, the client can be sent to pick it up and pay the fee.

> *.4 printing, reproductions, plots, standard form documents;* **(1.3.9.2.3; minor modifications)**

Printing bills can be significant. If the architect has in-house printing capabilities, it should bill out that printing on the same basis as if the drawings were sent out to commercial printers.

> *.5 postage, handling and delivery;* **(1.3.9.2.3; minor modifications)**
>
> *.6 expense of overtime work requiring higher than regular rates if authorized in advance by the Owner;* **(1.3.9.2.4; no change)**
>
> *.7 renderings, models, mock-ups, professional photography, and presentation materials requested by the Owner;* **(1.3.9.2.5; added photography and presentation materials)**

These should be discussed with the owner at the outset. Normally, the architect does not provide these items in its basic services.

> *.8 Architect's Consultant's expense of professional liability insurance dedicated exclusively to this Project, or the expense of additional insurance coverage or limits if the Owner requests such insurance in excess of that normally carried by the Architect's consultants;* **(1.3.9.2.6; modifications)**

This provision is limited to the consultants of the architect. If the owner requests that they carry additional insurance, that is reimbursable. Prior forms of this document allowed the architect to be reimbursed if the owner requested additional insurance for the architect. Many architects today do not carry professional liability insurance of any kind. If the owner insists on that insurance, the cost of coverage in excess of what the architect normally carries may not be reimbursable. If the architect does not carry any insurance, then this provision should be modified to state that it carries no insurance. Since this sentence is somewhat ambiguous as to what happens if the owner wants additional insurance by the architect, this should be discussed prior to execution of this agreement. The owner should be aware of the policy limits and the length of time this amount of insurance will be carried past the termination of the project. Although most claims are made soon after the end of the project, claims can be made many years after final completion. Each state has different statutes of limitations and repose. The architect should seek legal and insurance counsel whenever an owner requests additional insurance.

> *.9 all taxes levied on professional services and on reimbursable expenses;* (**new**)

To the extent that there are taxes on professional services of the architect or its consultants, or on reimbursable expenses, such taxes shall be reimbursed.

> *.10 site office expenses; and* (**new**)

The architect will not normally have a site office. To the extent that the parties agree that the architect will have an office at the site, all expenses for that site office will be reimbursed.

> *.11 other similar Project-related expenditures.* (**1.3.9.2.8; no change**)

> *11.8.2 For Reimbursable Expenses the compensation shall be the expenses incurred by the Architect and the Architect's consultants plus _____ (_____) of the expenses incurred.* (**1.5.4; modifications**)

This figure would be something in excess of 100 percent. This reimbursable expense multiplier could be lowered if the owner agreed to advance a fund to the architect against which the architect could draw, so that the architect does not bankroll the project. The owner could also establish an account with the printer so that those bills would not go through the architect.

§ 2.32 Compensation for Use of Architect's Instruments of Service: ¶ 11.9

11.9 COMPENSATION FOR USE OF ARCHITECT'S INSTRUMENTS OF SERVICE

If the Owner terminates the Architect for its convenience under Section 9.5, or the Architect terminates this Agreement under Section 9.3, the Owner shall pay a licensing fee as compensation for the Owner's continued use of the Architect's Instruments of Service solely for purposes of completing, using and maintaining the Project as follows: (**new**)

If the architect is terminated for the convenience of the owner per Section 9.5 or the architect terminates due to suspension of the project pursuant to Section 9.3, then the owner can only use the architect's drawings and specifications if this fee is paid. The use by the owner is limited to completing the project, and use and maintenance of the project. The owner is not authorized to build additional buildings, or to construct additions to this building using these Instruments of Service.

§ 2.33 Payments to the Architect: ¶ 11.10

11.10 PAYMENTS TO THE ARCHITECT

11.10.1 An initial payment of _____ ($_____) shall be made upon execution of this Agreement and is the minimum payment under this Agreement. It shall be credited to the Owner's account in the final invoice. (**1.5.7; modifications**)

This provision concerning crediting the initial payment to the final bill aids the architect in maintaining proper payments from the owner and avoids the problem of the architect's financing the project. In effect, the initial payment becomes like a rent security deposit. The architect's first bill would be for all time spent up to that first bill with no deduction for the initial payment. The initial payment is only applied towards the final bill.

11.10.2 Unless otherwise agreed, payments for services shall be made monthly in proportion to services performed. Payments are due and payable upon presentation of the Architect's invoice. Amounts unpaid _____ (_____) days after the invoice date shall bear interest at the rate entered below, or in the absence thereof at the legal rate prevailing from time to time at the principal place of business of the Architect.

(Insert rate of monthly or annual interest agreed upon.) (**1.5.7, 1.5.8; modifications**)

Billing frequency for additional services should be the same as for basic services. Unless the parties agree otherwise, the architect is entitled to payment on a monthly basis. Each item of billing should be separately listed, so there is no question of which are additional services.

In this provision, an interest rate should be inserted. Because many states have usury laws[294] that limit the interest rates that can be charged, legal advice should be obtained before entering an amount here. The legal rate referred to is provided by statute and is usually unrealistically low. One court has held that this is not the

---

[294] *See* Commerce, Crowdus & Canton v. DKS Constr., 776 S.W.2d 615 (Tex. App. 1989) (court discussed this provision and local usury laws).

maximum rate permitted by law but rather the statutory rate.[295] For a single-family home, with the homeowner as owner, a number of state and federal laws may also limit interest rates and require certain disclosures. The following can be inserted (typed in the space provided on AIA B101), subject to the above cautions:

> One percent per month. Payments are due thirty (30) days after the billing date shown on each invoice. All costs of collection, including reasonable attorneys' fees, shall be paid by the Owner.

The parties may want to include a clause that provides for payment of attorneys' fees in the event a lawsuit is brought to collect unpaid fees. The American Rule is that each party must bear its own attorneys' fees unless a statute or contract specifically provides for attorneys' fees. For a normal architectural project, there are no statutes in most states that permit recovery of attorneys' fees.[296]

Even when the parties to a construction contract have provided for attorney's fees, the courts will look carefully to the equity of the situation in determining whether such fees are due.[297]

11.10.3 The Owner shall not withhold amounts from the Architect's compensation to impose a penalty or liquidated damages on the Architect, or to offset sums requested by or paid to contractors for the cost of changes in the Work unless the Architect agrees or has been found liable for the amounts in a binding dispute resolution proceeding. (**1.3.9.1; substantial modifications**)

Unless the architect is found, by a court or arbitration, to be liable for particular problems with the work, the owner cannot withhold payments from the architect, even if the owner can withhold sums from the contractor.

11.10.4 Records of Reimbursable Expenses, expenses pertaining to Additional Services, and services performed on the basis of hourly rates shall be available to the Owner at mutually convenient times. (**1.3.9.3; modifications**)

---

[295] Jetty, Inc. v. Hall-McGuff Architects, 595 S.W.2d 918 (Tex. App. 1980). *See, e.g.,* Bolivar Insulation Co. v. R. Logsdon Builders, Inc., 929 S.W.2d 232 (Mo. Ct. App. 1996) (contract had provision for 18% per year interest; appellate court reversed trial court's limitation of interest to the statutory rate of 9%, finding the contractual provision valid.

[296] However, in the case of frivolous suits, Federal Rule of Civil Procedure 11 and many state statutes and rules, such as Rule 137 of the Illinois Supreme Court Rules, provide for payment of attorneys' fees if a party files an action, motion, or other paper without reasonable basis.

[297] In Willie's Constr. Co. v. Baker, 596 N.E.2d 958 (Ind. Ct. App. 1992), the contractor sued the owner for the balance of the cost of the house and homeowner counterclaimed. The court found that the contractor had breached the contract and awarded the owner the cost of repairs. The appellate court held that the contractor was not entitled to interest and attorney's fees on its unpaid bills because the amount awarded to the owners exceeded those costs.

In Marsh v. Loffler Hous. Corp., 102 Md. App. 116, 648 A.2d 1081, 1086 (1994), the contract contained this provision: "In the event that payment under this contract is enforced through legal action, or other collection action, homeowner agrees to pay contractor's costs and attorney's fees related to said action." The court held that attorneys' fees were properly awarded in arbitration.

If the contract is on a percentage or fixed-fee basis, the owner would be entitled to see only records that relate to reimbursable expenses or additional services, if any. Architects and all personnel should keep hourly records for time spent on each project.

§ 2.34   Article 12: Special Terms and Conditions

ARTICLE 12 SPECIAL TERMS AND CONDITIONS

Special terms and conditions that modify this Agreement are as follows: **(1.4.2; No change)**

Here, describe any riders that amend this document, or insert additional clauses that describe the services.

§ 2.35   Article 13: Scope of the Agreement

ARTICLE 13 SCOPE OF THE AGREEMENT

13.1 This Agreement represents the entire and integrated agreement between the Owner and the Architect and supersedes all prior negotiations, representations or agreements, either written or oral. This Agreement may be amended only by written instrument signed by both Owner and Architect. **(1.4.1; Deleted last sentence of prior paragraph)**

This is an *integration clause*, meaning that any agreements between the parties prior to the signing of the instant agreement are merged into the present agreement. If those prior agreements are inconsistent with the present agreement, they are void.[298]

---

[298] In American Demolition, Inc. v. Hapeville Hotel Ltd. P'ship, 413 S.E.2d 749 (Ga. 1991), the court stated that such a clause barred a fraud claim.

A Mississippi case, Godfrey, Bassett v. Huntington, 584 So. 2d 1254 (Miss. 1991), held an architect liable to a contractor for misrepresenting the terms of a contract. The court stated that if the contractor's recovery was contingent solely on this provision, he would lose. However, the court went beyond the contract. This case is instructive concerning the duties of the architect towards bidders. The architect had been hired by a school to prepare plans for the renovation of certain areas of the school. When the plans went out for bids, they included a requirement in the specifications that the bidders include a $9,000 contingency in their bids. Any unused portion would be credited to the owner. The architect then issued to all bidders an addendum that deleted the contingency. When the bids were opened, all bids exceeded the budget, and the architect was asked to reduce the scope of the work.

When the new plans were reissued, portions of the old plans and specifications were reused, including the section that required the $9,000 contingency. When one of the bidders noticed this, he called the senior partner at the architectural firm (with whom he was friendly) and asked whether the $9,000 contingency was intentionally included. The architect informed the contractor that an addendum was being mailed out to delete this contingency, just as had been done during the prior bidding. In fact, an addendum was sent out, but it did not delete the $9,000 contingency.

Under a prior version of the AIA documents, this provision prohibited the contractor from using parol evidence to show an oral representation by the owner and architect.[299]

Although the agreement states that it may be modified only by a written instrument, courts continually permit the introduction of evidence that a contract has been modified by a subsequent oral agreement.[300] For that reason, no part of the duties, scope, price, or other term of the contract should be orally changed. The prudent practice is to confirm each modification with a letter acknowledged by both parties. If a major modification occurs, a more formal memorial of it would be prudent. In either event, the documentation should emphasize that the modification is not effective until all required signatures are obtained.

A problem frequently occurs when the architect enters into more than one contract with the same owner. For instance, the architect may have one contract for the main building in a project and a separate contract for a separate building in the same project, or for certain interior work in the main building. The question then is whether the architect is entitled to compensation under both contracts. The simple solution to this problem is to specifically refer to the other contract in the present contract, then it is clear that there is more than one contract.

13.2 This Agreement is comprised of the following documents listed below: (**1.4.1; modifications**)

    .1   AIA Document B101-2007, Standard Form of Agreement Between Owner and Architect (**1.4.1.1; add reference to Digital Data; modifications**)

This is this document.

The contractor never saw the addendum, but he was the low bidder and was awarded the job. He signed the contract that incorporated the addendum. After the work was started, the contractor learned that the contingency had not been eliminated and that he was expected by the owner to have the contingency in place. At the end of the project, the contractor credited the owner the $9,000 and sued the architect to recover this amount.

The court found that it appeared that all parties had made an honest mistake. However, the contractor relied on the architect's representation that the contingency was not included. In the absence of this misrepresentation, the contractor would have read the addendum and learned the truth before submitting his bid. The contractor's negligence in failing to carefully read the contract documents was less than the architect's negligence in failing to make the planned corrections to the specifications. The contractor's negligence resulted directly from the architect's misrepresentation. The court held that the contractor was entitled to the $9,000 from the architect.

Cases in which this provision is referenced include: Warren Freedenfeld Assoc., v. McTigue (2007 WL 757874 (D.Mass., March 9, 2007) (copyright case—owner argued that he was a joint author); Town of Homer, Inc. v. General Design, Inc., 960 So. 2d 310 (La. Ct. App. 2007) (dispute over whether arbitration was required); Sears, Roebuck & Co. v. Enco Assocs., 370 N.Y.S.2d 338 (N.Y. Sup., 1975) (architect did not guarantee specific result).

[299] Hercules & Co. v. Shama Rest. Corp., 613 A.2d 916 (D.C. 1992) (arbitration in which the contractor was unhappy with the result).

[300] Somerset Cmty. Hosp. v. Mitchell & Assocs., Inc., 685 A.2d 141 (Pa. Super, 1996) (conduct of the parties orally modified agreement despite this language).

.2   AIA Document E201-2007, Digital Data Protocol Exhibit, if completed, or the following: **(new)**

With the widespread use of drawings and specifications in digital form, and contractors and owners requesting access to the architect's digital files, accommodations must be made with regard to digital files. See Section 1.6 of A201, which is a new section in 2007 regarding transmission of data in digital form. AIA Document E201-2007 covers procedures regarding the exchange of digital data.

.3   Other documents:

(List other documents, if any, including Exhibit A, Initial Information and additional scopes of service, if any, forming part of the Agreement.) **(1.4.1.3; add reference to Exhibit A)**

This might include proposals or other documents that are intended to be part of the owner-architect agreement. While Exhibit A is not required to be part of this agreement, it is not a good idea to omit this document. Unless Exhibit A is signed or initialed, or there is some other indication that the parties intend it to be binding, it does not become part of this agreement unless it is specifically referenced here.

EXHIBIT A

Initial Information for the following PROJECT:

Here the project name and location, as well as the identity of the owner and architect are inserted. See the discussion in § 2.3.

This Agreement is based on the following information.

(Note the disposition for the following items by inserting the requested information or a statement such as "not applicable," "unknown at time of execution" or "to be determined later by mutual agreement.") **(1.1.1; minor modifications)**

§ 2.36 Article A.1: Project Information

ARTICLE A.1 PROJECT INFORMATION

A.1.1 The Owner's Program for the Project:

(Identify documentation or state the manner in which the program will be developed.) **(1.1.2.3; minor modification)**

A program for even a very modest project may be many pages long. Here, insert a description of the program document that will be made an exhibit to the contract. If the program does not exist when this document is prepared, the parties should identify how the program will be developed and what these two parties will do to develop that document.

A.1.2 The Project's physical characteristics:

(Identify or describe, if appropriate, size, location, dimensions, or other pertinent information, such as geotechnical reports; site, boundary and topographic surveys; traffic and utility studies; availability of public and private utilities and services; legal description of the site; etc.) **(1.1.2.2; modifications)**

Reference can be made to a variety of documents, including soil and other geotechnical reports, environmental reports, and so on.

A.1.3 The Owner's budget for the Cost of the Work, as defined in Section 6.1:

(Provide total, and if known, a line item break down.)
(1.1.2.5; substantial modifications)

This is the construction cost and does not include costs of design professionals, interest, legal fees, or other costs associated with the project. This is the budget that the architect is concerned with and which constrains the architect's design. The provisions of Article 6 apply to this number.

To the extent that the owner has the experience to develop its budget, this number can be broken down into line items, such as costs for site work, the building, etc. In general, the more detailed this number is, the better.

A.1.4 The Owner's other anticipated scheduling information, if any, not provided in Section 1.2: **(new)**

Any other information related to scheduling should be inserted here.

A.1.5 The Owner intends the following procurement or delivery method for the Project:

(Identify method such as competitive bid, negotiated contract, or construction management.) **(1.1.2.7; minor modifications)**

This can make a substantial difference in the time required by the architect to perform the work. The method can be anything from the traditional design-bid-build method to a variety of other methods. The architect and owner should discuss the advantages and disadvantages of each method and insert the one that will work best. If the owner later decides to change the method, the architect may be entitled to an increased fee if the new method will result in more work.

A.1.6 Other Project information:

(Identify special characteristics or needs of the Project not provided elsewhere, such as environmentally responsible design or historic preservation requirements.) **(1.1.2.8; modifications)**

This might include any other special requirement for the project. One example listed is for a historic preservation project. In that case, special consultants may be required, as well as reports, specialty contractors, and so on. It is important to identify such characteristics up front.

§ 2.37   Article A.2: Project Team

ARTICLE A.2 PROJECT TEAM

A.2.1 *The Owner identifies the following representative in accordance with Section 5.3:*

(List name, address and other information.) **(1.1.3.1; minor modifications)**

This is the person who is authorized to make decisions on behalf of the owner (**see ¶** 5.3) and would likely have the primary interaction with the architect.

A.2.2 *The persons or entities, in addition to the Owner's representative, who are required to review the Architect's submittals to the Owner are as follows:*

(List name, address and other information.) **(1.1.3.2; modifications)**

If the owner wants persons other than the person listed in ¶ A.2.1 to review and authorize matters involving the owner, they should be listed here. This might arise, for instance, if the owner is a partnership or joint venture. Likewise, the owner's lender may want to review things prior to their approval.

A.2.3 *The Owner will retain the following consultants and contractors:*

(List discipline and, if known, identify them by name and address.) **(1.1.3.3; modifications)**

If the owner plans to retain consultants, they should be listed here. Sometimes, the owner may have engineers or specialty contractors that it has used in the past. The architect will want to know with whom it will be working. Also, if the owner already knows that the construction contract will be assigned to a particular contractor, that information should be inserted here.

A.2.4 *The Architect identifies the following representative in accordance with Section 2.3:*

(List name, address and other information.) **(1.1.3.4; modifications)**

This is the person at the architect's office who will interact with the client and contractor and who is authorized to make decisions and sign documents on behalf of the architect. (**See ¶** 1.2.2.3.)

A.2.5 *The Architect will retain the consultants identified in Sections A.2.5.1 and A.2.5.2.*

(List discipline and, if known, identify them by name and address.) **(1.1.3.5; modifications)**

A.2.5.1 Consultants retained under Basic Services:

    .1   Structural Engineer

    .2   Mechanical Engineer

    .3   Electrical Engineer

Here, the architect should identify all of the consultants the architect expects to retain. If the identity of a particular consultant is not known when this contract is signed, at least the discipline should be identified. For instance, an actual name of a firm can be inserted as the structural engineer, or, alternatively, "to be determined" could be inserted. If, for instance, there will not be an electrical engineer hired by the architect, "by owner" or "not required" could be inserted. Owners should review this section to make sure that the architect has included all disciplines required by the architect. If a particular discipline is not included here, the implication is that, if that discipline is required for the project, the owner will retain such a firm. **See ¶** A.2.3, in which the owner's consultants are listed.

A.2.5.2 Consultants retained under Additional Services: **(new)**

This section is to identify the type of consultant that, if hired by the architect, would be billed as an additional service under Section 11.4.

A.2.6 Other Initial Information on which the Agreement is based:
(Provide other Initial Information) **(1.1.4; minor modification)**

Here, any other information that would be important to memorialize should be recorded.

CHAPTER 3

ALTERNATIVE LANGUAGE FOR AIA DOCUMENT B101

§ 3.1   Introduction

The language in this chapter is offered as possible alternative language to AIA Document B101. The user is cautioned that some of these provisions are mutually incompatible, and each paragraph and each sentence should be carefully reviewed with legal counsel before use. Note, further, that some of these changes require changes to other documents. Finally, please note that no recommendations are being made or implied by this alternative language. Much of it is more favorable to the owner than the standard AIA language, while some is more favorable to the architect. Each party should consult with an attorney familiar with construction law in the appropriate jurisdiction before making any changes to the standard document, as such changes will often result in serious consequences to the parties. If in doubt, consideration should be given to using the AIA language as-is.

The recommended method of amending B101 is to use the AIA's Electronic Documents software and make the changes directly to the body of the document. Alternatively, attach separate written amendments that refer back to B101. This is normally done in riders to the agreement, or inserting language in Article 12. Another method is to graphically delete material and insert new material in the margins of the standard form. However, this method may create ambiguity, and it may result in rendering the stricken material illegible.

§ 3.2   Architect's Responsibilities: ¶ 2.2

The following alternatives represent two extreme positions for this clause. The AIA language is considered the most neutral provision.

A provision favorable to the architect:

2.2 The Architect's services shall be performed as expeditiously as is consistent with ordinary and reasonable care and with the orderly progress of the Project, subject to the reasonable interpretation of the Architect. The Architect may, from time to time, provide the Owner estimates of time in which the Architect may provide such services. However, the Owner understands and acknowledges that there are many circumstances beyond the Architect's control that will affect the Architect's services and the timing thereof. Therefore, Owner agrees that the Owner's sole remedy for any delay in the Architect's services shall consist of an extension of the time in which such services shall be completed and the Owner waives all other damages and causes of action related thereto.

A provision favorable to the owner:

2.2 The Architect acknowledges that the Owner is relying on the Architect's special skill and expertise in projects of the type herein. Therefore, the Architect's services shall be performed as expeditiously as is consistent with the highest standard of care and with the utmost diligence and the orderly progress of the Project. Architect acknowledges that it will furnish the most skilled personnel for the Project and will give the Project the highest priority. Architect further warrants that it is skilled and experienced in projects of the type herein; has experience with the designs, details, materials, procedures and methods intended for this Project; and has the capacity to meet all of the Owner's schedules. The Architect shall submit for the Owner's approval a schedule for the performance of the Architect's services which initially shall be consistent with the time periods established in Initial Information, and which shall be adjusted, if necessary and if approved by Owner, as the Project proceeds. This schedule shall include allowances for periods of time required for the Owner's review, for the performance of the Owner's consultants, and for approval of submissions by authorities having jurisdiction over the Project. Time limits established by this schedule approved by the Owner shall not be exceeded by the Architect. Upon approval of any such schedule submitted by Architect, such approved schedule shall be deemed a part of this Agreement. Architect also agrees that it will not be damaged by any delay caused by Owner or its consultants and shall not be entitled to any additional compensation for such delay, an extension of the time in which Architect is to provide its services being the sole remedy for any delay caused by other parties.

Other provisions related to the Architect's standard of performance follow:

All services to be performed by the Architect in respect of this Agreement shall be provided in a first-class manner [or "in a manner consistent with the degree of care and skill usually exercised by architects experienced in projects of similar scope"] and in accordance with standards of care and skill expected of architects experienced in the design of projects similar to the Project and under the direction of architects and engineers licensed and duly qualified in the jurisdiction in which the Project is located.

The Architect shall be responsible for the quality, technical accuracy, timely completion, and coordination of all plans, studies, designs, drawings, specifications, reports, and other services furnished by the Architect under this Agreement. The Architect shall, without additional compensation, correct or revise any errors, omissions, or other deficiencies in its plans, studies, designs, drawings, specifications, reports, and other services.

Another provision:

Architect shall at all times in performing its services under this Agreement exercise the professional standard of care and due diligence in a manner equivalent to other reputable architects performing similar services for projects of like size and kind. Such professional standard of care and due diligence shall include but not be limited to an obligation by Architect to cause all final documents, drawings and specifications prepared by Architect or its consultants to be in full compliance with laws, statutes, codes, ordinances, orders, rules and regulations applicable to Project.

Another provision:

Architect and its Consultants shall timely perform all Services under this Agreement in a skillful and competent manner in accordance with standards of nationally recognized architects engaged in the design of facilities similar to the Project. Architect shall be responsible to Owner for all reasonable damages due to Architect's or Architect's Consultants' failure to perform any or all Services under this Agreement in accordance with these standards. Neither review nor approval of any of Architect's or Architect's Consultants' work shall relieve Architect from its duty to adhere to these standards of professional care. Architect and its Consultants shall design the Project so that, when constructed in accordance with the Drawings and Specifications, the Project will operate in all material respects as a functional and efficient facility within the scope of the Project program.

Another provision:

Architect and its Consultants shall perform Services under this Agreement in a manner consistent with the degree of care and skill ordinarily exercised by professionals currently practicing in the same locality under similar conditions. No other representations, either express or implied, and no warranty of any sort is provided herein.

Another, more favorable to the owner:

The services of Architect and its Consultants, if any, shall be performed in accordance with and judged solely by the standard of care exercised by licensed members of their respective professions having substantial experience providing similar services on projects similar in type, magnitude and complexity to the Project that is the subject of this Agreement. The Architect shall be liable to the Owner for claims, liabilities, additional burdens, penalties, damages or third party claims (i.e. a Contractor claim against Owner), to the extent caused by errors or omissions that do not meet this standard of care.

§ 3.3 Conflicts of Interest: ¶ 2.4

Another version:

2.4 Architect and its consultants shall not have any employee that has a conflict of interest that may reasonably affect the Architect's or consultant's professional judgment in regard to the Project, unless such conflict is disclosed to the Owner and approved by the Owner in writing. It is the Architect's duty to enforce this provision with its consultants.

§ 3.4 Architect's Insurance: ¶ 2.5

The following language is a broad form of insurance/indemnification provision that favors an owner. The architect faced with such a provision should consult with its insurance carrier to ascertain whether its insurance will cover the indemnification provision in this section:

2.5.1 The Architect shall insure specifically the indemnity contained in sub paragraph 2.5.4(a) below and shall include the Indemnitees (as defined in 2.5.4(a) below) as additional insureds by causing amendatory riders or endorsements to be attached to the insurance policies described below in paragraphs 2.5.2(a), 2.5.2(b), and 2.5.2(c). The insurance coverage afforded under these policies shall be primary to any insurance carried independently by the Indemnitees. Said amendatory riders or endorsements shall indicate that as respects the Indemnitees, there shall be severability of interest under said insurance policies for all coverages provided under said insurance policies.

2.5.2 The Architect shall maintain, at its own expense, the following insurance coverages, insuring the Architect, its employees, agents, and designees, and the Indemnitees as required herein, which insurance shall be placed with insurance companies reasonably acceptable to the Owner and shall incorporate a provision requiring the giving of written notice to the Owner at least thirty

(30) days prior to the cancellation, nonrenewal, or material modifications of any such policies, as evidenced by return receipt of United States certified mail:

.1 Architect's Professional Liability Insurance in the amount of _____ ($ _____) (including contractual liability coverage with all coverage retroactive to the earlier of the date of this Agreement or the commencement of the Architect's services in relation to the Project), covering personal injury, bodily injury, and property damages, said coverage to be maintained for a period of three (3) years after the date of final payment hereunder.

.2 Comprehensive General Liability Insurance in the amount of _____ ($ _____), including coverage for blanket contractual liability, broad form property damage, and personal injury, political risk, and products/completed functions.

.3 Comprehensive Automobile Liability Insurance, including hired and non-owned vehicles, if any, in the amount of _____ ($ _____), covering personal injury, bodily injury, and property damage.

.4 Workers' Compensation Insurance in the amount of the statutory maximum with an employer's liability coverage of at least _____ ($ _____).

2.5.2.1 The Architect shall have the following endorsement added to its Comprehensive General Liability policy:

"It is hereby agreed and understood that the Indemnitee is named as an additional insured. The coverage afforded to the additional insured under this policy shall be primary insurance. If the additional insured has other insurance that is applicable to the loss, such other insurance shall be on an excess or contingent basis. The amount of the Company's liability under this policy shall not be reduced by the existence of such other insurance. It is further agreed that the coverage afforded to the additional insured shall not apply to the sole negligence of the additional insured."

All deductibles on any policy of insurance to be purchased by the Architect hereunder shall be borne by the Architect.

2.5.3 The Architect shall submit valid certificates in form and substance satisfactory to the Owner evidencing the effectiveness of the foregoing insurance policies, along with original copies of the amendatory riders to any such policies, to the Owner for the Owner's approval before the Architect commences rendering any services hereunder.

2.5.4.1 The Architect hereby agrees to indemnify, defend, and hold the Owner, and any subsidiary, parent, or affiliate corporations of the Owner, or other persons or entities designated by the Owner, and their respective directors, officers, agents, employees, and designees (collectively, the "Indemnitees") harmless from all losses, claims, liabilities, injuries, damages, and expenses, including attorneys' fees, that the Indemnitees may incur by reason of any injury or damage sustained to any person or property (including, but not limited to any one or more of the Indemnitees) arising out of or occurring in connection with the performance or lack of performance by the Architect of its duties and obligations under or pursuant to this Agreement, whether or not any other party contributes to such performance or lack of performance by the Architect. [The architect may wish to insert: "This paragraph shall be effective only to the extent that the Architect is adjudged by a court to be negligent hereunder."]

.2 The Architect hereby agrees to maintain the insurance described in Paragraph 2.5.2 hereof during the term hereof. If the Architect fails to furnish and maintain the insurance required by Paragraph 2.5.2, the Owner may purchase such insurance on behalf of the Architect, and the Architect shall pay the cost hereof to the Owner upon demand and shall furnish to the Owner any information needed to obtain such insurance.

Another provision for insurance:

1. The Architect agrees to maintain professional liability insurance in the amount of $ _____ with a deductible not to exceed $ _____ for a period from the date of this Agreement until thirty-six (36) months after Final Completion.

2. The Architect shall purchase and maintain throughout the duration of this Agreement workers' compensation insurance and employers liability insurance with a company to cover all employees engaged in services under the Agreement and in form satisfactory to the Owner in the maximum statutory liability amount to cover all employees engaged in work on the Project, naming the Owner as additional insured.

3. The Architect shall purchase and maintain throughout the duration of this Agreement general public liability insurance with a company and in form satisfactory to the Owner in the amount of $ _____ for each occurrence, naming the Owner as an additional insured. Said policies shall include contractual liability coverage and comprehensive automobile liability covering all owned, hired, and nonowned vehicles.

4. Each of the insurance policies described in paragraphs .1, .2, and .3 shall provide that insurance may not be canceled or nonrenewed without thirty (30) days' prior written notice to the Owner.

5. The Architect shall provide the Owner with evidence of the above insurance prior to execution of this Agreement. At the Owner's request, the Architect shall provide the Owner with full copies of the insurance policies required under paragraphs .1, .2, and .3.

Another insurance provision reads:

Architect accepts full responsibility for the maintenance and payment of all premiums on worker's compensation insurance and for all other sums that become due for unemployment insurance, health insurance, and/or any other insurance that may now or hereafter be required by any governmental entity with respect to persons employed by Architect for performance of services pursuant to this Agreement. Architect shall maintain at its own cost and expense policies of insurance against all liability arising out of the acts, neglect, errors, or omissions of Architect, its agents, and employees with respect to the Project. Architect shall cause its subcontractors and consultants likewise to maintain such insurance with respect to the services performed by them for the Project. All insurance policies so maintained shall be issued by insurance companies reasonably acceptable to Owner with one or more Certificates of Insurance evidencing such coverage. Each such Certificate shall provide that should the policy or policies described therein be canceled before the expiration date thereof, the issuing company shall endeavor to give to Owner by mail at least thirty (30) days' written notice prior to such cancellation. Architect agrees to furnish to Owner, at Owner's request, such information as may reasonably be requested to satisfy Owner of compliance with the provisions of this Paragraph. Notwithstanding the provisions of this Paragraph, Owner acknowledges that Architect's insurance coverage for the foregoing liabilities is limited to $500,000 per occurrence. Should Owner require Architect to procure additional or greater limits of coverage, Owner agrees to pay the costs of such additional or greater limits of coverage.

Other provisions for insurance:

A. CONTRACTUAL LIABILITY

The Comprehensive General Liability and Professional Liability coverage shall also protect the Architect for contractual liability in at least the same limits of liability against claims for errors, omissions, and negligent acts of the Architect and its employees or agents which may arise because of the indemnity or other contractual agreement contained within this Agreement.

B. CERTIFICATE OF INSURANCE

Prior to commencement of work under this Agreement, the Architect shall furnish to Owner certificates of insurance evidencing the above coverage by insurance companies authorized to do business under the laws of the State of _____(5). Said policies shall provide that they cannot be modified or canceled without at least thirty (30) days' prior written notice to the Owner. Owner reserves the right to require production of all insurance policies, or to require a statement from the Architect's insurance broker, attorney, or insurance representative that the policies provide the insurance required hereunder.

Another version:

2.5 INSURANCE. To protect against liability, loss and/or expense arising in connection with the performance of services described under this Agreement, the Architect shall obtain and maintain in force during the entire period of this Agreement without interruption, at its own expense, the following stated insurance from insurance companies authorized to do business in the State of _____, in a form and content satisfactory to the Owner, and rated "A-" or better with a financial size category of (a) Class X or larger where the applicable Construction Budget is $1,000,000 or greater; or (b) Class VII or larger where the applicable Construction Budget is under $1,000,000. All said ratings and financial size categories shall be as published by A.M. Best Company at the time this Agreement is executed. The Architect shall require all consultants to have and maintain similarly required policies. All of the following listed insurance coverages shall be provided by the Architect:

2.5.1 Architect's Professional Liability Insurance. The Architect shall maintain a policy on a claims-made basis, annual aggregate policy limit based on the following chart, unless modified in writing by the parties.

Construction Budget	Minimum Liability Coverage
$50,000,000 and above	$2,000,000 per claim, $4,000,000 aggregate
$25,000,000 and above, but under $50,000,000	$2,000,000 per claim, $2,000,000 aggregate
$1,500,000 and above but under $25,000,000	$1,000,000 per claim, $1,000,000 aggregate
Under $1,500,000	$ 500,000 per claim, $500,000 aggregate

2.5.2 The Owner reserves the right to require additional coverage from that stated in the chart herein above, at the Owner's expense for the additional coverage portion only. Owner also reserves the right to require project specific insurance, and if such right has been exercised it shall be indicated as an exhibit to this Agreement. Unless project specific insurance is required by the Owner, the coverage may be written under a practice policy with limits applicable to all projects undertaken by the firm but must be maintained in force for the discovery of claims for a period of three (3) years after the date final payment is made to the Architect under this Agreement. All policies provided by the Architect must contain a "retroactive" or "prior-acts" date which precedes the earlier of, the date of this Agreement or the commencement of the Architect's services hereunder. The Architect's policy must also include contractual liability coverage applicable to the indemnity provision of this Agreement for those portions of the indemnity provisions that are insured under the Architect's policy and in accordance with this Agreement, including the attachments hereto.

2.5.3 Commercial General Liability Insurance. Architect shall provide, at its own expense, Commercial General Liability Insurance, on an "occurrence basis," including insurance for premises and operations, independent consultants, projects/completed operations, and contractual liability coverage including specifically designating the indemnity provisions of this Agreement as an insured contract on the Certificate of Insurance. Such Commercial General

Liability Insurance must provide coverage for explosion, collapse and underground hazards. Insurance required by this paragraph shall provide for limits that are not less than the following:

$2,000,000 General Aggregate

$2,000,000 Products-Completed Operations Aggregate

$1,000,000 Personal and Advertising Injury

$1,000,000 Each Occurrence

$ 50,000 Fire Damage (any one fire)

$ 5,000 Medical Expense (any one person)

2.5.4 Workers' Compensation Insurance and Employers' Liability Insurance. Worker's Compensation Insurance shall cover full liability under the Worker's Compensation Laws of the jurisdiction in which the Project is located at the statutory limits required by said jurisdiction's laws. Employer's Liability Insurance shall provide the following limits of liability: $100,000 for each accident; $500,000 for Disease-Policy Limit; and $100,000 for Disease-Each Employee.

2.5.5 Automobile. Automobile liability insurance for claims arising from the ownership, maintenance, or use of a motor vehicle. The insurance shall cover all owned, non-owned, and hired automobiles used in connection with the work, with the following minimum limits of liability: $1,000,000—Combined Single Limit Bodily Injury and Property Damage Per Occurrence.

2.5.6 Valuable Papers and Records Coverage and Electronic Data Processing (Data and Media) Coverage. The Architect and all consultants of the Architect shall provide coverage for the physical loss of or destruction to their work product including drawings, specifications and electronic data and media.

2.5.7 Certificates. Before this Agreement is executed, the Architect shall submit certificates in form and substance satisfactory to the Owner as evidence of the insurance requirements of this Article. Such certificates shall contain provisions that no cancellation, or non-renewal shall become effective except upon thirty (30) days prior written notice by US Mail to Owner as evidenced by return receipt, certified mail sent to Owner. The Architect shall notify the Owner within thirty (30) days of any claim(s) against the Architect which singly or in the aggregate exceed 20% of the applicable required insured limits and the Architect shall, if requested by Owner, use its best efforts to reinstate the policy within the original limits and at a reasonable cost. The Owner shall be named as an insured party, as primary coverage and not contributing, on all the insurance policies required by this Article except the professional liability and workers' compensation policies. The Owner reserves the right to request the Architect to provide a loss report from its insurance carrier.

2.5.8 Maintain Throughout Agreement Term. The Architect agrees to maintain all insurance required under this Agreement during the required term. If the Architect fails to furnish and maintain said required insurance, the Owner may purchase such insurance on behalf of the Architect, and the Architect shall pay the cost thereof to the Owner upon demand and shall furnish to the Owner any information needed to obtain such insurance.

2.5.9 Waivers of Subrogation. All policies required, except Practice Professional Liability Insurance and Workers Compensation Insurance, shall be endorsed to include waivers of subrogation in favor of the Owner.

2.5.10 Excess Coverages. Any type of insurance or any increase of limits of liability not described in this Agreement which the Architect requires for its own protection or on account of any statute, rule or regulation, shall be its own responsibility and at its own expense.

2.5.11 Not Relieve Architect of Liability. The carrying of any insurance required by this Agreement shall in no way be interpreted as relieving the Architect of any other responsibility or liability under this Agreement or any applicable law, statute, rule, regulation or order.

2.5.12 Architect Compliance with Policies. Architect shall not violate or knowingly permit to be violated any of the provisions of the policies on insurance required under this Agreement.

§ 3.5   Architect's Basic Service: ¶ 3.1.1

Another provision:

3.1.1 The Architect's Basic Services shall consist of those services identified in this Agreement. Any services not specifically set forth in this Agreement, including engineering services, shall, if requested by the Owner, be provided as an Additional Service.

§ 3.6   Architect's Reliance on Owner-Furnished Information: ¶ 3.1.2

Because the architect will be relying on information furnished by the owner and consultants retained by the owner, the following provision gives the architect greater protection:

3.1.2 The Architect shall be entitled to rely on the accuracy and completeness of services and information furnished by the Owner. The Owner recognizes that it is impossible for the Architect to ensure the accuracy, completeness, and sufficiency of such information. The Architect shall, however, provide prompt written notice to the Owner if the Architect becomes aware of any errors, omissions, or inconsistencies in such services or information. The Owner agrees, to the fullest extent permitted by law, to indemnify and hold the Architect, the Architect's consultants, and others providing services through the Architect harmless from any claim, liability, or cost (including reasonable attorneys' fees and costs of defense) for injury or loss arising or allegedly arising from errors, omissions, or inaccuracies in documents or other information provided by the Owner to the Architect.

§ 3.7   Architect's Schedule: ¶ 3.1.3

This section relates to a schedule that the architect prepares for the services that the architect is required to provide. This schedule should, of course, permit the documents to be prepared in time to allow for bidding and construction of the project within the owner's overall schedule.

3.1.3 As soon as practicable after the date of this Agreement, the Architect shall submit for the Owner's approval a schedule for the performance of the Architect's services. The schedule shall cover the period of time from the date of the schedule until completion of the Construction Documents Phase. The schedule shall include allowances for periods of time required for the Owner's review and for the performance of the Owner's consultants. This schedule shall serve as a general guide for the Architect's services, but the parties acknowledge that the Architect's services will be influenced by many factors outside the Architect's control and, therefore, the Architect cannot guarantee that its services will be performed in substantial conformance with this schedule. The Architect will, however, exercise reasonable efforts to maintain the schedule.

§ 3.8   Architect's Review of Codes: ¶ 3.1.5

The following alternatives represent two extreme positions for this clause. The AIA language is considered the most neutral provision.

A provision favorable to the architect:

3.1.5 The Architect shall review laws, codes, and regulations applicable to the Architect's services. However, the Owner acknowledges that the Architect does not warrant that the Architect's analysis of such laws is accurate or will be followed by governmental authorities. Therefore, in the event that the Owner suffers any damages of any sort related to the Architect's failure to comply with any law, code or regulation, the Architect's liability to the Owner is limited to such damages or the Architect's fees paid by the Owner, whichever is less. To the extent the Owner is, or should be, aware of any law, code or regulation affecting the Project, the Owner shall examine the work of the Architect from time to time and promptly notify the Architect of any variation between such work and the laws, codes or regulations affecting the Project, and the Architect shall not be liable to the Owner for any damages resulting from any failure by the Owner to abide by this provision. The Architect shall respond in the design of the Project to requirements imposed by governmental authorities having jurisdiction over the Project.

A provision favorable to the owner:

3.1.5 The Architect represents that it is familiar with, and experienced in the interpretation and implementation of, laws, codes and regulations applicable to the Architect's services and the Project in general. Accordingly, the Architect shall be subject to the highest standard of care in its execution of the work of this Project and as applicable to such laws, codes and regulations. The Architect shall respond in the design of the Project to requirements imposed by governmental authorities having jurisdiction over the Project and shall comply with all directives of such authorities. Where necessary for the successful completion of the Project, the Architect shall meet with all appropriate governmental officials in the various design stages hereunder to apprise such officials of the specifics of the Project in order to avoid any deviations from such laws, codes and regulations and in order to expedite all permitting procedures. The Architect acknowledges that Owner is relying on the Architect's expertise in laws, codes and regulations concerning projects of this type. The Architect warrants that all work performed by the Architect, any consultants of the Architect, and any other party being coordinated by the Architect for this Project, shall fully comply with all such laws, codes and regulations. In the event that the Project fails to comply with any law, code or regulation, and such failure is not due to the Contractor's failure to comply with the Contract Documents, then the Architect shall be responsible to the Owner for any damages, including costs of replacement, lost income and all other direct and indirect costs associated with such failure.

Another provision favorable to the owner:

3.1.5 Compliance with Laws. Architect will give all notices and comply with all federal, state and local laws, ordinances, rules, regulations, codes, standards and orders governing or relating to the performance of the Services (collectively, "Laws") including, but not limited to, Laws regarding tax, social security, unemployment compensation, workers' compensation, equal employment opportunity, minority business enterprise, women's business enterprise, disadvantaged business enterprise, worker safety and health (including, without limitation, those established by or pursuant to the Occupational Safety and Health Act of 1970, as amended), accessibility (including, without limitation, those established by or pursuant to the FHA Act and Americans with Disabilities Act), and protection of the environment. Unless Owner otherwise agrees in writing, Architect will secure and pay for all permits, fees, taxes, licenses and inspections necessary for the proper execution and completion of the Services.

§ 3.9   Governmental Approvals: ¶ 3.1.6

Because no one can guarantee that approval of governmental agencies will be obtained, the architect may want to insert language to protect itself. The following additional language could be added to clarify this matter:

3.1.6.1 The Architect shall assist the Owner in connection with the Owner's responsibility for filing documents for zoning approvals and related matters. The Owner acknowledges, however, that the approval of governmental bodies having jurisdiction over the Project is based, in part, on nonobjective criteria. Such criteria include but are not limited to local political conditions, labor and union issues, community opposition, and similar matters. Therefore, the Architect cannot and does not warrant or guarantee any particular result or that the Project will receive the approvals that may be required. The Owner shall have no course of action against the Architect arising out of the failure of any governmental body to grant any approval, or for the revocation of any approval previously given.

Another version:

3.1.6 The Architect shall become familiar with community interests and standards that relate to the Project with the intent to design a structure that will not conflict with the interest of the community. To that end, the Architect will assist the Owner in preparing and filing documents necessary to obtain any necessary governmental approvals, in making reports and presentations to governmental bodies, agencies, community groups, and others with an interest in the Project. Any time spent by the Architect in such endeavors shall be reimbursed at the Architect's standard hourly rate set forth below.

§ 3.10   Existing Structures

An alternative clause that could be used here if the project includes work on an existing structure is:

3.1.7 The Architect shall not be responsible for verifying the condition of an existing structure, equipment, or appliance as part of Basic Services unless such verification can be made by simple visual observation. Any further investigation, if authorized or requested by the Owner, shall be provided as an Additional Service. If, after the Contract Documents are prepared, it appears from uncovering parts or portions of an existing structure that the plans and/or specifications must be altered to conform to previously hidden conditions, all such work shall be performed by the Architect as an Additional Service.

§ 3.11   Miscellaneous Provisions for Article 3.1

Other possible provisions for services in Article 3.1 are given below:

The Architect shall assist the Owner, as requested, in selecting finish materials and colors.

The Architect shall give full and prompt attention and shall recommend to the Owner methods for resolving any claims or controversies that arise during the course of construction of the Project. In the event of any proceeding to resolve any claim that involved any act or omission of the Architect, the Architect shall be present and shall participate in such proceedings.

The Architect shall give timely notice to the Owner for any meetings the Architect feels necessary in connection with this Project with tenants, utility companies, or city, state, or other regulatory agencies. Scheduling of such meetings is to be done by the Owner. In general, all contracts with such parties will be maintained by Owner.

The Architect shall cooperate with any interior designer or other professional consultant employed by the Owner in connection with the Project.

The Architect shall assist the Owner in any negotiations with governing authorities necessary to obtain temporary and permanent Certificates of Occupancy.

The Architect shall collaborate with the Owner in developing a construction testing program and in negotiating and awarding any professional service contract for such testing as required by the Owner.

The Architect shall provide any necessary assistance in the utilization of: any equipment or system such as initial start-up or testing, adjusting, and balancing, and preparation of operation and maintenance manuals in a form satisfactory to the Owner; the design parameters of the exterior enclosure, and the structural, electrical, vertical transportation, mechanical, heating, ventilation, air conditioning, building automation, building security, fire and life safety, and other systems of the Project; and orientation of personnel for operation and maintenance.

During the last week of the tenth (10th) month following the Completion Date, the Architect shall conduct inspections, witness tests, and conduct investigations of the Work in order to determine any discoverable deficiencies and defects in the work and to assist the Owner in obtaining the full benefits and privileges of all guarantees and warranties. The Architect shall file a written report with the Owner regarding the results of such inspections, witnessing, and investigations prior to the end of the eleventh (11th) calendar month following the Completion Date.

After the Completion Date, the Architect shall provide assistance to the Owner of up to 200 hours of time by the Architect or its Consultants in connection with troubleshooting, analysis, evaluation, and consultation in the utilization of any system or equipment installed in the Project.

The Architect shall incorporate a requirement within the plans and specifications that the Construction Manager and all subcontractors accurately and completely mark the sepias of the working drawings, and the large- and fullscale detail drawings, and the specifications to show field changes thereon and to describe in sufficient detail any deviation so as to evidence the "as-built" construction of the Improvements. The Architect shall review such drawings and specifications and promptly notify the Owner, the Construction Manager, and the applicable subcontractor of any deficiencies observed by the Architect.

Eleven months after Substantial Completion, the Architect shall participate in an inspection of the Project to identify any situation that may need further attention from the Owner or Contractor.

Upon completion of the Project and provided that construction phase professional services are included in this Agreement, one set of reproducible mylar record drawings shall be submitted to Owner. Prior to submittal of such reproducibles, Architect shall record all alterations that have been made in the Project during the construction phase so that the reproducibles will be a final record of the work as built. The Architect shall have the right to rely on information provided by the Contractor for this purpose, unless the Architect has actual knowledge that any such information is incorrect. Such reproducible drawings shall be labeled as "Record Drawings." Architect shall receive compensation for other supplemental forms of record drawings (e.g., AutoCAD diskettes or the like) requested by the Owner in addition to mylar as additional services and/or reimbursable expenses.

The Architect shall prepare the Contract Documents in accordance with all applicable codes, standards, and regulations. All Contract Documents shall provide that the Contractor shall comply with equal opportunity and prevailing wage requirements set forth by statute. No Contract Document shall contain an arbitration clause nor contain any interest clause for late payment unless said payment is made more than 60 days after approval of the Architect. All Contractors shall be required to furnish a 100% Performance, Labor, and Material Payment Bond with a corporate surety. All contracts shall be let only after an advertisement for public bid. The Architect shall not directly or indirectly own or control any Contractor.

The Architect agrees to cooperate with any consultant, construction supervisor, or superintendent retained by the Owner. The Architect agrees to complete and apply for any necessary building permit as part of its Basic Services.

The Architect shall be liable to Owner for reasonable expenses incurred by Owner, including court costs, as the result of the Architect's nonperformance or delay in the performance of the services required by the terms of this Agreement and to the extent not caused by persons or events beyond its control, including approval of governmental agencies and bodies having jurisdiction. In order for the Architect to complete its services within the time scheduled herein, the Architect, without additional compensation, may be required to increase the number of shifts, or overtime operations, days of work, or all of them.

Notice of Delay. Architect shall, immediately upon ascertainment, notify Owner of any delay in—(i) the preparation and/or production of any of Architect's documents hereunder, (ii) the performance by Contractor, (iii) Architect's Services, or (iv) in connection with any matter attended to by Architect or with which Architect is familiar (whether or not as the result of an act or omission of another) which would affect or delay the schedule. Architect shall consult and advise the Owner in connection with any such delay and its effect on the schedule and shall take such action on Owner's behalf as Owner may request in accordance with the terms and conditions of this Agreement.

Delays. If the work of Architect is delayed any time by reason of acts of God, war, civil commotion, riots, strikes, picketing or other labor disputes, damage to the Project by reason of fire or other casualty or other causes beyond the reasonable control of Architect (including failure of Owner or Contractor to respond timely) and not due to the willful or negligent act or omission, financial inability, or default of Architect, or events reasonably foreseeable to Architect then upon the written request of Architect to Owner the time for completion under the Design Schedule shall be appropriately extended by the number of working days of delay actually so caused. Provided, however, no such extension shall be made or allowed unless a written request therefor is made within ten (10) calendar days after the delay. In the case of a continuing cause of delay only one request shall be necessary, which request shall affirmatively state the delay is a continuing one and the reasons therefor. All delay requests or notices hereunder shall describe the nature of the delay and estimate the probable effect of such delay on the progress of the Work. The effect of any delay shall also be shown on the latest Design Schedule.

Commencement and Completion of Services. (a) Upon the request of Owner and prior to the commencement of the Services, Architect will prepare and deliver to Owner, for Owner's review and approval, a schedule for performance of the Services. The schedule will set forth the dates for the commencement, performance, sequencing and completion of the various stages of the Services and may be revised if and as required by the conditions of the Services, subject to Owner's prior written approval in accordance with this Agreement. (b) Upon execution of this Agreement and notification of authorization to proceed, Architect will promptly commence the Services and thereafter prosecute the Services diligently and continuously to completion, consistent with economy and the best interests of Owner, all in strict accordance with the terms of this Agreement. Upon written notification from Owner to suspend the Services, Architect shall immediately suspend the Services on any phase or item of the Services designated by Owner. Within a ninety (90) day period after any such suspension by Owner, Architect shall resume performance of the Services as directed by Owner. If there is a suspension of Services which extends beyond ninety (90) days, Architect shall resume performance of the Services as directed by Owner; provided, however, that Owner and Architect may negotiate such reasonably necessary modifications to the terms and conditions of this Agreement necessitated by such extension. All Services will be finally and fully completed in accordance with the time periods set forth in this Agreement. Architect understands and acknowledges that the time of completion takes into account the possibility of delays which may occur due to inclement weather, bidding or negotiations, market conditions, governmental action or inaction and all other facts or circumstances except those specifically listed elsewhere in this Agreement. (c) If completion of the Services is to be delayed for any cause beyond Architect's control that is not reasonably foreseeable or anticipated and is not due to any fault of Architect, including only (i) neglect, unreasonable delay or default of Owner, and (ii) damage or delay which arises by reason of fire, storm, lightning, flood, earthquake, tidal wave, surface or subsurface water, mob violence, vandalism, acts of war or other comparable casualty that affects Architect's performance of the Services or of changes in Laws or acts of any governmental authority having jurisdiction over the Services, then the time for completion will be extended for a period equivalent to the time lost by reason of any or all such causes. Notwithstanding the foregoing, as a condition precedent to any such extension, Architect shall deliver a written claim to Owner for such an extension within ten (10) business days after the occurrence of such cause specifying the nature of the occurrence and describing possible alternatives for making up the time, and the costs and other pertinent factors attributable to each alternative. (d) Notwithstanding any provision contained herein to the contrary, if Architect fails to (i) correct any defect in the Services, (ii) prosecute the Services diligently, (iii) cause payment to be made when due to any governmental authority, pension or other fund, or employee, (iv) comply with any Laws, or (v) otherwise breaches any terms or provisions of this Agreement, Owner, without prejudice to any right or remedy and after giving Architect seven (7) days' written notice, may terminate this Agreement and take over control of the Services unless the same shall have been fully remedied within such seven (7) day period.

§ 3.12   Schematic Design Phase Services: ¶ 3.2

If the owner wants the architect to provide programming, the following language could be used:

3.2.1 The Architect shall review with the Owner's designated personnel the owner's current program needs, current budget, and anticipated project schedule.

3.2.2 The Architect shall inventory and analyze the existing building, including mechanical/electrical systems and utilities, computer and communications facilities, furniture, and other systems, to develop facility design criteria.

3.2.2.1 The Architect shall submit the facility design criteria to the Owner's representatives for approval. Based on approved facility design criteria, the Architect shall develop preliminary design concepts for the existing building, including building systems, for review and approval by Owner. Such design concepts may be limited to block diagrams or bubble diagrams illustrating the relationships among the various elements.

3.2.2.2 The Architect shall prepare for approval by Owner a program analysis report establishing program needs, design concepts, building system description, project budget, and schedule.

3.2.3 Based on the approved program analysis report, the Architect shall prepare alternative approaches to the design and construction of the project. At the Owner's request, the Architect shall furnish up to three distinct designs for the Project. If the Owner desires more than three designs, additional designs shall be furnished by the Architect as an Additional Service.

Another version:

3.2 SCHEMATIC DESIGN PHASE

During the Schematic Design Phase, the Architect shall:

3.2.1 Review the program furnished by the Owner and provide architectural and engineering analysis of the Owner's requirements, access, zoning, planning, building code requirements, physical characteristics of the Site, traffic and utility requirements, and other information; and of applicable laws, statutes, ordinances, and regulations.

3.2.2 Based on the mutually agreed upon program and Budget Amount, prepare a preliminary study and report illustrating the scale and relationship of all components of the Project and possible future development of the Site, and prepare preliminary design sketches of the Project, outline the nature of the structure, exterior and basic building systems, which shall consist of suitable schematic drawings, layouts, concept drawings, flow diagrams, and such reports and studies as necessary for the review and written approval of the Owner.

3.2.3 Work and consult with the Owner and the Construction Manager on the preliminary engineering analysis for the various proposed systems of the Project to determine the most cost-effective, basic building systems, together with setting an achievable energy budget for the operation of the Project.

3.2.4 Furnish the Owner such copies as the Owner reasonably requests of the documents prepared pursuant to this ¶ 3.2.

Another:

3.2 SCHEMATIC DESIGN PHASE

The Architect shall provide the following services during the Schematic Design Phase:

3.2.1 Examine the site and the surrounding areas and community.

3.2.2 Review all applicable codes, ordinances, laws, rules, orders, regulations, statutes, and otherwise that may affect the Project; review the Owner's program for conformance with all such codes, ordinances, laws, rules, orders, regulations, statutes, and otherwise.

3.2.3 Review the availability of utilities, including but not limited to electrical, storm and sanitary drainage, and gas and water services.

3.2.4 Prepare alternative studies to determine density, scale, and relationship of the various elements on the site.

3.2.5 Make preliminary recommendations for structural, mechanical, and electrical systems for the Project.

3.2.6 Prepare an engineering analysis based on soil and site conditions to evaluate alternative foundation solutions.

3.2.7 Prepare an analysis of traffic flows and patterns and develop proposed traffic entrance and egress locations and parking designs; and prepare similar studies for pedestrian traffic.

3.2.8 Prepare schematic design drawings, including floor plans, elevations and sections showing floor-to-floor heights, wall sections, materials, types of vertical transportation, mechanical and structural systems; site plans and colored renderings; and site models to assist the Owner in understanding the Architect's design.

3.2.9 Prepare a preliminary estimate of construction costs.

3.2.10 Prepare environmental impact statement if required.

3.2.11 Prepare a time schedule for the work of the Architect, the Architect's consultants, bidding, and construction. When approved, this schedule shall govern the performance of the Architect. This schedule may be extended by approval of the Owner, but only if due to causes not in the control of the Architect.

3.2.12 Furnish the Owner with three copies of all documents prepared hereunder. Additional copies will be furnished by the Architect at cost.

Another alternative is:

3.2 SCHEMATIC DESIGN PHASE

During the Schematic Design Phase, the Architect shall:

3.2.1 Develop requirement guidelines for materials and finishes, security, and computer preparation. Develop detailed thermal, luminance, and acoustical performance requirements on a room-by-room basis to be used as performance specifications.

3.2.2 Develop the general description of facility program requirements, to include location criteria, parking and access, acceptable space arrangements and relationships, critical functional and aesthetic expectations, and the relationship with the design/build process of project advancement.

3.2.3 Prepare preliminary CPM schedule identifying major activities, including all Owner-required milestones and constraints. On a monthly basis, review preliminary CPM schedule and provide report to Owner.

3.2.4 Prepare budget cost estimate for anticipated furniture and shelving to meet program requirements from list compiled and provided by the Owner of recycled existing furnishing, and recycled and refinished existing furnishing, and new furnishing compiled from program data sheets. No specifications nor suggested manufacturers and their model numbers nor bidding documents are to be provided.

3.2.5 Present to the Owner the schematic design documents. Obtain approval of documents from Owner.

If the owner and architect agree that the architect will not furnish cost estimates, the parties should include language similar to the following:

[Insert the following provision in Article 2:]

The Owner agrees and acknowledges that the Architect will have no responsibility for providing any cost or budget estimates for the Work. The Architect shall not bear any responsibility or liability arising out of, or related to, any such estimates or construction costs. The Owner further acknowledges that the actual construction costs may vary substantially from the Owner's estimate or budget, but that the Architect cannot and does not warrant or guarantee in any manner whatsoever that these costs will be consistent with the Owner's estimate or budget. Any estimate prepared by the Architect, or any assistance given by the Architect in the preparation of such estimates by others does not operate as a waiver of this provision.

Note that the above modifications require that other provisions in the standard form will also change.

The owner may want to place additional responsibility for the budget on the architect in this phase with the following:

3.2.6 If the adjusted preliminary estimate of Construction Cost at this phase exceeds the preliminary construction cost budget previously established, Architect shall recommend to Owner items of possible cost reduction to the scope of the Project to bring it within such budget. The Owner may choose to adopt a new budget at this time, but this shall be done in writing. After Owner's written approval of these cost reductions, they will be incorporated into the design development phase.

To establish benchmarks for the architect's performance, the owner may want to add a paragraph such as this for each phase of the work:

[Replace Subparagraph 3.2.6 with:]

3.2.6 The Architect shall furnish the Schematic Design Documents and preliminary estimate of Construction Cost to Owner within 120 days after the date of this Agreement.

Here is another alternative:

3.2.1 The Architect shall provide Schematic Design Documents based on the mutually agreed-upon program, schedule, and budget for the Cost of the Work. The documents shall establish the conceptual design of the Project, generally illustrating the scale and relationship of the Project components. The Schematic Design Documents shall, at the Architect's discretion, include a conceptual site plan, if appropriate, and preliminary building plans, sections, and elevations. The Schematic Design Documents may also include study models, perspective sketches, electronic modeling, or combinations of these media. Preliminary selections of major building systems and construction materials shall be noted on the drawings or described in writing. The parties agree, however, that failure by the Architect to include any of the enumerated items shall not be considered a breach of this Agreement. If the Owner fails to timely object to the Schematic Design Documents submitted by the Architect, the Owner will be deemed to have accepted and approved these documents.

A more comprehensive alternative:

3.2 Schematic Design Phase. After receiving the information to be furnished by Owner pursuant to Section 2.2.1 hereof and after receiving Owner's written authorization to proceed, Architect shall proceed with the Schematic Design Phase in accordance with this Section 3.2.

3.2.1 Architect and Architect's consultants shall immediately review the program previously developed by or for Owner, and confirm its understanding of the program requirements with Owner. Architect shall work with Owner during the Schematic Design phase to further refine and define the program and Architect shall inspect the Project site and existing facilities, and provide a preliminary evaluation of the program and Project budget requirements, each in terms of the other. Owner shall furnish such existing information regarding utility services and site features, including existing construction, related to the Project as are available from Owner's records. Architect shall recommend which information should be relied upon and which should be subject to field verification. In addition, Architect shall determine, and notify Owner in writing, what additional information is required and the dates such information is needed to properly coordinate all phases of the Project. Architect shall review with Owner and Contractor site use and improvements, selection of materials, building systems, and equipment. There will be weekly meetings between Owner, Contractor and others during this Phase.

3.2.2 Based on Owner's program and Project Budget requirements, Architect shall prepare, for review and approval by Owner, Schematic Design Documents consisting of drawings and other documents illustrating the scale and relationship of project components. Architect shall first prepare drafts of Schematic Design Documents and review them with Owner's Project management team. The Schematic Design Documents shall include, without limitation, the following:

3.2.2.1 Plot plan indicating the proposed location of the building(s); major improvements such as proposed roads, boundary of the project site, parking areas, walks, plazas, and location of exterior utilities and service lines.

3.2.2.2 Floor plans showing all rooms and areas, entrances, exits, stairways, elevators, circulation corridors, toilet rooms, major mechanical and electrical areas. A tabulation of areas, including net and gross areas of the various parts of the Project shall be included.

3.2.2.3 Building elevations showing, by block outline and breaks, the various building masses and how they coincide with the floor plans, including colors and typical fenestration pattern.

3.2.2.4 Building sections showing floor to floor dimensions sufficient to indicate interface with existing structures.

3.2.2.5 Detailed code analysis including identifying building construction type, required egress units, occupancy, smoke and fire separations, maximum travel distances, and wall and building separations. This shall also include an analysis of zoning regulations applicable to the Project, including parking requirements, FAR limits, site coverage, building height and daylight planes.

3.2.2.6 A design schedule, addressing major elements of the design, including anticipated dates for the design reviews, Owner approvals and permit applications. The schedule shall identify any required phasing of the Work. The schedule shall also identify long lead items.

3.2.3 Architect shall propose and discuss with Owner a range of possible deduction alternatives which shall maximize program content and describe their impact on the Project sufficient to decrease the Statement of Probable Construction Cost by at least 10%.

3.2.4 Architect shall review with Owner alternative approaches for design and construction of the Project to permit Owner to determine the most economical design consistent with the requirements of the Project.

3.2.5 Architect shall make such changes and revisions in the Schematic Design Documents and provide such drawings, reproductions and supporting data as may be required or necessary to obtain the approval of Owner and remain within the Project Budget.

3.2.6 Architect shall, in a timely manner, provide architectural drawings, narrative description and other pertinent data to, and at the request of Owner, consult with appropriate governmental agencies (e.g. health, fire, building inspection) regarding compliance of the Schematic Design Documents with all Applicable Laws and shall make any supplementary or clarifying drawings or specifications as may be required in order to obtain all such governmental approvals or authorizations.

3.2.7 Architect and, if necessary, its Consultants, shall attend all meetings to obtain necessary Project approvals, including attendance at zoning boards, architectural review boards, planning boards or commissions, and other governmental meetings and hearings.

3.2.8 Architect shall involve Contractor whenever appropriate in the design process, and shall provide Schematic Design Documents for Contractor's review at intervals appropriate to the progress of the Schematic Design Phase.

3.2.9 Upon completion of the Schematic Design Phase, Architect shall provide the required number of sets of each of the drawings, a general narrative description of systems and materials and other documents approved by Owner in order for Contractor to prepare Contractor's Cost Estimate.

3.2.10 Architect shall obtain Owner's written approval of the Schematic Design Documents, including the drawings, systems checklist, general description, the tabulation of areas, and the final Design Schedule. Architect shall not proceed with the Design Development Phase until it has received Owner's written instructions to proceed.

3.2.11 Architect shall submit the required number of copies of each of the Schematic Design Documents and the tabulation of areas for approval by Owner.

3.2.12 Architect shall prepare the Schematic Design Documents based on the Preliminary Project Budget. Based upon the Contractor's Cost Estimate, the

Schematic Design Documents, and programmatic considerations, Owner shall establish and notify Architect in writing of the Project Budget which will govern the design of the Project up through and including the Bidding Negotiation Phase. Architect is responsible for providing a design which can be built within the Project Budget, as determined by construction bids received by Owner.

§ 3.13   Design Development Phase Services: ¶ 3.3

The following alternative language for ¶ 3.3 expands the architect's responsibility relative to cost evaluations or alternate designs created during the schematic design phase and during this phase.

3.3 DESIGN DEVELOPMENT PHASE

Upon receipt of Owner's written authorization to implement the documents presented in the Schematic Design Phase and to proceed with the Design Development Phase, the Architect shall:

3.3.1 Assist the Owner and the Construction Manager in developing and preparing detailed analysis of the long-term cost effectiveness of alternative design choices for the systems described in the Schematic Design Documents. The selection of any particular system shall not be considered as firm until the full interrelationship of all systems is fully approved and accepted by the Owner.

3.3.2 Prepare, from the approved Schematic Design Studies, the Design Development Documents consisting of design criteria, drawings, outline specifications, and other documents to establish and describe the size and character of the entire Project and as to architectural, structural, mechanical and electrical systems, materials, landscaping, and such other essentials as may be appropriate, and submit those documents for approval by Owner. The Design Development Documents shall be prepared after consultation with the Owner, Construction Manager, and such other consultants as may be retained by the Owner for the Project.

3.3.3 At such point during the Design Development Phase as the Architect deems appropriate, present for the Owner's approval floor plans and other necessary drawings and renderings depicting exterior treatment proposals and public areas for the Project. All proposals shall include the Architect's analysis of the cost of same.

3.3.4 Review any updated budget breakdown submitted by the Construction Manager or other parties retained by the Owner to indicate compliance with the Budget Amount through completion of the Design Development Phase and approve or reject same, indicating the reasons for any rejection.

3.3.5 Furnish the Owner such copies as the Owner reasonably requests of the documents prepared pursuant to this Paragraph 3.3.

Additional language might read as follows:

3.3.1.1 The documents to be submitted by the Architect for this phase shall include but not be limited to: plans of each floor; major building elevations; building sections; large-scale drawings of the building core, elevator, and escalator areas; mechanical areas; site plans; [revise to suit project]; and other drawings, as well as outline specifications and other documents.

3.3.3 If the adjusted preliminary estimate of Construction Cost at the end of this phase exceeds the preliminary Construction Cost budget previously established, Architect shall recommend to Owner items of possible cost reduction to the scope of the Project to bring it within such budget. The Owner may choose to adopt a new budget at this time, but this shall be done in writing. After Owner's written approval of these cost reductions, they will be incorporated into the Construction Documents phase.

Another alternative is as follows:

3.3.1 The Architect shall provide Design Development Documents based on the approved Schematic Design Documents and updated budget for the Cost of the Work. The Design Development Documents shall illustrate and describe the refinement of the design of the Project, establishing the scope, relationships, forms, size, and appearance of the Project by means of plans, sections and elevations, typical construction details, and equipment layouts. The Design Development Documents shall, at the Architect's option, include specifications that identify major materials and systems and establish in general their quality levels. The parties agree, however, that failure by the Architect to include any of the enumerated items shall not be considered a breach of this Agreement. If the Owner fails to timely object to the Design Development Documents submitted by the Architect, the Owner will be deemed to have accepted and approved these documents.

A more comprehensive alternative:

3.3 DESIGN DEVELOPMENT PHASE

3.3.1 After receiving written approval by Owner of the Schematic Design Documents and upon written authorization by Owner to proceed with the Design Development Phase, Architect shall develop the design of the Project in preparation for the Construction Documents Phase, fixing and describing in further detail the size and character of the entire Project as to architectural, structural, mechanical and electrical systems, space requirements, materials and such other elements as may be appropriate. The process shall be similar to the interactive process used in the Schematic Design Phase, involving weekly meetings between Owner, Architect and Contractor and meetings on special issues. The documents to be provided under this Phase shall incorporate those prepared by any Additional Consultants, if required, and shall include:

3.3.1.1 Architectural Drawings: (i) plot plan showing proposed roads, individual parking spaces, exterior utilities, sidewalks, other site improvements, grades, and drainage; (ii) floor plans, including roof, showing space assignments, sizes, and location of installed, fixed and movable equipment which affect the design of the spaces, and a tabulation of areas, including net and gross areas of various parts of the project. Floor plans should include doors, windows, fixed counters and shelving, utility systems outlets (electrical, telecommunications, mechanical, plumbing, computer, etc.) to facilitate furniture and equipment layout and interior design; (iii) building elevations indicating exterior design elements and features including fenestration, colors, materials, mechanical and electrical features appearing on walls, roofs, and adjacent areas; (iv) interior elevations to establish functional requirements, equipment, and systems locations, based on owner-approved system checklist; (v) typical building sections showing structural members, dimensions, accommodation of functional systems and other dimensions sufficient to indicate interface with existing structures; and (vi) typical wall sections sufficient to indicate materials, openings, and major features.

3.3.1.2 Structural Drawings: (i) plans and sections of sufficient clarity to show the extent and type of structures and foundations; (ii) details and notes to show that the structure conforms to the provisions of applicable laws and is otherwise sufficient; (iii) notes to indicate foundation and structural design complies with the requirements of soils analysis and applicable seismic requirements; (iv) notes on provisions to meet special requirements such as vibration and acoustical constraints; and (v) legible sheets showing the structural engineering calculations for all primary structural components of the Project.

3.3.1.3 Design Schedule updated monthly for any changes.

3.3.1.4 An updated code analysis, identifying any changes and ensuring compliance of the design with applicable code and zoning requirements.

3.3.2 Architect shall, in a timely manner, provide to Owner architectural drawings, narrative description, and other pertinent data prepared by Architect, and Architect and Owner shall review the documents with the governmental authorities having jurisdiction over the Project.

3.3.3 Architect shall involve Contractor whenever appropriate in the design process and shall provide Design Development Documents for Contractor's review, at intervals appropriate to the progress of the Design Development Phase. Architect shall review and comment on and work with Owner and Contractor in connection with Contractor's Cost Estimate.

3.3.4 Architect.shall provide Owner with monthly updates of the Design Schedule.

3.3.5 Upon completion of the Design Development Phase, Architect shall provide the required number of sets of drawings, outline specifications, and other documents for use by Owner and Contractor in preparing Contractor's Cost Estimate. If the Estimate exceeds the Project Budget, Owner may, in its discretion, require Architect to revise the design of the Project or applicable portion thereof so as to reduce the Project cost for the Work or applicable portion thereof to within the Project Budget therefor.

3.3.6 Architect shall submit the required number of copies of the Design Development Documents, the revised tabulation of areas for approval by Owner.

3.3.7 Architect shall not proceed to the Construction Documents Phase until Architect has secured Owner's written approval of the Design Development Documents, revised tabulation of areas, and the updated Design Schedule, and Owner's written instructions to so proceed.

§ 3.14   Construction Documents Phase Services: ¶ 3.4

The parties may want to insert specific provisions about what documents are to be provided at the end of the construction documents phase:

3.4.1.2 Upon completion of the Construction Documents Phase, the Architect shall deliver to the Owner as part of the contractual obligations unless otherwise noted: any data or field notes obtained pertaining to the Project, including data and field notes obtained from surveys and borings; one complete set of final approved Mylar tracings or Mylar polyester reproductions of contract plans required for the Project; one set of documents, other than plans, on size 8½ × 11 paper; ten sets of specifications printed on 8½ × 11 paper and suitably bound for bidding purposes; and copies of catalog information pertaining to any equipment incorporated in the Project.

A different version of language that could be substituted for ¶ 3.4 is:

3.4 CONSTRUCTION DOCUMENTS PHASE

Upon receiving the Owner's written authorization to implement the documents presented in the Design Development Phase and to proceed with the Construction Document Phase, the Architect shall:

3.4.1.1 Prepare, from the approved Design Development Documents, drawings and specifications in collaboration with the Construction Manager, setting forth in detail the requirements for the construction of the entire Project, including ceiling, lighting, and partition systems and common areas, landscaping and site work, and the placement and design of all utility services, together with necessary bidding information, and submit such documents for approval by the Owner. The Architect shall cooperate with the Owner and Construction

Manager in preparing bidding forms, instructions to bidders, and all Conditions of the Contract so as to design the project to meet the goals herein set forth and take advantage of all reasonable cost-saving suggestions.

3.4.1.2 Develop such Documents for competitive bidding at such point where bids can reasonably be expected to fall within ten percent (10%) of the estimated cost of the Project.

3.4.1.3 Throughout the Construction Documents Phase, provide the Owner with review and analysis of all costs submitted by the Owner and Construction Manager to achieve compliance with the Budget Amount, and review any list of proposed bidders submitted to the Architect and provide appropriate comments to the Owner with respect thereto.

3.4.1.4 Provide all requisite information and assist the Owner in filing the required documents for the approval of the Lender [or delete "Lender"] and all governmental authorities having jurisdiction over the Project, and shall be responsible for revising the drawings if necessary and obtaining all necessary approvals, permits, and certificates from those authorities on or before the date for completion of the Construction Document Phase as set forth in herein. All costs resulting from any such required revision are not reimbursable [or, All costs resulting from any such required revisions shall be considered Additional Services].

3.4.1.5 At the request of the Owner, assist the Owner's legal counsel in connection with its review of the Contract Documents for their legally related aspects.

3.4.1.6 Furnish the Owner such copies as the Owner reasonably requests of the documents prepared pursuant to this ¶ 3.4.1.

3.4.1.7 Furnish the Owner one (1) twenty-inch by thirty-inch (20 × 30 inch) professional color rendering of the approved exterior of the Project.

3.4.1.8 Before commencing construction, certify to the Owner and Lender that all plans, specifications, and drawings conform to all applicable governmental regulations, statutes, and ordinances; and that the improvements, when built in accordance therewith, shall likewise comply with all applicable governmental regulations, statutes, and ordinances. Said certification shall be in the form attached hereto as Exhibit "C" and by this reference incorporated herein.

Additional language might read as follows:

3.4.1.1 At the completion of the 50% and 90% Construction Documents phases, the Architect shall provide Owner with 5 sets of drawings and specifications for review and approval. Owner shall promptly review such drawings and specifications and return one marked-up set of drawings and specifications to the Architect for incorporation into the Architect's drawings and specifications.

3.4.1.2 Architect shall prepare construction cost estimates on a quantity or unit analysis basis, fully detailing unit prices and quantities in all divisions of work. Such construction cost estimates shall be initiated at the 50% stage and updated at the 90% stage.

3.4.1.3 If the Architect's construction cost estimates based on the construction documents exceed the latest approval construction cost budget, the Architect shall provide, either by using alternates to the bid or by reducing the scope of the Project, such cost savings as will satisfy Owner that the Project will be bid within the latest approved construction cost budget. All reductions or alternates shall be subject to approval in writing by Owner prior to bidding.

3.4.1.4 Based on the approved 90% construction documents, Architect shall prepare 100% completed construction documents ready for issuance to bidders as bid documents.

Another provision:

3.4.1.1 The Architect shall prepare and provide Construction Documents based on the approved Design Development Documents, or upon such other approval as the parties may deem necessary. The Construction Documents shall set forth in detail the requirements for construction of the Project so as to permit bidding and construction of the Work. The Construction Documents shall also be adequate for obtaining a permit for the Work from all required governmental bodies. The Construction Documents Phase shall not be deemed complete until necessary permits are obtained. The Construction Documents shall include Drawings and Specifications that establish in detail the quality levels of all materials and systems required for the Project.

3.4.1.2 During the development of the Construction Documents, the Architect shall prepare for the Owner bidding forms, instructions to bidders, the form of agreement between the Owner and the contractor, and such other documents as are reasonably necessary to obtain competitive bids for projects of this scope. As necessary, or as required by the Owner, the Architect shall prepare alternates for bidding.

3.4.1.3 The Architect shall furnish to the Owner sufficient copies of the Drawings and Specifications as may be required for bidding purposes and for permits. In addition, the Architect shall furnish the Owner with one set of Construction Documents in electronic form.

A more comprehensive alternative:

3.4.1 CONSTRUCTION DOCUMENTS PHASE

3.4.1.1 After receiving written approval by Owner of the Design Development Documents, and upon written authorization by Owner to proceed with the Construction Documents Phase, Architect shall provide the services and documents listed below in this Section 3.4.1. Architect and its Consultants, as appropriate, shall attend all regular and special meetings reasonably required by Owner, with Owner, Contractor and others to discuss and resolve specific issues. Regular Project meetings will be scheduled weekly and special meetings will be scheduled as needed in Owner's reasonable judgment.

3.4.1.2 The Construction Documents, which for purposes of this Agreement shall be defined as working drawings and specifications which set forth in detail the requirements for the construction of the Project, shall comply with all Applicable Laws, shall be sufficient for contractors to perform the Work without need for more drawings and details or change orders to correct or clarify Construction Documents, and shall include: (i) architectural drawings, details and specifications; (ii) structural plans, details, calculations and specifications; (iii) plans showing installation of major systems and equipment; (iv) door hardware and equipment specifications and schedules showing the sizes, locations and manufacturers of doors, hardware and equipment; (v) architectural specifications and finish schedules to set standards for the project and provide complete understanding by contractor, installers, fabricators and suppliers; (vi) detail drawings showing the design to be used in items such as special lighting, special partitions, cabinetwork, equipment, and for interior finishes such as wall coverings and floor coverings; (vii) Supplementary conditions and special conditions, as necessary; (viii) an updated code analysis, identifying any changes and ensuring compliance of the design with applicable code and zoning requirements; and (ix) an updated narrative sequence of operation for all building systems.

3.4.1.3 Prior to the completion of the Construction Documents Phase, Architect shall select finish materials and colors to be incorporated in the Work and shall prepare schedules of such materials and colors for Owner's approval. The schedule shall note any materials or finishes which are expected to involve extraordinary delays in delivery.

3.4.1.4 If so directed by Owner, Architect shall prepare separate packages of Construction Documents for site and/or foundation work so that such work may be commenced prior to other work.

3.4.1.5 At the time Construction Documents are 50% complete, Architect shall so notify Contractor and furnish documentation sufficient to allow Contractor to prepare Contractor's Cost Estimate.

3.4.1.6 The Plans and Specifications will include a requirement that Contractor and all subcontractors accurately and completely mark prints of the working drawings and specifications to show field changes therein. Architect shall immediately document and notify Owner of any field changes of which it is aware.

3.4.1.7 Architect shall provide Owner with monthly updates of the Design Schedule.

3.4.1.8 Architect shall provide, as an Additional Service, fully detailed and biddable alternatives as requested by Owner.

3.4.1.9 Architect, in a timely manner, shall provide architectural drawings, narrative description, and other pertinent data prepared by Architect to Owner, and shall file on behalf of Owner the required documents for design approval by the governmental authorities having jurisdiction over the Project. All Drawings and Specifications submitted for approval shall be stamped and certified by professionals licensed to practice in the state where the Project is located. Architect, as an Additional Service, shall appear at all meetings and hearings of governmental agencies necessary, in Owner's judgment, to obtain necessary approvals for the Project.

3.4.1.10 Architect shall assist Contractor and Owner in obtaining all required building permits and approvals and shall respond promptly and appropriately to all questions and comments, and make all required changes.

3.4.1.11 Architect shall consult and coordinate with Contractor and, as appropriate, other consultants on the Project team regarding any changes in requirements or in construction materials, systems or equipment as the Drawings and Specifications are developed. Final changes shall be reported so that Contractor can adjust its Contractor's Cost Estimate appropriately. Architect, as an Additional Service, shall review and comment on any procedures manual developed by Owner or Contractor.

3.4.1.12 Architect shall obtain Owner's written approval of each package of the Construction Documents, the final tabulation of areas and the updated Design Schedule. The approval process may take place in stages consistent with the division of the Work.

3.4.1.13 Architect shall provide Owner with the required number of sets of the Drawings and Specifications for review when the Construction Documents are 75% complete, and the required number of sets of the final Drawings and Specifications when the Construction Documents are 100% complete. Architect shall also provide a set of construction drawings in CAD form. Specific details of this requirement shall be obtained from Owner. Any additional sets shall be a Reimbursable Expense.

§ 3.15   Bidding—General: ¶ 3.5.1

The following revisions make it clear that the architect is not assuming responsibility for the contractor's financial status:

3.5.1.1 The Architect shall assist the Owner in establishing a list of prospective contractors. The Owner understands and acknowledges, however, that the Architect does not possess sufficient knowledge of the status of any prospective contractor so as to evaluate the financial condition of any such contractor and that the Owner is not relying on any evaluation by the Architect of the ability, financial or otherwise, of a contractor. Following the Owner's approval of the Construction Documents, the Architect shall assist the Owner in (1) obtaining either competitive bids or negotiated proposals; (2) confirming responsiveness of bids or proposals; (3) determining the successful bid or proposal, if any; and, (4) awarding and preparing contracts for construction. Architect shall assist the Owner in preparing such contracts, but the final responsibility for such contracts shall be on the Owner.

§ 3.16   Competitive Bidding: ¶ 3.5.2

This clause spells out that the architect is not to investigate the financial capability of the contractor:

3.5.2.4 The Owner shall make an independent investigation into the financial capability of the proposed contractor, and the Owner waives any claim against the Architect for any damages that may arise out of the bankruptcy or any other financial difficulties of the Contractor.

§ 3.17   Negotiated Proposals: ¶ 3.5.3

A more comprehensive alternative for negotiated proposals:

3.5.3 NEGOTIATED PROPOSALS

3.5.3.1 This Phase shall begin during the Construction Documents Phase so that bidding will not be delayed after Owner's approval of the Construction Documents.

3.5.3.2 As requested and authorized by Owner, Architect shall: (i) issue the required number of sets of the Construction Documents and bidding information for the purpose of obtaining bids and negotiating prices for subcontractors under the Construction Contract. Any additional sets shall be a Reimbursable Expense; (ii) include documents prepared by contractors or other consultants other than the Consultants retained by Architect, properly identified and signed, with the Construction Documents, provided that Architect shall have no responsibility for such other contractors' or consultants' work; (iii) provide for Owner's approval a list of potential bidders and Bidding Documents consisting of the Invitation to Bid, Instructions to Bidders and Proposal; (iv) assemble the complete bid package and, with Owner's approval, distribute it to the approved bidders; (v) After obtaining Owner's written approval, respond in writing to requests for clarifications of Architect's work and issue addenda and other pre-contract documents as requested; (vi) assist in the pre-bid conference and walk through, including giving technical narrative; (vii) provide Owner with a written analysis and recommendation of the bids and any alternatives included in each bid; and (viii) prepare and furnish to Owner the required number of stamped and signed copies of the Drawings and bound Contract Documents, including Specifications, fully prepared for execution.

3.5.3.3 Architect acknowledges that it is of primary concern to Owner that the Project be constructed within the Project Budget and agrees to design the Project so that the Guaranteed Maximum Cost will not exceed the Project Budget. If the lowest bona fide bid or negotiated proposal for any portion of the Work exceeds the Project Budget for such portion of the Work, then Owner may, in its reasonable discretion, require Architect to revise the Project so as to reduce the Project construction cost for such portion of the Work, in which case Architect shall, at Architect's expense, modify the Construction Documents, as so directed, in order to reduce the Project construction cost for such portion of the Work to within the Project Budget for such portion of the Work.

§ 3.18   General: ¶ 3.6.1

A disclaimer that the architect could place on each sheet of the record drawings might be worded as follows:

These record drawings have been prepared based, in part, on information prepared and submitted by others. The Architect is not responsible for their accuracy, nor for any errors or omissions that may be incorporated into these drawings as a result. Users of these documents must perform an independent verification of the information shown on these documents. ALL WARRANTIES ARE SPECIFICALLY DISCLAIMED.

The following language clarifies the fact that the architect is an independent consultant and should not mislead others about the extent of its authority:

3.6.1.4 The Architect is an independent contractor and in providing its services under this Agreement shall not represent to any third party that its authority is greater than that granted to it under the terms of this Agreement.

§ 3.19   Evaluations of the Work: ¶ 3.6.2

On some projects, the owner may not want to retain the architect for full construction administration services, or for any such services. This should be discouraged. In such event, however, the following provision should be added to protect the architect from inevitable claims:

3.6.2.6 The Owner acknowledges that, after careful deliberation, the Owner has chosen not to retain the Architect to provide complete Construction

Administration Services under this Agreement. The Owner understands that this decision will likely result in misinterpretations of the Contract Documents, in errors in implementing the Project, and in more construction defects than if the Architect renders normal services. Therefore, the Owner agrees to indemnify and hold the Architect harmless from any claims, suits, arbitrations, losses, and damages, including attorneys' fees, that may arise out of any errors, omissions, defects, or misinterpretations of any Contract Document that may have been prevented by having the Architect engaged to provide full Construction Administration Services hereunder. Any questions of whether any such error, defect, or otherwise could have been prevented shall be decided in favor of the Architect unless the evidence overwhelmingly shows that the error, defect, or otherwise would have occurred even with the Architect present. This indemnification shall apply to any suit, cause of action, arbitration, claim, or otherwise, whether brought by the Owner, any contractor, or any other third party, and shall apply in case of any damage of any sort, including death. This indemnification shall not apply in the event the Architect is judged to be solely negligent by a court of competent jurisdiction.

The owner can insert the following language regarding rejection of work by the architect:

3.6.2.2 Prior to the rejection of any work, the Architect shall submit to the Owner documentation supporting rejection of work performed. If the Owner is not satisfied with the specificity or adequacy of the documentation, the Owner may engage at its expense independent consultants or testing laboratories to confirm the Architect's conclusion. If the Architect requires special inspection or testing of the work, the Owner shall be notified of said special testing or inspection.

This language gives the sophisticated owner greater control over the project. It should not be used unless the owner actually intends to carefully review the documentation and is prepared to engage separate testing laboratories, especially because these actions can slow the progress of the work. The owner may also want to place additional responsibility on the architect to reject nonconforming work. This can be done by inserting in ¶ 3.6.2.2 "the responsibility and" before the word "authority" in the first sentence. This means that the architect must reject nonconforming work, even if it is better than specified.

Alternative language that does not require the architect to submit documentation when recommending a rejection of work is:

The Architect shall advise the Owner to reject work that does not conform to the Contract Documents. Whenever reasonable judgment would indicate a probability of a nonconforming or adverse circumstance, and in order to ensure the proper implementation of the intent of the Contract Documents, the Architect shall advise the Owner to require special inspection or testing of any work whether or not such work has been then fabricated, installed, or completed.

Additional language regarding field reports could be added:

3.6.2.6 The Architect shall submit periodic written reports, at least twice a month, on the progress of the work, manpower, and quality of the work. These reports shall detail the dates and times of each site visit by the Architect and/or its consultants, weather conditions, areas under construction, and other observations made. Copies of photographs taken shall be included in each report. These reports shall not relieve the architect of the responsibility to immediately notify the Owner of any observed material deficiencies in the Work.

§ 3.20   Certificates for Payment to Contractor: ¶ 3.6.3

The architect may want to insert the following language and make appropriate changes to this section to make the owner responsible for examining waivers of lien and related documents:

3.6.3.4 The Owner shall be responsible for reviewing all Applications for Payment by the Contractor and shall review all documentation, including waivers of lien, accompanying such Applications. The Architect's sole obligation relative to such Applications is to ascertain whether the Work has progressed to the point indicated in the Application.

Additional language inserts the following sentences at the end of Paragraph 3.6.3.1:

The Architect administratively shall obtain Contractor's mechanic's lien waivers and Contractor's sworn statements listing subcontractors and material suppliers before issuing Payment Certificates, and if such waivers or sworn statements cannot be obtained, then the Architect's Certificates shall be conditional upon the receipt of such waivers. In no event shall the Architect be responsible for verifying that mechanic's lien waivers or sworn statements are complete, sufficient, or valid.

Another version changes Subparagraph 3.6.3.1 by adding the following sentence:

No payment shall be certified to unless Contractor's Statement and Waivers of Lien for contractor, subcontractor, or material suppliers for the amounts required are reviewed and recommended for payment by the Architect as to the amount, not as to legal sufficiency. The Contractor shall be required to collect all such waivers for presentation to the Architect, all in accordance with the Contractor's Schedule of Values.

§ 3.21   Submittals: ¶ 3.6.4

Alternative language to that given in ¶ 3.6.4 for reviewing the contractor's submittals is given below:

3.6.4.1 The Architect shall review and approve in a timely manner, but in no event more than fourteen (14) days after submittal to the Architect, all shop drawings, samples, and submissions of the Construction Manager for conformance with the design concept of the Project, for compliance with the information given in the Contract Documents, for compatibility with adjacent and contiguous work, systems, and services and with limitations of space, weight, and services. Submissions that are not approved by the Architect are to be brought to the attention of the Owner concurrent with notification to the Construction Manager. The Architect shall inspect all mockups of any aspect of the Project when requested to do so by the Owner and shall report to the Owner in writing promptly whether the mockups are consistent with the design intent of the Project and advise the Owner of any of the Architect's other comments. The Architect shall visit manufacturing plants that are manufacturing components of the building as necessary to ensure that such components, when manufactured, will be consistent with the design intent and shall report to the Owner in writing promptly following any such visit. Travel expenses in connection therewith shall be reimbursable.

§ 3.22   Project Completion: ¶ 3.6.6

The architect may want to delete references to "inspections" entirely. This could be accomplished by modifying paragraph 3.6.6.1 to read as follows:

3.6.6.1 The Architect shall review the Work to determine the date or dates of Substantial Completion and the date of Final Completion, shall receive and forward to the Owner for the Owner's review and records written warranties and related documents required by the Contract Documents and assembled by the Contractor, and shall issue a final Certificate for Payment upon compliance with the requirements of the Contract Documents.

As an additional service, the parties may agree on the following:

3.7 Eight (8) months after the issuance of a Certificate of Substantial Completion, the Architect shall assist the Owner with an inspection of the Work, prepare a report of all observed defective materials, equipment, and workmanship that require corrective actions under any applicable warranties, and submit same to the Contractor(s).

§ 3.23   Owner: ¶ 5.1

The following is a provision more favorable to the owner:

5.1 Upon written request by the Architect, the Owner may, at its sole discretion, provide such information as may reasonably be required for the Architect to properly perform its duties under this Agreement. However, the failure by the Owner to furnish any information to the Architect shall not relieve the Architect of any liability hereunder, nor extend the time in which the Architect is to perform such duties, unless the Architect notifies the Owner in writing that such information is necessary and that the lack of such information may impede the progress of the Project.

§ 3.24   Owner's Budget: ¶ 5.2

5.2 The Owner may, in its sole discretion, from time to time update the budget for the Project, including that portion allocated for the Cost of the Work. The Owner shall not be responsible to the Architect for any damages, consequential or otherwise, for any failure to so notify the Architect. In the event that the Owner significantly increases or decreases the overall budget, the portion of the budget allocated for the Cost of the Work, or contingencies included in the overall budget or a portion of the budget, the Architect shall cooperate in all necessary revisions to the work of the Architect, including revisions to drawings, specifications, and other work of the Architect. Upon timely presentation of invoices reflecting any actual additional work required as a result of such change, the Architect shall be entitled to an equitable adjustment in its fee, as well as any reimbursable expenses arising therefrom.

§ 3.25   Owner's Representative: ¶ 5.3

If a designated representative is not used, the following language may be substituted for ¶ 5.3:

5.3 The Owner shall examine documents submitted by the Architect and shall render decisions pertaining thereto promptly, to avoid unreasonable delay in the progress of the Architect's services. Any approvals given by the Owner shall not relieve the Architect of any of its obligations hereunder.

Another provision:

5.3 Notwithstanding anything to the contrary contained in this Agreement, Owner's review and approval of any and all documents or other matters required herein shall be for the purpose of providing Architect with information and not for the purpose of determining the accuracy and completeness of such documents. Such review and approval by Owner shall in no way create any liability on the part of Owner (notwithstanding any professional skill and judgment possessed by Owner) for errors, inconsistencies or omissions in any approved documents, nor shall such review and approval alter Architect's responsibilities hereunder with respect to such documents.

§ 3.26   Additional Consultants: ¶ 5.6

The owner may want to consider using the following provision:

5.6 Upon the written request of the Architect, the Owner may, at its sole discretion, furnish the services of consultants other than those designated in Initial Information or authorize the Architect to furnish them as a Change in Services. The Owner shall only be required to pay for such additional consultants if such services were not reasonably foreseeable at the outset of the Project or were otherwise caused by events beyond the direct control of the Architect.

Another provision:

Owner reserves the right to retain other architects, engineers and consultants in connection with the Project. With respect to any portion of the design of the Project that is to be performed by any architect, engineer or other consultant retained by Owner pursuant to a separate agreement (any such other architect, engineer or consultant shall herein be referred to as "Separate Consultant"), Architect, as a Basic Service shall review and coordinate such designs with the Architect's Drawings and Specifications for conformance to the design, space limitations and performance criteria for the Project. Architect shall be entitled to rely on the accuracy and adequacy of the work of the Additional Consultants unless Architect has actual knowledge of any deficiency or error in such work, in which event Architect shall promptly notify Owner of same. Architect shall not be required to analyze or review the information prepared by the Separate Consultants unless specifically necessary for performance of Architect's services, or as stated above. Owner acknowledges that Architect shall not be liable for any errors or omissions in the services of the Additional Consultants, unless Architect has actual knowledge of such errors or omissions and fails to promptly notify Owner of such errors or omissions.

§ 3.27 Notice: ¶ 5.9

The following provision favorable to the owner may be added at the end of ¶ 5.9:

> but the Owner's failure or omission to do so shall not relieve the Architect of its responsibilities hereunder and the Owner shall have no duty of observation, inspection, or investigation.

Another provision to consider reads as follows:

> **5.9** In the event that the Owner becomes aware, or reasonably should have become aware in the exercise of ordinary care, of any fault or defect in the Project, including any errors, omissions, or inconsistencies in the Architect's Instruments of Service, the Owner shall promptly provide written notice to the Architect of same. Failure to do so shall constitute a waiver of any claim against the Architect that the giving of such notice could have cured.

Another provision:

> The Owner shall provide prompt written notice to the Architect if the Owner actually becomes aware of any fault or defect in the Project, including any errors, omissions or inconsistencies in the Architect's Instruments of Service, but the Owner has no architectural or construction expertise and is not obligated to investigate for any such faults or defects.

Another:

> The right of the Owner to perform plan checks, plan reviews, other reviews and/ or comment upon the work of the Architect, as well as any approval by the Owner, shall not be construed as relieving the Architect from its professional and legal responsibility for services required under this Agreement. No review by the Owner, approval or acceptance, or payment for any of the services required under this Agreement shall be construed to operate as a waiver by the Owner of any right under this Agreement or of any cause of action arising out of the performance or nonperformance of this Agreement, and the Architect shall be and remain liable to the Owner in accordance with applicable law for all damages to the Owner caused by the Architect's acts, errors and/or omissions.

§ 3.28 Owner's Communication: ¶ 5.10

Alternative language that gives the owner more direct communication may be used for ¶ 5.10:

> **5.10** Except as may otherwise be provided in the Contract Documents, the Owner and Contractor may communicate through the Architect.

Communications by and with the Architect's consultants shall be through the Architect. It is expressly understood, however, that the Owner may, at any time, directly communicate with the Contractor or any subcontractor. The Architect agrees to be the Owner's representative in any communications between the Contractor and the Architect's consultants.

Additional language changes Subparagraph 5.10 to read:

The Owner shall not issue directions to any Contractor contrary to any of the Construction Documents without prior consultation with the Architect. Because Architect shall not be informed of all communications from Owner to Contractor, the liability for said communications shall be the Owner's responsibility, not the Architect's, and Owner hereby agrees to accept full and complete responsibility for same.

§ 3.29   Article 6: Cost of the Work

The following are alternative clauses that can be substituted for the standard language:

6.1 The Cost of the Work shall be the total cost (as shown on the most recent Contractor's bid or Sworn Statement, whichever is later) of all elements of the Project designed or specified by the Architect. In the event bids have not been taken, this Cost shall be based on the most recent estimate prepared by the Architect, or the Owner's budget for the Cost of the Work, whichever is applicable.

Another is as follows:

6.1.1 The Cost of the Work shall be the most recent budget for the Cost of the Work approved by the Owner. If construction has commenced, the Cost of the Work shall be the construction cost accepted by the Owner, but only for those elements of the Project designed or specified by the Architect. Any revisions in price proposed by the Contractor shall not be considered as part of the Cost of the Work unless and until approved as a Change Order by the Owner.

6.1.2 The Cost of the Work shall include the cost at current market rates of labor and materials furnished by the Owner and any contractors hired by the Owner or others; equipment designed, specified, selected, or specially provided for by the Architect, including any items for which the Architect designs any structural, mechanical, or electrical components or considerations; the costs of management or supervision of construction or installation provided by a separate construction manager or contractor, plus a reasonable allowance for their overhead and profit. In the event that the Owner utilizes its own employees in the place of a contractor or construction manager, the reasonable costs of such employees shall be included as part of the Cost of the Work. In addition, a reasonable allowance for contingencies shall be included for market conditions at the time of bidding and for changes in the Work.

The following is another alternative for the latter clause:

6.1.2 The Cost of the Work shall include the cost at current market rates of labor and materials designed and specified by the Architect, but not including the cost of furniture, fixtures, or any items that are separately purchased by the Owner and not to be included on the Contractor's Sworn Statement, nor shall it include the costs of any construction manager hired by the Owner.

§ 3.30   Responsibility for Construction Costs: ¶ 6.6

The following are alternative provisions for specifying procedures when the bids come in higher than the budget or estimated cost:

6.6 If the lowest reasonable bid received from the Construction Manager exceeds the Budget Amount by more than ten percent (10%), the Architect shall revise the Construction Documents, rebid the General Construction, and perform all other services again as necessary to finally produce a contract with responsible contractors and for a cost that is within the Budget Amount. The only exception to this provision shall be if the increase is caused by the economic effect of extended delay in the Project, which arose through no fault of the Architect.

Another:

6.6 If the lowest bona fide bid or negotiated proposal by a responsible general contractor exceeds the last approved Statement of Probable Construction Costs, with later modifications approved by the Owner, if any, by more than ten percent (10%), then the Owner may elect to (i) accept such bid; or (ii) the Architect shall, at no additional cost or expense to the Owner, revise the Contract Documents in cooperation with the Owner so as to reduce the cost of the Project, and submit these revised Contract Documents for rebid so that the construction cost as rebid will not exceed such approved Statement of Probable Construction Costs by more than ten percent (10%).

Another provision:

6.6 The Architect understands and agrees that the budget for the Cost of the Work as approved in writing from time to time shall not be exceeded. In the event that such approved budget is exceeded by the lowest bona fide bid or negotiated proposal by more than _____ (_____ %) percent, the Owner may choose to do one of the following:

.1 give written approval of an increase in such budget;

.2 authorize rebidding or renegotiation of the Project within a reasonable time without additional compensation to the Architect;

.3 terminate the Project without additional compensation to the Architect; or

.4 cooperate in revising the Project scope and quality as required to reduce the Cost of the Work.

§ 3.31   Article 7: Instruments of Service

This section concerns the copyright in the drawings, specifications, and related documents that are authored by the architect and its consultants. Before making any of these changes, the parties should consult a knowledgeable attorney. An alternate to ¶ 7.2.1:

7.2.1 In the event of any unauthorized use, reuse or modification of the Architect's Drawings, Specifications, or other documents prepared by the Architect, whether such unauthorized use, reuse, or modification is made by the Owner, the Owner's employees, agents, consultants, contractors, or any third party whatsoever, the Owner agrees to indemnify and hold harmless the Architect, its officers, directors, and employees from and against any claims, suits, demands, losses, and expenses, including attorneys' fees, accruing or resulting to any and all persons, firms, or any other legal entity, on account of any damage or loss to property or persons, including death, arising out of such unauthorized use, reuse, or modification. The parties hereto intend to give this indemnification provision the broadest possible effect. This indemnification shall not be effective in the event of the Architect's sole negligence as determined by a court of competent jurisdiction.

Another alternate:

7.2.1 Owner's use on other projects, Owner's re-use, or Owner's modification of the Instruments of Service shall be at Owner's sole risk and without recourse against Architect, its consultants at any tier, and their principals, agents and employees. Owner shall hold harmless, indemnify and defend Architect, its consultants at any tier and their respective principals, agents and employees from and against any and all actions, claims, loss, or damages of any nature whatsoever to the extent related to and resulting from any said use, re-use, or modification of all or any portion of the Instruments of Service by or on behalf of Owner, or under any license issued by, through, or on behalf of Owner, irrespective of any actual or alleged fault on the part of the indemnitee(s). Under no circumstances shall Architect be indemnified for the use of the Instruments of Service for the Project that is the subject of this Agreement.

The owner may want to replace ¶ 7.2 with language such as this:

7.2 The Owner may use drawings, plans, and specifications without restrictions. This is not intended to create any rights by any other party in the drawings, plans, and specifications.

This or similar language should be of great concern to the architect, because an unauthorized reuse of the drawings can lead to unexpected situations that may create dangerous conditions and additional liability for the architect.

A better alternative if the architect agrees to give up ownership and copyright in the drawings is:

> **7.2** All plans, drawings, specifications, computations, sketches, data, surveys, models, photographs, renderings, and other like materials relating to the services ("Documents") shall become the property of the Owner at the conclusion of the project, or termination of the services of the Architect, whichever is earlier, and shall be delivered to the Owner clearly marked and identified and in good order. The Owner may use the Documents as it determines, but the Architect and the Architect's consultants shall incur no liability for the Owner's use of the Documents other than in connection with the Project, and the Owner hereby indemnifies and holds harmless the Architect and its consultants from any loss or damage, including attorneys' fees, incurred as a result of this provision.

The owner may want to strengthen its ability to use the drawings in the event of conflict with the architect, as follows:

> The Architect shall hold all copyrights in the Drawings, Specifications, and other documents prepared by the Architect for this Project. However, the Architect gives the Owner an irrevocable license to use the Drawings, Specifications, and other documents prepared by the Architect for completion of this Project in the event the Architect is terminated for any reason, as well as for any additions, alterations, or other work to the Project. This license is for the benefit of the Owner and its assigns and permits the Owner to retain other architects, engineers, and design professionals who may use the Drawings, Specifications, and other documents for such purposes.

> Drawings and specifications are, and shall remain, the property of the Owner, whether the Project for which they are made is executed or not. Such documents may be used by the Owner to construct one or more like projects without the approval of, or additional compensation to, the Architect. The Architect shall not be liable for injury or damage resulting from reuse of drawings and specifications if the Architect is not involved in the reuse project. Prior to the reuse of construction documents for a project in which the Architect is not also involved, the Owner will remove and obliterate from such documents all identification of the original Architect, including name, address, and professional seal and stamp.[1]

The following provisions give the copyright to the owner:

> (a) Architect agrees that any documents, models, renderings, and other materials to be furnished to Owner or anyone in connection with this Agreement by or on behalf of Architect, or by any consultants, contractors, or others retained by

---

[1] Adapted from Guillot-Vogt Assocs., Inc. v. Holly & Smith, 848 F. Supp. 682 (E.D. La. 1994).

Architect, and all modifications made by Architect to any documents supplied by Owner to Architect, and all written information, reports, studies, object or source codes, flow charts, diagrams, and other tangible material which have been created by Architect in order to provide services pursuant to this Agreement (collectively, the "Work Product") shall be the sole and exclusive property of Owner. Architect shall not be entitled to make any use of any of the Work Product whatsoever except as may be expressly permitted in or required by this Agreement. Architect shall deliver such Work Product (including those of persons engaged for special and consulting services) to Owner at the expiration or earlier termination of this Agreement or otherwise upon request of Owner. Architect, in consideration of Owner's execution of this Agreement and for other good and valuable consideration, the receipt and sufficiency of which are hereby acknowledged, hereby irrevocably grants, assigns, and transfers to Owner all of Architect's right, title, and interest of any kind in and to the Work Product and the copyright thereof, and in all renewals and extensions of the copyright that may be secured now or hereafter in force and effect in the United States of America or in any other country or countries.

(b) Owner may use all Work Product in whole or in part or in modified form in connection with this Agreement or otherwise as it shall determine in its discretion, without further employment of, or additional compensation to, Architect or any consultants retained by Architect. If (1) such Work Product is to be utilized by owner for any project subsequent to the termination of Architect's services under this Agreement, and (2) such Work Product is substantially modified by Owner in any material respect, Owner shall release, indemnify, and hold Architect harmless from and against any claim, loss, liability, damage, or expense sustained or incurred by Architect, to the extent that such claim, loss, liability, damage, or expense results from Owner's misapplication or modification of Architect's Work Product without Architect's involvement. If such Work Product is to be utilized by Owner on other projects, Owner shall release, indemnify, and hold harmless from and against any claim, loss, liability, damage, or expense sustained or incurred by Architect as a direct result of Owner's use of the Work Product. Notwithstanding anything to the contrary contained in this section, Owner shall in no event release, indemnify, and/or hold Architect harmless if the claim, loss, liability, damage, or expense results in whole or in part from the failure of Architect (or anyone for whom Architect may be liable under this Agreement) to comply with this Agreement in the preparation or publication of the applicable Work Product. Architect agrees to include in all of its contracts with architects, engineers, and other consultants or contractors a provision incorporating this section.

(c) Architect warrants that title to Work Products conveyed to Owner shall be delivered free and clear of all claims, liens, charges, encumbrances or security interests. Architect agrees to execute any documents reasonably requested by Owner in connection with the registration of patent and/or copyrights or any other statutory protection in such Work Product. All Work Product provided to Owner shall be marked as follows:

© 20 _____ by [Owner's name].
All rights reserved

Here is another provision favorable to the owner:

Architect and the Owner intend that, to the extent permitted by law, the Instruments of Service to be produced by Architect at the Owner's instance and expense under this Agreement are conclusively considered "works made for hire" within the meaning and purview of Section 101 of the United States Copyright Act, 17 U.S.C. §§ 101 et seq., and that the Owner will be the sole copyright owner of the Instruments of Service and of all aspects, elements and components of them in which copyright can subsist, and of all rights to apply for copyright registration or prosecute any claim of infringement.

To the extent that any Instruments of Service, or any part thereof, does not qualify as a "work made for hire," Architect hereby irrevocably grants, conveys, bargains, sells, assigns, transfers and delivers to the Owner, its successors and assigns, all right, title and interest in and to the copyrights and all U.S. and foreign copyright registrations, copyright applications and copyright renewals for them, and other intangible, intellectual property embodied in or pertaining to the Instruments of Service prepared for the Owner under this Agreement, and all goodwill relating to them, free and clear of any liens, claims or other encumbrances, to the fullest extent permitted by law. Architect will, and will cause all of its consultants and subcontractors, employees, agents and other persons within its control to, execute all documents and perform all acts that the Owner may reasonably request in order to assist the Owner in perfecting its rights in and to the copyrights relating to the Instruments of Service, at the sole expense of the Owner. Architect warrants to the Owner, its successors and assigns, that on the date of transfer, Architect is the lawful owner of good and marketable title in and to the copyrights for the Instruments of Service and has the legal rights to fully assign them. Architect further warrants that it has not assigned and will not assign any copyrights and that it has not granted and will not grant any licenses, exclusive or nonexclusive, to any other party, and that it is not a party to any other agreements or subject to any other restrictions with respect to the Instruments of Service. Architect warrants and represents that the Instruments of Service are complete, entire and comprehensive, and that the Instruments of Service constitute a work of original authorship, except to the extent such Instruments of Service incorporate work in the Public Domain or work licensed from other sources. To the extent any Instruments of Service, including but not limited to specifications, incorporate matters from other sources, Architect warrants that it has obtained proper licenses to incorporate such matters into the Instruments of Service.

Another provision:

Upon execution of this Agreement, the Architect grants to the Owner a nonexclusive license to reproduce the Architect's Instruments of Service solely for purposes of constructing, using and maintaining the Project, provided that the Owner shall comply with all obligations, including prompt payment of all sums when due, under this Agreement. The Architect shall obtain similar nonexclusive licenses from the Architect's consultants consistent with this Agreement.

In the event of termination of this Agreement prior to completion of this Project, Owner shall be entitled to use the Instruments of Service for completion of any Project on this Site only, subject to the terms, conditions and restrictions set forth in this Paragraph. Provided Owner has paid Architect for all sums due and owing up to the date of termination, Owner may use completed Instruments of Service that bear the Architect's professional seal and signature for completion of this Project. Owner may use in-progress or incomplete Instruments of Service for this Project only provided Owner removes the title block and name of Architect. Owner assumes all risk of use of incomplete or in-progress Instruments of Service and releases Architect from any liability for same. Further Owner agrees to defend, indemnify and hold harmless Architect and its Consultants from any claims arising out of Owner's use of the Instruments of Service on any other project or in a manner prohibited by this Agreement.

Another provision:

None of the licensed described in this Agreement authorize use of the instruments of Service for other projects. None of the licenses described in this Agreement may be assigned, delegated, sublicensed, pledged or otherwise transferred by the Owner to any party other than a lender providing financing for the Project without the prior written consent of the Architect. Any unauthorized use of the Instruments of Service shall be at the Owner's sole risk and without liability to the Architect and the Architect's consultants.

§ 3.32   Statutes of Limitations: ¶ 8.1.1

The parties can fix limitations periods other than those set by law. Thus, the following language more favorable to the architect could be substituted:

8.1.1 The parties agree that no action hereunder may be brought by either party more than one year after the earlier of the date of the Architect's Certificate of Substantial Completion, or the date of the Architect's last substantial work, except in the case of an action for nonpayment of any amounts due under this Agreement, in which case no action may be brought more than two years after payment is due.

Another:

8.1.1 The Owner and Architect shall commence all claims and causes of action, whether in contract, tort, or otherwise, against the other arising out of or related to this Agreement in accordance with the requirements of the method of binding dispute resolution selected in this Agreement within the period specified by applicable law, but in any case not more than two years *[or select another time limit]* after the date of Substantial Completion of the Work. The Owner and Architect waive all claims and causes of action not commenced in accordance with this Section 8.1.1

§ 3.33   Mediation: ¶ 8.2

The parties may want to attempt to resolve disputes through mediation, in which case the following provisions are useful:

8.2 MEDIATION

8.2.1 <u>Claims Subject to Mediation</u>: Any claim, dispute, or other matter in question arising out of or related to this Agreement shall be subject to mediation as a condition precedent to the institution of legal or equitable proceedings by either party.

8.2.2 <u>Mediator</u>: The mediation shall be before one disinterested mediator if one can be agreed upon; otherwise, a mediator will be selected by the Presiding Judge as hereinafter stated. This Agreement to Mediate shall be specifically enforceable.

8.2.3 <u>Appointment of Mediator</u>: Notice of the Demand for Mediation shall be given in writing to the other party to the Agreement within one hundred twenty (120) consecutive calendar days after the claim, dispute, or other matter in question has arisen. Such Notice of the Demand for Mediation shall specify the name and address of the person nominated to act as mediator of the dispute. If within fourteen (14) consecutive calendar days after service of such Notice, the other party shall agree to such person so nominated as mediator, then such person so selected shall become the mediator. If within said fourteen (14) consecutive calendar days after service of Notice on the other parties, there is no written objection made and served by the other party to the person nominated, then such person so nominated shall become the mediator. If within said fourteen (14) consecutive calendar days written objection is made and served by the other party to the person nominated and within said fourteen (14) consecutive calendar day period of time the parties hereto are unable to agree upon the appointment of one disinterested mediator, then the mediator shall be selected by the Presiding Judge of the Circuit Court of Jones County, Illinois, upon formal or informal application by either party hereto, and no party shall raise any objection or question as to the power and jurisdiction of the Presiding Judge to entertain the application and make the appointment. If both parties do not jointly make the application to said Presiding Judge, then the party making such application shall simultaneously serve a copy of the Notice of the application being filed with the Presiding Judge upon the other party to this Agreement. Either party may submit to the Presiding Judge a list of the names, addresses, and qualifications of no more than three disinterested persons who may be selected as a mediator. Such submission to the Presiding Judge shall take place no later than three (3) Court days after the application for appointment of a mediator has been filed with said Judge. The Presiding Judge may select as mediator the name of any person so nominated by a party hereto, or the Judge may select any other person as mediator.

8.2.4 <u>Mediation Sessions—Matters in Dispute:</u> The mediator so appointed, whether by agreement of the parties or by the Presiding Judge, shall fix the date, time, and place of each mediation session. Any party may be represented by a person of the party's choice. At least seven (7) consecutive calendar days prior to the first scheduled mediation session, each party shall provide the mediator and the other party with a brief statement of the nature of the dispute and a memorandum setting forth the position of the party with reference to such issues that need to be resolved. At the first mediation session, the parties will produce all information reasonably presented. The mediator may require any party to supplement such information.

8.2.5 <u>Authority of Mediator:</u> The mediator shall not have the authority to impose a settlement on the parties but will attempt to help the parties reach a satisfactory resolution of their dispute. The mediator is authorized to conduct joint and separate meetings with the parties and to make oral and written recommendations for settlement. The mediator is authorized to end the mediation whenever, in the judgment of the mediator, further efforts at mediation would not contribute to a resolution of the dispute between the parties.

8.2.6 <u>Use of Evidence:</u> The mediator shall not be compelled to divulge any records, reports, or other documents received by such mediator or to testify in regard to the mediation in any adversary proceeding or judicial forum. The parties shall not rely on or introduce as evidence in any judicial or other proceeding:

(a) views expressed or suggestions made by another party with respect to a possible settlement of the dispute;

(b) admissions made by another party in the course of the mediation proceedings;

(c) proposals made or views expressed by the mediator; or

(d) the fact that another party had or had not indicated willingness to accept a proposal for settlement made by the mediator.

There shall be no stenographic record of the mediation process.

8.2.7 <u>Termination of Mediation:</u> The mediation shall be terminated:

(a) by the execution of a Settlement Agreement by the parties;

(b) by a written Declaration of the mediator to the effect that further efforts at mediation are no longer worthwhile; or

(c) by a written Declaration of a party to the effect that the mediation proceedings are terminated.

8.2.8 <u>Rules:</u> The mediator shall have authority to make such other, further, and additional rules in the conduct of the mediation as the mediator decides necessary for the process to proceed in an orderly manner. However, unless otherwise agreed by all parties and the mediator, the mediator shall not have authority to modify or change the rules and principles contained herein as to the mediation process.

8.2.9 <u>Time Limit:</u> Notwithstanding any other provision contained herein, unless otherwise agreed to by the parties hereto, in writing, the mediation process as described herein shall cease sixty (60) consecutive calendar days after the appointment of the mediator.

8.2.10 <u>Costs, Expenses:</u> The expenses of witnesses, if any, for either side shall be paid by the party producing such witnesses. All other fees and expenses of the mediation shall be borne equally by the parties unless they agree otherwise.

8.2.11 <u>Venue:</u> Unless otherwise agreed by the parties, all mediation hearings shall take place in Jones County, Illinois.

8.2.12 <u>Governing Law:</u> All substantive and procedural matters involved in the mediation process shall be determined in accordance with the law of the State of Illinois.

§ 3.34   Arbitration: ¶ 8.3

The parties may want to provide for arbitration only when the amount in dispute is below a certain amount. The following provision can be tailored to specific requirements:

Delete Article 8.3 in its entirety and insert in lieu thereof:

8.3.1 All claims, disputes, and other matters in question involving amounts of less than $100,000 between any of the Architect, Construction Manager, Owner, Contractors, Surety, Subcontractors, or any Material Suppliers arising out of or relating to agreements to which two or more of said parties are bound or the Contract Documents or the breach thereof shall be decided by arbitration in accordance with the Construction Industry Arbitration Rules of the American Arbitration Association currently in effect, as modified herein, unless the parties mutually agree otherwise. Claims, disputes, and other matters in question involving amounts of $100,000 or more may be submitted to arbitration only with the Owner's written consent. The Owner, Architect, Construction Manager, Subcontractors, and Material Suppliers who have an interest in the dispute shall be joined as parties to the arbitration. The Owner's contract with the Architect and the Contractor's contract with the Subcontractors shall require such joinder. The arbitrator shall have authority to decide all issues between the parties including, but not limited to, claims for extras, delay and liquidated damages, matters involving defects in the Work, rights to payment, whether matters decided by the Architect involve artistic effect, and whether the necessary procedures for arbitration have been followed. The foregoing agreement to arbitrate and any other agreement to arbitrate with an additional person or persons duly consented to by the parties to the Owner-Contractor, Owner-Architect or Owner-Construction Manager Agreements shall be specifically enforceable under prevailing arbitration law. The award rendered by the arbitrator shall be final, and judgment may be entered upon it in accordance with applicable law in any court having jurisdiction thereof.

8.3.2 Notice of the demand for arbitration shall be filed in writing with the other party to the arbitration and with the American Arbitration Association. The demand for arbitration shall be made within a reasonable time after the claim, dispute, or other matter in question has arisen, and in no event shall it be made after the date when institution of legal or equitable proceedings based on such claim, dispute, or other matter in question would be barred by the applicable statute of limitation.

8.3.3 Unless otherwise agreed in writing, all parties shall carry on the work and perform their duties during any arbitration proceedings, and the Owner shall continue to make payments as required by agreements and the Contract Documents.

8.3.4 If any proceeding is brought to contest the right to arbitrate and it is determined that such right exists, the losing party shall pay all costs and attorneys' fees incurred by the prevailing party.

8.3.5 In addition to the other rules of the American Arbitration Association applicable to any arbitration hereunder, the following shall apply:

8.3.5.1 Promptly upon the filing of the arbitration, each party shall be required to set forth in writing and to serve upon each other party a detailed statement of its contentions of fact and law.

8.3.5.2 All parties to the arbitration shall be entitled to the discovery procedures and to the scope of discovery applicable to civil actions under [state] law, including the provisions of the Civil Practice Act and [state] Supreme Court Rules applicable to discovery. Such discovery shall be noticed, sought, and governed by those provisions of [state] law.

8.3.5.3 The arbitration shall be commenced and conducted as expeditiously as possible consistent with affording reasonable discovery as provided herein.

8.3.5.4 These additional rules shall be implemented and applied by the arbitrator(s).

8.3.6 In the event of any litigation or arbitration between the parties hereunder, all attorneys' fees and other costs shall be borne by the party determined to be at fault and, in the event that more than one party is determined to be at fault, shall be allocated equitably by the court or arbitrator.

Note that this language allows consolidation of the various claims and permits attorneys' fees to be recovered by the prevailing party.
A simplified version of the foregoing clause is as follows:

8.3.1 Claims, disputes, or other matters in question not exceeding $100,000 between the parties to this Agreement arising out of or relating to this Agreement or breach thereof shall be subject to and decided by arbitration in accordance with the Construction Industry Arbitration Rules of the American Arbitration Association currently in effect.

The 2007 arbitration language permits consolidation and joinder, so that an owner can force the architect and contractor to a single arbitration. This is not favorable to the architect, so the prior language of AIA documents might be used:

8.3.3 No arbitration arising out of or relating to this Agreement shall include, by consolidation or joinder or in any other manner, an additional person or entity not a party to this Agreement, except by written consent containing a specific reference to this Agreement and signed by the Owner, Architect, and any other person or entity sought to be joined. Consent to arbitration involving an additional person or entity shall not constitute consent to arbitration of any claim, dispute or other matter in question not described in the written consent or with a person or entity not named or described therein. The foregoing agreement to arbitrate and other agreements to arbitrate with an additional person or entity duly consented to by parties to this Agreement shall be specifically enforceable in accordance with applicable law in any court having jurisdiction thereof.

Another provision:

Notwithstanding anything contained in the Rules of the American Arbitration Association, or in applicable law or elsewhere, the parties and arbitrators shall be bound by the following:

No party shall be entitled to discovery.

The Arbitrator shall conduct a prehearing telephone conference with the parties at which schedules for the hearings shall be set.

The Arbitrators shall have no authority a) to order discovery, b) to award attorneys' fees, c) to set a site for hearings other than Chicago, Illinois, d) to award punitive damages, e) to compel anything in contravention of any agreement of the parties, or f) to enter an award totaling more than $50,000.

The sole remedy for any party hereunder shall be had in arbitration.

The sole reason for vacating the award of the Arbitrators shall be that the Arbitrators exceeded their authority.

Any petition to vacate an award of the Arbitrators shall be brought within 30 days of the date of the Award. Any petition to confirm an award of the Arbitrators shall be brought within five years of the date of the Award.

The parties can also limit arbitration to one or more matters. The following clause limits arbitration to fee disputes that involve $25,000 or less:

8.3.1 Fee disputes under $25,000 shall be decided by arbitration in accordance with the Construction Industry Arbitration Rules of the American Arbitration Association currently in effect.

The parties may want to require the payment of attorneys' fees to the winning party, in which case the following clause can be added:

The arbitrators shall award reasonable attorneys' fees to the prevailing party. Within three days after the close of the hearings, each party shall submit to the arbitrators a detailed fee petition, but no hearings shall be held on the fee petitions unless the arbitrators determine that such hearing is required. The arbitrators shall award attorneys' fees as seem just, but not to exceed the amount of the fee petition presented by the prevailing party. The arbitrators may determine that no fees are due either party.

Another method of resolving disputes is mediation. This is nonbinding, but its speed and relatively low cost make it attractive. A possible clause is as follows:

As a condition precedent to the filing of a lawsuit or other claim relating to or arising out of the Project, the Owner and Architect hereby agree to attempt to mediate the dispute with each other and any other interested parties in accordance with the Construction Mediation Rules of the American Arbitration Association. The Architect shall require a similar agreement from all of its consultants.

When the parties are in different locations, the following provision fixes the locale of the arbitration. Without such a provision, the first party to file an arbitration request has an advantage in fixing the location of the arbitration hearings. If there is a dispute, the American Arbitration Association has the authority to fix the locale under its rules. This provision avoids this problem:

Any arbitration hearing shall be heard at the offices of the AAA located in Chicago, Illinois. Any party requesting a site visit shall pay all costs of the AAA, including the arbitrator's fees, related to such site visit, unless both parties agree to a site visit.

Pursuant to the case of *Roadway Package Sys., Inc. v. Scott Kayser*, 257 F.3d 207 (3d Cir. 2001), the parties may want to "opt out" of the FAA regime:

Any arbitration hereunder shall be brought pursuant to the Illinois Arbitration Act (the "Act"), as well as the applicable Construction Industry Rules of the

American Arbitration Association. The parties hereby express their intent to "opt out" of the Federal Arbitration Act ("FAA"). To the extent any provision of the FAA conflicts with any provision of the Act, the Act shall control. This provision shall apply no matter where the arbitration is held or enforced. This provision supercedes Section 10.1.

In addition to the Federal Arbitration Act, the Revised Uniform Arbitration Act is a law that parties should be familiar with. A number of states have adopted this act and it is likely to be adopted in a majority of states by 2005. Parties may want to opt out of some of the questionable provisions of that act:

Notwithstanding the enacting into law of any act affecting arbitration hereunder by the applicable jurisdiction, the parties hereto agree to be bound solely by the arbitration laws in effect at the location of the Project as of the date of execution of this Agreement. Any provisions of any arbitration act that alter, add, decrease or otherwise affect the parties hereto shall not apply to any arbitration hereunder.

The American Arbitration Association (AAA) also regularly amends its Rules. For instance, the AAA recently adopted "Consumer Rules" whereby if one party to an arbitration is considered a "consumer" and the other party is a "business," then the business would pay the brunt of the fees involved in the arbitration. Consider the situation in which a small contractor agrees to build a home for a wealthy owner and a dispute develops. Under the AAA rules in effect in 2003 (since modified), the wealthy owner would pay no more than $375 in any arbitration (until the award), while the contractor might wind up fronting many tens of thousands of dollars in arbitrators' compensation, fees, and so on. These rules are binding even if not in effect at the time of execution of the contract, so long as they are in effect when the demand for arbitration is filed. Similar problems may arise in the future. Here is a possible way to avoid such problems:

8.3.6 The parties hereto agree that any arbitration shall be pursuant to the Construction Industry Arbitration Rules in effect at the time of execution of this Agreement, with the sole exception of the fee schedule. No additional procedures, rules or regulations shall be binding on these parties. Each party shall pay an equal share of administrative fees and arbitrator compensation. Should any party fail to timely pay such amounts, the arbitrator shall have the discretion to dismiss, with prejudice, any or all claims of the non-paying party and to enter an award, ex parte, in favor of the paying party, but only upon presentment of such evidence as in the arbitrator's sole discretion shall justify such award. Alternatively, the arbitrator shall have the discretion to enter such interim award as may do justice to the parties.

Counsel should carefully tailor this provision to take into account applicable law, since some provisions in the contract may not be waivable.

§ 3.35   Architect's Suspension of Services: ¶ 9.1

The architect should consider adding language giving it a right to retain all items in its possession until payment is made. The following language could be added:

> **9.1.1** In the event of any failure of payment by the Owner, the Architect shall have the right to retain any and all documents, drawings, specifications, models, surveys, reports, and similar items, whether prepared by the Architect or submitted to the Architect by others, until full payment is received. In such event, the Architect shall have no liability for any damages or losses that may result from the withholding of any such items.

§ 3.36   Calculation of Termination Expenses: ¶ 9.7

When the owner terminates the architect or the architect is retained for less than the normal full services, the architect is exposed to increased liability. The following additional provision could be used to address this issue:

> If the Owner terminates this Agreement for any reason before the completion of construction, or if the Owner retains the Architect for less than the normal full services, then the Owner agrees to indemnify and hold harmless the Architect from and against any and all claims, suits, demands, losses, and expenses, including attorneys' fees, accruing or resulting to any and all persons, firms, or any other legal entity, on account of any damage or loss to property or persons, including death, arising out of or related to the Work, except in the event the Architect is found to be solely liable for such losses or damages by a court of competent jurisdiction.

The owner may want to add this provision to avoid paying termination expenses to the architect if the bids are too high.

Add to end of Paragraph 9.7:

> In the event that the Project is abandoned because the lowest bona fide bid exceeds the Project budget, then no Termination Expenses of any sort shall be payable.

Other provisions:

> **Termination by Owner.** Owner shall have the right, at any time, acting in its sole discretion, with or without cause, to terminate Architect's rights under this Agreement by giving to Architect seven (7) days prior written notice. A termination effected under this Article 9 shall take effect at the conclusion of such seven (7) day period; provided, however, that Architect shall, upon receipt of such notice, immediately stop its work under this Agreement and deliver possession of the Drawings and Specifications and all other Architectural

Documents to Owner. Except as hereinafter set forth, within thirty (30) days after the date of such termination, Owner shall pay to Architect, and Architect shall be required to accept in full and final payment of any and all claims, including, but not limited to, loss of profits, job, administrative or company overhead, or any other claims whatsoever which it may have by reason of the work performed to the date of receipt of such notice and by reason of cancellation of the remaining work, an amount equal to (a) any unpaid services which have been completed by Architect, plus (b) a proportionate part of the fee for the Phase of Architect's services then in progress, which is the same proportion as the portion of the services then completed (as to that Phase) bears to the total services to have been completed (as to that Phase), plus (c) the amount, if any, of Reimbursable Expenses and/or fees for Additional Services incurred up to the date of receipt of such notice as provided above and not previously paid by Owner, less any payments made by Owner to Architect which are not reimbursable pursuant to, or in accordance with, this Agreement.

Delivery of Documents. In the event of any termination of this Agreement, for any reason, reproductions of all finished and unfinished documents, cost estimates, studies, surveys, drawings, maps, models, photographs and reports prepared by Architect in connection with the Project as part of its services under this Agreement shall be promptly delivered to Owner by Architect.

Survival. All representations made by Architect herein, together with any and all causes of action and other rights and remedies which Owner may have as a result of breach of any term, covenant or condition or representation contained in this Agreement, together with all obligations of Architect hereunder, shall survive any expiration or termination of Architect's rights under this Agreement. All rights and remedies of the parties hereunder are cumulative and the exercise of one or more of such rights and remedies shall not preclude the exercise of any other rights or remedies whether concurrently or sequentially.

§ 3.37   Article 10: Miscellaneous Provisions

Other miscellaneous provisions in Article 10 might include the following:

Owner's Approval. Whenever provision is made herein or in the Contract Documents for the approval or consent of the Owner, or that any matter be to the Owner's satisfaction, unless specifically stated to the contrary, such approval or consent shall be made by the Owner in its sole discretion and determination.

Personal Service Contract. This Agreement is entered into solely to provide for the design and administration of the Project and to define the rights, obligations, and liabilities of the parties hereto. This Agreement, and any document or agreement entered into in connection herewith, shall not be deemed to create any other relationship between the Architect and Owner other than as expressly provided herein. The Architect acknowledges that the Owner is not a partner or joint venturer of the Architect and that the Architect is not an employee or agent of the Owner.

Independent Contractor Provision. The Architect shall perform services as an independent contractor, and nothing contained herein shall be deemed to create any association, partnership, joint venture, or relationship of principal and agent or master and servant between the parties hereto or any affiliates or subsidiaries thereof, or to provide either party with the right, power, or authority, whether express or implied, to create any such duty or obligation on behalf of the other party. The Architect agrees that it will not hold itself out as an affiliate of or a partner, joint venturer, co-principal, or co-employer with Owner or any of its affiliates by reason of this Agreement, and that the Architect will not knowingly permit any of its employees, agents, or representatives to hold themselves out as, or claim to be, officers or employees of Owner or any of its affiliates by reason of this Agreement. In the event Owner is adjudicated to be a partner, joint venturer, co-principal, or co-employer of the Architect by reason of the Architect's failure to comply with this provision, the Architect shall indemnify and hold harmless Owner from and against any and all claims for loss, liability, or damages arising therefrom. Architect will be an independent contractor with respect to the Services and the Project, and neither Architect nor anyone employed by Architect will be deemed for any purpose to be the agent, employee, servant or representative of Owner in the performance of the Services. Architect acknowledges and agrees that Owner will have no direction or control over the means, methods, procedures or manner of performing the Services by Architect or any of its subcontractors, or any of their employees, vendors or suppliers.

Discounts, Rebates. The Architect shall accrue to the Owner all discounts, rebates and refunds obtained with respect to any reimbursable expense incurred in connection with the Project.

Judge of Performance. The Owner shall be the judge of performance by the Construction Manager. The Owner shall render approvals, rejections, decisions, and other communications necessary for the proper execution or progress of the Architect's services or the Work with reasonable promptness on written request of either the Architect or Construction Manager. The Architect shall, upon request, review and provide advice and interpretations to the Owner on claims or disputes arising between the Owner and the Construction Manager, based in whole or in part on the Contract Documents. The Owner shall render written decisions on all claims, disputes, and other matters in question between the Architect and the Construction Manager relating to the execution or progress of the work.

The above provision, which gives the owner more authority, changes the normal claims procedures, so extra caution must be exercised in its use.

Invalid Provisions. If any term or provision of this Agreement or the application thereof to any agency, person, firm, corporation, or circumstance shall, to any extent, be invalid or unenforceable, the remainder of the Agreement, or the application of such terms or provisions to agencies, persons, firms, corporations, or circumstances other than those to which it is held invalid or unenforceable, shall not be affected thereby, and each term or provision of this Agreement shall be valid and be enforced to the fullest extent permitted by law.

Attornment by Architect. The Architect agrees that notwithstanding a default by the Owner under the provisions of this Agreement that would give the Architect the right to terminate this Agreement, the Architect will continue to perform its obligations hereunder, on the same terms and conditions as are set forth herein, for and on account of the Lender if the Lender shall cure any such default by the Owner within fifteen (15) days after notice from the Architect to the Lender, and Architect shall agree in writing to perform all obligations of the Owner hereunder accruing from and after the date the Lender succeeds to the Owner's rights and obligations hereunder. If requested by the Lender, the Architect will execute a separate letter or other agreement with the Lender further evidencing its commitment to continue performance pursuant to this paragraph.

Lender's Rights. Whenever herein an approval, acceptance, direction, requirement, permission, designation, prescription, or other action by the Owner is required or permitted under the Contract Documents, such action may, at the option of the Owner and the Lender, be deemed to include and be conditioned upon the authorization of or joinder in such action by the Lender or an appropriate representative of the Lender, provided, however, no such action by the Lender shall abrogate any right granted to the Architect under the Contract Documents. The Architect shall prepare and submit all reports, certificates, and statements required by the Lender and shall perform such other actions as may be reasonably required of the Architect by the Lender.

These last two provisions give the owner's lender certain rights and may make termination of the contract more difficult if the owner defaults.

Notice of Injury. Should either party to this Agreement suffer injury or damage to person or property because of any act or omission of the other party, or of any of the other party's employees, agents, consultants, contractors, or others for whose acts it is legally liable, claim shall be made in writing to such other party within a reasonable time after the first observance of such injury or damage, but in no event later than 21 days after such first observance.

No Waiver. No action or failure to act by the Owner or Architect shall constitute a waiver of any right or duty afforded under this Agreement, nor shall any such action or failure to act constitute an approval of or acquiescence in any breach hereunder, except as may be specifically agreed in writing.

Approval by Owner of plans, studies, designs, specifications, reports, and incidental work furnished hereunder shall not in any way relieve the Architect of responsibility for the technical adequacy of its work. The Owner's approval or acceptance of, or payment for, any of the Architect's services shall not be construed to operate as a waiver of any rights under this Agreement or of any cause of action arising out of the performance of this Agreement.

Notices. Any notice required under this Agreement shall be deemed served when placed in a properly addressed envelope, postage prepaid, and placed in the U.S. mail.

An alternative notice provision:

Notice. Any written notice required hereunder shall be deemed properly given, delivered, and service thereof completed when said notice is deposited in any Post Office or Post Office Box in a post-paid envelope properly addressed, or when said notice is sent by telegram, or when said notice is delivered in person to the party to whom it is addressed or their authorized representatives. The addresses of the Owner and the Architect set forth in the beginning of this Agreement shall be deemed the place to which written notice to them shall be directed; provided, however, that any such party or parties may by written notice to the others given pursuant to this paragraph designate a different address to which notices to it shall be directed, or designate the name and address of another person, firm, or corporation to whom notices to it may be directed.

Notices. All notices required hereunder must be in writing, delivered in person to the individual or member of the firm or entity to whom it was intended, or if delivered at or sent by certified mail, return receipt requested, to the last known business address of the party to whom notice is being given.

Another notice provision:

Notices. All notices required hereunder shall be deemed delivered if served by personal delivery, by facsimile, or by certified mail, return receipt requested, delivered to the last known address of the person to whom it is intended.

No Limitation. The duties and obligations imposed upon the parties under this Agreement, and the rights and remedies available hereunder shall be in addition to, and not a limitation of, any duties imposed or available at law or in equity.

This last provision favors the owner. An alternative clause favoring the architect is as follows:

Limitation. The duties and obligations imposed upon the parties under this Agreement, and the rights and remedies available hereunder are specific, and are limited to the duties, obligations and remedies specifically set forth in this Agreement. The parties hereto do not intend to create any duties, obligations or remedies not specifically set forth herein.

In the event of a dispute between the architect and owner, attorneys' fees can be substantial. The parties may want to insert a provision whereby the prevailing party recovers such fees from the loser. Possible language:

Attorneys' Fees. In the event any legal or arbitration proceedings are commenced between the parties to this Agreement to enforce any part of this agreement, the prevailing party in such proceedings shall be entitled, in addition to any other relief granted in such proceedings, to reasonable attorneys' fees, which shall be determined by the court or other forum in such proceeding, or in a separate proceeding brought for such purpose.

Publicity. The Architect shall not use and shall keep its employee(s), agent(s), and/or subcontractor(s) from using the name and/or trademark/logo of Owner or any subsidiaries or affiliates thereof in any sales or marketing publication or advertisement, without prior written consent of the Owner.

Force Majeure. If either party is rendered unable, wholly or in part, to carry out its obligations under this Agreement by reason of Force Majeure, other than the obligation to make money payments, such party shall give to the other party prompt written notice thereof with reasonably full particulars; thereupon, the obligations of the party giving notice, so far as they were affected by the Force Majeure, shall be suspended during, but no longer than, the continuance of the Force Majeure, and said party shall use all possible diligence to remove the Force Majeure as quickly as possible.

Not withstanding the foregoing, if, after commencement, any project is suspended or abandoned in whole or in part for more than six (6) months, Architect shall be compensated for services performed up to and including the date it received written notice from Owner of such suspension or abandonment, together with reimbursable expenses then due and payable in connection with any project. Payment of any fee to Architect on account of the terminated or abandoned project shall be made on the basis of a mutual determination by Architect and Owner of the percentage of work completed and/or the number of hours expended as of the date of abandonment or termination. If any project is resumed after being suspended for more than six (6) months, then the portion of the fee payable to Architect for the remaining services necessary to complete the project may, upon request of Architect, be adjusted if necessary as of the date of the resumption of the project to compensate Architect for costs directly related to the delay in the project. Any adjustment hereunder shall be based on the type of project and the circumstances surrounding the delay in completion.

"Force Majeure" as used in this Article shall mean an act of God, strike, lockout, or other industrial disturbance, act of the public enemy, war, blockage, riots, lightning, fire, flood, explosion, failure to timely receive necessary government approvals, so long as there was no negligence on the part of the Architect, government restraint, and any other cause, whether of the kind specifically enumerated above or otherwise, which is not reasonably within the control of the party claiming suspension.

No Damages. (a) Owner shall not, under any circumstances, be liable for any damages (whether foreseen, unforeseen, actual, consequential, or otherwise) suffered by the Architect, its agents, or subcontractors (or anyone else for whom the Architect may be liable) arising from or in connection with any injury or damage suffered while on or around the specific project site or any portion thereof.

(b) Notwithstanding anything to the contrary contained elsewhere in this Agreement, in no event shall the Architect or any subcontractors claim or receive any consequential or other special damages, or lost profits on account of any claim submitted in connection with this Agreement, including, without limitation, expenses arising from Owner's performance or nonperformance of the terms of this Agreement, or otherwise, or claim damages for delay for any reason, for which the exclusive and sole remedy shall be an extension of the time for completion of the services, if such is warranted and permitted by Owner.

Liens. Architect will pay when due all claims for services, materials or labor incurred at Architect's request in the performance of this Contract. To the fullest extent permitted by law, and provided that Owner is not in default in the payment of compensation to Architect in accordance with the terms and conditions of this Contract, Architect will indemnify, defend and hold harmless Owner and the Project from and against any and all mechanics' liens or stop notices of any kind or character whatsoever that may be recorded, filed or served with respect to the Project by Architect or Architect's Consultants arising out of or in any manner connected with the performance of this Contract or any subcontract made pursuant to or in connection with the performance of this Contract. Architect will, at its own expense, defend any and all actions based upon such mechanics' liens or stop notices and will pay all charges of attorneys and all costs and other expenses arising therefrom. If Architect fails to defend any such action to which Owner is a party, Owner may defend itself with counsel of its choice, and Architect will indemnify Owner from and against all costs and fees incurred by Owner in such action. If any such lien or stop notice is recorded or served with respect to the Project or Architect's Consultants arising out of or in any manner connected with the performance of this Contract, Architect will, at its sole cost and expense, immediately record or file, or cause to be recorded or filed, in the office of the appropriate public official in which such lien or stop notice was recorded, or with person(s) on whom such notice was served, a bond executed by a good and sufficient surety, and approved by Owner, in a sum equal to one and one-half (1-1/2) times the amount of such lien or stop notice, which bond will guarantee the payment of any amounts that Architect's Consultants may recover on the lien or stop notice together with any attorneys' fees and costs of suit in the action, if any, that such Consultants may recover therein.

Architect's Taxes and Licenses. (a) Architect will be solely responsible for filing any reports and returns with federal and state tax and labor authorities required by reason of the employment of its employees and will withhold and pay all taxes, assessments or other amounts levied or required by reason of the employment of such employees including, without limitation, federal income taxes, taxes under the Federal Insurance Contribution Act, applicable state income taxes, state unemployment, disability or similar taxes and other taxes or assessments required by Laws.

(b) If Architect is a sole proprietor, Architect will be solely responsible for reporting net earnings, if any, from the performance of the Services as self-employment income on federal and state income tax returns when required, and paying federal and state income taxes and payroll taxes on such income under the Federal Insurance Contribution Act and any applicable Laws. Architect will indemnify, defend and hold harmless Owner and the Indemnified Parties (as hereinafter defined) from and against the failure to pay all federal, state and local taxes or contributions imposed or required under employment insurance, social security and income tax or other Laws with respect to Architect and Architect's employees engaged in the performance of the Services.

(c) Owner will have no obligation to withhold federal or state income taxes or payroll taxes under the Federal Insurance Contribution Act or under state or federal employment, disability or other Laws from amounts due Architect hereunder for the performance of the Services or to pay employer payroll taxes thereon.

(d) Architect will not be covered by any policy providing workers' compensation insurance obtained by Owner or otherwise be entitled to workers' compensation insurance or benefits obtained by Owner in connection with the performance of the Services.

(e) Architect will not be eligible for or allowed to participate in any retirement plan, group insurance policy providing life insurance, disability insurance or hospital or medical benefits, or any other plan providing benefits to employees of Owner with respect to performance of the Services.

(f) Architect will obtain and maintain in effect any business or Architect license required of persons in similar capacity whether self-employed or as an employee of others and will pay any business taxes or fees required in connection therewith.

No Personal Liability. Notwithstanding anything to the contrary contained within this Agreement, no shareholder, member, officer, director, employee, trustee, beneficiary or agent of Architect shall be personally liable, directly or indirectly, under or in connection with this Agreement, or any document, instrument or certificate securing or otherwise executed in connection with this Agreement, or any amendments or modifications to any of the foregoing made at any time or times, heretofore or hereafter; and the Owner and each of its successors and assigns waives and does hereby waive any such personal liability.

§ 3.38   Assignments: ¶ 10.3

When an owner assigns the agreement, the architect may incur additional costs. To prevent this, this language could be inserted:

> **10.3.7.9** In the event of any assignment of this Agreement by the Owner, the Architect shall be compensated for any costs, including reasonable attorneys' fees, incurred by the Architect as a result of such assignment.

The architect may intend to form a corporation after the project begins. To allow an assignment to a corporation, the following language could be used:

> The Architect shall have the right to assign its rights and interests in this Agreement to any corporation, professional or otherwise, of which the Architect is the principal shareholder (or director, etc.) and which is authorized under the laws of the state of _____ () [the state where the project is located] to transact business and to perform architectural services. Notice of such assignment may be given by the Architect to the Owner at any time during the term of this Agreement.

Frequently, an owner wants to assign the contract to an entity, such as a land trust or partnership to be formed at a later date. In such a case, the "owner" would change in name only, with the same people dealing with the architect. The agreement should state this intention of the parties so that the transition will be easy. This language could be used:

> Notwithstanding the foregoing, the parties agree that this Agreement may be assigned by the Owner to a [state] Limited Partnership to be formed at a later date, said Limited Partnership to have the Owner as a general partner.

> In case of the death or disability of one or more but not all of the principals and/or partners of the Architect, the rights and duties of the Architect shall, at the election of the Owner, devolve upon the survivor or survivors of them who shall be obligated hereunder.

Another provision:

> **Assignment.** Owner is retaining Architect in reliance upon the special expertise and skills of Architect. Architect will not assign its rights or interests under this Contract without the prior written consent of Owner, which consent may be withheld in Owner's sole and absolute discretion. Any assignment or delegation of rights, duties or obligations hereunder made by Architect without the prior written consent of Owner will be void and of no effect, but will constitute a default by Architect hereunder. Subject to the foregoing, this Contract will be binding upon, and inure to the benefit of, the parties hereto and their respective successors, heirs, administrators and permitted assigns. In the event Owner consents to an assignment by Architect of its rights and interests pursuant to

this Contract, such consent will not relieve or excuse Architect of or from any of its obligations arising under this Contract unless such written consent expressly provides. Owner may assign Owner's rights or interest in this Contract at any time without Architect's consent.

Parties can use contracts as collateral for a loan. If the architect does this, the owner may wind up dealing with a bank. To avoid this, the owner may want to insert the following language:

The Architect shall not assign any monies due under this Agreement without the Owner's prior written consent.

Owner shall have the right without the Architect's consent to sell, assign, or transfer its rights, title, and interest in, to, and under this Agreement to any parent, subsidiary, affiliate, or controlled corporation or corporations of Owner, or to any successor to Owner by consolidation, merger, or other corporate action, or to a corporation or other business entity to which Owner may sell all or substantially all of its assets. Except as aforesaid, neither party shall assign or transfer its interest in this Agreement, or any right or obligation under it, by operation of law or otherwise, to any entity without the other party's prior written consent. Any attempted assignment without the consent of the other party will be void.

The following is a typical collateral assignment currently in use in the Chicago area. Note that the title is somewhat innocuous. There are better collateral assignments in use, but this one is shown as an example of what is frequently encountered. The author definitely does not recommend it for the reasons discussed following the form.

CONSENT AGREEMENT

THIS CONSENT AGREEMENT is entered into as of this _____(3) day of _____(3) 20_____(3) by _____(3) ("Architect"), to and for the benefit of ABC BANK, a corporation organized under the laws of the United States of America ("Lender").

RECITALS:

A. Concurrently herewith, the Lender has agreed to make a $ _____(5) construction loan (the "Loan") to _____(5) , not personally, but solely as Trustee under Trust Agreement dated _____(5) , 20_____(5) , and known as trust No. _____(5) (the "Trust") and its sole beneficiary, _____ (5) , a _____(5) corporation (the "Beneficiary") (the Trust and the Beneficiary are hereinafter jointly referred to solely as "Borrower").

B. The Trust owns title to the real estate (the "Property") located in _____(5), _____(5) County, [state], and legally described in Exhibit A attached hereto.

C. _____(5) , as owner, has entered into an agreement (the "Contract") with the Architect dated _____(5) , 20_____(5) , pursuant to which the Architect has agreed to perform architectural design and supervision services in connection with certain site improvements to be constructed on the Property.

D. As a condition to the Lender's making the loan to the Borrower, the Lender has required that: (i) the Borrower collaterally assign the Contract to the Lender pursuant to an assignment (the "Assignment") in the form of Exhibit B attached hereto; and (ii) the Architect consent to the Assignment in favor of the Lender and agree that the Architect will continue to perform its duties and obligations under the Contract in the event of a default by the Borrower under the Assignment or the other documents and instruments evidencing or securing the Loan (collectively, the "Loan Documents").

E. The Architect will benefit if the Loan is made to the Borrower, because a portion of the proceeds of the Loan will be used to pay amounts owing to the Architect under the Contract.

NOW, THEREFORE, in consideration of the foregoing and other good and valuable consideration, the receipt and sufficiency of which are hereby acknowledged, the Architect hereby agrees as follows:

1. The Architect hereby consents to the Assignment and agrees, that (a) upon the occurrence of any event of default by the Trust and/or the Beneficiary in their respective obligations under the Loan Documents, which default is not cured within any applicable grace period expressly provided for in any such Loan Document, and (b) the giving of written notice from the Lender to the Architect of such default, the Lender shall have the right to use all plans and specifications prepared by the Architect, and the Architect will perform all of its duties and obligations under the Contract for the benefit of the Lender regardless of any default by either the Borrower or any other party under the Contract.

2. The Architect hereby acknowledges that the Assignment and the exercise by the Lender of any of its rights and remedies thereunder shall not render the Lender liable to the Architect for any obligations of either the Borrower or any other party under the Contract.

3. The Architect certifies that no party is in default under the Contract as of the date hereof and there are not any circumstances which would constitute a default upon the passage of time, the giving of notice, or both. The Architect agrees not to enter into any amendment or modification of the Contract without the prior written consent of the Lender.

4. The Architect agrees to give prompt written notice to the Lender of any default or breach by the Borrower of any of its obligations or duties under the Contract, and that the Lender shall have an opportunity to remedy or cure such breach within thirty (30) days after receipt of such notice thereof.

5. This Agreement shall be binding upon and inure to the benefit of the Lender, Architect, and their respective successors and assigns.

IN WITNESS WHEREOF, this Agreement has been executed and delivered as of the date first above written.

This consent agreement has a number of problems for the architect. In ¶ C, the agreement refers to "supervision." Many lenders and owners make this mistake. The word "observation" should be substituted. Paragraph E states that the architect benefits from this collateral assignment. This is certainly questionable, because the lender in ¶ 2 states that it will not be liable under the owner-architect agreement. This is an attempt to show consideration, a legal term that means "something of value." Consideration is required for both parties to a contract because without it there is no valid contract, although in this author's opinion there is no consideration for most of these collateral assignments.

Paragraph 1 states that the architect will perform, regardless of default by the owner. This contradicts ¶ 9.1 of AIA Document B101, which permits suspension of the architect's services in the event of failure to make payments in a timely manner. Paragraph 2 states that the lender is not obligated to make payments to the architect, presumably even after the assignment has taken place. At the least, the lender should be responsible for payments after the date of assignment, when the architect's services would directly benefit the lender. Paragraph 3 requires the architect to give the lender written notice before modifying or amending the owner-architect agreement. It is unclear whether this would include additional services. Paragraph 4 requires prompt notice to the lender of a default by the owner, and 30 days for the owner to cure, which presumably means lack of payment to the architect, which again contradicts ¶ 9.1 of B101. This provision is unclear as to whether the architect can suspend performance pursuant to ¶ 9.1 of B101.

It should be clear that any agreement such as this should be carefully reviewed before signing.

§ 3.39 Submittal of Certifications: ¶ 10.4

The following additional language may be helpful here:

10.4 The Owner agrees that the Architect shall not be required to execute any certification that extends its duties or liability beyond that contemplated by this Agreement. In the event that the Architect executes any such certificate that has the effect of extending the Architect's liability, the Owner acknowledges that such execution by the Architect shall not operate as a waiver of this provision, but shall be considered a mistake of fact or law, and the Owner shall indemnify and hold the Architect harmless from any damages or liability of any nature whatsoever, including attorneys' fees, that are in any manner related to the execution of such certification.

§ 3.40 Hazardous Materials: ¶ 10.6

A more comprehensive treatment of the issue of hazardous materials might include the following language:

10.6 The Owner acknowledges and agrees that the Architect is unable to obtain professional liability or other insurance for claims arising out of or related to the performance or failure to properly perform professional services related to the investigation, detection, abatement, replacement, removal, or detoxification of asbestos, asbestos-related materials, PCB, or other hazardous materials, products, or substances that are or may be toxic or dangerous. Accordingly, the Owner hereby agrees that no claim, cause of action, suit, arbitration, or otherwise, whether for breach of contract, negligence, indemnity, or any other cause of action, will be brought by the Owner against the Architect, its employees, directors, officers, agents, or their respective assigns arising out of the presence of any such materials, or out of the failure to detect such materials, or out of the improper removal or treatment of any such materials. The Owner further agrees to hold the Architect harmless from any causes of action, claims, or arbitrations accruing to any person, firm, or other legal entity arising out of or related to any injury, death, property damage, or other claim, related to the presence of any such toxic or other hazardous material. The only event in which this indemnification shall not apply is if the Architect is found to be solely responsible for any such damages or losses by a court of competent jurisdiction.

Another provision:

10.6 The Architect shall immediately notify the Owner and Contractor of any hazardous materials or toxic substances in any form at the Project site of which the Architect becomes aware, including the location thereof. However, the Architect shall have no responsibility for such hazardous materials or toxic substances beyond such notification.

Another provision that provides that the architect will be paid its additional costs relating to hazardous materials:

10.6.1 In the event of the discovery of any hazardous materials or toxic substances, the Architect shall be paid its additional costs and fees relating thereto as Additional Services.

§ 3.41 Architect's Promotional Materials: ¶ 10.7

The owner may prefer to add a clause such as this to ¶ 10.7:

The Architect shall in each instance obtain the prior written approval of the Owner concerning exact text and timing of news releases, articles, brochures, advertisements, prepared speeches, and other information releases concerning this Agreement or the Project.

In the alternative, the owner may prefer to substitute a stronger clause concerning the release of project information:

10.7 The Architect shall not divulge information concerning this project to anyone (including, without limitation, information in applications for permits, variances, etc.) without the Owner's prior written consent. The Architect shall obtain a similar agreement from firms, consultants, and others employed by it. The Owner reserves the right to release all information as well as to time its release, form, and content. This requirement shall survive the expiration of the contract.

§ 3.42   Confidentiality: ¶ 10.8

The owner may want to substitute this provision in place of ¶ 10.8:

10.8 The Architect acknowledges that certain of the Owner's valuable, confidential, and proprietary information may come into the Architect's possession. Accordingly, the Architect agrees to hold all information it obtains from or about the Owner in strictest confidence, not to use such information other than for the performance of the services, and to cause any of its employees or consultants to whom such information is transmitted to be bound to the same obligation of confidentiality to which the Architect is bound. The Architect shall not communicate the Owner's information in any form to any third party without the Owner's prior written consent. In the event of any violation of this provision, the Owner shall be entitled to preliminary and permanent injunctive relief as well as an equitable accounting of all profits or benefits arising out of such violation, which remedy shall be in addition to any other rights or remedies to which the Owner may be entitled.

§ 3.43   Article 11: Compensation

Additional language:

Payments to the Architect during the Design Development, Construction Documents, and Bidding or Negotiation Phases shall be made on a monthly basis upon billing submitted by the Architect based on percentage of work completed in each Phase. Payments to the Architect for the Construction Phase shall be made in equal monthly installments based on the projected period of construction. Payment for the Project Turnover and Closeout Phase shall be made upon final payment to the Contractor after issuance of the Final Certificate of Payment, or 60 days after building occupancy, whichever occurs first.

Other provisions:

Architect shall submit an itemized monthly statement of services. Architect's statement of services must state the phase and be in sufficient detail to allow Owner to monitor the activities of Architect to ensure that Architect is properly working toward the same objective as Owner and that Architect has not underestimated the expense of accomplishing each phase. The invoice should also specify the subparts of each phase which have been completed at that point.

For any work billed on an hourly basis, Architect will provide Owner with the hourly rate for each person who will perform work for Owner and obtain Owner's approval of the scope of and final charges for such services before beginning work. The rate will be the lowest charged by such persons for similar work performed for any other client. All additional services and reimbursable expenses shall be preapproved by Owner before Architect commences such work.

Copies of the records of reimbursable expenses, expenses pertaining to additional services, and services performed on the basis of direct personnel expense shall be delivered to Owner or Owner's authorized representative with each of Architect's statements of services.

Language favorable to an owner is:

11.4.1 Engineering consultants shall be paid at the hourly fixed rates agreed to by the Owner but not in excess of the above set forth rates applicable to Architects.

§ 3.44 Reimbursable Expenses: ¶ 11.8

The owner may want to insist that local travel be included in basic services. A typical provision might read:

Included in basic services are costs of travel within the [city] Metropolitan area.

The following language in favor of the owner can be substituted for ¶¶ 11.8.1.1 through 11.8.1.7:

11.8.1.2 The Owner shall, in addition to the amounts of compensation described herein, reimburse the Architect for the following and only the following costs and expenses. All other costs, expenses, or charges, including but not limited to all compensation and benefits paid to Architect's employees, incurred by the Architect in connection with the Project, shall be paid by the Architect without reimbursement from the Owner, and the Architect hereby indemnifies and agrees to hold the Owner harmless from and against payment of same:

11.8.1.2.1 Expenses of transportation and living when traveling beyond fifty (50) miles of the site when specifically authorized in writing by the Owner in advance, and long-distance calls and telegrams incurred in connection with the Project.

11.8.1.2.2 Expense of reproduction, postage, and handling of working drawings and specifications delivered to contractors, the Lender, Owner, and others at the Owner's direction. All other expenses for reproduction, for permits, for the Owner's review and approval, for the Architect's internal purposes, or for any other purpose whatsoever are to be at the Architect's sole expense. Expenses of printing services shall be reimbursable by the Owner only in the event the Owner has specifically authorized them in advance.

11.8.1.2.3 Expense of renderings or models for the Owner's use in excess of those provided herein.

11.8.1.3 Records of all reimbursable expenses and expenses pertaining to any change in services on the Project and for services performed on the basis of flat rates shall be kept on a generally recognized accounting basis and shall be available to the Owner or its authorized representative during business hours at the Architect's office, and copies thereof shall be made and presented to the Owner with each request for payment.

Other provisions change ¶ 11.8.1 to read:

Eligible reimbursable expenses include, but are not limited to: site surveying, subsurface borings, site construction testing, building construction testing and quality control, topographic survey, renderings, detailed models (when specifically requested by Owner), environmental impact statements, highly specialized consultants (when specifically authorized by Owner), overtime work and messenger services (when specifically requested by the Owner), and bidding documents. Noneligible expenses which are not reimbursable include but are not limited to: consultants hired by the Architect, meals, lodging, mileage (except for out-of-state meals, lodging, and mileage), postage and handling of submittals and correspondence, handling of bid documents, telephone expenses (except for long distance telephone calls), as-built drawings. The Architect shall employ at its own expense such structural, elevator, mechanical, electrical, civil engineers, and such other consultants as it deems necessary for its effective performance of services herein described. The names of said consultants shall be submitted to the Owner. The Architect shall be responsible for the services of and payment to its consultants. The Architect shall perform all of its services in conformity with the standards of the reasonable care and skill of the profession. The Architect shall be responsible for the performance of consultants or persons retained by the Architect as if performed by it, but the Architect shall not be responsible for the performance of consultants or persons retained or employed by the Owner or others, or consultants Owner directs to be retained by the Architect not related to design or construction services. Owner's right to review the work of the Architect, as hereinafter provided, shall not be construed as relieving the Architect from its professional and legal responsibility consistent with the services required under this Agreement.

§ 3.45 Payments to the Architect: ¶ 11.10

Subparagraph 11.10 is changed to read:

11.10 Payments due the Architect shall be made within 30 days after such Invoice is submitted to the Owner for payment, provided that said Invoice is correct. Interest on late payment shall commence 30 days after the correct Invoice is submitted and shall be at the rate of twelve percent (12%) per annum. An invoice shall be considered submitted when delivered or on the third day after deposit in the U.S. Mail. If the amount of any invoice is disputed by the Owner, the Owner shall pay the undisputed portion pursuant to this provision.

§ 3.46 Indemnification Clauses

An indemnification clause is a contract by which one engages to save another party from a legal consequence of the conduct of one of the parties or of some other person.[2] It is a unilateral agreement whereby one person pays the costs of another under certain circumstances. These circumstances should be clearly spelled out, or the indemnified party runs the risk that it will not be indemnified. Normally, a provision for attorneys' fees in an indemnity clause does not cover a contract action between the parties to the contract. Many states have laws that limit the effect of indemnification clauses. See discussion in § **4.28.**

In the following clause, the owner indemnifies the architect:

12.1.1 To the fullest extent permitted by law, the Owner shall waive any right of contribution and shall indemnify and hold harmless the Architect and its agents, employees, [and "officers and directors," in the case of architectural corporations] and consultants from and against any and all claims, damages, losses, and expenses, including any attorneys' fees, arising out of, resulting from, or in connection with the performance of the Work described herein, but only if such claim, damage, loss, or expense is caused in whole or in part by the Owner, its employees, agents, officers, directors, or any other party directly or indirectly employed by any of them or any party for whose acts any of them may be liable, regardless of whether or not it is caused by a party indemnified hereunder. Such obligation shall not be construed to reduce or negate any

---

[2] Myers Bldg. Indus., Ltd. v. Interface Tech., Inc., 17 Cal. Rptr. 2d 242 (Ct. App. 1993).

other right or obligation of indemnification that would otherwise exist as to any party hereto. This indemnification shall not apply to the liability of the indemnitee arising out of its own negligence. This indemnification shall not be limited in any way because of any limitation on damages, compensation, or benefits under any statute, law, or governmental requirement of any sort.

12.1.2 The following shall be included within the definition of "expenses" herein: (a) any time expended by the indemnified party or its employees, agents, officers and directors at their usual and customary billing rates, as well as all out-of-pocket expenses such as long-distance telephone calls, costs of reproduction, expenses of travel and lodging; (b) all costs and expenses of experts, consultants, engineers, and any other party retained by the indemnified party reasonably required to defend the claim; (c) all costs, including reasonable attorneys' fees, incurred in bringing any action to enforce the provisions of this indemnification. The following shall be included within the definition of "action" herein: any case brought in any state or federal court, any arbitration, any mediation, and any similar forum for resolution of any dispute herein, and shall also include any counterclaim or third-party action in any such forum.[3]

In the provision below, the architect and owner indemnify each other:

12.1.1 The Architect and the Owner each hereby agree to defend, indemnify, and save harmless the other party, its officers, servants, and employees, from and against any and all liability, loss, damage, cost, and expense (including attorneys' fees and accountants' fees) caused by an error, omission, or negligent act of the indemnifying party in the performance of services under this Agreement. The Architect and the Owner each agree to promptly serve notice on the other party of any claims arising hereunder, and shall cooperate in the defense of any such claims. In any and all claims asserted by any employee of the Architect against any indemnified party, the indemnification obligation shall not be limited in any way by any limitation on the amount or type of damages, compensation, or benefits payable by or for the Architect or any of the Architect's employees under workers' compensation acts, disability benefit acts, or other employee benefit acts. The acceptance by the Owner or its representatives of any certification of insurance providing for coverage other than as required in this Agreement to be furnished by the Architect shall in no event be deemed a waiver of any of the provisions of this indemnity provision. None of the foregoing provisions shall deprive the Owner of any action, right, or remedy otherwise available to the Owner at common law.

The following language indemnifies the owner:

12.1.1 To the extent permitted by law, the Architect, on behalf of itself and its agents (all of said parties are herein sometimes collectively referred to as the "Indemnitors"), waives any right of contribution against and shall indemnify, protect, defend, save, and hold the Owner, all entities related to the Owner, all

---

[3] *See generally* Mareci v. George Sollitt Constr. Co., 73 Ill. App. 3d 418, 392 N.E.2d 225 (1979).

principals of the Owner or its related entities, their respective agents, employees, partners, and anyone else acting for or on behalf of any of them (all of said parties are herein collectively referred to as the "Indemnitees") harmless from and against all liability, damage, loss, claims, demands, actions, and expenses, including but not limited to attorneys' fees of any nature whatsoever that arise out of or are connected with or are claimed to arise out of or be connected with: (i) the negligent performance of work to be performed by the Architect hereunder (or any negligent act or omission of the Indemnitors), or (ii) the failure of any Indemnitor to comply with the laws, statutes, ordinances, or regulations of any governmental or quasi-governmental authority.

[Here the architect would want to insert an indemnification clause from the owner for any loss because of the negligence of others.]

12.1.2 Without limiting the generality of the foregoing, the indemnity hereinabove set forth shall include all liability, damages, loss, claims, demands, and actions on account of personal injury, death, or property loss to any Indemnitee, any of Indemnitee's employees, agents, licensees, or invitees, or to any other persons, whether based on or claimed to be based on statutory, contractual, tort, or other liability of any indemnitor or any other persons. Without limiting the generality of the foregoing, the liability, damage, loss, claims, demands, and actions indemnified against shall include all liability, damage, loss, claims, demands, and actions for trademark, copyright, or patent infringement, for unfair competition or infringement of any other so-called "intangible" property right, for defamation, false arrest, malicious prosecution, or any other infringement of personal or property rights of any kind whatever or which arise out of failure of the Indemnitors to discharge the duties specified herein. The promise of indemnification in this paragraph shall not be construed to indemnify any Indemnitee for any loss or damage attributable to the negligent acts or omissions of such Indemnitee. Liability hereunder shall not extend to the liability of the Construction Manager or any Separate Construction Manager, their agents, or employees arising pursuant to the Contract Documents. Any Indemnitee shall be entitled to recover all costs and expenses, including attorneys' fees, from any Indemnitor whom the Indemnitee has had to compel by legal process to abide by the terms of this provision.

The following provision requires that both the owner and contractor indemnify the architect. Under this provision, the architect can insist that the contract documents contain this language:

12.1.1 The Owner will indemnify and hold harmless the Architect, its agents, employees, and consultants from and against any and all claims, losses, and expenses, including reasonable attorneys' fees, arising out of, in connection with, or resulting from the performance of the Work or of any professional services by any other person or entity with whom the Owner contracts for the Project; and from and against all claims, damages, losses, and expenses, including but not limited to attorneys' fees, arising out of, in connection with, or resulting from the performance (or failure to perform) of the Work where there has been a deviation from any Contract Document not approved by the

Architect or where there has been a failure to follow any written recommendation of the Architect, provided that this indemnity shall not apply to any claims, losses, or expenses arising out of the negligence of the Architect. In the event the Architect or any other party indemnified hereunder is required to incur expenses, including attorneys' fees, to enforce this indemnity obligation, such expenses shall be paid by the Owner.

12.1.2 The Owner shall require as one of the terms of its agreement with any Construction Manager, General Contractor, Subcontractors, Trade Contractors, or others with whom it has any agreement to perform any work on the Project (collectively hereinafter referred to as "Contractor") that:

12.1.2.1 The Contractor shall obtain Comprehensive General Liability insurance with broad form Property Damage coverage and a contractual liability endorsement. The Owner shall also make the following indemnity a part of its agreement with any Contractor and specify that the contractual liability endorsement required hereunder fall within the Contractor's coverage;

12.1.2.2 The Contractor shall name the Architect as an additional insured on the Contractor's Comprehensive General Liability insurance policy. The additional insured endorsement shall state (a) that the coverage afforded the Architect shall be primary insurance and include a Waiver of Subrogation in favor of the Architect (without any obligation on the part of the Architect to pay premiums or otherwise to the concerned carrier) with respect to claims arising out of operations performed by or on behalf of the Contractor, (b) that if the Architect has other insurance that is applicable to a loss, such other insurance shall be on an excess or contingent basis, and (c) that the amount of the insurance company's liability under this insurance policy shall not be reduced by the existence of such other insurance;

12.1.2.3 In the event of a claim or demand by the Contractor (arising from any reason whatsoever related to this Agreement) against the Architect, the Contractor is limited to a maximum total recovery of one-half of the amount of fees that the Architect has collected from the Owner for the Project hereunder, or the sum of Fifty Thousand Dollars ($50,000.00), whichever is less; all contractors shall include such a provision in their agreements with subcontractors and all agreements between the Owner and Contractor and Subcontractors shall state that the Architect is the third-party beneficiary of this damage limitation provision; and

12.1.2.4 The Contractor waives its right to make a claim against the Architect after more than three years from the date of Substantial Completion have elapsed, and the Architect is the third-party beneficiary of this limitation provision.

Another method of limiting some of the architect's liability is the following provision:

12.1.1 In any action of any nature, without limitation, by the Owner, its assignees, or successors against the Architect, its agents, employees, [and "officers and directors" in the case of architectural corporations] and consultants, any recovery by such Owner shall be limited to the contract price listed in this Agreement [If no single fee is listed, then "to the amount billed, or due and owing to the Architect, whichever is greatest, at the time such action is commenced."].

This type of clause is useful in smaller projects and when the architect does not carry errors and omissions coverage.[4] It does not cover the architect if someone other than the owner files an action.

---

[4] In W. William Graham, Inc. v. City of Cave City, 289 Ark. 104, 709 S.W.2d 94 (1986), the court construed the following clause:

> The Owner agrees to limit the Engineer's liability to the Owner and to all Construction Contractors and Subcontractors on the Project, due to the Engineer's professional negligent acts, errors or omissions, such that the total aggregate liability of the Engineer to those named shall not exceed Fifty Thousand Dollars ($50,000.00) or the Engineer's total fee for services rendered on this project, whichever is greater.

The court found that the clause did not cover actions for breach of contract and the damages that flowed from such a breach. Therefore, the limit does not apply.

California has allowed such a limitation of liability when the parties have had the opportunity to accept, reject, or modify the provision. Markborough Cal. v. Superior Court, 227 Cal. App. 3d 705, 277 Cal. Rptr. 919 (1991).

In Leon's Bakery, Inc. v. Grinnell Corp., 990 F.2d 44 (2d Cir. 1993), a business whose property was damaged by a fire brought an action against the company that manufactured and installed the fire protection system. The claim was barred by a limitation of liability provision that stated as follows:

LIMITATIONS OF LIABILITY

> In no event shall Seller be liable for special or consequential damages and Seller's liability on any claim whether or not based in contract or in tort or occasioned by Seller's active or passive negligence for loss or liability arising out of or connected with this contract, or any obligation resulting therefore, or from the manufacture, fabrication, sale, delivery, installation, or use of any materials covered by this contract, shall be limited to that set forth in the paragraph entitled "Warranty."

WARRANTY

> Seller agrees that for a period of one (1) year after completion of said installation it will, at its expense, repair or replace any defective materials or workmanship supplied or performed by Seller. Upon completion of the installation, the system will be turned over to the Purchaser fully inspected, tested and in operative condition. As it is thereafter the responsibility of the Purchaser to maintain it in operative condition, it is understood that the Seller does not guarantee the operation of the system. Seller further warrants the products of other manufacturers supplied hereunder, to the extent of the warranty of the respective manufacturer. ALL OTHER EXPRESS OR IMPLIED WARRANTIES OF MERCHANTABILITY OR FITNESS OR OTHERWISE ARE HEREBY EXCLUDED.

In Georgetown Steel Corp. v. Union Carbide Corp., 806 F. Supp. 74 (D.S.C. 1992), the engineer had the following limitation of liability provision in its contract:

> WARRANTY AND LIMITATION OF LIABILITY The only warranty or guarantee made by Law Engineering Testing Company in connection with the services performed hereunder, is that we will use that degree of care and skill ordinarily exercised under similar conditions by reputable members of our profession practicing in the same or similar locality. No other warranty, expressed or implied is made or intended by our proposal for consulting services or by our furnishing oral or written reports.

The following is an alternative clause that limits the architect's liability:

12.1.1 The Architect shall exercise that degree of skill and care ordinarily exercised under similar circumstances by other architects. However, the Architect makes no express or implied warranties by its provision of services under

> Our liability for any damage on account of any error, omission, or other professional negligence will be limited to a sum not to exceed $50,000 or our fee, whichever is greater. In the event the client does not wish to limit our professional liability to this sum, we agree to waive this limitation upon receiving client's request, and agreement by the client to pay additional consideration of 4% of our total fee or $200.00 whichever is greater.

In Valhal Corp. v. Sullivan Assocs., Inc., 44 F.3d 195, 205-06 (3d Cir. 1995), the architect had the following provision in the contract:

> The OWNER agrees to limit the Design Professional's liability to the OWNER and to all construction Contractors and Subcontractors on the project, due to the Design Professional's professional negligent acts, errors or omissions, such that the total aggregate liability of each Design Professional shall not exceed $50,000 or the Design Professional's total fee for services rendered on this project. Should the OWNER find the above terms unacceptable, an equitable surcharge to absorb the Architect's increase in insurance premiums will be negotiated.

In Pitts v. Watkins, 905 So. 2d 553 (Miss. 2005), involving a home inspection agreement, the Mississippi Supreme Court held the limitation of liability provision and arbitration provision unconscionable. The agreement required arbitration of any claim by the homeowner, but the inspector could bring a court action. The inspector's liability was capped at his fee—$265. "The limitation of liability clause, when paired with the arbitration clause, effectively denies the plaintiff of an adequate remedy and is further evidence of substantive unconscionability."

In Bolingbrook Hotel Corp., Inc. v. Linday, Pope, Brayfield & Assocs., Inc., 2005 WL 1226058 (N.D. Ill. Apr. 20, 2005), the owner sued the architect for improperly designing the plumbing pipes, resulting in inadequate water pressure. The contract contained this indemnification clause:

> In any project there will be ambiguities, inconsistencies, errors, and omissions in the Construction Documents. These conditions may result in changes to the Contract Work, Amount, and Time. To the extent of a Contingency Reserve of 5% of the Construction Cost, the Owner agrees to indemnify and hold the Architect harmless with respect to all claims, awards, or legal actions arising out of any suit concerning the aforementioned changes.

The architect argued that this provision limits claims against it by the owner. The owner argued that this language did not apply to design flaws, but rather to errors such as typos and other minor items. The court agreed with the owner, holding that the language indemnifies the architect to the extent that errors in the construction documents that result in changes to the contract work, amount, or time cause damages greater than five percent of the construction cost. Thus, it only covered errors that lead to changes to the contract work, amount, or time, none of which applied to the errors alleged by the complaint. This indemnification provision did not protect the architect.

The court found that this clause was not against public policy, stating that

> an architectural firm and real estate developer have attempted to allocate risks between themselves in such a way that neither is relieved from liability for its own negligence. We see no reason to hold that the policy enunciated in section 491 [68 Pa. Cons. Stat. Ann. § 491] precludes them from doing so.

Because the limit of the architect's liability was $50,000, the federal court lacked diversity jurisdiction. 28 U.S.C. § 1332(a) requires that the amount exceed $50,000. The *Valhal* case was followed in Marbro, Inc. v. Borough of Tinton Falls, 688 A.2d 159 (N.J. Super. Ct. Law Div. 1996).

this Agreement. The Owner and Architect have discussed their respective risks, rewards, and benefits of the Project and the Architect's total fee for services, and they have allocated the risks such that if the Owner makes a claim against the Architect, the Owner is limited in the amount which it may recover as "Damages" to the lesser of the amount of fees that the Architect has collected from the Owner hereunder or Fifty Thousand Dollars ($50,000). This provision shall be applicable to the fullest extent permitted by law. The term "Damages" as used herein includes but is not limited to any type of damages that are or could be awarded by any court or arbitration panel, such as, by way of general example, tort damages, contract damages, strict liability damages, liquidated damages, quantum meruit damages, and/or punitive damages. The Owner will make no claim against the Architect more than two years after the date of Substantial Completion of the Project.

The architect may wish to insert a clause that requires the owner to reimburse the architect for attorneys' fees except in the event the architect is found negligent:

12.1.1 The Owner shall promptly reimburse the Architect for all costs and expenses, including attorneys' fees, incurred by the Architect in defending any claim, in any court or in any arbitration proceeding, whether before or after substantial completion, arising out of or in connection with the work described herein. This indemnification shall not apply in the event the Architect is found to be negligent in the performance of any duty hereunder.

Another provision:

12.1.1 The Architect states that its consultants, subcontractors, agents, employees, and officers shall, to the best of the Architect's knowledge, information, and belief, possess the experience, knowledge, and character necessary to qualify them individually for the particular duties they perform in connection with the project and shall promptly, upon notice or discovery, make necessary revisions or corrections of errors, ambiguities, or omissions in its drawings and specifications for the project without additional compensation. Acceptance of the Architect's drawings and specifications by Owner shall not relieve the Architect of responsibility for subsequent corrections of its errors or omissions or for the clarification of any such ambiguities in the drawings and specifications.

12.1.2 The Architect hereby agrees to defend, indemnify, keep and save harmless Owner, its respective board members, officers, agents, and employees, in both individual and official capacities, against all suits, claims, damages, losses, and expenses, including attorneys' fees, which are the result of an error, omission, or negligent act of the Architect or any of its employees or agents arising out of or resulting from the performance of service under this Agreement, except where such is the result of the active negligence of the party seeking to be indemnified. The Owner hereby agrees to defend, indemnify, keep and save harmless Architect, its respective officers, agents, and employees, in both individual and official capacities, against all suits, claims, damages, losses, and expenses, including attorneys' fees, which are the result of an error, omission, or negligent act of the Owner or any of its employees or agents arising out of or resulting from the performance of service under this Agreement, except where such is due to the active negligence of the party seeking to be indemnified. This Paragraph is applicable to the full extent as allowed by the laws of the state of _____(5) and not beyond any extent which would render this provision void or unenforceable.

CHAPTER 4

AIA DOCUMENT A201 GENERAL CONDITIONS OF THE CONTRACT FOR CONSTRUCTION[1]

§ 4.1 Introduction

These General Conditions form the basic document of the contract for construction, along with the drawings and specifications and the agreement between the owner and contractor. This document provides the framework within which the parties operate and defines the relative duties and obligations of the parties. An understanding of this particular document is essential to any analysis of a construction project, because it has been developed by the industry itself. Although it cannot cover every situation, it does provide the foundation upon which any construction contract can be built.

This document has been a construction industry standard since it was first published in 1911 and is the most widely used document of its kind. The 2007 version (the 16th edition) cannot be mixed with related documents of a different generation. Thus, only the 2007 version of AIA Document A101, Owner-Contractor Agreement, can be used with it, and not any prior version.[2] The same holds true for the other related documents. Contact the national AIA office for a list of available documents. A number of revisions were made to the 2007 version of the General Conditions from the 1997 version. At the end of each paragraph of the 2007 version, which appears in italics, is a boldface reference to the 1997 paragraph, if any, and a short indication of the revision made to that paragraph, if any.

The provisions of the General Conditions can and should be modified to suit each particular project. Such modifications generally appear in the Supplementary Conditions, which are project-specific. AIA Document 511, Guide for Supplementary Conditions, provides alternative language for commonly encountered situations and should be consulted when preparing this document. Additional modifications of an administrative nature, such as shop drawing procedures, are usually placed in Division One of the specifications.

CAUTION: The AIA has made these documents available in electronic format, so that changes can be made by computer directly on the document. This results in a single document incorporating all changes. The next-best method of amending A201 is to attach separate written amendments that refer back to A201. This is normally done in supplementary general conditions. An alternative method is to graphically delete material and insert new material in the margins of the standard form. However, this method may create ambiguity, and it may result in the stricken material's being rendered illegible, and is the least preferred method.

---

[2] An example of problems caused by mixing documents is found in Eis Group/Cornwall Hill v. Rinaldi Constr., 154 A.D.2d 429, 546 N.Y.S.2d 105 (1989), in which the owner-contractor agreement was a 1987 AIA form that incorporated the 1987 version of A201. However, the 1976 version of A201 was actually attached to the contract, along with various modifications that included striking the arbitration clause. The court found an ambiguity (obviously) and declined to order arbitration.

The cited language of this document is as published by AIA in late 2007 in the Electronic Documents. There appear to be some minor inconsistencies between this document and others that are probably attributable to scrivenors' errors. AIA often corrects such errors in intervening years without any notice and without any way to determine exactly when a particular version of a document was actually released.

It is illegal to make any copies of this document in violation of the AIA copyright or to reproduce this document by computerized means, other than as specifically licensed by the AIA.

§ 4.2 Article 1: General Provisions

This article defines the basic elements of the construction contract, for example, what specifically is meant by contract documents, project, work, and drawings. It also includes some guidance concerning contract intent, execution, and interpretation, as well as provisions for copyright protection.

§ 4.3 Basic Definitions: ¶ 1.1

1.1 BASIC DEFINITIONS

1.1.1 The Contract Documents

The Contract Documents are enumerated in the Agreement between the Owner and Contractor (hereinafter the Agreement), and consist of the Agreement, Conditions of the Contract (General, Supplementary and other Conditions), Drawings, Specifications, Addenda issued prior to execution of the Contract, other documents listed in the Agreement and Modifications issued after execution of the Contract. A Modification is (1) a written amendment to the Contract signed by both parties, (2) a Change Order, (3) a Construction Change Directive or (4) a written order for a minor change in the Work issued by the Architect. Unless specifically enumerated in the Agreement, the Contract Documents do not include the advertisement or invitation to bid, Instructions to Bidders, sample forms, other information furnished by the Owner in anticipation of receiving bids or proposals, the Contractor's bid or proposal, or portions of Addenda relating to bidding requirements. (**1.1.1; reorganize paragraph, add "other information furnished . . .")**

These are the documents that form the contract between the owner and contractor for the construction work involved. The owner-architect agreement is not a part of the contract documents; neither are the bidding documents. Many architects and owners include the bid form and invitation to bid as part of the contract documents. This should be done with care, with a provision for inconsistencies between these bid documents and the others. Other documents that could be included as contract documents are contractor's proposals, shop drawings, and product data (in which case ¶ 3.12.4 must be deleted or modified). The contract documents may only be changed by one of the listed modifications and must be in writing. It is possible for the parties to agree to oral modifications, either specifically or by a course of conduct. This should be avoided. All modifications should be in writing to avoid later problems that could result in litigation.[3]

---

[3] In Campos Constr. Co. v. Creighton Saint Joseph Reg'l Health Care Sys., 1999 Neb. App. LEXIS 328 (Dec. 7, 1999), there was an issue as to whether a contract had been formed. The contractor submitted a bid in response to an invitation to bid. The architect then informed the contractor that it was the low bidder and was being awarded the contract. Later, the architect sent the contractor several copies of an owner/contractor agreement with directions to sign the copies and return them. The contractor did this, but the owner never forwarded a signed copy of the agreement or a formal notice to proceed to the contractor. Meanwhile, the contractor began entering into various contracts with subcontractors and suppliers while communicating with the architect about project progress, plan clarifications, and material substitutions. The architect responded to many of these requests and shop drawing submittals. Some time later, the owner informed the contractor that the project was being canceled for lack of funding. The trial court found that there was no contract. The appellate court reviewed the contract documents to determine exactly what the owner had to do to accept the contractor's bid, which constituted an offer. The only provision that applied was ¶ 1.5.1 (1997 version, now deleted), which required both parties to sign the Contract Documents. However, this provision was expressly deleted by the Supplementary General Conditions. The court held that the absence of a signature is not necessarily fatal to the contract claim. The contractor's position was that the architect accepted the offer as the owner's agent. The appellate court remanded the case to the trial court for a determination as to whether there was a contract.

In Cumberland Cas. & Sur. Co. v. Nkwazi, L.L.C., 2003 WL 21354608 (Tex. App.—Austin, June 12, 2003), the issue was whether A201 was a part of the contract documents. After problems developed between the owner and contractor, the owner declared the contractor in default and filed a claim against the performance bond. The bonding company refused coverage, arguing that its performance was excused because the owner materially altered the bonded contract by not hiring an architect to inspect the contractor's work, leading to a substantial overpayment for work the contractor had either improperly performed or failed to perform. The owner countered that its contract did not require it to hire an architect to inspect the work. The issue hinged on what constituted the contract. According to the owner, the bid proposal was the only contract between the parties. The bid proposal contained the following language:

> Bidder has carefully examined the form of contract, instructions to bidders, profiles, grades, specifications and the plans therein . . . and will do all the work and furnish all the material called for in the contract DRAWINGS and specifications. . . . In the event of the award of a contract to the undersigned, the undersigned will execute same on [an AIA] Standard Form Construction Contract and make bond for the full amount of the contract, to secure proper compliance with the terms and provisions of the contract. . . . The work proposed to be done shall be accepted when fully complied and finished to the entire satisfaction of the Architect and the Owner.

However, no AIA document was ever executed. The bonding company argued that the appropriate document was A101, which incorporated A201, which in turn required that the owner hire an architect to monitor construction. Since there was no evidence of another contract and because the owner never misrepresented that an AIA contract had been signed, the court held against the bonding company.

In Robert J. Denley Co., Inc., v. Neal Smith Constr. Co., 2007 WL 1153121 (Tenn. Ct. App. Apr. 19, 2007), the A101 was executed, but in a dispute over whether arbitration was required, the trial court refused to order arbitration on the basis that A201 was not incorporated. The appellate court reversed, finding that A201 was incorporated and there was no ambiguity. The court also rejected a fraudulent inducement argument on this issue.

1.1.2 The Contract

The Contract Documents form the Contract for Construction. The Contract represents the entire and integrated agreement between the parties hereto and supersedes prior negotiations, representations or agreements, either written or oral. The Contract may be amended or modified only by a Modification. The Contract Documents shall not be construed to create a contractual relationship of any kind (1) between the Contractor and the Architect or the Architect's consultants, (2) between the Owner and a Subcontractor or a Sub-subcontractor, (3) between the Owner and the Architect or the Architect's consultants or (4) between any persons or entities other than the Owner and the Contractor. The Architect shall, however, be entitled to performance and enforcement of obligations under the Contract intended to facilitate performance of the Architect's duties. **(1.1.2; add reference to Architect's consultants)**

All of the contract documents listed in ¶ 1.1.1 form the construction contract between the owner and contractor. Any documents or oral agreements that are not included in the contract documents are not part of the contract.[4] Negotiations, verbal representations, letters, prior drafts, and other documents that predate the contract documents are, by virtue of this paragraph, not a part of the contract documents.

In McCarl's, Inc. v. Beaver Falls Mun. Auth., 847 A.2d 180 (Pa. Cmwlth. 2004), when a dispute arose, the parties entered into a letter settlement agreement, and work continued. Subsequently, the owner refused to make further payments to the contractor and the contractor filed suit. The owner demanded arbitration. The contractor argued that the letter agreement superseded and was substituted for the original agreement that contained the arbitration clause. The court held for arbitration. The original contract stated that all modifications issued subsequent to the contract were to be considered part of the contract. In order to cancel the arbitration provisions of the original contract, a letter agreement must expressly cancel or otherwise nullify the arbitration provisions.

In Twin Oaks at Southwood, LLC v. Summit Constructors, Inc., 941 So. 2d 1263 (Fla. Dist. Ct. App. 1st Dist. 2006), the parties entered into a "Project Completion Agreement" several months after the completion date contemplated by the contract. This document contained an integration clause, but also stated that the "terms and conditions of the Contract shall remain in full force and effect." The contractor filed suit and the owner filed counterclaims. The contractor then filed a motion to dismiss, asserting that the counterclaims arose under the contract, which required arbitration, but the complaint was filed pursuant to the PCA which did not require arbitration. The owner then asserted that arbitration had been waived by both parties. The appellate court held that there was no waiver of arbitration, that the PCA met the AIA definition of a modification under this provision and was therefore subject to arbitration.

In Supreme Indus. v. Town of Bloomfield, 2007 WL 901805 (Conn. Super. Ct. Mar. 8, 2007), there was no signed agreement, although the parties had heavily negotiated the terms of A201. The court found that, by conduct, the parties had agreed to the terms of A201.

[4] In Employers Mut. Cas. Co. v. Collins & Aikman Floor Coverings, Inc., 2004 U.S. Dist. LEXIS 7193 (S.D. Iowa Feb. 19, 2004), the contract documents apparently included both A201 and an acknowledgment to a purchase order for carpeting, which included another integration clause with reference to certain warranties. Summary judgment was denied: "there cannot be two completely integrated contracts for the sale of the same carpet. That there are two candidates with different and conflicting warranty provisions suggests a genuine issue of material fact about which governs and whether either was truly, mutually intended to be completely integrated."

Courts have upheld the provision that no contractual relation is established between the architect or owner and any subcontractor.[5] A Connecticut court has

---

[5] Alvord & Swift v. Stewart M. Muller Constr. Co., 46 N.Y.2d 276, 385 N.E.2d 1238, 413 N.Y.S.2d 309 (1978). A subcontractor brought an action against the owner and architect for damages resulting from delays incurred in performance of the subcontract. Based on this clause, the subcontractor's only recourse was against the contractor, who was insolvent. A count for intentional interference with contractual relationship also failed. *See also* Hatzel & Buehler, Inc. v. Orange & Rockland Util., Inc., 1992 WL 391154 (D. Del. Dec. 14, 1992). In Cullum Mech. Constr., Inc. v. South Carolina Baptist Hosp., 520 S.E.2d 809 (S.C. 1999), the mechanical subcontractor was not fully paid for its work and sued the architect, arguing that the architect owed it a professional duty distinct from any contractual duty. According to the subcontractor, this purported duty was breached by the architect's certification of several payment requests without obtaining written documentation that the subcontractor had been paid. The court rejected this argument, finding that this provision, as well as a similar provision in the owner-architect agreement, did not support a legal duty on the part of the architect to assure payment to subcontractors.

This appellate decision was reversed in Cullum Mech. Constr., Inc. v. South Carolina Baptist Hosp., 344 S.C. 426, 544 S.E.2d 838 (2001). The court held that whether or not a design professional owes a duty depends on the facts and circumstances of each case, therefore, summary judgment was improper. Saying that "generally, an architect does not have a duty to assure payment to subcontractors," the court went on to state that "special conditions in these contract documents may have given rise to a special relationship with subcontractors, and therefore a duty of care."

Although the peculiar facts of this case undoubtedly influenced the reversal of this case, the appellate court made the correct finding. The AIA documents do not create any "special relationship" between the architect and any subcontractor. Only the actions or inactions by the architect may have given rise to such a relationship. Presumably, the court believed, the architect may have created a duty by leading the subcontractor astray. This is contrary to the intent of the AIA documents, which make it quite clear that there is no such duty. Only by pleading and showing actual intent to deceive a subcontractor by the architect should the subcontractor be able to make out a case of breach of duty against the architect. *Cullum* was cited in Eaton Corp v. Trane Carolina Plains, 350 F. Supp. 2d 699 (D.S.C. 2004) (buyer brought an action against seller of air conditioner unit to recover for damages caused by fire allegedly originating in the unit. Court discussed the economic loss rule in the context of this case).

There is no contractual relationship between the architect and the general contractor. Mountains Comty. Hosp. Dist. v. Superior Court of the County of San Bernardino, 2003 WL 21766472 (Cal. Ct. App. 4th Dist. July 31, 2003).

In James Stewart Polshek v. Bergen Iron Works, 142 N.J. Super. 516, 362 A.2d 63 (1976), an earlier version of this provision was cited by the court in allowing an architect to avoid arbitration with the contractor.

In Hogan v. Postin, 695 P.2d 1042 (Wyo. 1985), the architect ordered a subcontractor to change a sill height. The subcontractor sued the architect for payment, and the court held that the architect did not owe the subcontractor any contractual duty. The court based its decision on agency theory: the architect was an agent of the owner. Presumably, the architect would have been liable to the owner for any payment that the owner made to the contractor.

In Morse/Diesel, Inc. v. Trinity Indus., Inc., 859 F.2d 242 (2d Cir. 1988), *rev'g* 655 F. Supp. 346 (S.D.N.Y. 1987) (*citing* Widett v. United States Fidelity & Guar. Co., 815 F.2d 885 (2d Cir. 1987) (architects are not liable either in tort or contract absent privity under New York law)), the court held that damages are not available in a suit for negligent performance of a contractual duty when only economic injury is alleged.

This provision (and others) in A201 indicates an intention by the contractor to hold the owner rather than the architect liable for any economic damages arising from the plans and specifications.

held that this provision bars mandatory arbitration by subcontractors against the owner.[6]

A third party, such as a member of the public, must show that the parties to the contract intended to confer a benefit on that third party in order to obtain recovery.[7] This provision is intended to specifically negate such an intention.[8] The parties

---

Floor Craft Floor Covering, Inc. v. Parma Cmty. Gen. Hosp. Ass'n, 54 Ohio St. 3d 1, 560 N.E.2d 206 (1990); Tomb & Assocs. v. Wagner, 82 Ohio App. 3d 363, 612 N.E.2d 468 (1992).

In Clevecon, Inc. v. Northeast Ohio Reg'l Sewer Dist., 90 Ohio App. 3d 215, 628 N.E.2d 143 (1993), the general contractor sued the architect for malpractice to recover delay damages. The jury found the architect negligent in the preparation and drafting of the plans and specifications and in the administration of the project. It also found that the architect had breached warranties in the preparation and drafting of the plans and specifications. In affirming the trial court judgment, the appellate court found that the architect had exercised a substantial amount of control over the project, giving orders to the contractor during the construction. This was unlike the *Floor Craft* case, in which the architect had not exercised control at the job site. *Clevecon* was criticized in National Steel Erection, Inc. v. J.A. Jones Const. Co, 899 F. Supp. 268 (N.D.W.Va. 1995), and distinguished in Mosser Constr., Inc. v. Western Waterproofing, Inc., 2006 Ohio 2637 (Ohio Ct. App. 2006) (contractor and its insurer sued architect to recover costs of rebuilding defective trench drain, alleging implied indemnification and negligence. The court held that the architect lacked authority to make any major changes on the project. Contractor was actively negligent, thereby barring indemnification from the architect). *See also* Resurgence Prop., Inc. v. W.E. O'Neil Constr. Co., No. 92C6618, 1995 U.S. Dist. LEXIS 4939 (N.D. Ill. Apr. 14, 1995). *But see* Oldenburg v. Hageman, 159 Ill. App. 3d 631, 512 N.E.2d 718 (1987) (court denied claim by subcontractor against architect based on negligent misrepresentation when the damages were economic). In R.H. Macy & Co. v. Williams Tile & Terrazzo, 585 F. Supp. 175 (N.D. Ga. 1984), the court found that an architect owed no duty to a tile installer and supplier because there was no privity.

An interesting case is Pierce Assocs. v. Nemours Found., 865 F.2d 530 (3d Cir. 1988), in which an owner sued a subcontractor on a third-party beneficiary claim. The owner argued that, despite the presence of a similar clause, the General Conditions contained many provisions indicating that the owner is to be benefited and, further, this paragraph mentions only the contract documents as not creating a third-party beneficiary. There is nothing to prohibit the subcontract from creating this right. The federal court rejected this argument, especially because the General Conditions were incorporated into the subcontract by reference. *See also* Turner Constr., Inc. v. American States Ins. Co., 579 A.2d 915 (Pa. Super. Ct. 1990). Brick Constr. Corp. v. CEI Dev. Corp., 710 N.E.2d 1006 (Mass. 1999), wherein the court held that the subcontractor did not have a right to recover money from the owner based on its AIA subcontract that stated that the subcontractor had the benefit of all rights, remedies, and redress against the contractor that the contractor had against the owner.

[6] Wesleyan Univ. v. Rissil Constr. Assocs., 1 Conn. App. 351, 472 A.2d 23 (1984).

[7] A "third party beneficiary." *See* Richmond Shopping Ctr. v. Wiley N. Jackson Co., 220 Va. 135, 255 S.E.2d 518 (1979). A property owner sought recovery against the contractor on a highway construction contract. The contract did not intend to confer on any third party a status as third-party beneficiary, and recovery was denied. However, in Gaunt & Hayes, Inc. v. Moritz Corp., 138 Ill. App. 3d 356, 485 N.E.2d 1123 (1985), a grocery store operator was permitted to recover against a highway contractor for loss of access. The case was premised on the contractor's negligent performance of the work.

In F.O. Bailey Co. v. Ledgewood, Inc., 603 A.2d 466 (Me. 1992), the standard AIA contracts did not confer third-party beneficiary status on a condominium purchaser in his action against the contractor. *See also* Burns v. Black & Veach Architects, Inc., 854 S.W.2d 450 (Mo. 1993); Fisher v. M. Spinelli & Sons Co., Inc., 1999 WL 165674 (Mass. Super. 1999).

[8] In Board of Managers of the Arches at Cobble Hill Condominium v. Hicks & Warren, LLC, 836 N.Y.S.2d 497 (Sup. 2007), a condominium association sued the contractor, arguing that it was the

should carefully avoid any action that might indicate that a contractual relationship was intended between the architect and any contractor, or between the owner and any subcontractor. Thus, the owner should not direct the work of any subcontractor, and the architect should realize that its contract is only with the owner and not with the contractor. The architect, however, is entitled to limited performance of the contract as a third-party beneficiary.[9] The contractor may want to delete the last sentence because its contract is only with the owner and not the architect. This would also resolve any possible ambiguity related to a contractor's claim against the architect, with the contractor claiming that this provision does create contractual obligations between the architect and contractor.

Interestingly, this provision also states that the contract documents do not create a contractual relationship between the owner and architect. That relationship is defined exclusively by the owner-architect agreement. Thus, any language in this document does not modify the owner-architect agreement, even if the provisions conflict. Because the general conditions are normally created after the owner-architect agreement, the owner and architect should take care to coordinate those documents. The standard AIA documents are coordinated, but revisions to them may not be.

Note that the parties can, despite the express language of this paragraph, orally amend the contract documents. One court discussed the parol evidence rule as banning prior or contemporaneous agreements, but not subsequent agreements, stating:

> This rule of law applies notwithstanding any contractual provisions precluding oral modifications to a contract. . . . Such a "no oral modification" clause in a written contract may be waived or modified in the same way in which any other provision of a written agreement may be waived or modified, including a change in the provisions of the written agreement by the course of conduct of the parties.[10]

Because this modification might be inadvertent, the parties should take great care that anything that might be taken to modify the contract documents be in writing.

---

third-party beneficiary of the contract. Based on this provision, the court dismissed the case as to the contractor. The court found that this specific disclaimer trumped general provisions in the A201 that might be read to benefit other parties, such as condominium purchasers.

[9] Bates & Rogers Constr. v. North Shore Sanitary Dist., 109 Ill. 2d 225, 486 N.E.2d 902 (1985), aff'g 128 Ill. App. 3d 962, 471 N.E.2d 915 (1984). The contract contained a no-damages-for-delay clause. The contractor sought damages from the owner and architect for a delay. The architect was a third-party beneficiary of this contract and the exculpatory clause was a defense for him. In Temple Sinai-Suburban Reform Temple v. Richmond, 112 R.I. 234, 308 A.2d 508 (1973), the court found that the architect may be able to maintain a third-party action against a brick manufacturer, citing numerous cases in support of the proposition that privity was no longer a bar to such an action.

[10] J.A. Moore Constr. Co. v. Sussex Assocs., Ltd., 688 F. Supp. 982 (D. Del. 1988). See also Bouten Constr. Co. v. M&L Land Co., 125 Idaho 957, 877 P.2d 928 (1994) (court allowed oral modifications because the owner did not follow procedures set forth in contract).

1.1.3 The Work

The term "Work" means the construction and services required by the Contract Documents, whether completed or partially completed, and includes all other labor, materials, equipment and services provided or to be provided by the Contractor to fulfill the Contractor's obligations. The Work may constitute the whole or a part of the Project. (**1.1.3; no change**)

The *Work* is the object of the particular contract documents[11] and can be a portion of a larger *Project.* The definition has been revised from the 1976 edition to be more inclusive, which may possibly result in an ambiguity: the Work appears to include "services" of the contractor, including items such as scaffolding and shoring that are not to be incorporated into the project. However, a reasonable interpretation of the entire A201 results in the conclusion that the Work does not include the contractor's tools or other equipment belonging to the contractor or other items that are not to be incorporated into the completed project. It also does not include the process itself, only the finished "product."[12] Otherwise, the architect's duty to observe the Work under ¶ 4.2.2 would be in conflict with the balance of that paragraph, which states that the architect will not have control over construction procedures. Thus, scaffolding and similar items are not Work under A201. Even though the services of the contractor are included in the Work, as stated in ¶ 1.1.3, the architect's duty does not extend to observation of the work procedures.[13] For instance, the Work may include services such as a carpenter's hammering a nail. It is clear from the entire document that the architect has no duty to observe the hammering process, only the result of that process.

---

[11] For a case that refers to this definition of Work (in contrast to work), *see* Gilbane Bldg. Co. v. Nemours Found., 666 F. Supp. 649 (D. Del. 1985). Another case explained this provision to include either the construction project itself or the labor and materials required therefor. S.S.D.W. Co. v. Brisk Waterproofing Co., 153 A.D.2d 476, 544 N.Y.S.2d 139 (1989), *aff'd,* 76 N.Y.2d 228, 556 N.E.2d 1097, 557 N.Y.S.2d 290 (1990).

In Eldeco, Inc., v. Charleston County Sch. Dist., 642 S.E.2d 726 (S.C. 2007), a subcontractor complained that it was not awarded all of the work. This other work was not included in the original work, but was added pursuant to construction change directives and change orders. Pursuant to this definition and ¶ 6.1.1, which allows an owner to award work to other contractors, the court found against the subcontractor.

[12] In Willis Realty v. Cimino Constr., 623 A.2d 1287 (Me. 1993), the construction contract involved an addition attached to an existing building at its back wall. That back wall was therefore an integral part of the project and of the "Work."

[13] In Hanna v. Huer, Johns, Neel, Rivers & Webb, 233 Kan. 206, 662 P.2d 243, 249 (1983), the court held that this definition of Work did not include safety precautions:

Such a construction is obviously contrary to the clear wording of the contract itself and, if correct, would require that the ultimate responsibility for compliance with all of the provisions of the contract between the contractor and the owner rests upon the architect. We do not construe the contract so broadly as to the term "Work" as used in the contract and as applied to the architect's duties.

The sale of goods is the primary objective of some projects, with services being only incidental. In that case, the Uniform Commercial Code may apply.[14]

There is some question as to whether the work continues beyond final completion,[15] although in the context of arbitration, courts generally are liberal about permitting arbitration over matters beyond the end of the project.[16]

Note that the definition of Work also includes the work of subcontractors, sub-subcontractors, and others hired by the contractor, as well as all materials furnished by suppliers and installed in the completed project. It is not clear from the language that off-site work is included, although if off-site work will be incorporated into the final building, a reasonable interpretation of this provision is that it be included (**see** ¶ 9.3.2). This provision also included items to be furnished by the contractor to fulfil the contractor's obligations. This could be interpreted very broadly.

---

[14] According to J. Lee Gregory, Inc. v. Scandinavian House Ltd. P'ship, 209 Ga. App. 285, 433 S.E.2d 687 (1993), the Uniform Commercial Code applied to a contract to furnish and install windows in an apartment house. There was a question as to the formation of the contract. Because the sale of the windows was the predominant purpose of the transaction, with the installation of the windows a secondary purpose, the Code applied. *See also* D.N. Garner Co., Inc., v. Georgia Palm Beach Aluminum Window Corp, 504 S.E.2d 70 (Ga. Ct. App. 1998); Schenectady Steel Co., Inc., v. Bruno Trimpoli Gen. Const. Co., Inc., 350 N.Y.S.2d 920 (A.D.3d Dep't 1974) (contract to furnish and erect structural steel was a contract for services rather than a contract for the sale of goods, so that the UCC did not apply); G-W-L, Inc., v. Robichaux, 643 S.W.2d 392 (Tex. 1982) (contract for construction of house involved labor and performance of work; sales did not apply).

[15] In Automobile Ins. Co. v. United H.R.B. Gen. Contractors, Inc., 876 S.W.2d 791, 794 (Mo. Ct. App. 1994), the court held that the waiver of subrogation (¶ 11.3.7) ended upon final payment to the contractor. A fire occurred after completion of the building, causing substantial damage. The owner's insurer then sued the contractor, alleging faulty installation of the electrical system. The contractor defended based on the waiver of subrogation provision. The contractor argued that the definition of work included the completed building, demonstrating that the agreement as to waiver of claims extended beyond final payment. The court, however stated:

> [W]e agree that "work" includes the completed structure, but only for the time interval between the completion of the building and final payment. We find that the completed structure is no longer "work" after final payment is made and therefore the waiver of claims only applies to the completed structure up to the time of final payment.

The waiver of subrogation was only effective so long as the contractor had an insurable interest in the property. The court found that the contractor's interest in the project terminated on final payment.

In a case where fire damaged the building a year after construction was completed, the waiver of subrogation found in ¶ 11.3.7 applied because this definition of "the Work" was "the construction and services required by the Contract Documents, whether completed or partially completed." Colonial Prop. Realty Ltd. P'ship v. Lowder Constr. Co., Inc., 256 Ga. App. 106, 567 S.E.2d 389 (2002).

[16] In Contract Constr., Inc. v. Power Tech. Ctr. Ltd. P'ship, 100 Md. App. 173, 640 A.2d 251 (1994), one of the issues was whether the arbitration provisions covered only "the Work." The court examined this provision, as well as ¶¶ 3.2.3 (now 3.2.4) and 3.7.4 (now 3.7.3), and determined that arbitration could resolve issues relating to the indemnification provision of A201, ¶ 3.18. The arbitration provision was not limited to the Work.

In one case, the contractor was found to have contracted to furnish the design for a prestressed concrete system. The owner wanted to save money by not having the architect do this work.[17] It was the contractor and not the architect who was responsible for the defective design.

Because the work is only the specific work for which the contractor is hired, damage to work performed under another portion of the project is not recoverable under the provisions of ¶ 12.2.2, which provides that the contractor must correct work within one year.[18]

1.1.4 The Project

The Project is the total construction of which the Work performed under the Contract Documents may be the whole or a part and which may include construction by the Owner and by separate contractors. **(1.1.4; no change)**

The *Project* thus encompasses all work, either by this contractor, the owner, or other contractors required for completion. For instance, the Project may be a new university, whereas the Work may only be the construction of a dormitory.

1.1.5 The Drawings

The Drawings are the graphic and pictorial portions of the Contract Documents showing the design, location and dimensions of the Work, generally including plans, elevations, sections, details, schedules and diagrams. **(1.1.5; no change)**

The *Drawings* are sometimes referred to as the plans or blueprints, but they also consist of drawings of details, elevations, and other drawings. They also include drawings forming part of any addenda or other modifications to the contract. Some drawings might even be found in the specifications. Note that shop drawings are not included (**see** ¶ 3.12.4). Some owners may want to include shop drawings in this definition. Be extremely careful in adding shop drawings because they are produced by the contractor, whereas all the other contract documents are produced by the architect or owner. Note also that the shop drawings are produced after the contract has been signed, so that they are more like addenda or contract modifications. In that case, they should be approved in writing by the owner before being made a part of the contract documents.

The list of drawings in ¶ 1.1.5 is merely suggestive and is not meant to indicate that each of the items, such as schedules and diagrams, is required. The architect determines what method is best suited to the project to convey the required information to the contractor.

The drawings need to be specifically enumerated in order to avoid disputes. This is normally done in AIA Document A101, ¶ 9.1.5, or in a similar AIA document, where each drawing is identified by number and date. Many disputes revolve

---

[17] In Stevens Constr. Corp. v. Carolina Corp., 63 Wis. 2d 342, 217 N.W.2d 291 (1974), the prestressed concrete system was designed by the concrete subcontractor. Unfortunately, an error was made and the structure was designed to hold 120 pounds per square foot less than it should have.

[18] Idaho State Univ. v. Mitchell, 97 Idaho 724, 552 P.2d 776 (1976).

around the issue of the exact scope of the contractor's work, since there may be many revisions to each drawing and it is sometimes not clear exactly which drawings or which versions of particular drawings form the contract.

1.1.6 The Specifications

The Specifications are that portion of the Contract Documents consisting of the written requirements for materials, equipment, systems, standards and workmanship for the Work, and performance of related services. **(1.1.6; no change)**

The *Specifications* are the written documents that provide the contractor with various items of information, such as acceptable manufacturers of components, standards to be adhered to, cleanup, and many other facets of the project that are inappropriate on the drawings. On some smaller projects, the specifications can be found on the drawings.

The specifications are usually divided into sections, according to industry practice. Most architects and specifiers follow this practice, and contractors are familiar with it. In 2004, the Construction Specifications Institute issued a revised version of MasterFormat™, which is considered the specifications-writing standard for most commercial building design and construction projects in North America. For instance, Section 09 24 23 is for Portland Cement Stucco, while Section 09 24 33 is for Portland Cement Parging.

The drawings and specifications should be complementary and must be coordinated. Under the standard provisions, there is no order of precedence between the drawings and specifications (**see ¶** 1.2.1). In case of discrepancy between the drawings and specifications, one is not automatically assumed more correct than the other. Some architects may want to establish an order of precedence in the specifications, however, there is no substitute for careful coordination. These drawings and specifications are the contract documents that tell the contractor how the parts go together. Other contract documents, such as A201, give legal and procedural information.

Architects often use specialists in writing the specifications. Many of these are members of the Construction Specifications Institute and follow formats designed by that institute or the AIA.

In a federal case, the contractor argued that the specifications were "performance" specifications as opposed to "design" specifications. The court explained the distinction:

> Performance specifications set forth an objective or standard to be achieved, and the successful bidder is expected to exercise his ingenuity in achieving that objective or standard of performance, selecting the means and assuming a corresponding responsibility for that selection. . . . Design specifications, on the other hand, describe in precise detail the materials to be employed and the manner in which the work is to be performed. The contractor has no discretion to deviate from the specifications, but is required to follow them as one would a road map. Detailed design specifications contain an implied warranty that if they are followed, an acceptable result will be produced.[19]

---

[19] Blake Constr. Co. v. United States, 987 F.2d 743, 745 (Fed. Cir. 1993). Note that, in the context of performance versus design specifications, the word "specifications" also applies to the drawings.

The specifications should be specifically listed in AIA Document A101, ¶ 9.1.4, or similar document.

1.1.7 Instruments of Service

Instruments of Service are representations, in any medium of expression now known or later developed, of the tangible and intangible creative work performed by the Architect and the Architect's consultants under their respective professional services agreements. Instruments of Service may include, without limitation, studies, surveys, models, sketches, drawings, specifications, and other similar materials. **(1.6.1; revised; see also 1.5.1)**

This provision defines the term, "Instruments of Service." Note the language in the first sentence that these Instruments of Service can be in any medium, including any form of electronic media, and any media that may be developed in the future. The documents created under this concept are subject to copyright protection as set forth in the owner-architect agreement. Section 1.5 of A201 discusses copyright issues insofar as the contractor is concerned. The standard form documents created by the AIA, such as the A201, are not Instruments of Service, while drawings and specifications are.

1.1.8 Initial Decision Maker

The Initial Decision Maker is the person identified in the Agreement to render initial decisions on Claims in accordance with Section 15.2 and certify termination of the Agreement under Section 14.2.2. **(New)**

Note that the Initial Decision Maker may be the architect, or another party chosen by the parties to this agreement. In prior versions of A201, this person was the architect. This person could even be an employee of the owner or the contractor, although that would not be a good idea. Section 6.1 of AIA Document A101 states that the Initial Decision Maker may not be a party to the contract, which presumably would also disqualify the employees and agents of the owner and contractor. Since the architect is not an agent of the owner, that provision does not conflict with making the architect the default initial decision maker. This person will initially make decisions on claims. If either party is dissatisfied with the decision by this person, the mediation provision may be invoked, followed by either litigation or arbitration if mediation is unsuccessful, depending on which final method of dispute resolution was chosen by the parties.

In rejecting his claim, the court in *Blake* determined that the contractor was focusing on the wrong argument. The mere fact that a specification cannot be followed precisely does not, in and of itself, indicate that it is "performance" and not "design." Further, the distinction between the two is not absolute.

Spearin (United States v. Spearin, 248 U.S. 132 (1918)) type warranties apply to design specifications, but not performance specifications. "Design specifications explicitly state how the contract is to be performed and permit no deviations. Performance specifications, on the other hand, specify the result to be obtained, and leave it to the contractor to determine how to achieve those results." Stuyvesant Dredging Co., v. United States, 67 Fed. Cl. 362 (2005).

In concept, this Initial Decision Maker operates like a Dispute Resolution Board,[20] and the provisions of A201 can be adjusted to provide for a Dispute Resolution Board on larger projects. Historically, Dispute Resolution Boards have been comprised of three impartial and experienced persons who usually meet prior to the start of construction and periodically thereafter and become familiar with the project. They are on call whenever a dispute arises.

§ 4.4   Execution, Correlation, and Intent: ¶¶ 1.2 through 1.5

1.2 CORRELATION AND INTENT OF THE CONTRACT DOCUMENTS

1.2.1 The intent of the Contract Documents is to include all items necessary for the proper execution and completion of the Work by the Contractor. The Contract Documents are complementary, and what is required by one shall be as binding as if required by all; performance by the Contractor shall be required only to the extent consistent with the Contract Documents and reasonably inferable from them as being necessary to produce the indicated results. (**1.2.1; no change**)

All items reasonably inferable as required for the execution of the work by the contractor are included in the contract. This reasonable inferability appears to establish a standard of ordinary care for the contractor. The owner may want to change that to a higher standard of care. Not every detail can be shown on the drawings, so the contractor must use common sense in determining what is required. Good workmanship is also inferred,[21] as well as the use of proper materials.[22]

In a California case, the court construed an earlier version of A201 to find that the project manual was a part of the Contract Documents, holding that, per this provision, it is an "item necessary for the proper execution and completion" of the project.[23] This was supported by the fact that the project manual appeared to address all substantive aspects of the project.

---

[20] *See, e.g.*, John Carlo, Inc., v. Greater Orlando Aviation Auth., 2007 WL 430647 (M.D. Fla. Feb. 3, 2007), finding that submission of a claim to the DRB was a condition precedent to suit.

[21] In Robert G. Regan Co. v. Fiocchi, 44 Ill. App. 2d 336, 194 N.E.2d 665 (1963), the masonry subcontractor did not install wall ties per the contract documents. When the walls bulged, he was held liable for failing to follow the drawings and specifications. The court held that opinions that the walls would have bulged even if the drawings had been followed were immaterial. "When defendants departed from this specification they did so at their peril, and all attempted excuses for non-compliance became immaterial."

In Georgetown Township High Sch. Dist. No. 218 v. Hardy, 38 Ill. App. 3d 722, 349 N.E.2d 88 (1976), the court followed *Regan* and found for a contractor who had followed the drawings, finding that the contractor had no duty to construct a building that would withstand any given wind pressures. The "contractor is not liable for damages if he (1) performs his work in accordance with the plans and specifications furnished by the owner, and (2) does so in a workman-like manner."

[22] Cincinnati Gas & Elec. Co. v. General Elec. Co., 656 F. Supp. 49 (S.D. Ohio 1986).

Note that the last clause states that the contractor is to produce the "indicated results." This means that only what is shown on the actual drawings, specifications, and other Contract Documents, or is reasonably inferable from them, is required. The parties should be careful to list all of the Contract Documents in the appropriate location, such as in AIA Document A101, Section 9.1.

[23] Myers Bldg. Indus., Ltd. v. Interface Tech., Inc., 17 Cal. Rptr. 2d 242 (Ct. App. 1993).

There is no order of precedence in the documents, although many architects include language in the supplemental conditions that one document will have precedence over another in case of discrepancies.[24]

Note that the Work is also described in the Owner-Contractor Agreement. In the case of AIA Document A101, that is Article 2 (see **Chapter 10**). If not everything shown on the drawings and specifications is the Work, the parties should define what the Work is very carefully.[25]

> *1.2.2 Organization of the Specifications into divisions, sections and articles, and arrangement of Drawings shall not control the Contractor in dividing the Work among Subcontractors or in establishing the extent of Work to be performed by any trade.* (**1.2.2; no change**)

Many contractors depend on the architect's division of the work when submitting documents to the subcontractors. For instance, the plumbing contractor may only be given plumbing drawings. If a particular item is shown on the architectural drawings, the plumber might omit it. This paragraph places the burden on the contractor to examine the contract documents and to properly allocate the work among all the subcontractors. The responsibility for dividing the work among the various subcontractors rests with the general contractor, and not the architect. For that reason, the architect should not designate any particular trade to perform certain work. For instance, the use of language like "masonry anchors to be welded by structural steel contractor" should be avoided. The contractor may wish to have the masonry contractor do this work. Worse, another portion of the contract documents may call for a different contractor to do this work, with the result that this work is not in any contractor's bid. The bottom line is that the designer should show the work in its finished layout, and the contractor should be responsible for assigning the work to the various trades.

---

[24] In Mpact Constr. Group, LLC v. Superior Concrete Constructors, Inc., 802 N.E.2d 901 (Ind. 2004), the court held that the second sentence pertained to the work and did not operate to incorporate all of the general conditions into the subcontracts (which were not standard AIA form subcontracts). The contractor has a duty to immediately contact the architect or owner if it notices any discrepancies. However, the contractor is entitled to a set of contract documents that is reasonably free from discrepancies. These documents should be coordinated, so that the specifications describe the same piece of equipment that is shown on the drawings.

[25] In Taylor v. Allegretto, 112 N.M. 410, 816 P.2d 479 (1991), the owner claimed the work included construction of the shell of a three-unit medical complex and completion of the interior of Unit 2. The appellate court held, however, that because Work was defined in the Owner-Contractor Agreement as "Unit #2 As per plans and specifications," the contractor was only required to work on Unit 2 and not anything else:

> The clause making all contract documents [complementary] can only be reasonably read to require that the documents complement each other insofar as they are consistent. Where the language of the agreement, particularly typed-in language on a printed form contract, specifically indicates that the parties intended to limit the scope of the work to be performed, it is simply not "consistent" to expand the agreement because the plans as originally drawn included additional work.

1.2.3 Unless otherwise stated in the Contract Documents, words that have well-known technical or construction industry meanings are used in the Contract Documents in accordance with such recognized meanings. **(1.2.3; no change)**

Sometimes a word has a different meaning in the construction industry than in everyday life. This provision assures that those technical meanings will prevail. If a particular term is not meant in this technical sense, it must be clearly so indicated.[26]

1.3 CAPITALIZATION

Terms capitalized in these General Conditions include those that are (1) specifically defined, (2) the titles of numbered articles or (3) the titles of other documents published by the American Institute of Architects. **(1.3.1; minor change to subpart (2))**

This is a housekeeping provision. Certain words in this document are capitalized. For instance, the word "Work" is specifically defined in ¶ 1.1.3. Whenever this word is capitalized it takes on the defined meaning. If any additional terms are defined by the owner or contractor, they should follow this convention. AIA Document B101 refers to "terms" in ¶ 10.2. This refers to all defined capitalized terms such as "Work" and may also refer to other terms, such as "consequential damages."

1.4 INTERPRETATION

In the interest of brevity the Contract Documents frequently omit modifying words such as "all" and "any" and articles such as "the" and "an," but the fact that a modifier or an article is absent from one statement and appears in another is not intended to affect the interpretation of either statement. **(1.4.1; no change)**

This is another housekeeping provision.

§ 4.5   Ownership and Use of Instruments of Service: ¶ 1.5

1.5 OWNERSHIP AND USE OF DRAWINGS, SPECIFICATIONS AND OTHER INSTRUMENTS OF SERVICE

1.5.1 The Architect and the Architect's consultants shall be deemed the authors and owners of their respective Instruments of Service, including the Drawings and Specifications, and will retain all common law, statutory and other reserved rights, including copyrights. The Contractor, Subcontractors, Sub-subcontractors, and material or equipment suppliers shall not own or claim a copyright in the Instruments of Service. Submittal or distribution to meet official regulatory requirements or for other purposes in connection with this Project is not to be construed as publication in derogation of the Architect's or Architect's consultants' reserved rights. **(1.6.1; revisions)**

---

[26] In Ostrow Elec. Co. v. J.L. Marshall & Sons, Inc., 798 N.E.2d 310 (Mass. App. Ct. 2003), this provision was cited when the electrical contractor stated that he knew from his own experience that the term "outlet boxes" included speaker backboxes.

1.5.2 The Contractor, Subcontractors, Sub-subcontractors and material or equipment suppliers are authorized to use and reproduce the Instruments of Service provided to them solely and exclusively for execution of the Work. All copies made under this authorization shall bear the copyright notice, if any, shown on the Instruments of Service. The Contractor, Subcontractors, Sub-subcontractors, and material or equipment suppliers may not use the Instruments of Service on other projects or for additions to this Project outside the scope of the Work without the specific written consent of the Owner, Architect and the Architect's consultants.
(1.6.1; revisions; delete reference to return of drawings at end of project)

These paragraphs explain how the contractor is to use the documents, including any electronic documents, in terms of ownership and copyright. One provision is that the architect's copyright notice must be attached to any reproductions of the documents by any contractor on the project in order to prevent unauthorized use of the documents. This is applicable, for instance, when a contractor prepares shop drawings. If the contractor copies a portion of any drawing of the architect or the architect's consultant, it must place the architect's copyright notice on the relevant shop drawing. If the architect or consultant failed to place a copyright notice on their drawings, the contractor does not have to place a notice on the shop drawings, but this does not give the contractor any additional rights in the drawings. Note that the contractors, subcontractors, and suppliers are "authorized" to use the drawings but not "licensed." This implies a lesser legal right. Thus, the contractor cannot sublicense the documents. The only such permission is granted by this document, and it is given to the general contractor and all lower-tier subcontractors.

Copyrights are governed almost exclusively by the Copyright Act of 1976.[27] The Act was amended by the Architectural Works Copyright Protection Act ("AWCPA"), effective as of November 30, 1990. The Act applies to "any architectural work that, on the date of the enactment of this Act, is unconstructed and embodied in unpublished plans or drawings." Under the AWCPA, an *architectural work* is "the design of a building as embodied in any tangible medium of expression, including a building, architectural plans, or drawings. The work includes the overall form as well as the arrangement and composition of spaces and elements in the design, but it does not include individual standard features." In other words, architects are now protected from someone's copying the building itself and not only from physically copying just the drawings. The term *individual standard features* presumably refers to details that are commonly used, such as typical flashing details, standard doors and windows, toilets, and the like. The copyright protection does include the arrangement of such standard details into a new and original building shape. The AWCPA applies to *buildings,* meaning habitable structures. Non-habitable structures, such as bridges, are not covered by the AWCPA. Buildings existing prior to the effective date of the AWCPA are not covered, and photographing the building from an area accessible to the public is not prohibited. Also, the owners of such buildings may alter or demolish them without

---

[27] 17 U.S.C. § 101.

consent of the architect, so long as the architect's plans are only used pursuant to a license.

Unless the architect specifically transfers the copyright to another party, the architect keeps the copyright to the drawings and specifications. To protect itself, the architect should comply with the notice and registration provisions of the Act.[28] Under the notice provisions, all publicly distributed copies of the drawings or specifications should contain three elements: first, the word "copyright," the abbreviation "Copr," or the symbol ©; second, the year of the first publication of the work; and third, the name or a recognizable abbreviation of the copyright owner for identification. Although a notice is no longer required on documents, it is still a very good idea to include a copyright notice on every document created by the architect.

Drawings or specifications need not be registered unless an action for infringement is planned, in which case registration is mandated to obtain certain remedies. Such registration can be made at any time on forms furnished by the Register of Copyrights. Note, however, in the event of a possible infringement that there are filing time limits. If the architect learns of an infringement, it should immediately consult its attorney.

Until passage of the AWCPA, copyright protection only prevented the copying of the drawings or specifications.[29] It did not prevent another person from using the

---

[28] For an interesting case decided before the 1976 Act, *see* Nucor Corp. v. Tennessee Forging Steel Serv., Inc., 476 F.2d 386 (8th Cir. 1973), in which a senior officer of a joist manufacturer took a set of plans for a joist plant from his company. After he became the president of another joist manufacturing company, he had them slightly modified, and a new plant was constructed based on these drawings. When the first company learned of this, it initiated the suit to enjoin the use of the plans, based on infringement of common law copyright and improper disclosure of confidential information. Although the plant was finished, recovery was allowed against the defendant.

[29] In Guillot-Vogt Assocs. v. Holly & Smith, 848 F. Supp. 682 (E.D. La. 1994), the architect on a state project hired a firm to design the electrical and mechanical portions of the project. The architect then advised the state that, because of failing health, he wanted his contract terminated. The architect had been paid in full for his work, but he had failed to fully pay his subcontractor for the engineering work. When the replacement architect wanted the subcontractor to continue its work, the subcontractor refused because both the state and the second architect refused to pay the balance owed for the work previously done. The subcontractor then filed for copyright registration for its drawings and filed suit for copyright infringement.

The court reviewed the 1990 Act and stated: "The Court interprets the 1990 Amendment primarily as providing previously lacking copyright protection to physical architectural works, not drawings of such works. Indeed, drawings, whether of architectural works, or otherwise, have long enjoyed copyright protection."

The state also argued, based on 17 U.S.C. § 120(b), that copyright protection was lacking because the project involved an alteration to an existing structure. The court held that this section is not intended to nullify copyright protection in drawings. The drawings in question were not a building, which is the subject of that section of the Act.

Finally, the state argued that the drawings were not copyrightable because they were utilitarian or mechanical in nature. This was rejected on the basis that there was no showing that the drawings in question were the only meaningful way to depict such articles. In Hunt v. Pasternak, 179 F.3d 683 (9th Cir. 1999), the Ninth Circuit overruled the trial court, finding that the AWCPA allowed a

ideas shown in them.[30] Also, if another person examined the building, she could build another structure just like it as long as she did not copy the drawings or specifications. For new buildings subject to this Act, that is no longer true, as buildings now have the same protection from copying as drawings. See Article 7 of AIA Document B101 for additional considerations for the architect.

Architects who are registering their copyrights are advised to register the drawings in two separate copyright applications: once as "technical drawings" and second as "architectural works." In this way, the architect will obtain the maximum protection for the drawings.

A contractor who uses a set of drawings for another project without permission faces liability under this provision. In a Georgia case in which there was apparently no contract between the owner and architect, the architect had prepared a set of drawings for the owner.[31] Later, the architect learned that the owner was

---

valid copyright in an architectural work even when the work has not been constructed. The owner had hired an architect to design a pizza restaurant. Dissatisfied with the plans, the owner then hired the plaintiff architect, who prepared a new set of plans. Again dissatisfied, the owner hired a third architect, who prepared plans presumably based on those prepared by the second architect. The second architect obtained a copyright registration for the architectural work embodied in his plans and filed suit for copyright infringement against the owner and third architect. It appears that there was no separate registration of the drawings as technical drawings, only as an architectural work. This appeared to confuse the trial court, which dismissed the action, believing that it was based on alleged infringement of the plans, not the architectural work. The Ninth Circuit, however, read the complaint as alleging infringement of the design of the architectural work, not of a separate copyright in the plans or drawings themselves.

[30] Ideas are not protected by copyright. However, the expression of an idea, such as a drawing or a building that is constructed, is protected.

[31] G.E.C. Corp. v. Levy, 126 Ga. App. 604, 191 S.E.2d 461 (1972).

In an important decision that stripped the architect of copyright protection, the court in I.A.E. v. Shaver, 74 F.3d 768 (7th Cir. 1996), held that the architect had given an implied license to use his drawings. The architect was hired by an engineer to design an air cargo building for an airport. Believing that this was a contract for the schematic design phase of the project and that he would be retained to perform the remaining architectural services, the architect agreed to accept $10,000 for this first phase. The letter form of contract made no mention of copyright or license. After he prepared several schemes (with copyright notices) and one was chosen by the owner, the architect was terminated from the project by the engineer, who had contracted with another architect to do the remaining work. The first architect was paid only $5,000 of the fee. When he learned of the second architect's use of his drawings, he registered his drawings with the Copyright Office and notified all parties that his drawing could not be used. The engineer then brought a declaratory judgment action against the first architect, contending that he had a right to use the drawings.

The court agreed with the engineer, finding that the architect knew that his drawings would be used for the construction of the building. The fact that he was not paid the contract sum made no difference in a copyright action. The architect's attempted revocation of any implied license was ineffective. The court held that such an implied license could not be revoked. The architect also argued that the license was an exclusive implied license which required a writing to be effective. The court rejected this argument on the grounds that the architect could have licensed this particular design to any number of others.

The result of *Shaver* is that design professionals should not consider that they have any action under the Copyright Act against anyone other than a total stranger. In the *Shaver* case, the court had held that the architect had given an implied license to everyone that had anything to do with the

proceeding with the construction without paying the architect for the plans. The architect sued to recover his fees. The owner evidently claimed that the drawings that were used for the construction were not the same, although the trial court found that by overlaying the two sets of drawings, the plans were identical. The architect prevailed.

Approval of a site plan by a governmental agency does not invalidate a copyright. In one case, an architect prepared a subdivision plat that was approved by the municipality.[32] The architect subsequently registered the copyright and filed

---

project, including the second architect of whom Shaver was not even aware. The mere act of delivering drawings constituted the grant of an implied license.

Note that this document relies on the owner-architect agreement to provide contractors with permission to use the Instruments of Service (the drawings and specifications), since the owner of the copyright, the architect, is not a party to this agreement. The exception to this is where an owner has obtained the copyright from the architect by way of assignment.

In Eales v. Envtl. Lifestyles, Inc., 958 F.2d 876 (9th Cir. 1992), an architect had prepared plans for several homes in Arizona for a developer. The developer originally hired a California architect to prepare plans, but as the construction began, the developer became concerned that the plans were not suitable for Arizona. The developer then consulted with a local architect, who reported that the original plans should be changed. At that time, the foundations for the homes had been completed. The new architect did not use the old plans but had to design the homes based on the existing foundation locations and other constraints. Apparently, an AIA owner-architect contract was used, because the architect had retained ownership of the plans. New permits were issued and construction resumed. However, financial problems again stopped construction, and the developer went out of business. One of the lots had been sold to a buyer who wanted a certain model to be built on that site. The buyer contacted one of the people who had worked for the developer to construct the home because the developer was no longer in business. Nobody contacted the architect to ask permission to use the plans.

The architect found out that her plans were being used to construct the home, and she sent a notice to the contractor stating that her plans were being used without permission and demanding her usual fee of $4.00 per square foot. No response was made, so she registered her plans and then filed suit in federal court, claiming copyright infringement.

The court awarded the architect not only her normal fee but also an amount equivalent to the contractor's profit on the home, for a total of over $57,000. The defendants had argued that the plans fell under the useful article exception to the copyright code, but that argument was summarily rejected. The next argument was that the plans did not substantially differ from the earlier drawings prepared by the California architect. Because the second set of plans was created without consulting the prior drawings, and the plans were, in fact, different, this argument was likewise rejected. The court found that the plans were entitled to copyright protection and that they constituted an expression of an idea. An expression is subject to copyright, but an idea is not. The infringer in this case copied the plans and not the idea. The Shaver case was distinguished in Saxelbye Architects, Inc. v. First Citizens Bank & Trust Co., 129 F.3d 117, 44 U.S.P.Q.2d 1634 (4th Cir. 1997), wherein it was found that the implied-license exception to the requirement of a writing is a limited one. The case was remanded to the circuit court for a determination of factual issues regarding the scope of the architect's contract.

See Hunt v. Pasternak, 179 F.3d 683 (9th Cir. 1999), in which the Ninth Circuit clarified its decision in the Eales case relating to whether an architectural work can be embodied in plans or drawings as well as a building.

[32] Del Madera Prop. v. Rhodes & Gardner, Inc., 637 F. Supp. 262 (N.D. Cal. 1985). See also Ocean Atlantic Woodland Corp., v. DRH Cambridge Homes, Inc., 2004 WL 2203423 (N.D. Ill. Sept. 29, 2004).

suit for infringement against the successor owner of the subdivision. The owner argued that once the municipality had approved the plat, it was not copyrightable. The court found the copyright valid.

§ 4.6   Transmittal of Digital Data: ¶ 1.6

1.6 TRANSMISSION OF DATA IN DIGITAL FORM

If the parties intend to transmit Instruments of Service or any other information or documentation in digital form, they shall endeavor to establish necessary protocols governing such transmissions, unless otherwise already provided in the Agreement or the Contract Documents. **(new)**

This new section is intended to cover the issues related to the transfer of computer files such as CAD drawings to the contractor and subcontractors. Among the issues that concern the architect in allowing such transfers are prevention of use of the drawings on other projects, possible errors created by the transmission itself, use of the wrong version of a drawing, and the continuation of mistakes in the architect's drawings onto the shop drawings. For instance, subcontractors normally prepare shop drawings from the drawings created by the architect and its consultants. In the preparation of these shop drawings, there is an opportunity to catch and correct errors because the subcontractor is preparing new drawings. If, on the other hand, the subcontractor is using the architect's electronic files to create the shop drawings, there is a much greater likelihood that the architect's mistakes will not be caught in this process.

Note that as of 2007, the AIA has developed two new documents related to digital data. The first is AIA Document E201-2007, Digital Data Protocol Exhibit. This document is intended to cover procedures regarding the exchange of digital data. A second document, C107-2007, Digital Data Licensing Agreement, is for use when the parties do not already have an AIA agreement in place, such as when the architect sends digital documents to a subcontractor.

§ 4.7   Article 2: Owner

The information and services required of the owner are specified in this article, as are the owner's rights to both stop the work and perform the work in certain situations of default by the contractor.

2.1 GENERAL

2.1.1 *The Owner is the person or entity identified as such in the Agreement and is referred to throughout the Contract Documents as if singular in number. The Owner shall designate in writing a representative who shall have express authority to bind the Owner with respect to all matters requiring the Owner's approval or authorization. Except as otherwise provided in Section 4.2.1, the Architect does not have such authority. The term "Owner" means the Owner or the Owner's authorized representative.* **(2.1.1; no change)**

The *Owner* may be a tenant or other person who does not actually own title to the property. All parties who deal directly with the owner should have knowledge of the owner's actual situation.[33] See additional comments about owner in **Chapter 2**. The owner or its representatives must not hinder or delay the contractor.[34]

The owner is required to designate a person who will have authority to make decisions and sign documents on the owner's behalf. Unless the owner specifically designates the architect to have this authority, the architect has no authority to bind the owner. The architect is not the owner's agent, except to the very limited extent that authority granted by this document and the owner-architect agreement. This means that the architect cannot authorize changes to the project, other than minor changes that do not affect cost or time. **See ¶** 7.4. Only the owner or the owner's representative can do this. The owner's representative is normally designated in AIA Document A101, ¶ 8.3.

> *2.1.2 The Owner shall furnish to the Contractor within fifteen days after receipt of a written request, information necessary and relevant for the Contractor to evaluate, give notice of or enforce mechanic's lien rights. Such information shall include a correct statement of the record legal title to the property on which the Project is located, usually referred to as the site, and the Owner's interest therein. (2.1.2; no change)*

The contractor may request that the owner provide a legal description of the site as well as the status of ownership of the property. This request must be in writing, and the owner must furnish the information within 15 days of the request. The contractor should request this at the time of execution of the agreement.[35] It would be unwise for a contractor to rely on this provision when preparing a mechanic's lien, because the owner would be reluctant to help a contractor file such a lien. Because time requirements related to mechanic's liens are usually very strict, the contractor should know the requirements in its jurisdiction and obtain this information well before the due dates. Another problem with this provision is that there is no indication as to how much information is required or how often the contractor

---

[33] *See, e.g.,* Collins Dozer Serv., Inc. v. Gibbs, 502 So. 2d 1174 (La. Ct. App. 1987) (contractor who entered into an agreement with a lessee to clear some land was denied access and sued owner; court held that lessee had no authority to bind owner to contract).

[34] Lester N. Johnson Co. v. City of Spokane, 22 Wash. App. 265, 588 P.2d 1214 (1979). The owner's activity on an adjacent site hindered the contractor's work. The duty of the owner not to interfere is implied in every construction contract. The contractor asked for damages and the owner defended on the basis that the contract limited the contractor's remedy to an extension of time. The court held that the owner's interference was outside the scope of the contract (was not reasonably foreseeable), and the owner was liable for damages to the contractor.

[35] In Universal Contracting Corp. v. Aug, 2004 WL 3015325 (Ohio Ct. App. 1st Dist. Dec. 30, 2004), the construction manager brought an action against the executive director of the owner after the contractor had obtained a substantial arbitration award against the owner. The executive director had negligently informed the contractor that sufficient funding was available for the project when it was not. After a jury finding in favor of the contractor, the trial court granted judgment in favor of the executive director, and the appellate court affirmed.

can ask for this information, although it implies that the contractor can make such a request as often as it wants. A likely standard is one of reasonableness, If the contractor has reason to believe that the owner is having financial difficulties, for instance, it would be reasonable to request this information for lien purposes. This would be similar to the requirements of ¶ 2.2.1.

§ 4.8   Information and Services Required of Owner: ¶ 2.2

2.2 INFORMATION AND SERVICES REQUIRED OF THE OWNER

2.2.1 Prior to commencement of the Work the Contractor may request in writing that the Owner provide reasonable evidence that the Owner has made financial arrangements to fulfill the Owner's obligations under the Contract. Thereafter, the Contractor may request such evidence if (1) the Owner fails to make payments to the Contractor as the Contract Documents require; (2) a change in the Work materially changes the Contract Sum; or (3) the Contractor identifies in writing a reasonable concern regarding the Owner's ability to make payment when due. The Owner shall furnish such evidence as a condition precedent to commencement or continuation of the Work or the portion of the Work affected by a material change. After the Owner furnishes the evidence, the Owner shall not materially vary such financial arrangements without prior notice to the Contractor. (**2.2.1; substantial changes**)

This provision requires the owner to furnish evidence that it has the capability of paying the contractor for the project. Unless this information is furnished upon request, the contractor may refuse to proceed with the project.[36] The wording "thereafter" implies that the contractor can request this information at any time (although the 2007 revision permits the contractor to make this request without any reason only before construction starts and thereafter only if any of the three listed conditions are met) and that there is no waiver by starting the work without making such a request. If the information is not promptly provided, the contractor can refuse to continue with the work. If the work is later resumed, the contractor may be entitled to a change order on account of the owner's delay in furnishing this information. The contractor may also stop work for a period of 30 days and thereafter terminate if the owner refuses to comply (¶ 14.1.1.4). Note that the

---

[36] In Architectural Sys., Inc. v. Gilbane Bldg. Co., 779 F. Supp. 820 (D. Md. 1991), the subcontractor sued the general contractor after the owner became insolvent and before the subcontractor was fully paid. The general contract containing a prior version of this clause was incorporated by reference into the agreement between the general contractor and subcontractor. The subcontractor alleged that the general contractor breached its duty to the subcontractor by failing to provide it with information concerning the owner's financial condition in time to permit it to discontinue work on the project and mitigate its damages. The court dismissed the complaint, finding that there was no affirmative duty to disclose. This language says that the contractor "may" request such information. Therefore, unless the contractor makes such a request, there is no duty to disclose any information. The subcontractor's claims of fraudulent misrepresentation were likewise dismissed. In Enterprise Capital, Inc. v. San-Gra Corp., 284 F. Supp. 2d 166 (D. Mass. 2003), the contractor requested this information but the owner did not provide it. This case did not turn on this issue, but the decision suggests that it might have been significant.

owner cannot materially change the financial arrangements that relate to the money to be paid to the contractor. The owner might not want to furnish this information more than at the start of the work (or at all), and it may want to revise this provision to reflect that. The prohibition against materially varying the financial information may cause concern to the lender. The implication is that the contractor will have veto power over any changes to the loan between the owner and lender. Allowing the contractor to have this power would not be acceptable to the owner and lender. The parties may want to substitute language that gives the contractor an assurance that there are sufficient funds available to pay the costs of construction and that the contractor is entitled to rely on such an assurance.

> *2.2.2 Except for permits and fees that are the responsibility of the Contractor under the Contract Documents, including those required under Section 3.7.1, the Owner shall secure and pay for necessary approvals, easements, assessments and charges required for construction, use or occupancy of permanent structures or for permanent changes in existing facilities.* (**2.2.2; minor modifications**)

The contractor is responsible for obtaining the normal building permit (**see** ¶ 3.7.1). The owner, however, is responsible for any zoning variations, governmental approvals, public or private easements necessary for access to the site or adjacent properties or for utilities, any approvals required by the lender, and any other approvals from any other governmental agencies, including environmental studies and similar reports and analyses. The parties may want to specifically identify all such approvals to avoid any questions as to responsibility. Note that the responsibilities of the parties as to water, electrical, and other utilities are covered in ¶ 3.4.1.

> *2.2.3 The Owner shall furnish surveys describing physical characteristics, legal limitations and utility locations for the site of the Project, and a legal description of the site. The Contractor shall be entitled to rely on the accuracy of information furnished by the Owner but shall exercise proper precautions relating to the safe performance of the Work.* (**2.2.2; no change**)

This information is required by the contractor to situate the construction on the site, to locate utilities, and to establish building lines. The owner may want to include language requiring the contractor to verify the accuracy of the furnished information and to report back to the owner within a certain period of time in the event of any discrepancies. The contractor is entitled to rely on the accuracy of this information.

In an Alabama case, the owner furnished a report to a contractor that was found to be defective.[37] When the contractor sued the owner after incurring additional

---

[37] Berkel & Co. Contractors v. Providence Hosp., 454 So. 2d 496 (Ala. 1984).

In Green Constr. Co. v. Kansas Power & Light Co., 1 F.3d 1005, 1009 (10th Cir. 1993), the court stated that "when a contract contains a site inspection clause, it places a duty on the contractor to exercise professional skill in inspecting the site and estimating the cost of the work." In that case, the contractor had argued that information furnished to bidders concerning subsurface conditions created an implied warranty as to those conditions. The court refused to find an implied

costs because it had relied on the report, the court held that the owner's disclaimer was effective. The disclaimer put the contractor on notice that he was not to rely on the report and must obtain his own report at his own expense. Other courts have held that an owner has a duty to bidders to furnish information that is not misleading and have permitted recovery against an owner when a contractor was damaged.[38] This provision, along with other similar provisions in the General Conditions, shows that any duties relating to the architectural plans, and any alleged defects in those plans, arose from the obligations contained in the contract, and that a claim relating to the plans should be for breach of contract, not negligence.[39]

> **2.2.4** *The Owner shall furnish information or services required of the Owner by the Contract Documents with reasonable promptness. The Owner shall also furnish any other information or services under the Owner's control and relevant to the Contractor's performance of the Work with reasonable promptness after receiving the Contractor's written request for such information or services.* (**2.2.4; minor revisions**)

The owner must cooperate with the contractor and furnish all documents, information, services, or other items under the owner's control as expeditiously as possible,[40] but only if that information is required by the contract documents. Note that there is no requirement to furnish information that is not specifically

---

warranty, stating that "an owner does not create an implied warranty by providing some soil information but instructing the contractor that the information may not be complete and that an independent site and soil investigation is required." The contractor also argued that the extra work required to remedy the soil situation was a material change in the scope of the project which necessitated additional compensation. The contractor relied on a changes clause. The court also rejected this on the basis that the changes clause applies only to amendments to the project design, not to difficulties in performance because of unforeseen conditions.

[38] *E.g.,* Jacksonville Port Auth. v. Parkhill-Goodloe, 362 So. 2d 1009, 1011 (Fla. Dist. Ct. App. 1978). The owner had furnished bidders with boring reports that did not indicate any significant rock. After work started, significant amounts of rock were discovered, causing the contractor delays and increased costs. The owner then refused to pay the additional costs, relying on the following contractual language:

> Boring information shown on the Drawings or furnished with the specifications, or both, is not guaranteed to be more than a general indication of the materials likely to be found adjacent to holes bored at the site of the work approximately at the locations indicated. Bidders shall examine boring records and interpret the subsoil investigation and other preliminary data for himself, and shall base his bid on his own opinion of the conditions likely to be encountered.

This language was found to constitute a guarantee that the boring information gave a general indication of the materials likely to be found in the stated locations. When it turned out that the conditions differed, the owner had breached that guarantee and was liable to the contractor for damages. The court stated that the owner had a duty not to mislead bidders and had breached that duty. *See also* Taylor v. DeLosso, 725 A.2d 51 (N.J. Super. A.D. 1999).

[39] Cameo Homes, Inc. v. Kraus-Anderson Const. Co., 2003 WL 22867640 (D. Minn. 2003) *aff'd,* 394 F.3d 1084 (C.A.8 (Minn.) 2005) (contractor could not pursue negligence claim against owner, and its contract claims were barred because it had failed to provide notice to the architect).

[40] Osolo Sch. Bldgs., Inc. v. Thorleif Larsen & Son, Inc., 473 N.E.2d 643 (Ind. Ct. App. 1985).

called for in the contract documents, even if the contractor might need it, unless the contractor specifically requests it. If the owner breaches this provision, the contractor may be entitled to extensions of time or to damages for delay. The information, specifically called for, is required to be furnished without any notice by the contractor. Any other information the contractor wants must be specifically requested in writing.

One issue may arise when the owner has information "relevant to the Contractor's performance of the Work" and the contractor does not know this. This provision implies that the owner has no obligation to furnish that information unless it is requested in writing. If the contractor has no knowledge, how can it request that information? The better view might be that the owner has an obligation to provide the contractor with any relevant information in the owner's control, whether or not it is requested.

> **2.2.5** *Unless otherwise provided in the Contract Documents, the Owner shall furnish to the Contractor one copy of the Contract Documents for purposes of making reproductions pursuant to Section 1.5.2.* (**2.2.5; changed from a number of copies to "one copy"**)

The contractor is entitled to at least one copy of the contract documents without additional charge. The contractor will then need to make enough additional copies as may be required for its own use and the use of the various subcontractors and suppliers. One way of handling this is to furnish the contractor with a set of reproducible drawings to make its own copies of the drawings and specifications. Another is to provide electronic copies to the contractor.

§ 4.9   Owner's Right to Stop the Work: ¶ 2.3

2.3 OWNER'S RIGHT TO STOP THE WORK

> *If the Contractor fails to correct Work that is not in accordance with the requirements of the Contract Documents as required by Section 12.2 or repeatedly fails to carry out Work in accordance with the Contract Documents, the Owner may issue a written order to the Contractor to stop the Work, or any portion thereof, until the cause for such order has been eliminated; however, the right of the Owner to stop the Work shall not give rise to a duty on the part of the Owner to exercise this right for the benefit of the Contractor or any other person or entity, except to the extent required by Section 6.1.3.* (**2.3.1; change "persistently" to "repeatedly"**)

The architect has no independent right to stop the work. If the owner asks the architect to stop the work because the contractor is in breach of the contract, the architect must request a written order from the owner. The right to stop the work confers on the party that has that right a certain responsibility over safety at the job site. Some statutes place liability for construction accidents on the party "in charge" of the work. The right to stop the work is one of the primary criteria in determining whether a party has liability under such acts. Paragraph 2.3 states that, although the owner has the right to stop the work, it has no duty to do so for the benefit of any party except itself. The contractor is still responsible for safety under other provisions of A201. Paragraph 6.1.3 relates to work by the owner's own forces, for which the owner has responsibility for safety (¶ 6.1.4 and Article 10).

The owner, under this provision, can stop the work whenever the contractor fails to correct work that does not conform to the contract documents or fails to carry out the work in accordance with the contract documents. Thus, if the work is defective or the contractor is persistently late, this provision applies. Although there are no time limits given, other provisions of A201 would suggest that the contractor be given seven days' notice to come into compliance. Also, the deviations would need to be more than minor items (unless there are a large number of such minor items) before the owner could stop the work. This paragraph is not an excuse for an owner to stop the work unless there has been a substantial breach of the contract documents. The owner may want to delete the word "repeatedly" in order to invoke this provision in the case of any breach of the contract documents. Note that this provision is in addition to the owner's rights under Article 14.

§ 4.10   Owner's Right to Carry Out the Work: ¶ 2.4

2.4 OWNER'S RIGHT TO CARRY OUT THE WORK

If the Contractor defaults or neglects to carry out the Work in accordance with the Contract Documents and fails within a ten-day period after receipt of written notice from the Owner to commence and continue correction of such default or neglect with diligence and promptness, the Owner may, without prejudice to other remedies the Owner may have, correct such deficiencies. In such case an appropriate Change Order shall be issued deducting from payments then or thereafter due the Contractor the reasonable cost of correcting such deficiencies, including Owner's expenses and compensation for the Architect's additional services made necessary by such default, neglect or failure. Such action by the Owner and amounts charged to the Contractor are both subject to prior approval of the Architect. If payments then or thereafter due the Contractor are not sufficient to cover such amounts, the Contractor shall pay the difference to the Owner. **(2.4.1; changed seven-day period to ten days, eliminated second seven-day period)**

This paragraph provides a specific procedure the owner must follow to do some or all of the work of the contractor if the contractor fails to properly perform under the agreement. It also affords the contractor the opportunity to cure the problem. First, the owner must give the contractor written notice that the contractor must correct any defaults or neglect of the work within ten days. Note that, pursuant to ¶ 8.1.4, these are calendar days, not business days. Within that ten-day period, the contractor must commence and continue to correct deficiencies (but not necessarily finish such corrective work). If the contractor fails to do so or makes only a half-hearted attempt, the owner can take corrective action. The architect is required to approve these actions. If the architect does not do so, the owner can follow the claims procedure set forth in Section 15.1, starting with a submission to the Initial Decision Maker, and possibly followed by mediation. Note that under Section 15.2 an "Initial Decision Maker" will be appointed who may or may not be the architect. Under this provision, the architect must give the approval required by this paragraph, even if the Initial Decision Maker is a different person. The owner can also request a change order to cover the work performed by the owner, as well as additional architect's fees and related reasonable costs. Because a change order must be signed by the contractor as well as the owner and architect, it is quite unlikely that the contractor would do so. In that event, the owner and architect would likely prepare a construction change directive. The owner could also insert a provision that in the event this provision takes effect, the contractor will be deemed to have signed the change order, whether or not it has actually done so. The owner can also pursue other remedies against the contractor, such as termination. In the event that there is a bond, the owner should contact the contractor's surety concerning the situation.

This paragraph does not limit the deficiencies to major ones. The architect, in approving this remedy, must exercise its best judgment, in consultation with the owner. In some cases, the best remedy may be to let the contractor finish the work and withhold payment for minor items that are then completed by the owner. However, even minor items can become major if there are many of them and the contractor repeatedly fails to correct them. If the owner fails to follow the procedure, it may not be able to claim damages caused by the contractor's default.[41]

The owner may want the right to audit the contractor's books and records, particularly if the contract is a cost-plus contract or involves shared savings.

§ 4.11   Article 3: Contractor

Article 3 defines the contractor's responsibilities.

3.1 GENERAL

3.1.1 The Contractor is the person or entity identified as such in the Agreement and is referred to throughout the Contract Documents as if singular in number. The Contractor shall be lawfully licensed, if required in the jurisdiction where the Project is located. The Contractor shall designate in writing a representative who shall have express authority to bind the Contractor with respect to all matters under this

---

[41] Environmental Safety & Control Corp. v. Board of Educ., 580 N.Y.S.2d 595 (App. Div. 1992); State Sur. Co. v. Lamb Constr. Co., 625 P.2d 184 (Wyo. 1981); Bouchard v. Boyer, 1999 Conn. Super. LEXIS 1285 (May 17, 1999); A. Prete & Son Constr. v. Town of Madison, No. CV 91-03103073-S, 1994 Conn. Super. LEXIS 2532 (Oct. 4, 1994). In Sullivan v. George A. Hormel and Co., 303 N.W.2d 476 (Neb. 1981), the court held that this, and other provisions of A201, meant that the relationship of owner and independent contractor was not converted to that of master and servant in the context of a case where a worker was injured.

Contract. The term "Contractor" means the Contractor or the Contractor's autho-
rized representative. (**3.1.1; add second and third sentences**)

The project team should be identified so that a clear chain of command for all
parties is identified. This could be done in a separate memorandum. In general, the
contractor is not an "agent" of the owner but is an "independent contractor."[42]
This distinction has certain legal ramifications. The contractor is the party that
executes the owner-contractor agreement and is not necessarily the low bidder,
even if the owner has prequalified the bidders.

Note that this paragraph requires the contractor to be licensed if the laws that
apply to the project require that contractors be licensed. A failure to be properly
licensed may, in fact, render the contract void, with the contractor unable to collect
its fees or enforce any lien rights.[43]

3.1.2 The Contractor shall perform the Work in accordance with the Contract Docu-
ments. (**3.1.2; no change**)

The contractor must do the work in accordance with the contract documents.
The 1987 version included all approved shop drawings and other submittals.
Beginning with the 1997 version, shop drawings are not contract documents.
Thus, the contractor need not and, indeed, must not (**see** ¶ 3.12.8) comply with
approved shop drawings or submittals to the extent they are inconsistent with the
contract documents. Any such deviations would be ineffective without an appropriate change order. If the owner wants to make sure the contractor performs work
shown on shop drawings and that work is not shown on the contract documents, the
appropriate procedure would be to execute a change order.

The owner is entitled to strict compliance with the plans and specifications and
to recover damages sustained by reason of the contractor's failure to comply.[44] The
contractor is entitled to a reasonable opportunity to perform the contract without

---

[42] Fandrich v. Allstate Ins. Co., 25 Ill. App. 3d 301, 322 N.E.2d 843 (1975) (interpreted in the context
of a Structural Work Act case; if contractor had been an agent, owner would have been responsible
for worker's injury); Oldham & Worth, Inc. v. Bratton, 263 N.C. 307, 139 S.E.2d 653 (1965) (in
action by subcontractor against owner to recover for materials furnished, general contractor was
not an agent for the owner but an independent contractor); Lasky v. Realty Dev. Co., LLC, 2006
WL 1113510 (Mich. App. 2006).

[43] Midland Fire Prot., Inc., v. Clancy & Theys Const. Co., 623 S.E.2d 369 (N.C. App. 2006); Mortise v.
55 Liberty Owners Corp., 477 N.Y.S.2d 2 (App. Div. 1st Dep't 1984) (construction company's
failure to obtain license from city to engage in the business of home improvement rendered
contracts for building conversion unenforceable); Alexander v. Neal, 110 N.W.2d 797 (Mich.
1961) (unlicensed contractors cannot maintain action for recovery for services rendered pursuant
to statute providing for licensing of residential builders); Hydrotech Sys., Ltd., v. Oasis Waterpark,
803 P.2d 370 (Cal. 1991) (unlicensed out-of-state subcontractor could not recover because of
failure to substantially comply with licensing statute); State v. Wilkinson, 39 P.3d 1131 (Ariz.
2000) (unlicensed contractor charged with a misdemeanor for acting as a contractor without a
license required to pay restitution to homeowners for amounts paid by owners).

[44] Corbetta Constr. Co. v. Lake County Pub. Bldg. Comm'n, 64 Ill. App. 3d 313, 381 N.E.2d 758
(1978). In State v. Wolfenbarger & McCulley, P.A., 236 Kan. 183, 690 P.2d 380 (1984), the

interference or obstruction by the owner.[45] If the contractor has followed the contract documents and damages result because the contract documents were faulty, the owner cannot recover from the contractor if the contractor has pointed out the problem.[46] Note that courts have held that a builder is not required to perform perfectly but that it is held to a duty of substantial performance in a workmanlike manner.[47]

---

architect was held to be negligent when the contractor substantially followed the drawings. The architect had designed air intake louvers with 40% free air space. The contractor installed some with less free air space, and the building was damaged when snow entered the plenum space. The contractor demonstrated that even the proper louvers had a snow problem. The fault was the design, and not the contractor's slight deviation.

In Robert G. Regan Co. v. Fiocchi, 44 Ill. App. 2d 336, 194 N.E.2d 665 (1963), the masonry subcontractor did not install wall ties per the contract documents. When the walls bulged, he was held liable for failing to follow the drawings and specifications. The court held that opinions that the walls would have bulged even if the drawings had been followed were immaterial. "When defendants departed from this specification they did so at their peril, and all attempted excuses for non-compliance became immaterial."

In Georgetown Township High Sch. Dist. No. 218 v. Hardy, 38 Ill. App. 3d 722, 349 N.E.2d 88 (1976), the court followed *Regan* and found for a contractor who had followed the drawings, finding that the contractor had no duty to construct a building that would withstand any given wind pressures. The "contractor is not liable for damages if he (1) performs his work in accordance with the plans and specifications furnished by the owner, and (2) does so in a workman-like manner."

[45] Fehlhaber v. State, 65 A.D. 119, 410 N.Y.S.2d 920 (1978).

[46] J.R. Graham v. Randolph County Bd. of Educ., 25 N.C. App. 163, 212 S.E.2d 542 (1975). An architect specified an improper sealant, and the sealant sub-subcontractor called this to the contractor's attention. The contractor was not liable to the sub because the contractor advised the architect who had the final decision on the use of the sealant. *See also* Lawrence Dev. Corp. v. Jobin Waterproofing, Inc., 588 N.Y.S.2d 422 (App. Div. 1992) (general contractor had no control over or ability to change sealant specification, therefore, it owed no duty of care to sub).

The contractor may be able to avoid liability for deviating from the contract documents if it can demonstrate that the plans and specifications were faulty and that damage would have resulted from following the contract documents. City of Charlotte v. Skidmore, Owings & Merrill, 407 S.E.2d 571 (N.C. 1991).

[47] Mayfield v. Swafford, 106 Ill. App. 3d 610, 435 N.E.2d 953 (1983) (failing to perform in workmanlike manner constitutes breach of contract; if owner receives substantial performance, it must pay contractor the contract sum less the cost of correcting any deficiencies). *See also* Weidner v. Szostek, 245 Ill. App. 3d 487, 614 N.E.2d 879 (1993); J.R. Sinnott Carpentry, Inc. v. Phillips, 110 Ill. App. 3d 632, 443 N.E.2d 597 (1982); Watson Lumber v. Guennwig, 79 Ill. App. 2d 377, 226 N.E.2d 270 (1967); Cleveland Neighborhood Health Servs., Inc. v. St. Clair Builders, Inc., 64 Ohio App. 3d 639, 582 N.E.2d 640 (1989) ("Where the party obligated to perform under the contract makes an honest effort to do so, and there is no willful omission on its part, substantial performance is all that is required to entitle the party to payment under the contract").

In Lange Indus. v. Hallam Grain Co., 244 Neb. 465, 507 N.W.2d 465, 473 (1993), the court stated that:

in a building contract, substantial performance is shown when all the essential elements necessary for the full accomplishment of the purposes of the contract have been performed with such an approximation to complete performance that the owner obtains substantially what is called for by the contract. . . . In building and construction contracts, in the absence of an express agreement to the contrary, the law implies that the building will be erected in a

Contractors also owe a duty to third parties. This duty has been stated thus: "Contractors owe a duty to exercise the care required of their profession to those with whom they are not in privity when injury to those third persons is foreseeable."[48] But note that under ¶ 1.1.2, the contract documents do not create any contractual relationships other than between the owner and contractor.

3.1.3 The Contractor shall not be relieved of obligations to perform the Work in accordance with the Contract Documents either by activities or duties of the Architect in the Architect's administration of the Contract, or by tests, inspections or approvals required or performed by persons or entities other than the Contractor. **(3.1.3; add "entities")**

Unless there is direct interference by the architect not based on a legal right or duty, the contractor must perform.[49] The architect has a duty to cooperate with the contractor, but the architect must also ascertain that the owner is getting what it paid for. Thus, the architect can order tests (¶ 13.5) without relieving the contractor of its duty to properly perform the work, and the contractor must cooperate with those tests. The architect enjoys a privilege to interfere with the contract between the owner and contractor, so long as this is done without malice and in the architect's role as looking out for the interests of the owner.[50] The contractor

---

reasonably good and workmanlike manner and will be reasonably fit for the intended purpose.

    A contractor normally does not owe a fiduciary duty to an owner. Munn v. Thornton, 956 P.2d 1213 (Alaska 1998); Ginley v. E.B. Mahoney Builders, Inc., 2005 WL 27534 (E.D. Pa. Jan. 5, 2005). An exception is where a state construction trust fund statute creates a duty to hold funds in trust. *E.g.*, In re Bolger, 351 B.R. 165 (Bankr. N.D. Okla. 2006).

[48] Honey v. Barnes Hosp., 708 S.W.2d 686 (Mo. Ct. App. 1986). This duty ends upon the owner's acceptance, except in the case of latent defects.

[49] In Gordon J. Phillips, Inc. v. Concrete Materials, Inc., 590 N.Y.S.2d 344 (App. Div. 1992), the dissent correctly observed that this provision requires the contractor to do the work in accordance with the contract documents, whether or not the architect or owner advises the contractor of a problem. In that case, the asphalt failed to meet specifications. The contractor argued that the owner had, in the past, stopped work when its testing indicated that materials did not satisfy the specifications. The majority stated that this created a triable issue of fact.

    This might well create an issue of waiver, because the owner cannot sit by and allow the contractor to install something improperly, all the while knowing that the material will be rejected. An owner has a duty to mitigate a contractor's damages.

    See also Taber Partners I v. Insurance Co. of Am., Inc., 875 F. Supp. 81 (D.P.R. 1995).

[50] Waldinger Corp. v. CRS Group Eng'rs, Inc., 775 F.2d 781 (7th Cir. 1985), George A. Fuller Co. v. Chicago Coll. of Osteopathic Med., 719 F.2d 1326 (7th Cir. 1983) ("we conclude that Illinois would allow an architect a conditional privilege to interfere with the construction contract of its principal."); Geolar, Inc. v. Gilbert/Commonwealth, Inc., 874 P.2d 937 (Alaska 1994) (agent-engineer was only privileged to the extent that he acted in his principal's best interest, if he acted in spite or malice, then interference with the contract was not protected); Dehnert v. Arrow Sprinklers, Inc., 705 P.2d 846 (Wyo. 1985) ("An architect who acts within the scope of his contractual obligations to the owner will not be liable for advising the owner to terminate a contractor's performance unless the architect acts with malice or bad faith.").

cannot deny the architect access to the work under this provision and ¶ 3.16.1. The contractor remains responsible for safety, even if instructions are given by the architect or others that may create unsafe conditions.[51]

§ 4.12   Review of Contract Documents and Field Conditions by Contractor: ¶ 3.2

3.2 REVIEW OF CONTRACT DOCUMENTS AND FIELD CONDITIONS BY CONTRACTOR

3.2.1 Execution of the Contract by the Contractor is a representation that the Contractor has visited the site, become generally familiar with local conditions under which the Work is to be performed and correlated personal observations with requirements of the Contract Documents. **(1.5.2; no change)**

This is an acknowledgment by the contractor that it is familiar with the site conditions and has visited the site. The contractor is charged with such knowledge of the site as would be obtainable without extensive investigation. Thus, by this clause alone, the contractor would not be responsible for subsoil conditions but would be held to have knowledge of an open stream on the site. It would be good practice for the contractor to take a number of photographs of the site before any work, so that any extras caused by site conditions (**see** ¶ 3.7.4) can be more easily substantiated.

The contractor also is charged with knowledge of access to the site (if reasonably observable) and general governmental regulations for construction at the site. If any site access restrictions are not readily observable, such as special security requirements, these restrictions must be made known to the contractor as part of the bid documents. Otherwise, the contractor would be entitled to a change order. Likewise, if areas of the site are appropriated by others, the contractor may lose storage areas, resulting in additional costs that would be passed through to the owner, unless these conditions are related to the contractor for inclusion in its bid.

The contractor might want to more clearly set forth the site conditions found at the site visit. The conditions that the contractor is relying upon might include site access, labor conditions, and other conditions unique to this particular site. The description of the site conditions could be included as a rider to A101.

In a federal case, the court stated that "when a contract contains a site inspection clause, it places a duty on the contractor to exercise professional skill in inspecting the site and estimating the cost of the work."[52] In that case, the contractor had argued that information furnished to bidders concerning subsurface conditions created an implied warranty as to those conditions. The court refused to find an implied warranty, stating that "an owner does not create an implied warranty by providing some soil information but instructing the contractor that the information may not be complete and that an independent site and soil investigation is

---

[51] Koller v. Liberty Mut. Ins. Co., 541 N.W.2d 838 (Wis. Ct. App. 1995).

[52] Green Constr. Co. v. Kansas Power & Light Co., 1 F.3d 1005, 1009 (10th Cir. 1993).

required." The contractor also argued that the extra work required to remedy the soil situation was a material change in the scope of the project that necessitated additional compensation. The contractor relied on a changes clause. The court also rejected this on the basis that a changes clause applies only to amendments to the project design, not to difficulties in performance because of unforeseen conditions.

This provision has been held to charge the contractor with knowledge of site conditions so that a duty to warn of an unsafe condition by another party was not necessary.[53]

3.2.2 Because the Contract Documents are complementary, the Contractor shall, before starting each portion of the Work, carefully study and compare the various Contract Documents relative to that portion of the Work, as well as the information furnished by the Owner pursuant to Section 2.2.3, shall take field measurements of any existing conditions related to that portion of the Work and shall observe any conditions at the site affecting it. These obligations are for the purpose of facilitating coordination and construction by the Contractor and are not for the purpose of discovering errors, omissions, or inconsistencies in the Contract Documents; however, the Contractor shall promptly report to the Architect any errors, inconsistencies or omissions discovered by or made known to the Contractor as a request for information in such form as the Architect may require. It is recognized that the Contractor's review is made in the Contractor's capacity as a contractor and not as a licensed design professional, unless otherwise specifically provided in the Contract Documents. (**3.2.1; minor modifications; add "coordination"; add last sentence**)

The contractor is responsible for reviewing the contract documents and for comparing them with the surveys and other information furnished by the owner pursuant to ¶ 2.2. According to this paragraph, this is done to permit the contractor to become familiar with all of the contract documents, which are complementary (**see** ¶ 1.2.1). The idea is that the contractor goes to the site, carefully reviews the drawings, specifications, and other documents, and relies on its experience and other information before starting the project. The contractor should not then request an extra because an item is shown on the drawings and not the specifications, or vice versa. Note that this might mean multiple reviews, "before starting each portion of the Work."

It is usually difficult to prove what the contractor did or did not know. If an error is one that a reasonably prudent contractor should have discovered, the contractor should be charged with knowledge of the error. On the other hand, in the event of such an error, an argument can be made that the architect should have caught that error unless the error is of a type that a reasonable architect would not have caught. Expert testimony would likely be required to address this issue. The contractor is entitled to payment if it builds the project according to the contract documents, even if they contain errors.[54] On the other hand, if the contractor does not comply

---

[53] Jones v. James Reeves Contractors, Inc., 701 So. 2d 774 (Miss. 1997).

[54] J.D. Hedin Constr. Co. v. United States, 347 F.2d 235 (Ct. Cl. 1965); Blecick v. School Dist. No. 18, 2 Ariz. App. 115, 406 P.2d 750 (1965); Mayville-Portland Sch. v. C.L. Linfoot, 261 N.W.2d

with the plans and specifications provided by the owner, notwithstanding the fact
that they are defective, the contractor proceeds at its peril, assuming the risk of any

---

907 (N.D. 1978). The leading case is United States v. Spearin, 248 U.S. 132 (1918), in which the
Court stated: "But if the contractor is bound to build according to plans and specifications prepared
by the owner, the contractor will not be responsible for the consequences of defects in the plans and
specifications."

In McDermott v. Tendun Constructors, 211 N.J. Super. 196, 511 A.2d 690 (1986), the contractor was not liable in a wrongful death action because he followed the plans and had no
discretion in doing so. "This court has recognized that, in a products liability action, where a
defendant has had no discretion and has 'strictly adhered to the plans and specifications owned and
provided by the Government,' there can be no liability imposed." This defense was based on
governmental immunity. The architect was not entitled to this immunity because he was not
strictly bound by government specifications or demands.

In Wilkinson v. Landreneau, 525 So. 2d 617 (La. Ct. App. 1988), on an oral contract, a
contractor improperly built a chimney. He did not follow the plans and specifications. Even though
the court found that the plans did not show every detail, he was liable because he did not follow the
plans in other respects.

In Board of Educ. v. Mars Assocs., 133 A.D.2d 800, 520 N.Y.S.2d 181 (1987), the court said if
the contractors "followed the architects' plans and specifications and exercised reasonable care
and skill in the performance of their work, they will not be responsible for damages which occurred
as a result of defects in the architects' plans and specifications."

The court in Community Heating & Plumbing Co. v. Kelso, 987 F.2d 1575 (Fed. Cir. 1993),
reviewed a claim by a contractor concerning an ambiguity. A number of principles were enunciated:

Contracts are not necessarily rendered ambiguous by the mere fact that the parties disagree
as to the meaning of their provisions. A contract is ambiguous if it is susceptible of two
different and reasonable interpretations, each of which is found to be consistent with the
contract language. If a contract contains a patent ambiguity, the contractor is under a duty to
inquire and must seek clarification of the proper contract interpretation. . . . This policy,
known as the patent ambiguity doctrine, was established to prevent contractors from taking
advantage of the government, protect other bidders by assuring that all bidders bid on the
same specifications, and materially aid the administration of government contracts by
requiring that ambiguities be raised before the contract is bid, thus avoiding costly litigation
after the fact.

When a contractor incurs additional expenses when attempting to perform a contract pursuant
to defective specifications, the contractor is entitled to recover its extra costs thereafter incurred by
reason of the breach of the implied warranty. Fairbanks N. Star Borough v. Kandik Constr., Inc.,
795 P.2d 793 (Alaska 1990).

The case of Pete Wing Contracting, Inc. v. Port Conneaut Investors Ltd. P'ship, 1995 Ohio
App. LEXIS 4341 (Sept. 29, 1995), illustrates what can happen when the owner is too cheap to hire
the architect for the normal construction phase services. The architect issued a certificate of
substantial completion to the contractor, but the owner refused to accept it or release the retainage.
The problem was that the original bid plans called for the ceiling height in the loft area to be seven
feet, eight inches. When the construction plans were issued, the height was apparently given as 78"
"instead of 7"8". The architect testified that it was understandable that the contractor interpreted
the drawings in that manner, especially since the owner reduced the architect's scope of services
during construction, leaving the contractor to deal directly with the owner instead of the architect.
Thus, the ceiling height problem was not the contractor's fault, and the contractor attained
substantial completion of the project. The owner assumed the risk that the drawings were not
correct or in compliance with local, state, or federal codes, according to the court.

deviations from the plans and guaranteeing the suitability of the work.[55] The costs for such errors, inconsistencies, or omissions are to be allocated between the contractor and architect (and perhaps the owner). If the contractor gives proper notification of the error, it will not be liable unless it has contributed to the error. The owner may want to add language that includes the duty to verify the documents against all applicable codes, ordinances, regulations, and so on, but this would require modification of ¶ 3.2.3, which states the opposite.

The owner implicitly warrants that the plans and specifications are adequate and suitable for the work to be performed by the contractor.[56] If the plans are

---

Employers Mut. Cas. Co. v. Collins & Aikman Floor Coverings, Inc., 2004 WL 840561 (S.D. Iowa Feb. 13, 2004), discussed *Spearin,* but did not follow it. The project specifications for carpeting obtained from the manufacturer did not include anything suggesting that the carpet could be used under rolling chairs without mats. When the carpet began to wear significantly under the chairs, the owner sued. The carpet manufacturer defended on the basis of the *Spearin* doctrine. The court, finding that there was a question of fact whether the carpet manufacturer was verbally informed of the requirement, refused summary judgment. *See also* Travelers Indem. Co. v. S.M. Wilson & Co., 2005 WL 2234582 (E.D. Mo. Sept. 14, 2005).

[55] Burke County Pub. Sch. v. Juno Constr. Corp., 50 N.C. App. 238, 273 S.E.2d 504 (1981).

[56] United States v. Spearin, 248 U.S. 132 (1918); J.D. Hedin Constr. Co. v. United States, 347 F.2d 235 (Ct. Cl. 1965); Chaney Bldg. Co. v. City of Tucson, 148 Ariz. 571, 716 P.2d 28 (1986); Housing Auth. v. E.W. Johnson Constr. Co., 264 Ark. 523, 573 S.W.2d 316 (1978); Electronics Group v. Central Roofing, 164 Ill. App. 3d 915, 518 N.E.2d 369 (1987); North County Sch. Dist. v. Fidelity & Deposit Co., 539 S.W.2d 469 (Mo. 1976); Mayville-Portland Sch. v. C.L. Linfoot, 261 N.W.2d 907 (N.D. 1978); Metropolitan Sewerage Comm'n v. R.W. Constr., 72 Wis. 2d 365, 241 N.W.2d 371 (1976).

In Hunt v. Blasius, 74 Ill. 2d 203, 384 N.E.2d 368 (1978), the Illinois Supreme Court stated:

An independent contractor owes no duty to third persons to judge the plans, specifications or instructions which he has merely contracted to follow. If the contractor carefully carries out the specifications provided him, he is justified in relying upon the adequacy of the specifications unless they are so obviously dangerous that no competent contractor would follow them.

Hunt was cited in Soave v. National Velour Corp., 863 A.2d 186 (R.I. 2004). In that case, a pedestrian fell from a retaining wall, allegedly because there was no guardrail. The contractor constructed the wall in accordance with the plans and specifications and had relinquished control of the site prior to the accident. The court found no duty on the part of the contractor towards the plaintiff.

In Hubbard Constr. Co. v. Orlando/Orange County Expressway Auth., 633 So. 2d 1154 (Fla. Dist. Ct. App. 1994), the work included more than half a million cubic yards of embankment. The owner was to provide on-site sources for the embankment material, and the contractor was to compact this material in 12-inch layers. Each layer was to be compacted to a density of at least 100% of the maximum density as determined by the Standard Proctor test. After the contractor experienced extreme difficulty in satisfying the test, it learned that the owner, through its representative, had been applying the wrong test. Rather than the Standard Proctor test required by the contract, the more severe Modified Proctor standard had been applied. The error was compounded by the fact that the contractor was still held to a 100% maximum density requirement, although the industry standard required only 95% maximum density when the more stringent Modified Proctor was used. Although the problem was corrected, the contractor had already installed 80% of the embankment and had sustained substantial damages. The appellate court held that the owner had a nondelegable contractual obligation to apply the appropriate standard.

deficient, the contractor may be entitled to an extension of the contract time and damages.[57]

The contractor must take field measurements and compare them to the requirements of the contract documents. This is done for the purpose of "facilitating construction" and not for the purpose of discovering errors in the drawings. The type of field measurements depends on the type of project. For instance, in a remodeling project, the contractor takes a great deal of field measurements, whereas on new construction, field measurements generally are restricted to verification of survey and other site information. The contractor should also observe conditions at the site and surrounding locale and take into account any other information it knows. This includes information from prior jobs in the area, information about the availability of materials and services, and so on. The contractor cannot stand idly by and permit the project to suffer, if it has knowledge that would affect the work.[58]

3.2.3 The Contractor is not required to ascertain that the Contract Documents are in accordance with applicable laws, statutes, ordinances, codes, rules and regulations, or lawful orders of public authorities, but the Contractor shall promptly report to the Architect any nonconformity discovered by or made known to the Contractor as a request for information in such form as the Architect may require. **(3.2.2; modifications; added RFIs)**

In Alamo Cmty. Coll. Dist. v. Browning Constr. Co., 131 S.W.3d 146 (Tex. Ct. App. 2004), the appellate court refused to overturn a jury verdict in favor of a contractor who had been awarded damages due in part to the owner's failure or refusal to correct design errors. The owner argued that such damages are unenforceable because the contract did not impose a duty on the owner for such design errors. The court held that the instant contract provision did impose a duty on the owner for design errors.

When the owner disclaims warranties for its estimates, the risk is on the contractor. Brown Bros. v. Metropolitan Gov't, 877 S.W.2d 745, 746 (Tenn. Ct. App. 1993) (citing *Spearin*). The owner had provided its own estimates of earthwork quantities to be removed. When the actual quantities far exceeded the estimates, the contractor sought an extra on the basis of an implied warranty of accuracy. The court rejected this based on the contract language that "bidders shall rely exclusively upon their own estimates, investigation and other data which are necessary for full and complete information upon which the proposal may be based."

[57] J.D. Hedin Constr. Co. v. United States, 347 F.2d 235 (Ct. Cl. 1965); Housing Auth. v. E.W. Johnson Constr. Co., 264 Ark. 523, 573 S.W.2d 316 (1978).

[58] In Gillingham Constr., Inc. v. Newby-Wiggins Constr., Inc., 142 Idaho 15, 121 P.3d 946 (2005), a subcontractor sought additional compensation for dirt removal not reflected in elevations shown on plans and specifications. Apparently, the existing site elevations were much higher than shown on the drawings. The subcontractor requested additional compensation for the extra work, plus the time its equipment sat idle and the costs of moving its equipment back and forth onto the site. Under an earlier version of A201, this contract language required the subcontractor to check the site plan documents for obvious errors and discrepancies. There was a separate duty to go out and conduct field verification measurements prior to starting the work. Because the subcontractor breached this duty to verify field conditions, the subcontractor was not entitled to indemnification from the owner related to the accuracy of the plans.

This paragraph states what should be obvious: that it is the architect's job to design the project in conformance with codes, and not the contractor's. If, however, the contractor has actual knowledge that the design violates any code or law, it has a duty to inform the architect. Note, however, the interplay of Section 3.12.10, which requires the contractor to provide design services for "design-build" portions of the project. This would typically occur where the mechanical, electrical, or curtainwall systems are to be designed by the contractor's subcontractors. In that case, the architect would furnish performance criteria and the contractor would be responsible for meeting applicable codes and other laws relative to that portion of the work.

3.2.4 If the Contractor believes that additional cost or time is involved because of clarifications or instructions the Architect issues in response to the Contractor's notices or requests for information pursuant to Sections 3.2.2 or 3.2.3, the Contractor shall make Claims as provided in Article 15. If the Contractor fails to perform the obligations of Sections 3.2.2 or 3.2.3, the Contractor shall pay such costs and damages to the Owner as would have been avoided if the Contractor had performed such obligations. If the Contractor performs those obligations, the Contractor shall not be liable to the Owner or Architect for damages resulting from errors, inconsistencies or omissions in the Contract Documents for differences between field measurements or conditions and the Contract Documents, or for nonconformities of the Contract Documents to applicable laws, statutes, ordinances, codes, rules and regulations, and lawful orders of public authorities. **(3.2.3; minor modifications; add last clause regarding nonconformities)**

This reinforces the provisions of ¶ 3.2.2. The contractor must follow the claims procedure if a clarification or instruction by the architect would result in additional cost or time. If the contractor fails to point out errors in the contract documents and should have done so, the contractor will bear the costs resulting from its failure to do so. If the contractor does so and clarifications are provided by the architect, it is the responsibility of the contractor to determine whether the clarification will result in a change in the cost or time of the project. If so, the contractor must promptly make a claim. Note that the contractor is not responsible for discrepancies between actual field measurements or conditions and those indicated on the drawings, unless the contractor knows of such differences and fails to inform the architect.

§ 4.13 Supervision and Construction Procedures: ¶ 3.3

3.3 SUPERVISION AND CONSTRUCTION PROCEDURES

3.3.1 The Contractor shall supervise and direct the Work, using the Contractor's best skill and attention. The Contractor shall be solely responsible for, and have control over, construction means, methods, techniques, sequences and procedures and for coordinating all portions of the Work under the Contract, unless the Contract Documents give other specific instructions concerning these matters. If the Contract

Documents give specific instructions concerning construction means, methods, techniques, sequences or procedures, the Contractor shall evaluate the jobsite safety thereof and, except as stated below, shall be fully and solely responsible for the jobsite safety of such means, methods, techniques, sequences or procedures. If the Contractor determines that such means, methods, techniques, sequences or procedures may not be safe, the Contractor shall give timely written notice to the Owner and Architect and shall not proceed with that portion of the Work without further written instructions from the Architect. If the Contractor is then instructed to proceed with the required means, methods, techniques, sequences or procedures without acceptance of changes proposed by the Contractor, the Owner shall be solely responsible for any loss or damage arising solely from those Owner required means, methods, techniques, sequences or procedures. **(3.3.1; change last sentence)**

The contractor, not the architect, supervises the work,[59] including that of the subcontractors. This includes safety procedures (**see** Article 10). This could be varied by specific instructions in the contract documents. Thus, the architect or owner could direct the contractor as to specific sequences or other matters, but in that event, they would take on additional liability as to such items. Except under the most exceptional circumstances, architects should never include such instructions. Sometimes such instructions are buried in specifications, either intentionally or inadvertently. This provision requires the contractor to review any such instructions and object in writing if the contractor cannot warrant such work or the procedure is otherwise objectionable to the contractor. The contractor may want to provide that if unsafe conditions exist, the contractor may suspend work until such conditions are remedied and such suspension of work will not be considered a breach of the contract. The language that the contractor is "solely responsible" has been used by a developer to avoid liability for a worker's injuries.[60]

---

[59] In Slifer v. Wheeler & Lewis, 39 Colo. App. 269, 567 P.2d 388 (1977), the architect was liable for the injuries of workers because his contract required supervision. One portion of the owner-contractor contract read as follows: "The Contractor shall do the work herein contemplated in strict obedience to the directions which may be given . . . by the Owner through the Architect or his representatives."

In Marshall v. Allegheny County Port Auth., 106 Pa. Commw. 131, 525 A.2d 857 (1987), the engineer was not responsible for a worker's injuries, based on the following language in the owner-contractor agreement that placed sole responsibility for safety on the contractor: "The Contractor shall supervise and direct the Work, using his best skill and attention. He shall be solely responsible for all construction means, methods, techniques, sequences and procedures and for coordinating all portions of the Work under the Contract."

See also Burns v. Black & Veach Architects, Inc., 854 S.W.2d 450 (Mo. 1993); Dillard v. Shaughnessy, Fickel & Scott Architects, 864 S.W.2d 368 (Mo. Ct. App. 1993); Fisher v. M. Spinelli & Sons Co., Inc., 1999 WL 165674 (Mass. Super. 1999).

[60] Bryant v. Village Ctrs., Inc., 167 Ga. App. 220, 305 S.E.2d 907 (1983).

In Francavilla v. Nagar Constr. Co., 151 A.D.2d 282, 542 N.Y.S.2d 557 (1989), the New York court found the owner not liable for injuries sustained by a subcontractor's employee, because there was no evidence that the owner was negligent and this provision (quoting the 1976 version of A201), as well as ¶ 10.2, made the contractor rather than the owner responsible for safety at the job site.

Although the contractor is the superintendent for the construction, the contractor is not the owner's agent and cannot bind the owner to extras without the owner's consent.[61]

3.3.2 The Contractor shall be responsible to the Owner for acts and omissions of the Contractor's employees, Subcontractors and their agents and employees, and other persons or entities performing portions of the Work for or on behalf of the Contractor or any of its Subcontractors. (**3.3.2; no change**)

If a subcontractor, sub-subcontractor, or material supplier makes an error or omission, the contractor is responsible. This applies to all parties that have contracts with the contractor or the contractor's agents, employees, or their subcontractors. Not covered here are persons employed directly by the owner, other contractors employed by the owner, or the architect or its consultants. The contractor is responsible for the work and for seeing that it is completed in accordance with the contract documents.[62] If any of the subcontractors fails to complete its

---

In Hawthorne v. Summit Steel, Inc., 2003 WL 23009254 (Del. Super. July 14, 2003), the injured worker sued a number of parties, including the construction manager. The construction manager had hired a general contractor to perform the actual work. Because the contract between the owner and construction manager contained this language, summary judgment for the construction manager was not granted on the basis that safety had been delegated to another party.

In Cochran v. George Sollitt Constr. Co., 358 Ill. App. 3d 865, 832 N.E.2d 355 (Ill. App. Ct. 1st Dist. 2005), a new employee of the subcontractor was injured when he climbed a ladder that was supported by a sheet of plywood that sat on two milk crates. When he shifted his weight, the ladder fell, injuring him. Summary judgment in favor of the general contractor was appropriate because the general contractor did not exercise sufficient control over the subcontractor's work. *See also* Andrews v. DT Const., Inc., 205 S.W.3d 4 (Tex. App.-Eastland, 2006); Dilaveris v. W.T. Rich Co., Inc., 653 N.E.2d 1134 (Mass. App. Ct. 1995). *But see*, Lee Lewis Const., Inc., v. Harrison, 70 S.W.3d 778 (Tex. 2001) (general contractor retained control over safety, resulting in liability to construction worker); Point East Condominium Owner's Assn. v. Cedar House Assoc., 663 N.E.2d 343 (Ohio Ct. App. 8th Dist. 1995) (general contractor responsible for workmanship of its subcontractor); Mellon v. Shriners Hospitals for Children, 2007 WL 1977968 (Mass. Super. June 26, 2007) (architect had duty to injured worker despite this language).

In Townsend v. Muckleshoot Indian Tribe, 2007 WL 316504 (Wash. App. Div. 1, Feb. 5, 2007), the owner was not liable to an injured worker because the owner did not retain control pursuant to this provision. *See also* Neiman-Marcus Group, Inc. v. Dufour, 601 S.E.2d 375 (Ga. Ct. App. 2004); Perrit v. Bernhard Mech. Contractors, Inc., 669 So. 2d 599 (La. Ct. App. 1996).

This language also convinced a federal court that an engineer was not liable under OSHA because the jobsite was not a place of employment with regards to the engineer. Reich v. Simpson, Gumpertz & Heger, Inc., 3 F.3d 1 (1st Cir. 1993).

[61] Castle Concrete Co. v. Fleetwood Assocs., 13 Ill. 2d 289, 268 N.E.2d 474 (1971). In Kelsey Lane Homeowners Ass'n v. Kelsey Lane Co., Inc., 103 P.3d 1256 (Wash. Ct. App. Div. 1 2005), the court held that this provision meant that, while the owner has some control over the contractor, the contractor was not the owner's agent and any knowledge of construction defects that the contractor had was not imputed to the owner.

[62] In Delly v. Lehtonen, 21 Ohio App. 3d 90, 486 N.E.2d 251 (1984), the court upheld a homeowner's suit against the general contractor based on the subcontractor's improper installation of drain tile, which resulted in water damage in the basement. The contractor chose the subcontractor and was responsible for the poor workmanship of the sub.

portion of the work, the contractor must complete such work. This also applies to situations in which a subcontractor walks off the job or becomes insolvent.

This provision, read with Article 10 (safety), imposes on the contractor a duty of reasonable care for the benefit of all employees on the work and all other persons who may be affected by the work.[63]

3.3.3 The Contractor shall be responsible for inspection of portions of Work already performed to determine that such portions are in proper condition to receive subsequent Work. (**3.3.3; no change**)

This restates the industry practice. It primarily relates to the coordination between subcontractors. The contractor must inspect all work to ensure that one subcontractor does not slow the work or ask for extras because a prior subcontractor has failed to properly perform that portion of the work. For instance, the concrete subcontractor must install a level floor for the tile contractor. These things are

---

In Shaw v. Bridges-Gallagher, Inc., 174 Ill. App. 3d 680, 528 N.E.2d 1349 (1988), the court found that the owner did not waive defective performance by accepting an inferior roof. The defendant's estoppel argument was rejected. The general contractor was responsible for the work of the roofing subcontractor.

Pursuant to these provisions, the court in Rivnor Prop. v. Herbert O Donnell, Inc., 633 So. 2d 735, 744 (La. Ct. App. 1994), held against the general contractor:

O'Donnell was charged by contract with the sole responsibility for all construction means, methods, techniques, sequences, and procedures and for coordinating all portions of the work under the contract. Thus, his duty to the owner was to conduct periodic inspections as needed to assure all work was performed properly, resulting in a building free from defects.

In Point East Condominium v. Cedar House, 663 N.E.2d 343 (Ohio Ct. App. 1995), the contractor was responsible for the work of his subcontractor under these provisions. In Mosser Constr., Inc. v. Western Waterproofing, Inc., 2006 WL 1944934 (Ohio Ct. App. 6th Dist. 2006) the contractor and its insurer sued the architect to recover costs of rebuilding a defective trench drain, alleging implied indemnification and negligence. The contractor was actively negligent, based on its contractual provisions, including an earlier version of this paragraph, thereby barring indemnification from the architect.

[63] Ramon v. Glenroy Constr. Co., 609 N.E.2d 1123 (Ind. Ct. App. 1993). In Kleeman v. Fragman Constr. Co., 91 Ill. App. 3d 455, 414 N.E.2d 1064 (1981), in which the owner and the contractor brought an action against the subcontractor-employer of an injured worker, the court upheld the trial court's determination that the subcontractor could not introduce this language into evidence to show the authority of the parties over the work. The court stated that "this provision only relates to how the owners and general contractor would apportion possible technical liability under the Structural Work Act for the transgressions of the subcontractors, and not to the actual authority of the parties to control any portions of the work on the construction project."

This provision, along with Article 10, does not impose a contractual duty of care on the general contractor with regard to the safety of an employee of a subcontractor. Deleon v. DSD Dev., Inc., 2006 WL 2506743 (Tex. App.-Hous., 1st Dist. Aug. 31, 2006). Article 10 did not require the general contractor to control the means, methods, or details of how the subcontractor performed the work. The court found that the provisions of A201 governed only the relationship between the owner and general contractor and served to allocate the rights and responsibilities between them exclusively. It found that the purpose of these provisions to be to insulate the owner from tort liability and to create a duty of fiscal responsibility flowing from the general contractor to the owner for claims arising from the construction project.

the responsibility of the general contractor and they are among the primary duties for which the general earns its fee. The contractor must also inspect the work that was previously installed by other parties if its work is to interface with that prior work. If, in the example, the concrete subcontractor had been hired by the owner, the general contractor (or its tile subcontractor) must inspect that floor to ensure that it is properly installed. If the general does not perform the inspection and verify that it was properly done, the general would not be entitled to a change order if the tile has to be redone.

§ 4.14   Labor and Materials: ¶ 3.4

3.4 LABOR AND MATERIALS

3.4.1 Unless otherwise provided in the Contract Documents, the Contractor shall provide and pay for labor, materials, equipment, tools, construction equipment and machinery, water, heat, utilities, transportation, and other facilities and services necessary for proper execution and completion of the Work, whether temporary or permanent and whether or not incorporated or to be incorporated in the Work. **(3.4.1; no change)**

The contractor is responsible for all means required to carry out the work, including temporary utilities. The specifications should address the issue of temporary facilities. This paragraph is not meant to define the Work (**see** ¶ 1.1.3), and the listed items are not necessarily part of the Work.

A Massachusetts court found this provision not to be for the benefit of a lender who had cross-claimed against the general contractor.[64] The court also found that the contractor's obligations would not run to second-tier suppliers. One court has held that when the contract was primarily for the furnishing of goods, the UCC applied.[65]

3.4.2 Except in the case of minor changes in the Work authorized by the Architect in accordance with Sections 3.12.8 or 7.4 the Contractor may make substitutions only with the consent of the Owner, after evaluation by the Architect and in accordance with a Change Order or Construction Change Directive. **(3.4.2; reworded; add reference to minor changes in the work)**

This provision reinforces the requirement that the contractor must comply with the contract documents unless a change order is issued. (Of course, a change order is also a part of the contract documents once it is properly executed. **See** ¶ 1.1.1.) Thus, the contractor cannot simply substitute a specified product with something

---

[64] New England Concrete Pipe Corp. v. D/C Sys., Inc., 495 F. Supp. 1334 (D. Mass. 1980).

[65] J. Lee Gregory, Inc. v. Scandinavian House, Ltd. P'ship, 209 Ga. App. 285, 433 S.E.2d 687 (1993) (in deciding a question as to the formation of the contract to furnish and install windows in apartment building, because the sale of the windows was the predominant purpose of the transaction, with the installation of the windows secondary, UCC applied).

that the contractor believes to be equal unless the owner approves it with a change order. Simply having the architect "approve" a substitution is no longer sufficient to protect the contractor (**see ¶** 3.12.8). If substitutions that have not been properly approved by the Owner are later discovered, the Contractor can still be required to replace the improper substitution. Section 9.10.4.2 states that the Owner does not waive such claims by making final payment. Section 12.2.1 requires the Contractor to correct nonconforming work even if discovered after substantial completion.

3.4.3 The Contractor shall enforce strict discipline and good order among the Contractor's employees and other persons carrying out the Work. The Contractor shall not permit employment of unfit persons or persons not properly skilled in tasks assigned to them. (**3.4.3; minor modifications**)

This is part of the contractor's duty relative to the means, methods, techniques, sequences, procedures, and safety.

§ 4.15   Warranty: ¶ 3.5

3.5 WARRANTY

The Contractor warrants to the Owner and Architect that materials and equipment furnished under the Contract will be of good quality and new unless the Contract Documents require or permit otherwise. The Contractor further warrants, that the Work will conform to the requirements of the Contract Documents and will be free from defects, except for those inherent in the quality of the Work the Contract Documents require or permit. Work, materials, or equipment not conforming to these requirements, may be considered defective. The Contractor's warranty excludes remedy for damage or defect caused by abuse, alterations to the Work not executed by the Contractor, improper or insufficient maintenance, improper operation, or normal wear and tear and normal usage. If required by the Architect, the Contractor shall furnish satisfactory evidence as to the kind and quality of materials and equipment. (**3.5.1; modifications**)

This warranty is different from the one-year warranty of ¶ 12.2.2. One court has interpreted a similar warranty as warranting not the adequacy of the design or specifications but only that the material used would be new and installed in accordance with the manufacturer's recommendations.[66] This warranty is covered by each

---

[66] C.R. Perry Constr., Inc. v. C.B. Gibson & Assocs., 523 So. 2d 1221 (Fla. Dist. Ct. App. 1988). In Travelers Indem. Co. v. S.M. Wilson & Co., 2005 WL 2234582 (E.D. Mo. Sept. 14, 2005), the plaintiff argued that the contractor breached this warranty provision by using a cabinet that was not of good quality and not free of defects. The cabinet was an electrical cabinet installed in a new electronics store, and had been specified. Shortly after the store opened, a meltdown occurred in the cabinet where the lug assembly and the neutral conductor wire were connected to the neutral bus bar with screws, resulting in the destruction of the cabinet and damage to other property, including consumer electronic equipment in the store. The court found, as a matter of law, that the

state's statute of limitations or repose,[67] subject to ¶ 13.7, and a breach of this provision is a breach of contract.[68] The specifications also can provide for special warranties relative to specific items or systems of the work, such as roofs, exterior

---

contractor could not be held liable for damage caused by a latent defect in a product that was specified by the project owner, and cited the *Spearin* case.

[67] A statute of limitations operates to cut off claims after a certain period of time. Each state has various statutes of limitations to cover different types of situations, such as torts, breach of contract, and so forth. These limits are not consistent among the states. A typical statute of limitations involving construction activities might be four years from the accrual (start) of the action. Courts often impose a "discovery rule" to determine the start of the statute of limitations for a particular event. This means that the statute of limitations does not even start to run until a party knows or should know that some wrong was done. For example, if a roof starts to leak seven years after construction of a project is completed, a court might find that the discovery rule sets the accrual date at seven years after completion, assuming that the owner had no other reason to know that there was a design or construction defect. If the applicable statute of limitations is four years, the owner then has four more years after discovery to file suit against the architect or contractor. This might create situations where an owner does not discover a defect for decades after construction. To avoid exposing a party to liability forever, legislatures have enacted "statues of repose" which set an outside limit for discovering problems. A typical statute of repose might be ten years. If no problem is discovered within this period, liability ends and thereafter no suit can be brought. Note, however, that, in most jurisdictions, if a problem is discovered within the statute of repose period, the party then has the entire discovery period within which to bring suit. For example, if the roof leaks at nine years, the owner would have until year thirteen to bring suit against the owner and architect. If, however, the problem is first discovered ten years and one day after completion, no action can be brought.

This situation can be greatly simplified if the parties agree to a specific period within which suit can be brought (although such an agreement will not affect persons not parties to the agreement). The AIA did this in the 1997 version of A201 at Section 13.7. This has been eliminated in the current version except for a ten-year limitation.

[68] Waterfront Marine Constr. Co. v. North End 49ers, 251 Va. 417, 468 S.E.2d 894 (1996). In Graoch Assocs. #5 Ltd. P'ship v. Titan Constr. Corp., 109 P.3d 830 (Wash. Ct. App. 2005), the parties used the 1976 version of AIA Document A201. The owner filed suit against the general contractor relating to various defects. The contractor, in turn, sued several subcontractors who defended on the basis that this warranty and the one-year warranty provision of A201 (§ 12.2 of A201-2007), incorporated into the subcontract, barred the general contractors' breach of contract claim as a matter of law. The court stated:

> [T]his provision explicitly states that any one-year warranty required by the contract documents relates only to the obligation to correct the work, and not to the obligation to comply with the contractual requirement that all work be of "good quality, free from faults and defects and in conformance with the Contract Documents." . . . We conclude the one-year limited warranty does not bar Titan's claim against Purcell for breach of contract and the trial court erred in ruling as a matter of law that Purcell's warranty was the exclusive remedy for the breach of the contract claims.

Groach was followed in 1000 Va. Ltd. P'ship v. Vertecs Corp., 127 Wash. App. 899, 112 P.3d 1276 (2005), holding that the one-year warranty was not exclusive. "The establishment of the time periods noted in Subparagraph 13.2.2 or such longer period of time as may be prescribed by . . . any warranty required by the Contract Documents, relates only to the specific obligations of the Contractor to correct the Work, and has no relationship to the time within which the Contractor's obligation to comply with the Contract Documents may be . . . enforced, nor to

wall components, and so on. If the contractor has made substitutions that are not authorized by change orders or otherwise, a default can be declared and damages sought. The contractor should, therefore, not make any unpermitted substitutions without written authorization by the architect or owner. Also, any work that does not conform to the requirements of the contract documents can be declared defective.[69]

Note that this warranty provision lets the owner decide whether to declare such nonconforming work defective. The owner may want to change the word "may" to "shall." If the architect observes that any work does not conform to the contract documents (even if of a "better" quality), the architect should notify the owner. Only the owner can accept nonconforming work (¶ 12.3.1).

The contractor may want to limit its liability by limiting this warranty to a specific period of time, such as one year after substantial completion. This can result in a situation in which latent defects may not appear until a year has passed. For instance, if the contractor has omitted some insulation in the building, the owner and architect may not discover this omission for more than one year after substantial completion. For that reason, the time limitation is not a good idea for the owner. The contractor may also want to include language waiving warranties. To be effective, waiver language must be in conspicuous type (boldface) and expressly waive warranties. The owner may want to insert the words "or Owner" after "Architect" in the last sentence.

Even in the absence of this warranty provision, courts may find a warranty by a contractor. One court stated this warranty thus: "The general rule is that a contractor or builder impliedly warrants that the work he undertakes will be done in a good and workmanlike manner and will be reasonably fit for the intended purpose."[70]

---

the time within which proceedings may be commenced to establish . . . liability with respect to the Contractor's obligations other than specifically to correct the Work."

[69] In Shaw v. Bridges-Gallagher, Inc., 174 Ill. App. 3d 680, 528 N.E.2d 1349 (1988), the court found that the owner did not waive defective performance by accepting an inferior roof. The defendant's estoppel argument was rejected. The general contractor was responsible for the work of the roofing subcontractor.

Smith v. Erftmier, 210 Neb. 486, 315 N.W.2d 445 (1982), involved a provision under an earlier edition of AIA A201. The contractor argued that the owner had accepted the work. The court ruled that this was a latent defect and not waived by final payment. The contractor also argued that he properly constructed the footing in question according to local custom and usage. This was rejected because he had not properly pleaded this defense.

This provision was discussed in Trustees of Ind. Univ. v. Aetna Cas. & Sur. Co., 920 F.2d 429 (7th Cir. 1990), in which the owner had a series of problems with masonry.

This paragraph, along with a prior version of ¶ 12.2, prevented a contractor and its surety from arguing that the owner's acceptance of the work barred the owner's claim for damages at a later date. Brouillette v. Consolidated Constr. Co., 422 So. 2d 176 (La. Ct. App. 1982). Note, however, that ¶ 4.3.5 preserves certain claims of the owner after final payment.

In Garden City Osteopathic Hosp. v. HBE Corp., 55 F.3d 1126 (6th Cir. 1995), the statute of limitations was not a bar to an action against a contractor because there was fraudulent concealment.

[70] Carroll-Boone Water Dist. v. M.&P. Equip. Co., 280 Ark. 560, 661 S.W.2d 345, 353 (1983). In Korte Constr. Co. v. Deaconess Manor Ass'n, 927 S.W.2d 395 (Mo. Ct. App. 1996), the court held

One case held that the risk of loss is on the contractor before final acceptance of the work.[71] That means that if something is damaged before final acceptance, the contractor must repair or replace it.

There are actually three separate warranties contained in this provision. The first warranty is that materials and equipment will be of good quality and new. Issues may arise as to whether this warranty covers owner-furnished materials or equipment and whether the contractor also warrants that these materials and equipment are suitable for the intended purpose. A court likely will find that the contractor does not give a warranty of suitability where particular materials or equipment is specified and that owner-furnished materials are likewise not warranted by the contractor. However, where the contractor proposes substitutions, an appropriate warranty may be found.

Second, the contractor warrants that the work will be free of defects. The interesting twist on this is the limitation that the warranty applies only to defects "not inherent in the quality required or permitted." The issue then becomes to what degree a particular item has inherent defects. For example, if a specified wood veneer has inherent knots, there would be no breach of this warranty if there were knots in the wood veneer. Essentially, the specifications set forth the specific materials and, by implication, the defects to be expected in those materials or equipment. Thus, this warranty is not meant to guarantee perfection but only that defects not ordinarily expected to be found in such materials or equipment be furnished by the contractor.

Third, the contractor warrants that the work will conform to the requirements of the Contract Documents. In other words, the contractor must furnish and install precisely what is shown on the drawings and specifications. If the contractor fails to do so, the work can be declared defective.

§ 4.16   Taxes: ¶ 3.6

3.6 TAXES

3.6.1 The Contractor shall pay sales, consumer, use and similar taxes for the Work provided by the Contractor that are legally enacted when bids are received or negotiations concluded, whether or not yet effective or merely scheduled to go into effect. **(3.6.1; minor modification)**

---

that this section was an express provision governing the degree of skill and competence with which the work will be performed, thereby preventing the imposition of an implied warranty.

[71] Mayville-Portland Sch. v. C.L. Linfoot, 261 N.W.2d 907 (N.D. 1978). The contractor installed an underground storage tank for a school district in December, with the understanding that the tank would be partially re-excavated the following spring to complete the work. When the tank was uncovered the following June, it was discovered that it was severely damaged. The court held that the contractor was responsible for the damage, even if it were not specifically determined that the contractor was negligent, because under the AIA provision, the risk of loss is on the contractor until final acceptance.

This does not include new taxes that are not enacted at the time of bidding or execution of the owner-contractor agreement. It does cover taxes that are enacted and scheduled to go into effect during the construction. However, implicit in this paragraph is that the contractor must pay any increased taxes for whatever reason if the construction is delayed due to the contractor's fault.[72]

§ 4.17   Permits, Fees, Notices, and Compliance with Laws: ¶ 3.7

3.7 PERMITS, FEES, NOTICES AND COMPLIANCE WITH LAWS

3.7.1 Unless otherwise provided in the Contract Documents, the Contractor shall secure and pay for the building permit as well as for other permits, fees, licenses, and inspections by government agencies necessary for proper execution and completion of the Work that are customarily secured after execution of the Contract and legally required at the time bids are received or negotiations concluded. **(3.7.1; minor modifications)**

The contractor is required to obtain all necessary permits for the work.[73] It is required to be aware of all permit requirements and to figure such costs into its bids, including any bonding requirements of the local municipality. If a new permit regulation is enacted after the bidding, the owner would be responsible for the increased cost. The owner may want to delete that provision so as to require the contractor to be responsible, even if laws and ordinances are changed during the project. Of course, the contractor would probably add a contingency to its fee to cover this situation. If the owner is to obtain any permits, this should be specifically stated. **See ¶** 2.2.2 for the owner's responsibility in this regard.

If the owner undertakes to obtain permits even when it is the contractor's obligation to do so, the owner cannot later complain that the contractor failed to obtain the permits.[74]

---

[72] In ESI Cos., Inc. v. Fulton County, 609 S.E.2d 126 (Ga. Ct. App. 2004), the instructions to bidders required that taxes be excluded from the bid. Apparently, the AIA language relating to the contractor's obligation to pay for taxes was deleted. The contractor assumed that it would not have to pay sales taxes. However, Georgia required the contractor to pay sales taxes, despite the fact that a governmental entity was the ultimate user of the project. The contractor filed suit to recover the cost of the taxes. The court held that the contractor should have been aware of the state law requiring it to pay taxes. "Plainly, ESI made a mistake of law in believing that it could obtain a certificate of exemption for sales and use tax from Fulton County that would relieve it as the consumer of the State of Georgia sales and use tax. A contract will not be reformed in equity where one party makes a mistake of law through ignorance or neglect." This incredible decision is unlikely to be followed by more enlightened courts.

[73] In Western Reserve Transit Auth. v. B&B Constr. Co., 1996 Ohio App. LEXIS 143 (Ohio Ct. App. Jan. 16, 1996), the court found that the usual responsibility for obtaining a permit is on the contractor. Evidently, this case did not involve a standard A201.

[74] In Douglas Nw., Inc. v. Bill O'Brien & Sons Constr., Inc., 64 Wash. App. 661, 828 P.2d 565 (1992), the contract between the general contractor and subcontractor required the sub to secure permits. However, the general never mentioned, relied upon, or sought to enforce that provision.

In a Pennsylvania case, the owner was responsible for obtaining a permit for the installation of a swimming pool.[75] The contractor stated it was never his practice to check whether a permit was required or had been obtained. When the town notified the owner that the location of the pool violated the local ordinance and had to be removed, litigation resulted. The court held that the contract was not illegal but that the contractor had failed to mitigate damages by not checking into the permit issue. Because a permit is so essential to almost every construction project, it would seem foolish for a contractor to start work without evidence that a permit was issued.

> ***3.7.2*** *The Contractor shall comply with and give notices required by applicable laws, statutes, ordinances, codes, rules and regulations, and lawful orders of public authorities applicable to performance of the Work.* **(3.7.2; minor modifications)**

The contractor is responsible for compliance with all local and federal rules and regulations of any sort. In addition to OSHA (Occupational Health and Safety Administration) and other federal and state requirements, the contractor must cooperate with local building inspectors and is charged with knowledge of their procedures. At least one court has held that when a municipal ordinance is applicable to a contract, it is by operation of law an implied term of that contract.[76]

> ***3.7.3*** *If the Contractor performs Work knowing it to be contrary to applicable laws, statutes, ordinances, codes, rules and regulations, or lawful orders of public authorities, the Contractor shall assume appropriate responsibility for such Work and shall bear the costs attributable to correction.* **(3.7.4; minor modifications; deleted notice to owner and architect)**

Attributable costs include the costs of changing the work, the architect's fees related to correction of the problem, and other related costs. It probably does not include any attorneys' fees unless the parties modify this provision to include attorneys' fees.

Most architects will add language to the plans or specifications to the effect that the contractor is to construct the building in accordance with all ordinances, building codes, laws, and the like. Such language may conflict with ¶ 3.2.3. The question then becomes what the words "appropriate responsibility" in this section

---

In fact, the sub was assured that the permit issues would be taken care of. The sub was thus not in breach of the contract for failing to secure permits. If the contractor properly attempts to secure permits, but is unable to do so for reasons beyond his control, the contractor will be entitled to an extension of time. Bouchard v. Boyer, 1999 Conn. Super. LEXIS 1285 (May 17, 1999).

[75] Contractor Indus. v. Zerr, 241 Pa. Super. 92, 359 A.2d 803 (1976).

[76] Gutowski v. Crystal Homes, Inc., 26 Ill. App. 2d 269, 167 N.E.2d 422 (1960). The contractor had built the house too close to a lot line, in violation of the zoning ordinance. The owner was awarded damages for the decrease in value of the house. The contractor was obligated to construct the house in compliance with the ordinance, and its failure to do so was a breach of the contract. Apparently, no architect or AIA documents were involved.

mean when work is constructed in violation of some code.[77] Another issue may arise when the plans do not comply with code requirements and there is a substantial delay in obtaining a permit. Both the owner and the contractor may be damaged by such a delay. If so, the contractor may be entitled to extra time per ¶¶ 8.3.1 and 8.3.2.

> *3.7.4 Concealed or Unknown Conditions. If the Contractor encounters conditions at the site that are (1) subsurface or otherwise concealed physical conditions that differ materially from those indicated in the Contract Documents or (2) unknown physical conditions of an unusual nature, that differ materially from those ordinarily found to exist and generally recognized as inherent in construction activities of the character provided for in the Contract Documents, the Contractor shall promptly provide notice to the Owner and the Architect before conditions are disturbed and in no event later than 21 days after first observance of the conditions. The Architect will promptly investigate such conditions and, if the Architect determines that they differ materially and cause an increase or decrease in the Contractor's cost of, or time required for, performance of any part of the Work, will recommend an equitable adjustment in the Contract Sum or Contract Time, or both. If the Architect determines that the conditions at the site are not materially different from those indicated in the Contract Documents and that no change in the terms of the Contract is justified, the Architect shall promptly notify the Owner and Contractor in writing, stating the reasons. If either party disputes the Architect's determination or recommendation, that party may proceed as provided in Article 15.* **(4.3.4; changed notice requirement; require determination of architect; modified claims language)**

Paragraph 3.7.4 shifts the financial risk of concealed or unknown site conditions from the contractor to the owner. Two types of changed conditions are generally recognized: *Type I* occurs when the actual condition differs materially from that shown on the contract documents; *Type II* occurs when the actual conditions differ materially from those ordinarily found to exist or expected in work of the character shown in the contract documents. Type I involves a comparison of the contract documents with the conditions encountered, whereas Type II relates to the normal expectations of the contractor. The AIA language covers both types. Note that the concealed conditions need not be subsurface. As long as they are concealed and are not readily discoverable, this clause applies.

---

[77] The language of this section was discussed in Great N. Ins. Co. v. RLJ Plumbing & Heating, Inc., 2006 WL 1526066 (D. Conn. May 25, 2006), where the court held for the architect in finding that the contractor was responsible for a construction defect. The drawings for the new home established the location of the bathroom and plumbing fixtures in the second floor bathroom but did not detail how the piping to the fixtures would be run or make any provisions for freeze protection. The architect never inspected this portion of the work. The pipes later froze, causing extensive damage. According to the court, "because it was not [the architect's] responsibility under the contract to inspect the 'construction means, methods, techniques, sequences and procedures' used in the plumbing of the second floor bathroom, it was not in a position to prevent the harm that allegedly occurred in the instant case." In reviewing all of the contracts, including this provision, the court added that "in addition, this responsibility was accounted for within the contract scheme, and it fell on the general contractor . . ."

The Type I clause is also meant to cover the situation in which the contractor relies on the construction documents in preparing the bid. When the contract documents require the contractor to make its own investigation, such as soil borings, the protection afforded by this clause may not fully apply.[78]

---

[78] In Joseph F. Trionfo & Sons v. Board of Educ., 41 Md. App. 103, 395 A.2d 1207 (1979), the contract contained the following provision:

> No extra or additional compensation for excavation will be paid under this contract for work included in Bid Proposal at time of bidding. Subsurface Soil Data: Data concerning subsurface materials or conditions which is based upon soundings, test pits, or test borings, has been obtained by Architect for his own use in designing project. Its accuracy or completeness is not guaranteed by Owner or Architect and in no event is it to be considered as part of contract plans or specifications. Contractor must assume all responsibility in excavating for this project and shall not rely on subsurface information obtained from Architect, or indirectly from Owner. Bidders shall make their own investigation of existing subsurface conditions; neither Owner or Architect will be responsible in any way for additional compensation for excavation work performed under the Contract due to Contractor's assumptions based on sub-soil data prepared solely for Architect's use.

The test boring data that the architect had obtained was not provided to bidders with the specifications or other contract documents, but bidders could obtain access to these data after executing the following written request:

> Please forward copies of test boring data sheets for the subject property. The contracting firm herein named releases the owner and Architect from any responsibility or obligation as to its accuracy or completeness or for any additional compensation for work performed under the contract due to assumptions based on use of such furnished information.

During the excavation work, additional rocks were encountered, and the general contractor was required to compensate the excavation subcontractor. The contractor then sued the owner, claiming that inaccurate and misleading representations concerning the nature of the subsurface conditions at the site were made. The court held that the contract provision, along with the explicit waiver, should have caused the contractor to not rely on this data. The contractor's argument that he did not have sufficient time to conduct his own investigation was denied.

In Central Penn Indus., Inc., v. Commonwealth, 25 Pa. Commw. 25, 358 A.2d 445 (1976), the contract contained this language:

> The contractor further covenants and warrants that he has had sufficient time to examine the site of the work; that he has examined the site of the work; that he has had sufficient time to examine the site of the work to determine the character of the subsurface material and conditions to be encountered; that he is fully aware and knows of the character of the subsurface material and conditions to be encountered; and that he has based the within contract prices on his own independent examination and investigation of the site, subsurface materials, and conditions and has not relied on any subsurface information furnished to him by the Commonwealth of Pennsylvania, Department of Highways.

The court denied the contractor's claim for extra compensation, stating that insufficiency of time allowed for investigation by bidders, standing alone, will not support such a claim due to unanticipated subsoil conditions.

In Al Johnson Constr. Co. v. Missouri Pac. R.R., 426 F. Supp. 639 (E.D. Ark. 1976), the court held that the contractor justifiably relied on test data when a changed conditions clause, like ¶ 3.7.4, was included in the contract. The bidders did not have to include a contingency element in their bids.

In Condon-Cummingham, Inc. v. Day, 22 Ohio Misc. 71, 258 N.E.2d 264 (1969), the contractor justifiably relied on test data when no specific disclaimer, exculpatory clauses, or releases were involved.

Without such a changed conditions clause, the risk of uncertainty of subsurface conditions is placed on the contractor.[79]

One court used this clause to find that a contractor was entitled to additional compensation when the site conditions for Phase 2 of the project were not as the specifications for Phase 1 required.[80] The contractor had relied on those specifications in preparing his bid and incurred additional costs to bring the site up to the condition it should have been in if the prior specifications had been followed. Another court stated that "there is a distinction between extra work and work provided for in the contract. Extra work is work arising outside and independent

---

In Davidson & Jones v. County of New Hanover, 41 N.C. App. 661, 255 S.E.2d 580 (1979), the court found that an architect and soil testing engineer could be liable to a contractor for negligently misrepresenting subsurface soil conditions when the contractor relied on the soil report to its detriment. The report did not have the exculpatory language of the *Trionfo* case above. Also, the damages in this case were for economic damages. The *Moorman* (Economic Loss) doctrine (Moorman Mfg. Co. v. National Tank Co., 91 Ill. 2d 69, 435 N.E.2d 443 (1982)) would seem to hold contrary to the finding of this case.

In Jacksonville Port Auth. v. Parkhill-Goodloe, 362 So. 2d 1009 (Fla. Dist. Ct. App. 1978), the owner had furnished bidders boring reports that did not indicate any significant rock. After work started, significant amounts of rock were discovered, causing delays to the contractor and increased costs. The owner then refused to pay the additional costs, relying on the following contractual language:

Boring information shown on the Drawings or furnished with the specifications, or both, is not guaranteed to be more than a general indication of the materials likely to be found adjacent to holes bored at the site of the work approximately at the locations indicated. Bidders shall examine boring records and interpret the subsoil investigations and other preliminary data for himself, and shall base his bid on his own opinion of the conditions likely to be encountered [sic].

This language was found to constitute a guarantee that the boring information gave a general indication of the materials likely to be found in the stated locations. When it turned out that the conditions differed, the owner had breached that guarantee and was liable to the contractor for damages. The court stated that the owner had a duty not to mislead bidders, and had breached that duty.

In A. Amorello & Sons, Inc. v. Beacon Constr. Co., 422 N.E.2d 467 (Mass. App. Ct. 1981), the subcontractor was entitled to an increase in its contract sum for the increased cost of removal of ledge in excess of that shown in the borings report, based on this provision.

[79] Pinkerton & Laws Co. v. Roadway Express, Inc., 650 F. Supp. 1138 (N.D. Ga. 1986) (*citing* Flippin Materials Co. v. United States, 312 F.2d 408 (Ct. Cl. 1963)); Anderson v. Golden, 569 F. Supp. 122 (S.D. Ga. 1982); Eastern Tunneling Corp. v. Southgate Sanitation Dist., 487 F. Supp. 109 (D. Colo. 1979); Jahncke Serv., Inc. v. Department of Transp., 172 Ga. App. 215, 322 S.E.2d 505 (1984); Commonwealth v. Osage Co., 24 Pa. Commw. 276, 355 A.2d 845 (1976).

In Edward Fitzpatrick v. Suffolk County, 138 A.D.2d 446, 525 N.Y.S.2d 863 (1988), the court ruled that in the absence of privity of contract, the engineer was not under a duty to disclose to the contractor that certain subsurface conditions were materially different from those stated in the contract specifications.

In American Demolition, Inc. v. Hapeville Hotel Ltd. P'ship, 413 S.E.2d 749 (Ga. 1991), the parties deleted the changed conditions clause and did not substitute another. The risk was on the contractor.

[80] Moorhead Constr. Co. v. City of Grand Forks, 508 F.2d 1008 (8th Cir. 1975).

of the contract, something not required in its performance."[81] In that case, the contractor was denied additional costs when unanticipated rock was encountered in some excavation work.

An interesting case[82] involved a contractor who discovered an unknown artesian water condition that differed materially from the conditions shown in the drawings and indicated in the specifications (Type I). The extreme water conditions caused one problem after another, and the owner terminated the contractor and refused to pay for the contractor's additional work. The court stated that the contractor is held to the standard of what a reasonable contractor should have anticipated on the project, but the contractor is not bound to make a scientifically educated and skeptical analysis of the contract documents and the general situation. He was not required to consult scholarly treatises or review geological and hydrological information. The contractor was entitled to compensation for the unknown water conditions.

Notice of the condition must be given within 21 days of first observing the condition. Failure to do so may invalidate the contractor's claim.[83]

The contractor must make an initial site inspection (**see** ¶ 3.2), and it will be allowed to use this clause only if the contractor could not have reasonably discovered the condition by inspection or by reviewing the contract documents. Note that, without a changed conditions clause, the entire risk is placed on the contractor,[84] who would likely add a very substantial contingency onto the bid to cover the risk.

Both the owner and contractor may take advantage of the changed conditions clause by notifying the other party within 21 days of observing the condition. The condition must not be disturbed, although this is not always possible. If the condition is disturbed, work should stop and the architect should be immediately notified. The architect will then determine whether the condition is a changed condition. If it is, the architect may recommend either an increase or decrease in either the contract sum, or the contract time, or both. Note that if the conditions are changed "materially," then a change order is required. If the architect determines

---

[81] Brown-McKee, Inc. v. Western Beef, Inc., 538 S.W.2d 840, 844 (Tex. 1976).

[82] Metropolitan Sewerage Comm'n v. R.W. Constr., 72 Wis. 2d 365, 241 N.W.2d 371 (1976).

[83] Buckley & Co. v. City of N.Y., 121 A.D.2d 933, 505 N.Y.S.2d 140 (1986); Naclerio Contracting Co. v. EPA, 113 A.D.2d 707, 493 N.Y.S.2d 159 (1985). *But see* Cure v. City of Jefferson, 396 N.W.2d 727 (Mo. 1965) (permitted adjustments in accordance with unit prices).

In Davidson & Jones, Inc. v. North Carolina Dep't of Admin., 315 N.C. 144, 337 S.E.2d 463 (1985), the contractor failed to stop the work before the claim was processed. The contractor was to excavate some 800 cubic yards of rock at a unit price of $55 per yard. The project was substantially delayed because the contractor wound up removing 3714 cubic yards. The court permitted recovery for extra duration-related expenses because there was no work that was to be affected and no requirement to stop the work.

In Galin Corp. v. MCI Telecomm. Corp., 12 F.3d 465 (5th Cir. 1994), the contract required the contractor to notify the owner within five days of any event that would give rise to a claim. The contractor argued that this provision did not apply to extra work that was not anticipated and beyond the scope of the original contract. The court rejected this argument because the contractor failed to comply with the notice provisions.

[84] United States v. Spearin, 248 U.S. 132 (1918).

that the condition does not represent a change from the contract documents, or from what the contractor reasonably had a right to expect at the site, the architect must notify the owner and contractor in writing and explain the reasons. If either the owner or contractor is dissatisfied with the architect's determination, it must initiate a claim within 21 days after the architect's written decision. At this point, a changed condition is treated like any other claim.[85]

> *3.7.5 If, in the course of the Work, the Contractor encounters human remains or recognizes the existence of burial markers, archaeological sites or wetlands not indicated in the Contract Documents, the Contractor shall immediately suspend any operations that would affect them and shall notify the Owner and Architect. Upon receipt of such notice, the Owner shall promptly take any action necessary to obtain governmental authorization required to resume the operations. The Contractor shall continue to suspend such operations until otherwise instructed by the Owner but shall continue with all other operations that do not affect those remains or features. Requests for adjustments in the Contract Sum and Contract Time arising from the existence of such remains or features may be made as provided in Article 15.*
> **(New)**

This is a new provision for 2007 that requires the contractor to suspend operations in areas affected by the discovery of the listed items. Other areas of the project that would not affect these items would not be subject to this suspension of work. Of course, any such suspension, even if it affects only a small part of the site will result in the contractor having to alter the sequence of operations and probably affect scheduling of labor and materials, thereby necessitating a change in the contract time and/or cost.

§ 4.18   Allowances: ¶ 3.8

Allowances are set forth in the contract documents by the architect. They generally pertain to items that are undecided at the time the contract documents are prepared. For instance, the contract documents may specify an allowance of $15 per yard for carpeting of a type and color to be selected. If the owner subsequently selects carpeting that costs $20 per yard, the contractor is entitled to a change order for the additional $5 per yard.

3.8 ALLOWANCES

> *3.8.1 The Contractor shall include in the Contract Sum all allowances stated in the Contract Documents. Items covered by allowances shall be supplied for such amounts and by such persons or entities as the Owner may direct, but the Contractor*

---

[85] In Millgard Corp. v. McKee/Mays, 831 F.2d 88 (5th Cir. 1987), the subcontractor put in its claim for extra money for a concealed condition. This was denied much later, and the sub brought suit. One of the issues in the case was when the cause of action accrued and the statute of limitations began to run. The federal appellate court ruled that the limitations period began on the date the general contractor rejected the request for extras.

shall not be required to employ persons or entities to whom the Contractor has reasonable objection. (**3.8.1; no change**)

The contract sum includes all allowances that are stated in the drawings, specifications, or elsewhere. If an item that should have been listed as an allowance is omitted, but is shown on the plans or specifications, the Contractor must include the cost of that item in the contract sum. If it is not possible to place a cost on that item because the Contract Documents are incomplete, then the Contractor should either ask for clarification before submitting a price or make an assumption and clearly note that in the submission to the Owner. In the above example, if the contractor determines that 1,000 yards of carpeting will be required on the project, the contractor must include $15,000 for the carpeting in the amount bid. Then, if the carpeting costs $20,000, the contractor will be entitled to an additional $5,000. Unless the contractor has a reasonable objection, the owner can direct who will furnish and install the allowance items.

3.8.2 Unless otherwise provided in the Contract Documents:

> *.1   allowances shall cover the cost to the Contractor of materials and equipment delivered at the site and all required taxes, less applicable trade discounts;* (**3.8.2.1; no change**)

The allowance amounts given by the architect are for the item itself, and do not include the contractor's overhead, profit, costs of transport to the site, or costs of installation. However, the contractor must include these costs in its bid. If the cost of the item is in excess of the allowance, the contractor should not add any amounts for additional installation, handling, or other costs unless it can demonstrate that there is some justification because of some particularity of the item selected (**see ¶** 3.8.2.3).

> *.2   Contractor's costs for unloading and handling at the site, labor, installation costs, overhead, profit and other expenses contemplated for stated allowance amounts shall be included in the Contract Sum but not in the allowances; and* (**3.8.2.2; no change**)

These costs are not part of the allowance (but are included in the Contract Sum), since the Contractor should be able to determine them no matter what particular item is later selected. For instance, the cost of unloading carpeting from the delivery truck should be the same whether the carpet is an expensive wool carpet or an inexpensive synthetic blend.

> *.3   whenever costs are more than or less than allowances, the Contract Sum shall be adjusted accordingly by Change Order. The amount of the Change Order shall reflect (1) the difference between actual costs and the allowances under Section 3.8.2.1 and (2) changes in Contractor's costs under Section 3.8.2.2.* (**3.8.2.3; no change**)

Costs for allowance items should be substantiated by the contractor at the time of making requests for payment. This substantiation should include the actual costs that either have been paid by the Contractor or will be paid, and include any rebates that the Contractor may be entitled to. Note that if a change order is required, the time limits related to changes must be observed.

3.8.3 Materials and equipment under an allowance shall be selected by the Owner with reasonable promptness. (**3.8.3; modified "promptness" provision**)

If the owner materially delays selection of these items, the contractor would be entitled to a change order.

§ 4.19   Superintendent: ¶ 3.9

3.9 SUPERINTENDENT

3.9.1 The Contractor shall employ a competent superintendent and necessary assistants who shall be in attendance at the Project site during performance of the Work. The superintendent shall represent the Contractor, and communications given to the superintendent shall be as binding as if given to the Contractor. (**3.9.1; deleted last two sentences of prior section**)

This superintendent is normally a full-time field representative of the contractor during the construction. If less than a full-time superintendent is acceptable, this should be specifically stated. On larger projects, *necessary assistants* may be required to help the superintendent. This should be discussed between the parties prior to the start of construction. This also reinforces the idea that the architect is not a superintendent of the work. All communications of any importance whatsoever should be in writing. All meetings should be documented to provide a record for future use. Regular project meetings are extremely important and should include the architect, owner, and principal contractors.

3.9.2 The Contractor, as soon as practicable after award of the Contract, shall furnish in writing to the Owner through the Architect the name and qualifications of a proposed superintendent. The Architect may reply within 14 days to the Contractor in writing stating (1) whether the Owner or the Architect has reasonable objection to the proposed superintendent or (2) that the Architect requires additional time to review. Failure of the Architect to reply within the 14 day period shall constitute notice of no reasonable objection. (**new**)

This gives the Owner and Architect a chance to review the qualifications of the proposed superintendent and, if necessary, to object to that particular person. The success or failure of a construction project can often be traced to the quality of the superintendent who is on the job on a day-to-day basis and who makes the majority of the decisions on behalf of the Contractor and coordinates the work of the various subcontractors and suppliers. The Owner is entitled to the services of a capable and experienced superintendent. The 14-day time period gives the Owner and Architect a chance to check the qualifications and past experience on other projects of the proposed superintendent.

> *3.9.3 The Contractor shall not employ a proposed superintendent to whom the Owner or Architect has made reasonable and timely objection. The Contractor shall not change the superintendent without the Owner's consent, which shall not unreasonably be withheld or delayed.* **(new)**

This provision merely states what has been normal practice in the construction industry. The Owner and/or Architect should be able to substantiate a reasonable reason for objecting to the proposed superintendent, but "soft" reasons, such as the superintendent's reputation of not getting along with subcontractors, should be reason enough for an objection.

§ 4.20   Contractor's Construction Schedules: ¶ 3.10

3.10 CONTRACTOR'S CONSTRUCTION SCHEDULES

> *3.10.1 The Contractor, promptly after being awarded the Contract, shall prepare and submit for the Owner's and Architect's information a Contractor's construction schedule for the Work. The schedule shall not exceed time limits current under the Contract Documents, shall be revised at appropriate intervals as required by the conditions of the Work and Project, shall be related to the entire Project to the extent required by the Contract Documents, and shall provide for expeditious and practicable execution of the Work.* **(3.10.1; no change)**

This paragraph states that this schedule is for the information of the owner and architect.[86] The contractor is responsible for the scheduling of construction. If the owner or architect wants a detailed or specific type of schedule,[87] that should be spelled out in the specifications or Supplementary Conditions. There are several reasons for requiring the contractor to prepare a schedule: to ascertain that the contractor has a sound plan for execution of the work; to coordinate owner-furnished equipment; to assure that required milestones are incorporated into the schedule; and to assist the architect in determining the contractor's progress in making progress payments. The architect needs this schedule to determine whether the contractor is able to timely perform the work, in order to determine whether or not to certify payments. Under Section 9.5, the architect may withhold certification for a number of reasons, including "reasonable evidence that the Work will not be completed within the Contract Time, and that the unpaid balance would not be

---

[86] A schedule may provide the basis for a delay claim by a subcontractor. Williams Enter., Inc., v. Striat Mfg. & Welding, Inc., 728 F. Supp. 12 (D. D.C. 1990).

[87] For instance, a critical path method "is a construction schedule showing the separate elements of the work (electrical, plumbing, etc.), the period of time and sequence for the performance of each. An item that could delay the entire project is on the 'critical path'." Broadway Maint. Corp. v. Rutgers, 90 N.J. 253, 447 A.2d 906 (1982).

adequate to cover actual or liquidated damages for the anticipated delay. . . ." If the contractor falls behind the schedule, this may be evidence that there is an underlying problem, such as inadequate cash flow, and that the subcontractors are not being paid on time. The architect would generally not know this in the absence of a detailed construction schedule. If the contractor fails to follow this schedule, it will be in breach of the contract under Section 3.10.3 ("The Contractor shall perform the Work in general accordance with the most recent schedules submitted to the Owner and Architect.") The use of the word "shall" makes this a mandatory provision. If read in concert with Section 8.2.1, the "time is of the essence" provision, any delay from the most recent schedule would constitute a material breach of the contract and may be grounds for termination of the contractor.

Note that the schedules under this section do not require approval by either the owner or architect. This provision could be amended to require such approval. The owner and architect may also want the contractor to submit the schedule and any updates in electronic format, or use a specific program for such schedules.

3.10.2 The Contractor shall prepare a submittal schedule, promptly after being awarded the Contract and thereafter as necessary to maintain a current submittal schedule, and shall submit the schedule(s) for the Architect's approval. The Architect's approval shall not unreasonably be delayed or withheld. The submittal schedule shall (1) be coordinated with the Contractor's construction schedule, and (2) allow the Architect reasonable time to review submittals. If the Contractor fails to submit a submittal schedule, the Contractor shall not be entitled to any increase in Contract Sum or extension of Contract Time based on the time required for review of submittals. **(3.10.2; substantial revisions)**

The schedule in ¶ 3.10.1 is an overall schedule for the progress of the Work. *This* schedule sets forth when each required submittal (shop drawing, sample, etc.) will be provided to the architect for review.

The architect needs this information for its own scheduling so that it can provide fast turn-around times for shop drawings and other submittals. The owner may want to require a much more detailed monthly report from the contractor, giving details as to progress and other information that could be useful to the owner, architect, and lender. Note that this section gives the architect the right to approve this schedule. This is so because this directly affects the work the architect is doing. Each submittal is directed to the architect for approval. Submittals can be voluminous, perhaps running into hundreds of drawings for a single submittal. If the contractor gives the architect a dozen or more huge submittals all at the same time, it may make it impossible for the architect to properly respond to such submittals in a timely manner. An unscrupulous contractor may even use this technique to assert a delay caused by an architect's failure to process shop drawings, and ask for an extra to the contract.

If the submittals are properly spaced out, the architect will be in a much better position to respond to each one. The architect may even want to include a submission schedule in the specifications, thereby eliminating any complaints by the contractor that it was required to make submissions in an inefficient manner.

The schedule of submittals also must be coordinated with the construction schedule. This means that both schedules should be submitted at an early date, and the architect may be able to reject a submittal schedule if a later construction schedule does not coordinate with the submittal schedule.

If claims are made, particularly where time is important, regarding issues involving acceleration, liquidated damages, and the like, having access to accurate construction schedules becomes particularly important. This is where having the schedules in electronic format can make it much easier to analyze such a claim. The best time to obtain these documents and their backup information is before any claim is made. Documents generated after a claim should be viewed with suspicion. Other sources of project documentation are also important in this regard. Just because the contractor says that a conduit was installed by a certain date does not mean that it actually was installed. Regular project photographs or videos can be used to verify such claims. For instance, if the contractor alleges that it had to remove and replace the conduit because of a design error involving something above the conduit, it would be helpful to verify that the conduit was actually in place as claimed. Such photographic records will also be helpful to the owner in cases where adherence to the construction schedule is in question.

3.10.3 The Contractor shall perform the Work in general accordance with the most recent schedules submitted to the Owner and Architect. (**3.10.3; no change**)

Significant deviations from the most recent schedule may place the contractor in default. This would occur if the schedule included milestones (certain critical dates) that the contractor is to meet. The contractor may want to exclude such milestones in order to increase its flexibility. The owner may insist on them, particularly if interim dates are important, such as when a tenant must move in before final completion. A minor deviation from the schedule would not be a default unless the completion date would not be met as a result.

§ 4.21   Documents and Samples at the Site: ¶ 3.11

3.11 DOCUMENTS AND SAMPLES AT THE SITE

The Contractor shall maintain at the site for the Owner one copy of the Drawings, Specifications, Addenda, Change Orders and other Modifications, in good order and marked currently to indicate field changes and selections made during construction, and one copy of approved Shop Drawings, Product Data, Samples and similar required submittals. These shall be available to the Architect and shall be delivered to the Architect for submittal to the Owner upon completion of the Work as a record of the Work as constructed. (**3.11.1; minor modifications**)

These are for the use of the architect and owner during site visits. The municipality may also require that a copy of the permit set be kept at the jobsite, but the permit set is normally not kept up to date with revisions to the drawings and specifications. This provision requires that the most recent and complete set be at the site, in addition to any other sets of drawings that may be required by the municipality. The complete set of these documents is turned over to the owner at the completion of the project. The contractor must record all field changes on these documents. A field change is one that does not affect cost or time. An example is routing a pipe or duct a different way than the drawings indicate because of a space difficulty. Changes that do impact time or money require change orders, which are also included in this provision. The marked up set is often referred to as the "record set" or "as-built drawings."

§ 4.22   Shop Drawings, Product Data, and Samples: ¶ 3.12

It should be noted that the plans and specifications can never be totally complete. Not every detail can be illustrated and not every problem can be anticipated. For this reason, shop drawings are provided by the contractor, suppliers, or subcontractors, and the architect is available during the construction phase to resolve problems. The contractor is also required to use reasonable discretion in executing the work. The contractor is responsible for verifying the availability of materials and products used on the project. This is done by submitting product data that demonstrate the specific product that will meet each requirement of the contract documents. Samples are intended to show the specific item that will be used. Examples of samples would include bricks, carpeting, and other items whose color, texture, and other characteristics are important.

3.12 SHOP DRAWINGS, PRODUCT DATA AND SAMPLES

3.12.1 Shop Drawings are drawings, diagrams, schedules and other data specially prepared for the Work by the Contractor or a Subcontractor, Sub-subcontractor manufacturer, supplier or distributor to illustrate some portion of the Work. **(3.12.1; no change)**

3.12.2 Product Data are illustrations, standard schedules, performance charts, instructions, brochures, diagrams and other information furnished by the Contractor to illustrate materials or equipment for some portion of the Work. **(3.12.2; no change)**

3.12.3 Samples are physical examples that illustrate materials, equipment or workmanship and establish standards by which the Work will be judged. **(3.12.3; no change)**

Samples are normally delivered to the architect's office or to the job site. Samples also include mockups, such as samples of brick walls. These should be left at the job site until that particular component is well along or finished.

3.12.4 Shop Drawings, Product Data, Samples and similar submittals are not Contract Documents. Their purpose is to demonstrate the way by which the Contractor proposes to conform to the information given and the design concept expressed in the Contract Documents for those portions of the Work for which the Contract

Documents require submittals. Review by the Architect is subject to the limitations of Section 4.2.7. Informational submittals upon which the Architect is not expected to take responsive action may be so identified in the Contract Documents. Submittals that are not required by the Contract Documents may be returned by the Architect without action. **(3.12.4; minor modifications)**

This provision clarifies the fact that such submissions are not a part of the contract documents and do not amend or change the contract documents. Previously, some courts have held that shop drawings formed part of the contract.[88] If the contract documents are amended to make shop drawings part of the contract documents, this paragraph must be modified. Extreme caution must be observed in making this change because shop drawings then become the only contract documents not prepared by either the owner or architect.

The contract documents should indicate which submittals are required, including informational submittals. If the contractor makes any submittals that are not required by the contract documents, the architect may simply return them without comment.

3.12.5 The Contractor shall review for compliance with the Contract Documents, approve and submit to the Architect Shop Drawings, Product Data, Samples and similar submittals required by the Contract Documents in accordance with the submittal schedule approved by the Architect or, in the absence of an approved submittal schedule, with reasonable promptness and in such sequence as to cause no delay in the Work or in the activities of the Owner or separate contractors. **(3.12.5; minor modifications; delete reference to architect returning unwanted submittals)**

This provision requires the contractor to review and approve shop drawings and other submittals before sending them to the architect for review. In many cases, the contractor can forward submittals that are not completely approved if the discrepancies are minor. However, if the submittals do not reasonably comply with the contract documents, the contractor should reject them and not send them on to the architect. If the contractor does not indicate on the submittals that it has reviewed and approved them, they may simply be returned without action. This would likely delay the project, for which the contractor bears the responsibility.

3.12.6 By submitting Shop Drawings, Product Data, Samples and similar submittals, the Contractor represents to the Owner and Architect that the Contractor has (1) reviewed and approved them, (2) determined and verified materials, field measurements and field construction criteria related thereto, or will do so and (3) checked and coordinated the information contained within such submittals with the requirements of the Work and of the Contract Documents. **(3.12.6; minor modifications)**

By this representation, the contractor affirmatively states that it has reviewed these submittals and found that the measurements and related conditions are

---

[88] Cast-crete Corp. v. West Baro Corp., 339 So. 2d 413 (La. Ct. App. 1976).

proper. The contractor cannot later claim an extra if a field condition differs from that shown on a shop drawing. In regard to submittals and product literature, this representation means that the product is suitable for installation at the site and has been so confirmed by the contractor.

3.12.7 The Contractor shall perform no portion of the Work for which the Contract Documents require submittal and review of Shop Drawings, Product Data, Samples or similar submittals until the respective submittal has been approved by the Architect. (**3.12.7; no change**)

If the contractor starts any portion of the work before submitting and receiving shop drawings required by the contract documents, the contractor assumes the risk of error. This may constitute a substantial breach of contract, particularly if it is done repeatedly or results in a major problem. Note that the contractor is not relieved of responsibility if the submittals are approved, because the contract documents always control. Deviations from the contract documents can only be accomplished by change order, construction change directive, or minor change. On the other hand, if the architect fails to require from the contractor submittals that should have been required and the project is thereby delayed, the architect may be responsible for any resulting delays if the contractor fails to furnish the submittals in a timely manner (the contractor may even be entitled to a change order in this event). The architect can protect itself in such case by notifying the contractor and owner of a lack of timely submittals.

3.12.8 The Work shall be in accordance with approved submittals except that the Contractor shall not be relieved of responsibility for deviations from requirements of the Contract Documents by the Architect's approval of Shop Drawings, Product Data, Samples or similar submittals unless the Contractor has specifically informed the Architect in writing of such deviation at the time of submittal and (1) the Architect has given written approval to the specific deviation as a minor change in the Work, or (2) a Change Order or Construction Change Directive has been issued authorizing the deviation. The Contractor shall not be relieved of responsibility for errors or omissions in Shop Drawings, Product Data, Samples or similar submittals by the Architect's approval thereof. (**3.12.8; no change**)

Once a shop drawing or submittal has been approved, the work must be in accordance with the approved submittal. However, if there is any deviation between the approved submittal and the contract documents, then the contract documents control and the contractor would not be entitled to a change order. Even if the architect approves a shop drawing, unless the contractor has specifically pointed out deviations from the contract documents in writing, the deviations are not approved and can later be rejected. If the contractor wants to make a change, there must be a written change (**see** Article 7). The primary responsibility for shop drawings and submittals always remains that of the contractor. For that reason, such submittals are not contract documents and do not modify the contract documents (**see ¶** 3.12.4).

3.12.9 The Contractor shall direct specific attention, in writing or on resubmitted Shop Drawings, Product Data, Samples or similar submittals, to revisions other than those requested by the Architect on previous submittals. In the absence of such written notice the Architect's approval of a resubmission shall not apply to such revisions. (**3.12.9; no change**)

If a submission has been returned by the architect for corrections, the contractor must notify the architect of specific things that are revised on resubmissions other than what the architect previously noted. The architect retains a copy of the prior submission with its notations and usually checks only the noted items upon resubmission. If the contractor does not inform the architect of specific other revisions, those other items will probably not be reviewed. Because the architect's approval of submittals that do not conform to the contract documents is ineffective to bind the owner (¶ 4.2.7), the last sentence is actually redundant. The contractor must request a written change in order to be assured that there will not be any problem with a deviation from the contract documents (**see** ¶ 3.12.8).

3.12.10 The Contractor shall not be required to provide professional services that constitute the practice of architecture or engineering unless such services are specifically required by the Contract Documents for a portion of the Work or unless the Contractor needs to provide such services in order to carry out the Contractor's responsibilities for construction means, methods, techniques, sequences and procedures. The Contractor shall not be required to provide professional services in violation of applicable law. If professional design services or certifications by a design professional related to systems, materials or equipment are specifically required of the Contractor by the Contract Documents, the Owner and the Architect will specify all performance and design criteria that such services must satisfy. The Contractor shall cause such services or certifications to be provided by a properly licensed design professional, whose signature and seal shall appear on all drawings, calculations, specifications, certifications, Shop Drawings and other submittals prepared by such professional. Shop Drawings and other submittals related to the Work designed or certified by such professional, if prepared by others, shall bear such professional's written approval when submitted to the Architect. The Owner and the Architect shall be entitled to rely upon the adequacy, accuracy and completeness of the services, certifications and approvals performed or provided by such design professionals, provided the Owner and Architect have specified to the Contractor all performance and design criteria that such services must satisfy. Pursuant to this Section 3.12.10, the Architect will review, approve or take other appropriate action on submittals only for the limited purpose of checking for conformance with information given and the design concept expressed in the Contract Documents. The Contractor shall not be responsible for the adequacy of the performance and design criteria specified in the Contract Documents. (**3.12.10; minor revisions**)

The contractor is not required to provide design services—architecture or engineering—except as specifically required by the contract documents or required for construction procedures. The architect has no responsibility for construction procedures, so if engineering is required, it is the contractor's responsibility.

For instance, if sophisticated scaffolding requires engineering, that is the contractor's job and not that of the architect. The contractor is not required to provide any design services if such services would violate any law.

This paragraph permits the architect to rely on submissions of engineers' or other design professionals' certifications required by the contract documents—design delegation.[89] These can be professional certifications of performance criteria of materials, systems, and equipment. If such a submittal is made, the architect need not check any of the calculations or other information submitted by the engineer. If these design services are required of the contractor by the contract documents, the contract documents also must specify the criteria that are to be met. If the contractor meets the criteria, then the contract requirements are fulfilled, even if the system does not work. It is up to the architect to verify that the criteria will be suitable for the project.

Design delegation is often required on larger projects because architects simply do not have expertise in all areas. For instance, very few architects have the knowledge to properly detail a curtain wall system. The architect typically designs the aesthetics of the curtain wall and specifies the criteria that the system must meet, such as wind loading, water infiltration, and so forth. The manufacturer then designs the actual system, often constructing a mock-up for testing. In such a case, as long as the system meets the established criteria and aesthetics, the contractor and manufacturer will have met their contractual obligations. The contractor should carefully review the drawings and specifications to determine the extent to which such designs are required. The contractor's bid would, of course, include the costs of such design as well as the manufacturing and installation of the systems.

In order to minimize its exposure to design liability on those projects where there is design delegation, the contractor will want to take at least the following steps:

Carefully review the entire set of Contract Documents to determine whether all design criteria are adequately spelled out and seem reasonable;

Review the licensing laws of the state where the project is located to determine the impact of those laws on the delegated design process;

Investigate and obtain, if necessary, additional insurance that will protect the contractor from design-related liability.

§ 4.23   Use of Site: ¶ 3.13

3.13 USE OF SITE

The Contractor shall confine operations at the site to areas permitted by applicable laws, statutes, ordinances, codes, rules and regulations, and lawful orders of public

---

[89] The case General Bldg. Contractors of N.Y. State, Inc. v. New York State Educ. Dep't, 175 Misc. 2d 922, 670 N.Y.S.2d 697 (Sup. Ct. 1997), discusses design delegation in some detail in reference to New York rules governing such delegation.

authorities and the Contract Documents and shall not unreasonably encumber the site with materials or equipment. **(3.13.1; minor modifications)**

The contractor is required to be reasonable in the use of the site.[90] It may use no more of the site than is necessary for the work and cannot store materials for extended periods of time. If the work is for an occupied building, detailed provisions for use of the site should be included in the Contract Documents to minimize the inconvenience to the owner and tenants. If the drawings show specific areas that the Contractor may use during construction operations, that does not mean that the Owner or Architect are involved in the means or methods of construction, they are merely defining specific areas that are or are not available to the contractor for use.

§ 4.24   Cutting and Patching: ¶ 3.14

3.14 CUTTING AND PATCHING

3.14.1 The Contractor shall be responsible for cutting, fitting or patching required to complete the Work or to make its parts fit together properly. All areas requiring cutting, fitting and patching shall be restored to the condition existing prior to the cutting, fitting and patching, unless otherwise required by the Contract Documents. **(3.14.1; modified)**

This applies primarily to remodeling or otherwise working with existing construction, because this should be obvious for new construction. The contractor must install the work according to the contract documents and may have to cut existing walls, beams, or other conditions to incorporate the new work and then patch these things to properly complete the installation. Cutting and patching are required whether specifically called out in the drawings or not. This provision also implicitly requires that any patching be done so that it matches the surrounding materials and has a finished appearance, unless in an unexposed area. If a substantial amount of patching work is expected, the specifications should spell out the requirements for cutting and patching.

3.14.2 The Contractor shall not damage or endanger a portion of the Work or fully or partially completed construction of the Owner or separate contractors by cutting, patching or otherwise altering such construction, or by excavation. The Contractor shall not cut or otherwise alter such construction by the Owner or a separate contractor except with written consent of the Owner and of such separate contractor; such consent shall not be unreasonably withheld. The Contractor shall not unreasonably withhold from the Owner or a separate contractor the Contractor's consent to cutting or otherwise altering the Work. **(3.14.2; no change)**

---

[90] In Chadwick v. CSI, Ltd., 629 A.2d 820, 827 (N.H. 1993), the court stated that the term *site* referred "to the place where the Work is taking place, regardless of whether that place constitutes an entire parcel of property or just a portion thereof."

This is not in conflict with the prior provision. It merely means that the contractor must obtain written permission from the owner or other contractor before cutting or patching or otherwise altering such work. The contractor must also give reasonable approval if another contractor wants to do the same to its work. Of course, if another contractor delays or interferes with its work, the contractor may be entitled to a change order.

§ 4.25   Cleaning Up: ¶ 3.15

3.15 CLEANING UP

3.15.1 The Contractor shall keep the premises and surrounding area free from accumulation of waste materials or rubbish caused by operations under the Contract. At completion of the Work, the Contractor shall remove waste materials, rubbish, the Contractor's tools, construction equipment, machinery and surplus materials from and about the Project. **(3.15.1; minor modification)**

This has always been implicit in construction contracts. There may also be local laws concerning keeping the streets clean and requiring the Contractor to do certain things to insure that, all of which would be included in the basic cost of the project. The contractor must be reasonable in keeping the premises neat. The owner may want to expand this paragraph to make the contractor responsible for weekly or more frequent cleaning of the premises.

3.15.2 If the Contractor fails to clean up as provided in the Contract Documents, the Owner may do so and Owner shall be entitled to reimbursement from the Contractor. **(3.15.2; minor modification)**

If the owner finds that the contractor is failing to clean up pursuant to the contract documents, she should first make a written demand on the contractor to perform the proper cleanups. If the contractor continues to fail to perform, the owner can then do this work and charge the contractor for the cost. Note that this language says that the cost "shall" be charged to the contractor. While this may imply that a change order is not necessary, the better practice would seem to be for the owner to follow the change order and/or claims procedure for this item.

§ 4.26   Access to Work: ¶ 3.16

3.16 ACCESS TO WORK

The Contractor shall provide the Owner and Architect access to the Work in preparation and progress wherever located. **(3.16.1; no change)**

The contractor cannot deny the owner or architect access to the site and must permit access to components of the work if they are under construction elsewhere.

§ 4.27   Royalties, Patents, and Copyrights: ¶ 3.17

3.17 ROYALTIES, PATENTS AND COPYRIGHTS

The Contractor shall pay all royalties and license fees. The Contractor shall defend suits or claims for infringement of copyrights and patent rights and shall hold the Owner and Architect harmless from loss on account thereof, but shall not be responsible for such defense or loss when a particular design, process or product of a particular manufacturer or manufacturers is required by the Contract Documents or where the copyright violations are contained in Drawings, Specifications or other documents prepared by the Owner or Architect. However, if the Contractor has reason to believe that the required design, process or product is an infringement of a copyright or a patent, the Contractor shall be responsible for such loss unless such information is promptly furnished to the Architect. **(3.17.1; no change)**

If the contract documents call for a particular design, process, or product that requires the payment of a royalty or license fee, the contractor must pay the cost. The contractor should include this cost in its bid if it is known during the bidding. If the need for this fee is not discovered until later, the owner must pay this fee.

There are three requirements before a patent can be issued: the object must be novel, utilitarian, and nonobvious.[91] It is almost impossible to obtain a patent for a building in its entirety because of the requirements of novelty and nonobviousness. Components can, however, be patented under the proper circumstances.[92] Patent law is a specialty area, and the advice of a patent attorney should be sought if any questions arise in this area.

Section 3.17 places the burden of defense of claims for patent and copyright infringement on the contractor unless the problem is in the drawings or other contract documents. See the discussion in ¶ 1.5 in **§ 4.5.** If the problem is known to the contractor at the outset and the contractor fails to report the problem, then the contractor will bear the appropriate costs. Note that this situation may violate the anti-indemnification statute of some states (**see § 4.28**) unless this provision is read to limit such damages to the costs that could have been avoided if the contractor had given prompt notification.

§ 4.28   Indemnification: ¶ 3.18

3.18 INDEMNIFICATION

3.18.1 *To the fullest extent permitted by law the Contractor shall indemnify and hold harmless the Owner, Architect, Architect's consultants, and agents and employees of any of them from and against claims, damages, losses and expenses, including but not limited to attorneys' fees, arising out of or resulting from performance of the*

---

[91] 35 U.S.C. §§ 101-103.

[92] *See, e.g.,* Blumcraft of Pittsburgh v. United States, 372 F.2d 1014 (Ct. Cl. 1967) (handrail found to be patentable and action for infringement was upheld).

Work, provided that such claim, damage, loss or expense is attributable to bodily injury, sickness, disease or death, or to injury to or destruction of tangible property (other than the Work itself), but only to the extent caused by the negligent acts or omissions of the Contractor, a Subcontractor, anyone directly or indirectly employed by them or anyone for whose acts they may be liable, regardless of whether or not such claim, damage, loss or expense is caused in part by a party indemnified here-under. Such obligation shall not be construed to negate, abridge, or reduce other rights or obligations of indemnity which would otherwise exist as to a party or person described in this Section 3.18. **(3.18.1; deleted reference to Project Management Protective insurance)**

Indemnification clauses[93] (sometimes called hold harmless clauses) call upon one party to pay legal and other expenses of another party. This is a way of shifting the risk on a project. In a construction project, the greatest risk is typically based on the work of the contractor. The contractor, in its bid, anticipates the cost of such risk-shifting, and by undertaking to be solely responsible for the safety of the workers, the public, and the work during construction, the contractor attempts to keep injuries and damages to a minimum. A Massachusetts case involved an example of an indemnity clause.[94] A gas pipeline ruptured during construction,

---

[93] This provision has been held not to violate a Maryland law that would prohibit a clause that attempted to indemnify a party for its own negligence. Mason v. Callas Contractors, Inc., 494 F. Supp. 782 (D. Md. 1980). *See also* Cuhaci & Peterson Architects, Inc. v. Huber Constr. Co., 516 So. 2d 1096 (Fla. Dist. Ct. App. 1987).

The court in International Paper Co. v. Corporex Constructors, Inc., 385 S.E.2d 553 (N.C. 1989), reviewed a similar indemnity provision and determined that it violated North Carolina law. However, if the language "regardless of whether or not it is caused in part by a party indemnified hereunder" were deleted, the provision would comply with the law. Because the plaintiff was not negligent, the provision was effective to indemnify the plaintiff. *See also* Northwinds, Inc. v. Phillips Petroleum Co., 779 P.2d 753 (Wyo. 1989), for a careful analysis of the wording used here and in similar provisions.

In Burns v. DeWitt & Assocs., 826 S.W.2d 884 (Mo. 1992), this clause was upheld and the contractor had to pay sanctions to the architect for the appeal. But in Cabo Const., Inc., v. R S Clark Const., Inc., 227 S.W.3d 314 (Tex. App.-Hous. [1 Dist.], 2007), a Texas court held that this provision does not indemnify a party from its own negligence.

The broad arbitration provision found in the AIA contracts applies to indemnification provisions in a construction contract, such as this one. Contract Constr., Inc. v. Power Tech. Ctr. Ltd. P'ship, 100 Md. App. 173, 640 A.2d 251 (1994). *But see* Lopez v. 14th St. Dev., LLC, 835 N.Y.S.2d 186 (App. Div. 1st Dep't 2007), interpreting the 1997 version of A201, where the owner brought an action for indemnification against the contractor after being sued by an injured employee of the contractor. The contractor sought to stay the action pending arbitration. The court reversed, finding that the indemnity claim arose long after work was completed and final payment was made. Finding that the AIA document was "not a model of clarity," the court reviewed the provisions that require the architect to review claims as a condition precedent to arbitration. There was, however, no further provision in the document subjecting claims that arise after completion of the work to arbitration.

[94] Shea v. Bay State Gas Co., 383 Mass. 218, 418 N.E.2d 597, 602 (1981). This same language was the subject of Kelly v. Dimeo, Inc., 31 Mass. App. Ct. 626, 581 N.E.2d 1316 (1991), upholding the indemnity. *See also* Whittle v. Pagani Bros. Constr. Co., 422 N.E.2d 779 (Mass. 1981) (subcontractor required to indemnify contractor).

injuring members of the public. The contractor was sued and he, in turn, filed a third-party action against the engineers, claiming that the engineers negligently supervised, tested, and inspected the work. The court found that the indemnification clause was effective to shift the risk of loss and the burden of defending the lawsuits to the contractor: "in this case the indemnity clause as we have interpreted it places full responsibility for the proper conduct of the work on the party primarily responsible for it, thus giving the contractor a necessary and desirable incentive to assure both proper performance and appropriate safety measures during the construction process."

Some jurisdictions require that an indemnification provision must clearly and unequivocally state that a party may be indemnified for its own negligence before the provision will be effective to indemnify the party for its own negligence. In other words, the indemnification must say exactly what it means.[95]

---

[95] In Washington Elementary Sch. Dist. No. 6 v. Baglino Corp., 169 Ariz. 58, 817 P.2d 3 (1991), the court examined the language "regardless of whether or not such claim, damage, loss or expense is caused in part by a party indemnified hereunder." The court found that this language clearly and unequivocally protected the owner against its own active negligence. *See also* Moore Heating & Plumbing v. Huber, Hunt & Nichols, 583 N.E.2d 142 (Ind. Ct. App. 1991); Oster v. Medtronic, Inc., 428 N.W.2d 116 (Minn. 1988); Howe v. Lever Bros., 851 S.W.2d 769 (Mo. 1993); City of Pittsburgh v. American Asbestos, 629 A.2d 265 (Pa. Commw. Ct. 1993); Gunka v. Consol. Papers, Inc., 508 N.W.2d 426 (Wis. Ct. App. 1993).

In Facilities Dev. Corp. v. Miletta, 180 A.D.2d 97, 584 N.Y.S.2d 491, 494 (1992), the court stated that "[it] is the general rule that an indemnification agreement between sophisticated business entities will be construed as intending to indemnify either party for its own wrongdoing only when the language in the agreement clearly connotes an intent to provide for such indemnification." *See also* Mautz v. J.P. Patti Co., 298 N.J. Super. 13, 688 A.2d 1088 (1997). *Mautz* was cited in the case of Hagerman Constr. Corp. v. Long Elec. Co., 741 N.E.2d 390, 393 (Ind. Ct. App. 2000). The employee of a subcontractor was injured on the jobsite. Relying on the indemnification language of A401-1987 (very similar to the current language), the general contractor sought to be indemnified by the subcontractor. The court applied an extensive analysis of the language. Under Indiana law, a party may contract to indemnify another for the other's own language, but only if the party knowingly and willingly agrees to such indemnification. The court then applied a two-part test to determine whether the language indicates that a party has knowingly and willingly accepted the indemnification burden.

First, the clause must expressly state in clear and unequivocal terms that negligence is an area of application where the indemnitor has agreed to indemnify the indemnitee. This first part was met by the AIA language, as it:

expressly defines negligence as an area of application in clear and unequivocal terms. The clause speaks of claims, damages, losses and expenses attributable to bodily injury, sickness, disease or death, and injury to or destruction of property, as well as negligent acts or omissions. These words, taken in this context, are the language of negligence, and, as such, clearly and unequivocally demonstrate that the indemnification clause applies to negligence.

Next, the court sought to determine whether the clause expressly states, in clear and unequivocal terms, that it applies to indemnify the general contractor for its own negligence. The court concluded that it did not. The court held that:

the indemnification clause does not expressly state, in clear and unequivocal terms, that it applies to indemnify Hagerman for its own negligence. The clause explicitly indemnifies Hagerman for the acts of the sub-contractor, Long, and its subsubcontractors, employees and

Indemnifications and exculpatory clauses may not be effective when there is active interference by the owner.[96] In a Massachusetts case, the owner decided to

---

anyone for whom it may be liable, but it does not explicitly state that Long must indemnify Hagerman for its own negligent acts. Further, the phrase "but only to the extent" clearly limits Long's obligation to indemnify Hagerman only to the extent that Long, its sub-subcontractors, employees, and anyone for whom it may be liable are negligent. Otherwise, the clause contains no clear statement that would give the contractors notice of the harsh burden that complete indemnification would impose.

With regard to Hagerman's contentions, we conclude that inclusion of the phrase "to the fullest extent permitted by law" is not necessarily inconsistent with the use of the phrase "but only to the extent." The phrase "to the fullest extent permitted by law" is a preservation clause that preserves Hagerman's rights under the law to the extent that Long and/or its sub-subcontractors, etc. are negligent. In other words, Hagerman may pursue its rights to the fullest extent of the law as long as, and to the measure of, Long's negligence. Moreover, we think it also implausible to state that the phrase "regardless of whether or not such claim, damage, loss or expense is caused in part by a party indemnified hereunder" contradicts the other language of the clause that would limit Long's liability to Hagerman. Simply put, based upon this phrase, Long may not disregard its duty to indemnify Hagerman for Long's negligence merely because Hagerman may also be negligent under the circumstances. Therefore, the indemnification clause does not require Long to indemnify Hagerman for Hagerman's own negligence. However, the clause does require Long to indemnify Hagerman for Long's negligence.

See also Pitt v. Tyree Org., Ltd., 90 S.W.3d 244 (Tenn. Ct. App. 2002); Ryan v. United States, 233 F. Supp. 2d 668 (D. N.J. 2002); East-Harding, Inc. v. Horace A. Piazza & Assocs., 80 Ark. App. 143, 91 S.W.3d 547 (2002); In Burns v. DeWitt & Assocs., 826 S.W.2d 884 (Mo. 1992), this clause was upheld and the contractor had to pay sanctions to the architect for the appeal; Cabo Const., Inc., v. R S Clark Const., Inc., 227 S.W.2d 314 (Tex. App.-Hous. [1 Dist.], 2007), involving a subcontractor.

In Camp, Dresser & McKee, Inc. v. Paul N. Howard Co., 853 So. 2d 1072 (Fla. Dist. Ct. App. 2003), where the indemnification language was identical to A201, the court held that this language clearly expresses the parties' intent that the engineer may be indemnified by the contractor even if the engineer is sued for its own wrongful conduct.

The United States Supreme Court cited the AIA language as an example of an indemnification clause that makes specific reference to the effect of the negligence of the indemnitee. United States v. M. O. Seckinger, 397 U.S. 203 (1970). The *Seckinger* case was cited in Rhoades v. United States, 986 F. Supp. 859 (D. Del. 1997), in a review of indemnification clauses that attempted to indemnify the government from its own negligence.

[96] In Coatesville Contractors v. Borough of Ridley, 509 Pa. 552, 506 A.2d 862 (1986), the contractor signed a contract for dredging work on a municipal lake after inspecting the lake and finding it was drained. The specifications stated that the lake would remain drained until the work was completed. When the contractor went to start the work, the lake was filled, and the contractor encountered additional costs in performing the dredging work when the municipality failed to drain the lake. The contractor's claim for additional compensation for the extra work was upheld, even though the contract had an exculpatory clause, because the municipality had interfered with the work. The owner's notification to the contractor to proceed with the work when it was known that the lake was filled with water invalidated the exculpatory provisions of the contract.

In Franklin Contracting Co. v. State, 144 N.J. Super. 402, 365 A.2d 952 (1976), the contractor recovered in the face of a no-damage-for-delay clause. The state had failed to provide access to a right of way. The contractor had asked the state whether access would be a problem and was assured that access was available. The state had acted innocently.

"take away" the installation of a vertical lift door from the general contractor.[97] The contract was given by the owner to another contractor. A third party was injured when the door fell on him. When the plaintiff sued the owner, the owner sought indemnification from the general contractor under this clause. The court held that there could be no indemnification by the general contractor when the owner had taken that work away from the general contractor. The general contractor could not be charged with any negligent act or omission on the part of the substitute contractor.

The attorneys' fee provision in this paragraph has been held to apply only when third parties bring an action, not when the architect or owner brings an action against the contractor.[98] In a New York case, the contractor argued that because

---

[97] Urbanati v. Simplex Wire & Cable Co., 10 Mass. App. Ct. 881, 409 N.E.2d 243 (1980).

[98] Campbell v. Southern Roof Applicators, Inc., 406 So. 2d 910 (Ala. 1981).

In Myers Bldg. Indus., Ltd. v. Interface Tech., Inc., 17 Cal. Rptr. 2d 242 (Ct. App. 1993), the court stated that:

> a clause which contains the words "indemnify" and "hold harmless" is an indemnity clause which generally obligates the indemnitor to reimburse the indemnitee for any damages the indemnitee becomes obligated to pay third persons. . . . Indemnification agreements ordinarily relate to third-party claims.
>
> The very essence of an indemnity agreement is that one party hold the other harmless from losses resulting from certain specified circumstances. . . . Indemnification agreements are intended to be unilateral agreements.

The court held that attorneys' fees could not be awarded because this was an action for breach of contract by the contractor against the owner. This indemnification agreement is not a provision for attorneys' fees in an action to enforce the contract. It encompasses only losses attributable to bodily injury or injury to tangible property, not contract damages.

In Lagerstrom v. Beers Constr. Co., 157 Ga. App. 396, 277 S.E.2d 765 (1981), the structural engineer sought recovery from the contractor for his attorneys' fees incurred in defending a suit arising from a scaffolding collapse. The earlier version of this indemnification provision did not include "Architect's consultants." The court found that the engineer was not an agent of the architect and was not entitled to indemnification. Under the newer language, the outcome of the case would have been different.

In Adena Corp. v. Sunset View Ltd., 2001 Ohio App. LEXIS 1726 (April 10, 2001), *appeal denied*, 754 N.E.2d 259 (2001), the contractor filed suit against the owner. The owner moved for a stay and to compel arbitration. The arbitrator granted the owner a substantial award, which was confirmed by the trial court. The contractor then moved to vacate the award on the basis that the arbitrator had exceeded his authority. Apparently, the arbitrator had included attorneys' fees in the award to the owner. Because the arbitrator is authorized to award attorneys' fees in at least some incidences (here, the court cited the indemnification provision of ¶ 3.18.1 that permits attorneys' fees, although that was not the basis in this case), the arbitrator did not exceed his powers in rendering such an award.

Pinkert v. Olivieri, 2001 WL 641737 (D. Del. May 24, 2001) (the attorneys' fees provision in this section operates only as to third-party indemnification).

In Nusbaum v. City of Kansas City, 100 S.W.3d 101 (Mo. 2003), the court held that this language did not permit recovery of all attorneys' fees:

> While the indemnification provision at issue in the present case provides for the recovery of legal expenses, including attorneys' fees incurred in the defense of a claim, nothing in the indemnification provision suggests that it provides for the recovery of legal expenses incurred in establishing the right to indemnity.

the complaint alleged negligence by the architect, the exclusion provision of this paragraph negated the indemnification.[99] The court found that, because the architect had not been found guilty of negligence by the court, the mere fact that negligence had been alleged does not destroy the indemnification.

Some contracts have a no-damage-for-delay clause that is somewhat similar in that the contractor is not entitled to any damages resulting from delays to the project.[100] Another variation is a liquidated damages clause that assesses a per day charge against the contractor if the project is not substantially completed by a particular date.[101]

---

[99] Estate of Nasser v. Port Auth., 155 A.D.2d 250, 546 N.Y.S.2d 626 (1989). In Pugh v. Prairie Constr. Co., Inc., 602 N.W.2d 805 (Iowa 1999), an injured worker sued the general contractor and two architectural firms. Following a settlement, one of the architects brought an indemnity claim against the general contractor and the subcontractor that employed the worker under this provision. The jury in the trial court had previously found that the injured worker was 100% at fault for his injuries. The appellate court held that there were two reasons why the architect was not entitled to indemnification. First, the negligence of the employee cannot be imputed to the employer. The court held that the AIA language does not require this. Second, the language of the indemnification agreement is construed against the architect as drafter of the language. Because the allegation against the architect in the original complaint was for an allegedly defective design, the AIA language would have to require indemnification for the architect's sole negligence in clear and unequivocal language, which it did not.

[100] Annotation, *Validity, Construction, and Application of "No Damage" Clause with Respect to Delay in Construction Contract,* 10 A.L.R.2d 801 (1950); Buckley & Co. v. State, 140 N.J. Super. 289, 356 A.2d 56 (1975) ("a no-damage provision ought not be construed as exculpating a contractee from that liability unless the intention to do so is clear"; courts should look for ways to avoid result of this clause); Kiewit Constr. Co. v. Capital Elec. Constr. Co., Inc., 2005 WL 2563042 (D. Neb. Oct. 12, 2005).

[101] J.R. Stevenson Corp. v. County of Westchester, 113 A.D.2d 918, 493 N.Y.S.2d 819 (1985). The contract contained a no-damages-for-delay clause, a hold harmless clause, and a liquidated damages clause fixing damages at $300 per day. The owner filed suit against the contractor and asked for actual damages, which exceeded the $300 per day amount. The court held that "a liquidated damages clause which is reasonable precludes any recovery for actual damages. This is so even though the stipulated sum may be less than the actual damages sustained by the injured party." The court further held that a no-damages-for-delay clause will not bar an action based on delays or obstructions that were not within the contemplation of the parties when the contract was executed, or that resulted from wilful or grossly negligent acts of the other party.

In Twin River Constr. Co. v. Public Water Dist., 653 S.W.2d 682 (Mo. Ct. App. 1983), the court held that an owner could recover both liquidated damages and actual damages, as long as they are not duplicative.

In Gutowski v. Crystal Homes, Inc., 26 Ill. App. 2d 269, 167 N.E.2d 422 (1960), the contract contained a liquidated damage clause limiting recovery to $1,000. The contractor had built the house too close to a lot line, in violation of the zoning ordinance. The owner was awarded damages for the decrease in value of the house that exceeded the liquidated damages clause because "it is inconceivable that this clause was meant to cover every possible default or failure by either party to the agreement."

The court in Hartford Elec. Applicators, Inc. v. Alden, 169 Conn. 177, 363 A.2d 135 (1975), stated that "the majority of jurisdictions in this country hold that where there are delays attributable to both parties the liquidated damages clause based upon the contract date is abrogated; the contract is to be performed in a reasonable time."

In V.L. Nicholson Co. v. Transcon Inv. & Fin. Ltd., 595 S.W.2d 474 (Tenn. 1980), the court stated: "a party will not be allowed, however, to recover liquidated damages where he is responsible for or has contributed to the delay or nonperformance alleged as breach." *Accord* City of Whitehall v. Southern Mech. Contracting, Inc., 269 Ark. 563, 599 S.W.2d 430 (Ct. App. 1980).

In Unicon Mgmt. Corp. v. City of Chicago, 404 F.2d 627 (7th Cir. 1968), the contract had a no-damage-for-delay clause that included delay caused by the owner (with the possible exception of bad faith by the owner). The applicable provisions stated:

(A) Anything contained in the conditions or specifications document to the contrary notwithstanding, it is agreed that the following provisions relative to unavoidable delay shall be controlling:

"1. Should the Contractor be obstructed or delayed in the commencement, prosecution or completion of the work hereunder by any act or delay of the City or by unavoidable acts or delays on the part of transportation companies in transporting, switching or delivering material for said work, or by acts of public authorities, or by riot, insurrection, war, pestilence, fire, lightning, earthquake, cyclone, or through any delays or default of other parties under contract with said City or due to unavoidable delays in obtaining the specified materials, priorities, allocations, or prohibitions established by the United States of America, or due to strikes that cause unavoidable delays in performing said work, or to other unavoidable causes (except wet or cold weather), and provided the Purchasing Agent shall determine that any of such causes or delays were entirely beyond the control of the Contractor, then the times herein fixed for the completion of said work shall be extended for a period equivalent to the time lost by reason of any of the aforesaid causes or delays mentioned herein. No such allowance of time shall be made, however, unless notice in writing of a claim therefor is presented to the Purchasing Agent before the 30th day of each succeeding month of all delays occurring within the preceding month, and the Contractor shall satisfy the Purchasing Agent that the delays so claimed are unavoidable and substantial, and could not be reasonably anticipated or adequately guarded against, and said allowance of time is authorized in writing by the Mayor, Comptroller and Purchasing Agent.

2. It is further expressly understood and agreed that the Contractor shall not be entitled to any damages or compensation from the City, or be reimbursed for any losses, on account of any delay or delays resulting from any of the causes aforesaid. The Commissioner and the Purchasing Agent shall determine the number of days, if any, that the Contractor has been so delayed and his decision shall be final and binding upon other parties to this contract upon authorization in writing as hereinbefore set forth.

In Buckley & Co. v. City of N.Y., 121 A.D.2d 933, 505 N.Y.S.2d 140 (1986), the clause was:

The Contractor agrees to make no claim for damages for delay in the performance of this contract occasioned by any act or omission to act of the City or any of its representatives, and agrees that any such claim shall be fully compensated for by an extension of time to complete performance of the work as provided herein.

The court found that this provision creates a bar to any claim for delay damages when the delay was within the contemplation of the contracting parties.

In Blau Mech. Corp. v. City of N.Y., 158 A.D.2d 373, 551 N.Y.S.2d 228 (1990), the contractor was denied recovery for unforeseen subsurface conditions, changes to the contract, and delays caused by intrusion onto the work site of a local community group.

In State Highway Admin. v. Greiner Eng'g Sciences, Inc., 83 Md. App. 621, 577 A.2d 363 (1990), the contract between the owner and engineer read:

The Consultant agrees to prosecute the work continuously and diligently and no charges or claims for damages shall be made by him for any delays or hindrances, from any cause

Note that loss of use has been excluded from the indemnity provision since the 1997 version because of the waiver of consequential damages found in ¶ 15.1.6. One issue that may arise relates to delay damages, which are permitted under ¶ 8.3. Does a delay result in loss of use? This paragraph states that indemnification is not available for loss of use, but damages are. A clarification of the parties' intent would be wise. See also ¶ 11.3.3, which also contains a waiver related to loss of use.

This indemnity provision was examined in an Idaho case[102] involving a broken water line that flooded a sports arena. According to that case, ¶ 3.18.1 requires that

---

whatsoever during the progress of any portion of the services specified in this Agreement. Such delays or hindrances, if any, may be compensated for by an extension of time for such reasonable period as the Department may decide. Time extensions will be granted only for excusable delays such as delays beyond the control and without the fault or negligence of the consultant.

The court held that enforcement of this no-damage-for-delay clause was not unconscionable.

Department of Transp. v. Interstate Contractors Supply, 130 Pa. Commw. 334, 568 A.2d 294 (1990), analyzed whether a liquidated damages clause was an unenforceable penalty, holding it was not in that case.

[102] Idaho State Univ. v. Mitchell, 97 Idaho 724, 552 P.2d 776 (1976). In Crowinshield/Old Town Cmty. Urban Redevelopment Corp. v. Campeion Roofing & Waterproofing, Inc., 719 N.E.2d 89 (Ohio 1998), the court found no duty to indemnify under this provision. The project consisted of the conversion of several commercial buildings into residential units. Shortly after completion, the roof started leaking, leading to protracted litigation between the owner, architect, general contractor, and roofing contractor. Well into the litigation, the architect sought indemnification for its attorneys' fees from the roofing contractor under this provision. The court found that all claims for damages in the litigation, out of which the architect's claim for attorneys' fees arose, related to the faulty construction of the roof. This constituted "the Work" as defined in ¶ 1.1.3 of A201 and constituted an exception (the language in this ¶ 3.18.1 is "(*other than the Work itself*)") to the indemnification provision.

In American Realty and Constr., Inc. v. Duffy, 2002 WL 31117254 (Cal. Ct. App. 1st Dist. Sept. 25, 2002), the plaintiff was a real estate agent who sold a lot to a buyer. An affiliate of the real estate agent, the owner of the lot, contracted with a general contractor, Kilkee, to construct a home for the buyer. After construction was completed, the sale was closed. The buyer subsequently discovered that the foundation was defective and sued several parties, including the real estate agent. Following a settlement, the real estate agent sued the general contractor for indemnification under this provision. The appellate court found that this provision required the general contractor to indemnify the architect and owner for third-party bodily injury and property damage claims, but not to indemnify the various parties for claims of defective workmanship. Based on the fact that the damages were the result of defective construction and also on the grounds that the real estate agent was not provided with any warranties, the complaint was dismissed.

In Chester Upland Sch. Dist. v. Meloney, 901 A.2d 1055, 2006 WL 1644666 (Pa. Super. Ct. June 15, 2006), the architect specified a particular absorption chiller, which the contractor furnished and installed. The chiller malfunctioned and could not be repaired. The owner sued the architect and contractor and the architect sued the contractor for indemnification pursuant to this clause and several other indemnification clauses in the contract. The matter was settled by the owner but the architect pursued indemnification for its legal fees and expenses against the contractor. In ruling for the contractor, the court considered this indemnification language: "there are three important aspects to be noted: (1) the provision bars claims for injuries to or destruction of 'the Work' itself; (2) Contractor is only liable if the claim arises from either its negligence or

two things be proven to establish liability: first, that damage or injury occur to property other than the "Work" itself, and, second, that the damage be caused by a negligent act or omission of the contractor or someone working for the contractor. Thus, this paragraph requires that negligence of the contractor or its agents be established.[103]

Another court has held that the contractor has no indemnification obligation when the architect makes a "voluntary" settlement payment.[104] In a North Carolina case, the contractor performed repairs to a water cooler and lines leading to it.[105] However, the contractor had installed a defective pressure reducing

---

omission, regardless of whether it is only wholly or partially responsible; and (3) the provision specifically limits Contractor's liability to Architect for claims against Architect arising from its performance of various duties." This indemnification clause makes the contractor liable for claims for damage to tangible property provided that the tangible property is not the Work itself, so the contractor in this case was not liable because the chiller was the Work itself.

[103] Sears, Roebuck & Co. v. Shamrock Constr. Co., 441 So. 2d 379 (La. Ct. App. 1983). In Franklin St. Dev. Ltd. v. Bilow & Goldberg & Assocs., 1997 U.S. Dist. LEXIS 20580 (S.D. N.Y. 1997), the court held that this indemnification language applies only to claims for damages arising out of personal injuries or injury to property and not to mere economic loss.

[104] Best Prods. Co. v. A.F. Callan Co., 1997 U.S. Dist. LEXIS 1914 (E.D. Pa. Feb. 26, 1997). This case should not be relied on, because it is at odds with the intent of this paragraph. The court held that the architect was not liable for the underlying loss and its payment was "voluntary" and not subject to indemnification. This defies common sense since few people enter into a settlement as pure volunteers.

[105] Haywood County Consol. Sch. Sys. v. United States Fidelity & Guar. Co., 43 N.C. App. 71, 257 S.E.2d 670 (1979).

In Goffredo v. Bay St. Landing Assocs., 578 N.Y.S.2d 662 (App. Div. 1992), the court refused to dismiss the contractual indemnity count because there was a fact question as to whether the insurance was sufficient to cover the damages.

In Meyers v. Burger King Corp., 638 So. 2d 369, 380 81 (La. Ct. App. 1994), the contract contained an indemnification clause stating that "the contractor shall indemnify, defend and save and hold harmless the Owner from those claims set forth in the above paragraph captioned Insurance, including reasonable attorneys fees." The insurance provision had this language:

The contractor shall maintain insurance to protect the Owner and Himself from claims which may arise from the Contractor's operations, whether such operations be by himself, any sub-contractor, anyone directly or indirectly employed by any of them, or by whose acts they may be liable. This insurance shall be written for not less than limits of liability required by law, or $1,000,000 aggregate (single limit applicable to bodily injury and property damages combined) whichever is greater.

The court rejected the argument that this provision limited the indemnity to $1 million, especially because the contractor had purchased higher insurance limits. The indemnity clause required the contractor to fully indemnify the owner for all damages.

In Bentley Koepke, Inc. v. Jeffrey Allen Corp., 1998 Ohio App. LEXIS 684 (Feb. 27, 1998), the decedent's administrator sued the landscape architect and others. The architect sought to have the contractor's insurance carrier indemnify and defend it pursuant to this provision and ¶ 11.1.1.7 of the General Conditions. The architect claimed it was the intended third-party beneficiary of the contractor's insurance policy. The policy, however, did not name the architect, and the court held that the architect was not a third-party beneficiary. Presumably, the architect could still be indemnified by the contractor, but it would have been wise to obtain an insurance certificate that specifically named the architect and owner as insured in this situation.

valve. As a result of the installation of the defective valve, the supply tube to the water cooler blew out of a valve fitting, causing water damage to a newly constructed wooden gym floor. The contract had required a performance bond and insurance. Prior to the damage, the contractor went into bankruptcy and the surety paid the cost of repairing the defective valve. However, the surety argued that, because of ¶ 11.1.1 of the General Conditions, the insurance carrier should pay for the cost of the gym floor repair. Unfortunately, the insurance was canceled before the damage, and the owner learned of the cancellation only after the damage. The court held that the contractor's obligation to carry insurance did not remove his liability under this section. The surety was required to pay for the damage.

In a Georgia case, the indemnification clause allowed an architect to maintain an indemnification action against a contractor after the architect had been sued by the owner for a defective plumbing installation.[106] Even without specific indemnification clauses like ¶ 3.18.1, courts have found contractors liable in an indemnity action brought by an architect when the contractor was primarily responsible for injuries sustained by workers.[107]

In a Wyoming case, the contractor sought indemnification against a subcontractor for amounts paid to the subcontractor's injured employee.[108] The contract between the general and the subcontractor incorporated the 1976 version of A201. However, the subcontract itself contained a slightly different indemnification provision. A201 provided for indemnity for "all claims, damages, losses and expenses," but the subcontract indemnity clause was limited to "acts or omissions of the Subcontractor." The court construed these provisions strictly against the indemnitee, finding that the flow-down provision was not a clear and unequivocal agreement to indemnify the general contractor on account of its own negligence.

The architect, and not the contractor, is primarily responsible for the drawings and specifications. In a Florida case,[109] a contractor was found not liable for leaking caused by a latent defect of the particular brick specified.[110]

---

[106] Jova/Daniels v. B&W Mech. Contractors, 167 Ga. App. 649, 307 S.E.2d 97 (1983).

[107] Fidelity & Cas. Co. v. J.A. Jones Constr. Co., 200 F. Supp. 264 (E.D. Ark. 1961).

    In Aetna Cas. & Sur. Co. v. Lumbermens Mut. Cas. Co., 136 A.D.2d 246, 527 N.Y.S.2d 143 (1988), the court held that this express indemnity provision could co-exist with common-law indemnity, in finding that a contractor was entitled to coverage under its insurance policy. In North American Site Developers, Inc., v. MRP Site Developers, Inc., 63 Mass. App. Ct. 529, 827 N.E.2d 251 (2005), the court construed the phrase "but only to the extent" as limiting the contractor's indemnity obligation to losses caused by its own conduct.

[108] Wyoming Johnson, Inc. v. Stag Indus. Inc., 662 P.2d 96, 99 (Wyo. 1983).

[109] Wood-Hopkins Contracting Co. v. Masonry Contractors, Inc., 235 So. 2d 548 (Fla. Ct. App. 1970).

[110] See Fanning & Doorley Constr. Co. v. Geigy Chem. Corp., 305 F. Supp. 650 (D. R.I. 1969) (contractor not liable for damage caused by product that proved defective in this application because only that product was specified in contract). In Travelers Indem. Co. v. S.M. Wilson & Co., 2005 WL 2234582 (E.D. Mo. Sept. 14, 2005), the plaintiff argued that the contractor breached the warranty provision found in ¶ 3.5.1 of A201 by using a cabinet that was not of good quality and not free of defects. The cabinet, an electrical cabinet installed in a new electronics store, had been specified. Shortly after the store opened, a meltdown occurred in the

In an Alabama case, an architect specified a material called "Boncoat" for the exterior of a building.[111] The material leaked, but the court held that the contractor was responsible pursuant to a specific guarantee provision in the contract.

A Tennessee court held that this indemnification provision allowed an owner to recover attorneys' fees from the contractor, even though the evidence indicated that the owner was at least partially to blame for the delays that resulted in litigation. Most courts, however, will not allow attorneys' fees to a party that is at least partly at fault.[112]

The laws of each state should be reviewed to determine the extent to which an indemnification clause like ¶ 3.18.1 is valid. In many states, such as Illinois, a party cannot be indemnified for its own negligence, and any provision that attempts to do so is void.[113]

---

cabinet where the lug assembly and the neutral conductor wire were connected to the neutral bus bar with screws, destroying the cabinet and damaging other property, including consumer electronic equipment in the store. The court found, as a matter of law, that the contractor could not be held liable for damage caused by a latent defect in a product that was specified by the project owner, citing the *Spearin* case. Because the contractor was not negligent as a matter of law, it was not obligated to indemnify on the basis of this provision.

[111] United States Fidelity & Guar. Co. v. Jacksonville, 357 So. 2d 952 (Ala. 1978). This case was poorly reasoned in rejecting the logic of two cases cited in the opinion, which involved specified materials that were defective, and in which the contractor was found not at fault. The *Jacksonville* case involved a material specified that did not perform. There was no inquiry as to whether the material was inherently defective, and thus the damage was either the fault of the architect or manufacturer, or whether it was improperly installed. The case states that there were numerous meetings to discuss the problem, suggesting a defective product. In that case, the contractor should not be at fault for simply following the contractual provisions.

[112] *E.g.,* Airline Constr., Inc. v. Barr, 807 S.W.2d 247 (Tenn. Ct. App. 1990).

[113] Corrau Constr. Co. v. Curtis, 94 Nev. 569, 584 P.2d 1303 (1978), discussed whether an indemnification clause like ¶ 3.18.1 was void in the context of an injured employee of a contractor who sued the owner. Because the owner was free of negligence, he could be indemnified by the contractor.

In Inman v. Binghamton Hous. Auth., 3 N.Y.2d 137, 143 N.E.2d 895, 164 N.Y.S.2d 699 (1957), the architect and contractor were not required to indemnify the owner when the owner was actively negligent. Further, the court construed the contract as not requiring indemnification "for an accident that occurred some six years after the job had been completed and accepted and which did not arise from any defect in workmanship or in any material used."

In Bartak v. Bell-Gallyardt & Wells, Inc., 473 F. Supp. 737 (D.S.D. 1979), *rev'd and remanded,* 629 F.2d 523 (8th Cir. 1980), this indemnification provision did not protect the architect when the architects were partially negligent in the preparation of the drawings. The architect was not indemnified, even for his partial negligence. In St. Joseph Hosp. v. Corbetta Constr. Co., 21 Ill. App. 3d 925, 316 N.E.2d 51 (1974), the architect was not indemnified under this provision when he specified a material that did not meet code for flame spread rating.

For a case that states that a party is able to indemnify itself for its own negligence, *see* Southwestern Bell Tel. Co. v. J.A. Tobin Constr., 536 S.W.2d 881 (Mo. 1976). This case, however, found against indemnification because the language was too vague.

In LTV Steel Co. v. Northwest Eng'g & Constr., Inc., 845 F. Supp. 1295 (N.D. Ind. 1994), interpreting the Indiana construction indemnity statute, one of the issues was whether this maintenance contract fell within the definition of construction. The court held that it did.

In Sherman v. DeMaria Bldg. Co., 203 Mich. App. 593, 513 N.W.2d 187 (1994), the Michigan anti-indemnification statute did not bar indemnification when the construction manager who was

This provision attempts to circumvent this by limiting the contractor's liability to situations in which the contractor and/or its subcontractors are negligent. Thus, neither the architect nor the owner is indemnified for its own negligence.

3.18.2 In claims against any person or entity indemnified under this Section 3.18 by an employee of the Contractor, a Subcontractor, anyone directly or indirectly employed by them or anyone for whose acts they may be liable, the indemnification obligation under Section 3.18.1 shall not be limited by a limitation on amount or type of damages, compensation or benefits payable by or for the Contractor or a

---

seeking indemnification was not solely negligent. The complaint by the injured worker sought damages from a number of parties. This was apparently enough to convince the court that the statute did not apply and indemnification was proper.

In Dillingham v. CH2M Hill N.W., 873 P.2d 1271 (Alaska 1994), the contract contained the following indemnification provision:

> That the OWNER agrees to limit the ENGINEER'S liability to the OWNER and to all construction Contractors, Subcontractors, material suppliers, and all others associated with the PROJECT, due to the ENGINEER'S sole negligent acts, errors, or omissions, such that the total aggregate liability of the ENGINEER to all those named shall not exceed Fifty Thousand Dollars ($50,000.00) or the ENGINEER'S total compensation for services rendered on the portion(s) of the PROJECT resulting in the negligent acts, errors, or omissions, whichever is greater.

The court found that this language violated the Alaska anti-indemnification statute and was void. *See* Glendale Constr. Servs., Inc. v. Accurate Air Sys., Inc., 902 S.W.2d 536 (Tex. Ct. App. 1995).

In National Hydro Sys. v. Mortenson, 529 N.W.2d 690 (Minn. 1995), the supreme court of Minnesota found that a provision similar to that in *Glendale* was not against public policy because the provision was equivocal. The indemnification was not unequivocal because of the language, "which arise[s] out of or result[s] from performance of the work." Presumably, the court referred to an older version of the AIA document, because it stated that the language differed from the AIA language. Current AIA language is substantially the same as the quoted language that this court construed does not violate public policy.

In USX Corp. v. Liberty Mut. Ins. Co., 269 Ill. App. 3d 233, 645 N.E.2d 396 (1994), the court held that a subcontractor's agreement for self-insurance violated the Illinois anti-indemnification statute and was unenforceable. This decision stands for the proposition that in states with strict anti-indemnification acts, contractors must obtain insurance instead of being self-insured. Corrau Constr. Co. v. Curtis, 94 Nev. 569, 584 P.2d 1303 (1978), discussed whether an indemnification clause like ¶ 3.18.1 was void in the context of an injured employee of a contractor who sued the owner. Because the owner was free of negligence, he could be indemnified by the contractor.

In Hagerman Constr., Inc. v. Copeland, 697 N.E.2d 948 (Ind. Ct. App. 1998), *further proceedings,* 1998 Ind. App. LEXIS 2046 (4th Dist. Nov. 25, 1998), the subcontractor did not have to indemnify the general contractor under this provision where the jury found the subcontractor to be 0% at fault.

In Cafferky v. John T. Callahan & Sons, Inc., 2001 WL 1772003 (Mass. Super. Oct. 30, 2001), the language "to the fullest extent permitted by law" meant that the subcontractor's obligations under this provision are limited to an injury resulting from the negligence of the subcontractor.

In Nusbaum v. City of Kansas City, 100 S.W.3d 101 (Mo. 2003), the phrase "to the extent caused" expresses an intention to limit the indemnitor's liability to the portion of fault attributed to the indemnitor.

Subcontractor under workers' compensation acts, disability benefit acts or other employee benefit acts. (**3.18.2; no change**)

Some states limit the liability of a party whose worker is injured to the amount of workers' compensation benefits. Thus, a contractor who is primarily at fault for its employee's injury may escape liability except for the relatively small cost of workers' compensation. This paragraph is a waiver of that limitation.[114]

---

[114] In Illinois, for instance, the limitation can be waived in writing. Braye v. Archer-Daniels-Midland Co., 676 N.E.2d 1295, 175 Ill. 2d 201 (1997). This provision should do this, although it may be better to include state-specific language in the supplementary conditions. In a recent Illinois case, Christy-Foltz, Inc. v. Safety Mut. Cas., 722 N.E.2d 1206 (Ill. App. Ct. 4th Dist. 2000), an employee of Christy-Foltz sustained injuries while working on a construction project in which Christy-Foltz was engaged as a subcontractor to the general contractor, LISI. The employee filed an action against LISI, alleging negligence and violation of the Illinois Structural Work Act. LISI, in turn, brought a contribution action against Christy-Foltz. Christy-Foltz raised the defense that the Illinois Workers' Compensation Act limits the recovery against it to the amount under that Act. LISI then raised the waiver of this limit found in the subcontractor agreement (the language was similar to the AIA language in ¶ 3.18.2) and the court agreed, finding that Christy-Foltz had contractually waived this limit.

Next, Christy-Foltz brought a declaratory judgment action against its own insurance company, Safety National. Apparently, Safety National had refused any coverage beyond the limits under the Workers' Compensation Act, even though it appeared now that there was substantial liability above that amount (because of the waiver discussed above). The insurance policy had a clause that excluded coverage for "any loss or claim expenses voluntarily assumed by Employer under any contract or agreement, expressed or implied." Safety National argued that, by signing the contract that contained the waiver, Christy-Foltz had voluntarily assumed a loss under contract, triggering the exclusion.

The court agreed with the insurance carrier, finding that the subcontractor had contracted to do something for which coverage was excluded by the insurance policy. The subcontractor had argued that the waiver was not a loss "voluntarily assumed" under the contract as defined in the insurance policy exclusion, but rather an agreement to waive an affirmative defense. The court found this argument unpersuasive. The court noted that an insurance policy is a contract and must be interpreted in accordance with the rules of contract construction, just like any other contract. The court found the language in the insurance policy clear and unambiguous. That language excluded from coverage any loss voluntarily assumed under contract, which is what happened here. Thus, the subcontractor found itself without coverage for any loss in excess of the statutory limits under the Workers' Compensation Act. The subcontractor would have to pay those excess amounts out of its own pockets.

Christy-Foltz was discussed, and partly overruled, in Virginia Surety v. Northern Ins. Co. of N. Y., 866 N.E.2d 149 (Ill. 2007). The case involved a declaratory judgment action over whether a subcontractor's CGL carrier owed a duty to defend and indemnify the subcontractor against the general contractor's contribution claim arising out of an injury to the subcontractor's employee. The court held that the AIA provision waived the workers compensation limitation, but does not shift liability. Rather, the employer remains liable by not asserting an affirmative defense.

In Jackson v. Northeast United, 186 Misc. 2d 259, 718 N.Y.S.2d 564 (2000), the issue was whether there was a written agreement between the parties so as to preserve a third-party claim for contribution or indemnification against an injured worker's employer. The parties had normally used an AIA 401-1987 form in their course of dealings. In this case, the contract was not signed. Because there was no dispute that the terms and conditions found in the AIA form were binding on the parties, there was a "written contract" within the meaning of the statute.

In Pena v. Chateau Woodmere Corp., 759 N.Y.S.2d 451 (App. Div. 1st Dep't 2003), the issue was whether the indemnification provision of the AIA contract was enforceable in light of the local Workers' Compensation Law, which provides that the employer's liability is limited to that law, unless a contract for the work is entered into after an accident and the contract is intended to apply retroactively. A worker employed by Bernini was injured on November 13, 1998. He sued the owner, who, in turn, sued Bernini. The contract between the owner and Bernini was signed on December 10, 1998, but the space for the date of commencement on the AIA contract indicated November 10, 1998, three days prior to the accident. Because the contract was intended to apply retroactively, the local law did not prohibit enforcement of the indemnification provisions.

In Smay v. E.R. Stuebner, Inc., 864 A.2d 1266 (Pa. Super. Ct. 2004), an employee of the general contractor was injured and filed a Workers' Compensation claim against the contractor and a lawsuit against the owner and architect. The owner and architect brought an action against the contractor based on the indemnification provision of A201-1987. The contractor sought to arbitrate these claims and the owner and architect opposed the request. The appellate court reversed the trial court and granted arbitration, finding that the standard AIA non-consolidation language had been replaced by this provision that allowed the owner to consent to the architect's participation: "No arbitration arising out of or relating to the Contract Documents shall include by consolidation or joinder or in any other manner, the Architect, the Architect's employees or consultants, except by written consent of the Owner." The court failed to explain why the architect should be bound by this provision when the architect opposed arbitration and was not a party to the agreement between the owner and contractor.

In Estate of Robert P. Willis II v. Kiferbaum Constr. Corp., 830 N.E.2d 636 (Ill. App. Ct. 1st Dist. 2005), the language of A201 constituted a waiver of Illinois Worker's Compensation Act limits, but the subcontract, which did not contain the words "shall not be limited," did not.

In Merck & Co., Inc. v. Transcontinental Cas. Co., 2006 WL 2691987 (Pa. Com. Pl. Sept. 19, 2006), the indemnification language read as follows:

"11.1 . . . [Lorenzon] hereby assumes the entire responsibility and liability for any and all damage . . . and injury . . . to all persons, whether or not employees of [Lorenzon] . . . occurring in connection with (I) the Work . . . or (iv) any occurrence which happens in or about the area where the Work is being performed by [Lorenzon]. . . .

11.2 . . . [S]hould any such damage or injury referred to in Paragraph 11.1 be sustained . . . by Owner, Architect/Engineer, or Contractor, or should any claim for such damage or injury be made or asserted against any of them, whether or not such claim is based upon Owner's Architect/Engineer's or Contractor's alleged . . . negligence . . . [Lorenzon] shall indemnify and hold harmless Owner, Architect/Engineer and Contractor . . . against any such damages injuries, and claims . . . and against . . . all other loss, cost, expense and liability, including . . . legal fees . . . and [Lorenzon] agrees to assume . . . any action at law . . . which may be brought against any Indemnitee . . . by reason of such damage, injury or claim and to pay on behalf of every Indemnitee . . . the amount of any judgment decree, award, or order that may be entered against each said Indemnitee. . . ."

This language required the subcontractor, Lorenzon, to indemnify the parties after an employee of Lorenzon slipped and fell in the work area.

In U.S. Nat' Leasing, LLC v. Frazier Constr., Inc., 2006 WL 2766076 (Cal. App. 3d Dist. Sept. 27, 2006), involving indemnification language similar to the AIA language, Donald Frazier, the owner of Frazier Construction, fell to his death on the jobsite. The issue was whether this indemnification language applied. The parties had entered into an initial contract using an AIA document. Subsequent contracts were memorialized simply by a notice to proceed. The court held that each notice incorporated the AIA provisions. The decedent's estate also argued that he was not an employee, so that this provision did not apply. The court held that the language here does not limit coverage to employees, and that Frazier was an employee at the time of the accident.

§ 4.29   Article 4: Architect

This article describes the architect's authorities and duties in administering the contract. It designates the architect as the owner's representative and sets forth the authority adhering to this role. The architect's authority to resolve claims and disputes is also specified.

§ 4.30   General: ¶ 4.1

4.1 GENERAL

4.1.1 *The Owner shall retain an architect lawfully licensed to practice architecture or an entity lawfully practicing architecture in the jurisdiction where the Project is located. That person or entity is identified as the Architect in the Agreement and is referred to throughout the Contract Documents as if singular in number.* **(4.1.1; revisions)**

The Owner must use a person or firm that is actually licensed as an architect in the state where the project is located as the "Architect." This does not necessarily mean that a representative of the Architect who goes to the jobsite and works on the project necessarily needs to be licensed so long as that person works under the direct supervision and control of a properly licensed architect. Most architectural firms employ recent graduates of architectural schools and others who prepare drawings and work on the project, but who are not licensed. As long as a properly licensed architect supervises these people, this condition is met.

In a Louisiana case, the architect became ill during construction and the owner substituted a nonarchitect in his place.[115] Because of this, the contractor was delayed. The court found that the owner had breached this provision of the contract.

At least one court has found that an architect could be liable to a contractor for negligent misrepresentation.[116] The architect was aware that certain subsurface debris was located on a site but failed to disclose this on his drawings. In reliance on the drawings, the contractor submitted a low bid. After the contractor discovered

---

[115] E.B. Ludwig Steel v. C.J. Waddell, 534 So. 2d 1364 (La. Ct. App. 1988). In Manalili v. Commercial Mowing and Grading, 442 So. 2d 411 (Fla. App. 2d Dist. 1983), the owners were entitled to submit their dispute to arbitration despite the fact that they failed to first present the dispute to an architect where no architect was involved in the project. Instead, an engineer was presented with the dispute and the engineer found that the contractor had not satisfactorily performed.

[116] Gulf Contracting v. Bibb County, 795 F.2d 980 (11th Cir. 1986). Lack of privity did not deter the court. Privity is not required to support an action for negligent misrepresentation, although liability is limited to a foreseeable person for whom the information was intended, such as the contractor. *Accord* AAA Excavating v. Francis Constr., 678 S.W.2d 889 (Mo. Ct. App. 1984). Contra, SME Indus., Inc., v. Thompson, Ventulett, Stainback & Assocs., Inc., 28 P.3d 669 (Utah 2001).

the debris, additional work was required to remove it, causing extra expense to the contractor.

This language has been used to permit an individual architect to invoke the contract's arbitration provision despite the fact that the contract was between the owner and the architect's firm and not with him individually.[117]

> **4.1.2** *Duties, responsibilities and limitations of authority of the Architect as set forth in the Contract Documents shall not be restricted, modified or extended without written consent of the Owner, Contractor and Architect. Consent shall not be unreasonably withheld.* **(4.1.2; no change)**

As with many other provisions of the agreement that require modifications to be in writing, if the parties orally modify the agreement, the courts will uphold the modification. In order for the provision of this paragraph to remain effective, the parties must insist on written confirmation of any changes. Further, the parties must not act in a manner that is inconsistent with the authority created by the contract. The contractor's consent to such a modification can be effective without a written agreement, but the better practice is to obtain the contractor's written consent in the event the architect's duties are modified.[118] Such a change may affect the contractor's method of operation on the project and, therefore, result in a change order. The owner may want to delete the word "Contractor" in the first sentence.

> **4.1.3** *If the employment of the Architect is terminated, the Owner shall employ a successor architect as to whom the Contractor has no reasonable objection and whose status under the Contract Documents shall be that of the Architect.* **(4.1.3; minor modifications)**

In a Louisiana case, the architect was effectively terminated because of an illness and the owner substituted a nonarchitect in his place.[119] This constituted a breach of the contract, because the contractor had no say in the matter and suffered damages as a result. The owner may want to delete the words "as to whom the Contractor has no reasonable objection" in order to have more flexibility.

---

[117] Giller v. Cafeteria of South Beach, Ltd., LLP, 967 So. 2d 240 (Fla. Dist. Ct. App. 2007). In that case, the owner sued a principal of the architectural firm with which it had a contract for professional malpractice. There was no contract between the owner and the individual. The architect sought to compel arbitration and the appellate court agreed that the arbitration provision of A201 applied. The court reviewed this section of A201 and found it provided the proper definition of "architect" for the owner-architect agreement. That definition included the individual architect. This applied because of the language of ¶ 10.2 in B101 references definitions of A201, and that definition included the individual architect.

[118] In Windwood Dev. LLC v. Pelley, 2003 WL 2136949 (Mich. App. June 12, 2003), the engineer was not liable in negligence when the contractor followed the engineer's verbal instructions to use inferior fill materials. This provision meant that the engineer's duties could not be expanded to make the engineer responsible for the contractor's acts or omissions absent a change in the engineer's duties per the contract.

[119] E.B. Ludwig Steel v. C.J. Waddell, 534 So. 2d 1364 (La. Ct. App. 1988).

§ 4.31   Administration of the Contract: ¶ 4.2

4.2 ADMINISTRATION OF THE CONTRACT

4.2.1 *The Architect will provide administration of the Contract as described in the Contract Documents, and will be an Owner's representative during construction, until the date the Architect issues the final Certificate For Payment. The Architect will have authority to act on behalf of the Owner only to the extent provided in the Contract Documents.* **(4.2.1; delete reference to the one-year corrections period)**

The relationship between the owner and architect has evolved over the past few decades. Years ago, the architect was an agent of the owner under standard AIA agreements and normally in full charge of the construction, with a broad range of authority over the project and able to bind the owner. In more recent years, the AIA documents have limited this agency relationship so that the current contract makes the architect a *representative* of the owner and not an agent. The discussion below cites a number of cases involving older versions of AIA contracts, where non-AIA contracts were utilized, and other unusual circumstances, so that the architect was often considered an agent of the owner.

The architect is the owner's agent (representative) during the construction phase.[120] This provision was inserted as a result of a number of court cases that imposed liability against architects in favor of other parties, such as contractors. This provision means that the architect is acting for the owner and the owner's benefit. In a Texas case[121] the court held against the contractor on the basis that the architect was the agent of the owner.

An agent[122] has different legal obligations and duties from an independent contractor.[123] An agent acts on behalf of the owner and in her interest. An agent has certain authority to act on behalf of the owner, and can bind the

---

[120] Because this agency is so limited, it is probably best to not refer to the architect as the agent of the owner, but merely the owner's "representative." This is the approach taken by the AIA. Of course, the actual authority of the architect is derived not from this document, but from the owner-architect agreement. This is not a problem if that agreement is also a standard AIA document of the same vintage and with changes that mirror those found in the A201.

[121] Bernard Johnson, Inc. v. Continental Constructors, Inc., 630 S.W.2d 365 (Tex. App. 1982).

[122] An *agency relationship* is a fiduciary relationship that results from the consent by one person, the principal, to another, the agent, for the agent to act on the principal's behalf and subject to the principal's control. Tri-City Constr. Co. v. A.C. Kirkwood & Assocs., 738 S.W.2d 925 (Mo. Ct. App. 1987). In that case, the architect was an agent of the owner, and a release that the owner and contractor had signed that included the owner's agents was sufficient to release the architect from liability to the contractor.

Under the current AIA agreements, the architect is not a fiduciary to the owner. Parkman & Weston Assocs., Ltd., v. Ebenezer African Methodist Episcopal Church, 2003 WL 22287358 (N.D. Ill. Sept. 30, 2003).

[123] An *independent contractor* is one who contracts with another to do something but is not controlled by the other nor subject to the other's control with respect to physical conduct in the performance of the undertaking. Tri-City Constr. Co. v. A.C. Kirkwood & Assocs., 738 S.W.2d 925 (Mo. Ct. App. 1987).

owner to obligations, just as if the owner itself had acted. An independent contractor, on the other hand, acts on its own behalf. The independent contractor is hired to perform a particular duty and cannot generally bind the owner to any obligations. Note that the term *independent contractor* is used here in its legal sense; that person or entity does not need to be a contractor in the construction sense. For instance, a freelance drafter working for an architect can be an independent contractor (as opposed to an employee) if she generally directs herself in the particular task for which she is hired. These terms can get murky in the construction setting, particularly when the project is not standard, such as design-build or other variation. The distinction is important, because only an agent can bind the owner. If a person directs another party to perform work, the owner will have to take responsibility if the person was an agent, but not if the person was an independent contractor. Thus, all parties to a construction project should know who is an agent and who is not, particularly in nonstandard settings.

This agency is not unlimited, however, and can be limited by contract. Paragraph 4.2.1 provides that the architect has authority only to the extent specified in the contract documents. If the architect attempts to exceed that authority, the contractor can and should seek clarification from the owner in writing. In a Texas case, a subcontractor relied on representations by the architect's inspector in performing additional work.[124] The court held that the inspector did not have apparent authority to bind the owner and found against the subcontractor.

One court has held that, as the owner's agent, the architect can direct the contractor to make changes in the schedule and scope of the work, authorize additional costs, waive bidding requirements, and waive the requirements of written change orders.[125] This is, of course, not recommended and is contrary to other provisions of A201. Under the provisions of A201, the architect cannot change the contract sum or contract time without the owner's approval. Thus, under A201, the architect is not an agent in the usual, legal sense, primarily because the architect has no (or extremely limited) authority to bind the owner to changes to the contract documents.

Other courts have held that even under prior versions of A201 that gave the architect general supervisory authority, the architect was not the owner's agent for all purposes.[126] Thus, the architect had only the authority it was given in the

---

[124] Marble Falls Hous. Auth. v. McKinley, 474 S.W.2d 292 (Tex. 1971). In another Texas case, Alamo Cmty. Coll. Dist. v. Browning Constr. Co., 131 S.W.3d 146 (Tex. App. 2004), the court held that the architect was the owner's agent, based on various authorized duties of the architect identified in the contracts.

In Hunts Point Multi-Service Ctr., Inc., v. Terra Firma Constr. Mgmt., 2003 (NY Slip Op 51280U (N.Y. Misc. 2003), the contractor argued that the architect waived certain liquidated damages claims. The court held that the architect was not an agent of the owner for all purposes. The certification of final payment by the architect did not constitute a waiver by the owner.

[125] Love v. Double "AA" Constructors, Inc., 117 Ariz. 41, 570 P.2d 812 (Ct. App. 1977). This was a unique case in which the owner was unavailable during construction. The architect authorized changes based on an extremely tight time schedule.

[126] Kirk Reid Co. v. Fine, 205 Va. 778, 139 S.E.2d 829 (1965).

owner-architect and owner-contractor contracts, which did not include authority to make alterations in the plans and specifications.[127]

In one case involving an architect who wrote letters to the contractor's surety stating that the contractor was negligent in performing the work, the architect was not liable for defamation.[128]

As the owner's agent, he was privileged to make such statements if they were not made in bad faith and were within his duty to the owner. An architect also has a conditional privilege with regard to the owner's contract with the general contractor.[129] This privilege gives the architect the ability to advise the owner about the contractor without fear of lawsuits by the contractor, but only so long as the architect gives that advice in good faith and not with the intent to harm the contractor. This privilege is available to protect the architect even under current AIA agreements where the architect is something less than a full agent, but with some responsibility to help protect the owner. In order to properly accomplish that, the architect must be free to inquire into what the contractor and others are doing and, if necessary, interfere with such contracts in an effort to assist the owner.

The architect has a duty to the owner during the construction phase to report any serious problem with the design before construction has been completed if the architect reasonably should know of the problem.[130]

---

[127] *Id.;* Norair Eng'g v. St. Joseph's Hosp., 147 Ga. App. 595, 249 S.E.2d 642 (1978) (*citing* Mallard, Stacy & Co. v. Moody & Brewster, 105 Ga. 104, 31 S.E. 45 (1898), and Cannon v. Hunt, 113 Ga. 501, 38 S.E. 983 (1901)).

[128] Alfred A. Altimont v. Chatelain, Samperton, 374 A.2d 284 (D.C. 1977).

In Quality Granite Constr. Co. v. Hurst-Rosche Eng'rs, Inc., 632 N.E.2d 1139 (Ill. App. Ct. 1994), the general contractor brought an action against the engineer for libel. The engineer had certified the construction as 100% complete and satisfactory. Some months later, when the contractor was attempting to obtain additional compensation under its contract, the engineer sent a letter stating that the contractor "may be considered in default" because of "the contractor's failure to complete the project in a timely manner, substandard workmanship, reluctance to complete punch list items and inability to interpret the contract documents, plans and specifications as bid." A jury verdict in favor of the contractor was affirmed.

In another defamation case, Metzler Contracting Co., LLC v. Stephens, 2007 WL 1977732 (D. Haw., July 3, 2007), the contractor filed suit against the owner, claiming defamation. The owner demanded arbitration and the contractor argued that defamation was not subject to arbitration. The court ordered arbitration of the defamation dispute.

[129] Certified Mech. Contractors, Inc. v. Wight & Co., 162 Ill. App. 3d 391, 515 N.E.2d 1047 (1987) (*citing* Santucci Constr. Co. v. Baxter & Woodman, Inc., 151 Ill. App. 3d 547, 502 N.E.2d 1134 (1986); Waldinger Corp. v. CRS Group Eng'rs, Inc., 775 F.2d 781 (7th Cir. 1985); George A. Fuller Co. v. Chicago College of Osteopathic Med., 719 F.2d 1326 (7th Cir. 1983)) ("It has been held that an architect or engineer has a conditional privilege to interfere with the construction contract of its principal."). This must not be done in bad faith: "If an architect induces a breach of contract, not to further its principal's best interest, but with the intent to harm the other party to its principal's contract or to further its personal goals, the architect is liable for tortious interference with the contract." *Id.* at 1054.

[130] Comptroller v. King, 217 Va. 751, 232 S.E.2d 895 (1977). The architect should have discovered that the plans were faulty in that they did not provide for proper drainage of water in a project with limestone and flagstone exterior areas. After completion, these areas deteriorated because of

If the standard owner-architect agreement is modified to add or delete duties of the architect, A201 should be amended to alert the contractor of the architect's actual duties on the project (**see ¶** 4.1.2). A general contractor generally owes no fiduciary duty to the owner.[131]

> *4.2.2 The Architect will visit the site at intervals appropriate to the stage of construction, or as otherwise agreed with the Owner, to become generally familiar with the progress and quality of the portion of the Work completed, and to determine in general if the Work observed is being performed in a manner indicating that the Work, when fully completed, will be in accordance with the Contract Documents. However, the Architect will not be required to make exhaustive or continuous on-site inspections to check the quality or quantity of the Work. The Architect will not have control over, charge of, or responsibility for, the construction means, methods, techniques, sequences or procedures, or for the safety precautions and programs in connection with the Work, since these are solely the Contractor's rights and responsibilities under the Contract Documents, except as provided in Section 3.3.1.* (**4.2.2; modifications**)

This paragraph provides that the architect will review completed work as it is being constructed and will determine whether the work will conform to the contract documents when fully completed. This is a further attempt to distance the architect from control over the work and from any liability regarding safety on the job. The owner-architect agreement can be modified to provide for specific numbers of visits, but that is not necessary with the A201 form.

Paragraph 4.2.2 does not require the architect to observe the methods of construction. The architect's only duty is to determine if the completed work complies with the construction documents, and to determine if the uncompleted work, when fully completed, will conform to the contract documents. Although it may appear that this paragraph requires the architect to observe the means and methods of the construction in order to determine whether the completed work will conform, the paragraph suggests no such thing. Although the architect may, on occasion, observe the contractor in the performance of the work and will, therefore, observe the means and methods, there is no requirement and no duty to make such observations. The intent is clearly to avoid any duty to observe means and methods of

---

standing water. Although the statute of limitations had run for improperly preparing the drawings, it had not run for negligence in observing the construction. The owner can reasonably expect the architect to disclose defects observed during construction, or that reasonably should be observed, whether caused by errors in design or deviations from plans and specifications.

[131] In Eastover Ridge v. Metric Constructors, 139 N.C. App. 360, 533 S.E.2d 827 (2000), the general conditions contained the following language: "The Contractor accepts the relationship of trust and confidence established by this Agreement and covenants with the Owner to cooperate with the architect and utilize the Contractor's best skill, efforts and judgment in furthering the interests of the Owner. . . ." Based on Paragraphs 4.2.1, 4.2.2, 4.2.5, 4.2.6, 4.2.9, and 4.2.11 of A201-1987, the court held that, as a matter of law, the architect's constant, close involvement in the project as set forth in the AIA language belies any claim that a fiduciary relationship existed between the contractor and owner.

construction. Thus, the architect will not be required to report any irregular means and methods of construction, unless it is clear that the end product will deviate from the contract documents.

Paragraph 4.2.2 has been read as requiring only general supervision as opposed to the more detailed inspection services of a clerk of the works.[132] This is a factor in determining whether the architect "is in charge of the work" for purposes of determining liability to injured workers under statutes like the Illinois Structural Work Act.[133] The contract must be examined in order to answer the question of whether an architect has breached a duty to supervise.[134] Note that the AIA contracts do not state any duty to supervise by the architect.[135] The architect has no liability to a contractor's surety when the contractor fails to perform.[136]

---

[132] Moundsview Indep. Sch. Dist. No. 621 v. Buetow & Assocs., 253 N.W.2d 836 (Minn. 1977). In Central Sch. Dist. No. 2 v. Flintkote Co., 56 A.D.2d 642, 391 N.Y.S.2d 887, 888 (1977), the owner employed a clerk of the works. Under this provision, the architect could not rely solely on the assumption that the clerk of the works was properly inspecting the work of the roofing subcontractor.

If an architect can escape the obligation and supervisory duties it contracted to perform merely by accepting, at face value and without verification, the approval of the clerk of the works as to the progress of the work, the owner would be deprived of the professional judgment it had the right to expect. The owner's retainer of a clerk of the works for full-time, on-site services constituted a protection that is an addition to and not a substitute for the contractual and professional obligations of the architect.

[133] Ill. Rev. Stat. ch. 48, para. 60 (repealed in 1995). *See* Fruzyna v. Walter C. Carlson Assocs., Inc., 78 Ill. App. 3d 1050, 398 N.E.2d 60 (1979). In Jaroszewicz v. Facilities Dev. Corp., 115 A.D.2d 159, 495 N.Y.S.2d 498 (1985), the architect was responsible for providing "supervision and inspection" and employed a clerk of the works. There was no liability by the architect when a worker was electrocuted because there was no active malfeasance or right to control the manner in which the construction was performed.

See also Diomar v. Landmark Assocs., 81 Ill. App. 3d 1135, 401 N.E.2d 1287 (1980) (based on ¶ 4.2.2 and other provisions of newer versions of AIA documents, architect found not to be in charge of the work for Structural Work Act purposes).

[134] Moundsview Indep. Sch. Dist. No. 621 v. Buetow, 253 N.W.2d 836 (Minn. 1977). In Slifer v. Wheeler & Lewis, 39 Colo. App. 269, 567 P.2d 388 (1977), the architect was liable for the injuries of workers because his contract required supervision. In Great N. Ins. Co. v. RLJ Plumbing & Heating, Inc., 2006 WL 1526066 (D. Conn. May 25, 2006), frozen water pipes burst in a newly constructed house. In finding in favor of the architect, the court reviewed these contract provisions and stated that the architect "had a duty to inspect the construction to ensure that it conformed to the contract documents but was not charged with inspecting the workmanship of the general contractor or its subcontractors to ensure that the work was performed according to specific standards and municipal codes. Moreover, responsibility for the quality of the workmanship clearly fell on the general contractor. There was, therefore, no duty on [the architect] to inspect the insulation of the plumbing in the second floor bathroom. . . ."

[135] South Burlington Sch. Dist. v. Calcagni-Frazier-Zajchowski Architects, Inc., 138 Vt. 33, 410 A.2d 1359 (1980) (in school district's suit for leaking roof brought against architect and others, court held architect did not have duty to supervise project construction).

[136] J&J Elec., Inc. v. Gilbert H. Moen Co., 9 Wash. App. 954, 516 P.2d 217 (1974). *But see* Westerhold v. Carroll, 419 S.W.2d 73 (Mo. 1967) (architect owes duty to surety to correctly certify applications for payment). Following *Westerhold,* the court in Aetna Ins. Co. v. Hellmuth, Obata & Kassabaum, 392 F.2d 472 (8th Cir. 1968), allowed a surety to maintain an action against an architect for negligent supervision when the architect is obligated by the contract to supervise the construction. The AIA contracts do not provide for such supervision.

Other courts have stated that the right to stop the work under a general supervision or inspection clause is not the same as a duty when that duty is not contemplated by the contract or the parties.[137]

If the architect fails to guard the owner against defects and deficiencies, it may
be liable to the owner for damages,[138] particularly if the architect knows that the

---

[137] Reber v. Chandler High Sch. Dist. No. 202, 13 Ariz. App. 133, 474 P.2d 852 (1970). In Wheeler &
Lewis v. Slifer, 195 Colo. 291, 577 P.2d 1092 (1978), an architect designed a high school pursuant
to a contract. A worker was injured when a section of the roof collapsed because of insufficient
bracing and shoring of the roof. The worker sued the architect, contending that a contractual
provision permitting the architect to stop the work imposed a duty on the architect "to see that
reasonable precautions were taken in protecting the workmen on the site from unsafe conditions."
The appellate court held for the worker and the Colorado Supreme Court reversed, finding that
there was no duty to stop the work for the safety of the workers. The contract contemplated that
the contractor would be in charge of safety. The power to stop the work was to ensure that the
work conformed to the contract documents, not to ensure safety. The court specifically disagreed
with Miller v. DeWitt, 37 Ill. 2d 273, 226 N.E.2d 630 (1967). *See* Walker v. Wittenberg, Delony &
Davidson, Inc., 241 Ark. 525, 412 S.W.2d 621 (1966). In Emberton v. State Farm Mut. Auto. Ins.
Co., 44 Ill. App. 3d 839, 358 N.E.2d 1254 (1976), *rev'd*, 71 Ill. 2d 111, 373 N.E.2d 1348 (1978),
further proceedings, 85 Ill. App. 3d 247, 406 N.E.2d 219 (1980), the court found the architect in
charge of the work for Structural Work Act purposes. This case involved an earlier version of the
AIA documents in which the architect had authority to stop the work.

[138] Campbell County Bd. of Educ. v. Brownlee-Kesterson, 677 S.W.2d 457 (Tenn. Ct. App. 1984). In
Roland A. Wilson v. Forty-O-Four Grand Corp., 246 N.W.2d 922, 924-25 (Iowa 1976), the
architect discovered that the windows of a building leaked before final completion. It was agreed
that the contractor was to caulk all the windows. The architect then made a final inspection but did
not test the windows, and he notified the owner that the construction work passed the final
inspection. The owner paid the contractor and later discovered that the windows still leaked.
The owner refused to make a final payment to the architect, and the architect sued to collect his
fees. The court found that the architect should not have approved final payment to the contractor
without a water test on the windows to see if the leaking problem had been corrected by caulking.
The court stated: "This contract imposed two principal duties on plaintiff. One was to prepare
plans and specifications. The other was to see that the plans and specifications were carried out by
the contractor to achieve the result planned for."

In Westerhold v. Carroll, 419 S.W.2d 73 (Mo. 1967), the architect was contractually obligated
to inspect and supervise performance of the work. The architect improperly certified work,
resulting in overpayment to the contractor. The contractor defaulted, and the surety was required
to complete the project, with insufficient funds left to complete the work. The architect argued
that his negligence was not the proximate cause of the loss, but the default of the contractor was an
intervening cause. The court stated that the purpose of the inspection and certification was to
protect the owner and surety against payments to the contractor for work not performed and
materials not delivered in case of default.

In Lee County v. Southern Water Contractors, Inc., 298 So. 2d 518 (Fla. Dist. Ct. App. 1974),
the court found that an engineer had improperly supervised and inspected the laying of a pipeline.
The pipeline was later damaged because it was covered by only two feet of backfill instead of six.
The owner claimed against the supervising engineer for the resulting damages. Although the
engineer is not a guarantor of the contractor's work, if he fails to use reasonable care to prevent
material deviations from the plans and specifications and to prevent substandard workmanship, he
will be liable to the owner for the defects which could have been eliminated if he had properly
performed his work.

construction is not according to the contract documents and fails to notify the owner of the deficiency.[139] Note that under ¶ 12.3.1 of A201, only the owner,

---

In Shepard v. City of Palatka, 414 So. 2d 1077, 1078-79 (Fla. Dist. Ct. App. 1981), the contractor had used the wrong type of wallboard and the veneer plaster soon started to peel. The appellate court held for the architect on the basis of the AIA provision. It was clear that the architect had made periodic inspections of the site:

> In a proper factual situation, where it is demonstrated that the architect ignored his contractual duty to make periodic visits to the site, liability could possibly lie regardless of such exonerating language. Such was not the case here. . . . While he admitted that he had not discovered the misuse of the wallboard, the contract clearly protected him because it imposed no duty upon him to discover the omission of the contractor and clearly absolved him of liability if there were, in fact, such omissions.

See also First Nat'l Bank v. Cann, 503 F. Supp. 419 (N.D. Ohio 1980); Skidmore, Owings & Merrill v. Connecticut Gen. Life Ins. Co., 25 Conn. Supp. 76, 197 A.2d 83 (1963); Willner v. Woodward, 107 Ga. App. 876, 132 S.E.2d 132 (1959).

In Gables CVF, Inc. v. Bahr, Vermeer & Haecker Architect, Ltd., 244 Neb. 346, 506 N.W.2d 706, 709 (1993), the architect agreed to provide a set of "builders plans," an abbreviated set of drawings and specifications that cost the owner less than a full set of documents. The architect also agreed to be paid $125 per trip for "observation of construction" of the project. At substantial completion, the architect certified that the project was constructed "generally in accordance with the intent of the drawings and specifications." However, there were, in fact, some deviations, including different materials used in the decks and balconies of the building, lack of a required waterproof membrane, and studs spaced at 24 rather than 16 inches on center. The court found that the architect was not entitled to summary judgment in his favor and might be liable to the owner for failing to detect these problems, despite the architect's argument that it should not be liable for damages resulting from unreported deviations from the plans. The issue in such cases generally becomes one of whether the architect could reasonably have discovered the defects. If he could have done so, he has a duty to report this to the owner.

[139] In Brown v. McBro Planning & Dev. Co., 660 F. Supp. 1333 (D. V.I. 1987), the court held that the owner's acceptance of the project did not shift liability when a hospital technician injured his knee in a fall on the hospital floor. The owner's acceptance of full responsibility and control of the hospital was not a superseding cause of the injury. The architect was liable because he was aware of the defect early on and failed to order the contractor to correct the defect and continued to certify payments.

In Dan Cowling & Assocs. v. Board of Educ., 273 Ark. 214, 618 S.W.2d 158 (1981), the court held that the architect breached his duty to the owner by approving a wall that was in obviously defective condition.

In City of Houma v. Mun. & Indus. Pipe Serv., Inc., 884 F.2d 886 (5th Cir. 1989), the engineer was found liable to the owner for failing to detect incomplete work. The engineer had full-time personnel on the project, although they were poorly trained and "employed no reasonable, analytical method calculated to spot performance deficiencies." The engineer was not, however, liable for the contractor's criminal fraud.

In Board of Educ. v. Sargent, Webster, 146 A.D.2d 190, 539 N.Y.S.2d 814 (1989), the architect was liable because he knew that the roofing system was not installed according to the specifications. The court stated:

> When, as a result of periodic inspection, an architect discovers defects in the progress of the work which the owner, if notified, could have taken steps to ameliorate, the imposition of liability upon the architect for failure to notify would be based on a breach of his own contractual duty, and not as a guarantor of the contractor's performance.

and not the architect, can accept nonconforming work. The architect, of course, will not be able to observe all of the work at all times, particularly on projects where the architect comes to the site once a week or once a month. On such site visits, the architect is primarily looking to see how much construction has taken place since the last visit in order to certify payment to the contractor. This usually leaves little time to actually observe the work in any detail. The owner should, therefore, expect that the architect will not be able to observe most of the defects that may have been created by the contractors, particularly those that may have been covered up before the architect's visit. If an owner wants a greater degree of site observation by the architect, the owner should hire the architect to provide one or more full time field personnel, who will have a better chance of detecting construction defects.

It is a useful practice for the architect to prepare reports of each site visit, indicating the date and time of the visit, who was observed to be on the project, what work was being performed at the time, and any comments as to the quality of completed work. Comments should not be made as to the contractor's methods of construction or equipment. These reports should be forwarded to the owner and contractor.

4.2.3 On the basis of the site visits, the Architect will keep the Owner reasonably informed about the progress and quality of the portion of the Work completed, and report to the Owner (1) known deviations from the Contract Documents and from the most recent construction schedule submitted by the Contractor, and (2) defects and deficiencies observed in the Work. The Architect will not be responsible for the Contractor's failure to perform the Work in accordance with the requirements of the Contract Documents. The Architect will not have control over or charge of and will not be responsible for acts or omissions of the Contractor, Subcontractors, or their agents or employees, or any other persons or entities performing portions of the Work. (**4.2.3; added first sentence**)

These last two paragraphs are among the most important in the General Conditions for the protection of the architect. They reinforce the concept that the contractor is solely responsible for construction means and methods and for safety.

In U.R.S. Co. v. Gulfport-Biloxi Airport Auth., 544 So. 2d 824 (Miss. 1989), the architect was liable when it failed to notify the owner that the roof was defective. The architect had employed a full-time field representative, but that representative was not competent as an inspector of the roof work. Perhaps most damaging was the fact that the improper roofing work was pointed out to the field representative and to the architect's office on a number of occasions. The architect certified the work anyway. The architect was also found liable to the surety because of improper certifications for payment.

In Watson, Watson, Rutland v. Board of Educ., 559 So. 2d 168 (Ala. 1990), the court stated:

We conclude that, although the Architect had a duty under the contract to inspect, exhaustive, continuous on-site inspections were not required. We also hold, however that an architect has a legal duty, under such an agreement, to notify the owner of a known defect. Furthermore, an architect cannot close his eyes on the construction site and refuse to engage in any inspection procedure whatsoever and then disclaim liability for construction defects that even the most perfunctory monitoring could have prevented.

The contract provides that the architect will not have any control of or charge over the work. In many cases involving injuries to workers, this language has been held to exculpate the architect from injuries to such third parties.[140]

An Illinois case[141] involved a former mental patient who visited the construction site and made his way to the sixth floor, from which he either fell or jumped to his death. His wife sued the hospital, the architect, and several contractors under the Illinois Structural Work Act.[142] The court found that the architect had no duty to this person, his duties being defined by the contract. In cases involving damage to property, this clause has relieved the architect from liability for a contractor's negligence.[143]

---

[140] Jaroszewicz v. Facilities Dev. Corp., 115 A.D.2d 159, 495 N.Y.S.2d 498 (1985). The plaintiff filed suit against the architect under common law negligence theory for the death of a maintenance worker after completion of construction. Although the architect had contracted to provide "supervision and inspection" of the project and had employed a clerk of the works, there was no liability to third parties absent proof of active malfeasance that contributed to the accident. *See* Wheeler & Lewis v. Slifer, 195 Colo. 291, 577 P.2d 1092 (1978) (no duty on part of architect toward injured worker, even though architect had right to stop the work); Diomar v. Landmark Assocs., 81 Ill. App. 3d 1135, 401 N.E.2d 1287 (1980); Fruzyna v. Walter C. Carlson Assocs., 78 Ill. App. 3d 1050, 398 N.E.2d 60 (1979); Dillard v. Shaughnessy, Fickel & Scott Architects, 864 S.W.2d 368 (Mo. Ct. App. 1993); Fox v. Jenny Eng'g Corp., 122 A.D.2d 532, 505 N.Y.S.2d 270 (1986); Kerr v. Rochester Gas & Elec. Corp., 113 A.D.2d 412, 496 N.Y.S.2d 880 (1985) (engineer hired to supervise the work was not liable to injured worker and was not agent of owner).

In Hanna v. Huer, Johns, Neel, Rivers & Webb, 233 Kan. 206, 662 P.2d 243 (1983), the court held that the architect was not liable to a subcontractor's injured employee because of the general contractor's numerous safety violations. The employees argued that safety precautions came under the definition of Work in ¶ 1.1.3. The court rejected that argument, finding it contrary to the clear provisions of A201. Of interest is the fact that the architect had only an oral agreement with the owner. However, it was clear that A201 specified the duties of the architect.

See also Case v. Midwest Mech. Contractors, 876 S.W.2d 51, 53 (Mo. Ct. App. 1994) (architect not liable to injured worker even though architect had duty to review safety program of contractors; "It is a far cry from being responsible to review the safety programs of others to specifically assuming responsibility for safety precautions to be taken."); Romero v. Parkhill, Smith & Cooper, Inc., 881 S.W.2d 522 (Tex. App. 1994).

[141] Kelly v. Northwest Cmty. Hosp., 66 Ill. App. 3d 679, 384 N.E.2d 102 (1978).

[142] Ill. Rev. Stat. ch. 48, para. 60 (repealed in Feb. 1995).

[143] Jewish Bd. of Guardians v. Grumman Allied Indus., 96 A.D.2d 465, 464 N.Y.S.2d 778 (1983) (general contractor had duty to protect prefab units that were damaged by a severe rainstorm on-site during construction, and architect had no duty to supervise and no responsibility for this damage); Moundsview Indep. Sch. Dist. No. 621 v. Buetow & Assocs., 253 N.W.2d 836 (Minn. 1977). An architect's drawings and specifications required that wooden plates be fastened to the concrete walls of the building with washers and nuts attached to one-half-inch studs. During a severe windstorm, a portion of the roof blew off and it was discovered that the washers and nuts were missing. The architect had made 90 visits to the construction site but had not observed the omission of the washers and nuts. The court stated:

it is apparent that by the plain language of the contract an architect is exculpated from any liability occasioned by the acts or omissions of a contractor. The language of the contract is unambiguous. The failure of a contractor to follow the plans and specifications caused the mishap. By virtue of the aforementioned contractual provisions, Buetow is absolved from

In agreements in which the architect undertakes to supervise the construction, the architect may be liable if the construction is improperly performed.[144] Other courts have found that an architect had breached an implied warranty to supervise the construction.[145]

Note that, under the AIA documents, architects do not supervise construction. The contractor is charged with supervision of construction. Supervision connotes control over the activities of the workers and subcontractors. Architects have no such control. In fact, just the opposite is true. The AIA has taken great pains to state in various documents that the architect has no control over construction means, methods, and techniques. Of course, if non-AIA documents are used, the architect might undertake to supervise construction, but that would be counter to the custom and practice in the industry. While architects do occasionally go to the jobsite, such site visits are primarily for the purpose of observing the work that has been constructed and the amount of materials on site so as to properly certify payment to the contractor. These visits are not for the purpose of observing methods of construction or to supervise anyone.

One court has stated that "supervision" does not connote a higher standard of duty than "inspection."[146] In that case, the specifications prepared by the architect for a factory required the architect to oversee the placing of backfill against the foundation walls by calling for the architect to inspect and approve a number of items, including removal of debris, reinforcing splices, all areas to receive

---

any liability, as a matter of law, for a contractor's failure to fasten the roof to the building with washers and nuts.

See also, Hernandez v. Racanelli Constr. Co., Inc., 33 A.D.3d 536, 823 N.Y.S.2d 377 (App. Div. 1st Dep't 2006), where the architect was completely relieved of responsibility for construction defects and safety precautions under its contract with the owner, thus avoiding responsibility to the demolition contractor where a plywood fence collapsed during demolition, injuring passers-by.

In Weill Constr. Co. v. Thibodeaux, 491 So. 2d 166 (La. Ct. App. 1986), a skating rink was damaged by water that entered the foundation after construction. The architect was found to have a limited duty that did not include construction supervision based on a contractual provision modeled after the standard AIA clause.

[144] Corbetta Constr. v. Lake County Pub. Bldg. Comm'n, 64 Ill. App. 3d 313, 381 N.E.2d 758 (1978). This case also stated that an owner may have a cause of action against both an architect and a contractor if both faulty construction and faulty design appear in the same project. The contractor cannot interpose the defense of improper design if the contractor does not follow the drawings of specifications.

[145] *See, e.g.,* Tamarac Dev. v. Delamater, Freund, 234 Kan. 618, 675 P.2d 361 (1984). The developer alleged that the architect orally agreed to supervise the grading construction and to check the grades of a trailer park on completion. The grades were later found to be improper, causing drainage problems. A written contract also existed, but no breach of that was alleged. The court held that an action in express contract, implied warranty, or negligence could be maintained. The court stated that although the work of some professionals, such as lawyers or doctors, is not an exact science, that of an architect and engineer is, giving rise to an implied warranty. "A person who contracts with an architect or engineer for a building of a certain size and elevation has a right to expect an exact result." Although this is true in the context of supervising grading, it is not true of many architectural or engineering projects.

[146] Dickerson Constr. Co. v. Process Eng'g, 341 So. 2d 646 (Miss. 1977).

concrete, and pipe lines. A soils report had indicated a layer of expansive clay from five to seven feet below grade. The engineer had recommended that surface grading be kept to a minimum; that piers be sunk to at least 20 feet below grade; that voids filled with vermiculite be placed below the grade beams to allow for swelling; that soil under the floor slab be of a select clay; and that drainage on the east side of the building provide for rapid drainage. Soon after the building was occupied, severe cracking of the slab started; the roof leaked, walls cracked, and other problems developed, to the point that the building became unsafe. It was found that the water table had risen and the clay had expanded, causing the cracking. Much testimony was offered as to the reason for the water. The architect contended that the owner allowed chemicals to corrode the plastic piping under the slab, causing leaking that saturated the soil. The owner offered evidence that the recommendations of the soils engineer had not been followed. The architect was held liable because he did not inspect the construction properly, and the contractor was found liable for not following the plans and specifications.

The court stated "the requirements that the [architect] be present at certain stages of construction and to certify progress payments and to make a final inspection and certification means that the [architect] will use reasonable care to see that the Contractor constructs the building in substantial compliance with the plans and specifications."[147] The key here was probably the fact that the specifications were quite specific about the approval required by the architect at certain points of the construction of the foundation.

One court has construed the general duty of the architect under an earlier version of ¶ 4.2.3 in this way:

> As we view the matter, the primary object of this provision was to impose the duty or obligation on the architects to insure to the owner that before final acceptance of the work the building would be completed in accordance with the plans and specifications; and to insure this result the architects were to make "frequent visits to the work

---

[147] *Id.* at 652. *Accord* Skidmore, Owings & Merrill v. Connecticut Gen. Life Ins. Co., 25 Conn. Supp. 76, 197 A.2d 83 (1963); Lee County v. Southern Water Contractors, Inc., 298 So. 2d 518 (Fla. Dist. Ct. App. 1974); Willner v. Woodward, 107 Ga. App. 876, 132 S.E.2d 132 (1959); Roland A. Wilson v. Forty-O-Four Grand Corp., 246 N.W.2d 922 (Iowa 1976); Westerhold v. Carroll, 419 S.W.2d 73 (Mo. 1967).

The court in Gables CVF, Inc. v. Bahr, Vermeer & Haecker Architect, Ltd., 244 Neb. 346, 506 N.W.2d 706 (1993), found an ambiguity existed as to what the architect's duties were in observing the construction and determined that an architect was not entitled to summary judgment. One of the partners of the owner was also the general contractor. All of the provisions in the construction phase of the 1977 version of B141 except the comparable version of this one were stricken. The architect was to be paid $125 per trip for observing construction, and the owner alleged that the architect failed to notify the owner of defects in the work. The court stated that the language of this provision provides that

the architect is not the insurer or guarantor of work completed by the contractor. However, such language does not absolve the architect from liability for a breach of the architect's duty, if one exists, to inform the owner of deviations from the building plans when the architect has agreed to make periodic observations.

site" during the progress of the work. Under the contract they as architects had no duty to supervise the contractor's method of performing his contract unless such power was provided for in the specifications. Their duty to the owner was to see that before final acceptance of the work the plans and specifications had been complied with, that proper materials had been used, and generally that the owner secured the building it had contracted for.[148]

A New York court held that when an engineer had contractual authority to stop work if the contractor failed to correct conditions that were unsafe for workers, the engineer had no duty to the workers.[149] The engineer in this case had no authority to supervise or control the injured worker or direct construction procedures or safety measures. The worker was not permitted to recover from the engineer on a negligence theory. A federal court has held that the construction site was not a "place of employment" which the structural engineer had a duty to protect under OSHA.[150]

One case has held that if the design professional allows the contractor to deviate from the drawings and specifications and that deviation results in injury to third parties, ¶ 4.2.3 will not protect the designer.[151] Such a holding, however, is contrary to the terms of A201, since such a deviation cannot be authorized by the architect absent the owner's approval via a change order.

4.2.4 Communications Facilitating Contract Administration. Except as otherwise provided in the Contract Documents or when direct communications have been specially authorized, the Owner and Contractor shall endeavor to communicate with each other through the Architect about matters arising out of or relating to the Contract. Communications by and with the Architect's consultants shall be through the Architect. Communications by and with Subcontractors and material suppliers shall be through the Contractor. Communications by and with separate contractors shall be through the Owner. **(4.2.4; no change)**

This paragraph provides for orderly communications on the project. Unless the architect is kept informed of such communications, the probability of errors

---

[148] Day v. National U.S. Radiator Corp., 241 La. 288, 128 So. 2d 660, 666 (1961).

[149] Fox v. Jenny Eng'g Corp., 122 A.D.2d 532, 505 N.Y.S.2d 270 (1986). *See also* Jaroszewicz v. Facilities Dev. Corp., 115 A.D.2d 159, 495 N.Y.S.2d 498 (1985) (worker was electrocuted while working on a parking lot lighting system when a short circuit in a line was probably caused by post-installation excavation work; architect had no duty or right to control the manner in which the work was performed, and there was no proof that the architect engaged in any active malfeasance that contributed to the accident). New York also will not impose liability on an engineer who is engaged to assure compliance with construction plans and specifications for an injury sustained by a worker, unless the engineer commits an affirmative act or negligence or such liability is imposed by a clear contractual provision. Brooks v. A. Gatty Serv. Co., 27 A.D.2d 553, 511 N.Y.S.2d 642 (1987).

[150] Reich v. Simpson, Gumpertz & Heger, Inc., 3 F.3d 1 (1st Cir. 1993).

[151] Campbell v. The Daimler Group, Inc., 115 Ohio App. 3d 783, 686 N.E.2d 337 (1996) (engineer who approved substitution of expansion anchor bolts in place of originally specified imbedded anchor bolts was not protected by this provision).

and increased costs is heightened. The paragraph also recognizes the realities of the construction site by allowing communications to go through those parties most responsible for them, such as contractor-subcontractor communications. One exception to this flow of information is found in ¶ 9.6.3, relating to the furnishing of information by the architect to the subcontractor related to payments.

> *4.2.5 Based on the Architect's evaluations of the Contractor's Applications for Payment, the Architect will review and certify the amounts due the Contractor and will issue Certificates for Payment in such amounts.* **(4.2.5; no change)**

The architect has a duty to exercise reasonable care to see that the work is done in a proper manner with proper materials, and to certify payment to the contractor only upon verification that the contractor has performed the work as stated in the application for payment.[152] The language should be corrected to read: "Based on the Architect's evaluations of the *Work and of the* Contractor's . . ." in order to make this clear.

The architect's duties under this provision are a matter of judgment and do not constitute a guarantee.[153] The architect must be more than a rubber stamp for the contractor's applications and must take an active role in observing the quantity and quality of the work completed to date. The architect is also entitled to reasonably rely on the contractor's application for payment and is not required to independently check the information in the application. Thus, if the contractor states that certain materials have been paid for, the architect is not expected to know whether that statement is true or not. What the architect must do is to compare the application with its site observations and then certify that the application correctly shows the amount of work completed to date. Anything beyond that is beyond the architect's knowledge and responsibility. These certificates may be withdrawn at a later date if the work does not conform to the contract documents.

An interesting case concerning certification of payouts is *Northwest Bank v. Garrison.*[154] The owner decided to build a custom home and hired a contractor. The contractor convinced the owner that he could save $60,000 by not hiring an architect. All references to the term "architect" were deleted from the construction

---

[152] Skidmore, Owings & Merrill v. Connecticut Gen. Life Ins. Co., 25 Conn. Supp. 76, 197 A.2d 83 (1963); Lee County v. Southern Water Contractors, Inc., 298 So. 2d 518 (Fla. Dist. Ct. App. 1974); Willner v. Woodward, 107 Ga. App. 876, 132 S.E.2d 132 (1959); Roland A. Wilson v. Forty-O-Four Grand Corp., 246 N.W.2d 922 (Iowa 1976); Westerhold v. Carroll, 419 S.W.2d 73 (Mo. 1967).

[153] *See* Roland A. Wilson v. Forty-O-Four Grand Corp., 246 N.W.2d 922 (Iowa 1976) (court found architect had duty to inspect the work and not to certify completion of defective or incomplete work; "This obligation does not make the architect a guarantor of the contractor's work. Instead, it requires the architect to exercise reasonable care in supervision and inspection of the work to protect the owner against payment of money to the contractor for work not performed or materials not delivered."). *See also* Lee County v. Southern Water Contractors, Inc., 298 So. 2d 518 (Fla. Dist. Ct. App. 1974).

[154] 874 S.W.2d 278 (Tex. App. 1994).

contract and the term "lending institution" was substituted in its place (without the bank's approval). Apparently, the bank performed some normal lender's inspections during the first part of the construction. After problems with the construction developed, the bank refused to fund further draws. In the foreclosure action on the mortgage, the owner defended, arguing that the bank had undertaken a common law duty to perform inspections. As evidence, the owner submitted a G702 form that had been prepared by the owner and was signed by the bank in the space reserved for the architect. Because there was a dispute over this issue, summary judgment was not proper, and the case was remanded to the trial court for further evidence on the subject of whether the bank owed the owner a duty.

4.2.6 The Architect has authority to reject Work that does not conform to the Contract Documents. Whenever the Architect considers it necessary or advisable, the Architect will have authority to require inspection or testing of the Work in accordance with Sections 13.5.2 and 13.5.3, whether or not such Work is fabricated, installed or completed. However, neither this authority of the Architect nor a decision made in good faith either to exercise or not to exercise such authority shall give rise to a duty or responsibility of the Architect to the Contractor, Subcontractors, material and equipment suppliers, their agents or employees, or other persons or entities performing portions of the Work. **(4.2.6; minor revisions)**

Note that there is no time specified in this paragraph. Because the architect makes periodic visits to the site (¶ 4.2.2), it can reject nonconforming work as the project proceeds. Failure to reject work does not necessarily result in a waiver of the nonconforming work.[155]

When rejecting work, the architect should exercise caution to avoid giving the contractor any direction regarding methods of construction. The 1967 version of AIA Document B131 included the architect's authority to stop the work. This was a basis upon which many courts found the architect liable for injuries sustained by workers.[156] Under more recent versions, the architect can only reject work that does not conform to the contract documents.[157] The architect has no authority as to the method of construction, only as to the finished product. The architect also cannot accept nonconforming work but can only recommend to the owner that it be accepted.

---

[155] Stevens Constr. Corp. v. Carolina Corp., 63 Wis. 2d 342, 217 N.W.2d 291 (1974) (failure to object does not constitute an express or implied waiver if the defect is not obvious).

[156] *See, e.g.,* Emberton v. State Farm Mut. Auto. Ins. Co., 44 Ill. App. 3d 839, 358 N.E.2d 1254 (1976), *rev'd,* 71 Ill. 2d 111, 373 N.E.2d 1348 (1978), *further proceedings,* 85 Ill. App. 3d 247, 406 N.E.2d 219 (1980) (court found architect in charge of work for Structural Work Act purposes) (This case involved an earlier version of the AIA documents in which the architect had authority to stop the work.); Fruzyna v. Walter C. Carlson Assocs., 78 Ill. App. 3d 1050, 398 N.E.2d 60 (1979).

[157] *See* Diomar v. Landmark Assocs., 81 Ill. App. 3d 1135, 401 N.E.2d 1287 (1980) (based on ¶ 4.2.6 and other provisions of newer versions of AIA documents, architect was found not to be in charge of the work for Structural Work Act purposes).

Generally, the architect does not owe any contractual duties to the contractor or to others on the project.[158] However, architects have been held to owe a duty to the contractor to avoid negligently causing extra expenses for the contractor in the completion of a construction project.[159]

4.2.7 The Architect will review and approve or take other appropriate action upon the Contractor's submittals such as Shop Drawings, Product Data and Samples, but only for the limited purpose of checking for conformance with information given and the design concept expressed in the Contract Documents. The Architect's action will be taken in accordance with the submittal schedule approved by the Architect or, in the absence of an approved submittal schedule, with reasonable promptness while allowing sufficient time in the Architect's professional judgment to permit adequate review. Review of such submittals is not conducted for the purpose of determining the accuracy and completeness of other details such as dimensions and quantities, or for substantiating instructions for installation or performance of equipment or systems, all of which remain the responsibility of the Contractor as required by the Contract Documents. The Architect's review of the Contractor's submittals shall not relieve the Contractor of the obligations under Sections 3.3, 3.5 and 3.12. The Architect's review shall not constitute approval of safety precautions or, unless otherwise specifically stated by the Architect, of any construction means, methods, techniques, sequences or procedures. The Architect's approval of a specific item shall not indicate approval of an assembly of which the item is a component. **(4.2.7; requires contractor to follow submittal schedule; minor modifications)**

Shop drawings[160] are drawings, diagrams, schedules, and other data specially prepared for the work by the contractor, subcontractor, sub-subcontractor, manufacturer, supplier, or distributor that show some portion of the work. The shop drawings are not contract documents. **See ¶** 3.12.4. Normally, these are much more detailed than the architect's drawings and often provide a last check before materials are fabricated and substantial money is spent. However, not all materials or items incorporated into a project are shown on shop drawings. Some architects list those items for which shop drawings are required. It would be virtually impossible for most projects to be built without some shop drawings. The contractor takes a substantial risk if it does not submit most of these to the architect for review.

The procedure for shop drawing review is essentially the following. First, the supplier or subcontractor prepares the shop drawing based on the architect's contract documents and submits this to the contractor. Second, the contractor checks

---

[158] Chicago College of Osteopathic Med. v. George A. Fuller Co., 719 F.2d 1335 (7th Cir. 1983). *See also* AIA Document A201, ¶ 1.1.2.

[159] *See, e.g.,* Forte Bros., Inc., v. National Amusements, Inc., 525 A.2d 1301 (R.I. 1987); Detweiler Bros., Inc., v. John Graham & Co, 412 F. Supp. 416 (E.D. Wash. 1976). The cases generally turn on whether a particular state recognizes the Economic Loss Rule and if the damages to the contractor are solely economic. In states that do not recognize this Rule, contractors are generally allowed to sue architects in tort despite a lack of privity.

[160] An excellent article about shop drawings is Stubbs, *Shop Drawings: Minding Someone Else's Business,* Architecture (Jan. 1987).

this against the contract documents and against other shop drawings that affect the same portion of the work. This is then forwarded to the architect, along with notations as to any corrections required, only if such corrections are minor. If corrections are major, the shop drawing should be returned by the contractor to the subcontractor or supplier. Some architects insist that the contractor forward only an approved shop drawing or that the drawing contain a notation that the contractor has reviewed the shop drawing and has found no errors. The architect then checks this for conformance with the design intent and the contract documents, notes on each drawing the state of approval, as well as any comments or notations, and returns one or more copies to the contractor. The contractor then notes the approval status and returns one set to the supplier or subcontractor. Unless the drawing is marked "approved" or "approved as noted" or some variation of these words, the entire process is repeated with the revised shop drawings.

The architect's review of the shop drawings is "only for the limited purpose of checking for conformance with information given and the design concept expressed in the Contract Documents." Thus, the architect does not check for quantities, field dimensions, techniques of erection, and so on. The architect should refuse to approve shop drawings that contain statements to the effect that approval of the shop drawing shifts responsibility for dimensions, quantities, or other such items to the architect. Also, the architect should refuse to review shop drawings that have not been reviewed by the contractor. Even if the architect approves shop drawings, the contractor is not relieved of responsibility for errors or omissions in those shop drawings.[161] **See** ¶ 3.12.8. Some courts have held architects liable for failing to detect errors in shop drawings.[162] An architect may be liable to a

---

[161] In Acmat Corp. v. Daniel O'Connell's Sons, 17 Mass. App. Ct. 44, 455 N.E.2d 652 (1983), *appeal denied*, 459 N.E.2d 824 (1984), the contractor submitted a brochure for "Spraydon," an acoustical coating material that the subcontractor was proposing as a substitute for the originally specified material, "Cafco Soundshield." Pursuant to the contract documents, the contractor later submitted a sample of the product, which the architect rejected as not meeting certain requirements. In the resultant lawsuit, the architect's decision was upheld because the contract gave him broad rights to make final and conclusive decisions related to the work. Under the current AIA Documents, the contractor would have the right to litigation or arbitration.

In Aetna Cas. & Sur. Co. v. Leo A. Daly Co., 870 F. Supp. 925 (S.D. Iowa 1994), the court found that the architect could limit his liability contractually so long as third parties were not injured. In this case, the architect, Daly, was contractually relieved of the obligation to review the contractor's proposed changes in the details of the building. When a sprinkler pipe froze and broke, each party blamed the other. The architect's original design for the area had been revised at the suggestion of the contractor. The court held that the architect had no duty to evaluate the performance of a proposed design substitution.

In Fauss Constr., Inc. v. City of Hooper, 197 Neb. 398, 249 N.W.2d 478 (1977), the drawings required solid core doors. The contractor submitted shop drawings that showed particle core doors. The fire marshal inspected the doors and rejected them. The contractor then demanded an extra for the change. The court rejected the claim. Even though the architect had approved the shop drawing (although that may have been an oversight), the architect had no authority to change the contract documents. No change order had been requested or issued, and this was not an oral modification of the written contract.

[162] *E.g.,* Jaeger v. Henningson, Durham & Richardson, Inc., 714 F.2d 773 (8th Cir. 1983). Here, the shop drawings provided for 14-gauge steel stair pans instead of 10-gauge as shown on the

contractor for delays caused by the architect in approving shop drawings if the delays are the result of bad faith by the architect.[163]

In a Missouri case,[164] the structural engineer who reviewed shop drawings for the walkways that collapsed at the Kansas City Hyatt Hotel, killing over 100 people, asserted that the connections that failed were to be designed by the steel fabricator. The drawings failed to indicate this, and the court found the engineer responsible for the design of these connections. This case appears to make the design professional responsible for shop drawings to a greater degree than the AIA documents and custom provide. The court stated, "Custom, practice, or 'bottom line' necessity cannot alter that responsibility."[165]

If the contractor submits unwanted shop drawings, the architect can simply return them. According to ¶ 3.12.4, "Submittals that are not required by the Contract Documents may be returned by the Architect without action."

When the contractor submits alternative materials for the architect's approval, the architect must review the characteristics of the materials to verify that they comply with applicable building codes.[166]

Some contractors simply reuse the contract documents as their shop drawings. This is not a good procedure, because one of the purposes of shop drawings is to provide an additional check for accuracy and to spot things that were overlooked in the original drawings. Only when the contractor does new shop drawings can there be an independent check of the work.

It would be wise to have a meeting before the start of construction with the architect, contractor, and major subs to review the procedure for shop drawing submittals, number of copies required, turn-around time expected, and the extent of information expected on the submittals.

If the architect takes a particularly long time to process shop drawings that result from a change order, the contractor can refuse to accept the change order. Otherwise, the contractor may be barred from asking for additional time or money.[167]

4.2.8 The Architect will prepare Change Orders and Construction Change Directives, and may authorize minor changes in the Work as provided in Section 7.4. The Architect will investigate and make determinations and recommendations regarding concealed and unknown conditions as provided in Section 3.7.4. **(4.2.8; added last sentence)**

---

drawings. The architect did not detect the discrepancy and the contractor did not call it to the attention of the architect. The pan collapsed, and two workers were injured during construction. The architect was held negligent in supervision (as opposed to negligent in preparation of plans).

[163] Prichard Bros. v. Grady Co., 428 N.W.2d 391 (Minn. 1988), *further proceedings,* 436 N.W.2d 460 (Minn. Ct. App. 1989).

[164] Duncan v. Missouri Bd. for Architects, 744 S.W.2d 524 (Mo. Ct. App. 1988).

[165] *Id.* at 537.

[166] St. Joseph Hosp. v. Corbetta Constr. Co., 21 Ill. App. 3d 925, 316 N.E.2d 51 (1974).

[167] Huber, Hunt & Nichols, Inc. v. Moore, 67 Cal. App. 3d 278, 136 Cal. Rptr. 603 (1977).

It is important to obtain written verification from the contractor that the minor change will not result in any increased cost or extension of time. Note that the architect cannot approve changes involving changes in time or money without the owner's approval. **See** ¶ 7.1. Thus, if the architect approves what she believes to be minor, the contractor cannot use this approval to argue for a change involving time or money.

> *4.2.9 The Architect will conduct inspections to determine the date or dates of Substantial Completion and the date of final completion; issue Certificates of Substantial Completion pursuant to Section 9.8; receive and forward to the Owner, for the Owner's review and records, written warranties and related documents required by the Contract and assembled by the Contractor pursuant to Section 9.10; and issue a final Certificate for Payment pursuant to Section 9.10.* **(4.2.9; minor modifications)**

These two inspections are the only ones referred to in the General Conditions; all others are observations. The difference is one of thoroughness. An *inspection* connotes that the architect has done a more thorough review of the work than mere observation. The architect should be careful when using this terminology in letters and reports of the architect. The warranties required in the contract documents are forwarded to the owner for review. The architect does not furnish legal advice, so the owner must obtain its own legal advice as to whether the warranties conform to the requirements of the contract.

The architect has no authority to accept work that is incomplete or deviates from the contract documents.[168] It cannot change the specifications or interpolate something into the contract that is not justified by any fair interpretation of its terms.[169] The architect is not liable to the contractor for failing to issue a certificate of substantial completion,[170] absent bad faith.

A New York architect issued a final certificate for payment one week after notifying the general contractor that roof repairs were required because of leakage.[171] After several years, the owner sought damages from the architect and demanded arbitration with the architect based on the roof problem. The architect sought to stay the proceeding because the three-year statute of limitation had run. The court found in favor of the architect, denying the owner's argument that a longer statute should apply because issuance of the certificate amounted to fraud. There was no attempt to conceal the facts, and the leakage was not caused by the issuance of the certificate.

---

[168] Hurley v. Kiona-Benton Sch. Dist. No. 27, 124 Wash. 537, 215 P. 21 (1923) (all parties knew construction was not completed, and the architect's issuance of final certificate constituted fraud and was not binding on the owner).

[169] Northwestern Marble & Tile Co. v. Megrath, 72 Wash. 441, 130 P. 484 (1913).

[170] Blecick v. School Dist. No. 18, 2 Ariz. App. 115, 406 P.2d 750 (1965). This issue was decided on the basis of the architect's role as arbitrator.

[171] Naetzker v. Brocton Cent. Sch. Dist., 50 A.D.2d 142, 376 N.Y.S.2d 300 (1975).

4.2.10 If the Owner and Architect agree, the Architect will provide one or more project representatives to assist in carrying out the Architect's responsibilities at the site. The duties, responsibilities and limitations of authority of such project representatives shall be as set forth in an exhibit to be incorporated in the Contract Documents. **(4.2.10; no change)**

AIA Document B352 can be used when a project representative is provided. Although the owner and architect may agree to provide a full-time field representative, the architect is still only responsible for its own documents. The contractor remains liable for the end product and, if the building deviates from the plans and specifications, the contractor and not the architect would be liable.[172]

The probability of detecting deviations from the contract documents increases when the architect has one or more project representatives.

4.2.11 The Architect will interpret and decide matters concerning performance under, and requirements of, the Contract Documents on written request of either the Owner or Contractor. The Architect's response to such requests will be made in writing within any time limits agreed upon or otherwise with reasonable promptness. **(4.2.11; deleted last sentence of prior version)**

Any provision of the contract documents is subject to interpretation by the architect.[173] If a request for an interpretation is made, the parties can agree to a time limit within which the architect will make the interpretation. If no such agreement is made, the contractor may be entitled to additional time and money after a reasonable time without an interpretation by the architect, if the contractor can substantiate such additional costs or time. The 1997 version of A201 provided 15 days as such a "reasonable" time.

4.2.12 Interpretations and decisions of the Architect will be consistent with the intent of and reasonably inferable from the Contract Documents and will be in writing or in

---

[172] Cincinnati Riverfront Coliseum v. McNulty Co., 28 Ohio St. 3d 333, 504 N.E.2d 415 (1986). Deviations from the design must be material and must be the proximate cause of the damages for the architect to be absolved from liability.

The deviation must not be the fault of the designer, otherwise the designer will be held liable. In Campbell v. The Daimler Group, Inc., 686 N.E. 2d 337 (Ohio App. 10th Dist., 1996), the court distinguished the *Cincinnati Riverfront* case (where the deviations were not created by the architect) from a situation in which the engineer approved the substitution of expansion anchor bolts in place of the originally specified imbedded anchor bolts. When the structure collapsed, the substituted bolts were the suspected cause of the accident. The designer is not exculpated in that situation by virtue of the standard AIA disclaimer language in ¶ 4.2.3.

[173] In Fru-Con Constr. Co. v. Southwestern Redevelopment Corp., 908 S.W.2d 741 (Mo. Ct. App. 1995), the contract was amended to allow arbitration only for claims that were less than $200,000. The contractor submitted a claim consisting of numerous claims, and the architect determined that only one of these claims exceeded the limit. The contractor filed suit, and the owner sought to arbitrate, asking the court to stay the lawsuit pending the arbitration. The trial court declined to stay the proceeding. The appellate court reversed, saying that the architect had the power to make the initial determination under the contract. Since the architect determined that all the claims except one were subject to arbitration, the court proceeding should have been stayed.

the form of drawings. When making such interpretations and decisions, the Architect will endeavor to secure faithful performance by both Owner and Contractor, will not show partiality to either, and will not be liable for results of interpretations or decisions rendered in good faith. **(4.2.12; delete "initial" decision)**

If requested by either the owner or contractor, the architect is required to interpret the contract documents. The architect's determination will be given great weight by the courts.[174]

---

[174] Bolton Corp. v. T.A. Loving Co., 94 N.C. App. 392, 380 S.E.2d 796 (1989) ("the architect's determination is prima facie correct, and the burden is upon the other parties to show fraud or mistake.").

In Mayfair Constr. v. Waveland Assocs., 249 Ill. App. 3d 188, 619 N.E.2d 144 (1993), the contract between the owner and architect specifically excluded the provision that would have required the architect to be the judge of performance on the project and to issue decisions on disputes arising between the owner and contractor. The General Conditions, on the other hand, contained this provision. The architect pointed out to the owner that there were significant conflicts between the construction contract and the architect's contract, but these did not get resolved. When a dispute arose between the owner and contractor, the contractor attempted to submit the dispute to the architect. The owner threatened the architect with a lawsuit if the architect attempted to resolve the dispute, and the architect then refused to be involved in the dispute. The dispute was then submitted to the trial court, which found that the owner breached its agreement with the contractor by refusing to allow the architect to initially resolve the dispute. The contract between the owner and contractor required that the architect be the initial judge of performance. It made no difference to the contractor that the owner-architect agreement conflicted with these provisions.

In James A. Cummings, Inc. v. Young, 589 So. 2d 950 (Fla. 1992), the architect had authority to determine the value of work in dispute between the general contractor and the subcontractor. The court stated that "when parties to a contract agree by its express terms to be bound to the determination made by an architect, that agreement is binding upon the parties."

In Airline Constr., Inc. v. Barr, 807 S.W.2d 247 (Tenn. Ct. App. 1990), the court held that a court is not bound by the architect's determination and not limited to the architect's findings of deficiencies in the project.

In Fontaine Bros. v. City of Springfield, 35 Mass. App. Ct. 155, 617 N.E.2d 1002 (1993), the arbitration provisions had been deleted from the owner-contractor agreement. This left the architect as the sole arbiter of disputes, and his decision was final and binding. The architect in that case had ruled that the contractor was not entitled to extra compensation for certain work. The contractor brought a court action and this court held that the owner was entitled to a directed verdict based on the architect's decision.

In A. Prete & Son Constr. v. Town of Madison, No. CV 91-03103073-S, 1994 Conn. Super. LEXIS 2532 (Oct. 4, 1994), the owner failed to obtain the architect's disapproval of a gymnasium floor before rejecting the floor. This was one provision cited by the court in holding for the contractor. The owner, the court said, should have followed the procedures in the AIA documents, which required the architect to make a finding that the contractor failed to comply with the contract documents.

In Ostrow Elec. Co. v. J.L. Marshall & Sons, Inc., 798 N.E.2d 310 (Mass. App. Ct. 2003), the court cited *Fontaine* for the proposition that contract terms are not changed when the architect decides in the course of a job what a specification means or whether one specification trumps another. The subcontractor had argued that the architect's decision had no rational basis and was therefore arbitrary and capricious. The court held that the architect's resolution did not constitute bad faith.

These provisions are widely misunderstood by owners and contractors. They believe that because the architect is paid by the owner, she will be partial to the owner. Owners sometimes resent an architect siding with a contractor. However, architects have traditionally attempted to settle construction disputes at an early stage. It is to everyone's best interest to resolve such disputes quickly and get on with the work. If either party is dissatisfied, that party can then follow the claims procedure in Article 15.

An error on the part of the architect in making such a determination will not cause the determination to be overturned unless the error is so gross as to amount to fraud or bad faith.[175] Thus, when an architect was a shareholder in the development entity and had promised that the project would be completed for a certain price, that architect could not arbitrate the question of extras by a contractor in an impartial manner.[176]

The architect enjoys a quasi-judicial immunity in making such determinations so that the architect will not fear lawsuits by one of the parties.[177] This immunity is only for acts performed in this judicial role, but not for acts performed as an architect.[178] Thus, in a case in which a contractor sued an architect because of errors in drawings and specifications that resulted in increased costs to the contractor, the immunity did not apply.[179] The architect also cannot give a decision contrary to a custom or trade usage that has been read into the contract or made a part of the contract.[180] The immunity has been held to apply to the issuance of a final certificate.[181]

---

[175] Ballou v. Basic Constr. Co., 407 F.2d 1137 (4th Cir. 1969); Roberts v. Security Trust & Sav. Bank, 196 Cal. 557, 238 P. 673 (1925); M&L Bldg. Corp. v. Housing Auth., 35 Conn. App. 379, 646 A.2d 244 (1994); Fontaine Bros. v. City of Springfield, 35 Mass. App. Ct. 155, 617 N.E.2d 1002 (1993); Prichard Bros. v. Grady Co., 407 N.W.2d 324 (Minn. Ct. App. 1987), rev'd, 428 N.W.2d 391 (Minn. 1988), further proceedings, 436 N.W.2d 460 (Minn. Ct. App. 1989); Meco Sys., Inc. v. Dancing Bear Entm't, Inc., 948 S.W.2d 185 (Mo. Ct. App. 1997); Odell v. Colomor Irrigation & Land Co., 34 N.M. 277, 280 P. 398 (1929); Helmer-Cronin Constr., Inc. v. Central Sch. Dist. No. 1, 51 A.D.2d 1085, 381 N.Y.S.2d 347 (1976).

[176] Manett, Seastrunk & Buckner v. Terminal Bldg. Corp., 120 Tex. 374, 39 S.W.2d 1 (Civ. App. 1931). In Steinberg v. Fleischer, 706 S.W.2d 901 (Mo. Ct. App. 1986), the court stated that if an architect is an undisclosed co-owner of a project, the provisions for arbitration by the architect would be a nullity.

[177] Craviolini v. Scholer & Fuller Associated Architects, 89 Ariz. 24, 357 P.2d 611 (1960); Lundgren v. Freeman, 307 F.2d 104 (9th Cir. 1962) (architects are immune only if they are acting in a judicial capacity within their jurisdiction).

[178] Id. In Britt Constr., Inc. v. Magazzine Clean LLC, 2006 WL 240619 (Va. Cir. Ct. Feb. 1, 2006), the contractor asserted a claim for tortious interference with contract against the architect, based on this provision. The court found that the language "not to show partiality to either [the owner or contractor]" when making interpretations and initial decisions meant that the architect's claim of privilege could not stand.

[179] Donnelly Constr. Co. v. Oberg/Hunt/Gilleland, 139 Ariz. 184, 677 P.2d 1292 (1984).

[180] John W. Johnson, Inc. v. J.A. Jones Constr. Co., 369 F. Supp. 484 (E.D. Va. 1973) (decision of architect was in excess of his authority, and such decision was not valid).

[181] Blecick v. School Dist. No. 18, 2 Ariz. App. 115, 406 P.2d 750 (1965).

The jurisdiction of architects to interpret contract documents and resolve disputes and claims between owner and contractor is restricted by the work in progress limitation. Disputes submitted to the architect after the operational phase of the project are not subject to the architect's jurisdiction. Thus, the architect's jurisdiction has been held to end upon substantial completion[182] or upon termination of the construction project.[183] One court has even held that the architect's jurisdiction ends when the particular work ends, not when the project as a whole ends.[184] If an architect is asked to decide a dispute between the owner and contractor, it should obtain a waiver in advance from both parties in order to be protected from the possibility that its quasi-judicial immunity had ended earlier or did not apply to the particular question.

At least one court has found that language like that in ¶ 4.2.12 does not mean that the architect is to act as an arbitrator between the contractor and owner, but rather that the determination of the architect would be a precondition to the contractor's right to payment.[185]

---

[182] County of Rockland v. Primiano Constr. Co., 51 N.Y.2d 1, 409 N.E.2d 951, 431 N.Y.S.2d 478 (1980). A dispute arose between the owner and contractor two years after substantial completion of the project and occupation of the building. The contractor demanded arbitration and insisted that the dispute must first be submitted to the architect. The court rejected this contention, because the architect's jurisdiction to hear such claims is limited to the operational phases of the construction. *Accord* Pigott Constr. Int'l, Ltd. v. Rochester Inst. of Tech., 84 A.D. 679, 446 N.Y.S.2d 632 (1981); Lopez v. 14th Street Dev., LLC, 835 N.Y.S.2d 186 (App. Div. 1st Dep't 2007) (also stating that the "A201 is not a model of clarity.")

[183] Liebhafsky v. Comstruct Assocs., 62 N.Y.2d 439, 466 N.E.2d 844, 478 N.Y.S.2d 252 (1984). An owner terminated the construction contract, and the contractor thereafter demanded arbitration in seeking amounts still owed. The owner sought to stay the arbitration on the basis that the contractor had not submitted the dispute to the architect. The courts found that the operation phases had ceased after the owner had terminated the contract and the contractor ceased work on the project, therefore submission to the architect was no longer a condition precedent to arbitration. The architect's responsibility to initially mediate disputes between the owner and contractor ends upon termination of the contract, whether because of substantial completion or because the owner terminated it.

[184] J.A. Sullivan Corp., v. Commonwealth, 397 Mass. 789, 494 N.E.2d 374 (1986). A contractor had included unit costs for removing a ledge under the trench and open-cut methods, the trench method being more expensive. Neither method was specified in the contract documents. A ledge was encountered and the contractor notified the architect and commenced removal operations using the trench method. The architect observed the progress of this work on several occasions. After all ledge operations had been completed but before work on the entire project was completed, the contractor submitted a request for extras for the more costly method. The architect rejected this in part, saying that the less costly method would have sufficed. The contractor sued the owner for the balance. The court found that the architect's authority to arbitrate ended when the "relevant" work was completed, not necessarily when the project as a whole was completed.

This decision could be reconciled with others on an estoppel basis, because the architect was told in advance what the contractor was about to do and observed the work being done. It would be unfair for the architect, as arbitrator, to subsequently reject the additional costs when unit prices had been submitted.

[185] Harry Skolnick & Sons v. Heyman, 7 Conn. App. 175, 508 A.2d 64 (1986). A contractor claimed certain payments from the owner. Pursuant to the contract, the architect determined that the

An architect or engineer owes a duty to a contractor to avoid negligently causing extra expenses for the contractor in the completion of a construction project.[186]

---

contractor was owed over $240,000. The contractor then sought to confirm the architect's decision as an arbitration award. The court found that this was not an agreement to arbitrate but was merely a condition precedent to the plaintiff's right to payment. The contractor would then have to bring a plenary action to obtain payment.

[186] A.R. Moyer, Inc. v. Graham, 285 So. 2d 397 (Fla. 1973) ("A third party general contractor, who may foreseeably be injured or sustained an economic loss proximately caused by the negligent performance of a contractual duty of an architect, has a cause of action against the alleged negligent architect, notwithstanding absence of privity."); Navajo Circle, Inc. v. Development Concepts, 373 So. 2d 689 (Fla. Dist. Ct. App. 1979); Case Prestress v. Chicago Coll. of Osteopathic Med., 118 Ill. App. 3d 782, 455 N.E.2d 811 (1983); Normoyle-Berg & Assocs. v. Village of Deer Creek, 39 Ill. App. 3d 744, 350 N.E.2d 559 (1976); Forte Bros. v. National Amusements, Inc., 525 A.2d 1301 (R.I. 1987); J&J Elec., Inc. v. Gilbert H. Moen Co., 9 Wash. App. 954, 516 P.2d 217 (1974).

In Spancrete, Inc. v. Ronald E. Frazier & Assocs., 630 So. 2d 1197 (Fla. Dist. Ct. App. 1994), the architect owed no duty to a subcontractor because the architect had no right to stop the work, unlike the A.R. Moyer case, which was distinguished by the court.

In Fleischer v. Hellmuth, Obata & Kassabaum, 870 S.W.2d 832 (Mo. Ct. App. 1993), the court held that an architect does not owe a duty to the construction manager, rejecting *A.R. Moyer* and similar cases.

In Engineered Refrigeration Sys., Inc. v. Albertson's Inc., 1996 U.S. Dist. LEXIS 9090 (N.D. Fla. May 29, 1996), the court reviewed *A.R. Moyer* and *Spancrete* and determined that an architect and the construction manager had no power to stop the work. Thus, a negligence action against them was barred by the economic loss rule.

In A.E. Inv. Corp. v. Link Builders, Inc., 62 Wis. 2d 479, 214 N.W.2d 764 (1974), privity was not necessary for a foreseeable tenant to maintain an action for negligence against an architect. In Bacco Constr. Co. v. American Colloid Co., 148 Mich. App. 397, 384 N.W.2d 427 (1986), the court stated:

it is certainly foreseeable that an engineer's failure to make proper calculations and specifications for a construction job may create a risk of harm to the third-party contractor who is responsible for applying those specifications to the job itself. The risk of harm would include the financial hardship created by having to cure the defects which may very well not be caused by the contractor.

In Morse/Diesel, Inc. v. Trinity Indus., Inc., 859 F.2d 242 (2d Cir. 1988), *rev'g* 655 F. Supp. 346 (S.D. N.Y. 1987) (*citing* Widett v. United States Fidelity & Guar. Co., 815 F.2d 885 (2d Cir. 1987) (architects not liable either in tort or contract absent privity under New York law) the court held that damages are not available in a suit for negligent performance of a contractual duty when only economic injury is alleged.

But see the cases arising out of Moorman Mfg. Co. v. National Tank Co., 91 Ill. 2d 69, 435 N.E.2d 443 (1982), which have held that there is no tort recovery for economic damages.

Moorman was applied to architects in a case in which the architect's negligence caused only economic damages. 2314 Lincoln Park W. Condominium Ass'n v. Mann, Gin, Ebel & Frazier, Ltd., 136 Ill. 2d 302, 555 N.E.2d 346 (1990).

In American Towers Owners Ass'n, v. CCI Mech., 930 P.2d 1182 (Utah 1996), a condominium association sued the designers and contractors of a large condominium complex, alleging design and construction defects in the plumbing and mechanical systems. The court found that the economic loss rule applied. The association argued that there was damage to other property as a result of the leaking pipes. This included damage to walls, wall coverings, carpeting, wall

This duty is a duty of care in the design and administration of the project.[187] Although one court[188] used the term "supervising engineer," the negligent items included failing to give timely responses to requests for instructions and failing to complete details of plans. In one case, the owner had a contract with the general contractor containing a no-damage-for-delay clause.[189] That clause defeated a delay claim by the contractor against the owner, but it did not protect the engineer against the contractor's claim for delay resulting from the engineer's negligent administration of its duties under the owner-engineer contract.

4.2.13 The Architect's decisions on matters relating to aesthetic effect will be final if consistent with the intent expressed in the Contract Documents. **(4.2.13; no change)**

These decisions are not subject to litigation, arbitration, or mediation, as are other interpretations of the architect.[190] It is not improper that the representative of one of the contracting parties is the one designated as the mediator whose decision

---

hangings, curtains, and other furnishings. The court found that the property in this case was the entire complex itself that was constructed under one general contract. What the unit owners purchased was a finished product, their dwellings, and not the individual components of the dwellings.

In Federal Ins. Co. v. Grunau Project Dev., Inc., 721 N.W.2d 157 2006 WL 1528948 (Wis. Ct. App. June 6, 2006), the economic loss doctrine applied. During construction, a load-bearing wall was demolished, which in turn caused the building to partially collapse, resulting in substantial damage. The plaintiff argued that the "damage to other property" exception applied because the existing roof and flooring were damaged as a result of the wall collapse. This contention was rejected on the basis of the "integrated system limitation." This provides that the damage-to-other-property exception does not apply to escape the application of the economic loss doctrine when the other property is a component part of the renovated building.

[187] E.C. Ernst Inc. v. Manhattan Constr. Co., 551 F.2d 1026 (5th Cir. 1977); Bates & Rogers Constr. v. North Shore, 92 Ill. App. 3d 90, 414 N.E.2d 1274 (1980); W.H. Lyman Constr. Co. v. Village of Gurnee, 84 Ill. App. 3d 28, 403 N.E.2d 1325 (1980); Normoyle-Berg Assoc. v. Village of Deer Creek, 39 Ill. App. 3d 744, 350 N.E.2d 559 (1976). *See also* Annotation, *Tort Liability of Project Architect for Economic Damages Suffered by Contractor,* 65 A.L.R.3d 249, 256-59 (1975).

[188] Normoyle-Berg & Assoc. v. Village of Deer Creek, 39 Ill. App. 3d 744, 350 N.E.2d 559 (1976).

[189] Bates & Rogers Constr. v. North Shore, 92 Ill. App. 3d 90, 414 N.E.2d 1274 (1980).

[190] In Baker v. Keller Constr. Corp., 219 So. 2d 569 (La. 1969), this clause was held to not bar a claim by the contractor. The architect stated that a terrazzo floor was artistically unacceptable and recommended that $750 be deducted from the contract sum. The court held this to be arbitrary, partly because the project had received an award that encompassed the terrazzo work.

In Mississippi Coast Coliseum Comm'n v. Stuart Constr. Co., 417 So. 2d 541 (Miss. 1982), the architect recommended that the owner accept a discolored concrete wall, subject to certain guarantees from the contractor. The owner rejected the recommendation and withheld $75,000 from the contractor. The court held that this clause was not applicable because the architect had only made a qualified recommendation, not a final determination.

In NSC Contractors v. Borders, 317 Md. 394, 564 A.2d 408 (1989), the architect rejected discolored bricks and refused to certify payment to the contractor. The court found that, although the architect's decision was final as to the artistic effect of the bricks, the contractor could litigate the amount withheld because of the brick problem.

See Bolton Corp. v. T.A. Loving Co., 94 N.C. App. 392, 380 S.E.2d 796 (1989), for a case interpreting a similar provision.

is to be recognized as final and binding.[191] One court has held that the architect's decision is not the same as an arbitrator's decision.[192] This provision contemplates a controversy between the owner and contractor that is actually submitted to the architect for decision. Presumably, both the owner and the contractor would be given an opportunity to present their cases. Whether or not this decision is consistent with the intent expressed in the Contract Documents would be subject to a separate dispute process. To the extent the architect's decision affected the cost to either the owner or contractor, those parties could make a claim under Article 15 to allocate the costs, but not to challenge the interpretation itself.

4.2.14 The Architect will review and respond to requests for information about the Contract Documents. The Architect's response to such requests will be made in writing within any time limits agreed upon or otherwise with reasonable promptness. If appropriate, the Architect will prepare and issue supplemental Drawings and Specifications in response to the requests for information. (**new**)

A request for information is commonly referred to as an RFI. Although an RFI can be made orally, the better practice for the contractor is to submit them in writing. This is normally a request for clarification. For instance, a note might be ambiguous, or one item might conflict with information given elsewhere on the contract documents. The parties should understand that most sets of drawings and plans are one-of-a-kind, and that, therefore, some degree of errors and inconsistencies will be inevitable. These will usually require some exchange of information between the contractor and architect to clarify what the architect intended. On larger projects it is common to maintain a log of all RFIs, indicating the date requested, who is in charge of answering the item, the current status, and the date an answer is submitted. Such a log will be useful if the contractor has complaints about inadequate plans and specifications. These will form the basis for requesting change orders. The architect may also want additional fees from the owner if the architect believes that the contractor is issuing too many RFIs that could be answered merely by looking at the drawings. Note that an RFI might trigger the time limit for making claims, since the existence of an RFI is proof that the contractor has knowledge of a particular condition. The contractor, on the other hand, may claim that the RFI constitutes the required written 21-day notice of ¶ 15.1.2.

It might be a good idea for the parties to agree on what a reasonable time limit is for the architect to respond to an RFI. This might be done at an initial project meeting, memorialized by a written memorandum.

If the RFI is legitimate, the architect may respond in one of several ways, including letters, drawings, specifications, and so forth. If the architect's response requires the contractor to change anything, the contractor may be entitled to a change order, depending on whether or not the contractor should have anticipated the resolution.

---

[191] Joseph F. Trionfo & Sons v. Ernest B. LaRosa, 38 Md. App. 598, 381 A.2d 727 (1978).

[192] Martel v. Bulotti, 65 P.3d 192 (Idaho 2003).

§ 4.32 Article 5: Subcontractors

This article discusses subcontracts and subcontractual relations, such as procedures for changing subcontractors and the conditions to which subcontractors are legally bound.

§ 4.33 Definitions: ¶ 5.1

5.1 DEFINITIONS

5.1.1 A Subcontractor is a person or entity who has a direct contract with the Contractor to perform a portion of the Work at the site. The term "Subcontractor" is referred to throughout the Contract Documents as if singular in number and means a Subcontractor or an authorized representative of the Subcontractor. The term "Subcontractor" does not include a separate contractor or subcontractors of a separate contractor. (**5.1.1; no change**)

This provision could include architects or others who are not normally thought of as subcontractors if they have a contract with the contractor for work at the site. It is limited to entities that perform work at the site and does not include material suppliers or others who work off the site.

5.1.2 A Sub-subcontractor is a person or entity who has a direct or indirect contract with a Subcontractor to perform a portion of the Work at the site. The term "Sub-subcontractor" is referred to throughout the Contract Documents as if singular in number and means a Sub-subcontractor or an authorized representative of the Sub-subcontractor. (**5.1.2; no change**)

Again, this does not include material suppliers or others who do not actually perform work at the site.[193]

§ 4.34 Award of Subcontracts: ¶ 5.2

5.2 AWARD OF SUBCONTRACTS AND OTHER CONTRACTS FOR PORTIONS OF THE WORK

5.2.1 Unless otherwise stated in the Contract Documents or the bidding requirements, the Contractor, as soon as practicable after award of the Contract, shall furnish in writing to the Owner through the Architect the names of persons or entities

---

[193] In Aetna Cas. & Sur. v. Canam Steel, 794 P.2d 1077 (Colo. Ct. App. 1990), a roof collapsed after substantial construction. The owner's builder's risk insurer then sued the steel supplier. The court held that the supplier was not a subcontractor under A201, and therefore, the waiver provision of ¶ 11.3.7 did not apply. Waco Scaffolding Co. v. National Union Fire Ins. Co. of Pittsburgh, 1999 Ohio App. LEXIS 5058 (Oct. 28, 1999).

In Vulcraft v. Midtown Bus. Park, Ltd., 110 N.M. 761, 800 P.2d 195 (1990), the court analyzed whether a supplier of materials was a subcontractor or a materialman. This case gives an overview of two different lines of cases that distinguish these terms.

(including those who are to furnish materials or equipment fabricated to a special design) proposed for each principal portion of the Work. The Architect may reply within 14 days to the Contractor in writing stating (1) whether the Owner or the Architect has reasonable objection to any such proposed person or entity or (2) that the Architect requires additional time to review. Failure of the Owner or Architect to reply within the 14 day period shall constitute notice of no reasonable objection. **(5.2.1; add "14 day" in place of "promptly"; add additional review time)**

Prior to the 2007 version of A201, which gives a 14-day period for an objection to a subcontractor, the objection had to be "prompt." The architect has a qualified privilege to advise the owner to reject unsuitable contractors or subcontractors.[194]

5.2.2 The Contractor shall not contract with a proposed person or entity to whom the Owner or Architect has made reasonable and timely objection. The Contractor shall not be required to contract with anyone to whom the Contractor has made reasonable objection. **(5.2.2; no change)**

If the bidding documents require the contractor to contract with a particular subcontractor and the contractor submits a bid, it would be deemed to have waived

---

[194] In Kecko Piping Co. v. Town of Monroe, 172 Conn. 197, 374 A.2d 179 (1977), the architect requested information from a subcontractor for the apparent low bidder. The sub failed to respond. Other information indicated that the sub was overextended and did not have sufficient experience with jobs of that size. The architect advised the owner to reject the subcontractor, which was done. The subcontractor sued the owner and architect for interference with its contract. The court held for the defendants. The architect was under an obligation to advise the owner as to the suitability of contractors and subcontractors and accordingly enjoyed a qualified privilege.

Under this provision, the owner has a right to reject proposed subcontractors, even if such a subcontractor was on a list of prequalified subcontractors. In North Am. Mech., Inc. v. Diocese of Madison, 600 N.W.2d 55 (Wis. Ct. App. 1999), the owner, with the assistance of the architect, had prepared a list of prequalified contractors who were to bid on a project. The plaintiff was an HVAC subcontractor who was on the prequalified list and who submitted the lowest bid for that work. However, the owner decided to object to the use of the plaintiff because that firm was a non-union shop and the owner wanted to use only union labor. The plaintiff sued the owner, alleging misrepresentation and promissory estoppel. Plaintiff's argument was that, by prequalifying it, the owner represented that if plaintiff successfully competed for the subcontract, the owner would not disqualify plaintiff from receiving the contract. Based on the Project Manual, which contained a standard A201, including the language here, and the bidding documents, the court ruled against the contractor. Although in this case the subcontractor was rejected by the owner even before the contract was signed, this language put the contractor on notice that the owner had the right to reject any contractor. Both the Invitation to Bid and the Instructions to Bidders included provisions giving the owner the right to reject the bids of prequalified subcontractors and prequalified general contractors. The court further held that the criteria by which the owner would evaluate the bids of the contractors is not a fact basic to the transaction which the owner had a duty to disclose to the contractors. There was no duty on the part of the owner to disclose its preference for union contractors.

In Gaglioti Contracting, Inc. v. City of Hoboken, 307 N.J. Super. 421, 704 A.2d 1301, 1305 (1997), the court found that this provision, allowing the contractor to submit names of subcontractors after the award of bids, "fosters an opportunity for bid shopping, an atmosphere ripe for corruption, and benefits only the general contractor who receives the award and not the public entity." In this case, there was another provision requiring submission of subcontractor names along with the bid.

the provision in the second sentence.[195] If the owner or architect objects to a subcontractor in a timely manner, the contractor must obtain a substitute. If this is done, the originally proposed subcontractor has no legal action against the contractor, owner, or architect unless the contractor has entered into a contract with the subcontractor (which would be very unwise for the contractor to do, given this provision), in which case the subcontractor would only have an action against the general contractor.

5.2.3 If the Owner or Architect has reasonable objection to a person or entity proposed by the Contractor, the Contractor shall propose another to whom the Owner or Architect has no reasonable objection. If the proposed but rejected Subcontractor was reasonably capable of performing the Work, the Contract Sum and Contract Time shall be increased or decreased by the difference, if any, occasioned by such change, and an appropriate Change Order shall be issued before commencement of the substitute Subcontractor's Work. However, no increase in the Contract Sum or Contract Time shall be allowed for such change unless the Contractor has acted promptly and responsively in submitting names as required. **(5.2.3; no change)**

This allows the contractor to recover any increased cost if the substituted subcontractor bids a higher amount than the original subcontractor. If the new subcontractor bids a lower amount, the owner is entitled to a credit, but only if the original subcontractor was reasonably capable of performing the work. Presumably, the burden is on the contractor to show this qualification. Note that if there is to be a change in the contract sum because of a substitution of subcontractors, the change order must be issued prior to the start of that subcontractor's work. The contractor would need to present proper evidence of such price increases, including the relevant subcontractor proposals.

The contractor always proposes the names of the subcontractors in order to retain control and responsibility for each such subcontractor, with the owner and architect then having this veto power. The owner might want to add language that permits the use of the sum shown on the schedule of values submitted by the contractor instead of the actual subcontract price. This may prevent a situation in which the original subcontractor has an unusually low price and still appears qualified.

5.2.4 The Contractor shall not substitute a Subcontractor, person or entity previously selected if the Owner or Architect makes reasonable objection to such substitution. **(5.2.4; minor revision)**

The owner may want to obtain copies of the subcontracts, although contractors generally are reluctant to give the owner this information. This is useful if the owner later wants to take over certain subcontracts in the event of the contractor's default (**see ¶ 5.4**).

---

[195] Norair Eng'g v. St. Joseph's Hosp., 147 Ga. App. 595, 249 S.E.2d 642 (1978).

§ 4.35   Subcontractual Relations: ¶ 5.3

5.3 SUBCONTRACTUAL RELATIONS

By appropriate agreement, written where legally required for validity, the Contractor shall require each Subcontractor, to the extent of the Work to be performed by the Subcontractor, to be bound to the Contractor by terms of the Contract Documents, and to assume toward the Contractor all the obligations and responsibilities, including the responsibility for safety of the Subcontractor's Work, which the Contractor, by these Documents, assumes toward the Owner and Architect. Each subcontract agreement shall preserve and protect the rights of the Owner and Architect under the Contract Documents with respect to the Work to be performed by the Subcontractor so that subcontracting thereof will not prejudice such rights, and shall allow to the Subcontractor, unless specifically provided otherwise in the subcontract agreement, the benefit of all rights, remedies and redress against the Contractor that the Contractor, by the Contract Documents, has against the Owner. Where appropriate, the Contractor shall require each Subcontractor to enter into similar agreements with Sub-subcontractors. The Contractor shall make available to each proposed Subcontractor, prior to the execution of the subcontract agreement, copies of the Contract Documents to which the Subcontractor will be bound, and, upon written request of the Subcontractor, identify to the Subcontractor terms and conditions of the proposed subcontract agreement that may be at variance with the Contract Documents. Subcontractors will similarly make copies of applicable portions of such documents available to their respective proposed Sub-subcontractors. **(5.3.1; no change)**

All the contract documents are binding on all the subcontractors. The contractor should indicate this in all subcontracts—something unlikely if a subcontract is merely the subcontractor's proposal, or a verbal contract. This is sometimes known as a *flowdown provision*. The contractor has the right to insert additional terms into the subcontracts that do not prejudice the rights of the owner or architect. These could be items such as no-lien provisions or payment contingent upon receipt of payment from the owner.

Paragraph 5.3 has been held not to require each subcontractor to read the entire General Conditions, other than those portions that concern the specific subcontractor, and does not constitute a waiver of lien rights when read in conjunction with ¶ 9.3.3.[196] One court allowed a contractor to compel a subcontractor to arbitrate a claim because the 1970 version of A201 was deemed to bind the subcontractor to the provisions of A201, even though he did not sign the General Conditions, and even though the contractor-subcontractor agreement did not have an arbitration clause.[197]

---

[196] Luczak Bros. v. Generes, 116 Ill. App. 3d 286, 451 N.E.2d 1267 (1983) (the recorded construction contract, which incorporated A201, did not constitute a waiver of the subcontractors' mechanic's lien rights).

[197] Vespe Contracting Co. v. Anvan Corp., 399 F. Supp. 516 (E.D. Pa. 1975), *aff'd*, 433 F. Supp. 1226 (E.D. Pa. 1977).

In Massachusetts Elec. Sys., Inc. v. R. W. Granger & Sons, Inc., 32 Mass. App. Ct. 982, 594 N.E.2d 545 (1992), the subcontractor was bound by the arbitration clause in the 1976 version of A201 because the prime contract was incorporated into the subcontract.

The warranty and indemnity provisions of the 1976 A201 were incorporated into the subcontract in McDevitt & St. Co. v. K-C Air Conditioning Serv., Inc., 418 S.E.2d 87 (Ga. Ct. App. 1992).

In Wasserstein v. Kovatch, 261 N.J. Super. 277, 618 A.2d 886, 89 (1993), the court ordered consolidation under AIA Document A107 (1987). In a very questionable decision, the court allowed consolidation between the owner, the general contractor, and various subcontractors. The owners opposed the motion by the other parties to arbitrate, but the court held that, by virtue of the general contractor's joining in the motion, all the various contractors had consented to arbitration. The court also relied on a provision in the General Conditions, similar but not identical to ¶ 5.3 (1997 version), that allowed the subcontractors "the benefit of all rights and remedies" available to the contractor by virtue of the contract. The better view is that this sort of consolidation is simply not allowed under A201.

See also Wasserstein v. Kovatch, 261 N.J. Super. 277, 618 A.2d 886 (1993) (based on provision that gives subcontractor the "benefit of all rights, remedies and redress" found in contract documents; integration clause of contract was held to do only with scope of work, not procedural basis for claim by subcontractor). But in Williams Tile & Marble Co. v. Ra-Lin & Assocs., 426 S.E.2d 598 (Ga. Ct. App. 1992), the court found that the architect's plans and specifications were incorporated into the subcontract solely to establish the materials and manner of installation, not for the purpose of arbitration. It is unclear whether AIA documents were used in this case.

In 3A Indus., Inc. v. Turner Constr. Co., 71 Wash. App. 407, 869 P.2d 65 (1993), the subcontractor was bound to arbitrate because the general conditions were incorporated by reference. The subcontractor argued that it should not be held to arbitrate a claim based on a Little Miller Act. After reviewing several federal cases decided under the federal Miller Act, the court held that this case involved a dispute resolution clause, as opposed to an arbitration clause. The subcontractor had specifically agreed that it had the same remedies as the general contractor, namely, arbitration.

In Lord & Son Constr., Inc. v. Roberts Elec. Contractors, Inc., 624 So. 2d 376 (Fla. Dist. Ct. App. 1993), the appellate court overturned the trial court's determination that only the portion of the subcontract requiring the furnishing and installation of the items listed in the subcontract was governed by the arbitration provision of the general contract. The general contractor sued the subcontractor for damages caused by the subcontractor's delay, and the sub wanted to arbitrate. Apparently the subcontract was not an AIA form but referenced the contract documents which did, in turn, reference A201.

In Point E. Condo. Owner's Assn. v. Cedar House Assoc., 663 N.E.2d 343 (Ohio Ct. App. 1995), the contractor was responsible for the work of his subcontractor under these provisions.

In Premier Elec. Constr. Co. v. American Nat'l Bank, 658 N.E.2d 877 (Ill. App. 1st Dist. 1999), the court held that only those general contract provisions that relate to the subcontractor's work are incorporated into the subcontract.

In Mpact Constr. Group, LLC v. Superior Concrete Constructors, Inc., 802 N.E.2d 901 (Ind. 2004), where the subcontracts did not include arbitration provisions, this language was held to place the burden on the general contractor to obtain arbitration agreements from the subcontractors. The subcontractors were not required to arbitrate.

In Tribble & Stephens Co., v. RGM Constructors, L.P., 154 S.W.3d 639 (Tex. Ct. App. 2005), the court held that the flow-down provision in the subcontract is limited to the performance of the subcontractor's work; therefore, it could not conclude as a matter of law that the subcontractor had agreed to be bound by the condition precedent provision of A201 (following *Mpact*).

The term "unless specifically provided otherwise in the subcontract agreement" was discussed in Technosteel, LLC v. Beers Constr. Co., 271 F.3d 151 (4th Cir. 2001). There, the subcontract contained a provision requiring litigation, rather than arbitration. The court rejected an argument that the parties had intended to delete this provision but failed to do so.

§ 4.36   Contingent Assignment of Subcontract: ¶ 5.4

5.4 CONTINGENT ASSIGNMENT OF SUBCONTRACTS

5.4.1 Each subcontract agreement for a portion of the Work is assigned by the Contractor to the Owner provided that:

> *.1   assignment is effective only after termination of the Contract by the Owner for cause pursuant to Section 14.2 and only for those subcontract agreements that the Owner accepts by notifying the Subcontractor and Contractor in writing; and*

> *.2   assignment is subject to the prior rights of the surety, if any, obligated under bond relating to the Contract.*

The assignment does not become effective unless and until the owner accepts the assignment. Thus, the owner can determine which, if any, of the existing subcontractors it wants to keep on the project.

When the Owner accepts the assignment of a subcontract agreement, the Owner assumes the Contractor's rights and obligations under the subcontract. (**5.4; add final paragraph**)

This last paragraph means that the owner "steps into the shoes" of the general contractor insofar as the rights and obligations toward the subcontractors that the owner chooses to accept. If a subcontractor is owed money by the general contractor for that project, for instance, the owner will now owe that money. This, of course, does not extend to money owed by the general for other projects.

This section is inserted for the owner's convenience if the owner terminates the contract with the general contractor. Because A201 is incorporated into all the subcontracts (**see** ¶ 5.3), each subcontractor agrees to this contingent assignment when it executes the subcontract. Note the importance of the contractor including a flowdown provision in the subcontracts. Otherwise, a subcontractor may refuse to accept such an assignment. In the event of termination, the owner can pick and choose which subcontracts to affirm by notifying each such subcontractor in writing, as well as the contractor. The owner should also notify rejected subcontractors that they are to perform no more work at the site. If there is a surety, the surety would probably exercise its rights to complete the work, making it unnecessary for the owner to use this provision.

5.4.2 Upon such assignment, if the Work has been suspended for more than 30 days, the Subcontractor's compensation shall be equitably adjusted for increases in cost resulting from the suspension. (**5.4.2; no change**)

When the general contractor is terminated, there is usually some delay. This provision allows the subcontractors who remain on the project to obtain additional compensation if the work is suspended for more than 30 days. The owner might want to delete this provision or increase the time limits. Note that this provision does not permit the subcontractor to renegotiate the contract but merely allows for additional costs caused by delay.

5.4.3 Upon such assignment to the Owner under this Section 5.4, the Owner may further assign the subcontract to a successor contractor or other entity. If the Owner assigns the subcontract to a successor contractor or other entity, the Owner shall nevertheless remain legally responsible for all of the successor contractor's obligations under the subcontract. **(new)**

The owner may want to hire a replacement general contractor. In that case, the owner can assign those subcontractors it wants to keep to the replacement contractor. Naturally, the replacement contractor will want to approve any such subcontractors. This provision also makes the owner the guarantor of the obligations of the replacement contractor towards these subcontractors for the Project.

§ 4.37   Article 6: Construction by Owner or by Separate Contractors

6.1 OWNER'S RIGHT TO PERFORM CONSTRUCTION AND TO AWARD SEPARATE CONTRACTS

6.1.1 The Owner reserves the right to perform construction or operations related to the Project with the Owner's own forces, and to award separate contracts in connection with other portions of the Project or other construction or operations on the site under Conditions of the Contract identical or substantially similar to these including those portions related to insurance and waiver of subrogation. If the Contractor claims that delay or additional cost is involved because of such action by the Owner, the Contractor shall make such Claim as provided in Article 15. **(6.1.1; change reference)**

Often, this means that the construction process is a *multiple prime contracting* arrangement, whereby the owner hires several general contractors to perform parts of the work. The owner stands to avoid paying some profit and overhead that would have been paid to a single general contractor, but also has additional risk in the coordination of such multiple prime contractors. The owner also has a duty to do this coordination (¶ 6.1.3).[198]

If the owner intends to award such other contracts,[199] this should be disclosed in the bidding documents or contract documents. The owner always has the right to enter into such other contracts, but if that has any adverse impact on the contractor, the contractor may make a claim for extras. A Tennessee court has interpreted this provision of the 1976 version of A201 to mean that one prime contractor was an

---

[198] Amp-Rite Elec. Co., Inc., v. Wheaton Sanitary Dist., 580 N.E.2d 622 (Ill. App. Ct. 2d Dist., 1991).

[199] In Eldeco, Inc., v. Charleston County Sch. Dist., 642 S.E.2d 726 (S.C. 2007), a subcontractor complained that it was not awarded all of the Work. This other work was not included in the original work, but was added pursuant to construction change directives and change orders. Pursuant to this section and the definition of "Work" at ¶ 1.1.3, the court found against the subcontractor.

intended (third-party) beneficiary of a construction contract between the owner and another prime contractor.[200]

> *6.1.2 When separate contracts are awarded for different portions of the Project or other construction or operations on the site, the term "Contractor" in the Contract Documents in each case shall mean the Contractor who executes each separate Owner-Contractor Agreement.* **(6.1.2; no change)**

> *6.1.3 The Owner shall provide for coordination of the activities of the Owner's own forces and of each separate contractor with the Work of the Contractor, who shall cooperate with them. The Contractor shall participate with other separate contractors and the Owner in reviewing their construction schedules. The Contractor shall make any revisions to the construction schedule deemed necessary after a joint review and mutual agreement. The construction schedules shall then constitute the schedules to be used by the Contractor, separate contractors and the Owner until subsequently revised.* **(6.1.3; minor modification)**

The various contractors must cooperate with one another. If meetings are required to coordinate schedules, the contractor may be entitled to an extra pursuant to ¶ 6.1.1. The construction schedule referred to here is identified in ¶ 3.10.1.

> *6.1.4 Unless otherwise provided in the Contract Documents, when the Owner performs construction or operations related to the Project with the Owner's own forces, the Owner shall be deemed to be subject to the same obligations and to have the same rights that apply to the Contractor under the Conditions of the Contract, including, without excluding others, those stated in Article 3, this Article 6 and Articles 10, 11 and 12.* **(6.1.4; no change)**

---

[200] Moore Constr. v. Clarksville Dep't of Elec., 707 S.W.2d 1 (Tenn. Ct. App. 1985). The court reviewed a number of decisions and stated:

> Unless the construction contract provides otherwise, prime contractors on construction projects involving multiple prime contractors will be considered to be intended or third party beneficiaries of the contracts between the project's owner and other prime contractors. They have been permitted to recover when the courts have found that their fellow prime contractor assumed an obligation of the owner to them (the "duty owed" test) or that their fellow contractor assumed an independent duty to them in their own contract with the owner (the "intent to benefit" test). The courts have generally relied upon the following factors to support a prime contractor's third party claim: (1) the construction contracts contain substantially the same language; (2) all contracts provide that time is of the essence; (3) all contracts provide for prompt performance and completion; (4) each contract recognizes other contractors' rights to performance; (5) each contract contains a non-interference provision; and (6) each contract obligates the prime contractor to pay for the damage it may cause to the work, materials, or equipment of other contractors working on the project.

See also Barth Elec. Co. v. Traylor Bros., 553 N.E.2d 504 (Ind. Ct. App. 1990).

In Leon Williams Gen. Contractor, Inc. v. Hyatt, 2002 WL 192548 (Tenn. Ct. App. Feb. 7, 2002), the court held that this provision, along with ¶ 6.1.4, did not give a separate contractor the right to arbitrate against the owner under the first contractor's agreement.

The owner is bound to the same obligations as the contractor under A201 when the owner is performing work on the project. While the contractor is expected to cooperate with the owner's separate contractors, the contractor will not be liable to the owner if the owner's contractors delay the project or otherwise damage the owner. Even where there are provisions requiring the contractor to coordinate the work of the owner's contractors, the contractor will not be liable for the fault of those contractors, since there is no contract between those parties. Under ¶ 3.3.2, the contractor is responsible for its own subcontractors.

§ 4.38   Mutual Responsibility: ¶ 6.2

6.2 MUTUAL RESPONSIBILITY

6.2.1 The Contractor shall afford the Owner and separate contractors reasonable opportunity for introduction and storage of their materials and equipment and performance of their activities, and shall connect and coordinate the Contractor's construction and operations with theirs as required by the Contract Documents. **(6.2.1; no change)**

All dealings between the various contractors and the owner are subject to a standard of reasonableness. Thus, the contractor must be reasonable in permitting the owner and other contractors to have access to the site and to store material at the site. The contractor must also be reasonable in coordinating its work with that of the others. The owner and the other contractors, in turn, must be reasonable in their demands on this contractor.

6.2.2 If part of the Contractor's Work depends for proper execution or results upon construction or operations by the Owner or a separate contractor, the Contractor shall, prior to proceeding with that portion of the Work, promptly report to the Architect apparent discrepancies or defects in such other construction that would render it unsuitable for such proper execution and results. Failure of the Contractor so to report shall constitute an acknowledgment that the Owner's or separate contractor's completed or partially completed construction is fit and proper to receive the Contractor's Work, except as to defects not then reasonably discoverable. **(6.2.2; no change)**

If the contractor does not promptly report an error or defect in the work of another contractor or the owner's own forces wherever the contractor's work abuts or joins that other work, the contractor is deemed to have accepted such other work and waive any right to receive additional compensation. This is true so long as the errors or deficiencies are reasonably discoverable or "apparent." If, on the other hand, the defects are latent—that is, something that a reasonably competent contractor would not discover—then there is no such waiver. In one case, a contractor who notified the owner that Phase 1 work was not as the specifications required and upon which he had submitted a bid for Phase 2 work was allowed his additional costs of correction.[201] If the contractor is delayed because of improper work by another contractor, it should promptly request a change order for additional time and an increase in the contract sum, if appropriate.

> *6.2.3 The Contractor shall reimburse the Owner for costs the Owner incurs that are payable to a separate contractor because of the Contractor's delays, improperly timed activities or defective construction. The Owner shall be responsible to the Contractor for costs the Contractor incurs because of a separate contractor's delays, improperly timed activities, damage to the Work or defective construction.* (**6.2.3; minor revisions**)

If the owner is required to pay any costs to a separate contractor because this contractor delayed the work or because of another fault of this contractor, the owner is entitled to be reimbursed by this contractor. The reverse is also true: if this contractor is delayed by another contractor on this project, it is entitled to be reimbursed by the owner.

There may be a question as to whether this provision, allowing for delay damages, conflicts with the waiver of consequential damages found at ¶ 15.1.6. See also ¶ 8.3.3, allowing delay damages in some cases. Presumably, costs paid to other contractors are considered "direct" as opposed to "consequential" damages.[202] However, if the contractor is delayed as a result of delays by a separate contractor, "costs incurred by the Contractor" are recoverable. Do these costs include consequential damages subject to the waiver contained in ¶ 15.1.6? Presumably, reading A201 as a whole, consequential damages are not due the contractor in this situation.

> *6.2.4 The Contractor shall promptly remedy damage the Contractor wrongfully causes to completed or partially completed construction or to property of the Owner, separate contractors as provided in Section 10.2.5.* (**6.2.4; no change**)

If the contractor damages work performed by another contractor or the owner, or to property of one of them, the contractor is responsible for such costs. The owner might want to expand this paragraph to provide for full indemnification of the owner and architect if the contractor causes damage. While ¶ 3.18.1 arguably does this, there may be an issue as to whether such property is part of the Work. The owner might also want to delete the word "wrongfully" in order to require the contractor to pay for such damages whether wrongfully caused or not.[203] See also ¶ 9.5.1.5, wherein certification can be withheld for this reason.

> *6.2.5 The Owner and each separate contractor shall have the same responsibilities for cutting and patching as are described for the Contractor in Section 3.14.* (**6.2.5; no change**)

---

[201] Moorhead Constr. Co. v. City of Grand Forks, 508 F.2d 1008 (8th Cir. 1975).

[202] See the discussion of consequential damages at **§ 4.85**.

[203] For a case that discusses a similar but nonstandard clause, *see* Gilbane Bldg. Co. v. Nemours Found., 666 F. Supp. 649 (D. Del. 1985).

§ 4.39   Owner's Right to Clean Up: ¶ 6.3

6.3 OWNER'S RIGHT TO CLEAN UP

If a dispute arises among the Contractor, separate contractors and the Owner as to the responsibility under their respective contracts for maintaining the premises and surrounding area free from waste materials and rubbish, the Owner may clean up and the Architect will allocate the cost among those responsible. **(6.3.1; no change)**

If such an event occurs, it would be wise to notify the various parties that the architect will be making this determination and give everybody a final opportunity to settle the matter. Once the architect makes the determination, it will be subject to the dispute resolution procedures of Article 15.

§ 4.40   Article 7: Changes in the Work

Change orders and change order directives are the main focus of this article, which stipulates that changes to the work be specified in writing. However, courts frequently uphold certain types of oral agreements to change the work.

7.1 GENERAL

7.1.1 *Changes in the Work may be accomplished after execution of the Contract, and without invalidating the Contract, by Change Order, Construction Change Directive or order for a minor change in the Work, subject to the limitations stated in this Article 7 and elsewhere in the Contract Documents.* **(7.1.1; no change)**

The requirement that change orders be in writing is primarily for the benefit of the owner. The owner is put on notice that a claim is being made, thereby giving it an opportunity to take appropriate corrective action or to prepare a proper response to the claim.[204] This provision also protects the contractor, particularly when the

---

[204] Ida Grove Roofing & Improvement, Inc. v. City of Storm Lake, 378 N.W.2d 313 (Iowa Ct. App. 1985); Moore Constr. v. Clarksville Dep't of Elec., 707 S.W.2d 1 (Tenn. Ct. App. 1985).

In Whitfield Constr. Co. v. Commercial Dev. Corp., 392 F. Supp. 982 (D. V.I. 1975), the court stated:

As a general rule, the provision in a private building or construction contract that alterations or extras must be ordered in writing is valid and binding upon the parties, and therefore, so long as such a provision remains in effect no recovery can be had for alterations or extras done without a written order in compliance therewith.

In Castle Concrete Co. v. Fleetwood Assocs., Inc., 13 Ill. 2d 289, 268 N.E.2d 474 (1971), the court stated the elements that a contractor must show before he is entitled to an extra: (a) the work was outside the scope of the contract; (b) the extra items were ordered by the owner; (c) the owner agreed to pay extra, either by words or conduct; (d) the extras were not furnished by the contractor as his voluntary act; and (e) the extra items were not rendered necessary by any fault of the contractor.

In Chiappisi v. Granger Contracting Co., 352 Mass. 174, 223 N.E.2d 924 (1967), the subcontractor scaled the architect's drawing and miscalculated the amount of spray insulation required.

work involves public projects. In one case, the contract required that changes be approved in writing by the county commissioners.[205] The court held that the contractor could not recover the cost of extra work because the director of public works, instead of the commissioners, had authorized the extras in writing.

For protection, the contractor can refuse to perform the extra work unless properly authorized in writing.[206] However, if the parties orally agree to change orders, the courts will uphold those change orders,[207] but it must be the owner and not the

---

He asked for an extra at the conclusion of the work. The court held that the extra was not permitted because the subcontractor failed to follow the claims procedures.

[205] Burke v. Allegheny County, 336 Pa. 411, 9 A.2d 396 (1939), *cited in* Dick Corp. v. State Pub. Sch. Bldg. Auth., 27 Pa. Commw. 498, 365 A.2d 663 (1976).

[206] Dick Corp. v. State Pub. Sch. Bldg. Auth., 27 Pa. Commw. 498, 365 A.2d 663 (1976); J.M. Humphries Constr. Co. v. City of Memphis, 623 S.W.2d 276 (Tenn. Ct. App. 1981).

In Naek Constr. Co., Inc. v. PAG Charles Street Ltd. P'ship, 2004 WL 2757623 (Conn. Super. Nov. 3, 2004), the contract modified the standard AIA language to make the contractor responsible for all change orders unless the owner agreed to pay. Although the contractor performed in excess of $1 million in extra work, the court held that it was not entitled to be paid for such work.

[207] In Custom Builders, Inc. v. Clemons, 52 Ill. App. 3d 399, 365 N.E.2d (1977), the court stated: "Generally, a condition that all extra work be performed only upon written orders can be waived orally by the owner, but before the contractor is entitled to compensation for such extras, the waiver must be proved by clear and convincing evidence." *Accord* Delta Constr., Inc. v. Dressier, 64 Ill. App. 3d 867, 381 N.E.2d 1023 (1978).

In Burdette v. Lascola, 40 Md. App. 720, 395 A.2d 169 (1979), the issue was whether an oral change order was material. If material, the surety would have been discharged and the owner would have been without a remedy after the contractor defaulted. Most surety bonds require that the surety be notified of any substantial changes to the contract.

See also Arc & Gas Welder Assoc. v. Green Fuel Economizer Co., 285 F.2d 863 (4th Cir. 1960); Harrington v. McCarthy, 91 Idaho 307, 420 P.2d 790 (1966); Joray Mason Contractors v. Four J's Constr., 61 Ill. App. 3d 410, 378 N.E.2d 328 (1978); Coonrod & Walz Constr. Co. v. Motel Enter., Inc., 217 Kan. 63, 535 P.2d 971 (1975); Wisinger v. Casten, 550 So. 2d 685 (La. Ct. App. 1989); Eastline Corp. v. Marion Apartments, Ltd., 524 So. 2d. 582 (Miss. 1988); Lazer Constr. Co. v. Long, 296 S.C. 127, 370 S.E.2d 900 (1988); City of Mound Bayou v. Roy Collins Constr. Co., 499 So. 2d 1354 (Miss. 1986); Union Bldg. Corp. v. J&J Bldg. & Maintenance, 578 S.W.2d 519 (Tex. App. 1979).

In Forest Constr. v. Farrell-Cheek Steel Co., 484 So. 2d 40 (Fla. Dist. Ct. App. 1986), a unit price on one change order established the price of a subsequently unsigned change order.

In Moore Constr. v. Clarksville Dep't of Elec., 707 S.W.2d 1 (Tenn. Ct. App. 1985), the court stated:

The course of dealing between the parties can also amount to a waiver where the conduct of the parties makes it clear that they did not intend to rely strictly upon a contract's written notice requirement and that adherence to such a requirement would serve no useful purpose. Copco Steel & Eng'g Co. v. United States, 341 F.2d 590 (Ct. Cl. 1965), and Willey v. Terry & Wright, Inc., 421 S.W.2d 362 (Ky. Ct. App. 1967). Thus, an owner's consideration of a claim on its merits without invoking a formal written notice requirement has been held to amount to the waiver of the requirement thereby preventing the owner from asserting this claim at a later time. Blount Bros. v. United States, 424 F.2d 1074 (Ct. Cl. 1970); Morrison-Knudsen Co. v. United States, 397 F.2d 826 (Ct. Cl. 1968). Once a party has waived the requirement with regard to a particular matter, it cannot revoke its waiver, in whole or in part, at its convenience.

architect who authorizes oral change orders.[208] Note that an architect generally has no authority to unilaterally alter plans or specifications once the construction contract has been signed.[209]

In Devenow v. St. Peter, 134 Vt. 245, 356 A.2d 502 (1976), the court held that the extra work was done pursuant to a separate oral agreement in order to get around the requirement for written change order.

In V.L. Nicholson Co. v. Transcon Inv. & Fin. Ltd., 595 S.W.2d 474 (Tenn. 1980), the owner authorized a developer to act for it in the construction of a low-rent housing complex. When the contractor did additional work for changes that the developer was aware of, the owner was bound to pay, even though the owner did not sign the change orders, based on an implied contract theory. The developer had before it the proposed written change orders that detailed the work and took no steps to stop the additional work.

In Care Sys., Inc. v. Laramee, 155 A.D.2d 770, 547 N.Y.S.2d 471 (1989), the court stated:

When an owner knowingly receives and accepts the benefits of extra work orally directed by himself and his agents, that owner is equitably bound to pay the reasonable value thereof, notwithstanding the provisions of his contract that any extra work must be supported by a written authorization signed by the owner; such conduct constitutes a waiver of that requirement.

In Turner Excavating Co., Inc. v. GWX Ltd. P'ship, 2002 WL 4603 (Minn. Ct. App. Jan. 2, 2002), the contractor argued that the owner waived the written notice requirement by establishing a practice that was inconsistent with this provision. In this case, the record established that written change orders were prepared and signed after work was done. This did not establish that the owner waived the contractual provisions requiring written change orders. More proof would be required for the contractor to prevail. Creative Marine, Inc. v. Zaccai, 2006 WL 2678307 (Mass. Super. Aug. 29, 2006).

[208] Whitfield Constr. Co. v. Commercial Dev. Corp., 392 F. Supp. 982 (D. V.I. 1975) (*citing* McNulty v. Keyser Office Bldg. Co., 112 Md. 638, 76 A. 1113 (1910)) ("In the absence of express authority, an architect, as such, has no power to waive or modify a stipulation requiring a written order for alterations or extras.").

In Citizens Nat'l Bank v. L.L. Glascock, Inc., 243 So. 2d 67 (Miss. 1971), under the 1963 version of A201, the contractor could not recover without the owner's signature on a change order even though the architect originally told the contractor that a change order would be approved.

In Cape Fear Elec. Co. v. Star News Newspapers, Inc., 22 N.C. App. 519, 207 S.E.2d 323, *cert. denied,* 285 N.C. 757, 209 S.E.2d 280 (1974), the court held that there was no evidence that the owner had appointed the engineer its agent. The contractor had installed more expensive conduit at the insistence of the engineer, but no change order was ever issued. The contractor could not recover for the additional work. *But see* J.R. Graham v. Randolph County Bd. of Educ., 25 N.C. App. 163, 212 S.E.2d 542 (1975).

[209] Alexander v. Gerald E. Morrisey, Inc., 137 Vt. 20, 399 A.2d 503 (1979). The contractor submitted to the architect a cut sheet that showed the type of insulation proposed. The contract documents indicated R-19 insulation for the ceiling areas. The submittal indicated that the material proposed had a thermal resistance of R-19 for air conditioning purposes but only R-15 for heating. The architect responded with a notation that "R19-ceilings; R-11 walls o.k." Several years later, the owner complained that his heating bills were high. Subsequent investigation showed that the insulation was inferior to that originally specified. The architect and contractor were both liable for the error.

In Dehnert v. Arrow Sprinklers, Inc., 705 P.2d 846 (Wyo. 1985), the contractor sought to substitute a different sprinkler head from the one specified. At bid opening, the contractor

In a Missouri case, the court held that a drilling subcontractor was entitled to additional compensation for removal of unforeseen obstructions.[210] The contract sum was a fixed price, plus a unit price for the removal of obstructions at $800 per yard. When the contractor encountered obstructions and did not obtain a change order, the owner refused to pay. The court held that a change order was not required, because there was neither a change in the work or an adjustment in the contract sum. The contractor had notified the architect that he was encountering obstructions. The architect was presumably aware that additional compensation would be requested. The court stated that no written notice was required each time rock was encountered because such a requirement would necessitate many delays, and the work was clearly within the terms of the contract. To avoid any ambiguity, the contract documents in that case could have specified written notice if the contractor was to submit a claim for the removal of such obstructions.

One court held that an owner who took possession of a building acquiesced to the changes made by the contractor.[211] Under ¶ 9.10.4, occupancy, whether partial or complete, should not operate as an acceptance of non-conforming work.

One issue that arises on many projects is whether proposed, but not yet approved, change orders should appear on the contractor's payment requests, schedule of values, and similar documents. From the viewpoint of the lender and title company, all proposed or probable change orders should be shown so that the lender and title company can see where the project is headed financially. Unexecuted change requests can pose a major problem on a project, especially where there is little or no contingency. From the owner's perspective, such proposed change requests should not appear, since they may never become approved change orders. If there is a lender on the project, the owner would also be at risk of having the lender declare the loan out of balance, thereby jeopardizing the progress of the work due to lack of further financing by the lender. These are issues that the parties often struggle with, and competent counsel is important in such situations.

---

informed the owner that his bid was based on the substitute head and later presented the architect with information concerning the new head. The architect verbally approved the substitution. Later, the architect rejected the heads and demanded that the contractor install the ones originally specified. The owner terminated the contract after the contractor refused to make the substitution. The court held that the architect is not liable to the contractor for damages when he acts within the scope of his contractual obligations unless he acts with bad faith or malice. The contractor had not obtained a written change order, and the court found that the documentation the contractor furnished the architect was insufficient for the purpose of initiating a change order because it did not show that the quality of the proposed head equaled or exceeded that of the specified head. The contractor also failed to provide cost quotations or substantiate in writing its reasons for requesting the change. The architect's written approval of the sprinkler layout was not effective to authorize the contractor to make the substitution.

See also Kirk Reid Co. v. Fine, 205 Va. 778, 139 S.E.2d 829 (1965) (architect had "general supervision of the work" but had no authority to make alterations in plans and specifications nor to bind owner with respect thereto except as stated in the contract).

[210] Hayes Drilling, Inc. v. Curtiss-Manes Constr., 715 S.W.2d 295 (Mo. Ct. App. 1986).

[211] Shimek v. Vogel, 105 N.W.2d 677 (N.D. 1960).

Of course, once a change order is accepted and signed, it usually changes the contract sum and this must be reflected on all future payment applications.

7.1.2 A Change Order shall be based upon agreement among the Owner, Contractor and Architect; a Construction Change Directive requires agreement by the Owner and Architect and may or may not be agreed to by the Contractor; an order for a minor change in the Work may be issued by the Architect alone. **(7.1.2; no change)**

In each case, the architect must reasonably agree to the change. However, because the construction contract is between the owner and contractor, they could agree to changes without the architect by amending the contract, either orally or in writing, or even by conduct. Thus, if the owner and contractor agree to a change order, the change order is immediately effective, notwithstanding the lack of the architect's approval. This would, in most cases, be detrimental to the project but would be valid despite this provision.

The contractor must perform additional work that is within the general scope of the contract, although it must be paid for such additional work.

7.1.3 Changes in the Work shall be performed under applicable provisions of the Contract Documents, and the Contractor shall proceed promptly, unless otherwise provided in the Change Order, Construction Change Directive or order for a minor change in the Work. **(7.1.3; no change)**

Unless the change has a different time schedule, the contractor must promptly perform the additional work, even if it does not know the cost of the additional work or does not agree to the proposed change to the contract sum or contract time.

§ 4.41   Change Orders: ¶ 7.2

7.2 CHANGE ORDERS

7.2.1 A Change Order is a written instrument prepared by the Architect and signed by the Owner, Contractor and Architect, stating their agreement upon all of the following:

> *.1   the change in the Work;*
>
> *.2   the amount of the adjustment, if any, in the Contract Sum; and*
>
> *.3   the extent of the adjustment, if any, in the Contract Time.* **(7.2.1; minor modification)**

In a Utah case, the owner inserted a provision in a change order whereby he reserved the right to assert a claim against the contractor based on a delay.[212] If, the

---

[212] John Price Assocs. v. Davis, 588 P.2d 713 (Utah 1978). The contractor sued the owner on a promissory note. The owner defended and counterclaimed, saying that the project was late resulting in damages to the owner. Paragraph 4.3.5 of A201 waives claims by the owner related

contract time is extended in a change order, the owner cannot later assert that it was damaged by a delay of the contractor, because a change order was signed by all parties and constitutes a modification of the contract (¶ 1.1.2).[213] Normally, a contractor would refuse to sign a reservation of right in a change order, resulting in a construction change directive and subsequent litigation or arbitration.

In a Missouri case, change orders were submitted to the owner and the Farmer's Home Administration (FmHA), which provided financing for the public project.[214] The contract did not address the fact that a third party (the FmHA) was approving change orders concerning time extensions. When the owner later refused to comply with a change order granting the contractor a time extension that previously had been approved by the owner and contractor, but not FmHA, the court held that parties may make their performance conditional on the determination of a third party.

A change order must cover each of three items in ¶ 7.2.1.[215] If there is no change in the contract time or amount, it should be noted that no change is intended. The change order is all-inclusive. That is, a change order must indicate the change in contract amount, including the contractor's overhead and profit, so that the contractor cannot later request additional sums for a prior change order because overhead, profit, or similar items were not included. If additional contract time is indicated on the change order, the change order must include all additional costs, if any, associated with this additional time.

The prudent architect and owner will verify that the appropriate blanks in the change order form are filled in. For instance, the space for a change in time should have a "0" inserted instead of being left blank if no extension of time is anticipated for that change order. If either the contract time or the contract amount change

---

to delays, and the court held for the contractor. The reservation of rights by the owner in the change order was not similarly reserved in later change orders, in a certificate of substantial completion, or in making final payment. If the owner had reserved rights in making final payment, the result might have been different.

[213] J.A. Jones Constr. Co. v. Greenbrier Shopping Ctr., 332 F. Supp. 1336 (N.D. Ga. 1971) (change order was an accord and satisfaction and extinguished claims of the parties).

In Airline Constr., Inc. v. Barr, 807 S.W.2d 247 (Tenn. Ct. App. 1990), the appellate court held that the trial court could not rewrite the parties' contract. In that case, the owner and contractor had signed a change order for an underground sprinkler system. The trial court changed the amount of the change order, on the basis that it was already included in the original contract. On appeal, the court found no evidence that the change order has been procured by coercion, deceit, or fraud. Because the contract provided a means (change orders) whereby the parties could amend their contract, the courts could not interfere without evidence that the change order was improperly obtained.

[214] Twin River Constr. Co. v. Public Water Dist., 653 S.W.2d 682 (Mo. Ct. App. 1983) (unit price contract for construction of extensions to an existing water system; court held unit prices were valid in different geographic locations involved in the contract).

Another Missouri case, Midwest Materials Co. v. Village Dev. Co., 806 S.W.2d 477 (Mo. Ct. App. 1991), discusses various provisions of the General Conditions related to change orders.

[215] The AIA has a standard change order form, AIA Document G701.

spaces are left blank, the assumption must be that no change of that particular item was intended.

§ 4.42   Construction Change Directives: ¶ 7.3

7.3 CONSTRUCTION CHANGE DIRECTIVES

7.3.1 A Construction Change Directive is a written order prepared by the Architect and signed by the Owner and Architect, directing a change in the Work prior to agreement on adjustment, if any, in the Contract Sum or Contract Time, or both. The Owner may by Construction Change Directive, without invalidating the Contract, order changes in the Work within the general scope of the Contract consisting of additions, deletions or other revisions, the Contract Sum and Contract Time being adjusted accordingly. **(7.3.1; no change)**

Although a contractor must agree to a change order, a construction change directive is an order to the contractor to proceed immediately with the work without necessarily the contractor's agreement nor any agreement to the change in price or time. If the change is within the general scope of the project, the contractor must proceed. If the contractor refuses to proceed with a construction change directive, it runs the risk of being in default under the contract.

However, if the change is a major or *cardinal change,*[216] the contractor must agree to the change in order for it to be effective.[217] In some cases, it may be difficult to determine when such a change is beyond the scope of the contract.[218] The doctrine of *abandonment* is often used interchangeably with *cardinal change.*[219] This involves situations where the number of changes on a project is so excessive that they are beyond the reasonable expectation of the contractor, who

---

[216] A *cardinal change* is "an alteration in the work so drastic that it effectively requires the contractor to perform duties materially different from those originally bargained for." Allied Materials & Equip. Co., Inc. v. United States, 215 Ct. Cl. 406, 569 F.2d 562 (1978). *See also* Edward R. Marden Corp. v. U.S., 442 F.2d 364 (Ct. Cl. 1971) ("The cardinal change doctrine is not a rigid one. Its purpose is to provide a breach remedy for contractors who are directed by the Government to perform work which is not within the general scope of the contract. In other words, a cardinal change is one which, because it fundamentally alters the contractural undertaking of the contractor, is not comprehended by the normal Changes clause."); Merrill Contractors, Inc. v. GST Telecom, Inc., 2003 WL 464069 (Wash. App. Div. 2 Feb. 25, 2003).

  Not all states have adopted the cardinal change rule. Mellon Stuart Constr. v. Metropolitan Water Reclamation Dist. of Chicago, 1995 WL 124133 (N.D. Ill. March 20, 1995) (construing Illinois law), although older Illinois cases seem to take the opposite view. The County of Cook v. Henry Harms, 108 Ill. 151 (1883).

[217] *See* Daugherty Co. v. Kimberly-Clark Corp., 14 Cal. App. 3d 151, 92 Cal. Rptr. 120 (1971) ("abandonment of the contract can occur in instances where the scope of the work when undertaken greatly exceeds that called for under the contract.").

[218] Wunderlich Contracting Co. v. U.S., 351 F.2d 956 (Ct. Cl. 1965).

[219] *See* L.K. Comstock & Co., Inc. v. Becon Constr. Co., Inc., 932 F. Supp. 906 (E.D. Ky. 1993) ("Both abandonment and cardinal change may properly be utilized to establish a basis for recovery outside the original contract in cases where the contractual obligations of a construction

cannot reasonably recover the overall costs of the changes through the contract's change order procedures. The term *cardinal change* had its origin in governmental contracts, but can also apply to private contracts.[220] *Abandonment* is a similar concept, wherein a contract is abandoned expressly or implicitly by the parties acting inconsistently with its terms.[221] California draws a sharp distinction between these two doctrines, wherein the parties fail to follow the changes procedures in the contract under the abandonment concept, while the procedures are followed under the cardinal change concept.[222]

The architect prepares the construction change directive and, if a change in the contract amount is contemplated by the directive, the architect should include a proposed method of setting the cost change, based on ¶ 7.3.3. If no method is included, the procedure of ¶ 7.3.7 is used.

7.3.2 *A Construction Change Directive shall be used in the absence of total agreement on the terms of a Change Order.* **(7.3.2; no change)**

If the parties cannot agree on the terms of a change order, the work can be accomplished using the construction change directive. In many cases, the only difference between the two procedures is the lack of the contractor's agreement to the proposed terms of the change order.

7.3.3 *If the Construction Change Directive provides for an adjustment to the Contract Sum, the adjustment shall be based on one of the following methods:*

　.1　*mutual acceptance of a lump sum properly itemized and supported by sufficient substantiating data to permit evaluation;*

　.2　*unit prices stated in the Contract Documents or subsequently agreed upon;*

　.3　*cost to be determined in a manner agreed upon by the parties and a mutually acceptable fixed or percentage fee; or* **(7.3.3; no change)**

---

contractor vary materially from the original expectations of the parties regarding the scope and manner of work.")

[220] L.K. Comstock & Co., Inc. v. Becon Const. Co., Inc., 932 F. Supp. 906 (E.D. Ky. 1993) ("A cardinal change is a breach. It occurs when the government effects an alteration in the work so drastic that it effectively requires the contractor to perform duties materially different from those originally bargained for. By definition, then a cardinal change is so profound that it is not redressable under the contract, and thus renders the government in breach. . . .

This broad principle has been recognized by state courts as well as federal tribunals, and in private as well as public contract settings.")

[221] L.K. Comstock & Co., Inc. v. Becon Const. Co., Inc., 932 F.Supp. 906 (E.D. Ky. 1993); C. Norman Peterson Co. v. Container Corp. of Am., 172 Cal. App. 3d 628, 218 Cal. Rptr. 592 (Cal. App. 1st Dist. 1985) ("when an owner imposes upon the contractor an excessive number of changes such that it can fairly be said that the scope of the work under the original contract has been altered, an abandonment of contract properly may be found."); Schwartz v. Shelby Constr. Co., 338 S.W.2d 781 (Mo. 1960).

[222] Amelco Elec. v. City of Thousand Oaks, 27 Cal. 4th 228, 38 P.3d 1120 (Cal. 2002).

If the parties can agree, the construction change directive then becomes a change order.

> .4   *as provided in Section 7.3.7.* **(7.3.3.4; change reference)**

The contractor must keep accurate accounting records of all costs incurred for any changes. If the contractor disagrees with the proposed method of determining the change in the contract amount, it must submit a detailed breakdown of such costs, as outlined in ¶ 7.3.7.

> ***7.3.4*** *If unit prices are stated in the Contract Documents or subsequently agreed upon, and if quantities originally contemplated are materially changed in a proposed Change Order or Construction Change Directive so that application of such unit prices to quantities of Work proposed will cause substantial inequity to the Owner or Contractor, the applicable unit prices shall be equitably adjusted.* **(4.3.9; no change)**

Changes in quantities are matters of judgment. The intent is that if the quantities upon which unit prices are based are significantly changed, the unit prices will be adjusted accordingly. In a Florida case, the construction contract included furnishing one ton of asphalt for $180.[223] Subsequently, the owner and contractor signed a written change order for an additional 130 tons of asphalt for the same $180 per ton. When the owner then asked for an additional 381 tons of asphalt, the contractor requested the same price. The owner refused to pay, but the court found for the contractor because the additional asphalt was of the same nature and character as the work provided for in the first change order, and was therefore chargeable at the same rate. If the owner had not signed the first change order, the court would have charged the owner with a "reasonable" amount (presumably at a rate less than $180 per ton). The first change order, however, established the price for like quantities, which this was.

> ***7.3.5*** *Upon receipt of a Construction Change Directive, the Contractor shall promptly proceed with the change in the Work involved and advise the Architect of the Contractor's agreement or disagreement with the method, if any, provided in the Construction Change Directive for determining the proposed adjustment in the Contract Sum or Contract Time.* **(7.3.4; no change)**

The contractor must promptly proceed with the work indicated in the construction change directive (unless the architect specifies another time frame) and notify the architect in writing if it disagrees with the proposed method of determining the adjustment in the contract price or time. The architect may, instead of listing a

---

[223] Forest Constr. v. Farrell-Cheek Steel Co., 484 So. 2d 40 (Fla. Dist. Ct. App. 1986). This provision was referred to as the "equitable adjustment clause" in Bellvue Tech. Tower, LLC v. DPR Constr., Inc., 125 Wash. App. 1003 (2005). The original contract called for the removal of 500 cubic yards of contaminated soil, while the final amount was 24,555 cubic yards. The court held that the contractor was not entitled to an equitable adjustment because it failed to comply with the 21-day notice provision of A201.

method of determining the contract cost or time, simply indicate the proposed change in the contract cost or time. In that case, the contractor can disagree with the proposed figure, and it must do so in writing as promptly as possible. No time limit is stated, but the contractor should be able to do this within a matter of a few days. If the contractor does not give written notice of disagreement to the architect, the contractor will be held to have agreed with the stated amount or method of determining the amount.

> *7.3.6 A Construction Change Directive signed by the Contractor indicates the agreement of the Contractor therewith, including adjustment in Contract Sum and Contract Time or the method for determining them. Such agreement shall be effective immediately and shall be recorded as a Change Order.* (**7.3.5; no change**)

This simply states that if the contractor agrees to a construction change directive, including the time and cost change, or at least the methodology of determining these numbers, the construction change directive is transformed into a change order.

> *7.3.7 If the Contractor does not respond promptly or disagrees with the method for adjustment in the Contract Sum, the Architect shall determine the method and the adjustment on the basis of reasonable expenditures and savings of those performing the Work attributable to the change, including, in case of an increase in the Contract Sum, an amount for overhead and profit as set forth in the Agreement, or if no such amount is set forth in the Agreement, a reasonable amount. In such case, and also under Section 7.3.3.3, the Contractor shall keep and present, in such form as the Architect may prescribe, an itemized accounting together with appropriate supporting data. Unless otherwise provided in the Contract Documents, costs for the purposes of this Section 7.3.7 shall be limited to the following:* (**7.3.6; minor modification**)

The architect will make a determination of the amount of the change in cost or time if the parties do not agree,[224] based on the contractor's submitted data as

---

[224] The 1970 version of A201 contained this provision: "If the Owner and Contractor cannot agree on the amount of the adjustment in the contract sum, it shall be determined by the Architect." In Elec-Trol, Inc. v. C.J. Kern Contractors, Inc., 54 N.C. App. 626, 284 S.E.2d 119 (1981), *appeal denied,* 305 N.C. 299, 290 S.E.2d 701 (1982), the court found that under the 1970 provision, the architect would determine the amount of claims for additional cost if the owner and contractor could not agree, and that such a determination was final and binding on the parties, absent bad faith or failure to exercise honest judgment by the architect. This language was also at issue in Cal Wadsworth Constr. v. City of St. George, 865 P.2d 1373, 1376 (Utah Ct. App. 1993). The owner, a municipality, had requested bids from contractors. After the low bid was disqualified, the plaintiff's bid was accepted, with the qualification that it be accepted subject to negotiation of the price. No formal contract was ever executed, and the parties never reached agreement on the cost of the project. The contractor sued for breach of contract, arguing that this clause, which permitted the owner to order changes to the contract without invalidating the contract, indicates an intent to accept the contract. The court held that "[e]ven when we consider the standard AIA provision that the City may unilaterally issue a change directive for changes within the general scope of the contract, we nevertheless conclude that the City's response contained material reservations or

indicated below. If either the contractor or owner disagree with the architect's determination, the procedure in Article 15 must be used to resolve the dispute. This provision also encourages the parties to spell out a formula for calculating overhead and profit, otherwise, the architect will determine what a reasonable amount for these items will be.

> *.1   costs of labor, including social security, old age and unemployment insurance, fringe benefits required by agreement or custom, and workers' compensation insurance;* **(7.3.6.1; no change)**

These costs are usually easily ascertainable.

> *.2   costs of materials, supplies and equipment, including cost of transportation, whether incorporated or consumed;* **(7.3.6.2; no change)**

The contractor should have invoices, cancelled checks, mileage logs, and so on.

> *.3   rental costs of machinery and equipment, exclusive of hand tools, whether rented from the Contractor or others;* **(7.3.6.3; no change)**

This could be based on actual invoices or on standard rates for equipment rental in the location of the project.

---

conditions, and was therefore not within the general scope of the contract." Thus, the owner's acceptance of the contractor's offer was not a true acceptance, and no contract was ever formed.

In RaDec Constr., Inc. v. School Dist. No. 17, 248 Neb. 338, 535 N.W.2d 408 (1995), the contract contained the following provision:

12.1.4 . . . The Contractor, provided he receives a written order signed by the Owner, shall promptly proceed with the Work involved. The cost of such Work shall then be determined by the Architect on the basis of the reasonable expenditures and savings of those performing the Work attributable to the change, including, in the case of an increase in the Contract Sum, a reasonable allowance for overhead and profit. In such case . . . the Contractor shall keep and present, in such form as the Architect may prescribe, an itemized accounting together with appropriate supporting data for inclusion in a Change Order. Unless otherwise provided in the Contract Documents, cost shall be limited to the following: cost of materials, including sales tax and cost of delivery; cost of labor, including social security, old age and unemployment insurance, and fringe benefits required by agreement or custom; workers' or workmen's compensation insurance; bond premiums; rental value of equipment and machinery; and the additional costs of supervision and field office personnel directly attributable to the change. Pending final determination of cost to the Owner, payments on account shall be made on the Architect's Certificate for Payment. The amount of credit to be allowed by the Contractor to the Owner for any deletion or change which results in a net decrease in the Contract Sum will be the amount of the actual net cost as confirmed by the Architect. When both additions and credits covering related Work or substitutions are involved in any one change, the allowance for overhead and profit shall be figured on the basis of the net increase, if any, with respect to that change.

After the contract was signed, the architect revised the earthworking plan so that no fill would be required and asked the contractor to provide a suitable credit. The credit proposed by the contractor was then rejected, and the architect came up with his own figure. The court found that the architect's determination of the amount of the credit was patently erroneous and therefore legally equivalent to bad faith, so the court could reverse the architect's determination.

.4   *costs of premiums for all bonds and insurance, permit fees, and sales, use or similar taxes related to the Work; and* **(7.3.6.4; no change)**

The contractor should have invoices or other forms of proof for all such items.

.5   *additional costs of supervision and field office personnel directly attributable to the change.* **(7.3.6.5; no change)**

All of the above costs are costs that are directly attributable to the change. The contractor must be able to substantiate these with time tickets.

7.3.8 The amount of credit to be allowed by the Contractor to the Owner for a deletion or change that results in a net decrease in the Contract Sum shall be actual net cost as confirmed by the Architect. When both additions and credits covering related Work or substitutions are involved in a change, the allowance for overhead and profit shall be figured on the basis of net increase, if any, with respect to that change. **(7.3.7; minor modification)**

7.3.9 Pending final determination of the total cost of a Construction Change Directive to the Owner, the Contractor may request payment for Work completed under the Construction Change Directive in Applications for Payment. The Architect will make an interim determination for purposes of monthly certification for payment for those costs and certify for payment the amount that the Architect determines, in the Architect's professional judgment, to be reasonably justified. The Architect's interim determination of cost shall adjust the Contract Sum on the same basis as a Change Order, subject to the right of either party to disagree and assert a Claim in accordance with Article 15. **(7.3.8; include payments for Construction Change Directives; delete references to amounts in dispute; minor modifications)**

While the contractor is performing the work under a construction change directive, the contractor can ask for payment for work performed and materials installed, and the architect will make a determination as to whether the owner should pay such amounts. These interim determinations are subject to the Article 15 dispute resolution procedures, although in most cases, the entire construction change directive work will have been completed before that process is actually implemented. Once agreement has been reached on the changed cost, the contractor can substitute the actual agreed-upon cost. In the event of a net decrease in contract cost, there is no provision for decrease in overhead. If there is a net increase, the additional overhead allowed the contractor will be based only on the net increase. Assume that one part of the change involves a decrease of $1,000, while another part involves an increase of $1,500. The contractor will be allowed additional overhead based on $500, the net increase. Obviously, it would be to the contractor's advantage to separate out decreases to the contract amounts into separate change orders to maximize the overhead increases. The architect and owners would want to lump all revisions based on an overall change together. If the parties fail to agree on a change order following these processes, claims can be made as set forth in Article 15.

7.3.10 When the Owner and Contractor agree with a determination made by the Architect concerning the adjustments in the Contract Sum and Contract Time, or otherwise reach agreement upon the adjustments, such agreement shall be effective immediately and the Architect will prepare a Change Order. Change Orders may be issued for all or any part of a Construction Change Directive. **(7.3.9; minor revision; add last sentence)**

This merely restates that whenever agreement is reached as to the terms of a construction change directive, it becomes a change order. If no agreement is reached, the contractor can follow the procedures found in Article 15. The last sentence states that a change order may be issued for only part of a construction change directive. This might apply where there is agreement as to part of the construction change directive, but another part is disputed and might only be resolved by the claims procedure. In the meantime, however, the contractor will want to be fully paid for the part that is not in dispute. This provides for such an occurrence.

§ 4.43 Minor Changes in the Work: ¶ 7.4

7.4 MINOR CHANGES IN THE WORK

The Architect has authority to order minor changes in the Work not involving adjustment in the Contract Sum or extension of the Contract Time and not inconsistent with the intent of the Contract Documents. Such changes will be effected by written order signed by the Architect and shall be binding on the Owner and Contractor. **(7.4.1; minor revisions)**

A minor change in the work is defined as a change that does not involve either a change in the contract sum or the time to complete the work. An example of this might be a change in a wall color before the contractor has purchased the paint.

If the contractor believes that it is entitled to an adjustment in the contract sum or the contract time, the contractor can follow the procedure for resolving disputes found in Article 15.[225] Because of this possibility, the architect would be wise to not unilaterally order a minor change unless the contractor confirms in advance that the price and time will not change. The best practice is to not use a minor change at all, and use the change order procedure for such changes, with the form indicating what the change is and include a "0" for time and cost change.

---

[225] This provision was discussed in Turner Excavating Co., Inc. v. GWX Ltd. P'ship, 2002 WL 4603 (Minn. Ct. App. Jan. 2, 2002). There, the contractor sought $150,049 for extra work and relied on directions from the engineer to "dig deeper." The court held that this provision, absent other facts, did not confer authority on the engineer to bind the owner.

§ 4.44   Article 8: Time

This article defines beginning and substantial completion dates for construction and addresses delays and time extensions.

8.1 DEFINITIONS

8.1.1 Unless otherwise provided, Contract Time is the period of time, including authorized adjustments, allotted in the Contract Documents for Substantial Completion of the Work. **(8.1.1; no change)**

Note that this requires only substantial completion (**see ¶** 9.8). Any changes in the contract time because of change orders or construction change directives will adjust this time. Thus, the contract time can only be adjusted with the consent of the owner, except through the arbitration or litigation procedure.[226]

8.1.2 The date of commencement of the Work is the date established in the Agreement. **(8.1.2; no change)**

Unless the owner directs the contractor to postpone the work, the time starts as of the date stated in the owner-contractor agreement, whether or not the contractor actually starts the work at that time. Note that the parties should insert a date of commencement in the owner-contractor agreement. See, e.g., AIA Document A101, ¶ 3.1. If this date is omitted, the courts would probably determine that the contractor will be given a reasonable amount of time from the date that the agreement is executed, perhaps 14 days.

8.1.3 The date of Substantial Completion is the date certified by the Architect in accordance with Section 9.8. **(8.1.3; no change)**

If the position of architect is vacant at the time of substantial completion, or if the architect refuses or neglects to issue the certificate of substantial completion,

---

[226] In Fluor Daniel Carribean, Inc., v. Humphreys (Cayman), Ltd., 2005 WL 1214278 (S.D. N.Y. May 23, 2005), ¶ 8.1.1 was modified to read as follows:

> 8.1.1 Time is of the essence in the performance of the Work by [Fluor] and its Subcontractors. . . . In the event [Fluor] fails to achieve Substantial Completion or complete the Work by the dates set forth in the Construction Schedule. . . . [Fluor] agrees to pay liquidated damages only as set forth in Paragraph 4.2 of the [Contract]. Under no circumstances, whether arising out of contract, tort (including negligence), strict liability, warranty or otherwise, shall [Fluor] be liable for special, indirect, incidental, exemplary or consequential damages of any nature, other than the aforementioned liquidated damages. Notwithstanding anything herein or in the [Contract] to the contrary, [Fluor's] aggregate liability for direct and liquidated (if any) damages shall in no case exceed 10% of the [C]ontract value, and [Humphreys] agrees to release [Fluor] for all damages in excess of 10% of the [C]ontract value.

The issue is whether this 10% restriction is a complete limit of Fluor's liability or only a limitation on damages related to work delays. Summary judgment was denied pending further factual determinations.

the date would probably be the date on which the owner occupies the property, or possibly the date the municipality issues a certificate of occupancy.

8.1.4 The term "day" as used in the Contract Documents shall mean calendar day unless otherwise specifically defined. **(8.1.4; no change)**

This is as opposed to business day or 24-hour periods. Thus, the period from 5:00 P.M. on one day to noon the next day constitutes two days. Note that weekends and holidays count as "days" for all purposes, so that the effective period of time in which to give a notice might be significantly shortened on holiday weekends.

§ 4.45   Progress and Completion: ¶ 8.2

8.2 PROGRESS AND COMPLETION

8.2.1 Time limits stated in the Contract Documents are of the essence of the Contract. By executing the Agreement the Contractor confirms that the Contract Time is a reasonable period for performing the Work. **(8.2.1; no change)**

The contractor here warrants that the time provided is reasonable. By implication, any delay could cause damages to the owner for which the contractor may be liable. The contractor should exercise care in agreeing to the allotted time. This provision is a time-is-of-the-essence clause, and it means that the contractor must strictly adhere to the time limits. If the contractor fails to perform within the stated time, the owner may be able to treat this failure as a total breach of contract.[227] Without a time-is-of-the-essence clause, the completion time is less important, and

---

[227] Monmouth Pub. Sch. Dist. 38 v. D.H. Rouse Co., 153 Ill. App. 3d 901, 506 N.E.2d 315 (1987) (construed ¶¶ 13.4.1 and 14.2.2 of A201, the 7-day notice provision for termination of the contractor).

In Vermont Marble Co. v. Baltimore Contractors, Inc., 520 F. Supp. 922 (D. D.C. 1981), under a time-is-of-the-essence clause, the subcontractor attempted to rescind the contract when it was delayed. The court ruled that the contract, read as a whole, did not give a right of rescission for delay to the sub. The proper remedy was to obtain delay damages. The parties had included the following clause in the contract:

[1] Should Subcontractor be obstructed or delayed in the commencement, prosecution or completion of the Work because of conditions attributable to Owner and which by the terms of the Prime Contract may be grounds for an extension of time or money damages, Subcontractor shall promptly make claim therefore in writing, and Contractor shall present said claim to the Owner. Contractor shall pay to Subcontractor whatever money damages or extensions of time allowed by Owner to Contractor under the Prime Contract for said delay claim. [2] Contractor for just cause shall have the right at any time to delay or suspend the commencement or execution of the whole or any part of the Work without compensation or obligation to Subcontractor other than to extend the time for completing the Work for a period equal to that of such time or suspension.

In Commonwealth Constr. Co. v. Cornerstone Fellowship Baptist Church, Inc., 2006 WL 2567916 (D. Super. Aug. 31, 2006), the owner's two-month delay in obtaining a necessary variance constituted a breach of the contract.

it is more difficult for an owner to obtain damages for delays. The owner can waive a time-is-of-the-essence clause by permitting the contractor to work past the scheduled completion date.[228] In one case, the contractor argued that the owner warranted that weather conditions prevailing during the construction would permit completion of construction.[229] The owner had furnished weather data to the contractor for the prior winter, and the contractor argued that this amounted to a representation that the weather would be no worse the following year. The contractor was substantially delayed by weather. The court refused to find that the owner had made such a warranty, and found that the risk of abnormal weather conditions was to be borne by the contractor.

> *8.2.2 The Contractor shall not knowingly, except by agreement or instruction of the Owner in writing, prematurely commence operations on the site or elsewhere prior to the effective date of insurance required by Article 11 to be furnished by the Contractor and Owner. The date of commencement of the Work shall not be changed by the effective date of such insurance.* **(8.2.2; deleted last sentence of prior version)**

Usually, the date of commencement is the date specified in the owner-contractor agreement. If another date is chosen, these provisions must be followed. This is for the protection of both the owner and contractor, because the insurance does not protect the parties before its effective date. The security interests referred to here relate to the interests of the owner's lenders and others who may want to have a priority over the contractor and others working at the project. If the owner permits the contractor to start construction before a lender's mortgage or other security interest is filed, the owner may be in default on its loan.

> *8.2.3 The Contractor shall proceed expeditiously with adequate forces and shall achieve Substantial Completion within the Contract Time.* **(8.2.3; no change)**

Based on the project schedule submitted by the contractor (¶ 3.10), the architect can withhold certification (¶ 9.5.1) if the contractor falls behind. The schedule prepared by the contractor in ¶ 3.10 is important here. If the contractor falls behind schedule, it is in default. If the owner has failed to require the schedule, there is no way to determine if the contractor is behind schedule until the end of the project.

---

[228] Baker Domes v. Wolfe, 403 N.W.2d 876 (Minn. Ct. App. 1987); Ryan v. Thurmond, 481 S.W.2d 199 (Tex. 1972). *See also* Herbert & Brooner Constr. Co. v. Golden, 499 S.W.2d 541 (Mo. 1973) (owner's waiver of time is of the essence clause does not relieve contractor from damages for the delay. Waiver of the time of performance is not the same as waiver of damages for failure to perform in time.).

[229] Associated Eng'rs & Contractors, Inc. v. State, 58 Haw. 187, 567 P.2d 397 (1977).

§ 4.46   Delays and Extensions of Time: ¶ 8.3

8.3 DELAYS AND EXTENSIONS OF TIME

8.3.1 If the Contractor is delayed at any time in the commencement or progress of the Work by an act or neglect of the Owner or Architect, or of an employee of either, or of a separate contractor employed by the Owner; or by changes ordered in the Work; or by labor disputes, fire, unusual delay in deliveries, unavoidable casualties or other causes beyond the Contractor's control; or by delay authorized by the Owner pending mediation and arbitration; or by other causes that the Architect determines may justify delay, then the Contract Time shall be extended by Change Order for such reasonable time as the Architect may determine. **(8.3.1; minor modifications)**

All of the listed causes are beyond the contractor's control. If any of these events occur, the architect must issue a change order to extend the contract time. If the contractor or owner refuses to sign such a change order, the parties should follow the dispute resolution procedure of Article 15. In some circumstances, the contractor may be entitled to additional costs because of such delays (**see** ¶ 8.3.2).[230] The delay can be either before the start of the project or during the project itself.

Some contracts provide for liquidated damages[231] if the project is not completed on time. Any extension of time delays the commencement of such liquidated

---

[230] *See* Nelse Mortensen & Co. v. Group Health Coop., 17 Wash. App. 703, 566 P.2d 560 (1977) (court interpreted an amended A201 relating to this and other time provisions).

In McGee Constr. Co. v. Neshobe Dev., Inc., 594 A.2d 415 (Vt. 1991), the court reviewed the three paragraphs of ¶ 8.3 and stated that "the avenue for recovery of delay damages under [¶ 8.3.3] was the bringing of a claim for additional cost in accordance with the procedure set forth in [¶ 4.3]." The court further stated that if the contractor had walked off the job without following the claims procedure, the contractor would have been in material breach of the contract. *See also* Airline Constr. Inc., v. Barr, 807 S.W.2d 247 (Tenn. Ct. App. 1990).

In Foley Co. v. Walnut Assocs., 597 S.W.2d 685 (Mo. Ct. App. 1980), the court held that this provision exonerates the contractor from delays beyond its control, including labor disputes. The fact that the architect did not issue a change order did not affect the contractor's right to an extension of time.

See also, Murdock & Sons v. Goheen General Constr., Inc., 461 F.3d 837 (7th Cir. 2006).

In Bouchard v. Boyer, 1999 Conn. Super. LEXIS 1285 (May 17, 1999), the court found that the contractor should have been granted a change order for an extension of time when the contractor was unable to timely secure a building permit for reasons beyond the control of the contractor.

In Watson Elec. v. City of Winston-Salem, 109 N.C. App. 194, 426 S.E.2d 420 (1993), the court found a similar provision ambiguous when read in conjunction with a no-damage-for-delay clause in the contract.

[231] "Liquidated damages clauses are essentially artificial damages agreed to at the time of contracting, and these clauses are enforceable if actual damages are difficult to ascertain, and if the liquidated damages provision is a reasonable estimate of the damages which would actually result from a breach of the contract. However, if both of these criteria are not met, then the liquidated damages clause is unenforceable since, in that situation, the provision would actually be a penalty." Calumet Constr. Corp., v. Metropolitan Sanitary Dist. of Greater Chicago, 533 N.E.2d 453 (Ill. App. Ct. 1st Dist. 1988).

damages.[232] In the past, courts did not favor liquidated damages clauses.[233] The more modern view, favored by a majority of jurisdictions, is to allow the contracting parties to address damages in the contract unless the damages are obviously unreasonable.[234] If the contract contains a valid liquidated damages clause, the injured party is entitled to the liquidated damages, irrespective of whether or not actual damages were incurred.[235]

8.3.2 Claims relating to time shall be made in accordance with applicable provisions of Article 15. (**8.3.2; change reference**)

As with most other claims under the contract documents, the parties must first submit the claim to the architect and then to litigation or arbitration, depending on what method of dispute resolution has been selected by the parties. The contractor must submit its claim for an extension of time within 21 days of the occurrence of the event that triggers the claim, otherwise the claim is waived.[236] One case[237] held

---

[232] In Arrowhead, Inc. v. Safeway Stores, Inc., 179 Mont. 510, 587 P.2d 411 (1978), a contractor claimed delays of 144 days. The court found that the owner properly allowed the contractor an additional 51 days before assessing liquidated damages.

[233] General Ins. Co. v. Commerce Hyatt House, 5 Cal. App. 3d 460, 85 Cal. Rptr. 317 (1970). There, the court stated:

"It is well established that where the owners seek liquidated damages pursuant to the provisions of a contract, they must show that they have strictly complied with all requisites to the enforcement of that contractual provision. An owner whose acts have contributed substantially to the delayed performance of a construction contract may not recover liquidated damages on the basis of such delay." Liquidated damages are a penalty not favored in equity and should be enforced only after he who seeks to enforce them has shown that he has strictly complied with the contractual requisite to such enforcement." [Citations omitted]

In Centex-Rodgers Constr. Co. v. McCann Steel Co., 426 S.E.2d 596 (Ga. Ct. App. 1992), the court found that the general contractor was not limited to the liquidated damages provision against a subcontractor because the sub had breached a material term of its contract. The general could recover its actual damages, which were apparently greater than the liquidated damages amount.

[234] Calumet Constr. Corp., v. Metropolitan Sanitary Dist. of Greater Chicago, 533 N.E.2d 453 (Ill. App. Ct. 1st Dist. 1988) ("While liquidated damages provisions have traditionally been subject to a court's scrutiny because of their inherent speculative nature, liquidated damages provisions have been recognized more recently as appropriate in circumstances where the complexity of contractual relationships make damages difficult to determine, since reasonably related, agreed upon liquidated damage amounts are easy to apply in such situations and satisfy the needs of the parties.")

[235] Southwest Eng'g Co., v. U.S., 341 F.2d 998 (8th Cir. 1965) (Government entitled to liquidated damages even if there were no actual damages). In Worthington Corp. v. Consolidated Aluminum Corp., 544 F.2d 227 (5th Cir. 1976), the owner sought some $4 million from the contractor. The contract, however, set the liquidated damages at $500 per day, with a maximum of $100,000. The owner was not entitled to more than this maximum.

[236] State Sur. Co. v. Lamb Constr. Co., 625 P.2d 184 (Wyo. 1981). In American Nat'l Elec. Corp. v. Poythress Commercial Contractors, Inc., 604 S.E.2d 315 (N.C. Ct. App. 2004), the subcontract incorporated A201. The subcontractor was aware in April 2000 that its work was being delayed, but did not notify the general contractor of its delay claim in writing until a letter dated 20 September 2000. This lack of timely notice defeated the claim.

[237] Steinberg v. Fleischer, 706 S.W.2d 901 (Mo. Ct. App. 1986).

that such a submittal to the architect was not necessary when the architect was an owner of the project. A California court, interpreting an earlier version of A201 which allowed only seven days to make such a claim for additional time, held the provision was inapplicable when the owner caused some of the delays.[238]

8.3.3 This Section 8.3 does not preclude recovery of damages for delay by either party under other provisions of the Contract Documents. **(8.3.3; no change)**

One court interpreted this provision related to delay damages to be different from other claims under the contract.[239] If the contract has a liquidated damages clause, the contractor may be liable to the owner for delay, even if it was caused by weather or other things beyond the contractor's control.[240] The owner can waive or extend the time within which a building contract is to be performed by the contractor.[241]

---

[238] General Ins. Co. v. Commerce Hyatt House, 5 Cal. App. 3d 460, 85 Cal. Rptr. 317 (1970).

[239] Osolo Sch. Bldgs., Inc. v. Thorleif Larsen & Son, Inc., 473 N.E.2d 643 (Ind. Ct. App. 1985). The court made this analysis:

> Additional costs arise when construction is begun in a timely fashion and extra work is necessitated by changed specifications or unforeseen conditions. While the parties would seek to avoid additional costs, such costs can be contemplated by the parties as they occur frequently in construction situations. On the other hand, delay damages are extra costs that arise solely as a result of delay by the owner, contractor, or subcontractor. Party-caused delays are not contemplated by the parties, and the damages resulting therefrom should not be treated as additional costs. Delays are breaches of implied or express contractual provisions, whereas additional costs do not stem from any such breach. Therefore, in the absence of express language in the construction contract equating the treatment of claims for additional costs and delay damages, they will not be treated the same.

> The *Osolo* case was discussed in Starks Mech., Inc. v. New Albany-Floyd County Consolidated Sch. Corp., 854 N.E.2d 936 (Ind. App. 2006). The school filed a declaratory judgment action against the mechanical and plumbing contractor, asserting that the contractor failed to provide timely notice of a claim. In this case, the contract contained express language equating the treatment of claims for delay damages to the treatment of all other claims, unlike the *Osolo* case. Here, the general conditions required notice by the contractor within 14 days, as well as weekly updates of the claim and estimated associated costs to the construction manager for the duration of the delay. The contractor failed to do this, and the court held in favor of the owner.

[240] Associated Eng'rs & Contractors, Inc. v. State, 58 Haw. 187, 567 P.2d 397 (1977).

> In Twin River Constr. Co. v. Public Water Dist., 653 S.W.2d 682 (Mo. Ct. App. 1983), the court held that an owner could recover both liquidated damages and actual damages, as long as they are not duplicative.

[241] Ryan v. Thurmond, 481 S.W.2d 199 (Tex. 1972). The court said:

> The waiver or extension may be implied as well as express. What acts or omissions will constitute a waiver, depends on the nature and the circumstances of each case. If, after the time for completion of the work has expired, and the owner assents to the continuance of the work without objection to the delay, he will be deemed to have waived the provision as to time of performance. [Citation omitted.] The effect of an extension or waiver of time for performance is merely to substitute a new time for the old. It does not affect the other provisions of the contract.

In an Illinois case, the contract had a no-damages-for-delay clause,[242] which courts

---

[242] Bates & Rogers Constr. v. North Shore Sanitary Dist., 109 Ill. 2d 225, 486 N.E.2d 902 (1985), *aff'g* 128 Ill. App. 3d 962, 471 N.E.2d 915 (1984). The clause read:

> The Contractor agrees to make no claim for damages for delay in the performance of this Contract occasioned by any act or omission to act of the District or any of its representatives, or because of any injunctions which may be brought against the District or its representatives, and agrees that any such claim shall be fully compensated for by an extension of time to complete performance of the work as provided herein.

The contractor bid on the basis of this exculpatory clause, and its bid should have reflected the possibility that delays could result. The architect was a third-party beneficiary of this contract and the exculpatory clause was a defense for him. This clause also bound the subcontractors because the main contract required that subcontracts include similar language.

In J.R. Stevenson Corp. v. County of Westchester, 113 A.D.2d 918, 493 N.Y.S.2d 819 (1985), the contract contained a no-damages-for-delay clause, a hold harmless clause, and a liquidated damages clause fixing damages at $300 per day. The owner filed suit against the contractor and asked for actual damages, which exceeded the $300 per day amount. The court held that "a liquidated damages clause which is reasonable precludes any recovery for actual damages. This is so even though the stipulated sum may be less than the actual damages sustained by the injured party." The court further held that a no-damages-for-delay clause does not bar an action based on delays or obstructions that were not within the contemplation of the parties when the contract was executed, or that resulted from willful or grossly negligent acts of the other party.

The court in Hartford Elec. Applicators, Inc. v. Alden, 169 Conn. 177, 363 A.2d 135 (1975), stated that "the majority of jurisdictions in this country hold that where there are delays attributable to both parties the liquidated damages clause based upon the contract date is abrogated; the contract is to be performed in a reasonable time."

In Carrabine Constr. Co. v. Chrysler Realty Corp., 250 Ohio St. 3d 222, 495 N.E.2d 952 (1986), the contract contained the following no-damages-for-delay clause:

> The Contractor shall have no claim against the Owner for an increase in the contract price or a payment or allowance of any kind based on any damage, loss or additional expense the Contractor may suffer as a result of any delays in prosecuting or completing the work under the contract, whether such delays are caused by the circumstances set forth in the preceding paragraph or by any other circumstances. It is understood that the contractor assumes all risks of delays in prosecuting or completing the work under the contract.

A zoning ordinance was required to permit the project to proceed, and a delay resulted. The contractor was not allowed to recover damages because of this delay. *Carrabine* was cited in Dugan & Meyers Constr. Co., Inc., v. Ohio Dep't of Admin. Serv., 864 N.E.2d 68 (Ohio 2007), wherein the court held that the *Spearin* Doctrine does not allow the contractor to collect delay damages in the face of a no-damages-for-delay clause, and discussed cumulative impact claims.

In Unicon Mgmt. Corp. v. City of Chicago, 404 F.2d 627 (7th Cir. 1968), the contract had a no-damage-for-delay clause that included delay caused by the owner (with the possible exception of bad faith by the owner).

In Ace Stone, Inc. v. Township of Wayne, 47 N.J. 431, 221 A.2d 515 (1966), the contract contained this clause: "No claim for damages or any claim other than for an extension of time as herein provided shall be made or asserted against the Owner by reason of the delays herein-before mentioned." The court stated that when the parties enter into a construction contract with a customary no-damage clause they clearly contemplate that the contractor himself will bear the risks of the ordinary and usual types of delay incident to the progress and completion of the work. *See also* Vermont Marble Co. v. Baltimore Contractors, Inc., 520 F. Supp. 922 (D. D.C. 1981).

consistently uphold.[243] Typical no-damages-for-delay clauses allow for additional

---

In Watson Elec. v. City of Winston-Salem, 109 N.C. App. 194, 426 S.E.2d 420, 422 (1993), the contract contained this language:

> If the contractor is delayed by the Owner or Architect or any Agent or employee of either, the Contractor's sole and exclusive remedy for the delay shall be the right to a time extension for completion of the Contract and not damages.

The contractor in this case was claiming damages not for the delay but for the architect's refusal to grant it the time extension. The court found this provision ambiguous when read with this version of ¶ 8.3.1.

[243] Peter Kiewit Sons' Co. v. Iowa S. Utils. Co., 355 F. Supp. 376 (S.D. Iowa 1973); Herlihy Mid-Continent Co. v. Sanitary Dist., 390 Ill. 160, 60 N.E.2d 882 (1945); Annotation, *Validity, Construction, and Application of "No Damage" Clause with Respect to Delay in Construction Contract,* 10 A.L.R.2d 801 (1950).

In Kalisch-Jarcho, Inc. v. City of N.Y., 58 N.Y.2d 377, 448 N.E.2d 413, 461 N.Y.S.2d 746 (1983), a no-damage-for-delay clause was enforceable, even for unreasonable delay, unless the owner was guilty of gross negligence. If the owner were guilty of gross negligence, even if the gross negligence was within the contemplation of the parties to the contract, the exculpatory clause would be void as being against public policy. Accord Williams & Sons Erectors, Inc. v. South Carolina Steel, 983 F.2d 1176 (2d Cir. 1993).

In Nix, Inc. v. City of Columbus, 111 Ohio App. 133, 171 N.E.2d 197 (1959), the city and contractor signed a road contract containing a no-damage-for-delay clause. Unbeknownst to either party, the city had not secured a right of way, resulting in a substantial delay in construction and damages to the contractor. The court held that, because these particular damages were not within the contemplation of the parties at the time the contract was made, the contractor was entitled to damages.

In Buckley & Co. v. State, 140 N.J. Super. 289, 356 A.2d 56 (1975), the court stated that "a no-damage provision ought not be construed as exculpating a contractee from that liability unless the intention to do so is clear." The courts should look for ways to avoid the result of this clause, according to the opinion.

In Earthbank Co. v. City of N.Y., 145 Misc. 2d 937, 549 N.Y.S.2d 314 (Sup. Ct. 1989), the owner failed to obtain a permit required by law. The contractor was entitled to damages despite this clause (identical to the clause in *Kalisch-Jerico*).

Spearin v. City of N.Y., 160 A.D.2d 263, 553 N.Y.S.2d 372 (1990), held that when delay is caused by the owner's bad faith or willful, malicious, or grossly negligent conduct, there is no waiver of delay damages. *Accord* Blau Mech. Corp. v. City of N.Y., 158 A.D.2d 373, 551 N.Y.S.2d 228 (1990).

In Broadway Maint. Corp. v. Rutgers, 90 N.J. 253, 447 A.2d 906 (1982), the contract contained the following clause:

> If the Contractor is delayed in completion of the work by any act or neglect of the Owner, Architect, or of any other Contractor employed by the Owner, or by changes ordered in the work, or by strikes, lockouts, fire, unusual delay by common carriers, unavoidable casualties, or any cause beyond the Contractor's control or by any cause which the Architect shall decide to justify the delay, then for all such delays and suspensions the Contractor shall be allowed one day additional to the time limitations herein stated for each and every day of such delay so caused in the completion of the work, the same to be ascertained solely by the Architect, and a similar allowance of extra time will be made for such other delays as the Architect may find to have been caused by the Owner.

> No such extension of time shall be made for any one or more delays unless within three (3) days after the beginning of such delays a written request for additional time shall be filed with the Architect. In case of a continuing cause of delay, only one request is necessary.

No claim for damages or any claim other than for extensions of time as herein provided shall be made or asserted against the Owner by reason of any of the delays herein mentioned.

Anything contained in the Contract to the contrary notwithstanding the Contractor shall not be entitled to damages or to extra compensation by reason of delays occasioned by proceedings to review the awarding of the Contract to the Contractor or to review the awarding of any other Contract to any other Contractor.

However, in John E. Green Plumbing & Heating Co. v. Turner Constr. Co., 742 F.2d 965 (6th Cir. 1984), the court held that a similar no-damage-for-delay clause bars only delay damages, as opposed to other kinds of damages, such as damages for hindering work on the project. It defined delay damages as simply the cost of idle workers. Although the decision says it follows other decisions in this area, the reasoning is poor, and it is at odds with the majority of decisions.

In Phoenix Contractors, Inc. v. General Motors Corp., 135 Mich. App. 787, 355 N.W.2d 673 (1984), the owner contracted directly with various contractors and gave instructions to one of them to not let the other contractors keep it from completing its work on time. Phoenix was hindered by this contractor, because Phoenix could not perform its work while the other contractor was working. Phoenix did manage to complete its work on time, although it expended considerably more money to do so. The court held that, despite the no-damage-for-delay clause, the plaintiff could proceed with its action against the owner.

In Chicago Coll. of Osteopathic Med. v. George A. Fuller Co., 776 F.2d 198 (7th Cir. 1985), the subcontract had this provision:

> The Subcontractor expressly agrees not to make, and hereby waives, any claim for damages on account of any delay, obstruction or hindrance from any cause whatsoever . . . and agrees that its sole right and remedy in the case of any delay, obstruction or hindrance shall be an extension of the time fixed for completion of the Work.

However, the no-damage-for-delay clause in *Chicago College* was held to be waived (despite the existence of a clause prohibiting oral waivers) where the general repeatedly promised the subcontractor that the general would be personally liable to the owner for the delays and the general stated that the delay waiver clause would not apply to the extraordinary delays involved in this case.

In Marriott Corp. v. Dasta Constr. Co., 26 F.3d 1057, 1065-67 (11th Cir. 1994), the contract contained this provision:

> TIME IS OF THE ESSENCE OF THIS AGREEMENT. The Owner may sustain financial loss if the project or any part thereof is delayed because the Contractor fails to perform any part of the work in accordance with the contract documents, including, without limitation, a failure to comply with the schedule for this project, or any revision thereof, established by Owner. The Contractor shall begin the work at the time directed by the Owner and perform its obligations under this agreement with diligence and sufficient manpower to maintain the progress of the work as scheduled by Owner, without delaying other contractors or areas of work. At the request of the Owner, the Contractor shall perform certain parts of the work before other parts, add extra manpower, or order overtime labor in order to comply with the schedule (or any revision thereof), all without any increase in the contract sum.
>
> If the Contractor is delayed at any time in the progress of the Work by any act or neglect of Owner or by any contractor employed by Owner, or by changes ordered in the scope of the Work, or by fire, adverse weather conditions not reasonably anticipated, or any other causes beyond the control of the Contractor, then the required completion date or duration set forth in the progress schedule shall be extended by the amount of time that the Contractor shall have been delayed thereby. However, to the fullest extent permitted by Law, Owner and Marriott Corporation and their agents and employees shall not be held responsible for any loss or damage sustained by Contractor, or additional costs incurred by Contractor, through delay caused by Owner or Marriott Corporation, or their agents or employees, or any other

time, but no additional money. In that case, if the contractor is not granted additional time, the contractor may be entitled to recover money damages resulting from the delay.[244]

Note that in ¶ 15.1.6, claims for consequential damages are waived. By including this provision, it appears that delay claims are not considered consequential damages.

§ 4.47 Article 9: Payments and Completion

Article 9 covers the conditions under which progress payments and the final payment to the contractor are to be made. The article specifies procedures and time periods for the contractor to apply for payments and for the architect and owner to certify the work and make payments.

9.1 CONTRACT SUM

The Contract Sum is stated in the Agreement and, including authorized adjustments, is the total amount payable by the Owner to the Contractor for performance of the Work under the Contract Documents. (**9.1.1; no change**)

The authorized adjustments referred to are change orders only. Construction change directives are effective to change the contract sum only when they are converted into a change order. If the contract specifies a guaranteed maximum price, the contractor is limited to that amount plus signed change orders.[245]

Contractor or Subcontractor, or by abnormal weather conditions, or by any other cause, and Contractor agrees that the sole right and remedy therefor shall be an extension of time.

The contractor was unable to recover more than $2 million in damages for delays on a fast-track project because of this and other contract provisions. One of these required that any request for an extension of time be submitted in writing within seven days. The contractor only gave oral warnings to the owner. The court did not find this sufficient.

In Brown Bros. v. Metropolitan Gov't, 877 S.W.2d 745, 746 (Tenn. Ct. App. 1993), the contractor was delayed by the utility company. The court rejected a claim for extra costs because the contract specifically anticipated such delays.

See Blake Constr. Co., Inc./Poole & Kent v. Upper Occoquan Sewage Auth., 266 Va. 564, 587 S.E.2d 711 (2003), for a no-damages-for-delay clause held to violate a Virginia state statute.

[244] U.S., for Use and Benefit of Pertun Constr. Co., v. Harvesters Group, 918 F.2d 915 (11th Cir. 1990) (involving a Miller Act claim).

[245] In Gilbert v. Powell, 165 Ga. App. 504, 301 S.E.2d 683, 684 (1983), the contract provided that "the maximum cost to the Owner, including the Cost of the Work and the Contractor's Fee, is guaranteed not to exceed the sum of $21,500." When the contractor brought an action against the owner for extra work, the court found that the contractor was limited to the guaranteed maximum amount, because the contract provided a procedure for the payment of changes.

§ 4.48   Schedule of Values: ¶ 9.2

9.2 SCHEDULE OF VALUES

Where the Contract is based on a stipulated sum or Guaranteed Maximum Price, the Contractor shall submit to the Architect before the first Application for Payment, a schedule of values allocating the entire Contract Sum to the various portions of the Work and prepared in such form and supported by such data to substantiate its accuracy as the Architect may require. This schedule, unless objected to by the Architect, shall be used as a basis for reviewing the Contractor's Applications for Payment. **(9.2.1; add first clause; minor modifications)**

One of the principal purposes for the architect's careful review of these items is that the contractor should not be paid more than it is entitled to at any stage of the work. If the contractor is unable or unwilling to complete the project, there should be enough money left to complete the project. The architect should advise the contractor at the outset of the project as to the degree of detail required in the schedule of values. The contractor may wish to submit a schedule of values in a form different from that required by the architect. In that event, the contractor should submit the schedule before executing the contract. Otherwise, the architect is entitled to a reasonably detailed schedule of values.

§ 4.49   Applications for Payment: ¶ 9.3

9.3 APPLICATIONS FOR PAYMENT

9.3.1 At least ten days before the date established for each progress payment, the Contractor shall submit to the Architect an itemized Application for Payment prepared in accordance with the schedule of values, if required under Section 9.2, for completed portions of the Work. Such application shall be notarized, if required, and supported by such data substantiating the Contractor's right to payment as the Owner or Architect may require, such as copies of requisitions from Subcontractors and material suppliers, and shall reflect retainage if provided for in the Contract Documents. **(9.3.1; minor modification)**

This ten-day period may require revision based on the requirements of the owner's lender and title company, if any. In many states, owners require waivers of lien in proper form along with the application for payment.

The payment procedure used by A201 may run afoul of the requirements of the lien acts of several states, including Illinois. These states have some form of statutory disbursement procedure that does not square with the AIA procedure. For instance, the AIA procedure does not require the contractor to disclose the identity of every subcontractor or the actual amount of each subcontract. Counsel for owners should carefully review state law and make any necessary modifications to A201. Language requiring one such form is included here at § 5.53.

The owner might want to consider making payments directly to the subcontractors. Most title companies that provide construction services have procedures for this. The advantage for the owner is that it eliminates most claims by subcontractors. General contractors do not like this because it reduces their control over the subcontractors.

9.3.1.1 As provided in Section 7.3.9, such applications may include requests for payment on account of changes in the Work that have been properly authorized by Construction Change Directives, or by interim determinations of the Architect, but not yet included in Change Orders. **(9.3.1.1; minor modifications)**

This provision means that when the architect has established an amount, it may immediately be included in an application for payment and adjusted when the construction change directive is turned into a change order.[246] Lenders and title insurers will definitely want each payment application to include information about pending change orders. Of course, until a pending change order is actually approved by the owner and contractor, there is often a dispute about the cost involved. If the contractor's proposed cost is shown on payment requests, the construction loan may become "out of balance," which is not something the owner will want.

9.3.1.2 Applications for Payment shall not include requests for payment for portions of the Work for which the Contractor does not intend to pay a Subcontractor or material supplier, unless such Work has been performed by others whom the Contractor intends to pay. **(9.3.1.2; minor modification)**

If the contractor has a dispute with a subcontractor or material supplier and does not intend to pay all or part of the amount in dispute, the application for payment may show only the amount that the contractor actually intends to pay. This may, of course, result in problems in obtaining waivers of lien from the subcontractor or material supplier. The owner might want to allow for retentions (or *retainage*)[247] from the contractor.[248] The contractor may include in the application the amounts paid to "others" whom the contractor intends to pay. This means substitute

---

[246] In Framingham Heavy Equip. Co., Inc. v. John T. Callahan & Sons, Inc., 61 Mass. App. Ct. 171, 807 N.E.2d 851 (2004), the court discusses payments to a subcontractor under a construction change directive.

[247] This means that the contractor is paid less than the value of the work performed to date in order that the owner can have a margin of safety and to assure completion of the project. For instance, with a 10% retention, if the contractor states that it has completed $100,000 worth of work and materials and the architect certifies this as correct, the owner pays the contractor $90,000 at that time. If the contractor later defaults, the excess money can be used to complete the project. Of course, at final payment, the contractor will be paid all retained funds. The owner never wants to be in a position of having advanced the contractor more than the project is worth to that point.

[248] In Firemen's Ins. Co. v. State, 65 A.D.2d 241, 412 N.Y.S.2d 206 (1979), the state attempted to use retainage to pay for the contractor's unpaid unemployment insurance premiums. The surety was entitled to the balance of the retainage after it had completed the work for the defaulting contractor.

contractors who are called in to replace subcontractors whom the contractor does not intend to pay. Of course, the contractor is not entitled to double payment, even if more than one subcontractor does the same work.

> *9.3.2 Unless otherwise provided in the Contract Documents, payments shall be made on account of materials and equipment delivered and suitably stored at the site for subsequent incorporation in the Work. If approved in advance by the Owner, payment may similarly be made for materials and equipment suitably stored off the site at a location agreed upon in writing. Payment for materials and equipment stored on or off the site shall be conditioned upon compliance by the Contractor with procedures satisfactory to the Owner to establish the Owner's title to such materials and equipment or otherwise protect the Owner's interest, and shall include the costs of applicable insurance, storage and transportation to the site for such materials and equipment stored off the site.* **(9.3.2; no change)**

The contractor is entitled to include the amounts of materials delivered to the site for the project in its application for payment. The contractor cannot include materials that are stored elsewhere unless there is a prior written agreement and the contractor has provided the owner with suitable assurances that title to the materials is in the owner and that adequate insurance is in place. This appears to be in conflict with ¶ 11.3.1.4, which requires the owner to purchase insurance for work stored off the site and in transit. These provisions are harmonized by allowing the contractor to purchase "off-site insurance" and charging the owner for the cost.

> *9.3.3 The Contractor warrants that title to all Work covered by an Application for Payment will pass to the Owner no later than the time of payment. The Contractor further warrants that upon submittal of an Application for Payment all Work for which Certificates for Payment have been previously issued and payments received from the Owner shall, to the best of the Contractor's knowledge, information and belief, be free and clear of liens, claims, security interests or encumbrances in favor of the Contractor, Subcontractors, material suppliers, or other persons or entities making a claim by reason of having provided labor, materials and equipment relating to the Work.* **(9.3.3; no change)**

An Illinois case[249] interpreted the 1976 version of this clause as providing two promises: first, that the contractor warranted that the property would pass to the owner free of any liens when the contractor received payment. This was a guarantee that the subcontractors and material suppliers would be paid if the general contractor was paid. Second, the contractor warranted that the labor or materials used would not be subject to any security agreements. This provision, however, did not bar a lien claim by a subcontractor, even when a portion of the general contract incorporating A201 by reference was recorded.[250]

---

[249] Luczak Bros. v. Generes, 116 Ill. App. 3d 286, 451 N.E.2d 1267 (1983).

[250] The first three pages of the construction contract were recorded, incorporating A201. The owner alleged that, under Illinois lien laws, this constituted a waiver of the subcontractors' mechanic's

In a Virginia case, the court held that, under a similar provision, a subcontractor's creditors had no security interest in the subcontractor's goods and inventory when they were sold to the general contractor.[251]

In some states, the owner is permitted to insert a no lien provision in the contract documents. This can then be recorded in order to be effective against the subcontractors.[252]

This provision has been found by a Florida court as not meeting the specific requirements of a state statute that requires a contractor to give to the owner a statement at final payment to the effect that all lienors under the contractor's direct contract have been paid in full.[253]

§ 4.50   Certificates for Payment: ¶ 9.4

9.4 CERTIFICATES FOR PAYMENT

9.4.1 *The Architect will, within seven days after receipt of the Contractor's Application for Payment, either issue to the Owner a Certificate for Payment, with a copy to the Contractor, for such amount as the Architect determines is properly due, or notify the Contractor and Owner in writing of the Architect's reasons for withholding certification in whole or in part as provided in Section 9.5.1.* (**9.4.1; no change**)

The architect has a duty to the contractor to respond within the seven days by either approving the application for payment or issuing a written statement of rejection, stating the reasons for the rejection. These reasons must be clear and concise in order to enable the contractor to promptly correct the application for payment.[254]

---

lien rights. This assertion was rejected by the court. The owner's interpretation that ¶ 9.3.3 bars any lien by any subcontractor was rejected. The court found that the paragraph warrants that the property would pass to the owner free of any liens when the general contractor is paid, but not if full payment is not made; and it warrants that none of the labor or materials would be subject to any security agreements. It also was held that this provision merely gives a right against the general contractor but is not a waiver.

   In Southeastern Sav. & Loan Ass'n v. Rentenbach Constructors, Inc., 907 F.2d 1139 (4th Cir. 1990), the court held that this no lien provision applied only to the lower tier subcontractors and material suppliers, otherwise, it would be unenforceable under applicable North Carolina law.

[251] Graves Constr. Co. v. Rockingham Nat'l Bank, 220 Va. 844, 263 S.E.2d 408 (1980). *See also* Whirlpool Corp. v. Dailey Const., Inc., 429 S.E.2d 748 (N.C. App. 1993) (appliances incorporated into the project not subject to security interest, *citing Graves*).

[252] In Marriott Corp. v. Dasta Constr. Co., 26 F.3d 1057 (11th Cir. 1994), the contract contained a no lien provision:

   in the event that any such lien shall be filed, [Contractor] agrees to take all steps necessary and proper for the release and discharge of such lien . . . and in default of performing such obligation, agrees to reimburse the Owner, on demand, for all monies paid by Owner in releasing, satisfying, and discharging of such liens.

[253] McMahon Const. Co., Inc. v. Carol's Care Center, Inc., 460 So. 2d 1001 (Fla. App. 5th Dist. 1984).

[254] In General Ins. Co. v. K. Capolino Constr. Corp., 983 F. Supp. 403 (S.D.N.Y. Oct. 7, 1997), the court held that the architect's refusal to certify the amounts shown on the contractor's application

Note that the architect can reject the application for payment in part and certify the balance.

> ***9.4.2*** *The issuance of a Certificate for Payment will constitute a representation by the Architect to the Owner, based on the Architect's evaluation of the Work and the data comprising the Application for Payment, that, to the best of the Architect's knowledge, information and belief, the Work has progressed to the point indicated and that, the quality of the Work is in accordance with the Contract Documents. The foregoing representations are subject to an evaluation of the Work for conformance with the Contract Documents upon Substantial Completion, to results of subsequent tests and inspections, to correction of minor deviations from the Contract Documents prior to completion and to specific qualifications expressed by the Architect. The issuance of a Certificate for Payment will further constitute a representation that the Contractor is entitled to payment in the amount certified. However, the issuance of a Certificate for Payment will not be a representation that the Architect has (1) made exhaustive or continuous on-site inspections to check the quality or quantity of the Work, (2) reviewed construction means, methods, techniques, sequences or procedures, (3) reviewed copies of requisitions received from Subcontractors and material suppliers and other data requested by the Owner to substantiate the Contractor's right to payment, or (4) made examination to ascertain how or for what purpose the Contractor has used money previously paid on account of the Contract Sum.* **(9.4.2; minor modifications)**

The architect must carefully review the application for payment by the contractor to assure that the percentage of work completed is as the contractor stated.[255] The contractor might tend to overstate this percentage in order to obtain payment as soon as possible, but the owner would want to retain its funds for as long as possible. If the contractor defaults, the owner needs to have sufficient funds available to complete the project. The architect's certificate of payment can be revoked at a later date if the architect discovers that the work deviates from the contract documents (¶ 9.5.1).[256] Normally, the architect does not know how the contractor

---

for payment amounted to a breach of the owner-contractor contract because this provision was not followed.

In ATAP Const., Inc. v. Liberty Mut. Ins. Co., 1998 WL 341931 (E.D. Pa. June 25, 1998), the owner, a county, added a provision that allowed it to withhold payment.

[255] In Broadmoor/Roy Anderson Corp., v. XL Reinsurance Am., Inc., 2006 WL 2873473 (USDC, W.D. La. Oct. 4, 2006), the architect certified payment after an arbitration award and after the architect was no longer being paid. Apparently, the surety argued that such a late certification by an architect made the certification invalid and allowed the surety to escape its obligation to pay. The court rejected this argument. The architect had testified that, "When you're the architect, you're the architect forever." The delay in obtaining the certification was harmless and understandable.

[256] In Longview Constr. & Dev., Inc. v. Loggins Constr. Co., 523 S.W.2d 771 (Tex. 1975), the subcontractor argued that it was entitled to 98% of its fee because of the architect's certification. The work was improper, and the general contractor employed another subcontractor to complete the work for substantial additional money. The court held that the general contractor could recover from the sub because the architect's certificate of payment was not final and conclusive.

has spent the money that it was paid. If the contractor has submitted waivers of lien when required, it will be presumed that the money is being properly given to the various subcontractors and suppliers.[257]

§ 4.51   Decisions to Withhold Certification: ¶ 9.5

9.5 DECISIONS TO WITHHOLD CERTIFICATION

9.5.1 The Architect may withhold a Certificate for Payment in whole or in part, to the extent reasonably necessary to protect the Owner, if in the Architect's opinion the representations to the Owner required by Section 9.4.2 cannot be made. If the Architect is unable to certify payment in the amount of the Application, the Architect will notify the Contractor and Owner as provided in Section 9.4.1. If the Contractor and Architect cannot agree on a revised amount, the Architect will promptly issue a Certificate for Payment for the amount for which the Architect is able to make such representations to the Owner. The Architect may also withhold a Certificate for Payment or, because of subsequently discovered evidence, may nullify the whole or a part of a Certificate for Payment previously issued, to such extent as may be necessary in the Architect's opinion to protect the Owner from loss for which the Contractor is responsible, including loss resulting from acts and omissions described in Section 3.3.2, because of: **(9.5.1; no change)**

.1   defective Work not remedied; **(9.5.1.1; no change)**

This provision applies only when notice is given to the contractor to remedy defective work. In a Maryland case,[258] the architect refused to approve final payment because the contractor had not corrected defective bricks. The architect had

---

Taber Partners I v. Insurance Co. of Am., 875 F. Supp. 81 (D.P.R. 1995), held that, under this and other provisions, payment did not constitute acceptance of the contractor's work. In Hunts Point-Multi Service Ctr., Inc., v. Terra Firma Constr. Mgmt., 2003 NY Slip Op 51280U (N.Y. Misc. 2003), the architect did not waive a liquidated damages clause by certifying payment.

[257] Fabe v. WVP Corp., 760 S.W.2d 490 (Mo. App. E.D. 1988). For a case that reviewed the history of the architect's role in monitoring and approving construction work on behalf of the owner, *see* Sweeney Co. of Maryland v. Engineers-Constructors, Inc., 823 F.2d 805 (4th Cir. 1987).

[258] NSC Contractors v. Borders, 317 Md. 394, 564 A.2d 408 (1989).

In School Bd. v. Southeast Roofing, 532 So. 2d 1353 (Fla. Dist. Ct. App. 1988), the court construed an "AIA style construction contract." The architect had reduced the contractor's request, stating that there were deficiencies in the work. The owner refused to pay anything and the contractor filed suit. Unfortunately, the architect's findings were ambiguous and the case was sent to the lower court for further findings.

In A. Prete & Son Constr. v. Town of Madison, No. CV 91-03103073-S, 1994 Conn. Super. LEXIS 2532 (Oct. 4, 1994), the owner failed to obtain the architect's disapproval of a gymnasium floor before rejecting the floor. This was one provision cited by the court in holding for the contractor. The owner, the court said, should have followed the procedures in the AIA documents, which required the architect to make a finding that the contractor failed to comply with the contract documents.

This provision authorizes only the architect, and not the owner, to certify payment to the contractor. James Talcott Constr., Inc., v. P&D Land Enter., 141 P.3d 1200 (Mont. 2006). In this

determined that the aesthetic effect of the bricks was improper and demanded that the contractor replace the bricks. The court held that the contractor could not challenge the architect's decision as to aesthetic effect but could dispute the amount withheld.

Because withholding of funds could be subject to abuse by an owner, the contractor may want to insert language here that provides that there be a good-faith basis for believing that there are third-party claims and that if the contractor submits a reasonable bond, there would be no withholding.

> *.2*   *third party claims filed or reasonable evidence indicating probable filing of such claims unless security acceptable to the Owner is provided by the Contractor;* **(9.5.1.2; no change)**

This is not meant to establish any such third party as a beneficiary of the contract; it is meant only to protect the owner from mechanics' liens and similar claims. If the architect or owner reasonably believes that a mechanics' lien will be filed in the near future, this provision allows withholding certification.[259] The security referred to in this clause usually consists of a letter of credit, title indemnity, or similar security that will protect the owner from mechanics' liens.

> *.3*   *failure of the Contractor to make payments properly to Subcontractors or for labor, materials or equipment;* **(9.5.1.3; no change)**

In many cases, the owner or architect would learn of this by receiving notice of the filing of mechanics' liens or material suppliers' liens. If any such liens are filed or threatened, an amount sufficient to pay the disputed amount should be withheld from the contractor. Note that the general contractor is not an agent of the owner but an independent contractor, so that if the general contractor fails to pay a subcontractor for materials, the subcontractor usually does not have a cause of action against the owner.[260]

---

case, the architect had certified payment and the owner refused to pay the contractor. The court held that the contractor was not obligated to perform punch list and warranty work as a result.

    This provision was held to bar a promissory estoppel claim by a contractor. Greg Allen Const. Co., Inc., v. Estelle, 762 N.E.2d 760 (Ind. App. 2002).

[259] In Lewis-Brady Builders Supply, Inc., v. Bedros, 231 S.E.2d 199 (N.C. App. 1977), this provision allowed the architect to withhold a progress payment after the architect discovered the existence of a lien claim by a subcontractor.

[260] Oldham & Worth, Inc. v. Bratton, 263 N.C. 307, 139 S.E.2d 653 (1965) (subcontractor filed an action against an owner to recover for materials furnished to the job; general contractor held not an agent for the owner, but an independent contractor).

    In In re Modular Structures, Inc., 27 F.3d 72 (3d Cir. 1994), the court held that the general contractor was not entitled to final payment until it had paid its subcontractors, based in part on this provision.

    The contractor should be permitted to withhold funds from a subcontractor for good cause—for example, if the subcontractor is not performing. In such a case, the general contractor should be required to post reasonable security to protect the owner from the subcontractor's lien.

> *.4   reasonable evidence that the Work cannot be completed for the unpaid
>   balance of the Contract Sum;* **(9.5.1.4; no change)**

The architect should specify this, with as much detail as possible.[261]

> *.5   damage to the Owner or a separate contractor;* **(9.5.1.5; minor
>   modification)**

This means damage to the owner's property that is not the Work of this contractor. If the contractor damages this particular Work, it must correct it without additional cost. If the contractor shows no sign of repairing the damage, the architect should withhold certification.[262]

"A separate contractor" here refers to contractors hired by the owner, not the contractor's subcontractors.

> *.6   reasonable evidence that the Work will not be completed within the Contract
>   Time, and that the unpaid balance would not be adequate to cover actual or
>   liquidated damages for the anticipated delay; or* **(9.5.1.6; no change)**

This provision has been upheld by a Wyoming court.[263] The contractor will want the owner to furnish a detailed accounting that provides the basis for the conclusion that the work could not be completed in the Contract Time or that there are insufficient funds to pay for the damages resulting from delay. This will avoid a capricious decision by the owner that could seriously impact the contractor. See the discussion at § 4.46 about liquidated damages. If the owner is entitled to liquidated damages because the contractor failed to complete the work on time, the owner should withhold further payment to the contractor, otherwise there may be a waiver.[264]

> *.7   repeated failure to carry out the Work in accordance with the Contract
>   Documents.* **(9.5.1.7; change "persistent" to "repeated")**

---

[261] In General Ins. Co. v. K. Capolino Constr. Corp., 983 F. Supp. 403 (S.D.N.Y. Oct. 7, 1997), the court stated that if this provision was the reason for withholding money from the contractor, the architect should have stated that. Instead, the money was simply held back.

[262] In Manhattan Real Estate Partners, LLP v. Harry S. Peterson Co., Inc., 1992 WL 15130 (S.D.N.Y. Jan. 17, 1992), a subcontractor struck a parapet wall, causing substantial damage to the building. The owner hired a different contractor to repair the wall and withheld that amount of payment from the general contractor. Based on an earlier version of A201, the owner was not obligated to give the contractor notice and an opportunity to cure.

[263] State Sur. Co. v. Lamb Constr. Co., 625 P.2d 184 (Wyo. 1981). In Dove Bros., Inc., v. Horn, 1992 WL 117799 (Conn. Super. April 22, 1992), failure to complete the work on time was a breach of contract, under a different version of this provision. In James Talcott Constr., Inc., v. P&D Land Enter., 141 P.3d 1200 (2006), the architect certified payment and the owner failed to pay the contractor. The contractor then stopped work. This provision alone does not provide a basis for finding that the contractor was not entitled to payment.

[264] Centerre Trust Co. v. Continental Ins., 167 Ill. App. 3d 376, 521 N.E.2d 219 (1988). *See also* Polk County v. Widseth, Smith, Nolting, 2004 WL 2940847 (Minn. App. Dec. 21, 2004), following *Centerre*; County of Dauphin, Pa., v. Fidelity and Deposit Co. of Maryland, 770 F. Supp. 248 (M.D. Pa. 1991), analyzing *Centerre* and finding that a delay in completion of the work is immediately discoverable, unlike a defect in the work itself, which may take time to manifest itself.

If the architect cannot reasonably make the representations stated in ¶ 9.4.2, it should withhold certification. Because this will delay payment to the contractor, it is an important decision that normally requires review with the owner and contractor.

The contractor will want the owner to state precisely when and where the contractor failed to carry out the work. Presumably, the owner would have previously notified the contractor in writing of such problems.

9.5.2 When the above reasons for withholding certification are removed, certification will be made for amounts previously withheld. **(9.5.2; no change)**

This is usually accomplished with the contractor's submitting the next application for payment in the proper amounts.

9.5.3 If the Architect withholds certification for payment under Section 9.5.1.3, the Owner may, at its sole option, issue joint checks to the Contractor and any Subcontractor or material or equipment suppliers to whom the Contractor failed to make payment for Work properly performed and material or equipment suitably delivered. If the Owner makes payments by joint check, the Owner shall notify the Architect and the Architect will reflect such payment on the next certificate for payment. **(new)**

If the owner learns that the contractor is not properly paying the subcontractors or material suppliers, this provision will provide the legal basis to issue joint checks to the contractor and the subcontractors, thereby ensuring proper payments in the future. Often, this is done through a title company as a way to further protect the owner's interests.

§ 4.52   Progress Payments: ¶ 9.6

9.6 PROGRESS PAYMENTS

9.6.1 After the Architect has issued a Certificate for Payment, the Owner shall make payment in the manner and within the time provided in the Contract Documents, and shall so notify the Architect. **(9.6.1; no change)**

The owner must make payments pursuant to the contract documents and must notify the architect that payment was made. If the owner does not make payments or authorize its lender to make payments to the contractor after the contractor performs the work for which payment is due, the owner is in breach of the contract.[265] Certification by the architect is required before the contractor can recover payments from the owner unless the contractor can show that the architect acted in

---

[265] Brady Brick & Supply Co. v. Lotito, 43 Ill. App. 3d 69, 356 N.E.2d 1126 (1976); Baker Domes v. Wolfe, 403 N.W.2d 876 (Minn. Ct. App. 1987); Adams Builders & Contractors, Inc., v. York Saturn, Inc., 798 N.E.2d 586 (Mass. App. Ct. 2003) (owner was required to make payment to contractor where architect had certified payment. The contractor was entitled to suspend work, but failure to do so was not a waiver of this provision).

bad faith or failed to exercise honest judgment.[266] In a Florida case,[267] the contractor, by letter, directed the owner to make checks out jointly to the contractor and its surety. Later, the contractor revoked the letter, but the owner refused to make checks out only to the contractor, although there had been no breach of the contract. The court ruled that the owner had an obligation to pay the contractor according to the contract if there was no material breach.

The owner should carefully review the time periods within which it will make payments. If a lender and title company plan to review the contractor's application for payment, the time required for this review should be included in the time allowed for payment.

9.6.2 The Contractor shall pay each Subcontractor no later than seven days after receipt of payment from the Owner, the amount to which the Subcontractor is entitled, reflecting percentages actually retained from payments to the Contractor on account of the Subcontractor's portion of the Work. The Contractor shall, by appropriate agreement with each Subcontractor, require each Subcontractor to make payments to Sub-subcontractors in a similar manner. **(9.6.2; add "seven days"; modifications)**

The contractor must pay its subcontractors upon receiving payment from the owner. This does not establish payment from the owner as a precondition to payment to the subcontractors, although many contractors place such a precondition in their subcontracts. This provision could be read as a pure contingent payment provision, so that the subcontractor is only entitled to payment if and when the contractor is paid. However, the language is not that clearly a contingency. The better reading is that this language imposes a time restriction on the contractor as to when payment is due the subcontractor. If the contractor is never paid, the subcontractor is still due payment from the contractor after a reasonable period of time has elapsed. This portion of the clause must await interpretation by the courts of the various jurisdictions.[268] Note that the contractor cannot withhold more from the subcontractors than the owner withholds from the contractor. One court has held

---

[266] City of Mound Bayou v. Roy Collins Constr. Co., 499 So. 2d 1354 (Miss. 1986); J.R. Graham v. Randolph County Bd. of Educ., 25 N.C. App. 163, 212 S.E.2d 542 (1975); Elec-Trol, Inc., v. C.J. Kern Contractors, Inc., 284 S.E.2d 119 (N.C. App. 1981).

[267] Newkirk Constr. Corp. v. Gulf County, 366 So. 2d 813 (Fla. Dist. Ct. App. 1979). Although the letter asserted that there was an irrevocable assignment, the contract was not affected. The owner was under no obligation to anyone other than the contractor. When the owner continued to make checks out jointly to the contractor and surety and the surety refused to endorse the checks, the contractor had a claim against the owner. As long as the contractor has committed no vital breach, it is entitled to payment.

[268] In Galloway Corp. v. S.B. Ballard Constr. Co., 250 Va. 493, 464 S.E. 2d 349 (1995), the contractor modified the payment provision in each subcontract, so that language in the standard AIA subcontractor agreement was deleted. The contractor believed that this made the contract a pay when paid contract. The court disagreed, finding that there was a latent ambiguity in the contracts. Several subcontractors were not paid and sued the general contractor. The general's defense was

that this language created an express trust for the benefit of subcontractors.[269] This language does not confer third-party-beneficiary status on any other party.[270] This language also requires that the subcontractor be "entitled" to payment.[271]

> **9.6.3** *The Architect will, on request, furnish to a Subcontractor, if practicable, information regarding percentages of completion or amounts applied for by the Contractor and action taken thereon by the Architect and Owner on account of portions of the Work done by such Subcontractor.* **(9.6.3; no change)**

This is an exception to the provisions of ¶ 4.2.4, which provides that communications between the architect and subcontractor shall be through the contractor. A subcontractor may directly ask the architect for information concerning the payment applications and approval status relating to the subcontractor's work. This is an effective method for a subcontractor to determine if the contractor is dealing fairly with the sub, and the architect and owner should cooperate with the subcontractors to minimize the possibility of mechanics' liens.

> **9.6.4** *The Owner has the right to request written evidence from the Contractor that the Contractor has properly paid Subcontractors and material and equipment suppliers amounts paid by the Owner to the Contractor for subcontracted Work. If the Contractor fails to furnish such evidence within seven days, the Owner shall have the right to contact Subcontractors to ascertain whether they have been properly paid. Neither the Owner nor Architect shall have an obligation to pay or to see to the payment of money to a Subcontractor except as may otherwise be required by law.* **(9.6.4; add first two sentences)**

In most cases, owners will require the general contractor to submit waivers of lien with each request for payment, thus satisfying the first sentence. In situations where lien waivers have not been provided, the owner should make such a request

---

that he was not obligated to pay until the owner paid him. The court's discussion of latent ambiguities in construction contracts is interesting.

In the end, those contractors who testified that they understood that payment by the owner was a condition precedent to their payment were denied payment, because there was no ambiguity as to them. Those others who testified the opposite were entitled to payment notwithstanding lack of the owner's payment. In each case, the court could allow extrinsic evidence as to what the parties intended when the contract was signed. If there were no ambiguity, the contract would be enforced as written without evidence of what the parties meant.

In Framingham Heavy Equip. Co., Inc. v. John T. Callahan & Sons, Inc., 61 Mass. App. Ct. 171, 807 N.E.2d 851 (2004), the court held that this language did not set forth a clear precondition to payment to the subcontractor.

[269] Westview Inv., Ltd., v. US Bank Nat. Ass'n., 138 P.3d 638 (Wash. App. Div. 1 2006). In In re Concraft, 206 B.R. 551 (Bankr.D. Or. 1997), this provision created a constructive trust in the context of a bankruptcy proceeding. *But see* In re H. & A. Const. Co., Inc., 65 B.R. 213 (Bankr.D. Mass. 1986), holding to the contrary.

[270] Badiee v. Brighton Area Schools, 695 N.W.2d 521 (Mich. App., 2005).

[271] Continental Air Conditioning, Inc., v. Keller Const. Co., Ltd., 2003 WL 22120886 (Cal. App. 2d Dist. Sept. 15, 2003).

and, if the proper documentation is not provided within seven days, the owner should proceed to contact the various subcontractors and material suppliers to ascertain whether or not they have been paid. Otherwise, the owner may be at risk based on applicable lien laws.

Under those laws, the subcontractor has statutory rights to collect payments directly from the owner if statutory requirements are met. This clause is meant to again reinforce the concept that the subcontractors do not have any third-party beneficiary rights under the contract documents and must look solely to the contractor for payment, except insofar as lien and similar laws may apply. The last sentence does not operate to give rights in the progress payments to the subcontractors.[272]

> *9.6.5 Contractor payments to material and equipment suppliers shall be treated in a manner similar to that provided in Sections 9.6.2, 9.6.3 and 9.6.4.* (**9.6.5; minor modifications**)

Material suppliers are treated in the same manner as subcontractors.

> *9.6.6 A Certificate for Payment, a progress payment, or partial or entire use or occupancy of the Project by the Owner shall not constitute acceptance of Work not in accordance with the Contract Documents.* (**9.6.6; no change**)

Even though the owner uses the property or makes a partial payment, the owner will not be deemed to have accepted work that is not in conformance with the contract documents. Even when the owner makes final payment, there is no waiver of non-conforming work (**see ¶ 9.10.4.2**).[273] The owner may want to retain the option to make direct payments to the subcontractors under certain conditions.

> *9.6.7 Unless the Contractor provides the Owner with a payment bond in the full penal sum of the Contract Sum, payments received by the Contractor for Work properly performed by Subcontractors and suppliers shall be held by the Contractor for those Subcontractors or suppliers who performed Work or furnished materials, or both, under contract with the Contractor for which payment was made by the Owner. Nothing contained herein shall require money to be placed in a separate account*

---

[272] Capitol Indem. Corp., v. U.S., 41 F.3d 320 (7th Cir. 1994).

[273] But in Centerre Trust Co. v. Continental Ins., 167 Ill. App. 3d 376, 521 N.E.2d 219 (1988), there was a waiver of a right to liquidated damages, despite the AIA language.

In Taber Partners I v. Insurance Co. of Am., 875 F. Supp. 81 (D.P.R. 1995), the court held that, under this and other provisions, payment did not constitute acceptance of the contractor's work.

In Hunts Point Multi-Service Center, Inc. v. Terra Firma Constr. Mgmt., 2003 NY Slip Op 51280U (N.Y. Misc. 2003), the contractor argued that the architect waived certain liquidated damages claims. The court held that the architect was not an agent of the owner for all purposes. The certification of final payment by the architect did not constitute a waiver by the owner, but is simply the procedure of making payment with no concomitant waiver effect upon the owner. The fact that the architect certified final payment does not affect a waiver; RLI Insurance Company v. St. Patrick's Home for the Infirm and Aged, 452 F. Supp. 2d 484 (S.D.N.Y. 2006), citing *Hunts Point*.

*and not commingled with money of the Contractor, shall create any fiduciary
liability or tort liability on the part of the Contractor for breach of trust or shall
entitle any person or entity to an award of punitive damages against the Contractor
for breach of the requirements of this provision.* **(9.6.7; no change)**

This provision follows ¶ 9.6.2, which requires payment to the subcontractor
upon payment from the owner to the general contractor. Some courts have held that
this provision, along with ¶ 9.6.2, establishes a trust.[274] This paragraph allows the
contractor to keep all funds in one account instead of segregated accounts.[275]

§ 4.53 Failure of Payment: ¶ 9.7

9.7 FAILURE OF PAYMENT

*If the Architect does not issue a Certificate for Payment, through no fault of the
Contractor, within seven days after receipt of the Contractor's Application for Payment, or if the Owner does not pay the Contractor within seven days after the date
established in the Contract Documents the amount certified by the Architect or
awarded by binding dispute resolution, then the Contractor may, upon seven additional days' written notice to the Owner and Architect, stop the Work until payment
of the amount owing has been received. The Contract Time shall be extended appropriately and the Contract Sum shall be increased by the amount of the Contractor's
reasonable costs of shut-down, delay and start-up, plus interest as provided for in the
Contract Documents.* **(9.7.1; change reference from arbitration to binding dispute resolution)**

The contractor may stop performance of the work if the owner is at least seven
days late with payment and the contractor is not responsible for the delay in
payment. Also, the contractor must give an additional seven days' notice after
the initial seven-day lapse before stopping the work. If the contractor complies
with these requirements, it would be entitled to the extra costs associated with
demobilization and remobilization.

In a Texas case,[276] a contractor performed work for an owner on both a
commercial project and the owner's residence. Both projects were submitted
on the contractor's application for payment. The owner acted as supervising

---

[274] Westview Inv., Ltd., v. US Bank Nat. Ass'n., 138 P.3d 638 (Wash. App. Div. 1 2006).

[275] In Taylor Pipeline Const., Inc., v. Directional Road Boring, Inc., 438 F. Supp. 2d 696 (E.D. Tex.
2006), this language helped the contractor avoid a claim of conversion.

[276] Texas Bank & Trust Co. v. Campbell Bros., 569 S.W.2d 35 (Tex. 1978) (construing the 1967
version of AIA Documents A111 and A201 (1970)). The court also found that the contractor did
not have an independent duty to the lender as co-obligee on the performance bond. *Texas Bank*
was cited in Argee Corp. v. Solis, 932 S.W.2d 39 (Tex. App.-Beaumont, 1995) for the proposition
that where an owner fails to make timely payment to the contractor, the contractor's performance
is excused.

architect and approved payments. When one payment was overdue, the contractor complied with the notice requirements and stopped the work. The fact that both projects were included on one application (as were other costs that were project-related but should not have been included on the application) was not a defense for the owner when he knew of this arrangement and had previously approved such payments. The older version of the owner-contractor agreements provided that payment dates were "on or about" certain days of the month. The current AIA Document A101 (¶ 5.1.3) provides that "Provided that an Application for Payment is received by the Architect not later than the _____ day of a month, the Owner shall make payment of the certified amount to the Contractor not later than the _____ day of the _____ month." The fact that the contractor submitted his application one day late did not mean that he had to wait until the following month. The contractor may stop work pursuant to ¶ 9.7.

A Missouri court held that a contractor did not repudiate his contract with the owner when he walked off the project after payment was due and not paid, although the architect had certified payment.[277] Because the date for completion of the project had not come and because the contractor was still ready, willing, and able to perform his part of the contract (if paid), there was not a breach of the contract by the contractor.

In another Texas case, the contractor argued that the paragraph which provided that "if the Owner does not pay the Contractor within seven days after the date established in the Contract Documents the amount . . . awarded by arbitration, then the Contractor may . . . stop the work," allowed it to stop the work when the owner failed to pay an arbitration award within seven days.[278] The court stated that state statute permitted the owner to challenge the award within a 90-day period. Therefore, failure to pay the arbitration award within seven days did not constitute a covenant to pay an award within a specified time.

---

[277] Hart & Son Hauling, Inc. v. MacHaffie, 706 S.W.2d 586 (Mo. Ct. App. 1986). If the amount not paid is not significant, the contractor would be in breach by walking off the job. Stewart v. C & C Excavating & Const. Co., 877 F.2d 711 (8th Cir. 1989).

In Adams Builders & Contractors, Inc. v. York Saturn, Inc., 60 Mass. App. Ct. 1101, 798 N.E.2d 586 (2003), the court held that the contractor's option to stop the work was not an exclusive remedy.

Section 9.5 authorizes only the architect, and not the owner, to certify payment to the contractor. James Talcott Construction, Inc. v. P&D Land Enter., 141 P.3d 1200 (2006). In this case, the architect had certified payment and the owner refused to pay the contractor. The court held that the contractor was not obligated to perform punch list and warranty work as a result, pursuant to this provision.

[278] Snyder v. Eanes Indep. Sch. Dist., 860 S.W.2d 692 (Tex. App. 1993).

§ 4.54   Substantial Completion: ¶ 9.8

9.8 SUBSTANTIAL COMPLETION

9.8.1 Substantial Completion is the stage in the progress of the Work when the Work or designated portion thereof is sufficiently complete in accordance with the Contract Documents so that the Owner can occupy or utilize the Work for its intended use. **(9.8.1; no change)**

The date of substantial completion is an important one, on which the contractor is due its payment including any remaining retainage, less the value of punchlist items.[279] At this point, the risk of loss passes to the owner, who must insure the building and take other responsibilities for it, as set forth in the architect's certificate of substantial completion (**see** ¶ 9.8.4).[280] The contractor's warranties start to run from this date, and liquidated damages, if any, normally end on this date unless otherwise specified in the contract. Note that this date is different from the date, if any, on which a municipality issues a certificate of occupancy. The project may be substantially complete, but a certificate of occupancy may not be granted, and vice-versa.[281]

---

[279] Mayfield v. Swafford, 106 Ill. App. 3d 610, 435 N.E.2d 953 (1983). Failing to perform in a workmanlike manner constitutes a breach of the contract. If the owner receives substantial performance, it must pay the contractor the contract sum less the cost of correction of any deficiencies. *See also* J.R. Sinnott Carpentry, Inc. v. Phillips, 110 Ill. App. 3d 632, 443 N.E.2d 597 (1982); Watson Lumber v. Guennwig, 79 Ill. App. 2d 377, 226 N.E.2d 270 (1967); Cleveland Neighborhood Health Servs., Inc. v. St. Clair Builders, Inc., 64 Ohio App. 3d 639, 582 N.E.2d 640 (1989).

In Pete Wing Contracting, Inc. v. Port Conneaut Investors Ltd. P'ship, 1995 Ohio App. LEXIS 4341 (Sept. 29, 1995), the substantial completion provision was examined. The architect issued a certificate of substantial completion to the contractor, but the owner refused to accept it or release the retainage. The problem was that the original bid plans called for the ceiling height in the loft area to be seven feet, eight inches. When the construction plans were issued, the height was apparently given as 78 "instead of 7'8". The architect testified that it was understandable that the contractor interpreted the drawings in that manner, especially because the owner reduced the architect's scope of services during construction, leaving the contractor to deal directly with the owner instead of the architect. Thus, the ceiling height problem was not the contractor's fault, and the contractor had attained substantial completion of the project. In this case, it was apparent that the architect sided with the contractor, and the court was convinced that the owner brought about his own problems.

Without substantial completion, the contractor can only recover in quantum meruit. Stephenson v. Smith, 337 So. 2d 570 (La. Ct. App. 1976).

[280] Since ¶ 9.8.4 of A201 requires that the architect designate responsibility "for security, maintenance, heat, utilities, damage to the Work and insurance," what if the architect fails to issue the certificate, or fails to address these issues? What if the architect places all responsibility on the contractor when the contractor thought the owner would have such responsibility? The parties can proceed to the dispute resolution process found in Article 15, but this procedure opens the door to litigation. It would be better if this was addressed at the outset of the project.

[281] *See, e.g.,* Miller v. Bourgoin, 613 A.2d 292 (Conn. App. 1992) ("although the issuance of a certificate of occupancy may be evidence of substantial completion, it is not dispositive of the question.")

It is the architect who determines the date of substantial completion.[282] After this date, the contractor is entitled to its fee, less the value of any uncompleted work.[283]

---

[282] *But see* Holy Family Catholic Congregation v. Stubenrauch Assoc., 136 Wis. 2d 515, 402 N.W.2d 382 (Ct. App. 1987). The case involved an action by a church against the architect and contractor for a leaky roof. The question turned on the date of substantial completion and whether the statute of limitations had run. The court held that it was up to the court, and not the architect, to determine the date of substantial completion. The architect's certificate may be persuasive, but it is not determinative.

In a Louisiana case, Stephenson v. Smith, 337 So. 2d 570 (La. Ct. App. 1976), the court held that because the contractor had left more than 10% of the work undone, he had not substantially complied with the contract.

In American Prod. Co. v. Reynolds & Stone, 1998 Tex. App. LEXIS 7387 (Nov. 30, 1998), the owner added a new building and modified an existing structure at its facilities. The architect certified Phase I of the project as substantially complete on October 1, 1984. On October 7, 1994, the roof on the addition collapsed, apparently due to the omission of certain specified anchor bolts. Texas has a ten-year statute of repose (a statute of repose is similar to a statute of limitations in that it imposes an absolute time limit in which to bring an action). Because more than ten years had elapsed from the date of substantial completion, the court dismissed the action. The owner argued that substantial completion must be measured from the date the entire project is substantially complete, not just one phase. Because the construction contract expressly provided for phased construction, it was proper to have separate dates of substantial completion. Even though the architect had inadvertently left the date on the Certificate of Substantial Completion blank, there was other evidence indicating the correct date, and the architect filed an affidavit attesting to the correct date. The architect was assigned exclusive authority to determine the substantial completion date by the contract, and the court would not disturb that determination.

The substantial completion provision can be modified. In Brookridge Apartments, Ltd. v. Universal Constructors, Inc., 844 S.W.2d 637 (Tenn. Ct. App. 1992), the owner sued the architect and the contractor after two defective balconies at the apartment complex caused serious injuries four years after the initial construction. The issue was whether the trial court properly applied the four-year statute of limitations. The AIA contracts were modified by certain HUD amendments. The architect issued a certificate of substantial completion on June 11, 1986. However, the owner and the contractor had agreed that the date of substantial completion would be the date the HUD representative signed the final HUD Representative's Trip Report, which occurred on August 21, 1986. The suit filed August 20, 1990, was therefore timely filed.

In B.M. Co. v. Avery, 2001 WL 1658197 (Guam Terr. Dec. 2001), the court stated that the issuance of an occupancy permit is one factor indicating that a project is substantially complete.

[283] In Forrester v. Craddock, 51 Wash. 2d 315, 317 P.2d 1077 (1957), the court stated the rule as follows:

> Where the builder has substantially complied with his contract, the measure of damages to the owner would be what it would cost to complete the structure as contemplated by the contract. There is a substantial performance of a contract to construct a building where the variations from the specifications or contract are inadvertent and unimportant and may be remedied at relatively small expense and without material change of the building; but where it is necessary, in order to make the building comply with the contract, that the structure, in whole or in material part, must be changed, or there will be damage to parts of the building, or the expense of such repair will be great, then it cannot be said that there has been a substantial performance of the contract.

9.8.2 When the Contractor considers that the Work, or a portion thereof which the Owner agrees to accept separately, is substantially complete, the Contractor shall prepare and submit to the Architect a comprehensive list of items to be completed or corrected prior to final payment. Failure to include an item on such list does not alter the responsibility of the Contractor to complete all Work in accordance with the Contract Documents. **(9.8.2; no change)**

9.8.3 Upon receipt of the Contractor's list, the Architect will make an inspection to determine whether the Work or designated portion thereof is substantially complete. If the Architect's inspection discloses any item, whether or not included on the Contractor's list, which is not sufficiently complete in accordance with the Contract Documents so that the Owner can occupy or utilize the Work or designated portion thereof for its intended use, the Contractor shall, before issuance of the Certificate of Substantial Completion, complete or correct such item upon notification by the Architect. In such case, the Contractor shall then submit a request for another inspection by the Architect to determine Substantial Completion. **(9.8.2; no change)**

9.8.4 When the Work or designated portion thereof is substantially complete, the Architect will prepare a Certificate of Substantial Completion which shall establish the date of Substantial Completion, shall establish responsibilities of the Owner and Contractor for security, maintenance, heat, utilities, damage to the Work and insurance, and shall fix the time within which the Contractor shall finish all items on the list accompanying the Certificate. Warranties required by the Contract Documents shall commence on the date of Substantial Completion of the Work or designated portion thereof unless otherwise provided in the Certificate of Substantial Completion. **(9.8.4; no change)**

The contractor is due payment, less the amounts required to properly finish the work, upon substantial completion.[284] Under the terms of A201, the architect's certificate of substantial completion is required before the contractor is entitled to payment for substantial completion. *Substantial completion* means that despite deficiencies, a construction is fit for its intended purpose.[285] This is the time stated in the contract documents that effectively ends the project except for minor

---

Accord J.M. Beeson Co. v. Sartori, 553 So. 2d 180 (Fla. Dist. Ct. App. 1989); Mayfield v. Swafford, 106 Ill. App. 3d 610, 435 N.E.2d 953 (1983); J.R. Sinnott Carpentry, Inc. v. Phillips, 110 Ill. App. 3d 632, 443 N.E.2d 597 (1982); Watson Lumber v. Guennwig, 79 Ill. App. 2d 377, 226 N.E.2d 270 (1967); J&J Elec., Inc. v. Gilbert H. Moen Co., 9 Wash. App. 954, 516 P.2d 217 (1974); E.B. Ludwig Steel v. C.J. Waddell, 534 So. 2d 1364 (La. Ct. App. 1988).

For an extensive discussion of substantial completion, *see* Perini Corp. v. Greate Bay Hotel & Casino, Inc., 129 N.J. 479, 610 A.2d 364 (1992).

[284] Walter Lafaruge Real Estate, Inc., v. Raines, 420 So. 2d 1309 (La. Ct. App., 1982); Jim Arnott, Inc., v. L & E, Inc., 539 P.2d 1333 (Colo. App. 1975).

[285] Weill Constr. Co. v. Thibodeaux, 491 So. 2d 166 (La. Ct. App. 1986).

In Glacier Springs Property Owners Ass'n v. Glacier Springs Enters., Inc., 41 Wash. App. 829, 706 P.2d 652 (1985) (*citing* Patraka v. Armco Steel Co., 495 F. Supp. 1013 (M.D. Pa. 1980) ("substantial completion of construction occurs when the entire improvement, not merely a component part, may be used for its intended purpose.")), the court stated that substantial completion in that case occurred no later than the date the contractor billed for final payment. *Glacier Springs* apparently did not involve AIA documents.

completion items. Liquidated damages are usually computed as of the date of substantial completion.[286] Note that it is the contractor who initiates the process, although the architect determines whether the contractor has substantially completed the work.

If substantial completion has been obtained, the architect is under a legal duty to issue a certificate of substantial completion.[287]

In Moore's Builder & Contractor v. Hoffman, 409 N.W.2d 191 (Iowa Ct. App. 1987), the court stated that "substantial performance allows only the omissions or deviations from the contract that are inadvertent and unintentional, not the result of bad faith."

One court held that an architect's determination of the date of substantial completion is not necessarily final. Allen v. A&W Contractors, Inc., 433 So. 2d 839, 841 (La. Ct. App. 1983). The court found that an arbitrator had the power to fix a date of substantial completion different from the date the architect had determined. The fact that the contract provided that the architect will establish the date of substantial completion is "not sacrosanct if the facts show substantial completion at a date earlier than that certified by the owner's architect."

In Russo Farms, Inc. v. Vineland Bd. of Educ., 675 A.2d 1077 (N.J. 1996), the court followed the AIA definition as well as the following: "Substantial completion occurs when the architect certifies such to the owner and a certificate of occupancy is issued attesting to the building's fitness." Note that this definition is not accurate, because a certificate of occupancy does not trigger the date of substantial completion under the AIA documents. The court went on to state that at the time of substantial completion, all that remains is the punch list, which is a final list of small items requiring completion, or finishing, corrective, or remedial work.

Russo was examined in Daidone v. Buterick Bulkheading, Inc., 2006 WL 2346286 (N.J. Super. A.D. Aug. 15, 2006), wherein the owner filed suit against the foundation contractor and architect for the cost of repairing damage caused by settlement. The contractor completed its work and was paid in full by May 24, 1993. The architect completed its work on June 23, 1993. An occupancy certificate was granted on June 14, 1994. Plaintiff noticed the problem in 1999. In late 2001, plaintiff hired an expert and repairs were performed in July, 2002. Suit was not filed until June 2, 2004. The defendants moved to dismiss based on the ten-year statute of repose. The owner argued that the start of the statute of repose should commence on the date of issuance of the occupancy certificate, not the date of completion of a particular party's work. The court rejected that argument.

[286] Page v. Travis-Williamson County Water Control and Imp. Dist. No. 1, 367 S.W.2d 307 (Tex. 1963) (owner had taken possession and was using the property. There is no delay for which liquidated damages may be awarded. Owner is entitled to actual damages after substantial completion.); Hungerford Constr. Co., v. Florida Citrus Exposition, Inc., 410 F.2d 1229 (5th Cir. 1969); Stone v. City of Arcola, 536 N.E.2d 1329 (Ill. App. 4th Dist. 1989). In Ledbetter Bros., Inc., v. North Carolina Dep't of Transportation, 314 S.E.2d 761 (N.C. App. 1984), the liquidated damages clause read: "It is mutually recognized that time is an essential element of the contract, and that delay in completing the work will result in damages due to public inconvenience, obstruction to traffic, interference with business, and the increasing of engineering, inspection, and administrative costs to the Commission. It is therefore agreed that in view of the difficulty of making a precise determination of such damages, a sum of money in the amount stipulated in the contract will be charged against the Contractor for each calendar day that the work remains uncompleted after the expiration of the completion date, not as a penalty but as liquidated damages." The court held that this meant that liquidated damages terminated on final, not substantial, completion.

[287] Haugen v. Raupach, 260 P.2d 340 (Wash. 1953) ("If the architect is satisfied that there has been a substantial performance of the contract it then becomes his duty to issue the certificate of completion, and if he does not do so, his conduct is regarded as arbitrary and capricious. If the architect

The comprehensive list is usually called the *punchlist*. If an item does not conform to the contract documents and does not appear on the punchlist, the contractor is still responsible for it. Even final payment does not constitute a waiver of items that do not conform to the contract documents (**see** ¶ 9.10.4). The punchlist must be completed prior to the time the contractor is entitled to final payment.

If the work is not substantially complete, the contractor must proceed with the work and request a reinspection by the architect. When the architect agrees that the work is substantially complete, it will issue the certificate of substantial completion, with the final punchlist and a description of the other items mentioned attached to the certificate. Warranties start to run as of this date for that portion of the work that has been completed.

Even if the architect issues a certificate of substantial completion, the contractor is not relieved from liability if the building is not constructed according to the plans and specifications. This occurred in a Mississippi case, in which the court held that the fact that the certificate had been issued did not mean that the building had been properly built, or even that the architect was acting as arbiter between the owner and contractor for the purpose of making such a determination.[288]

An architect has been held immune from suit for refusing to issue a certificate of substantial completion based on his capacity as arbiter.[289] In one case, the "inspecting architect" was held liable to the owner for failing to notify the owner of deviations (including the use of smaller wires and the use of 100-amp instead of 125-amp breakers) from the electrical drawings by the contractor.[290]

Sometimes, for various reasons, there is no certificate of substantial completion and a court may be asked to determine whether the project is substantially complete. Often there are incomplete or defective items. Courts will look to several factors in making this determination, including (1) the extent of the defect or nonperformance; (2) the degree to which the purpose of the contract is defeated; (3) the ease of correction; and (4) the use or benefit to the owner of the work performed.[291]

9.8.5 The Certificate of Substantial Completion shall be submitted to the Owner and Contractor for their written acceptance of responsibilities assigned to them in such Certificate. Upon such acceptance and consent of surety, if any, the Owner shall make payment of retainage applying to such Work or designated portion thereof.

---

is in collusion with his principal, or yields to his opposition to the issuance of the certificate when such opposition is not justified, then in such cases the contractor has a legal excuse for not obtaining the certificate as a condition precedent to recovering on his contract.")

[288] May v. Ralph L. Dickerson Constr. Corp., 560 So. 2d 729 (Miss. 1990).

[289] Blecick v. School Dist. No. 18, 2 Ariz. App. 115, 406 P.2d 750 (1965) (contractor could recover from owner when a required certificate was unreasonably withheld). However, if the architect does not issue a certificate of substantial completion because the owner prevents it from doing so, the contractor is excused from obtaining the certificate. Haugen v. Raupach, 260 P.2d 340 (Wash. 1953).

[290] South Union v. George Parker & Assocs., 29 Ohio App. 3d 197, 504 N.E.2d 1131 (1985).

[291] Gibbens Pools, Inc., v. Corrington, 446 So. 2d 420 (La. App. 4th Cir. 1984).

Such payment shall be adjusted for Work that is incomplete or not in accordance with the requirements of the Contract Documents. **(9.8.5; no change)**

If a surety is involved in the project, its consent should be obtained before paying the retainage (**see AIA** Document G707A, Consent of Surety to Reduction in or Partial Release of Retainage). If the work is incomplete or improper, the owner may deduct the value of that work from the retainage and pay it at the time of final payment if the work is then completed.

§ 4.55   Partial Occupancy or Use: ¶ 9.9

9.9 PARTIAL OCCUPANCY OR USE

9.9.1 The Owner may occupy or use any completed or partially completed portion of the Work at any stage when such portion is designated by separate agreement with the Contractor, provided such occupancy or use is consented to by the insurer as required under Section 11.3.1.5 and authorized by public authorities having jurisdiction over the Project. Such partial occupancy or use may commence whether or not the portion is substantially complete, provided the Owner and Contractor have accepted in writing the responsibilities assigned to each of them for payments, retainage, if any, security, maintenance, heat, utilities, damage to the Work and insurance, and have agreed in writing concerning the period for correction of the Work and commencement of warranties required by the Contract Documents. When the Contractor considers a portion substantially complete, the Contractor shall prepare and submit a list to the Architect as provided under Section 9.8.2. Consent of the Contractor to partial occupancy or use shall not be unreasonably withheld. The stage of the progress of the Work shall be determined by written agreement between the Owner and Contractor or, if no agreement is reached, by decision of the Architect. **(9.9.1; change reference; minor modifications)**

When the owner wishes to occupy all or a portion of the project before final completion, the parties must agree and set forth the various provisions listed, including responsibility for security, utilities, and so forth. Although this has been a practice in the industry for some time, it has been on an informal basis, with little agreement as to allocation of risks between the owner and contractor. Further, unless the parties had specific written agreements related to partial occupancy, the contractor was frequently in a good position to argue that the owner had accepted the entire project by this partial occupancy. The ability to enforce the contract documents was thus compromised.

Among the important points to note are the following:

1. Permission must be obtained by appropriate governmental agencies and by the insurer.
2. Immediately before the partial occupancy, the owner, contractor, and architect must jointly inspect the area to be occupied. The condition of the area must be carefully documented, so that the contractor's obligation to complete the area

is clear as to each item. Note that the area to be occupied need not be substantially complete. This "inspection" is more thorough than the "observations" that the architect is required to make periodically as the project progresses. A201 requires only two other inspections: one at substantial completion and the second at final completion. If the owner decides to partially occupy one or more areas of the project, the architect will then be required to conduct more than these two inspections. This additional cost to the architect is not covered under either contingent or optional additional services. The architect should consider adding this to the list of additional services for which she will be paid in addition to the basic services.

3. The owner and contractor must agree, in writing, to their relative responsibilities as to security, maintenance, heat, utilities, damage to the work, and insurance. In most cases, the owner would assume these responsibilities for the portions of the project that are partially occupied. However, the parties could agree that the contractor would continue to provide insurance and other items until substantial completion.

4. The owner and contractor must also agree as to payments, including retainage. The contractor usually requests release of the retainage and wants to be substantially paid at this stage.

5. Agreement must be reached relative to the correction period. (The contractor is responsible, within one year after substantial completion, to correct work that does not conform to the contract documents. **See ¶** 12.2.2.) There also must be an agreement about any warranties required by the contract documents. The owner may want the correction period to commence upon substantial completion of the entire project, whereas the contractor may want it to start at the time of the partial occupancy. The specifications may have specific guarantee periods for particular items, such as elevators. Again, the contractor would want these periods to start immediately upon partial occupancy.

Paragraph 9.9.1 requires written agreement between the owner and contractor concerning these items. If they cannot agree, presumably the owner cannot start the partial occupancy. However, if the contractor is unreasonable, the owner may invoke the dispute resolution procedure in Article 15. If such partial occupancy is anticipated, the contract documents should address these items more fully.

Paragraph 9.9.1 specifically states that partial occupancy does not mean that work that does not conform to the contract documents has been accepted. The inspection that is conducted before the occupancy should stipulate all work that does not conform. Nonconforming work discovered later must be corrected by the contractor, or else the owner can obtain damages.

If partial occupancy delays the contractor or causes it additional cost, the contractor is entitled to a change order. If partial occupancy is anticipated, the owner might want to include a no-damage-for-delay clause related to the partial occupancy.

9.9.2 Immediately prior to such partial occupancy or use, the Owner, Contractor and Architect shall jointly inspect the area to be occupied or portion of the Work to be used in order to determine and record the condition of the Work. (**9.9.2; no change**)

This is to protect the contractor against claims by the owner that items were damaged before occupancy, when instead they could have been damaged by the owner's own forces.

9.9.3 Unless otherwise agreed upon, partial occupancy or use of a portion or portions of the Work shall not constitute acceptance of Work not complying with the requirements of the Contract Documents. (**9.9.3; no change**)

Even final payment does not constitute acceptance or waiver (**see** ¶ 9.10.4.2). That acceptance must be made in writing by the owner (**see** ¶ 12.3). One court has held that occupancy constituted acceptance of changes in the work by the contractor.[292] This provision should be read so that any occupancy by the owner does not waive any rights of the owner, and does not constitute an acceptance unless the owner knowingly accepts changes in the work.[293] The contractor may want to have the owner execute a letter of acceptance, specifying particular work that deviates from the contract documents.

§ 4.56   Final Completion and Final Payment: ¶ 9.10

The contract is not finished until the date of final completion. A New Jersey case[294] held that the risk of fire loss was upon the contractor until the project was fully completed, even though substantial completion had occurred.[295]

---

[292] Shimek v. Vogel, 105 N.W.2d 677 (N.D. 1960).

[293] In Mt. Hawley Ins. Co. v. Structure Tone, Inc., 2004 WL 2792039 (N.J. Super. July 13, 2004), a fire caused extensive damage to a building that was about 70% complete. The loss also included furniture, workstations, and computer equipment. The appellate court reversed the trial court's grant of summary judgment that dismissed all claims for subrogation, based on the waiver of subrogation provision (¶ 11.4.7 of A201). Responding to the "Work" versus non-Work argument, the contractor argued that, pursuant to this provision, which requires agreement of the owner and contractor before the owner may partially occupy the premises, failure to secure such agreement meant that the contractor had full control of the premises, thereby triggering the waiver of subrogation provision as to everything at that location, including computer equipment and the like. The appellate court sent this matter back to the trial court as a factual issue to be determined at a later date.

[294] Hartford Fire Inc. v. Riefolo Constr. Co., 161 N.J. Super. 99, 390 A.2d 1210 (1978).

[295] *See also* Brown v. McBro Planning & Dev. Co., 660 F. Supp. 1333 (D. V.I. 1987) (owner's acceptance of the project did not shift liability when a hospital technician injured his knee in a fall on the hospital floor. The owner's acceptance of full responsibility and control of the hospital was not a superseding cause of the injury.).

9.10 FINAL COMPLETION AND FINAL PAYMENT

9.10.1 *Upon receipt of the Contractor's written notice that the Work is ready for final inspection and acceptance and upon receipt of a final Application for Payment, the Architect will promptly make such inspection and, when the Architect finds the Work acceptable under the Contract Documents and the Contract fully performed, the Architect will promptly issue a final Certificate for Payment stating that to the best of the Architect's knowledge, information and belief, and on the basis of the Architect's on-site visits and inspections, the Work has been completed in accordance with terms and conditions of the Contract Documents and that the entire balance found to be due the Contractor and noted in the final Certificate is due and payable. The Architect's final Certificate for Payment will constitute a further representation that conditions listed in Section 9.10.2 as precedent to the Contractor's being entitled to final payment have been fulfilled.* **(9.10.1; written notice to come from "Contractor")**

Note that the work must be fully completed in order for the contractor to be entitled to final payment.[296] The contractor must initiate the architect's final

---

[296] *See* Ryan v. Thurmond, 481 S.W.2d 199 (Tex. 1972), which interpreted the following language (which is similar to AIA language): "Final payment shall be due Thirty days after (1) Substantial Completion of the work provided the work be then (2) fully completed and the contract (3) fully performed." The court held that the contractor was not entitled to payment because the contract was not completed. The contractor had failed to plead substantial performance, and the owner introduced evidence that the contract was not completed. *See also* RLI Ins. Co., v. MLK Ave. Redevelopment Corp., 925 So. 2d 914 (Ala. 2005).

In Martinson v. Brooks Equip. Leasing, Inc., 36 Wis. 207, 152 N.W.2d 849 (1967), the court construed a similar provision as requiring 100% completion. Substantial completion was not sufficient. The contractor had demanded full payment before completion of the project. The court held that this was a breach of contract.

In Laurel Race Course, Inc. v. Regal Constr. Co., 274 Md. 142, 333 A.2d 319 (1975), the contractor was not entitled to recover the balance of the contract because it had not obtained the engineer's "Final Certificate." "Where payments under a contract are due only when the certificate of an architect or engineer is issued, production of the certificate becomes a condition precedent to liability of the owner for materials and labor in the absence of fraud or bad faith." *See also* Russell H. Lankton Constr. Co. v. LaHood, 143 Ill. App. 3d 806, 493 N.E.2d 714 (1986) (architect's final certification required after arbitration award contingent on certificate); Bolton Corp., v. T.A. Loving Co., 380 S.E.2d 796 (N.C. App. 1989).

For an interesting case in which the architect had a 50% ownership in the general contracting firm, *see* Professional Builders, Inc. v. Sedan Floral, Inc., 819 P.2d 1254 (Kan. Ct. App. 1991). After the owner lost in an arbitration with the contractor, the owner sought to overturn the arbitration on the grounds of fraud in issuing the equivalent of a final certificate for payment. The court held that this was actually the issue before the arbitrator. The type of fraud necessary to overturn an arbitration award deals with fraud in the arbitration, not that which is totally outside the process of arbitration. The arbitration was upheld.

In In re Modular Structures, Inc., 27 F.3d 72 (3d Cir. 1994), the court held that the general contractor was not entitled to final payment until it had paid its subcontractors, based in part on this provision.

In Johnson City Cent. Sch. Dist. v. Fidelity & Deposit Co., 641 N.Y.S.2d 426 (App. Div. 1996), the dispute centered around the timing of filing the suit against the surety. The performance bond had a requirement that suit must be initiated within two years of the date on which final

inspection. Even if a final certificate for payment is obtained, the contractor is not relieved of liability for work that does not conform to the contract documents (**see** ¶ 9.10.4.2).[297]

Note, also, that even if the owner waives the requirement that the architect issue a formal final certificate for payment, that does not mean that the owner has waived its right to have the contract fully performed.[298] Some courts have held that issuance of the architect's final certificate was a required condition before the contractor was entitled to final payment.[299] The architect should not omit the language "to the best of the Architect's knowledge" from the certificate, because otherwise the certificate might be taken to guarantee the work.

> *9.10.2 Neither final payment nor any remaining retained percentage shall become due until the Contractor submits to the Architect (1) an affidavit that payrolls, bills for materials and equipment, and other indebtedness connected with the Work for which the Owner or the Owner's property might be responsible or encumbered (less amounts withheld by Owner) have been paid or otherwise satisfied, (2) a certificate evidencing that insurance required by the Contract Documents to remain in force after final payment is currently in effect and will not be canceled or allowed to expire until at least 30 days' prior written notice has been given to the Owner, (3) a written statement that the Contractor knows of no substantial reason that the insurance will not be renewable to cover the period required by the Contract Documents, (4) consent of surety, if any, to final payment and (5), if required by the Owner, other data establishing payment or satisfaction of obligations, such as receipts, releases and waivers of liens, claims, security interests or encumbrances arising out of the Contract, to the extent and in such form as may be designated by the Owner. If a Subcontractor refuses to furnish a release or waiver required by the Owner, the Contractor may furnish a bond satisfactory to the Owner to indemnify the Owner against such lien. If such lien remains unsatisfied after payments are made, the Contractor shall refund to the Owner all money that the Owner may be*

---

payment under the contract fell due. The architect's final certificate for payment was issued on December 21, 1992. Suit was filed on January 12, 1995. The court looked at this provision and Article 6 of A101 to determine whether the 30-day period meant that the two-year period began on December 21, 1992, or 30 days later. There was some degree of uncertainty as to which construction was more plausible. Holding that there is a general rule of liberal construction in favor of insureds, the court held that the action was not time-barred. *See also* American Motorists Inc. Co., v. Gottfurcht, 2004 WL 909799 (Cal. App. 2d Dist. April 29, 2004); Decca Design Build, Inc., v. American Auto. Ins. Co., 77 P.3d 1251 (Ariz. App. Div. 1 2003); Menorah Nursing Home, Inc., v. Zukov, 153 A.D.2d 13, 548 N.Y.S.2d 702 (App. Div. 2d Dep't 1989).

[297] Environmental Safety & Control Corp. v. Board of Educ., 580 N.Y.S.2d 595 (App. Div. 1992) ("[r]eceipt of the architects' final certificate does not foreclose the defendant, however, from asserting claims against the contractor.").

[298] Ryan v. Thurmond, 481 S.W.2d 199 (Tex. 1972).

[299] *See, e.g.,* I. Perlis & Sons v. Peacock Constr. Co., 222 Ga. 723, 152 S.E.2d 390 (1966).

In Pickett v. Chamblee Constr. Co., 124 Ga. App. 769, 186 S.E.2d 123 (1971), failing to obtain an architect's final certificate did not affect the contractor's right to final payment when there had been compliance. The owner had taken possession of the building.

compelled to pay in discharging such lien, including all costs and reasonable attorneys fees. **(9.10.2; no change)**

The contractor should submit final waivers of lien for itself and all subcontractors and material suppliers that have been listed on the contractor's statements. Most title companies have such forms. The contractor can bond over any unsatisfied liens and close out the project. In some cases, the owner may need to hire an attorney to clear up liens by subcontractors. The costs of paying off the subcontractors and other costs such as attorneys fees are to be paid by the contractor to the owner. These items are best addressed by the owner's counsel. If a contractor fails to furnish satisfactory evidence of payment to subcontractors and suppliers, the owner is entitled to withhold final payment.[300] The furnishing of the listed items is a condition precedent to the contractor's right to final payment from the owner.

Steffek v. Wichers, 211 Kan. 342, 507 P.2d 274 (1973), held that, by making changes and additions without resorting to or consulting with the architect and by disregarding the contract, the owner had waived the requirement of the architect's certificate for final payment.

In American Continental Life Ins. Co. v. Ranier Constr. Co., 125 Ariz. 53, 607 P.2d 372 (1980), the court held that the owner did not waive the requirement of a final certificate for payment. The contractor argued that both the owner and contractor had deviated from the formal requirements of the contract in several respects: change orders were not signed, informal extensions of time were granted, and so on. The court, however, said:

> the waiver of one right under a contract does not necessarily waive other rights under the contract. [Citation omitted.] Thus, even if American did waive other rights under the contract relating to change orders or extensions of time, that conduct does not manifest an intent to waive any right relating to payment for work.

In Kilianek v. Kim, 192 Ill. App. 3d 139, 548 N.E.2d 598 (1989), the appellate court overturned an arbitrator's award in favor of the contractor because the architect's final certificate had not been obtained. The court ruled that the arbitrator had exceeded his power because the owner's obligations had ended when the condition precedent (obtaining the certificate) had not been met. *See* Formigli Corp. v. Fox, 348 F. Supp. 629 (E.D. Pa. 1972), for a case in which consistent failure previously to obtain the architect's approval waived the requirement for final payment approval. Here, the owner had made 12 payments to the contractor without the architect's approval, thereby waiving the requirement.

In Redevelopment Auth. v. Fidelity & Deposit Co., 665 F.2d 470 (3d Cir. 1981), the court found that the architect was the representative of the owner charged with administrative responsibility regarding final completion. The architect's execution of a final certificate for payment constituted final settlement of the contract for purposes of the surety's contractual limitations period.

In Decca Design Build, Inc. v. American Auto. Ins. Co., 77 P.3d 1251 (Ariz. Ct. App. 2003), final completion had not been attained and the general contractor's action against a subcontractor's surety was not untimely. *Accord* American Motorists Ins. Co. v. Gottfurcht, 2004 Cal. App. Unpub. LEXIS 4258 (2d Dist. Apr. 29, 2004).

[300] Williard, Inc. v. Powertherm Corp., 497 Pa. 628, 444 A.2d 93 (1982). However, in Henrico Doctors' Hosp. & Diagnostic Clinic, Inc. v. Doyle & Russell, Inc., 221 Va. 710, 273 S.E.2d 547 (1981), these documents were not required after the time for filing liens had passed.

See also Hagerstown Elderly Assocs. v. Hagerstown Elderly Building Assocs., 368 Md. 351, 793 A.2d 579 (2002); Beard Family P'ship v. Commercial Indem. Ins. Co., 116 S.W.3d 839 (Tex. App.-Austin, 2003); Brown and Kerr, Inc., v. American Stores Prop., Inc., 715 N.E.2d 804 (Ill. App. 1st Dist. 1999); In re Modular Structures, Inc., 27 F.3d 72 (3d Cir. 1994).

9.10.3 If, after Substantial Completion of the Work, final completion thereof is materially delayed through no fault of the Contractor or by issuance of Change Orders affecting final completion, and the Architect so confirms, the Owner shall, upon application by the Contractor and certification by the Architect, and without terminating the Contract, make payment of the balance due for that portion of the Work fully completed and accepted. If the remaining balance for Work not fully completed or corrected is less than retainage stipulated in the Contract Documents, and if bonds have been furnished, the written consent of surety to payment of the balance due for that portion of the Work fully completed and accepted shall be submitted by the Contractor to the Architect prior to certification of such payment. Such payment shall be made under terms and conditions governing final payment, except that it shall not constitute a waiver of claims. **(9.10.3; no change)**

If the final completion of the project is delayed through no fault of the contractor or because of the issuance of change orders, the contractor is entitled to payment for the work performed, and the owner cannot delay such payment. The final payment referred to is the actual last payment and not any prior payment. Thus, if two payments are made after substantial completion, the first such payment does not operate as a waiver under ¶ 9.10.3. The owner might want to insert a statement requiring a certain retainage tied to the estimated cost of the punchlist items until final completion.

One issue is what happens if the architect does not "so confirm" the contractor's right to receive the stated payment. Presumably, the contractor can avail itself of the claims procedure in Article 15.

9.10.4 The making of final payment shall constitute a waiver of Claims by the Owner except those arising from:

> *.1   liens, Claims, security interests or encumbrances arising out of the Contract and unsettled;* **(9.10.4.1; no change)**

This provision was held to bar a claim when the second change order issued reserved a claim by the owner for the contractor's failure to complete the project on time. A subsequent change order failed to reserve this right, and the final payment authorization also did not reserve the claim.[301] It is important to note the reservation of any claims at the time of substantial and final completion.

> *.2   failure of the Work to comply with the requirements of the Contract Documents; or* **(9.10.4.2; no change)**

An owner demanded arbitration against the contractor, alleging faulty construction. The contractor sought to enjoin the arbitration, alleging that the final payment

---

In R.W. Grainger & Sons v. Nobel Ins. Co., 2005 WL 3729018 (Mass. Super. Dec. 20, 2005), a sub-subcontractor filed an action for injunction, requiring a response by the general contractor, who backcharged the subcontractor for its legal fees. The court held that this provision did not apply, since the action was not to secure the discharge of a lien.

[301] John Price Assocs. v. Davis, 588 P.2d 713 (Utah 1978); Fitzgerald v. Corbett, 793 P.2d 356 (Utah, 1990).

constituted a waiver. An earlier version of this clause was held to negate the contractor's argument, and the case was sent to arbitration.[302] This clause has been held to refer to defects in materials and workmanship, not failure to comply with time deadlines.[303]

> *.3*   *terms of special warranties required by the Contract Documents.* (**9.10.4.3; no change**)

When the owner makes final payment to the contractor, the owner waives all claims against the contractor or subcontractors except the listed exceptions.[304]

---

[302] Woodward Heating v. American Arbitration, 259 Pa. Super. 460, 393 A.2d 917 (1978).

An Illinois case, Village of Westfield v. Loitz Bros. Constr. Co., 165 Ill. App. 3d 338, 519 N.E.2d 37 (1988), involved an owner's claim that a contractor had accepted final payment from the owner and this acceptance operated as a release of all claims. Because the owner had stated that no final payment would be made until it received final waivers of lien from subcontractors and the final contractor's affidavit, and these documents were never given, there was no such release. *See also* Burke County Pub. Sch. v. Juno Constr. Corp. 50 N.C. App. 238, 273 S.E.2d 504(1981).

In Automobile Ins. Co. v. United H.R.B. Gen. Contractors, Inc., 876 S.W.2d 791 (Mo. Ct. App. 1994), the court held that the waiver of subrogation (¶ 11.3.7) ended upon final payment to the contractor. A fire occurred after completion of the building, causing substantial damage. The owner's insurer sued the contractor, alleging faulty installation of the electrical system. The contractor defended based on the waiver of subrogation provision. The court construed an earlier version of the AIA contract. The exceptions contained in ¶ 4.3.5.2 (in an earlier version of this document) were at odds with the contractor's contention that the owner's insurer could not bring an action against the contractor after final payment. Because the owner expressly reserved the right to bring an action against the contractor after final payment for defective work, the owner's insurer could bring an action for subrogation after final payment. The waiver of subrogation was only effective so long as the contractor had an insurable interest in the property. The court found that the contractor's interest in the project terminated on final payment.

In Warwick Township Water & Sewer Auth. v. Boucher & James, Inc., 2004 WL 557597 (Pa. Super. Mar. 23, 2004), the trial court refused to compel arbitration after the project was completed, based in part on the waiver provisions in the contract, similar to A201. The appellate court reversed and ordered arbitration, holding that claims relating to defects were not waived.

[303] Centerre Trust Co. v. Continental Ins., 167 Ill. App. 3d 376, 521 N.E.2d 219 (1988).

[304] In John Price Assocs. v. Davis, 588 P.2d 713 (Utah 1978), a contractor sued the owner on a promissory note. The owner defended and counterclaimed, saying that the project was late, resulting in damages to the owner. This provision of A201 waived claims by the owner related to delays, and the court held for the contractor.

Centerre Trust Co. v. Continental Ins., 167 Ill. App. 3d 376, 521 N.E.2d 219 (1988), involved an owner's action against the contractor for liquidated damages. The court held that the owner waived its right to liquidated damages arising from breach of the contract by making a final payment. *But see* Illinois State Toll Highway Auth. v. Gust K. Newberg, Inc., 531 N.E.2d 982 (Ill. App. 2d Dist. 1988), distinguishing *Centerre*.

In People ex rel. Skinner v. Graham, 170 Ill. App. 3d 417, 524 N.E.2d 642 (1988), the court interpreted this section as meaning that liability of a surety was extinguished 12 months after final payment when the contract contained a specific limitation period and the contractor was not notified within that period of unfinished or defective work in need of correction.

It is therefore important that the final inspection of the architect be thorough so that other claims can be asserted. Courts have held that payment and acceptance of improvements constituted a waiver of all damages for defects that were known to the owner or that were observable by a reasonable inspection.[305] Claims arising from unknown (latent) defects are not waived by final payment.[306] Some states have adopted the *accepted work rule doctrine,* which states that an independent contractor will not be liable to third parties for injuries that occur after the contractor has completed the work and the work has been turned over to and accepted by the employer.[307]

In David Co. v. Jim W. Miller Constr., Inc., 428 N.W.2d 590 (Minn. Ct. App. 1988), the court held that there was no waiver when there was a breach of contract by the contractor. This was the same as "failure of the work to comply with the requirements of the Contract Documents."

In Wilson Area Sch. Dist. v. Skepton, 860 A.2d 625 (Pa. 2004), the defendants were three contractors on a school project. Pursuant to their contract, these contractors paid some $120,000 in permit fees to the borough in which the school was located, but under protest. The contractors then filed actions against the borough, and the trial court found that the permit fees were grossly disproportionate. The Pennsylvania Supreme Court ultimately ruled that the borough must refund the excess fees to the contractors. Thereafter the school district filed an action to recover the fees, asserting that it had a superior interest in the fees. The court found that the permit fees were not a separate line item in the bids, but were absorbed and buried in the total or composite contract price. Pursuant to this provision of A201, the school district had waived all claims against the contractors by making final payment and by not reserving a right to claim an interest in the refunded permit fees.

[305] Town of Tonawanda v. Stapell, Mumm & Beals Corp., 240 A.D. 472, 270 N.Y.S. 377 (1934); Shaw v. Bridges-Gallagher, Inc., 174 Ill. App. 3d 680, 528 N.E.2d 1349 (1988); State v. Wilco Constr. Co., 393 So. 2d 885 (La. Ct. App. 1981).

In Eastover Corp. v. Martin Builders, 543 So. 2d 1358 (La. Ct. App. 1989), the court held that the architect knew or should have known that certain pipe hangers were improperly spaced, causing damage when the pipes failed. Because the architect was an agent or representative of the owner, and because it was not a latent defect, it was waived.

[306] Intaglio Serv. Corp. v. J.L. Williams & Co., 95 Ill. App. 3d 708, 420 N.E.2d 634 (1981); Centerre Trust Co. v. Continental Ins., 167 Ill. App.3d 376, 521 N.E.2d 219 (1988).

The contractor is also liable to third parties for latent defects, even when final acceptance of the project by the owner would normally end such liability. Honey v. Barnes Hosp., 708 S.W.2d 686 (Mo. Ct. App. 1986). Contra, Bruzga v. PMR Architects, P.C., 693 A.2d 401 (N.H. 1997), declining to hold contractor liable for suicide; R.W. Gast v. Shell Oil Co., 819 S.W.2d 367 (Mo. 1991) (cashier who was shot to death could not maintain action against general contractor who converted service bay into cashier's room).

[307] Harrington v. LaBelle's, Inc., 235 Mont. 80, 765 P.2d 732 (1988) (contractor who installed a "speed bump" was not liable when bicyclist was injured several months after owner accepted the work and made final payment).

In R. W. Gast v. Shell Oil Co., 819 S.W.2d 367 (Mo. 1991), the parents of a gas station cashier who was shot to death during a robbery filed a wrongful death action against the contractor who converted a service bay into a cashier's room. The contractor apparently did not deviate from the plans and specifications. The court found that the contractor owed no duty to the owner's employees with respect to the design of the modifications. Also, the specifications were not so imperfect or improper that the contractor should have realized that the work would make the structure inherently unsafe. Therefore, there was no duty to third persons after acceptance by the owner.

The term *special warranty* in ¶ 9.10.4.3 refers to warranties that are specific to the project.[308] In most cases, the specifications require particular warranties, such as roof warranties. These would fit the definition of special warranty.

In Pierce v. ALSC Architects, P.S., 52 Mont. 93, 890 P.2d 1254 (1995), the architect designed a remodeling project for a supermarket, which had a walk-in cooler located on the main floor immediately below the observation and storage room. The roof of the cooler was even with the security walkway and provided a floor for the storage room and a place for storage of seasonal displays used in the store. During the remodeling, the cooler was removed and replaced with a smaller walk-in freezer. A suspended ceiling was installed in the space between the new freezer and the observation walkway. It was agreed that the walkway was to be sealed off by removing the access door, using it in another location, and covering the opening with drywall. It was understood that if the walkway was not to be sealed off, guardrails, lighting, and walkway improvements would be required by code. During construction, the contractor had an extra door and used it where the access door was to go. The access door was not replaced or sealed. Later the plaintiff, a clerk at the supermarket, was asked by a customer to retrieve a poster from the area of the walkway. Unaware that the roof of the cooler had been replaced, he stepped through the access door. The area was unlighted and did not appear different than it had before the remodeling. The clerk fell through the ceiling and injured himself.

The architect claimed that the accepted work doctrine barred recovery. That doctrine holds that, once the owner accepts the work, the contractor or architect is free from liability to third parties. In rejecting this doctrine, the supreme court of Montana stated:

> This defense, as previously applied, has the undesirable effect of shifting responsibility for negligent acts or omissions from the negligent party to an innocent person who paid for the negligent party's services. Furthermore, the shifting of responsibility is based on the legal fiction that by accepting a contractor's work, the owner of property fully appreciates the nature of any defect or dangerous condition and assumes responsibility for it. In reality, the opposite is usually true. Contractors, whether they be building contractors, or architects, are hired for their expertise and knowledge. The reason they are paid for their services is that the average property owner does not have sufficient knowledge or expertise to design or construct real property improvements safely and soundly. The mere fact that expert testimony is required to establish professional negligence makes it clear that nonexperts are incapable of recognizing substandard performance on their own. How then can we logically conclude that simply because the professional has completed his or her services and the contractee has paid for those services, liability for the contractor's negligence should shift to the innocent and uninformed contractee? We cannot.

The court also found that the architect was guilty of negligence per se because the area in question was admittedly in violation of the building code.

See also Ogles v. E.A. Mann & Co., Inc., 625 S.E.2d 425 (Ga. App. 2005); Washington v. Qwest Communications Corp., 704 N.W.2d 542 (Neb. 2005).

The accepted work doctrine was rejected in Davis v. Baugh Indus. Contractors Inc., 150 P.3d 545 (Wash. 2007), finding that this doctrine was outmoded, incorrect, and harmful. *See, also,* Griggs v. Shamrock Bldg. Serv., Inc., 634 S.E.2d 635 (N.C. App. 2006) (there is an exception to the completed and accepted rule, where a contractor remains liable where the work completed and turned over to the owner was imminently dangerous to third persons).

[308] Hillcrest Country Club v. N.D. Judds Co., 236 Neb. 233, 461 N.W.2d 55 (1990). In that case, the special warranty was a specific 20-year roof warranty. The contractor argued that its liability was limited to one year under ¶ 12.2.2 and because the warranty was contained in its quotation to the owner rather than in the contract documents. The court rejected this contention on the basis that the quotation was made a part of the contract. Compare that situation with the current version of A201, which does not incorporate the bid documents (see ¶ 1.1.1). The *Hillcrest* warranty was

9.10.5 Acceptance of final payment by the Contractor, a Subcontractor or material supplier shall constitute a waiver of claims by that payee except those previously made in writing and identified by that payee as unsettled at the time of final Application for Payment. **(9.10.4; no change)**

If the contractor accepts final payment, it waives all claims except those specifically reserved in writing.[309] One court held that a payment to a contractor is not the final payment which triggers the running of a statute of limitations if the payment was merely an interim payment that happened to be the last payment made.[310]

§ 4.57   Article 10: Protection of Persons and Property

This article covers safety, both to people and things, on the job site, which is the contractor's responsibility. In addition to basic safety measures, the article includes procedures for encountering asbestos or PCB on the job site and for emergencies in general.

§ 4.58   Safety Precautions and Programs: ¶ 10.1

10.1 SAFETY PRECAUTIONS AND PROGRAMS

The Contractor shall be responsible for initiating, maintaining and supervising all safety precautions and programs in connection with the performance of the Contract. **(10.1.1; no change)**

---

for "future performance" (this relates to a statute of limitations defense—see U.C.C. § 2-725(2)) Joswick v. Chesapeake Mobile Homes, Inc., 747 A.2d 214 (Md. App. 2000).

[309] For a case that held that a claim had previously been made and the contractor did not waive the claim, *see* Cape Fear Elec. Co. v. Star News Newspapers, Inc., 22 N.C. App. 519, 207 S.E.2d 323, *cert. denied,* 285 N.C. 757, 209 S.E.2d 280 (1974).

In McKeny Constr. Co. v. Town of Rowlesburg, 187 W. Va. 521, 420 S.E.2d 281, 283 (1992), the contract contained this provision:

The acceptance by the Contractor of the Final Payment shall be and shall operate as a release to the Owner of all claims and of all liability to the Contractor for all things done or furnished in connection with this work and for every act and neglect of the Owner and others relating to or arising out of this work, excepting the Contractor's claims for interest upon the Final Payment, if this payment is improperly delayed. No payment, however final or otherwise, shall operate to release the Contractor or his sureties from any obligation under this Contract or the performance bond.

This provision also operated as a release by the contractor of claims against the engineer. There was no evidence that the owner ever waived any provision of the contract. *See also* Absher Constr. Co. v. Kent. Sch. Dist. No. 415, 77 Wash. App. 137, 890 P.2d 1071 (1995). In Everman's Elec. Co., Inc., v. Evan Johnson, 955 So. 2d 979 (Miss. App. 2007), the court rejected the argument by one prime contractor that this provision meant that another prime contractor had waived claims against it.

[310] Credit Gen. Ins. Co. v. Atlas Asphalt, Inc., 304 Ark. 522, 803 S.W.2d 903 (1991). The court rejected the argument that final payment could be made without regard to the retainage.

It is the contractor's responsibility, and not the architect's, to ensure that a safety program is in force at the site.[311] This duty, however, is no greater than the duty an

---

[311] Gero v. J.W.J. Realty, 757 A.2d 475 (Vt. 2000); Graham v. Freese & Nichols, Inc., 927 S.W.2d 294 (Tex. App.-Eastland 1996); Padgett v. CH2M Hill Southeast, Inc., 866 F. Supp. 563 (M.D. Ga. 1994). For a case in which an engineer undertook a safety program, *see* Caldwell v. Bechtel, 631 F.2d 989 (D.C. 1980). A worker contracted silicosis while performing "mucking" operations in a subway tunnel. The engineer had a contract to provide "safety engineering services" and owed the worker a duty to take reasonable steps to protect him from the silica dust in the tunnel.

This provision was examined in a case in which the architect was asked to interpret it in a situation of multiple prime contractors. The court held that consolidation of the subsequent arbitration proceeding was proper. Children's Hosp. v. American Arbitration Ass'n, 231 Pa. Super. 230, 331 A.2d 848 (1974).

In Lewis v. N.J. Riebe Enters., Inc., 170 Ariz. 384, 825 P.2d 5 (1992), the general contractor was found to have a duty toward the subcontractor regarding safety on the job. *See also* Burns v. Black & Veach Architects, Inc., 854 S.W.2d 450 (Mo. 1993); Fisher v. M. Spinelli & Sons Co., Inc., 1999 WL 165674 (Mass. Super. 1999).

Architects and engineers are subject to OSHA regulations if they undertake safety duties at a construction project. On the other hand, OSHA does not apply to design professionals if they do not have a duty to oversee safety. CH2M Hill, Inc. v. Alexis Herman, Secretary of Labor and Occupational Safety and Health Review Comm'n, 192 F.3d 711 (7th Cir. 1999). In 1977, the Milwaukee Metropolitan Sewerage District ("MMSD") undertook a $2.2 billion pollution abatement program that included construction of some 80 miles of sewer tunnels. CH2M Hill was the lead engineering consultant firm and Healy was the contractor for one of the tunnels, CT-7. In November 1988, the tunnel boring machine's methane monitor detected a high concentration of methane. The tunnel was evacuated, but the electrical power was not turned off. Contrary to Healy's evacuation plan, three supervisors reentered the tunnel only 17 minutes after the methane was detected. Apparently, one of the three attempted to operate the grout pump, which was not explosion-proof. The gas exploded, killing all three.

The engineer was cited for willful violation of OSHA standards. At the initial trial before an administrative law judge, the engineer was found not guilty because it was not engaged in construction work, but was engaged in "typical engineering services." This finding was reversed by a review panel. Ultimately, the case was brought before the Seventh Circuit, which reinstated the original finding that the engineer was not liable under the OSHA regulations. The court did not hold that an engineer could never be liable under OSHA, but rather, that each case must be examined to determine the scope of the engineer's duty on each particular construction project. The court found that OSHA was intended to apply to those employers who were best suited to alleviate hazards at the construction site.

With this intent in mind, the Seventh Circuit examined the contractual duties of the engineer, as well as how the project was actually run, to determine if this engineer was in a position to "alleviate hazards at the construction site." The starting point was the contract. The court stated that "contracts represent an agreed upon bargain in which the parties allocate responsibilities based on a variety of factors. To ignore the manner in which the parties distributed the burdens and benefits is contrary to our notion of contract law."

The engineer's contract forbade it from supervising and directing Healy or any other contractor. The engineer was to "administer the contracts in accordance with the requirements of the Contract Documents and act as the DISTRICT's representative during construction." The contract also stated that "the construction contractors shall remain responsible for construction means, methods, techniques, procedures and safety precautions on the construction portions of the PROGRAM." These provisions are similar to the language found in AIA Document B141-1997, at ¶ 2.6.2.1.

owner assumes towards third parties.[312] This provision also acts to insulate the owner from claims by employees of the contractor who are injured,[313] and contractors from claims by injured employees of subcontractors, where a similar provision requires the subcontractor to be in charge of safety for its work.[314]

Additional language in the engineer's contract referred to site visits and observations by the engineer during construction, which included this language: the visits "shall not relieve a construction contractor of its obligation to conduct comprehensive inspections of the work sufficient to ensure conformance with the intent of the construction contract documents, and shall not relieve a construction contractor of its responsibility for means, methods, techniques, sequences and procedures necessary for coordinating and completing all portions of the work under the construction contract and for all safety precautions incidental thereto."

The court also examined whether or not the engineer actually exercised substantial control or supervision so as to place it in charge of safety and concluded that there was no evidence of such control or supervision. While the engineers did enter the tunnel, and were further involved in issuing contract modifications with regard to discovery of methane gas in another tunnel, the owner retained the right to approve any modification dealing with safety. The engineer could not act on its own. The court then went on to state:

> In addition, the fact that Healy turned to CH2M Hill for advice does not indicate that CH2M Hill was acting as the de facto director of safety. Before it issued the clarification, CH2M Hill had to check with MMSD. From the record it is clear that CH2M Hill functioned as an intermediary between Healy and MMSD. MMSD directed the modifications; it ordered CH2M Hill to conduct an inquiry, reviewed CH2M Hill's suggestions and granted the final approval to the draft language, as well as the clarification of it requested by Healy. With this fuller picture, it would be disingenuous to say that the record supports the Commission or the Secretary's conclusion that CH2M Hill played a 'central role' as the 'nerve center' for developing and implementing safety practices for tunnel CT-7.

Under Article 10, the contractor has exclusive control over the construction site. In Fe' E. Clardy v. PCL Constr., 2001 Minn. App. LEXIS 225 (March 6, 2001), the appellate court affirmed the jury's finding that the contractor was responsible when a worker fell through an opening that had not been properly protected. *See also* Townsend v. Muckleshoot Indian Tribe, 2007 WL 316504 (Wash. Ct. App. Div. 1 Feb. 5, 2007); Gero v. J.W.J. Realty, 757 A.2d 475 (Vt. 2000).

In Hawthorne v. Summit Steel, Inc., 2003 WL 23009254 (Del. Super. July 14, 2003), an injured worker sued a number of parties, including the construction manager. The construction manager had hired a general contractor to perform the actual work. Because the contract between the owner and construction manager contained this language, summary judgment for the construction manager was not granted, on the ground that safety responsibilities had been delegated to another party.

[312] Jakubowski v. Alden-Bennett Const. Co., 763 N.E.2d 790 (Ill. App. 1st Dist. 2002) (trespasser injured on construction site. Contractor not liable. The contractor's duties towards the trespasser were determined by premises liability rules rather than the contractor's obligations under the contract). *Jakubowski* was followed in Smithey v. Stueve Const. Co., 2007 WL 172511 (D. S.D. 2007) (no tort duty for negligent performance of this contractual obligation. Instead, common law duty applies).

[313] Townsend v. Muckleshoot Indian Tribe, 2007 WL 316504 (Wash. App. Div. 1, February 5, 2007); Perrit v. Bernhard Mech. Contractors, Inc., 669 So. 2d 599 (La. Ct. App. 1996). Contra, Moe v. Eugene Zurbrugg Const. Co., 123 P.3d 338 (Or. Ct. App. 2005).

[314] Legros v. Lone Star Striping and Paving, LLC, 2005 WL 3359740 (Tex. App.-Houston [14 Dist.], Dec. 6, 2005); Fernandez v. Concrete Shell Structures, Inc., 2002 WL 1389828 (Cal. App. 1st Dist. June 27, 2002).

The issue in these cases is usually one of assumption of duty regarding safety. In other words, whose job is it to oversee safety? The architect should not include any specific directions concerning site safety in the contract documents, nor should the architect review the contractor's safety program. Doing so may create an ambiguity between such directions and the provisions found here, and may result in the architect being held liable for injuries at the jobsite.

The provisions of this document create duties between the owner and contractor, and, to some degree, with the architect. However, the duties of safety created herein do not extend to workers, according to a California court.[315]

§ 4.59   Safety of Persons and Property: ¶ 10.2

10.2 SAFETY OF PERSONS AND PROPERTY

10.2.1 The Contractor shall take reasonable precautions for safety of and shall provide reasonable protection to prevent damage, injury or loss to: (**10.2.1; no change**)

> *.1   employees on the Work and other persons who may be affected thereby;* (**10.2.1.1; no change**)

The contractor must take reasonable precautions to assure the safety of the workers. In one case, an engineer who undertook to provide "safety engineering services" owed a duty to a worker who contracted silicosis while working in a subway tunnel.[316] It is the contractor and not the architect who bears this responsibility for safety under the standard AIA agreements.[317] This provision has also defeated a claim by a worker against a developer.[318] In a Minnesota case, an

---

[315] Fernandez v. Concrete Shell Structures, Inc., 2002 WL 1389828 (Cal. App. 1st Dist. June 27, 2002), although California has a line of cases following Privette v. Superior Court, 854 P.2d 721 (Cal. 1993), which severely limited the right of an employee of an independent contractor to sue the employer's hirer for injuries suffered on the jobsite. This line of cases restricts the employee to the worker's compensation laws and do not rely on the provisions of the AIA language. Nevertheless, *Fernandez* provides an interesting analysis of the language of this AIA document and the duties it creates.

[316] Caldwell v. Bechtel, 631 F.2d 989 (D.C. 1980). An architect/construction manager who contracted to develop a project safety manual, review contractors' safety programs, and had authority to stop the work was not insulated from liability related to a worker's job site injuries by the exculpatory language that he would not be "responsible for construction means, methods, techniques, work sequences or procedures." Wenzel v. Boyles Galvanizing Co., 920 F.2d 778 (11th Cir. 1991).

[317] Waggoner v. W&W Steel Co., 657 P.2d 147 (Okla. 1983) (construction workers were injured as a result of unsafe construction practices). *See also* Burns v. Black & Veach Architects, Inc., 854 S.W.2d 450 (Mo. 1993); Fisher v. M. Spinelli & Sons Co., Inc., 1999 WL 165674 (Mass. Super. 1999).

[318] Bryant v. Village Centers, Inc., 167 Ga. App. 220, 305 S.E.2d 907 (1983); Francavilla v. Nagar Constr. Co., 151 A.D.2d 282, 542 N.Y.S.2d 557 (1989); Townsend v. Muckleshoot Indian Tribe, 2007 WL 316504 (Wash. App. Div. 1, February 5, 2007); Perrit v. Bernhard Mech. Contractors,

injured worker was not allowed to recover against an engineer who had loaned the worker a wrench.[319] The engineer had no duty to provide safe tools.

Paragraph 10.2.1, read with ¶ 3.3.2, imposes a duty of reasonable care upon the contractor for the benefit of all employees on the work and all other persons who may be affected by the work.[320]

Inc., 669 So. 2d 599 (La. Ct. App. 1996). Contra, Moe v. Eugene Zurbrugg Const. Co., 123 P.3d 338 (Or. Ct. App. 2005).

In Vickers v. Hanover Constr. Co., 875 P.2d 929 (Idaho 1994), the court held that the general contractor and owner owed no duty to the subcontractor's employee under the AIA language. The worker's employer was dismissed because of the exclusivity provision of the workers' compensation act. The court found that, although the general contractor had a right to inspect the construction site, exercise of that right was limited to inspection for the purpose of determining whether the subcontractor's work complied with the contract specifications. Caution should be taken in following this case because, in most jurisdictions, the general contractor has charge of safety at the job site and owes a duty to most, if not all, workers at the site.

[319] Graham v. Abe Mathews Eng'g, 358 N.W.2d 131 (Minn. Ct. App. 1984). In Lewis v. N.J. Riebe Enters., Inc., 170 Ariz. 384, 825 P.2d 5 (1992) (*citing* Restatement (Second) of Torts § 414, cmt. b (1965)), the court found that the general contractor was responsible for the safety of the subcontractors' workers:

> Because the general contractor coordinates the work of various subcontractors at a work site, it is in the best position to provide for the safety of all workers and to reduce the risk of injury at the work site. . . . Therefore, we hold that if a general contractor contractually assumes the responsibility for safety at a work site, it is liable for any injury resulting from its negligent exercise of that responsibility as long as the general contractor knew or by the exercise of reasonable care should have known that the subcontractors' work was being done in a dangerous manner, and had the opportunity to prevent it by exercising the power of control which he retained in himself.

Section 414 of the Restatement was also the basis for a negligence claim by an injured worker for a masonry subcontractor against the general contractor in Aguirre v. Turner Construction Company, 501 F.3d 825, (7th Cir. (Ill.) 2007). This case involved the "retained control" theory of negligence, wherein if the general retains some control over the manner in which the work is done, there might be liability. "It is not enough that the GC has merely a general right to order the work stopped or resumed, to inspect its progress or receive reports, to make suggestions or recommendations which need not necessarily be followed, or to prescribe alterations and deviations." In this case, the GC had an extensive safety program, thus retaining control. Unfortunately, this case suggests that contractors who do less to insure safety will have less liability if workers are injured.

[320] Ramon v. Glenroy Constr. Co., 609 N.E.2d 1123 (Ind. Ct. App. 1993).

In Eischeid v. Dover Constr., Inc., 217 F.R.D. 448 (N.D. Iowa 2003), a construction worker was injured when an unbraced wall collapsed under the force of a strong wind. The general contractor argued that it was not liable because its subcontractor, the masonry contractor, was responsible for bracing the wall. The court found that its duty under the contract was nondelegable; performance of that duty may be delegated to another, but such delegation of performance does not excuse the general contractor from liability for non-performance. The general contractor, as a matter of law, was subject to a nondelegable duty to provide a reasonably safe workplace on the basis of its "control of the job." The court found that the injured worker can sue for breach of the general contractor's duty as a third-party beneficiary.

In Cochran v. George Sollitt Constr. Co., 358 Ill. App. 3d 865, 832 N.E.2d 355 (1st Dist. 2005), a new employee of the subcontractor was injured when he climbed a ladder that was supported by a sheet of plywood that sat on two milk crates. When he shifted his weight, the ladder fell and he was injured. Summary judgment in favor of the general contractor was appropriate because the general contractor did not exercise sufficient control over the subcontractor's work.

*.2   the Work and materials and equipment to be incorporated therein, whether
       in storage on or off the site, under care, custody or control of the Contrac-
       tor or the Contractor's Subcontractors or Sub-subcontractors; and*
(10.2.1.2; no change)

This provision places the risk of loss on the contractor for the Work and all of the
materials that will ultimately be incorporated into the Work, whether or not such
materials are actually installed at the time of a loss. This also applies to equipment,
such as mechanical equipment, that is located off-site, but is to be incorporated into
the Work at a later date. This provision also applies to any such items if they are
under the care, custody, or control of this contractor, any lower tier contractor, or
anyone else in privity with this contractor. This provision does not apply to items
that will not be incorporated into the Work, such as tools or scaffolding.

This is, however, apparently not absolute. The contractor must take "reasonable" precautions to protect the work. What this means must be determined on a
case-by-case basis. Note that ¶ 11.3.1 requires the owner to provide insurance for
the work at the site.

*.3   other property at the site or adjacent thereto, such as trees, shrubs, lawns,
       walks, pavements, roadways, structures and utilities not designated for
       removal, relocation or replacement in the course of construction.*
(10.2.3; no change)

The contractor is responsible for reasonable safety precautions on the job site in
order to prevent damage to non-Work items at the site. This would include the
listed items, plus other materials or items belonging to the owner or to other
contractors, or to property owned by neighbors or other parties.

10.2.2 *The Contractor shall comply with and give notices required by applicable
laws, statutes, ordinances, codes, rules and regulations, and lawful orders of public
authorities bearing on safety of persons or property or their protection from damage,
injury or loss.* **(10.2.2; minor modifications)**

This paragraph applies to notices required by any governmental entity related to
safety. A prime example would be OSHA, which has various rules regarding
safety. Compliance with such rules rests with the contractor and not the architect
or owner.

These provisions do not impose a contractual duty of care on the general contractor with
regard to the safety of an employee of a subcontractor. *Deleon v. DSD Dev., Inc.,* 2006 WL
2506743 (Tex. App.-Hous., 1st Dist. Aug. 31, 2006). Article 10 did not require the general
contractor to control the means, methods, or details of how the subcontractor performed the
work. The court found that the provisions of A201 governed only the relationship between the
owner and general contractor and served to allocate the rights and responsibilities between them
exclusively. It found that the purpose of these provisions to be to insulate the owner from tort
liability and to create a duty of fiscal responsibility flowing from the general contractor to the
owner for claims arising from the construction project.

10.2.3 The Contractor shall erect and maintain, as required by existing conditions and performance of the Contract, reasonable safeguards for safety and protection, including posting danger signs and other warnings against hazards, promulgating safety regulations and notifying owners and users of adjacent sites and utilities. **(10.2.3; no change)**

This provision has protected a developer from a suit by an injured worker.[321] This paragraph requires the contractor to take "reasonable" safeguards regarding safety. The contractor must consider existing conditions and what the Contract Documents require, as well as the contractor's own means and methods to be used for the construction of the project. Only the contractor, as opposed to either the owner or architect, has the necessary expertise to properly perform this duty. A Connecticut court has examined this provision in addressing the question of whether this contractual obligation can overcome the exclusive remedies provision of a state worker's compensation law.[322]

10.2.4 When use or storage of explosives or other hazardous materials or equipment or unusual methods are necessary for execution of the Work, the Contractor shall exercise utmost care and carry on such activities under supervision of properly qualified personnel. **(10.2.4; no change)**

When using hazardous materials or methods, the contractor must use "utmost care." This means that the contractor will be liable for injuries or damage even if it is not negligent, if hazardous materials or methods are related to the damage.

10.2.5 The Contractor shall promptly remedy damage and loss (other than damage or loss insured under property insurance required by the Contract Documents) to property referred to in Sections 10.2.1.2 and 10.2.1.3 caused in whole or in part by the Contractor, a Subcontractor, a Sub-subcontractor, or anyone directly or indirectly employed by any of them, or by anyone for whose acts they may be liable and for which the Contractor is responsible under Sections 10.2.1.2 and 10.2.1.3, except damage or loss attributable to acts or omissions of the Owner or Architect or anyone directly or indirectly employed by either of them, or by anyone for whose acts either of them may be liable, and not attributable to the fault or negligence of the Contractor. The foregoing obligations of the Contractor are in addition to the Contractor's obligations under Section 3.18. **(10.2.5; no change)**

---

[321] Bryant v. Village Ctrs., Inc., 167 Ga. App. 220, 305 S.E.2d 907 (1983); Klein v. Cisco-Eagle, Inc., 855 So. 2d 844 (La. App. 2d Cir. 2003); Perrit v. Bernhard Mech. Contractors, Inc., 669 So. 2d 599 (La. App. 1st Cir. 1996); Townsend v. Muckleshoot Indian Tribe, 2007 WL 316504 (Wash. App. Div. 1, February 5, 2007).

[322] Riviera v. City of Meriden, 2006 WL 1000003 (Conn. Super. March 23, 2006). Where the contractor undertakes to perform the work in a "safe" manner, there is an independent duty that avoids the exclusive remedies provision. Britt v. Danziger Dev., 28 Conn. L. Rptr. 625, 2000 WL 1862651 (Conn. Super. Nov. 27, 2000).

Unless damage is caused by the owner or architect[323] or someone for whom the architect or owner are responsible, such as a separate contractor hired by the owner,[324] the contractor must promptly remedy damage and loss that occurs at the job site. This damage includes damage to adjacent property, including streets, sidewalks, shrubs, and so on.[325]

> *10.2.6 The Contractor shall designate a responsible member of the Contractor's organization at the site whose duty shall be the prevention of accidents. This person shall be the Contractor's superintendent unless otherwise designated by the Contractor in writing to the Owner and Architect.* **(10.2.6; no change)**

The contractor's superintendent is in charge of safety unless the contractor notifies the owner and architect in writing that another person is responsible for it. Note that the contractor must designate a safety person. This imposes a duty on the contractor relative to other persons at the job site, including subcontractors.[326]

---

[323] Sun Pipe Line Co., v. Conti Const. Co., Inc., 1989 WL 58679 (D. N.J. 1989).

[324] Barth Elec. Co., v. Traylor Bros., Inc., 553 N.E.2d 504 (Ind. App. 1st Dist. 1990) (given this language, "it is only reasonable to conclude that contracting parties intended each contractor involved in the project to benefit from timely, competent work of the other contractors or to be able to seek compensation from those contractors failing to complete their work in an appropriate manner.") This interpretation would seem to be at odds with the language of paragraph 1.1.2, which states that there are no third-party beneficiaries.

[325] In Watral & Sons, Inc., v. OC Riverhead 58, LLC, 34 A.D.3d 560, 824 N.Y.S.2d 392 (App. Div. 2d Dep't 2006), the court stated that this provision "clearly reflects an intent to require the plaintiff to indemnify the defendant for property damage caused by its work, even in circumstances where the plaintiff's negligence cannot be established . . ." The court compared Par. 3.18.1 with this one and found that "subparagraph 10.2.5 serves to broaden the plaintiff's liability under the common-law rules of implied indemnity by requiring it to remedy any damage or loss arising out of its work regardless of whether the plaintiff has been negligent." The court found that this paragraph requires the contractor to remedy any and all damage it causes to any property referred to in paragraph 10.2.1.3 for which it is responsible. However, the contractor is obligated only to take all reasonable precautions for the safety of, and provide all reasonable protections to prevent damage to, underground utilities. "Thus, it is only in the event that Watral's responsibility under clause 10.2.1.3 is engaged that Watral's duty to indemnify OC for any resulting property damage is triggered." *See also* Manhattan Real Estate Partners, ILP v. Harry S. Peterson Co., 1992 WL 15130 (S.D.N.Y. 1992).

[326] Lewis v. N.J. Riebe Enters., Inc., 170 Ariz. 384, 825 P.2d 5 (1992) (general contractor found to have a duty to subcontractor regarding safety on the job).

In Kilgore v. R.J. Kroner, Inc., 2002 WL 480944 (Del. Super. March 14, 2002), the general contractor did not have a duty to protect the employees of independent contractors. "A general contractor will normally be expected to oversee to some degree the safety precautions its independent contractors employ, but not every such exercise of oversight will necessarily lead to a finding of control or to a voluntary assumption of responsibility for workplace safety not otherwise required by contract." Here, the general conditions apparently were similar to A201.

In Martens v. MCL Constr. Corp., 2004 WL 369143 (Ill. App. 1st Dist., Feb. 27, 2004), the subcontractor had control over safety. The general contractor had only a general right to control construction means and methods. Although the general contractor had safety responsibilities, that did not equate to control over the specific work, particularly where the subcontractor maintained

10.2.7 The Contractor shall not permit any part of the construction or site to be loaded so as to cause damage or create an unsafe condition. **(10.2.7; minor modifications)**

This is meant to cover temporary storage of materials in portions of the building, although it also covers any loading, including machinery, equipment, and temporary and permanent items in the construction. This would not cover items shown on the contract documents in their final locations, because these are the responsibility of the architect and engineer. This would also apply to loads applied to the structure in moving heavy items to their final location. This would be part of the "means and methods" for which the contractor is responsible. Such temporary loads might significantly exceed the normal design loads of the structure, in which case the contractor should have a structural engineer design safe procedures for moving these heavy objects.

*10.2.8 **Injury or Damage to Person or Property.** If either party suffers injury or damage person or property because of an act or omission of the other party, or of others for whose acts such party is legally responsible, written notice of such injury or damage, whether or not insured, shall be given to the other party within a reasonable time not exceeding 21 days after discovery. The notice shall provide sufficient detail to enable the other party to investigate the matter.* **(4.3.8; minor modification)**

If no written notice of the injury is provided, no claim can be made. This provision was inserted to give a party time to investigate the situation and prepare a defense to the claim. Conditions that caused damage or injury are usually altered and are difficult to investigate after time has passed. The maximum amount of time that is considered reasonable is 21 days, although under the proper circumstances, this time period might be significantly less if there might be changes to the location of the injury or damage.

---

contractual control of the supervision and safety of its workers. The court stated that the central issue is retained control of the independent contractor's work, whether contractual, supervisory, operational, or some mix thereof.

However, in Stumpf v. Hagerman Constr. Corp., 863 N.E.2d 871 (Ind. App., 2007), the general contractor assumed a duty of care for the safety of its subcontractors' employees, where the language used was thus: "The Contractor shall take all necessary precautions for the safety of employees on the work, and shall comply with all applicable provisions of Federal, State, and Municipal safety laws and building codes to prevent accidents or injury to persons on, about or adjacent to the premises where the work is being performed. . . . Contractor shall designate a responsible member of its organization on the work, whose duty shall be the prevention of accidents."

The owner has no duty to warn, absent some superior knowledge. Under this provision, the owner has no responsibility for assuring safe work conditions. Kosan v. Pegasus Integrated Living Cen., 1993 WL 1156111 (Pa. Com. Pl. Feb. 1, 1993).

In a New York case, the court found that a predecessor to this provision did not create a waiver of the statutory time limit within which the contractor was to file claims (relying on ¶ 13.4).[327]

§ 4.60   Hazardous Materials: ¶ 10.3

10.3 HAZARDOUS MATERIALS

10.3.1 The Contractor is responsible for compliance with any requirements included in the Contract Documents regarding hazardous materials. If the Contractor encounters a hazardous material or substance not addressed in the Contract Documents and if reasonable precautions will be inadequate to prevent foreseeable bodily injury or death to persons resulting from a material or substance, including but not limited to asbestos or polychlorinated biphenyl (PCB), encountered on the site by the Contractor, the Contractor shall, upon recognizing the condition, immediately stop Work in the affected area and report the condition to the Owner and Architect in writing. (**10.3.1; added first sentence and first part of second sentence**)

If the Contract Documents include requirements concerning known hazardous materials, the contractor must comply with those requirements. Under the provisions of ¶ 3.2.2, the contractor should have investigated such requirements at the outset, and cannot later complain about them, unless there is something that a reasonably competent contractor could not have anticipated. The remainder of this paragraph deals with the situation where the contractor finds something at the jobsite that reasonably appears to be a hazardous material or substance, but was not addressed by the Contract Documents.

If the contractor encounters what it believes is asbestos, PCB, or other "hazardous materials," the contractor is required to stop the work in that area unless reasonable precautions would be sufficient to prevent injury. Thus, if the contractor can take reasonable precautions, stopping the work would not be an option. Even if it turns out that the material is safe, the contractor may still be entitled to a time extension or additional costs if it was reasonable to believe that the material was hazardous. If the parties know they will encounter such materials on a project, this provision could be changed to require the contractor to continue work on other parts of the project while the asbestos or other hazardous material is removed.

Note that the contractor can stop work only in the "affected area" and not necessarily in the entire project site. Thus, if hazardous materials are discovered to be in only one area of a larger project, the contractor would be required to continue work on the remainder of the site if it is reasonable to do so. Determining which areas are affected might create difficulties between the owner and the contractor. The contractor may want to expand the list of materials that constitute hazardous materials, possibly by defining hazardous materials to be those listed by

---

[327] Geneseo Cent. Sch. v. Perfetto & Whalen Constr. Corp., 53 N.Y.2d 306, 423 N.E.2d 1058, 441 N.Y.S.2d 1058 (1981), *rev'g* 434 N.Y.S.2d 502 (App. Div. 1980).

the United States Environmental Protection Agency or the state where the project is located, whichever list is more comprehensive.

> *10.3.2 Upon receipt of the Contractor's written notice the Owner shall obtain the services of a licensed laboratory to verify the presence or absence of the material or substance reported by the Contractor and, in the event such material or substance is found to be present, to cause it to be rendered harmless. Unless otherwise required by the Contract Documents, the Owner shall furnish in writing to the Contractor and Architect the names and qualifications of persons or entities who are to perform tests verifying the presence or absence of such material or substance or who are to perform the task of removal or safe containment of such material or substance. The Contractor and the Architect will promptly reply to the Owner in writing stating whether or not either has reasonable objection to the persons or entities proposed by the Owner. If either the Contractor or Architect has an objection to a person or entity proposed by the Owner, the Owner shall propose another to whom the Contractor and the Architect have no reasonable objection. When the material or substance has been rendered harmless, Work in the affected area shall resume upon written agreement of the Owner and Contractor. By Change Order, the Contract Time shall be extended appropriately and the Contract Sum shall be increased in the amount of the Contractor's reasonable additional costs of shut-down, delay and start-up.* **(10.1.2; add notice requirement in first sentence; add reference to change order in last sentence; minor modifications)**

Once the owner receives the written notice of the contractor pursuant to the prior paragraph concerning the presence of possible hazardous materials, the owner is required to retain a licensed laboratory to verify whether the suspected material is, indeed, a hazardous material. If the laboratory reports that it is not, then the contractor is informed of this fact and may proceed. The contractor may also be entitled to an extension of the contract time and, possibly, an increase in the contract price, assuming that the contractor acted reasonably.

If, on the other hand, the suspect material is hazardous, then the laboratory, or other entity hired by the owner, must neutralize the hazardous material, possibly by removing it from the site. The owner must also let both the contractor and architect know the identities of the entities who are testing the materials and who will be remediating the situation. Both the architect and contractor are then given a chance to object to these entities, in which case, the owner must propose alternate entities. Once the remediation has been completed, work will restart, and the contractor will be entitled to a change order for the cost of any additional costs reasonably incurred by the contractor as a result of the work stoppage and the restart, as well as an extension of time. This provision trumps the waiver of consequential damages found in ¶ 15.1.6.

> *10.3.3 To the fullest extent permitted by law, the Owner shall indemnify and hold harmless the Contractor, Subcontractors, Architect, Architect's consultants and agents and employees of any of them from and against claims, damages, losses and expenses, including but not limited to attorneys' fees, arising out of or resulting from performance of the Work in the affected area if in fact the material or substance*

presents the risk of bodily injury or death as described in Section 10.3.1 and has not been rendered harmless, provided that such claim, damage, loss or expense is attributable to bodily injury, sickness, disease or death, or to injury to or destruction of tangible property (other than the Work itself), except to the extent that such damage, loss or expense is due to the fault or negligence of the party seeking indemnity. **(10.3.3; reword last sentence as to exception)**

As with all indemnification clauses, legal counsel should be consulted in the state in which the project is to be constructed. Each state has a different set of rules concerning these clauses. Some states have laws regulating hazardous materials that might come into play in construction situations.

Under this paragraph, the owner will indemnify all of the listed parties if there has been bodily injury, sickness, disease, or death, or damage to property other than the Work. Thus, if the architect is sued by a worker for damage to the worker's truck arising from hazardous materials, the architect must be indemnified by the owner, unless the architect was negligent or otherwise at fault related to such injury.

10.3.4 The Owner shall not be responsible under this Section 10.3 for materials or substances the Contractor brings to the site unless such materials or substances were required by the Contract Documents. The Owner shall be responsible for materials or substances required by the Contract Documents, except to the extent of the Contractor's fault or negligence in the use and handling of such materials or substances. **(10.4; Add second sentence; minor modifications)**

Under this provision, the owner is not required to indemnify the contractor or architect for hazardous materials brought to the site by the contractor. The exception to this consists of those materials that are included or required by the contract documents. Thus, the owner would indemnify the contractor for damage or injury related to materials to be incorporated into the work but not materials used only in the construction process, such as gasoline to run excavating equipment. The owner remains responsible for any dangerous or hazardous materials required by the Contract Documents to be brought to the site, except to the extent that the contractor was careless in the handling of such materials. This provision would not absolve the architect from liability to the owner resulting from the specification of dangerous materials if it was not reasonable for the architect to specify such materials, since such liability would arise out of a different document—the owner-architect agreement.

10.3.5 The Contractor shall indemnify the Owner for the cost and expense the Owner incurs (1) for remediation of a material or substance the Contractor brings to the site and negligently handles, or (2) where the Contractor fails to perform its obligations under Section 10.3.1, except to the extent that the cost and expense are due to the Owner's fault or negligence. **(new)**

If the contractor is negligent in its handling of any material or substance brought to the site, the contractor must indemnify the owner if the owner incurs any cost or expense because of such negligence. Similarly, if the contractor recognizes a hazardous substance and fails to promptly notify the owner as stated in ¶ 10.3.1, the owner will be entitled to indemnification for costs reasonably incurred as a result of the contractor's failure. To the extent that the owner has been negligent, the contractor is not required to indemnify the owner.

> *10.3.6 If, without negligence on the part of the Contractor, the Contractor is held liable by a government agency for the cost of remediation of a hazardous material or substance solely by reason of performing Work as required by the Contract Documents, the Owner shall indemnify the Contractor for all cost and expense thereby incurred.* **(10.5; add reference to a government agency)**

This covers situations in which the contractor is fined for these activities and the contractor was not negligent. In that event, the owner bears the cost of the fines and other expenses.

§ 4.61   Emergencies: ¶ 10.4

10.4 EMERGENCIES

In an emergency affecting safety of persons or property, the Contractor shall act, at the Contractor's discretion, to prevent threatened damage, injury or loss. Additional compensation or extension of time claimed by the Contractor on account of an emergency shall be determined as provided in Article 15 and Article 7. **(10.6.1; revised reference)**

In one case, the contractor was almost finished with the construction when a flood caused substantial damage to the structure.[328] The court held that the remedial repairs performed by the contractor after the waters receded were not done in an emergency, thus the contractor was required to obtain a written change order to obtain payment for this work.

The contractor must act with reasonable prudence to minimize damages from emergencies. It may request a change order for the cost of any additional work, but the change order is subject to the normal review and approval process.

§ 4.62   Article 11: Insurance and Bonds

Because the architect is usually not qualified to give the owner advice concerning insurance, the owner should consult its own insurance counsel regarding the requirements of this article.

---

[328] Goin v. Board of Educ., 298 Ky. 645, 183 S.W.2d 819 (1944) (contractor was not entitled to additional money for work performed after a flood, although architect gave contractor assurances that he would be paid. The contractor was required to deliver a completed building, and he had to repair the flood damages.).

§ 4.63   Contractor's Liability Insurance: ¶ 11.1

11.1 CONTRACTOR'S LIABILITY INSURANCE

11.1.1 The Contractor shall purchase from and maintain in a company or companies lawfully authorized to do business in the jurisdiction in which the Project is located such insurance as will protect the Contractor from claims set forth below which may arise out of or result from the Contractor's operations and completed operations under the Contract and for which the Contractor may be legally liable, whether such operations be by the Contractor or by a Subcontractor or by anyone directly or indirectly employed by any of them, or by anyone for whose acts any of them may be liable: **(11.1.1; add reference to completed operations)**

> *.1   claims under workers' compensation, disability benefit and other similar employee benefit acts that are applicable to the Work to be performed;* **(11.1.1.1; no change)**

> *.2   claims for damages because of bodily injury, occupational sickness or disease, or death of the Contractor's employees;* **(11.1.1.2; no change)**

In an Illinois case, a subcontractor was required by the contract to obtain insurance against Structural Work Act claims.[329] The subcontractor had failed to obtain this coverage and the owner filed suit when an injured worker claimed against the owner. The provision was not void because of a state statute that voided indemnification provisions when a party sought indemnification against its own negligence. A promise to obtain insurance is not the same as a promise to indemnify. In another Illinois case, the owner required the contractor to provide insurance to replace an earlier indemnity provision that had been voided by law.[330] The owner could not subsequently sue the contractor under common law indemnity.

---

[329] Zettel v. Paschen Contractors, Inc., 100 Ill. App. 3d 614, 427 N.E.2d 189 (1981) The subcontract provided that:

> Subcontractor further agrees to cause contractual liability endorsements to be issued by the insurance companies and attached to the above-mentioned policies, to include under the coverage therein extended an obligation on the part of the insurers to insure against Subcontractor's contractual liability hereunder and to indemnify the Owner, Architect, and Contractor against loss, liability, costs, expenses, attorney's fees and court costs as provided in Section 10 hereof, and further agrees that said coverage shall be afforded therein against all claims arising out of the operation of any structural work law or law imposing liability arising out of the use of scaffolds, hoists, cranes, stays, ladders, supports, or other mechanical contrivances.

> *See also* Lulich v. Sherwin-Williams Co., 799 F. Supp. 64 (N.D. Ill. 1992) (following *Zettel*).
> In Hurlburt v. Northern States Power, 549 N.W.2d 919 (1996), this provision was incorporated into a subcontract but was modified to require indemnification only to the extent the injury was attributable to the subcontractor's negligence. When a worker was injured, the general contractor sought to enforce the indemnification clause against the subcontractor who was not negligent. The court held that there was no obligation to indemnify and that the insurance that the subcontractor purchased to fund the indemnity was not applicable to the claim.

[330] Vandygriff v. Commonwealth Edison Co., 87 Ill. App. 3d 374, 408 N.E.2d 1129 (1980) (insurance carrier could not sue contractor under its right of subrogation because it was an additional insured).

There is an interaction between this provision and the waiver of subrogation provision of ¶ 11.3.7. Under that provision, the parties waive claims against each other to the extent that there is insurance to cover damage to the Work. However, where the damage is to non-Work items, there is no waiver of subrogation. Instead, this provision requires the contractor to carry insurance to cover such damage.[331]

> *.3*   *claims for damages because of bodily injury, sickness or disease, or death of any person other than the Contractor's employees;* **(11.1.1.3; no change)**
>
> *.4*   *claims for damages insured by usual personal injury liability coverage;* **(11.1.1.4; no change)**
>
> *.5*   *claims for damages, other than to the Work itself, because of injury to or destruction of tangible property, including loss of use resulting therefrom;* **(11.1.1.5; no change)**

In a Washington case, the owner's microfiche records were damaged as a result of water seepage during construction.[332] The court stated that the question was whether the microfiche records were Work under A201. If they were Work, then the owner's insurance purchased in accordance with ¶ 11.3.1 would cover the loss. If they were not Work, the contractor was required to provide insurance against damage to them under this provision.[333]

Note that this provision is in apparent conflict with ¶ 11.3.3, which states that the owner is to purchase insurance to cover the owner's property for loss of use. In ¶ 11.1.1.5, the contractor is required to purchase insurance to cover the owner's non-Work property, including loss of use. This can only be reconciled by interpreting these provisions as requiring the contractor to insure the owner's non-Work property, including adjacent property of the owner that is not part of the Work, whereas the owner is to provide loss of use insurance for the owner's property that encompasses the Work. In other words, if the Work is damaged, the resulting loss of use is covered by the owner's insurance as in ¶ 11.3.3. If the owner does not carry such insurance, the owner will take the risk of such loss, per the waiver in ¶ 11.3.3.

---

[331] Knob Noster R-VIII Sch. Dist., v. Dankenbring, 220 S.W.3d 809 (Mo. App. W.D. 2007). In that case, the masonry subcontractor negligently applied an acid wash that damaged the owner's property that was not part of the Work. The subcontractor defended the subrogation action based on the waiver of subrogation provision. The court cited this paragraph in support of its holding against the subcontractor.

[332] Public Employees Mut. Ins. Co. v. Sellen Constr. Co., 48 Wash. App. 792, 740 P.2d 913 (1987).

[333] If damage results from construction defects, does the contractor's CGL policy cover such damage? Insurance carriers often take the position that defective work does not create an "occurrence" so as to trigger coverage. The Texas supreme court, in Lamar Homes, Inc., v. Mid-Continent Cas. Co., 242 S.W.3d 1 (Tex., 2007), held that there was an occurrence that triggered coverage.

This is in keeping with the waiver of consequential damages found at ¶ 15.1.6, which specifically waives damages for the owner's loss of use. Given this waiver, the owner should purchase insurance to cover this risk.[334]

> *.6*   *claims for damages because of bodily injury, death of a person or property damage arising out of ownership, maintenance or use of a motor vehicle;* (**11.1.1.6; no change**)
>
> *.7*   *claims for bodily injury or property damage arising out of completed operations; and* (**11.1.1.7; no change**)

Note that the contractor is required to purchase completed operations insurance coverage.

> *.8*   *claims involving contractual liability insurance applicable to the Contractor's obligations under Section 3.18.* (**11.1.1.8; no change**)

In a North Carolina case, the contractor had performed repairs to a water cooler and lines leading to it.[335] However, the contractor had installed a defective pressure reducing valve. As a result of the installation of the defective valve, the supply tube to the water cooler blew out of a valve fitting, causing water damage to a newly constructed, wooden gym floor. The contract had required a performance bond and insurance. Before the damage, the contractor went into bankruptcy, and the surety paid the cost of repairing the defective valve. However, the surety argued that, because of ¶ 11.1.1 of the General Conditions, the insurance carrier should pay for the cost of the gym floor repair. Unfortunately, the insurance was canceled before the damage, and the owner learned of the cancellation only after the damage. The court held that the contractor's obligation to carry insurance did not remove his liability under ¶ 3.18. The surety was required to pay for the damage.

Various courts have examined this provision with respect to the indemnification provisions found in ¶ 3.18 and have used it to require subcontractors to indemnify

---

[334] In Mu Chapter of the Sigma Pi Fraternity of the United States, Inc. v. Northeast Constr. Serv., Inc., 179 Misc. 2d 374, 684 N.Y.S.2d 872 (Sup. 1999), the building was destroyed by fire on the first day of construction. The owner failed to follow the provisions of ¶ 11.4.1 by not purchasing the required insurance and by not notifying the contractor of that fact. The court found that this provision is not an undertaking to insure the building against fire. Paragraph 11.4, on the other hand, assigns to the owner the risk of loss and the duty of insuring the building against fire.

[335] Haywood County Consol. Sch. Sys. v. United States Fidelity & Guar. Co., 43 N.C. App. 71, 257 S.E.2d 670 (1979).

In Bentley Koepke, Inc. v. Jeffrey Allen Corp., 1998 Ohio App. LEXIS 684 (Feb. 27, 1998), the decedent's administrator sued the landscape architect and others. The architect sought to have the contractor's insurance carrier indemnify and defend it pursuant to this provision. The architect claimed it was the intended third-party beneficiary of the contractor's insurance policy. The policy, however, did not name the architect, and the court held that the architect was not a third-party beneficiary. Presumably, the architect could still be indemnified by the contractor, but it would have been wise to obtain an insurance certificate that specifically named the architect and owner as insured in this situation.

the general contractor[336] and in construing various anti-indemnification statutes.[337]

11.1.2 The insurance required by Section 11.1.1 shall be written for not less than limits of liability specified in the Contract Documents or required by law, whichever coverage is greater. Coverages, whether written on an occurrence or claims-made basis, shall be maintained without interruption from the date of commencement of the Work until the date of final payment and termination of any coverage required to be maintained after final payment, and, with respect to the Contractor's completed operations coverage, until the expiration of the period for correction of Work or for such other period for maintenance of completed operations coverage as specified in the Contract Documents. **(11.1.2; add last clause; minor modifications)**

The owner must determine the limits of liability specified in the contract. *Occurrence-basis coverage* means that the insurance is effective if the event that causes the claim occurs during the policy period, whereas *claims-made coverage* is effective for claims that are made during the policy period, irrespective of when the actual damage took place. The contractor's completed operations coverage must remain in place for at least the period for correction of the Work unless the Contract Documents require a longer period of time.

11.1.3 Certificates of insurance acceptable to the Owner shall be filed with the Owner prior to commencement of the Work and thereafter upon renewal or replacement of each required policy of insurance. These certificates and the insurance

---

[336] Whittle v. Pagani Bros. Constr. Co., 422 N.E.2d 779 (Mass. 1981).

[337] In Lulich v. Sherwin-Williams Co., 799 F. Supp. 64, 69 (N.D. Ill. 1992), the court held that the indemnification provision (under an earlier version of A201) was void under the Illinois Indemnification Act. The court further found that this provision, which requires the contractor to obtain insurance to cover the indemnification provision, is likewise void and unenforceable. The court went on, however, to find that another provision that required the contractor to obtain comprehensive general liability insurance, including protective liability insurance, was not void. The court held that "an agreement requiring a contractor to provide insurance protecting the owner also protect[s] the interests of the construction worker and the general public by reserving a potential source of compensation for injured workers," and this was therefore not against public policy.

In Meyers v. Burger King Corp., 638 So. 2d 369, 380-81 (La. Ct. App. 1994), the contract contained an indemnification clause stating that "the contractor shall indemnify, defend and save and hold harmless the Owner from those claims set forth in the above paragraph captioned Insurance, including reasonable attorneys fees." The insurance provision had this language:

The contractor shall maintain insurance to protect the Owner and Himself from claims which may arise from the Contractor's operations, whether such operations be by himself, any sub-contractor, anyone directly or indirectly employed by any of them, or by whose acts they may be liable. This insurance shall be written for not less than limits of liability required by law, or $1,000,000 aggregate (single limit applicable to bodily injury and property damages combined) whichever is greater.

The court rejected the argument that this provision limited the indemnity to $1 million, especially because the contractor had purchased higher insurance limits. The indemnity clause required the contractor to fully indemnify the owner for all damages.

policies required by this Section 11.1 shall contain a provision that coverages afforded under the policies will not be canceled or allowed to expire until at least 30 days' prior written notice has been given to the Owner. An additional certificate evidencing continuation of liability coverage including coverage for completed operations shall be submitted with the final Application for Payment as required by Section 9.10.2 and thereafter upon renewal or replacement of such coverage until the expiration of the time required by Section 11.1.2. Information concerning reduction of coverage on account of revised limits or claims paid under the General Aggregate, or both, shall be furnished by the Contractor with reasonable promptness. **(11.1.3; add requirement that certificates of insurance must be filed with the owner at each renewal or replacement; delete reference to requirement that coverages remain in place after final payment; other revisions)**

The owner should insist that the contractor furnish these certificates, with a copy to the architect.[338] If the contractor does not do so, it will be in default. However, failure to require such certificates may constitute a waiver of this requirement.[339] The certificate should be examined to determine that the 30-day notice provision is included, affording the owner sufficient time to obtain alternative coverage. The owner might want to consider requiring the actual policies in order to determine whether the required coverage is provided. Note that the reporting requirement includes changes in coverage limitations resulting from claims paid under the policy.

***11.1.4** The Contractor shall cause the commercial liability coverage required by the Contract Documents to include (1) the Owner, the Architect and the Architect's Consultants as additional insureds for claims caused in whole or in part by the Contractor's negligent acts or omissions during the Contractor's operations; and (2) the Owner as an additional insured for claims caused in whole or in part by the Contractor's negligent acts or omissions during the Contractor's completed operations.* **(new)**

This provision requires that the contractor include the owner, architect, and architect's consultants as additional insureds on the contractor's commercial liability insurance policies. It has long been common practice for the contract documents to include a similar provision requiring the owner and architect to be named as additional insureds on the contractor's CGL policy. The 1997 version of A201 stated that there would not be any additional insureds.

---

[338] It is important to carefully examine these certificates. In Modern Builders, Inc., v. Alden-Conger Public School District #242, 2005 WL 2089195 (D. Minn. August 30, 2005), the roofing contractor started work on the school building. Shortly thereafter, a storm damaged the school. The contractor had neglected to request that the insurance carrier name the school as an additional insured on its insurance policies. The school was issued a certificate listing the school as the "Certificate Holder," however, nothing on the certificate indicated that the school was an additional insured. The court declined to find that the school was an additional insured under various equitable principles.

[339] Whalen v. K-Mart Corp., 519 N.E.2d 991 (Ill. App. 1st Dist., 1988).

§ 4.64   Owner's Liability Insurance: ¶ 11.2

11.2 OWNER'S LIABILITY INSURANCE

The Owner shall be responsible for purchasing and maintaining the Owner's usual liability insurance. (**11.2.1; no change**)

The owner should consult insurance and legal counsel to determine what coverage is required and whether it may be more advantageous to have the contractor obtain some of this insurance. Note the coverage that the contractor is required to purchase under ¶ 11.1.1. Some owners do not have any "usual" liability insurance, while others have different types of insurance, such as risk "pools" carried by governmental bodies. Note that the "usual" insurance refers to that particular owner, not what is normal or standard in the industry. The contractor should find out at the outset what types of insurance the owner will have for the project and coordinate this with the contractor's own coverages.

§ 4.65   Property Insurance: ¶ 11.3

11.3 PROPERTY INSURANCE

11.3.1 Unless otherwise provided, the Owner shall purchase and maintain, in a company or companies lawfully authorized to do business in the jurisdiction in which the Project is located, property insurance written on a builder's risk "all-risk" or equivalent policy form in the amount of the initial Contract Sum, plus value of subsequent Contract Modifications and cost of materials supplied or installed by others, comprising total value for the entire Project at the site on a replacement cost basis without optional deductibles. Such property insurance shall be maintained, unless otherwise provided in the Contract Documents or otherwise agreed in writing by all persons and entities who are beneficiaries of such insurance, until final payment has been made as provided in Section 9.10 or until no person or entity other than the Owner has an insurable interest in the property required by this Section 11.3 to be covered, whichever is later. This insurance shall include interests of the Owner, the Contractor, Subcontractors and Sub-subcontractors in the Project. (**11.4.1; minor modification; change reference**)

It is the responsibility of the owner, unless otherwise stated in the contract documents,[340] to provide coverage for the Project until it is completed. Prior to

---

[340] In Farmers Elevator & Mercantile Co. v. Farm Builders, Inc., 432 N.W.2d 864 (N.D. 1988), the Supplementary Conditions provided: "The Contractor shall purchase and maintain Property Insurance as called for in paragraph [11.3.1] of Article [11] of the General Conditions." The court ruled that this shifted the responsibility for obtaining property insurance to the contractor, and that the waiver of rights clause (similar to ¶ 13.3.7) was inapplicable.

    See also Monical v. State Farm Ins. Co., 211 Ill. App. 3d 215, 569 N.E.2d 1230 (1991), in which the court stated:

    When parties to a business transaction agree that insurance will be provided as a part of the bargain, the agreement must be interpreted as providing mutual exculpation to the

the 1997 version of A201, this provision read "the entire Work at the site." A number of cases, cited below, construed that version of this paragraph. There is, however, a possible quandary with this change of terminology. As identified in ¶ 1.1.4, the "Project" could be substantially larger than the "Work." For instance, if a hospital is adding a new wing to the existing building, the Work will be just the new wing, but not the existing parts of the building that will remain in place. However, the entire building, both old and new, might be considered the "Project." Under this analysis, the owner needs to carry a builder's risk policy on much more than simply the Work. This would then protect the contractor if the contractor damages adjacent structures that are part of the Project but not part of the Work.[341] The waiver of subrogation provision at ¶ 11.3.7 would prohibit the owner's insurer to look to the contractor for reimbursement for such damage. Note that a number of the cases interpreting the waiver of subrogation language turn on whether the damaged items constitute part of the Work or other property.

Complicating the analysis still further, this paragraph requires that the insurance be in the value of the cost of the Work, including the original contract sum, plus all Change Orders. In addition, the cost of materials supplied or installed by others is to be added to the value of this insurance. This would include work done by the owner's separate contractors or by the owner itself. The total of these numbers is more than just the "Work" as that is defined in A201. This leaves open the question of whether adjacent existing work (or new work constructed by others for the same owner) is considered part of the Project for purposes of the amount of insurance required. The intent seems to be that all of the Owner's property, including the Work and any adjacent property owned by the Owner, would be included.

If any portion of the Work (now Project) is damaged, destroyed, or stolen, the insurance will cover the cost of replacing such work, even if title to the affected work is still in the contractor or subcontractor. The phrase "the entire Work at the site" (the 1976 edition of A201) was reviewed by a Massachusetts court, which determined that the waiver clause of ¶ 11.3.7 applied even when the owner purchased insurance that covered more than the Work.[342]

---

bargaining parties. The parties are thus deemed to have agreed to look solely to the insurance in the event of loss, and not impose liability on the other.

[341] However, the court in Knob Noster R-VIII School Dist. v. Dankenbring, 220 S.W.3d 809 (Mo. App. W.D. 2007) stated that, by its very nature, Builder's Risk insurance is intended to cover only new work. It does not apply to the value of any existing structures that may be damaged during construction. In that case, the masonry subcontractor negligently applied an acid wash that damaged the owner's property that was not part of the Work (but was presumably part of the Project—this was not discussed by the court). The subcontractor defended the subrogation action based on the waiver of subrogation provision. The court held that subrogation was not waived.

[342] Haemonetics Corp. v. Brophy & Phillips Co., 23 Mass. App. Ct. 254, 501 N.E.2d 524 (1986). *See also* Willis Realty v. Cimino Constr., 623 A.2d 1287 (Me. 1993); Lloyd's Underwriters v. Craig & Rush, Inc., 32 Cal. Rptr. 2d 144 (Ct. App. 1994) (citing *Haemonetics*) (owner failed to purchase a separate builder's risk policy as required by this provision, choosing instead to rely on its existing all-risk property insurance, and also failed to add contractors as additional insureds).

In a Washington case, the owner's microfiche records were damaged as a result of water seepage during construction.[343] The court stated that the question was whether the microfiche records were Work under A201. If they were Work, then the owner's insurance purchased under this provision would cover the loss. If they were not Work, the contractor was required to provide insurance against damage to them under ¶ 11.1.1.5.

In a New York case, the contractor was hired to perform extensive renovations and improvements on an existing fraternity house at a cost of $664,200.[344] On the morning of April 28, 1994, work began with the removal of some iron work with acetylene torches. At noon, the crew took a lunch break. Returning a few minutes later, they found the building in flames. It was a total loss. The fraternity's insurer paid out $670,000 and then sued the contractor to recover that money. The fraternity also sued the contractor for more than $1.4 million based on cost of replacement plus consequential damages. The court found that this AIA language is designed to put in place a "wrap-up" plan whereby the owner, for the mutual benefit of all the participants in the construction project, would insure the entire bundle of property interests, regardless of any technical notions regarding the multiplicity of insurable interests that might be involved. Here, the fraternity opted not to purchase the insurance specified in A201 but instead relied upon an existing property policy with a limit of $670,000 plus an additional $200,000 installation floater for the first portion of the construction contract. This, the fraternity argued, satisfied its duty under the insurance provisions of A201. The court rejected this argument, finding that the fraternity's actions were not in compliance with the contract, and, having failed to notify the contractor in advance of its proposed actions, the fraternity would be held liable under ¶ 11.3.1.2 of A201, which requires the owner to pay all reasonable costs properly attributable to the owner's failure or neglect to purchase the proper insurance. This part of A201 assigns the risk of loss and the duty of insuring the building against fire to the owner. The court held that because the fraternity's claims involved losses that it should have insured against under the contract, the fraternity could not recover from the contractor. Furthermore, the waiver of consequential damages found at ¶ 11.3.7 prohibited recovery for loss of use. As to the subrogation action, the court

---

The phrase "the entire Work at the site" from the 1987 version of A201 (now replaced with "the entire Project at the site") was reviewed in Walker Engineering, Inc. v. Bracebridge Corp., 102 S.W.3d 837 (Tex. App. 2003). The court found that the waiver of subrogation language now found in ¶ 11.3.7 prohibited recovery from a subcontractor responsible for a construction loss from flooding. The court found that the AIA language does not revolve around the issue of whether the property damage is to the entire Work at the site or to property other than the Work. Rather, the question is whether, and to what extent, insurance is available. *See also* Western Washington Corp. of Seventh-Day Adventists v. Ferrellgas, Inc., 7 P.3d 861 (Wash. App. Div. 2, 2000); Church Mut. Ins. Co. v. Palmer Constr. Co., Inc., 2004 WL 1438177 (E.D. Pa. June 9, 2004), following *Haemonetics*.

[343] Public Employees Mut. Ins. Co. v. Sellen Constr. Co., 48 Wash. App. 792, 740 P.2d 913 (1987).

[344] Mu Chapter of the Sigma Pi Fraternity of the United States, Inc. v. Northeast Constr. Serv., Inc., 179 Misc. 2d 374, 684 N.Y.S.2d 872 (Sup. Ct. 1999).

held that the insurer was not entitled to sue the contractor unless losses were incurred that were beyond those that the fraternity promised to insure for the benefit of all.

In a Connecticut case, the town cancelled its all-risk policy some four months before the rupture of a heating coil in an air handling unit at a school caused property and water damage.[345] The contractor alleged that the town breached these provisions by failing to maintain the required insurance. The town argued that its obligation to maintain insurance "until final payment has been made" meant that its obligation ceased upon substantial completion. This contention was rejected by the court.

The architect is not listed in ¶ 11.3.1 as a party whose interests are covered. A North Carolina Court found that this helped to create an ambiguity in a situation in which the owner's insurance carrier filed suit against the architect for a construction collapse.[346] The architect defended on the basis of the mutual waiver clause now found in ¶ 8.1.2 of AIA Document B101. The court held that by requiring the architect to obtain professional liability insurance, the parties may not have meant the owner to waive rights of subrogation for property insurance purchased under this paragraph. The architect may want to insert "Architect" as one of the interests to be insured in the last sentence of this paragraph in order to avoid this situation.

> *11.3.1.1 Property insurance shall be on an "all-risk" or equivalent policy form and shall include, without limitation, insurance against the perils of fire (with extended coverage) and physical loss or damage including, without duplication of coverage, theft, vandalism, malicious mischief, collapse, earthquake, flood, windstorm, false-work, testing and startup, temporary buildings and debris removal including demolition occasioned by enforcement of any applicable legal requirements, and shall cover reasonable compensation for Architect's and Contractor's services and expenses required as a result of such insured loss.* **(11.4.1.1; no change)**

When there is no contractual provision, the risk of losses occurring during construction is on the builder.[347] Moreover, even if the contract requires the owner or the builder to carry insurance while the building is under construction, the risk of loss does not shift, absent a contrary contractual provision.[348] In a Florida case, the general contractor bore the risk of loss when the contract between the owner and contractor required the contractor to carry insurance against vandalism.[349]

---

[345] Haynes Constr. Co. v. Town of Newtown, 2005 WL 2857750 (Conn. Super. Oct. 14, 2005).

[346] St. Paul Fire & Marine v. Freeman-White, 322 N.C. 109, 366 S.E.2d 480 (1988).

[347] Hartford Fire Inc. v. Riefolo Constr. Co., 161 N.J. Super. 99, 390 A.2d 1210 (1978).

    Best Friends Pet Care, Inc. v. Design Learned, Inc., 2003 WL 21235341 (Conn. App. June 3, 2003) (examining this provision from a pre-1997 version of A201 in light of the waiver of subrogation, finding that loss of use of personal and real property is not covered by the waiver).

[348] *Id.*

[349] Fred McGilvray, Inc. v. International Builders, 354 So. 2d 103 (Fla. Dist. Ct. App. 1978). The action was by the subcontractor against the contractor. Vandalism occurred and the subcontract was silent as to who should bear the loss. The court ruled against the contractor.

11.3.1.2 If the Owner does not intend to purchase such property insurance required by the Contract and with all of the coverages in the amount described above, the Owner shall so inform the Contractor in writing prior to commencement of the Work. The Contractor may then effect insurance that will protect the interests of the Contractor, Subcontractors and Sub-subcontractors in the Work, and by appropriate Change Order the cost thereof shall be charged to the Owner. If the Contractor is damaged by the failure or neglect of the Owner to purchase or maintain insurance as described above, without so notifying the Contractor in writing, then the Owner shall bear all reasonable costs properly attributable thereto. **(11.4.1.2; no change)**

This paragraph places the risk of loss on the owner unless the owner informs the contractor in writing before the work commences that the owner does not intend to purchase such insurance. In that case, the contractor can purchase the insurance and, by change order, charge the owner for the cost of that insurance.[350]

11.3.1.3 If the property insurance requires deductibles, the Owner shall pay costs not covered because of such deductibles. **(11.4.1.3; no change)**

Note that this provision requires the owner and not the contractor to bear the costs that are not covered because of deductibles under the policy.

In The Gray Ins. Co. v. Old Tyme Builders, Inc., 2004 WL 691393 (La. App. 1st Cir. Apr. 2, 2004), the contractor improperly installed exterior stucco, causing water infiltration. This, in turn, damaged some of the interior furnishings. The contractor repaired the stucco, but the insurance carrier paid for the other damage, and then sought to recover for its payments. The carrier alleged that the waiver of subrogation did not apply since the damages were not caused by a "peril" within the meaning of this provision. The court rejected this argument, and the waiver of subrogation was held applicable.

In Haynes Constr. Co. v. Town of Newtown, 2005 WL 2857750 (Conn. Super. Oct. 14, 2005), the town cancelled its all-risk policy some four months before the rupture of a heating coil in an air handling unit at a school caused property and water damage. The contractor alleged that the town breached these provisions by failing to maintain the required insurance. The town argued that this particular incident was not included within the specific list of covered losses found in ¶ 11.4.1.1. The court held that the list was not intended to be an exclusive list. The waiver of subrogation provision also supported the contractor's contention "The conclusion that the rupture of the heating coil in this case falls within the all-risk coverage mandated by the contract thus furthers the contractual goal of resolving disputes by insurance payments rather than lawsuits."

If the required insurance is not obtained, there is a breach of contract, but this breach only affects who will pay for the damages and does not, by itself, entitle anyone to damages. Modern Builders, Inc., v. Alden-Conger Public Sch. Dist. #242, 2005 WL 2089195 (D. Minn. August 30, 2005).

[350] In Mu Chapter of the Sigma Pi Fraternity of the United States, Inc. v. Northeast Constr. Serv., Inc., 179 Misc. 2d 374, 684 N.Y.S.2d 872 (1999), the contractor was not informed that the owner was relying on an existing insurance policy. After a fire destroyed the building on the first day of construction, the owner and its insurance carrier sued the contractor. The court held that the provisions of this section barred the action. Having failed to notify the contractor in advance, the owner must bear all reasonable costs attributable to the failure.

Because such deductibles may be significant, the owner may want to delete this provision.[351] This provision applies only to an action against the "Contractor" and not any consultants. This is as opposed to the language found in ¶ 11.3.7, where rights are waived against a number of parties.

>*11.3.1.4 This property insurance shall cover portions of the Work stored off the site, and also portions of the Work in transit.* **(11.4.1.4; no change)**

If the contractor informs the owner that certain materials are stored off the site, the insurance will cover those items to the value agreed upon between the owner and contractor. It is up to the contractor, however, to initiate the approval and indicate the insurable value of those materials. This paragraph appears to be in conflict with ¶ 9.3.2, which apparently requires the contractor to insure work stored off the site and in transit. These provisions are harmonized by permitting the contractor to purchase off-site insurance and charge the owner for its cost.

>*11.3.1.5 Partial occupancy or use in accordance with Section 9.9 shall not commence until the insurance company or companies providing property insurance have consented to such partial occupancy or use by endorsement or otherwise. The Owner and the Contractor shall take reasonable steps to obtain consent of the insurance company or companies and shall, without mutual written consent, take no action with respect to partial occupancy or use that would cause cancellation, lapse or reduction of insurance.* **(11.4.1.5; no change)**

Some insurance policies may contain provisions that void or otherwise impede the insurance in case of partial use or occupancy of the project. If there is any question about whether insurance will apply after the owner has taken occupancy, the parties are required to obtain the consent of the insurance carrier. If such occupancy would result in any change or cancellation of the insurance, the parties must address this issue.

>*11.3.2 Boiler and Machinery Insurance. The Owner shall purchase and maintain boiler and machinery insurance required by the Contract Documents or by law, which shall specifically cover such insured objects during installation and until final acceptance by the Owner; this insurance shall include interests of the Owner, Contractor, Subcontractors and Sub-subcontractors in the Work, and the Owner and Contractor shall be named insureds.* **(11.4.2; no change)**

>*11.3.3 Loss of Use Insurance. The Owner, at the Owner's option, may purchase and maintain such insurance as will insure the Owner against loss of use of the Owner's property due to fire or other hazards, however caused. The Owner waives all rights of action against the Contractor for loss of use of the Owner's property, including consequential losses due to fire or other hazards however caused.* **(11.4.3; no change)**

---

[351] In Best Friends Pet Care, Inc. v. Design Learned, Inc., 77 Conn. App. 167, 823 A.2d 329 (Conn. App. 2003), the court looked at the pre-1997 version of A201 and held that the waiver of subrogation did not apply to loss of use where the damage was caused by a consultant to the construction manager.

If the work is damaged for any reason (presumably other than the contractor's negligence), the owner will have no right to recover damages from the contractor for the loss of use of the project. The owner may obtain insurance to cover this possibility. See the discussion at ¶ 11.1.1.5, § **4.63,** for an apparent conflict between these two provisions.

11.3.4 If the Contractor requests in writing that insurance for risks other than those described herein or other special causes of loss be included in the property insurance policy, the Owner shall, if possible, include such insurance, and the cost thereof shall be charged to the Contractor by appropriate Change Order. (**11.4.4; no change**)

Such coverage requested by the contractor may be obtained more cheaply by the owner as part of its insurance package than by the contractor. This gives the contractor the opportunity to take advantage of the savings.

11.3.5 If during the Project construction period the Owner insures properties, real or personal or both, at or adjacent to the site by property insurance under policies separate from those insuring the Project, or if after final payment property insurance is to be provided on the completed Project through a policy or policies other than those insuring the Project during the construction period, the Owner shall waive all rights in accordance with the terms of Section 11.3.7 for damages caused by fire or other causes of loss covered by this separate property insurance. All separate policies shall provide this waiver of subrogation by endorsement or otherwise. (**11.4.5; change reference**)

This relates to the waiver of subrogation discussed in ¶ 11.3.7.[352]

11.3.6 Before an exposure to loss may occur, the Owner shall file with the Contractor a copy of each policy that includes insurance coverages required by this Section 11.3. Each policy shall contain all generally applicable conditions, definitions, exclusions and endorsements related to this Project. Each policy shall contain a provision that the policy will not be canceled or allowed to expire, and that its limits will not be reduced, until at least 30 days' prior written notice has been given to the Contractor. (**11.4.6; change reference**)

Note that this paragraph requires the owner to furnish the contractor with copies of the actual policies, not merely certificates.

11.3.7 Waivers of Subrogation. The Owner and Contractor waive all rights against (1) each other and any of their subcontractors, sub-subcontractors, agents and employees, each of the other, and (2) the Architect, Architect's consultants, separate contractors described in Article 6, if any, and any of their subcontractors, sub-subcontractors, agents and employees, for damages caused by fire or other causes

---

[352] In Colonial Prop. Realty Ltd. P'ship v. Lowder Constr. Co., Inc., 256 Ga. App. 106, 567 S.E.2d 389 (2002), the court relied on this provision in finding a waiver of subrogation even though the fire that damaged the building occurred a year after construction was completed. *See also* Lumbermens Mut. Cas. Co. v. Grinnell Corp., 477 F. Supp. 2d 327 (D. Mass. 2007).

of loss to the extent covered by property insurance obtained pursuant to this Section 11.3 or other property insurance applicable to the Work, except such rights as they have to proceeds of such insurance held by the Owner as fiduciary. The Owner or Contractor, as appropriate, shall require of the Architect, Architect's consultants, separate contractors described in Article 6, if any, and the subcontractors, sub-subcontractors, agents and employees of any of them, by appropriate agreements, written where legally required for validity, similar waivers each in favor of other parties enumerated herein. The policies shall provide such waivers of subrogation by endorsement or otherwise. A waiver of subrogation shall be effective as to a person or entity even though that person or entity would otherwise have a duty of indemnification, contractual or otherwise, did not pay the insurance premium directly or indirectly, and whether or not the person or entity had an insurable interest in the property damaged. **(11.4.7; change reference)**

Without this provision, the insurance company that has paid off for a loss to the owner could then sue the contractor for the loss. The purpose of carrying insurance, of course, is to protect the parties from loss, not merely to shift the risk from one member of the construction team to another.[353] With the

---

[353] Haemonetics Corp. v. Brophy & Phillips Co., 23 Mass. App. Ct. 254, 501 N.E.2d 524 (1986) (*quoting* Tokio Marine & Fire Ins. Co. v. Employers Ins., 786 F.2d 101, 104 (2d Cir. 1986)) ("A waiver of subrogation is useful in such projects because it avoids disruptions and disputes among the parties to the project. It thus eliminates the need for lawsuits, and yet protects the contracting parties from loss by bringing all property damage under the all risks builder's property insurance.").

See also United States Fidelity & Guar. Co. v. Farrar's Plumbing & Heating Co., 158 Ariz. 354, 762 P.2d 641 (Ct. App. 1988); Trump-Equitable Fifth Ave. Co. v. H.R.H. Constr. Corp., 106 A.D.2d 242, 485 N.Y.S.2d 65 (1985); New York Bd. of Fire Underwriters v. Trans Urban Constr. Co., 91 A.D. 115, 458 N.Y.S.2d 216 (1983).

In Industrial Risk Insurers v. Oarlock Equip. Co., 576 So. 2d 652 (Ala. 1991), the court stated that the manufacturer of a tar kettle used by a roofing contractor was not covered by the waiver of subrogation clause.

In Touchet Valley Grain Growers, Inc. v. Opp & Seibold Gen. Constr., Inc., 119 Wash. 2d 334, 831 P.2d 724 (1992), the language of the subrogation clause was not from an AIA document: "Subrogation rights, if any, are expressly waived by each party to the extent of insurance coverage afforded on any claim, loss or casualty arising from or in connection with the Project." The court held that this provision, unlike the AIA provision, did not protect subcontractors when the subcontractors were not parties to the agreement and were not specifically mentioned.

In Reliance Ins. Co. v. Liberty Mut. Fire Ins. Co., 13 F.3d 982 (6th Cir. 1994), both the owner and the general contractor purchased builder's risk policies. During construction, a fire caused extensive damage to the building. One of the insurance carriers paid for the damage and sought contribution from the other carrier. The court refused to require contribution, stating that the carrier had merely fulfilled its obligations under its policy.

In Northern Adirondack Cent. Sch. Dist. v. L.H. LaPlante Co., 645 N.Y.S.2d 893 (App. Div. 1996), the court found that the parties had substituted another insurance article in place of this Article 11. Because the substitute article did not contain the waiver of subrogation clause, it had been deleted, despite other provisions that cross-referenced the original Article 11.

In Employers Mut. Cas. Co. v. A.C.C.T., Inc., 580 N.W.2d 490 (Minn. 1998), the court held the waiver of subrogation clause applied to both "Work" and "non- work" items. In this case, the owner utilized its existing all-risk policy to cover its obligations under A201. That policy covered property beyond that included in the scope of the contractor's work. The court found that if the

inclusion of this paragraph, the insurance company cannot sue any of the stated parties.[354]

---

owner purchases a separate all-risk policy specifically to cover the work of that contract, then only the "Work" is subject to the waiver. However, if the owner relies on an existing all-risk policy that is so broad as to cover more than the "Work," then the owner waives the right to sue for all damages as long as such damages are covered by the policy.

The *Employers Mut.* case was followed in Rahr Malting Co. v. Climatic Control Co., 150 F.3d 835 (8th Cir. 1998), in determining that the waiver of subrogation covered both "Work" and non-Work items where insurance covered both categories. *See also* Trinity Universal Ins. Co. v. Bill Cox Constr., Inc., 2001 WL 1161227 (Tex. App.-San Antonio, Oct. 3, 2001); Walker Eng'g, Inc. v. Bracebridge Corp., 102 S.W.3d 837 (Tex. App. 2003) (to the extent that the damaged property was covered by insurance, the waiver of subrogation applies). In Best Friends Pet Care, Inc. v. Design Learned, Inc., 2003 WL 21235341 (Conn. App. June 3, 2003), the waiver found in the owner-contractor agreement applied to the design consultant hired by the construction manager even though the manager-consultant contract did not contain the waiver of subrogation provision. Under the pre-1997 version, the waiver did not apply to consequential damages. *See* St. Paul Fire & Marine Ins. Co. v. Elkay Mfg. Co., 2003 WL 139775 (Del. Super. Jan. 17, 2003), for a survey of cases reviewing this provision.

In Nordaway Valley Bank v. E.L. Crawford Constr., Inc., 126 S.W.3d 820 (Mo. App. 2004), the court held that indemnification provisions are subject to harmonization with the waiver of subrogation. The indemnification clause refers to compensation and liability for losses not covered by the property insurance policy, that is, compensation and liability to third parties. The waiver of subrogation is enforceable to the extent that the property damage is covered by the contractually required property insurance.

See also Church Mut. Ins. Co. v. Palmer Constr. Co., Inc., 2004 WL 1438177 (E.D. Pa. June 9, 2004), *following Haemonetics;* Lumbermens Mut. Cas. Co. v. Grinnell Corp., 477 F. Supp. 2d 327 (D. Mass. 2007)

[354] In Atlantic Mut. Ins. Co. v. Metron, 83 F.2d 897 (7th Cir. 1996), the trial court was reversed on the issue of whether the general conditions were incorporated into the contract. The document was a standard AIA A101/CM. The boilerplate and instructions to that document indicate that it is to be used along with A201/CM. The federal appeals court found the contract ambiguous because the parties failed to actually list A201 in the place where it should have been listed. The matter was sent back to the district court for a factual determination whether the general conditions were actually incorporated.

In Adventists v. Ferrellgas, Inc., 102 Wn. App. 488, 7 P.3d 861 (2000), a church brought an action on behalf of its property insurer against a heating systems contractor and a propane supplier after a fire damaged the partially constructed church building. The church had hired a project manager/superintendent to run the project. The superintendent entered into a "Trade Contract" with the heating contractor. The supplier did not have such a trade contract, but merely a sales agreement drafted by the supplier. The issue became whether A201 was incorporated, thereby invoking the waiver of subrogation clause. The court held that the heating contract included A201 because the specifications given to that contractor incorporated the "Conditions of the Contract." However, the supplier's agreement did not reference A201. Therefore, a subrogation action could be maintained against the supplier, but not the heating contractor.

In Federal Ins. Co. v. Grunau Project Dev., Inc., 721 N.W.2d 157 2006 WL 1528948 (Wis. Ct. App. June 6, 2006), the owner entered into contracts with the architect and general contractor, but, according to practice, did not actually sign the contracts. Work progressed, with various change orders being signed and payments made. During construction, a load-bearing wall was demolished, which in turn caused a partial collapse of the building, resulting in substantial damage. The insurance carrier paid for the loss and had the owner sign a statement to the effect that there was never any AIA contract between the owner and general contractor. This statement was not

The insurance company should be made aware of this language at the time the insurance is purchased, because a special endorsement may be required. A New York court has differentiated a similar provision from an indemnification clause that would have been illegal.[355] Note that subcontractors are included in this waiver provision.[356] In a case in which the subcontractor neglected to purchase required insurance, it was not protected by the waiver provision.[357]

---

disclosed to the parties to the subsequent subrogation action. In upholding the trial court's ruling in favor of the defendants based on the waiver of subrogation provision, and finding that the statement was a "sham affidavit," the appellate court found that, based on the evidence, there was only one reasonable inference—that the owner, by its conduct, clearly intended to be bound by the AIA contracts. The fact that the owner never actually signed the contracts did not render the contracts ineffective or unenforceable.

In Travelers Indem. Co. v. Losco Group, Inc., 136 F. Supp. 2d 253 (S.D.N.Y. 2001), the owner's insurance carrier brought a subrogation action against the architect and others arising out of a construction accident. The issue was whether a count for gross negligence should be dismissed based on the waiver of subrogation language. The court held that when the losses are the result of gross negligence, a waiver clause (such as the waiver of subrogation) will not preclude recovery in tort or contract. Here, the insurance carrier had set forth sufficient allegations that, if proved, would constitute gross negligence.

[355] Board of Educ. v. Valden Assocs., 46 N.Y.2d 653, 389 N.E.2d 798, 416 N.Y.S.2d 202 (1979). In Summit Contractors, Inc. v. General Heating & Air Conditioning, Inc., 2004 WL 834017 (S.C. Apr. 19, 2004), the court found no ambiguity in the waiver of subrogation clause referring to indemnification.

[356] In Fortin v. Nebel Heating Corp., 12 Mass. App. Ct. 1006, 429 N.E.2d 363 (1981), the court held that the 1970 version of A201 did not include the subcontractor so that there was no waiver of subrogation as to the subcontractor. The court noted that the subsequent 1976 version of A201 did include subcontractors. If the later version had been used, the waiver would have applied. Note that this case shows that courts will look at later versions of AIA documents in interpreting them.

In Aetna Cas. & Sur. v. Canam Steel, 794 P.2d 1077 (Colo. Ct. App. 1990), a roof collapsed after substantial construction. The owner's builder's risk insurer then sued the steel supplier. The court held that the supplier was not a subcontractor under A201 and, therefore, the waiver provision did not apply. In Midwestern Indem. Co. v. Systems Builders, Inc., 801 N.E.2d 66 (Ind. Ct. App. 2004), the court held that the waiver of subrogation applied to subcontractors. *Midwestern* was followed in S.C. Nestel, Inc. v. Future Const., Inc., 836 N.E.2d 445 (Ind. Ct. App. 2005). The trial court had refused to follow the waiver of subrogation provision because the subcontractor had subcontracted some of its work without informing the general contractor, thereby breaching its contract. The appellate court held that it made no difference whether the theory of recovery is negligence or breach of contract; the waiver of subrogation provision bars recovery. The dissent in *Midwestern* was followed in Lumbermens Mut. Cas. Co v. Grinnell Corp., 477 F. Supp. 2d 327 (D. Mass. 2007) ("It does not follow that a general contractor and subcontractors should be released from post-construction liability under [§ 11.3.5] simply because the property owner shooses to purchase a post-construction insurance policy for its own benefit.")

[357] Executive Dev. Prop., Inc. v. Andrews Plumbing Co., 134 Ga. App. 618, 215 S.E. 318 (1975).

A subcontractor also is subject to the waiver of subrogation when A201 is incorporated into the subcontract by reference. Davlar Corp. v. Wm. H. McGee & Co., 62 Cal. Rptr. 2d 199 (Cal. Ct. App. 1997).

In Haynes Constr. Co. v. Town of Newtown, 2005 WL 2857750 (Conn. Super. Oct. 14, 2005), the town cancelled its all-risk policy some four months before a heating coil in an air handling unit at a school ruptured, causing property and water damage. The contractor alleged that the town breached these provisions by failing to maintain the required insurance. The town argued that this

In a Georgia case involving this provision, the insurer could not sue the contractor who purchased the insurance because the contractor had an "insurable interest" in the property.[358] A claim for damages to the property caused by fire as a result of the negligence of the contractor or subcontractor was waived by this provision.[359]

---

particular incident was not included within the specific list of covered losses found in ¶ 11.4.1.1. The court held that the list was not intended to be an exclusive list. The waiver of subrogation provision also supported the contractor's contention. "The conclusion that the rupture of the heating coil in this case falls within the all-risk coverage mandated by the contract thus furthers the contractual goal of resolving disputes by insurance payments rather than lawsuits."

[358] E.C. Long, Inc. v. Brennan's of Atlanta, 148 Ga. App. 749, 252 S.E.2d 642 (1979). The court construed the 1970 A201, which was similar to the 1987 version. The court stated that "an insurer may not insure an insured against a peril for a premium, and when the loss occurs—pay the insured, take subrogation, and then sue the insured on the basis that his negligence caused the damage."

In Island Villa Developers, Inc. v. Bonner Roofing & Sheet Metal Co., 175 Ga. App. 713, 334 S.E.2d 41 (1985), the owner's fire insurer paid a claim for fire damage caused by a subcontractor's negligence. The insurer could not recover from the subcontractor because of the waiver.

In Blue Cross of S. W. Va. v. McDevitt & Street Co., 234 Va. 191, 360 S.E.2d 825 (1987), the owner had occupied a building that was substantially complete. Water pipes burst, causing damage. The owner's insurer sued the architect and contractor, alleging that this waiver provision was not effective because the loss occurred after substantial completion and occupancy. The court held that this provision was effective until final payment. The court stated that,

read together, the insurance and waiver provisions in the three contract documents signify a consensus among the contracting parties to exempt themselves from liability inter se. Said differently, the contracts show that the parties agreed to shift the risk of loss from themselves to a commercial insurer by acquiring policies insuring their own interests and the interests of subcontractors and sub-subcontractors.

See South Tippecanoe Sch. Bldg. Corp. v. Shambaugh & Sons, 395 N.E.2d 320 (Ind. Ct. App. 1979), for a thorough analysis of these insurance provisions (the 1970 versions) and a number of cases about insurance. See also State v. United States Fidelity & Guar. Co., 577 So. 2d 1037 (La. 1991).

South Tippecanoe was followed in Home Ins. Co. v. Bauman, 684 N.E.2d 828 (Ill. Ct. App. 1997).

[359] Motorists Mut. Ins. Co. v. Jones, 9 Ohio Misc. 113, 223 N.E.2d 381 (1966) (earlier version of A201 did not operate as a waiver, and insurer was allowed to recover against contractor for fire damage).

In A.S.W. Allstate Painting v. Lexington Ins. Co., 94 F. Supp. 2d 782 (W.D. Tex. 2000), a fire caused substantial damage to a building under construction. The owner's insurer paid the owner and put the contractor on notice that it had a claim. The contractor brought a declaratory action against the insurer to bar any subrogation claim. The insurer then sought to compel arbitration of the claim. The court held that there was no agreement to arbitrate with the insurer by virtue of this provision and the nonconsolidation in arbitration provision (¶ 4.6.4 of A201).

In Midwestern Indem. Co. v. Systems Builders, Inc., 801 N.E.2d 66 (Ind. Ct. App. 2004), the owner hired Systems Builders as general contractor to construct an addition to a commercial building designed and manufactured by Varco-Pruden. The addition was completed in the summer of 1995. In January 1996, a snowstorm hit the area and the addition collapsed. The owner's insurance carrier paid for the loss and brought a subrogation action. Several counts were directed against Varco-Pruden. Four significant issues were raised on cross-motions for summary judgment, with the trial court ruling in favor of Varco-Pruden. The first issue was whether the waiver of subrogation language covers damage to the property insured under a policy issued after project completion. The waiver applied. There was an interesting dissent on this issue.

The second issue was whether a subcontractor is a third-party beneficiary of the contract in the context of the waiver of subrogation. Because subcontractors are specified in the AIA language, the court held that subcontractors are beneficiaries of the waiver of subrogation clause contained in the contract between the owner and general contractor.

The third issue was whether the waiver applied to negligence of the contractor. The AIA language indicates that the insurance coverage and waiver of subrogation rights pertain to damage caused by perils insured against, such as fire and collapse. It is the relationship between the damage and the perils insured against that controls the waiver of subrogation. Waiver is not dependent on what theory—contract, warranty, or negligence—might be asserted to seek recovery for the damages caused by the collapse. Thus the waiver applies to negligence.

The final issue was whether the waiver applied to the contents of the building. The court held that the waiver applied only to the value of the work performed under the contract, but did not extend to the contents.

In Town of Fairfield v. Commercial Roofing, 2003 WL 21404027 (Conn. Super. June 5, 2003), the owner sued several defendants, including the architect, for damages relating to a school re-roofing project that caused extensive water damage. The architect moved for summary judgment, claiming that the waiver of subrogation provision in A201 was applicable. The court agreed that the architect was an intended beneficiary of this provision. The real issue was to what extent the waiver applied. The claim presented by the owner to the insurance carrier was for $5 million. These parties then settled for $1.2 million. The difference resulted from the fact that the overall damages included closing the school, transporting the children to other schools, and other costs, including remediation of mold, which was subject to an exclusion in the insurance policy. The architect's argument, that ¶ 11.4.1 of A201 required the owner to purchase an "all risk" policy meant that all risks should have been covered, was rejected by the court: "However, despite its title, an 'all risk' policy does not insure against every conceivable risk. In general, an 'all risk' insurance policy insures against 'all fortuitous losses, unless specifically excluded, not resulting from the wilful misconduct or fraud of the insured.' " The architect failed to present any evidence that the owner's claims in the case were covered by the policy. Summary judgment was, therefore, denied as to costs beyond the $1.2 million.

In The Gray Ins. Co. v. Old Tyme Builders, Inc., 2004 WL 691393 (La. App. 1st Cir. Apr. 2, 2004), the contractor improperly installed exterior stucco, causing water infiltration, which in turn damaged some of the interior furnishings. The contractor repaired the stucco, but the insurance carrier paid for the other damage. The carrier then sought to recover for its payments for these damages. The carrier alleged that the waiver of subrogation did not apply since the damages were not caused by a "fire or other peril" (prior version of A201). The court rejected this argument, and the waiver of subrogation was held applicable.

In Ohio Cas. Ins. Co. v. Oakland Plumbing Co., 2005 WL 544185 (Mich. App. March 8, 2005), the insurance company argued that the waiver of subrogation did not apply where there was an allegation of gross negligence and wilful and wanton misconduct. A fire that had damaged two hotels was allegedly caused by one of the defendant's employees who used a propane torch to solder pipe connections in the hotels in close proximity to wood beams and framing, using a process called "brazing" or "pipe sweating," and who failed to follow safety precautions in violation of applicable fire codes and Occupational Safety and Health Administration (OSHA) regulations. Plaintiff argued: (1) that a party may not use a contract to shield itself from liability for gross negligence or wilful or wanton misconduct; (2) that a party may not use a contractual provision to shield itself from liability for violations of public safety regulations and codes; and (3) that, as a matter of public policy, the above rules prohibiting the contractual avoidance of liability for a party's own negligence should apply irrespective of the nature of the damages. The court examined the pleadings and concluded that, at best, the allegations were for ordinary negligence. Further, the presumption that arises from violation of a statute is ordinary negligence, not gross negligence. Thus, the waiver of subrogation provision applied to bar the action.

In Federal Ins. Co. v. CBT/Childs Bertman Tseckares, Inc., 22 Mass. L. Rep. 472 (Mass. Super. Ct. 2007), the waiver did not apply where there was a failure to comply with applicable statutes, regulations, codes, rules, industry standards, and/or ordinances of the state. Violation of a statutory duty resulted in loss of the protection afforded by the waiver.

A Maryland case involved an apartment building that was almost complete.[360] A skunk had made its way into an interior wall. The construction superintendent attempted to remove the skunk by inserting propane gas into the partition wall. An explosion occurred, damaging the building and, presumably, removing the skunk. The contractor was not liable because of the waiver provision of this paragraph.

---

[360] Brodsky v. Princemont Constr. Co., 30 Md. App. 569, 354 A.2d 440 (1976).

In a case where fire damaged the building a year after construction was completed, the waiver of subrogation applied because the definition of "the Work" was "the construction and services required by the Contract Documents, whether completed or partially completed." Colonial Prop. Realty Ltd. P'ship v. Lowder Constr. Co., Inc., 256 Ga. App. 106, 567 S.E.2d 389 (2002). The court also relied on ¶ 11.4.5, which provides for a waiver of subrogation rights after final payment.

Colonial Properties was followed in Midwestern Indem. Co. v. Systems Builders, Inc., 801 N.E.2d 66 (Ind. Ct. App. 2004), holding that the waiver of subrogation applied even to insurance purchased after completion of the project. There was an interesting dissent on this issue. In TX. C.C., Inc., v. Wilson/Barnes General Contractors, Inc., 233 S.W.3d 562 (Tex. App. -Dallas, 2007). The court looked to the phrase "other property insurance applicable to the Work," as indicating that the parties intended the waiver to apply if the owner obtained property insurance that covered any damage to the property resulting from fire and/or other perils, even after completion of the project.

In Royal Surplus Lines Ins. Co. v. Weis Builders, Inc., 2006 WL 897078 (W.D. Ky. Apr. 3, 2006), the parties used an AIA-A111, but deleted the standard waiver of subrogation language and replaced it with the following "waiver of claims" provision:

Notwithstanding anything in this Agreement to the contrary, Owner and Contractor hereby waive and release each other from any and all right of recovery, claim, action, or cause of action against (a) each other, their subcontractors, agents, officers and employees and (b) the Architect and its consultants, contractors, agents, officers, and employees, for any loss or damage that may occur to the Project, improvements to the Project, or personal property within the Project by reason of fire or the elements, accident, or other casualty, regardless of whether the negligence or fault of the other party or their agents, officers, employees or contractors caused such loss, to the extent the same is insured against under insurance policies carried by the waiving party (or required to be carried by such party). Owner and Contractor shall obtain a waiver of subrogation from the respective insurance companies which have issued policies of insurance covering all risk of direct physical loss, and to have the insurance policies endorsed, if necessary, to prevent the invalidation of the insurance coverages by reason of the mutual waivers. The Contractor shall cause each contract with each Subcontractor to contain a similar waiver in favor of the Owner, Contractor, Architect, and their respective subcontractors, agents, consultants, officers and employees.

The court held that, because the AIA language was deleted, incidents occurring after substantial completion and final payment were not covered by this waiver of claims provision.

In Church Mut. Ins. Co. v. Palmer Constr. Co., Inc., 153 Fed. Appx. 805 (2005), a church hired the contractor to construct an addition and renovation to an existing church structure. During construction, a fire caused extensive damage to the work, including the existing structure. An investigation revealed that the fire had been deliberately set by a trespasser to conceal evidence that he had been stealing tools from the contractor. The insurance carrier brought a subrogation action against the contractor, contending that the contractor breached the contract by failing to protect the property. The court held that the waiver of subrogation language barred the action. "We have held that the policy underlying AIA waiver clauses, such as the clause at issue here, is the avoidance of disputes among construction project participants, and is best effectuated by interpreting the clause as effectively abrogating any subrogation right of the owner's insurer against the contractor."

Paragraph 11.3.7 covers the Work but not adjacent property.[361] Thus, if damage occurs to adjacent property or other property that is not considered part of the

---

[361] In S.S.D.W. Co. v. Brisk Waterproofing Co., 153 A.D.2d 476, 544 N.Y.S.2d 139 (1989), aff'd, 76 N.Y.2d 228, 556 N.E.2d 1097, 557 N.Y.S.2d 290 (1990), the contractor was to provide corrective work on the exterior walls of an existing building. A fire started in the contractor's shack and caused extensive damage to the owner's property that was not part of the "Work." The waiver clause was held not to apply in this case. See also Town of Silverton v. Phoenix Heat Source Sys., Inc., 1997 Colo. App. LEXIS 8 (Jan. 9, 1997); Employers Mut. Cas. Co. v. A.C.C.T., Inc., 1997 Minn. App. LEXIS 1048 (Sept. 9, 1997).

Mu Chapter of the Sigma Pi Fraternity of the U.S., Inc. v. Northeast Constr. Servs., Inc., 179 Misc. 2d 374, 684 N.Y.S.2d 872 (Sup. Ct. 1999). American Home Ins. Co. v. Monsanto Enviro-Chem Sys., Inc., 16 Fed. Appx. 172, 2001 WL 878323 (4th Cir. (N.C.), Aug. 3, 2001); Allianz Ins. Co. of Canada v. Structural Tone, 2005 WL 2006701 (S.D.N.Y. Aug. 15, 2005); TX. C.C., Inc., v. Wilson/Barnes General Contractors, Inc., 233 S.W.3d 562 (Tex. App. -Dallas, 2007).

In Trinity Universal Ins. Co. v. Bill Cox Constr., Inc., 75 S.W.3d 6 (Tex. App.-San Antonio 2001), the court reviewed numerous decisions concerning the scope of the waiver. It found that there were two approaches to the question of when an insurer's subrogation rights are barred: one approach makes a distinction between work (see ¶ 1.1.3 of A201) and non-work property and limits the scope of the waiver to damages to the work; and the second approach draws no distinction between work and non-work, but instead, limits the scope of the waiver to the proceeds of the insurance provided under the contract between the owner and contractor.

Under the first approach, there is only one question: Was the work damaged? If the answer is yes, then the waiver applies. If not, then the waiver does not apply. Under the second approach, the scope of the waiver is limited to the proceeds of the insurance provided under the contract. If the loss was paid by a policy "applicable to the Work," then there is a waiver. The court found that the AIA language defines the waiver by the source of the insurance proceeds, not the property damaged.

The Trinity approach was also followed in Independent School Dist. 833 v. Bor-Son Constr., Inc., 631 N.W.2d 437 (Minn. Ct. App. 2001). There, the school district had pre-existing property insurance for which it purchased a builder's risk endorsement. After water damaged the buildings and caused substantial damage, the school's insurer brought a subrogation action against the contractor. The insurer claimed that it was not subject to the waiver because the damage was not work related. The court held that there was a waiver and that the only way an owner can preserve subrogation rights to sue for damage to non-work property is to purchase a separate policy, not supplement an existing policy.

In Behr v. Hook, 787 A.2d 499 (Vt. 2001), a newly constructed home was destroyed by fire. The waiver of subrogation applied even though there were allegations of gross negligence against the painting contractor for leaving a heater unattended and propped near a freshly painted porch. The owners also argued that, because the general contractor failed to obtain waivers from the subcontractor as required by the contract, the general contractor materially breached the contract, thereby preventing defendants from enforcing the waiver of subrogation clause. The court rejected this argument, holding that the contract does not make obtaining the waivers from subcontractors a condition precedent to application of the waiver.

In Travelers Indemnity Co. v. The Losco Group, Inc., 204 F. Supp. 2d 639 (S.D.N.Y. 2002), the court held that the waiver of subrogation can be defeated if gross negligence is proved.

In Mt. Hawley Ins. Co. v. Structure Tone, Inc., 2004 WL 2792039 (N.J. Super. July 13, 2004), a fire caused extensive damage to a building that was about 70 percent complete. The loss also included furniture, workstations, and computer equipment. The appellate court reversed the trial court's grant of summary judgment that dismissed all claims for subrogation. Responding to the "Work" versus non-Work argument, the contractor argued that, pursuant to ¶ 9.9.1 of A201, which requires agreement of the owner and contractor before the owner may partially occupy the premises, failure to secure such agreement meant that the contractor had full control of the

Work, as defined by A201, the waiver would not apply. Note that ¶ 11.3.3 includes a waiver by the owner related to loss of use of all of the owner's property, including non-Work property.

If the parties decide to change the provisions regarding who is to purchase insurance coverages, ¶ 11.3.7 must be reviewed.[362] In a Colorado case, the court held that the contractor did not waive his rights under this clause by beginning work before receiving copies of the insurance policies (which he was entitled to receive under that agreement).[363] The phrase "other property insurance applicable

---

premises, thereby triggering the waiver of subrogation provision as to everything at that location, including computer equipment and the like. The appellate court sent this matter back to the trial court as a factual issue to be determined at a later date.

But see Willis Realty v. Cimino Constr., 623 A.2d 1287 (Me. 1993), in which the court held that this clause applied. The construction contract involved an addition attached to an existing building at its back wall. That back wall was therefore an integral part of the project and of the Work.

Colonial Properties was followed in Midwestern Indem. Co. v. Systems Builders, Inc., 801 N.E.2d 66 (Ind. Ct. App. 2004), in which the court held that the waiver of subrogation was limited to the value of the work performed under the contract. The contents of the building were not subject to the waiver.

In Butler v. Mitchell-Hugeback, Inc., 895 S.W.2d 15 (Mo. 1995), the court held that the meaning of this provision is that the waiver of subrogation is effective only to the extent of the value of the work. The court pointed to ¶ 11.1.1.5, which requires the owner to insure against damages in excess of the value of the work itself.

[362] In City of Sioux Falls v. Henry Carlson Co., 258 N.W.2d 676 (S.D. 1977), the owner changed the 1970 version of A201 to provide that the contractor was to obtain the property insurance that the owner is normally required to carry. The waiver provision of ¶ 11.3.7 was, however, not changed. Although not required to do so, the owner did carry property insurance. When a fire damaged the property during construction, the owner's insurance paid, and the owner sued the contractor. The court found that there was an ambiguity in the contract. The owner had been fully compensated and could not recover from the contractor.

In Lloyd's Underwriters v. Craig & Rush, Inc., 32 Cal. Rptr. 2d 144, 146 (Ct. App. 1994), the owner failed to purchase a separate builder's risk policy as required by 11.3.1, choosing instead to rely on its existing all-risk property insurance. The owner also failed to add the contractors as additional insureds. The waiver applied, because the contract language provided that the waiver was in force "to the extent covered by property insurance."

Lloyd's was followed in St. Paul Mercury Ins. Co. v. Vahdani Constr. Co., Inc., 2003 WL 21300107 (Cal. App. 1 Dist., June 6, 2003). In this subrogation case, the carrier sued a subcontractor after the carrier had made payments for losses resulting from a fire. The insurance carrier argued that its insured, the owner, was not a party to the subcontract that incorporated the prime contract by reference. It also argued that the terms of ¶ 1.1.2 of A201 meant that the subcontractor could not enforce any provision of the prime contract. The court disagreed and held for the subcontractor.

[363] Richmond v. Grabowski, 781 P.2d 192 (Colo. Ct. App. 1989).

In Chadwick v. CSI, Ltd., 629 A.2d 820, 825 26 (N.H. 1993), the question was whether the parties intended to enter into a standard AIA contract that contained the waiver of subrogation clause. Following a fire that destroyed the building under construction, the contractor sued the owner for nonpayment. The owner counterclaimed, alleging the contractor's negligence resulted in the fire. The owner argued that the parties did not intend to waive subrogation rights. The jury, however, believed the contractor, finding that the standard language was part of the contract. On

to the Work" refers to any property insurance applicable to the work other than that procured under paragraph 11.3.1 (insurance required of the owner).[364]

> *11.3.8 A loss insured under the Owner's property insurance shall be adjusted by the Owner as fiduciary and made payable to the Owner as fiduciary for the insureds, as their interests may appear, subject to requirements of any applicable mortgagee clause and of Section 11.3.10. The Contractor shall pay Subcontractors their just shares of insurance proceeds received by the Contractor, and by appropriate agreements, written where legally required for validity, shall require Subcontractors to make payments to their Sub-subcontractors in similar manner.* (**11.4.8; revised reference; minor modification**)

If any property of the contractor, subcontractors, or others on the project is damaged and payment is received from the insurance company, the owner collects the money as a fiduciary for the contractor or others. The owner is like a trustee for

---

appeal, that finding was upheld. The court held that the waiver of subrogation provision was not a straight exculpatory provision:

> [Waiver of subrogation provisions] exist in the contract as part of a larger comprehensive approach to indemnifying the parties involved in the construction project, allocating the risks involved, and spreading the costs of different types of insurance. These paragraphs do not present the same concerns as naked exculpatory provisions. . . . The insurance provisions of the standard AIA contract are not designed to unilaterally relieve one party from the effects of its future negligence, thereby foreclosing another party's avenue of recovery. Instead, they work to ensure that injuries or damage incurred during the construction project are covered by the appropriate types and limits of insurance, and that the costs of that coverage are appropriately allocated among the parties.

In Rivnor Prop. v. Herbert O'Donnell, Inc., 633 So. 2d 735, 744 (La. Ct. App. 1994), the insurance policy excluded loss of use or occupancy. The waiver did not apply to such exclusions.

In Automobile Ins. Co. v. United H.R.B. Gen. Contractors, Inc., 876 S.W.2d 791 (Mo. Ct. App. 1994), the court held that the waiver ended upon final payment to the contractor. The court construed an earlier version of the AIA contract. The exceptions contained in ¶ 4.3.5 were at odds with the contractor's contention that the owner's insurer could not bring an action against the contractor after final payment. Because the owner expressly reserved the right to bring an action against the contractor after final payment for defective work, the owner's insurer could bring an action for subrogation after final payment. The waiver under this provision was only effective so long as the contractor had an insurable interest in the property. The court found that the contractor's interest in the project terminated on final payment.

In Butler v. Mitchell-Hugeback, Inc., 895 S.W.2d 15 (Mo. 1995), the court held that the waiver of subrogation is effective so long as insurance was required to be maintained. In this case, even though the contractor sent the owner an invoice stating that the work was 100 percent complete, there was no evidence that a final certificate of payment had been issued. Further, under ¶ 2.6.1 of B141-1997, the architect's services are not concluded until final payment to the contractor or 60 days after substantial completion, whichever occurs first. The building collapsed suddenly shortly after completion of construction, but within 60 days of substantial completion. The court held that insurance was required to be maintained during this time, so that the waiver of subrogation was effective to protect the architect.

[364] Town of Silverton v. Phoenix Heat Source System, Inc., 948 P.2d 9 (Colo. App. 1997); Blue Cross of Southwestern Virginia and Blue Shield of Southwestern Virginia v. McDevitt & Street Co., 360 S.E.2d 825 (Va. 1987).

the money that rightfully belongs to the other party. Any agreements made under this provision should be in writing.

> *11.3.9 If required in writing by a party in interest, the Owner as fiduciary shall, upon occurrence of an insured loss, give bond for proper performance of the Owner's duties. The cost of required bonds shall be charged against proceeds received as fiduciary. The Owner shall deposit in a separate account proceeds so received, which the Owner shall distribute in accordance with such agreement as the parties in interest may reach, or as determined in accordance with the method of binding dispute resolution selected in the Agreement between the Owner and Contractor. If after such loss no other special agreement is made and unless the Owner terminates the Contract for convenience, replacement of damaged property shall be performed by the Contractor after notification of a Change in the Work in accordance with Article 7.* (**11.4.9; change reference to arbitration**)

This requires the owner, if requested in writing by a party in interest (a contractor, subcontractor, or other party whose property has been damaged), to post a bond for its performance as a fiduciary. The owner collects the insurance proceeds for the benefit of one or more of the contractors or the owner. The bond protects the contractors, and the cost of the bond is deducted from the insurance proceeds. These provisions are subject to the claims procedure in Article 15. Change orders may be required when accomplishing the replacement of damaged work.

> *11.3.10 The Owner as fiduciary shall have power to adjust and settle a loss with insurers unless one of the parties in interest shall object in writing within five days after occurrence of loss to the Owner's exercise of this power; if such objection is made, the dispute shall be resolved in the manner selected by the Owner and Contractor as the method of binding dispute resolution in the Agreement. If the Owner and Contractor have selected arbitration as the method of binding dispute resolution, the Owner as fiduciary shall make settlement with insurers or, in the case of a dispute over distribution of insurance proceeds, in accordance with the directions of the arbitrators.* (**11.4.10; revise dispute resolution language**)

Unless objected to in writing, the owner will settle with the insurance company.

§ 4.66   Performance Bond and Payment Bond: ¶ 11.4

11.4 PERFORMANCE BOND AND PAYMENT BOND

11.4.1 The Owner shall have the right to require the Contractor to furnish bonds covering faithful performance of the Contract and payment of obligations arising thereunder as stipulated in bidding requirements or specifically required in the Contract Documents on the date of execution of the Contract. (**11.5.1; no change**)

The owner can require the contractor to furnish performance and payment bonds, but the contractor must be notified before execution of the contract so that the cost of such bonds can be included in the contract sum. In a Florida case, the owner notified the contractor that the owner had accepted the contractor's bid and requested the contractor to proceed with obtaining the bonds.[365] After the owner informed the contractor that the project was abandoned, the contractor sued for enforcement of the contract. The owner defended on the basis that the contractor had not furnished the bonds. The appellate court found for the contractor because the contract was formed when the bid was accepted and was breached by the owner by refusing to proceed with the contract.

Paragraph 11.4.1 does not require an owner to give notice to the contractor before filing a bond claim. The contractor had argued that the owner had a duty to notify the contractor prior to contacting the bonding company. The court rejected this argument, finding no contractual obligation to do so. There were other notice provisions in the contract documents, but none that supported the contractor's position.

In a Louisiana case, one issue was whether the two-year prescriptive period found in A311 applied.[366] The trial court found that it did not and that the 10-year statutory period applied because the surety was not a signatory to the contract. The appellate court reversed this holding.

This provision does not require a bond; that is left up to the owner to decide.[367]

11.4.2 *Upon the request of any person or entity appearing to be a potential beneficiary of bonds covering payment of obligations arising under the Contract, the Contractor shall promptly furnish a copy of the bonds or shall authorize a copy to be furnished.* **(11.5.2; minor modifications)**

The contractor must furnish copies of the payment bonds to anyone who appears to be a potential beneficiary of the bonds, upon request. This includes not only the owner but subcontractors,[368] lenders, and others.

---

[365] Terra Group, Inc. v. Sandefur Mgmt., Inc., 527 So. 2d 849 (Fla. Dist. Ct. App. 1988).

[366] National Tea Co. v. Plymouth Rubber Co., 663 So. 2d 801 (La. Ct. App. 1995).

[367] West Durham Lumber Co., v. Aetna Cas. & Sur. Co., 184 S.E.2d 399 (N.C. App. 1971). As a result, the subcontractor could not recover on the general contractor's bond.

[368] In Roulo v. Automobile Club, 24 Mich. App. 32,179 N.W.2d 712 (1970), aff'd, 386 Mich. 324, 192 N.W.2d 237 (1971), the court held that an owner cannot be held liable to a subcontractor for failing to require the general contractor to furnish proper bonds. The contract had required a labor and material bond and a performance bond in the amount of 100% of the contract price. The bond was not furnished, and the general contractor went bankrupt. There was no duty by the owner to provide the bonds.

In Gary Boren v. Thompson & Assocs., 999 P.2d 438 (Ok. 2000), the school board hired the architect for a school library project that required both performance and payment bonds by the contractor. After approving several payments to the contractor without any bonds, the architect sent a letter to the contractor reminding him that bonds were required. Thereafter, the architect received a copy of the performance bond, but not the payment bond. Later, a lawyer contacted the architect on behalf of a subcontractor, informing the architect of a problem with payments. At that time, the architect realized that he did not have a copy of a payment bond. Nevertheless, the architect continued certifying payments. After the contractor failed to make payment to the subcontractors, the subcontractors filed suit against the architect for negligence in certifying payments to the contractor in the absence of the statutorily required payment bond. The architect moved to dismiss on three bases: (1) no duty to the subcontractors, (2) no privity of contract, and

§ 4.67   Article 12: Uncovering and Correction of Work

Correction of nonconforming work is covered in Article 12. This article explains conditions and time periods under which the contractor must correct nonconforming work.

§ 4.68   Uncovering of Work: ¶ 12.1

12.1 UNCOVERING OF WORK

12.1.1 *If a portion of the Work is covered contrary to the Architect's request or to requirements specifically expressed in the Contract Documents, it must, if requested in writing by the Architect, be uncovered for the Architect's examination and be replaced at the Contractor's expense without change in the Contract Time.* **(12.1.1; change "required" to "requested")**

This provision covers the situation where there were specific instructions to the contractor to keep certain work uncovered but the contractor ignored those instructions. If the contract documents or the architect require that certain work be kept uncovered for "examination" by the architect, the contractor must abide by those instructions. If the contractor covers up such work contrary to those instructions, the architect can request, in writing, that the contractor uncover the work. If the covered work is contrary to the contract documents, the contractor must correct the work, plus pay for uncovering that work and all related costs. If the work is correct, the contractor must at its own expense correct any damage caused by uncovering the work. In either case, the Contract Time will not be changed.

12.1.2 *If a portion of the Work has been covered which the Architect has not specifically requested to examine prior to its being covered, the Architect may request to see such Work and it shall be uncovered by the Contractor. If such Work is in accordance with the Contract Documents, costs of uncovering and replacement shall, by appropriate Change Order, be at the Owner's expense. If such Work is not in accordance with the Contract Documents, such costs and the cost of correction shall be at the Contractor's expense unless the condition was caused by the Owner or*

---

(3) the subcontractors had constructive knowledge of the contractor's duty to procure the bonds, making the proximate cause of the loss their own negligence. The court denied the motion. At trial, the subcontractors obtained an award against the architect. Finding that the purpose of bonding statutes would not be served by protecting a private, for-profit company (the architect) engaged in the business of designing and overseeing public construction projects from potential liability. The damage arose from the architect's negligence in failing to ascertain that there was no payment bond and in making unauthorized payment to the contractor after he discovered that no payment bond existed. Further, once a public entity has contracted with a private party to oversee a construction project, subcontractors should be able to assume that the private party responsible for certifying payments has verified the existence of the bonds. The court stated that the architect ought to labor under a duty to subcontractors to refrain from paying the contractor without the required bond being secured. *See also* Hydro Turf, Inc., v. International Fidelity Ins. Co., 91 P.3d 667 (Okla. Civ. App. Div. 3 2004).

a separate contractor in which event the Owner shall be responsible for payment of such costs. **(12.1.2; add language including costs and costs of correction)**

In this situation, the architect has not specifically instructed the contractor not to cover a portion of the work and the architect suspects that the hidden work does not conform to the contract documents. The architect may request that the work be uncovered, and if it does not conform, the contractor must correct and replace the work at its own cost. The contractor must also pay for the costs of uncovering the work, as well as all associated costs. If the work does conform, the contractor is entitled to reimbursement by change order, as well as an extension of the contract time. For instance, if the architect suspects that certain plumbing work hidden inside a wall is defective, the architect can request the contractor to uncover the plumbing work for inspection. If it turns out that the work is, indeed, not in accordance with the Contract Documents, then the contractor must not only correct the plumbing work, but also pay for the cost of opening up the wall and the cost of patching or replacing the wall after the plumbing is corrected. Before making a demand under this provision, the architect should get the owner's approval.

§ 4.69   Correction of Work: ¶ 12.2

12.2 CORRECTION OF WORK

12.2.1 BEFORE OR AFTER SUBSTANTIAL COMPLETION

The Contractor shall promptly correct Work rejected by the Architect or failing to conform to the requirements of the Contract Documents, whether discovered before or after Substantial Completion and whether or not fabricated, installed or completed. Costs of correcting such rejected Work, including additional testing and inspections, the cost of uncovering and replacement, and compensation for the Architect's services and expenses made necessary thereby, shall be at the Contractor's expense. **(12.2.1.1; add language including cost of uncovering and replacing)**

The architect may reject (but has no authority to accept) work that is not in compliance with the contract documents at any time.[369] The contractor must correct nonconforming work at its own expense, and must pay for the architect's services related to this nonconforming work. The architect should point out non-conforming work as soon as it becomes aware of it in order to limit the contractor's

---

[369] In *Dehnert v. Arrow Sprinklers, Inc.,* 705 P.2d 846 (Wyo. 1985), the contractor sought to substitute a different sprinkler head from the one specified. At bid opening, the contractor informed the owner that his bid was based on the substitute head and later presented the architect with information concerning the new head. The architect verbally approved the substitution. Later, the architect rejected the heads and demanded that the contractor install the ones originally specified. The owner terminated the contract when the contractor refused to make the substitution. The court held that the architect was not liable to the contractor for damages when he acted within the scope of his contractual obligations unless he acted with bad faith or malice. The architect had a contractual duty to reject nonconforming work and generally to advise and protect the owner. Thus, the architect properly rejected the nonconforming sprinkler heads.

damages. However, if the architect does not notice that certain work is nonconforming until it is installed, the contractor is responsible for correction, including the cost of any damage to surrounding work. Note that this applies to work rejected both before and after substantial completion.

12.2.2 AFTER SUBSTANTIAL COMPLETION

12.2.2.1 In addition to the Contractor's obligations under Section 3.5, if, within one year after the date of Substantial Completion of the Work or designated portion thereof or after the date for commencement of warranties established under Section 9.9.1, or by terms of an applicable special warranty required by the Contract Documents, any of the Work is found to be not in accordance with the requirements of the Contract Documents, the Contractor shall correct it promptly after receipt of written notice from the Owner to do so unless the Owner has previously given the Contractor a written acceptance of such condition. The Owner shall give such notice promptly after discovery of the condition. During the one-year period for correction of Work, if the Owner fails to notify the Contractor and give the Contractor an opportunity to make the correction, the Owner waives the rights to require correction by the Contractor and to make a claim for breach of warranty. If the Contractor fails to correct nonconforming Work within a reasonable time during that period after receipt of notice from the Owner or Architect, the Owner may correct it in accordance with Section 2.4. (**12.2.2.1; no change**)

This is sometimes referred to as the "call-back warranty." If any work is discovered to be defective after substantial completion, the owner can give notice to the contractor within this one-year time period and the contractor must come back and repair or replace the defective work. If defects are discovered after this one-year period, the contractor may still be liable to the owner for monetary damages for improper work, but after the one-year period, the contractor does not have to actually correct the work (**see** ¶ 3.5 for the contractor's general warranty, which is governed by state law). This one-year provision relates solely to the contractor's duty to correct defects and is not the owner's exclusive remedy if problems with the work appear.[370]

---

[370] John W. Cowper Co. v. Buffalo Hotel Dev., 115 A.D.2d 346, 496 N.Y.S.2d 127 (1985); Carrols Equities Corp. v. Villnave, 57 A.D.2d 1044, 395 N.Y.S.2d 800 (1977). For a case that incorrectly stated that the two contractual remedies available were under this section and ¶ 3.18.1 (ignoring the provisions of ¶ 13.4.1), *see* Idaho State Univ. v. Mitchell, 97 Idaho 724, 552 P.2d 776 (1976). Philco Corp., v. Automatic Sprinkler Corp. of Am., 337 F.2d 405 (7th Cir. 1964) (apparently a non-AIA document with a one-year guarantee and one-year correction period, holding that the owner could not recover from the contractor after one year. Ambiguity in contract construed against drafter). *Philco* was distinguished in Smith v. Berwin Builders, Inc., 287 A.2d 693 (Del. Super. 1972).

In Beacon Plaza Shopping Ctr. v. Tri-Cities Constr. & Supply Co., 2 Mich. App. 415, 140 N.W.2d 531 (1966), the owner wanted to be the general contractor for the project. When the owner could not be bonded, he entered into an arrangement with one of the subcontractors to act as general contractor, although the owner still maintained control and was a subcontractor on the project. When damage occurred within a year of completion because of settlement, the owner sued the general contractor. The court ruled against the owner because the damage occurred as a result of the defective workmanship of the owner as subcontractor.

Of course, it would still be a good idea for the owner to give the contractor the chance to do such corrective work after the one-year period simply to avoid litigation, as it is probably much cheaper for the contractor to fix defective work rather than have the owner hire another contractor to deal with the problem, all of which would then be followed by litigation.

The term *"special warranty"* in the first sentence of ¶ 12.2.2.1 refers to warranties that are specific to the project.[371] In most cases, the specifications require particular warranties, such as roof warranties. These would then fit the definition of special warranty. These warranties should be obtained from the contractor before final payment is made. Note that the terms of special warranties survive final payment by the owner.[372]

When an error is attributable to the architect's design and not to a construction error, the contractor has been held not liable to correct such work.[373] The contractor must receive notice of the defect, and the statute of limitations with regard to the defect starts at the time the notice is given.

This language, that the contractor "shall correct" work that is not in accordance with the contract documents, makes this a warranty. Cause is not an issue, so long as the defective work is work performed by the contractor or its subcontractors. The only question is whether or not the work fails to conform to the requirements of the contract documents.[374] There is, of course, some proof required by an owner,

---

A Florida court held that work performed under this provision would not qualify as final furnishing of labor within the meaning of a Florida statute providing that a claim of lien may not be recorded later than 90 days after final furnishing of labor. Aronson v. Keating, 386 So. 2d 822 (Fla. 1980).

In Grass Range High School Dist. No. 27 v. Wallace Diteman, Inc., 465 P.2d 814 (Mont. 1970), the defects were noted prior to final payment and acceptance. The court found a waiver of a provision similar to this.

This paragraph, along with the prior version of ¶ 3.5.1, prevented a contractor and its surety from arguing that the owner's acceptance of the work barred the owner's claim for damages at a later date. Brouillette v. Consolidated Constr. Co., 422 So. 2d 176 (La. Ct. App. 1982). Note, however, that ¶ 4.3.5 preserves certain claims of the owner after final payment. *See also* Larkins v. Cage Contractors, Inc., 580 So. 2d 1068 (La. App. 4th Cir. 1991); Sanders v. Zeagler, 670 So. 2d 748 (La. App. 3 Cir., 1996); Fair Street LLC v. A. Secondino and Son, Inc., 2007 WL 1600012 (Conn. Super., May 18, 2007) (using language similar to A201).

[371] Hillcrest Country Club v. N.D. Judds Co., 236 Neb. 233, 461 N.W.2d 55 (1990). In that case, the term referred to a specific 20-year roof warranty. The contractor argued that its liability was limited to one year under this provision and because the warranty was contained in its quotation to the owner rather than in the contract documents. The court rejected this contention on the basis that the quotation was made a part of the contract. Compare that situation with the current version of A201, which does not incorporate the bid documents (**see** ¶ 1.1.1).

A performance bond is not a "special warranty" pursuant to this provision. County of Dauphin, Pa., v. Fidelity and Deposit Co. of Maryland, 770 F. Supp. 248 (M.D. Pa. 1991).

[372] A201, ¶ 9.10.4.1.

[373] Atlantic Nat'l Bank v. Modular Age, Inc., 363 So. 2d 1152 (Fla. 1978); J.R. Graham v. Randolph County Bd. of Educ., 25 N.C. App. 163, 212 S.E.2d 542 (1975).

[374] Nelson v. Marchand, 691 N.E.2d 1264 (Ind. App., 1998) ("When a homebuilder guarantees the quality of workmanship and materials, a homeowner need not prove why a particular system failed, only that it did fail").

particularly for projects where the work consists of remodeling of an existing structure, of causation.[375]

A breach of warranty is different than a breach of contract.[376]

The owner should give the contractor prompt written notice of an observed defect. This notice should be as specific as possible, and should give the contractor a reasonable time, given the circumstances, within which to respond, inspect the problem, and take corrective action. If the contractor fails to do so, it would be in breach of this provision of the contract. It would also be a repudiation of this warranty.

12.2.2.2 The one-year period for correction of Work shall be extended with respect to portions of Work first performed after Substantial Completion by the period of time between Substantial Completion and the actual completion of that portion of the Work **(12.2.2.2; change from performance to completion of that portion of the Work)**

If Work is first performed after substantial completion, then the one-year period for that particular portion of the Work starts when that segment of the Work is completed, and extends for one year beyond that time. The latest that the one-year period can start is the date of final completion.

12.2.2.3 The one-year period for correction of Work shall not be extended by corrective Work performed by the Contractor pursuant to this Section 12.2. **(12.2.2.3; no change)**

In an Illinois case, this one-year period (and the Illinois two-year statute of limitation in force at the time) was extended when the contractor made numerous attempts to correct leaking in a building over a four-year period.[377] In most cases, a single correction of work does not extend the one-year period for that item.

---

[375] In Cocke v. Odom, 385 So. 2d 1321 (Ala. Civ. App. 1980), the contractor was hired to repair an existing swimming pool. The contract contained a guarantee that "All material is guaranteed to be as specified. All work to be completed in a workmanlike manner according to standard practices." A few months after completion, the owner noticed that the pool began to lose water and notified the contractor. The contractor failed to inspect the pool or take any action. Because the owner failed to present evidence at trial that the loss was from leaks repaired by the contractor or that the leaks were the result of the contractor's failure to use adequate materials or make repairs in a workmanlike manner, the contractor prevailed.

[376] Nelson v. Marchand, 691 N.E.2d 1264 (Ind. App. 1998) ("Much of the confusion at trial and on appeal can be explained by Nelson's failure to distinguish between a breach of contract action and a breach of warranty action. Although closely related, the two actions are not identical. A warranty is a promise, usually collateral to the principal contract, although not necessarily so. . . . In this case, the contract consisted of a promise to build a home. The warranty was a subsequent promise to perform the first promise in a workmanlike manner.")

[377] Axia Inc. v. I.C. Harbour Constr. Co., 150 Ill. App. 3d 645, 501 N.E.2d 1399 (1986). The architect was allowed to assert the statute of limitations because he took no active role in the efforts to correct the problems.

12.2.3 The Contractor shall remove from the site portions of the Work that are not in accordance with the requirements of the Contract Documents and are neither corrected by the Contractor nor accepted by the Owner. **(12.2.3; minor modification)**

The owner (but not the architect) may accept nonconforming work and take a credit for the difference in value of such work compared to the required work (**see ¶** 12.3.1). If the owner chooses not to accept the nonconforming work, the contractor must either correct the work or replace it, in which case the contractor must promptly remove the nonconforming work.

12.2.4 The Contractor shall bear the cost of correcting destroyed or damaged construction, whether completed or partially completed, of the Owner or separate contractors caused by the Contractor's correction or removal of Work that is not in accordance with the requirements of the Contract Documents. **(12.2.4; minor modification)**

If the contractor, in the course of correcting work that does not conform to the Contract Documents, damages any other work of the owner, the contractor must pay the cost of such damage. Presumably, the insurance obtained by the contractor pursuant to ¶ 11.1.1 would cover damage to non-Work items.

12.2.5 Nothing contained in this Section 12.2 shall be construed to establish a period of limitation with respect to other obligations the Contractor has under the Contract Documents. Establishment of the one-year period for correction of Work as described in Section 12.2.2 relates only to the specific obligation of the Contractor to correct the Work, and has no relationship to the time within which the obligation to comply with the Contract Documents may be sought to be enforced, nor to the time within which proceedings may be commenced to establish the Contractor's liability with respect to the Contractor's obligations other than specifically to correct the Work. **(12.2.5; minor modifications)**

Section 12.2 relates only to the contractor's correction of the work. Subject to the applicable statute of limitations,[378] the owner can still sue the contractor for

---

[378] *See* Garden City Osteopathic Hosp. v. HBE Corp., 55 F.3d 1126 (6th Cir. 1995) (statute of limitations was not a bar to an action against contractor because there was fraudulent concealment).

   In Groach Assocs. #5 Ltd. P'ship v. Titan Constr. Corp., 109 P.3d 830 (Wash. App. 2005), the parties used the 1976 version of AIA Document A201. The owner filed suit against the general contractor relating to various defects. The contractor, in turn, sued several subcontractors who defended on the basis that the one-year warranty provision of A201, incorporated into the subcontract, barred the general contractors' breach of contract claim as a matter of law. The court stated that

   this provision explicitly states that any one-year warranty required by the contract documents relates only to the obligation to correct the work, and not to the obligation to comply with the contractual requirement that all work be of "good quality, free from faults and defects and in conformance with the Contract Documents." . . . We conclude the one-year limited warranty does not bar Titan's claim against Purcell for breach of contract and the trial court erred in ruling as a matter of law that Purcell's warranty was the exclusive remedy for the breach of the contract claims.

damages beyond the one-year period in this section. That suit may be based on the contract or the contractor's warranties found in ¶ 3.5.

§ 4.70   Acceptance of Nonconforming Work: ¶ 12.3

12.3 ACCEPTANCE OF NONCONFORMING WORK

If the Owner prefers to accept Work that is not in accordance with the requirements of the Contract Documents, the Owner may do so instead of requiring its removal and correction, in which case the Contract Sum will be reduced as appropriate and equitable. Such adjustment shall be effected whether or not final payment has been made. (**12.3.1; no change**)

The owner may choose to accept nonconforming work. In that case, the owner is entitled to a reduction in the contract sum for the reduced value of the project. It is possible that the nonconforming work is actually more valuable than the required work, in which case the owner would not be obligated to pay more unless the contractor had previously obtained the owner's approval (in which case it would not be nonconforming work). This can be accomplished by change order or construction change directive. Note that there is no provision for increasing the Contract Sum in the event that the nonconforming work is actually more valuable. If the owner first discovers the nonconforming work after final payment has been made to the contractor, this adjustment can still be made, as the provision of ¶ 9.10.4, relating to waivers of claims, would not apply.

§ 4.71   Article 13: Miscellaneous Provisions

Forum selection, successors to the contract, notice provisions, and interest on unpaid funds are among the miscellaneous topics covered in Article 13.

§ 4.72   Governing Law: ¶ 13.1

13.1 GOVERNING LAW

The Contract shall be governed by the law of the place where the Project is located, except, that if the parties have selected arbitration as the method of binding dispute

---

Graoch was followed in 1000 Va. Ltd. P'ship v. Vertecs Corp., 127 Wash. App. 899, 112 P.3d 1276 (2005), holding that the one-year warranty was not exclusive. "The establishment of the time periods noted in Subparagraph 13.2.2 or such longer period of time as may be prescribed by . . . any warranty required by the Contract Documents, relates only to the specific obligations of the Contractor to correct the Work, and has no relationship to the time within which the Contractor's obligation to comply with the Contract Documents may be . . . enforced, nor to the time within which proceedings may be commenced to establish . . . liability with respect to the Contractor's obligations other than specifically to correct the Work." *See also* Mount Calvary Baptist Church v. Williams Const. Co. of Port Allen, 2007 WL 2193771 (La. App. 1st Cir. Aug. 1, 2007).

resolution, the Federal Arbitration Act shall govern Section 15.4. (**13.1.1; add reference to the Federal Arbitration Act**)

This clause has been upheld on several occasions.[379] Frequently, the contractor and owner are in different states and the project is in yet a third state. The contract can contain a clause that selects the forum for litigation in advance. The parties thus choose where a case is to be tried and whose law is to be used. The language of this paragraph is not a forum selection clause and selects only which laws are to be used but not where the case is to be heard. Thus, if the contractor is headquartered in New York and the owner and project are in California, the owner can bring suit in California (although the laws of New York would apply), which can cause considerable inconvenience for the contractor and pressure to settle.

In a Florida case in which the parties contractually agreed that arbitration would be conducted in New York, the architect sought to compel arbitration.[380]

The court held that the arbitration clause was voidable by either party because it was a rejection of the Florida Arbitration Code. The architect sought to arbitrate in New York because the owners, residents of Florida, filed a Florida action against the architect. The outcome might have been different if the architect had originated the action in New York.

The forum should be a court that has some expertise in construction litigation, so as to offer the fairest treatment for the parties.

---

[379] In Alwynseal, Inc. v. Travelers Indem. Co., 61 A.D.2d 803, 402 N.Y.S.2d 33 (1978), a case brought by a subcontractor was dismissed based on this clause. The contract required that the case be brought in New Jersey because the project was located there. Although the subcontractor was based in New York, the most important issue was that the parties had agreed among themselves as to the forum.

In Roberts & Schaefer Co. v. Merit Contracting, Inc., 901 F. Supp. 1349, 1351 (N.D. Ill. 1995), the court considered the following forum selection clause: "This purchase order and all disputes between the parties hereto shall be governed by and construed according to the laws of Illinois whose Circuit Court shall have exclusive jurisdiction to determine all such issues." However, the general conditions that contained this forum selection clause were in a document that was not attached to the contract and was not signed. The court held that the clause was not incorporated into the contract, so the lawsuit could be removed to the federal court. The obvious lesson here is to make sure that all provisions of the contract are actually attached to or otherwise properly incorporated into the correct contract. A problem that sometimes arises is that a party prepares a "standard" purchase order or rider that references particular sections of standard AIA documents but then neglects to change that language when new editions of the AIA documents come out.

See New England Tel. & Tel. Co. v. Gourdeau Constr. Co., 419 Mass. 658, 647 N.E.2d 42 (1995), for a case that revolves around the issue of which state's statute of limitations would apply, citing the forum selection clause in the contract.

In C & L Enters. v. Citizen Band Potawatomi Indian Tribe, 121 S. Ct. 1589, 149 L. Ed. 2d 623 (2001), a federally recognized Indian Tribe entered into a contract with a contractor to install a roof on a Tribe-owned commercial structure located outside the Tribe's reservation. When a dispute arose, the contractor demanded arbitration and the Tribe refused to participate on the grounds of sovereign immunity. When the contractor sought to confirm the award, the case went all the way to the Supreme Court, which held that, because of the arbitration agreement and choice-of-law provision in the standard AIA agreement, the parties had waived sovereign immunity.

[380] Damora v. Stresscon Int'l, Inc., 324 So. 2d 80 (Fla. 1976).

Several states have recently passed laws setting the venue for construction projects. In Illinois, for instance, Public Act 92-0657, effective July 16, 2002, states:

A provision contained in or executed in connection with a building and construction contract to be performed in Illinois that makes the contract subject to the laws of another state or that requires any litigation, arbitration, or dispute resolution to take place in another state is against public policy. Such a provision is void and unenforceable.

While this is consistent with the A201 venue provision, it may not be in relation to the venue provision in the owner-architect agreement, such as AIA Document B101 (¶ 10.1). Note that the AIA clauses are not "forum selection" clauses, but merely state which state's laws apply. The Illinois act cited above actually requires both that the laws of Illinois apply and that hearings must be held in that state if the project is located there. Laws such as this will likely be challenged in the future as an impingement on the parties' ability to contract and the Commerce Clause.

If the parties have opted into arbitration as the dispute resolution procedure, the Federal Arbitration Act[381] will govern the arbitration, although the laws of the state where the project is located will also apply, except as to the arbitration law.

§ 4.73   Successors and Assigns: ¶ 13.2

13.2 SUCCESSORS AND ASSIGNS

13.2.1 The Owner and Contractor respectively bind themselves, their partners, successors, assigns and legal representatives to covenants, agreements and obligations contained in the Contract Documents. Except as provided in Section 13.2.2, neither party to the Contract shall assign the Contract as a whole without written consent of the other. If either party attempts to make such an assignment without such consent,

---

In Johnson County v. R.N. Rouse & Co., 331 N.C. 88, 414 S.E.2d 30 (1992), the contract between the owner and contractor contained the usual arbitration clause. In addition, it contained the following provision in the supplementary general conditions:

By executing a contract for the Project the Contractor agrees to submit itself to the jurisdiction of the courts of the State of North Carolina for all matters arising or to arise hereunder, including but not limited to performance of said contract and payment of all licenses and taxes of whatever nature applicable thereto.

When the contractor sought arbitration of a dispute, the owner convinced the trial and appellate courts that this provision required the contractor to go to court with any dispute and that the arbitration provision was no longer operative. The North Carolina Supreme Court reversed, finding that this provision was a consent to jurisdiction clause that did not conflict with the arbitration provision.

[381] 9 U.S.C. § 1, *et seq.* The states follow either the Uniform Arbitration Act or the Revised Uniform Arbitration Act. The federal act contains differences from these state acts which may have significant impacts on an arbitration.

that party shall nevertheless remain legally responsible for all obligations under the Contract. **(13.2.1; minor modification)**

The language of the first sentence of this provision does not confer personal responsibility on an agent of a disclosed principal in the absence of some other agreement to the contrary or other circumstances showing that the agent has expressly or impliedly incurred or intended to incur personal responsibility to the other party.[382]

If the owner agrees to an assignment of the contract by the contractor, the owner will be bound by the terms of the assignment.[383] If the contract is assigned by the owner, the assignee can bring an action against the contractor just as the original owner could.[384] In a Florida case, the contractor wrote a letter to the owner

---

[382] Folgers Architects, Ltd. v. Pavel, 2001 Neb. App. LEXIS 9 (Jan. 16, 2001).

[383] A.E. Finlay & Assocs. v. Hendrix, 271 S.C. 312, 247 S.E.2d 328 (1978). The contractor assigned the contract proceeds of a paving contract to his equipment supplier. The owner acknowledged the assignment in writing. The owner was bound by the assignment and was obligated to pay the assignee.

[384] Herlitz Constr. Co. v. Matherne, 476 So. 2d 1037 (La. 1985). An architect designed a hotel. During the construction, a discrepancy between the architectural and mechanical drawings was discovered. It was decided to follow the mechanical drawings and revise the architectural and electrical drawings, resulting in revisions to room layouts and problems with television cables. The lender then called the loan because the work was not done in accordance with the approved drawings, resulting in additional financing costs to the owner of about $700,000. The owner's assignee sued the architect, claiming negligence. The court held for the architect on the basis that it was the owner's obligation to notify the lender whenever substantial changes were made to the plans. There was no evidence that if the owner had done so in a timely manner that the bank would not have approved the changes. The owner's contributory negligence was imputed to the assignee.

In Mears Park Holding Corp. v. Morse/Diesel, Inc., 427 N.W.2d 281 (Minn. Ct. App. 1988), the lender was assigned a development project when the developer defaulted. The lender brought suit against the architect and a subcontractor. The court held that the assignment was not valid because of this clause.

Crown Oil & Wax Co. v. Glen Constr. Co., 320 Md. 546, 578 A.2d 1184 (1990), involved the owner's assignee's attempt to enforce the arbitration provision against the contractor. The court held that the broad arbitration clause required the parties to arbitrate, despite this language.

The court in Mississippi Bank v. Nickles & Wells Constr. Co., 421 So. 2d 1056 (Miss. 1982), held that a contractual provision prohibiting assignment was ineffectual if the assignment was for the purpose of creating a security interest under the Uniform Commercial Code. The Code specifically prohibits contractual restrictions on assignments under this circumstance.

In TRUST Atlanta, Inc. v. 1815 The Exchange, Inc., 220 Ga. App. 184, 469 S.E.2d 238 (1996), the court found a valid assignment after a default by the original owner. The lender took over the project and filed this action against the contractor. The court found that the "contract as a whole" had not been assigned, that the language did not expressly operate to release the nonconsenting party, and that the clause actually anticipated assignments. Note that other AIA contracts, such as B101, do not have the "as a whole" language, indicating an intent that part of the construction contract can be assigned without the contractor's consent.

In another case, Korte Constr. Co. v. Deaconess Manor Ass'n, 927 S.W.2d 395 (Mo. Ct. App. 1996), this provision was examined in the context of a demand for arbitration. The assignee of the owner attempted arbitration against the contractor, who resisted. The trial court found that the

advising the owner that the contractor was experiencing financial difficulties, and that checks should be made out jointly to the contractor and surety, based on an irrevocable assignment of its contract to the surety.[385] Later, the contractor revoked the letter, but the owner refused the revocation. The court held that the attempted assignment was invalid in that the contract was not modified or amended.

When the contract is assigned after completion, this nonassignability clause may be ineffective. Courts have held that a party can assign its rights to sue for damages in the face of such a clause.[386]

---

assignment was valid because less than the whole was assigned. The appellate court rejected this distinction as ascribing an extraordinary degree of carelessness to the drafting of this provision.

The court held that the language prohibited subletting of the contract as a whole without consent and also prohibited assignment of the contract without consent, in whole or in part. The court went on to examine a doctrine that validates certain assignments when a contract is no longer executory, notwithstanding the assignor's failure to comply with the terms of a non-assignment clause. Under that doctrine, if the contract were no longer executory, the assignment would be effective. The court held that this was the case here. According to this court, "contract rights are distinct from causes of action which accrue from the violations of such rights. Thus, the prohibitions in the contract against assignment of the contract without the consent of the other party [do] not bar assignment of causes of action accruing from breach of the contract."

In Wade M. Berry v. Absher Constr. Co., 1999 Wash. App. LEXIS 1853 (Oct. 29, 1999), a corporation entered into a contract with defendant. The corporation then merged into a partnership, with one partner assigning her interest to plaintiff, the other partner. Defendant argued that the anti-assignment provision (similar to the language here) prohibits the assignment, thus the case should be dismissed. The court held that the AIA language did not prohibit successors and the plaintiff was a successor. The court found that, under the facts of the case where there was a family-owned corporation and an assignment between spouses, there was no assignment in violation of the contract.

In TC Arrowpoint, L.P. v. Choate Constr. Co., 2006 WL 91767 (W.D.N.C. Jan. 13, 2006), the assignee of the original owner brought an arbitration action against the contractor. The contractor argued that the assignee was not a signatory to the agreement and could not enforce the arbitration agreement. The court disagreed, citing this provision.

In Folgers Architects, Ltd. v. Kerns, 9 Neb. App. 406, 612 N.W.2d 539 (2000), an assignment was proper when the only thing being assigned was a right to money damages for breach of contract under this provision.

In SME Indus., Inc. v. Thompson, Ventulett, Stainback & Assocs., Inc. 28 P.3d 669 (Utah 2001), the case was sent back to the trial court for a determination of whether the parties intended the anti-assignment clause, which was substantially similar to the AIA language, to prohibit only the assignment of the performance of the contract, or whether it also prohibited the assignment of a cause of action seeking money damages for breach of the contract after the contract had been fully performed.

[385] Newkirk Constr. Corp. v. Gulf County, 366 So. 2d 813 (Fla. Dist. Ct. App. 1979). Although the letter asserted that there was an irrevocable assignment, the contract was not affected. The owner was under no obligation to anyone other than the contractor. When the owner continued to make checks out jointly to the contractor and surety and the surety refused to endorse the checks, the contractor had a claim against the owner. As long as the contractor has committed no vital breach, he is entitled to payment.

[386] Ford v. Robertson, 739 S.W.2d 3 (Tenn. Ct. App. 1987). The owner sold an apartment complex after substantial completion and after the architect was fully paid. The new owner sued the architect for breach of contract, negligence, and others. The architect defended on the basis of the nonassignability language of the 1987 version of B141. The court held that the agreement

Frequently, an owner wants to assign the contract to an entity, such as a land trustee or partnership to be formed at a later date. The owner-contractor agreement should state this intention of the parties so that the transition will be easy.

Sometimes the effect of such an assignment is to assign the contract to a party that will not assume the original owner's obligation. This can occur when a lender asks for a collateral assignment of the agreement so that if the owner defaults under the loan agreement, the lender can take over the project. In such cases, the assignment sometimes provides that the lender does not have to pay the contractor. Such assignments should be closely reviewed by legal counsel. The last sentence of this paragraph states that, if there is an assignment without consent, the original party will still be liable under the contract. Except as noted above, such an assignment without consent would not be effective and no duties would be owed to the assignee.

> *13.2.2 The Owner may, without consent of the Contractor, assign the Contract to a lender providing construction financing for the Project, if the lender assumes the Owner's rights and obligations under the Contract Documents. The Contractor shall execute all consents reasonably required to facilitate such assignment.* **(13.2.2; minor modifications)**

This clause permits the owner to assign the contract to the lender without prior approval from the contractor. In that case, the lender assumes the owner's obligations, including the obligation to pay the contractor.

---

prohibited assigning an interest in the performance of the executory contract. The owner could assign his claim for damages for nonperformance, which was not prohibited by the agreement.

In Elzinga & Volkers, Inc. v. LSSC Corp., 838 F. Supp. 1306, 1313 14 (N.D. Ind. 1993), *additional proceedings,* 852 F. Supp. 681 (N.D. Ind. 1994), the owner had assigned its rights to arbitrate. The contractor attempted to block the arbitration based on the nonassignment language in this paragraph. The court found the assignment valid:

> First the court notes that the meaning of a prohibition on assignment is a matter of interpretation for the court. . . . In addition, because the clause is a restriction on alienation, it must be strictly construed against the party urging the restriction. . . . Here the nonassignment provision appears internally inconsistent: in the first sentence it states that each party is bound to the assigns of the other while in the next sentence it states that neither party may assign the contract without the consent of the other. However, it is important to note that consent is required only where "the contract as a whole" is assigned. Reading the two sentences together in order to give meaning to both leads to the interpretation that each party will be bound to the other's assignee for a partial assignment, even where consent is not given.

Next, case law of most states is in accord that a prohibition on assignment does not prohibit assignment of claims for money damages for nonperformance. . . . In the present situation, all the work had been performed and the money paid under the contract at the time of the assignment. The only right remaining under the construction contract was the right to sue for nonperformance or breach. Therefore this was the only right that was assigned to L & P and as such the assignment was not barred by the nonassignment clause.

Accord, A&B Prop., Inc., v. Dick Pacific Const. Co., Ltd., 140 P.3d 1033 (Haw. 2006); Smith v. Cumberland Group, Ltd., 687 A.2d 1167 (Pa. Super. 1997).

In order to be effective, the part of this provision that requires a lender to "step into the shoes" of the owner would require the lender's approval of this paragraph. Most lenders would not be willing to assume the owner's obligations toward the contractor. This would only occur if the owner is in default when, presumably, the contractor would be owed a substantial amount of money. If the lender then took over, the lender would have to pay the contractor out of its own funds. Because the lender is not a party to this agreement, the contractor would likely not be able to force payment from the lender. On the other hand, the lender could not enforce the contract absent payment of at least most of the overdue sums to the contractor. Most lenders will require a contractor to agree to a collateral assignment of the contract, with the lender not being obligated to pay the contractor any sums owed by the owner to the contractor prior to the time the lender took the project over. This, of course, is unfair to the contractor, particularly if that collateral assignment prohibits the filing of any mechanics liens. Contractors should not agree to such assignments.

§ 4.74   Written Notice: ¶ 13.3

13.3 WRITTEN NOTICE

Written notice shall be deemed to have been duly served if delivered in person to the individual, to a member of the firm or entity, or to an officer of the corporation for which it was intended; or, if delivered at or sent by registered or certified mail or by courier service providing proof of delivery to, the last business address known to the party giving notice. **(13.3.1; add reference to courier service)**

The parties may specify that notice must be given to particular persons at particular addresses.

This paragraph was cited in a Louisiana case[387] in which a contractor filed an arbitration demand against the owner. When the contractor petitioned the court to confirm the award in its favor, the owner defended on the grounds that it had not been properly served. The contractor had sent written notice to the owner's registered agent's last known address. The agent had moved but the contractor was not notified. The court confirmed the award.

§ 4.75   Rights and Remedies: ¶ 13.4

13.4 RIGHTS AND REMEDIES

13.4.1 Duties and obligations imposed by the Contract Documents and rights and remedies available thereunder shall be in addition to and not a limitation of duties, obligations, rights and remedies otherwise imposed or available by law. **(13.4.1; no change)**

---

[387] Commercial Renovations, Inc. v. Shoney's of Boutte, Inc., 797 So. 2d 183 (La. Ct. App. 2001).

This paragraph provides that the parties may take advantage of any common law or statutory rights that may be available in addition to those found in the contract documents.[388] In an Illinois case, the owner hired a contractor to repair a leaking roof in a school.[389] The completion date passed before the contractor even started the project. The owner then notified the contractor that he was being terminated, but the seven-day notice of ¶ 14.2.2 was not given. The contractor argued that he should have been given the seven-day notice and the opportunity to complete the work within those seven days. Based on ¶ 13.4.1, the owner prevailed. The owner can elect remedies other than those stated in the contract documents if the other

---

[388] This clause was interpreted in Harris v. Dyer, 292 Or. 233, 637 P.2d 918 (1981), in a case that involved award of attorneys' fees under a statute, after the matter was arbitrated. The court stated:

> We disregard "duties and obligations" as referring most likely to the substantive performance required of the respective parties. Their "rights and remedies" clearly extend to the contingency of nonperformance. One of the standard remedies that must have been contemplated in this provision is the security of a construction lien and its eventual foreclosure. . . . Article [13.4.1] provides that the rights and remedies under the contract (which includes the remedy of arbitration) shall not be a limitation of rights and remedies otherwise available by law. It does not seem the most likely reading of this reservation of rights that it meant to sacrifice attorney fees which the law allows when all phases of the foreclosure remedy are litigated in court.

Accord Morrison-Knudsen Co. v. Makahuena Corp., 5 Haw. App. 315, 690 P.2d 1310 (1984); Sentry Eng'g & Constr., Inc. v. Mariner's Cay Dev. Corp., 287 S.C. 346, 338 S.E.2d 631 (1985). Stiglich Constr., Inc. v. Larson, 621 N.W.2d 801 (Minn. Ct. App. 2001). In L.H. Morris Elec., Inc. v. Hyundai Semiconductor Am., Inc., 125 P.3d 1 (Or. Ct. App. 2005), this provision allowed attorney's fees in a lien foreclosure action. In Everman's Electric Co., Inc., v. Evan Johnson, 955 So. 2d 979 (Miss. App. 2007), one prime contractor sued another for causing delays and interference. The defendant argued that it was an agent for the owner, but this argument was unpersuasive. This paragraph preserved the plaintiff's claim against parties other than the owner.

In Beaver Constr. Co. v. Lakehouse, 742 So. 2d 159 (Ala. 1999), the owner argued that this provision allows the parties to choose rights and remedies they desire such as the right to a jury trial through litigation. The court rejected this argument and ordered arbitration.

[389] Monmouth Pub. Sch. Dist. 38 v. D.H. Rouse Co., 153 Ill. App. 3d 901, 506 N.E.2d 315 (1987).

In Ingrassia Constr. Co., Inc. v. Vernon Township Bd. of Ed., 784 A.2d 73 (N.J. Super. 2001), the owner terminated the contractor. Pursuant to ¶ 14.2.2, the architect certified that sufficient cause existed to terminate. However, the architect was licensed in Canada but not anywhere in the United States. That made the certification invalid. The court then examined the consequences that flow from a termination based on a defective architect's certificate. It held that an invalid certificate deprives the owner of any finality, presumption of correctness, or obligation of judicial deference that would otherwise attach to a proper architect's certification. This then left the parties to their common-law causes of action for breach of contract. In this case, the arbitration provisions were stricken from the contract. This meant that a proper architect's certificate would not be subject to arbitration. The court stated that "it appears that if the architect's certificate is given in good faith, without gross error, and complies with the procedural and substantive requirements of the contract, it is conclusive with respect to the facts stated therein." The contractor argued that the architect's certification is a condition precedent to the owner's right to terminate the contractor. The court disagreed with this assertion. The absence of this certificate may be a condition precedent to the owner's right to terminate under the contract, but not "to the owner's right to exercise its common-law right of termination subject to the normal and traditional burden of proof of material breach from which it would be largely exempted by a proper certificate."

remedies are available at common law.[390] This provision does not make arbitration optional if arbitration is selected as the dispute resolution procedure.[391]

In a New York case, the court relied on this provision in finding that a contractor had not filed a notice of claim within the statutory time limit.[392] The lower court had found that there was a waiver of the statute under a predecessor clause to ¶ 10.2.8, which allowed claims within a reasonable time. However, under ¶ 13.4, there is no waiver of the statutory time limit within which to file the claim.

13.4.2 No action or failure to act by the Owner, Architect or Contractor shall constitute a waiver of a right or duty afforded them under the Contract, nor shall such action or failure to act constitute approval of or acquiescence in a breach thereunder, except as may be specifically agreed in writing. (**13.4.2; no change**)

This provides that even if the parties waive a particular right or duty on one occasion, they do not waive it on any subsequent occasions. It refers to things that are in conflict with the requirements of the contract documents, so that a party that has done something once is not assured that it can do that thing again if it conflicts with the contract documents. However, final payment under ¶ 9.10.4 waives certain rights, regardless of this provision.[393] The parties to a contract can waive the requirement that everything must be in writing.[394]

§ 4.76   Tests and Inspections: ¶ 13.5

13.5 TESTS AND INSPECTIONS

13.5.1 Tests, inspections and approvals of portions of the Work shall be made as required by the Contract Documents and by applicable laws, statutes, ordinances, codes, rules and regulations or lawful orders of public authorities. Unless otherwise provided, the Contractor shall make arrangements for such tests, inspections and approvals with an independent testing laboratory or entity acceptable to the Owner,

---

[390] For a case that held that the only contractual remedies were under ¶¶ 3.18.1 and 12.2.2, *see* Idaho State Univ. v. Mithcell, 97 Idaho 724, 552 P.2d 776 (1976). In Centex-Rodgers Constr. Co. v. McCann Steel Co., 426 S.E.2d 596 (Ga. Ct. App. 1992), the court found that the general contractor was not limited to the liquidated damages provision against a subcontractor where the sub had breached a material term of its contract. The general could recover its actual damages, which were apparently greater than the liquidated damages amount.

[391] State ex rel. Skinner v. Lombard Co., 106 Ill. App. 3d 307, 436 N.E.2d 566 (1982) (this section "is a general provision designed to cover rights and remedies available under the law of a given jurisdiction which are not specifically provided for in the contract.").

[392] Geneseo Cent. Sch. v. Perfetto & Whalen Constr. Corp., 53 N.Y.2d 306, 423 N.E.2d 1058, 441 N.Y.S.2d 1058 (1981), *rev'g* 434 N.Y.S.2d 502 (App. Div. 1980).

[393] *See* Centerre Trust Co. v. Continental Ins., 167 Ill. App. 3d 376, 521 N.E.2d 219 (1988) (in owner's action against contractor for liquidated damages, court held that owner waived its right to liquidated damages arising from breach of the contract by making a final payment). *But see* Taber Partners I v. Insurance Co. of Am., 875 F. Supp. 81 (D.P.R. 1995) (payment did not constitute acceptance of contractor's work).

[394] Hempel v. Bragg, 856 S.W.2d 293 (Ark. 1993) (owner can waive requirement that changes must be authorized in writing, based on the general rule of law).

or with the appropriate public authority, and shall bear all related costs of tests, inspections and approvals. The Contractor shall give the Architect timely notice of when and where tests and inspections are to be made so that the Architect may be present for such procedures. The Owner shall bear costs of (1) tests, inspections or approvals that do not become requirements until after bids are received or negotiations concluded, and (2) tests, inspections or approvals where building codes or applicable laws or regulations prohibit the Owner from delegating their cost to the Contractor. **(13.5.1; modifications; add last clause)**

The contractor must pay for all tests, inspections, and approvals of the work unless they are first required after the bids are received, or if building codes or laws prohibit the owner from passing those costs on to the contractor.

In a Tennessee case, the court construed an earlier version of this provision, finding that the owner was responsible for the costs of consulting engineers who inspected the project when deficiencies were found.[395] The owner argued that the inspections were a reasonable and foreseeable result of the contractor's failure to perform the contract. The court held that, because the tests were not required by "the Contract Documents, laws, ordinances, rules, regulations or orders of any public authority," the specific provision stating that the owner would bear all other costs (now the last sentence of ¶ 13.5.2) applied, so the owner had to bear the costs of inspecting faulty work. This case is clearly contrary to the intent of A201 and should not be relied upon. In the current edition of A201, ¶ 13.5.3 would cover this situation. The tests must be conducted by an entity independent of the contractor, and the architect must be given notice.

13.5.2 If the Architect, Owner or public authorities having jurisdiction determine that portions of the Work require additional testing, inspection or approval not included under Section 13.5.1, the Architect will, upon written authorization from the Owner, instruct the Contractor to make arrangements for such additional testing, inspection or approval by an entity acceptable to the Owner, and the Contractor shall give timely notice to the Architect of when and where tests and inspections are to be made so that the Architect may be present for such procedures. Such costs, except as provided in Section 13.5.3, shall be at the Owner's expense. **(13.5.2; no change)**

This may occur if the quality of the work comes into question, such as when a test shows a deficiency in some of the work and similar work has already been

---

In Gilbert Moen Co. v. Spokane River Associates, 88 Wash. App. 1064, 1997 Wash. App. LEXIS 2123 (Dec. 23, 1997), the court found that this provision prevented a waiver. The contractor had argued that the parties had disregarded the formal process regarding written orders, timely notice of claims, and verbal changes. The court found that there was no evidence that the owner had waived the claim notice requirement in writing. This case should not be relied on since most courts would probably find that this provision can be waived orally.

In Barclay White Skanska, Inc. v. Battelle Memorial Institute, 2006 WL 950375 (D. Md. Apr. 12, 2006), this provision did not salvage the contractor's breach of contract claims that were filed beyond the 21-day time limitation. The court held that Section 4.3 of A201 meant that each party could waive claims if written notice was not timely given.

[395] Airline Constr., Inc. v. Barr, 807 S.W.2d 247 (Tenn. Ct. App. 1990).

completed. In that case, the owner would want the prior work tested to assure that it conforms to the contract documents and codes.[396]

> *13.5.3 If such procedures for testing, inspection or approval under Sections 13.5.1 and 13.5.2 reveal failure of the portions of the Work to comply with requirements established by the Contract Documents, all costs made necessary by such failure including those of repeated procedures and compensation for the Architect's services and expenses shall be at the Contractor's expense.* (**13.5.3; no change**)

If tests reveal that the tested work does not comply, the contractor must pay all associated costs. If the work does comply, the owner pays.

> *13.5.4 Required certificates of testing, inspection or approval shall, unless otherwise required by the Contract Documents, be secured by the Contractor and promptly delivered to the Architect.* (**13.5.4; no change**)

Often, the Contract Documents will require that certificates of testing be provided for the owner's records and the architect's review. Examples of such tests include concrete slump and welding tests. This provision requires that the contractor obtain those certificates and forward them to the architect.

> *13.5.5 If the Architect is to observe tests, inspections or approvals required by the Contract Documents, the Architect will do so promptly and, where practicable, at the normal place of testing.* (**13.5.5; no change**)

The architect should observe tests promptly, so as to cause no delay to the contractor. Some tests are done on site, while other tests, such as curtainwall tests, are done at a testing facility. The architect is to observe the tests wherever they are performed.

> *13.5.6 Tests or inspections conducted pursuant to the Contract Documents shall be made promptly to avoid unreasonable delay in the Work.* (**13.5.6; no change**)

If there is a delay caused by the owner or architect, the contractor is entitled to a change order for additional time.

§ 4.77 Interest: ¶ 13.6

13.6 INTEREST

> *Payments due and unpaid under the Contract Documents shall bear interest from the date payment is due at such rate as the parties may agree upon in writing or, in the absence thereof, at the legal rate prevailing from time to time at the place where the Project is located.* (**13.6.1**)

---

[396] In The Clark Constr. Group Inc., v. Allglass Systems, Inc., 2004 WL 1778862 (D. Md. Aug. 6, 2004), a similar provision was held to permit the owner the right to order "whatever testing it deemed appropriate."

Because of this provision, an interest rate should be inserted in the owner-contractor agreement. (See the alternative language for ¶ 8.2 in AIA Document A101 in **Chapter 11.**) Many states have usury laws[397] that limit the interest rates that can be charged, so legal advice should be obtained in entering any amount here. The "legal rate" referred to is provided by statute and is usually unrealistically low. If the project is for a single-family home, with the homeowner as owner, a number of state and federal laws may also limit interest rates and require certain disclosures.

§ 4.78   Time Limit on Claims: ¶ 13.7

All states fix the time period after which parties cannot file suit. These vary from state to state, and even within each state the period depends upon the type of action brought. For instance, the period within which the owner may bring an action against the contractor based on the contract usually differs from the period within which the owner must bring an action based on negligence. In any case, it is important for the parties and the court to be able to determine when the period starts so that the cutoff date can be calculated. That is the purpose of this section.

13.7 TIME LIMIT ON CLAIMS

The Owner and Contractor shall commence all claims and causes of action, whether in contract, tort, breach of warranty or otherwise, against the other arising out of or related to the Contract in accordance with the requirements of the final dispute resolution method selected in the Agreement within the period specified by applicable law, but in any case not more than 10 years after the date of Substantial Completion of the Work. The Owner and Contractor waive all claims and causes of action not commenced in accordance with this Section 13.7.1. (**13.7.1; substantial modifications; deleted reference to the start of the running of the statute of limitations; set outside limit for actions at 10 years**)

The paragraph states that the parties must look to applicable state law for the period in which any lawsuit, arbitration, claim, or cause of action must be commenced. It sets an outside time limit for this at ten years after the date of substantial completion.[398] In determining the proper time limit within which a suit or arbitration must be filed, several considerations must be given.

---

[397] *See* Commerce, Crowdus & Canton v. DKS Constr., 776 S.W.2d 615 (Tex. Ct. App. 1989) (court discussed this provision and the local usury laws).

[398] In Louisville/Jefferson County Metro Government v. HNTB Corporation, 2007 WL 1100743 (W.D. Ky. April 11, 2007), it was argued that there had not been substantial completion so as to trigger the running of the statute of limitations. The project was for a baseball field, and the plaintiff argued that the project had failed to meet the standards of the governing body for baseball. The court rejected this argument on the basis that baseball games were being played at the facility without apparent problems, so that it was sufficiently complete for the owner to occupy and use the facility. This was under an earlier version of A201, which actually set the start of the statute of limitations upon the date of substantial completion.

First, there is the applicable statute of limitations itself. Unless the contract says otherwise, this will be found in the laws of the state where the project is located. Most states have specific statutes of limitations for construction and design, which are often different time limits than simple breach of contract claims. Typically, construction statutes of limitation are in the range of two to four years.

Second, is the concept of "discovery," which is a judicially-created exception to the statute of limitations. This means that the statute of limitations does not even start to run until the problem is "discovered," meaning either that the party knew of the problem or should have known of the problem. Latent defects in construction usually mean that this consideration comes into play, since they are the types of defects that an owner may not learn about until years after completion of construction. Whether a party should have known of a problem can be a difficult question and often results in litigation over this issue. Once the owner or its representative learns of a problem, the statute of limitations starts to run.

Finally, the statute of repose comes into play. Statutes of repose were passed by the states as a way to limit the discovery exception. Without a statute of repose, a plaintiff might have an unlimited amount of time in which to sue if the problem is not discovered. A statute of repose places an outside limit in which a plaintiff can discover a defect and in which the statute of limitations starts. A typical statute of repose limit is ten years. If, within this time period, a plaintiff learns of a problem, the statute of limitations starts to run. If the problem is discovered after this repose time, the claim is cut off. Notice that a plaintiff will have the benefit of the full statute of limitations period following discovery of the problem, but only if the problem is discovered within the statute of repose period.

For example, if an owner discovers a roof leak nine years and eleven months after substantial completion, he will have a further four years to sue the contractor and architect, assuming the typical time limits. If he discovers the problem two months later—ten years and one month after substantial completion—the statute of repose would cut off the claim.

Under this document, there is an absolute ten-year cut-off of claims. However, state law might shorten that time if the plaintiff discovers the problem at an earlier date. For example, if the owner discovers a roof leak one year after substantial completion and the state's construction statute of limitations is four years, the claim is barred five years after substantial completion. Claims made after this ten-year period are waived.

§ 4.79 Article 14: Termination or Suspension of the Contract

This article provides for the contractor to terminate the contract and for the owner to terminate it for cause or to suspend it for convenience.

§ 4.80   Termination by the Contractor: ¶ 14.1

14.1 TERMINATION BY THE CONTRACTOR

14.1.1 The Contractor may terminate the Contract if the Work is stopped for a period of 30 consecutive days through no act or fault of the Contractor or a Subcontractor, Sub-subcontractor or their agents or employees or any other persons or entities performing portions of the Work under direct or indirect contract with the Contractor, for any of the following reasons: **(14.1.1; no change)**

The contractor may terminate under this provision only because of the following five reasons and not for any other reason. If the contractor or other listed party contributes to the occurrence of one of the following reasons, the contractor may not terminate.

> *.1   issuance of an order of a court or other public authority having jurisdiction that requires all Work to be stopped;* **(14.1.1.1; no change)**

This can refer to any order of a building or fire department, zoning official, the Environmental Protection Agency (EPA), Occupational Safety and Health Administration (OSHA), or other local, state, or federal agency. If OSHA, for instance, stops the work because the contractor is not in compliance with safety standards, the contractor cannot terminate, because it was at fault.

> *.2   an act of government, such as a declaration of national emergency, that requires all Work to be stopped;* **(14.1.1.2; no change)**

Strikes are not included, nor are material shortages for other than governmental acts. Paragraph 8.3.1 permits the contractor additional time in the event of strikes and other events.

> *.3   because the Architect has not issued a Certificate for Payment and has not notified the Contractor of the reason for withholding certification as provided in Section 9.4.1, or because the Owner has not made payment on a Certificate for Payment within the time stated in the Contract Documents; or* **(14.1.1.3; no change)**

The contractor may terminate the contract if the architect fails to issue the certificate and also fails to notify the contractor of the reason for that.[399]

> *.4   the Owner has failed to furnish to the Contractor promptly, upon the Contractor's request, reasonable evidence as required by Section 2.2.1.* **(14.1.1.4; no change)**

---

[399] In General Ins. Co. v. K. Capolino Constr. Corp., 983 F. Supp. 403 (S.D. N.Y. Oct. 7, 1997), the court held that the architect's refusal to certify the amounts shown on the contractor's application for payment amounted to a breach of the owner-contractor contract because this provision was not followed. The contractor was entitled to at least payment for the uncontested portions of the payment.

This paragraph relates to proof by the owner that it is financially capable of making payments to the contractor for the project. Note that the work must be stopped for 30 consecutive days before the contractor can terminate. Thus, the contractor, after making the demand under ¶ 2.2.1 and not receiving a prompt response, must stop the work for the 30 days and thereafter terminate.

14.1.2 The Contractor may terminate the Contract if, through no act or fault of the Contractor or a Subcontractor, Sub-subcontractor or their agents or employees or any other persons or entities performing portions of the Work under direct or indirect contract with the Contractor, repeated suspensions, delays or interruptions of the entire Work by the Owner as described in Section 14.3 constitute in the aggregate more than 100 percent of the total number of days scheduled for completion, or 120 days in any 365-day period, whichever is less. **(14.1.2; no change)**

The contractor may terminate the contract if there are delays to the project that total the lesser of: 120 calendar days in any 365-day period, or 100 percent of the scheduled work days. For instance, if the work is stopped (through no fault of the contractor) for a month, restarted, stopped for two months, restarted, and stopped again for a month, the contractor may terminate if the stoppages all occurred within a one-year window. Alternatively, if the work is scheduled to last 45 days, and, through no fault of the contractor, the work is at the 90-day point, the contractor can terminate. Note that all time is measured in calendar days. The owner should consider agreeing to change orders if such events occur, wherein the contractor would agree to adjust the Contract Time and negate this provision.

14.1.3 If one of the reasons described in Section 14.1.1 or 14.1.2 exists, the Contractor may, upon seven days' written notice to the Owner and Architect, terminate the Contract and recover from the Owner payment for Work executed, including reasonable overhead and profit, costs incurred by reason of such termination, and damages. **(14.1.3; modifications)**

The seven-day notice is in addition to the 30-day work stoppage of ¶ 14.1.1, or the time periods in ¶ 14.1.2. This provision does not allow for an opportunity to cure on the part of the owner. Once the time periods are passed, the contractor has an absolute right to terminate, but only after giving this notice.

14.1.4 If the Work is stopped for a period of 60 consecutive days through no act or fault of the Contractor or a Subcontractor or their agents or employees or any other persons performing portions of the Work under contract with the Contractor because the Owner has repeatedly failed to fulfill the Owner's obligations under the Contract Documents with respect to matters important to the progress of the Work, the Contractor may, upon seven additional days' written notice to the Owner and the Architect, terminate the Contract and recover from the Owner as provided in Section 14.1.3. **(14.1.3; changed "persistently" to "repeatedly")**

Under this provision, the contractor can terminate by providing the seven-day notice after the work has been stopped for 60 days as a result of the owner's fault.

This can include the owner's failure to give approvals, to provide necessary surveys or other documents, to provide access to the site, or any other duty of the owner or its agents. Under an earlier version of A201, a contractor was held in breach of the contract because of his failure to give a seven-day notice.[400] Note that ¶ 9.7 permits the contractor to suspend the work if timely payments are not made by the owner.

§ 4.81   Termination by the Owner for Cause: ¶ 14.2

14.2 TERMINATION BY THE OWNER FOR CAUSE 14.2.1

14.2.1 The Owner may terminate the Contract if the Contractor:

> *.1   repeatedly refuses or fails to supply enough properly skilled workers or proper materials;* **(14.2.1.1; eliminate "persistently")**

One court has found that a similar provision permits the owner to insist that the contractor add a second shift if it becomes evident that the work cannot be completed on time.[401]

> *.2   fails to make payment to subcontractors for materials or labor in accordance with the respective agreements between the Contractor and the Subcontractors;* **(14.2.1.2; no change)**

In a federal case, the court held that the general contractor was not entitled to final payment until it had paid its subcontractors, based in part on this provision.[402] The owner may receive notice of this directly from the subcontractors, or because liens have been filed, or because necessary lien waivers are improper or not furnished.

> *.3   repeatedly disregards applicable laws, statutes, ordinances, codes, rules and regulations, or lawful orders of a public authority; or* **(14.2.1.3; modifications)**

This usually is brought to the owner's or architect's attention by the building inspectors.

> *.4   otherwise is guilty of substantial breach of a provision of the Contract Documents.* **(14.2.1.4; no change)**

---

[400] Provident Wash. Ins. Co. v. Beck, 356 Mass. 739, 255 N.E.2d 600 (1970). The contractor argued that the owner was in default for not complying with the schedule of payments. The court found that by accepting late payments and continuing the work, the contractor waived those breaches. When the contractor terminated the contract on only three days' notice, the court held that it was the contractor who breached the contract, and the owner had no further obligation to make payments.

[401] Black Lake Pipe Line Co. v. Union Constr. Co., 538 S.W.2d 80 (Tex. 1976).

[402] In re Modular Structures, Inc., 27 F.3d 72 (3d Cir. 1994).

This could include the contractor's failure to construct according to the contract documents, to correct deficient work, to meet the construction schedule, or other reasons.

14.2.2 *When any of the above reasons exist, the Owner, upon certification by the Initial Decision Maker that sufficient cause exists to justify such action, may without prejudice to any other rights or remedies of the Owner and after giving the Contractor and the Contractor's surety, if any, seven days' written notice, terminate employment of the Contractor and may, subject to any prior rights of the surety:* **(14.2.2; changed architect to initial decision maker)**

The term *without prejudice to any other rights or remedies* includes a court action or arbitration to obtain damages from the contractor. The certification by the Initial Decision Maker should be in writing.[403] In one case, the court held that the owner's failure to give the surety the seven-day notice was an insignificant breach and the surety was liable on the bond when the contractor failed to perform.[404] Failure to obtain the certification is a breach that renders the termination invalid.[405] Of course, there may be extenuating circumstances that make it impossible or unreasonable to obtain this certification.

---

[403] In Ingrassia Constr. Co., Inc. v. Vernon Township Bd. of Ed., 784 A.2d 73 (N.J. Super. 2001), the owner terminated the contractor. Pursuant to this paragraph, the architect certified that sufficient cause existed to terminate. However, the architect was licensed in Canada but not anywhere in the United States. That made the certification invalid. The court then examined the consequences that flow from a termination based on a defective architect's certificate. It held that an invalid certificate deprives the owner of any finality, presumption of correctness, or obligation of judicial deference that would otherwise attach to a proper architect's certification. This then left the parties to their common-law causes of action for breach of contract. In this case, the arbitration provisions were stricken from the contract. This meant that a proper architect's certificate would not be subject to arbitration. The court stated that "it appears that if the architect's certificate is given in good faith, without gross error, and complies with the procedural and substantive requirements of the contract, it is conclusive with respect to the facts stated therein." The contractor argued that the architect's certification is a condition precedent to the owner's right to terminate the contractor. The court disagreed with this assertion. The absence of this certificate may be a condition precedent to the owner's right to terminate under the contract, but not "to the owner's right to exercise its common-law right of termination subject to the normal and traditional burden of proof of material breach from which it would be largely exempted by a proper certificate."

[404] Winston Corp. v. Continental Cas. Co., 508 F.2d 1298 (6th Cir. 1975).
   In Enterprise Capital, Inc., v. San-Gra Corp., 284 F. Supp. 2d 166 (D. Mass. 2003), the failure to provide the seven-day notice was a material breach of the construction contract, thereby excusing the surety's performance. The seven-day notice requirement exists precisely to provide the surety an opportunity to protect itself against loss by participating in the selection of the successor contractor to ensure that the lowest bidder is hired and damages are mitigated.

[405] Supreme Indus. v. Town of Bloomfield, 2007 WL 901805 (Conn. Super. March 8, 2007). The owner argued that it was not required to obtain the certification. The evidence showed that it attempted, but failed, to obtain the architect's certification. This indicated to the court that this was intended to be a material part of the contract. *See also* Gulf Ins. Co., v. Fidelity & Deposit Co. of Maryland, 2007 WL 2162885 (N.Y. Sup. July 20, 2007).

If a surety is involved, it should be notified immediately if the contractor is not properly performing. In an Illinois case, the owner hired a contractor to repair a leaking roof in a school.[406] The completion date passed before the contractor even started the project. The owner then notified the contractor that he was being terminated, but the seven-day notice required in ¶ 14.2.2 was not given. The contractor argued that he should have been given the seven-day notice and the opportunity to complete the work within those seven days. Based on ¶ 13.4.1, the owner prevailed. The owner can elect remedies other than those stated in the contract documents if other remedies are available at common law.

Although failure to provide notice of termination in compliance with the terms of the contract may be grounds for breach of contract, if the contractor is notified in substantial accord with the agreement and understands the circumstances, there will be no breach of the notice provision.[407]

> *.1   exclude the Contractor from the site and take possession of all materials,
> equipment, tools, and construction equipment and machinery thereon
> owned by the Contractor;* (**14.2.2.1; allow owner to exclude contractor
> from site**)

The owner is here given the right to take possession of all equipment and other items owned by the contractor. When the equipment is to be returned is not stated, although it is reasonable to presume that the owner may use all this equipment to complete the project and may incorporate materials into the project.

---

[406] Monmouth Pub. Sch. Dist. 38 v. D.H. Rouse Co., 153 Ill. App. 3d 901, 506 N.E.2d 315 (1987). *But see* Willms Trucking Co. v. JW Constr. Co., 442 S.E.2d 197 (S.C. Ct. App. 1994) (failure to give required seven-day notice was a material breach of the agreement).

Bouchard v. Boyer, 1999 Conn. Super. LEXIS 1285 (May 17, 1999). In 566 New Park Assocs., LLC v. Blardo, 906 A.2d 720 (Conn. App. 2006), the owner terminated the contractor by taking possession of the jobsite without giving the contractor the required notice.

In Blaine Econ. Dev. v. Royal Elec., 520 N.W.2d 473, 477 (Minn. Ct. App. 1994), the owner brought an action against a contractor for breach of contract. The contractor counterclaimed, alleging wrongful termination. The court found that the owner had failed to give the contractor the required seven-day notice to cure. Rejecting the owner's arguments that the contractor had been notified of the owner's unhappiness with its performance, the court held that the notice was required under the contract; it must fairly advise the contractor that "the inadequate performance was serious enough that, without prompt correction, the contract will be terminated."

In Gulf Ins. Co., v. Fidelity & Deposit Co. of Maryland, 2007 WL 2162885 (N.Y. Sup. 2007 WL 2162885, July 20, 2007), the termination was invalid and breached the contract because of a failure to obtain the architect's certification and provide the seven days' written notice.

[407] J.A. McDonald, Inc. v. Waste Systems Int'l, 189 F. Supp. 2d 174 (D. Vt. 2001).

In United Stone America, Inc. v. Frazier, Lamson, Budlong, P.C., 2002 WL 241370 (Conn. Super. Jan. 30, 2002), the architect was sued by the general contractor pursuant to an assignment by the owner. The contractor claimed that the architect tortiously interfered with the owner-contractor agreement by preparing a certificate that the contractor was in default, even though substantial completion had been achieved, as evidenced by the fact that the owner had moved in. The court held that this was an issue for the trial court to determine, with a finding of bad faith, malice, wantonness, or acting beyond the scope of authority a requirement before liability is imposed on the architect.

> *.2   accept assignment of subcontracts pursuant to Section 5.4; and* (**14.2.2.2; no change**)

Note that ¶ 5.4 allows the owner to pick and choose which subcontracts it wishes to accept. All others would not be allowed back on the site.

> *.3   finish the Work by whatever reasonable method the Owner may deem expedient. Upon written request of the Contractor, the Owner shall furnish to the Contractor a detailed accounting of the costs incurred by the Owner in finishing the Work.* (**14.2.2.3; require "written" request**)

The owner is permitted to choose any reasonable method for completing the work. It need not be the best method or the cheapest. The cost of completion is covered by ¶ 14.2.4.

Note that, upon request by the contractor in writing, the owner must provide a "detailed accounting" of the costs required to finish the work. This is a condition precedent to the owner's ability to recover these costs from or backcharge the costs to the contractor. The contractor should request that this information be in accordance with generally accepted accounting principles applied in the construction industry.

> *14.2.3 When the Owner terminates the Contract for one of the reasons stated in Section 14.2.1, the Contractor shall not be entitled to receive further payment until the Work is finished.* (**14.2.3; no change**)

When the contractor did not achieve substantial completion, the owner did not have to pay the contractor and could deduct the cost of completion and repair of defects, according to an Illinois case.[408]

> *14.2.4 If the unpaid balance of the Contract Sum exceeds costs of finishing the Work, including compensation for the Architect's services and expenses made necessary thereby, and other damages incurred by the Owner and not expressly waived, such excess shall be paid to the Contractor. If such costs and damages exceed the unpaid balance, the Contractor shall pay the difference to the Owner. The amount to be paid to the Contractor or Owner, as the case may be, shall be certified by the Initial Decision Maker, upon application, and this obligation for payment shall survive termination of the Contract.* (**14.2.4; change architect to initial decision maker**)

The owner can deduct all reasonable expenses incurred in completing the project after the contractor's termination. However, if any money is left over after such payments, the original contractor is entitled to those amounts. For instance, if the contract is for $100,000 and the contractor is terminated for cause and has been paid $80,000, the owner has $20,000 left to complete the project. If the total fees,

---

[408] J.R. Sinnott Carpentry, Inc. v. Phillips, 110 Ill. App. 3d 632, 443 N.E.2d 597 (1982) (owner's refusal to pay was not breach of the contract by owner without substantial completion).

In Ironbound Fin. Servs. v. Certified Contracting, Inc., 1997 U.S. Dist. LEXIS 2574 (Mar. 10, 1997), the assignee of the contractor was barred from payment under this provision.

including those of the replacement contractor and the architect, are less than $20,000, the original contractor is entitled to the difference.[409] If the cost is greater, the original contractor must pay the owner the difference. The Initial Decision Maker must certify the amounts due under this paragraph, either to the owner or the contractor, and payment is to be made even though the contract has been terminated. If the contract is a guaranteed maximum cost contract, this provision should be reviewed and changed.

§ 4.82   Suspension by the Owner for Convenience: ¶ 14.3

14.3 SUSPENSION BY THE OWNER FOR CONVENIENCE 14.3.1

14.3.1 The Owner may, without cause, order the Contractor in writing to suspend, delay or interrupt the Work in whole or in part for such period of time as the Owner may determine. (**14.3.1; no change**)

The owner need not specify any reason for suspending the work, but the order must be in writing. If these periods are extensive, the contractor may be able to terminate the contract pursuant to section 14.1.

14.3.2 The Contract Sum and Contract Time shall be adjusted for increases in the cost and time caused by suspension, delay or interruption as described in Section 14.3.1. Adjustment of the Contract Sum shall include profit. No adjustment shall be made to the extent: (**14.3.2; no change**)

---

[409] In Anagnostopoulos v. Union Turnpike Mgmt. Corp., 300 A.D.2d 393, 751 N.Y.S.2d 762 (2002), the contractor agreed to perform demolition and excavation work and construct a new house for the owner. The owner terminated the contractor during construction due to the contractor's failure to comply with the plans and specifications. The contractor filed a demand for arbitration, and the owners moved to stay the arbitration because the contractor had not fulfilled the condition precedent found in ¶ 4.4.1 of submitting its claim to the architect. The court agreed and stayed the arbitration without prejudice and until ripe pursuant to Section 14.2 of A201. The owners then hired a substitute contractor to complete the house. After the house was completed, the contractor tried to revive its former claim, but the AAA refused due to the prior stay. Once again, the contractor submitted the matter to the court and the owner objected, based on the contractor's failure to timely submit the issue to the architect. The trial court granted the motion to compel arbitration on the grounds of equity. The appellate court reversed:

> A court has the jurisdiction to determine whether contractual conditions precedent to arbitration have been fulfilled. . . . In this case, the Supreme Court should have denied the motion to compel arbitration based upon the respondent's failure to fulfill in a timely fashion the contractually mandated condition precedent. . . . Consensual arbitration is a creature of contract. . . . The Supreme Court had no authority to substitute its notions of equity for what the parties have agreed to in their contract.

Apparently, the trial court believed that the claim was not "ripe" until construction was completed pursuant to the language of this paragraph. It is unclear from the case what relief the contractor was seeking. If the relief consisted of the difference between what was paid and the contract amount, then the court was correct. If, however, the contractor merely sought its actual damages, then ripeness should not have been an issue.

This would be done by change order or construction change directive. At a minimum, the contractor would be given additional time equal to the delay to complete the work. The contractor may also be entitled to additional time caused by demobilization and remobilization. Note that the contractor would be entitled to an adjustment in profit on account of a suspension.

> *.1 that performance is, was or would have been so suspended, delayed or interrupted by another cause for which the Contractor is responsible; or* **(14.3.2.1; no change)**

If the contractor is partly responsible for the delay or suspension, no adjustment need be made.

> *.2 that an equitable adjustment is made or denied under another provision of the Contract.* **(14.3.2.2; no change)**

This means that if the contractor has filed a claim for the delay and that claim has been acted upon (under the procedure for claims), then no separate adjustment is made under this provision. Other provisions in the General Conditions may affect whether or not and to what extent the contractor is entitled to an adjustment.

§ 4.83 Termination by the Owner for Convenience: ¶ 14.4

14.4 TERMINATION BY THE OWNER FOR CONVENIENCE

14.4.1 The Owner may, at any time, terminate the Contract for the Owner's convenience and without cause. **(14.4.1; no change)**

The owner may terminate the contract for any reason, subject to payments to the contractor as described below.

14.4.2 Upon receipt of written notice from the Owner of such termination for the Owner's convenience, the Contractor shall: **(14.4.2; no change)**

> *.1 cease operations as directed by the Owner in the notice;* **(14.4.2.1; no change)**
>
> *.2 take actions necessary, or that the Owner may direct, for the protection and preservation of the Work; and* **(14.4.2.2; no change)**
>
> *.3 except for Work directed to be performed prior to the effective date of termination stated in the notice, terminate all existing Subcontracts and purchase orders and enter into no further Subcontracts and purchase orders.* **(14.4.2.3; no change)**

These provisions describe the steps that the contractor must take upon receiving the owner's notice to terminate for convenience.

14.4.3 In case of such termination for the Owner's convenience, the Contractor shall be entitled to receive payment for Work executed, and costs incurred by reason of such termination, along with reasonable overhead and profit on the Work not executed. **(14.4.3; no change)**

The contractor is entitled to costs for such termination, as well as overhead and profit on the work not performed.

This provision, allowing "reasonable overhead and profit on the Work not executed," appears to be in conflict with the waiver of consequential damages provision found in ¶ 15.1.6, even though that waiver provision states that it applies to this Article 14. Paragraph 15.1.6.2 permits the contractor to recover "anticipated profit arising directly from the Work." Thus, the contractor's anticipated profit is permitted by both provisions. However, ¶ 15.1.6.2 implies that the contractor's overhead expenses are waived as an element of damages, while ¶ 14.4.3 permits recovery of overhead for the uncompleted work. These apparent conflicts could be harmonized by allowing the contractor to recover its anticipated overhead costs for the uncompleted work as an exception to the waiver of consequential damages, but only in the situation where the owner terminates for convenience.

The word "may" in ¶ 6.1 of the 1997 version of A101 makes termination optional.[410] This does not appear in the 2007 version.

§ 4.84   Article 15: Claims and Disputes

In the 2007 version of A201, this Article has been substantially modified with the introduction of the Initial Decision Maker and making arbitration optional.

§ 4.85   Claim and Disputes: ¶ 15.1

15.1 CLAIMS

15.1.1 Definition. *A Claim is a demand or assertion by one of the parties seeking, as a matter of right, payment of money, or other relief with respect to the terms of the Contract. The term "Claim" also includes other disputes and matters in question between the Owner and Contractor arising out of or relating to the Contract. The responsibility to substantiate Claims shall rest with the party making the Claim.* **(4.3.1; delete reference to adjustment or interpretation of Contract terms; delete extension of time; delete reference to written notice)**

---

[410] BBC Co., Inc. v. Town of Easton, 842 N.E.2d, 2006 WL 266107 (Mass. App. Ct. 2006). The parties had amended A101 to provide that the contract was contingent upon obtaining certain permits and approvals. Incorrectly assuming that the permit process had been completed, the town issued the contractor a notice to proceed. Upon discovering to the contrary, the town informed the contractor that construction would have to cease. There was a delay of 14 months. Neither side terminated the contract in accordance with A201. The town paid the contractor its mobilization and site work costs. The contractor then sued the town for lost profits pursuant to ¶ 14.4.3. Based on the contingency provision, the court ruled against the contractor.

These claims are contested claims.[411] If the owner and contractor agree, they may modify the contract. If one party does not agree, the procedure outlined is followed.[412] In the prior version, there was a requirement that claims be made by written notice. This has not been eliminated, as the following paragraph still requires the written notice of claims by either the owner or contractor.

15.1.2 Notice of Claims. Claims by either the Owner or Contractor must be initiated by written notice to the other party and to the Initial Decision Maker with a copy sent to the Architect, if the Architect is not serving as the Initial Decision Maker. Claims by either party must be initiated within 21 days after occurrence of the event giving rise to such Claim or within 21 days after the claimant first recognizes the condition giving rise to the Claim, whichever is later. **(4.3.2; modifications)**

---

[411] In David R. Bain v. Joan Paciotti-Orr, 1999 Ohio App. LEXIS 5698 (Dec. 2, 1999), the court reviewed this language and found that a dispute about draw payments constituted a claim as defined here. In this case, a dispute arose between the owner and contractor regarding payment. The owner filed a demand for arbitration and hearings were held on that claim as well as the contractor's claim for contract damages. The arbitrators issued an award denying both claims. The contractor then filed a new action in court, seeking damages for unpaid work. The owner moved to dismiss on the basis of the arbitration award. The contractor argued that this was a new and different claim for damages. The appellate court rejected this argument, as these were the same issues that were litigated in the arbitration.

In Fe-Ri Constr., Inc. v. Intelligroup, Inc., 218 F. Supp. 2d 168 (D.P.R. 2002), the owner failed to pay the contractor. This was a "claim" under this provision. The contractor argued that this was an action for breach of contract and not a claim that required submission to the architect, followed by mediation and/or arbitration. This argument was rejected and the lawsuit dismissed without prejudice.

In Cameo Homes v. Kraus-Anderson Constr. Co., 394 F.3d 1084 (8th Cir. 2005), the court stated that "the general conditions distinguish claims from change orders. Change orders modify the terms of a contract, while claims seek relief owed 'as a matter of right' under the existing terms of a contract." This distinction is important since "written notice of a claim must be made within 21 days of an event or of the discovery of an event giving rise to the demand," and "failure to present a claim to the architect would preclude later litigation about it, for the architect's decision was a condition precedent to the right to litigate a dispute." The contractor presented evidence that the parties had changed the change order process. The court held that this was insufficient, since the change order process was distinct from the claims process and the contractor "has not shown that the parties understood that its submission of change order requests to [the construction manager] was effectively equivalent to submission of claims to the architect."

In Starks Mech., Inc. v. New Albany-Floyd County Consolidated Sch. Corp., 854 N.E.2d 936 (Ind. App. 2006), a school filed a declaratory judgment action against the mechanical and plumbing contractor, asserting that the contractor failed to provide timely notice of a claim. The contract contained express language equating the treatment of claims for delay damages to the treatment of all other claims. Here, the general conditions required notice by the contractor within 14 days, as well as weekly updates of the claim and estimated associated costs to the construction manager for the duration of the delay. Although the contractor knew of the claim near the beginning of the project, it waited until construction was almost complete before submitting a claim for more than $1.3 million. The court held in favor of the owner.

[412] In Fletcher v. Laguna Vista Corp., 275 So. 2d 579 (Fla. 1973), under the 1967 version of A201 in which the architect was authorized to determine the adjustment in the contract sum or contract time if the owner and contractor did not agree, the court held that an ambiguity existed and the contractor's claim was allowed.

This giving of this notice is a condition precedent to the contractor's ability to recover the costs of such work.[413] The notice requirement can be waived.[414]

A Missouri court interpreted the 1976 version of this provision in this way:

> It is incumbent upon the contractor, in order to avail of an extension of time . . . to lay claim therefor at the time and within the time and manner stipulated. Otherwise no such extension of time is available to him on the grounds of waiver . . . in the very face of a provision stipulating a precise time.[415]

---

[413] U.S. v. Centex Constr. Co., 638 F. Supp. 411 (W.D. Va. 1985); Mike M. Johnson, Inc., v. County of Spokane, 78 P.3d 161 (Wash. 2003); Westates Const. Co., v. City of Cheyenne, 775 P.2d 502 (Wyo. 1989).

[414] In E.C. Ernst, Inc., v. Koppers Co., Inc., 626 F.2d 324 (3d Cir. 1980), there was a 30-day notice requirement. The contractor was issued revised drawings that stated "approved for construction." In addition, the changes were "massive," leading the court to conclude that there was a waiver of the notice requirement.

[415] Steinberg v. Fleischer, 706 S.W.2d 901, 904-05 (Mo. Ct. App. 1986) (*quoting* Herbert & Brooner Constr. Co. v. Golden, 499 S.W.2d 541 (Mo. Ct. App. 1973)).

However, in Triangle Air Conditioning, Inc. v. Caswell County Bd. of Educ., 291 S.E.2d 808, 811 (N.C. Ct. App. 1982), there was a substantial delay in the construction which was not the fault of the contractor. The owner contended that the contractor should have given written notice of the delay and a claim within 20 days of the completion date specified in the contract, and that therefore the contractor was not entitled to delay damages. The court found that "the event which gave rise to the plaintiffs demand was the delay in the construction. It did not occur on a specific date." The court held that the contractor, who gave notice 49 days later, complied with the notice requirement. Note that in this case the court also found that the owner had waived the change order procedures by failing to issue the required change order for the delay. This case should be viewed skeptically, because the contract clearly requires the contractor to request a change order for an extension of time within the stated time limits. The contractor certainly knew on the contract completion date that the project was delayed and that a change order had to be requested. Unless the owner has waived the change order procedures, it seems only logical to require the contractor to make a claim no later than 21 days (under the current version of A201) after the contract completion date, or the date it first learns of the pending delay, whichever occurs first.

Waiver by the owner should not be inferred if the owner seems to be following the change order procedures. McKeny Constr. Co. v. Town of Rowlesburg, 187 W. Va. 521, 420 S.E.2d 281 (1992). When the owner considers claims without following the procedures strictly, an issue of fact is created as to whether there is a waiver. Grimm Constr. Co. v. Denver Bd. of Water Comm'rs, 835 P.2d 599 (Colo. Ct. App. 1992). For a case involving waiver in the context of an arbitration, *see* Southern Md. Hosp. Ctr. v. Edward M. Crough, Inc., 48 Md. App. 401, 427 A.2d 1051 (1981). In L. Loyer Constr. Co. v. Hartland Meadows, 2002 WL 347817 (Mich. Ct. App. Feb. 26, 2002), the owner sought damages from the contractor for improper grading. The contractor argued that the owner had failed to assert its claim within 21 days. The arbitrators found for the owner without giving any explanation for the award. The award was confirmed on the basis that the arbitrators could have found that this provision was waived.

In East 26th Street & Park Realty v. Shaw Indus., Inc., 6 Misc. 3d 1036A (N.Y. Sup. 2004), in a motion for summary judgment, the contractor argued that the owner's claim is barred for failing to make the claim within the 21-day limit. The court declined to enter summary judgment because the owner alleged that the contractor concealed the cause of the damage prior to the commencement of the action.

In Bellvue Tech. Tower, LLC v. DPR Constr., Inc., 125 Wash. App. 1003 (Div. 1, 2005), the contractor was not entitled to an equitable adjustment because it failed to comply with the 21-day notice provision of A201.

In one California case, the contractor was denied damages for the architect's delay in processing change orders.[416] His remedy was to refuse to accept the change orders after he was aware of the architect's delay. One court has held that the owner cannot waive the claims procedure without the architect's consent because the architect is a third-party beneficiary to the claim provisions.[417]

Paragraph 15.1.2 also states that claims must be filed with the other party, with a copy to the architect if the architect is not the Initial Decision Maker, as well as the Initial Decision Maker (who might or might not be the architect). Even if the architect is not the Initial Decision Maker, it would be prudent to furnish the architect with a copy of every claim. One major purpose of a timely claim is to permit the architect (or Initial Decision Maker) to investigate in a timely manner. If the architect is not promptly notified, conditions may change, thereby making it difficult or impossible for the architect to properly evaluate a claim. Even if the architect is not the Initial Decision Maker and is not required to investigate claims, it would seem very unwise to keep the architect out of the loop and merely present the architect with the results of the claim.

Note that the contractor need only make the claim itself in a timely manner.[418] There is no requirement that the amount of time or money involved be submitted at

---

In Kingsley Arms, Inc., v. Sano Rubin Constr. Co., Inc., 791 N.Y.S.2d 196 (2005), failure to strictly comply with the 21-day notice requirement was a waiver of the claim.

In R.W. Grainger & Sons v. Nobel Inst. Co., 2005 WL 3729018 (Mass. Super. Dec. 20, 2005), the court discusses a number of instances where the 21-day provision was violated.

In Barclay White Skanska, Inc. v. Battelle Mem'l Inst., 2006 WL 950375 (D. Md. Apr. 12, 2006), ¶ 13.4.2 did not salvage the contractor's breach of contract claims that were filed beyond the 21-day time limitation for claims. The court held that Section 4.3 of A201 meant that each party could waive claims if written notice was not timely given.

In Lake Eola Builders, LLC v. The Metropolitan at Lake Eola, LLC, 2006 WL 1360909 (MD. Fla. May 17, 2006), the defendant was not entitled to summary judgment because there was some evidence that the defendant waived its right to enforce the time limit on claims or that the parties orally agreed to waive the notice provisions of the contract.

See also T.W. Morton Builders v. von Buedingen, 450 S.E.2d 87 (S.C. Ct. App. 1994). In Carteret County v. United Contractors of Kinston, Inc., 462 S.E.2d 816 (N.C. Ct. App. 1995), the court found that the contractor waived its right to arbitrate the claim by failing to file its claim within the 21-day limit. The court stated that, because of the strong policy in favor of arbitration, the party opposing it must first demonstrate that the delay caused it prejudice. In this case, the owner did not claim any prejudice, so the arbitration was ordered.

[416] Huber, Hunt & Nichols, Inc. v. Moore, 67 Cal. App. 3d 278, 136 Cal. Rptr. 603 (1977).

[417] Skidmore, Owings & Merrill v. Intrawest 1, 1997 Wash. App. LEXIS 1505 (Sept. 8, 1997). The court held that a party must strictly comply with procedural claim requirements. The architect was a third-party beneficiary because this claim procedure gave the architect the opportunity to avoid litigation through dispute resolution.

[418] In County Commissioners of Caroline County, Maryland v. J. Roland Dashiell & Sons, Inc., 747 A.2d 600 (Md. App. 2000), the contractor notified the owner that a claim would be filed and that the contractor was working to put the claim together. The court held that this was insufficient to constitute a claim under the 1987 version of A201, and that the claim that was then filed beyond the 21-day window was untimely. The court also noted that the claim did not include "an estimate of cost and of probable effect of delay on progress of the Work" as required by ¶ 4.3.7.1. Finally, the contractor could not recover under a theory of unjust enrichment because there was a valid contract.

the time the claim is submitted. The claim is only a notice of the general nature of the claim, although it should be as detailed as circumstances will allow at the time. It is often the case that when events outside the contractor's control occur—for example, when there is a flood or other natural disaster—the contractor will know that a claim exists but may not know for some time just what the impact will be in terms of delay or cost. Paragraph 15.2.2 requires the Initial Decision Maker to take one of a number of actions following receipt of a claim, one of which is to request additional supporting data. The contractor should carefully document the reason for not being able to provide the necessary supporting information if there will be an extended delay. Failure by the contractor to furnish this information if it could have been timely furnished might be cause to deny a claim. The time delay and cost information should be furnished as soon as it is available.

15.1.3 Continuing Contract Performance. Pending final resolution of a Claim, except as otherwise agreed in writing or as provided in Section 9.7 and Article 14, the Contractor shall proceed diligently with performance of the Contract and the Owner shall continue to make payments in accordance with the Contract Documents. The Architect will prepare Change Orders and issue certificates for payment in accordance with the decisions of the Initial Decision Maker. (**4.3.3; add last sentence**)

This provision requires the work to continue and the owner to continue payments to the contractor pending the outcome of a claim, including mediation and litigation or arbitration. Otherwise, the potential damages of shutting down the project would be enormous.[419] The Initial Decision Maker is empowered to make "initial decisions" concerning claims. See ¶ 15.2.5. This means that the Initial Decision Maker could authorize partial payments for items in dispute in the claims submitted to that person, and the architect would issue change orders in compliance with those initial decisions.

15.1.4 Claims for Additional Cost. If the Contractor wishes to make a Claim for an increase in the Contract Sum, written notice as provided herein shall be given before proceeding to execute the Work. Prior notice is not required for Claims relating to an

---

Strand Hunt Constr., Inc. v. Lake Wash. Sch. Dist. No. 414, 2006 WL 2536315 (Wash. App. Div. 1, Sept. 5, 2006) involved a similar provision with a 14-day notice requirement.

[419] For a case that interpreted a similar clause, *see* Village of Cairo v. Bodine Contracting Co., 685 S.W.2d 253 (Mo. Ct. App. 1985). In McGee Constr. Co. v. Neshobe Dev., Inc., 594 A.2d 415 (Vt. 1991), the court stated that if the contractor had "walked off the job" without following the claims procedure, this paragraph would have placed the contractor in material breach of the contract. In Cleveland Wrecking Co. v. Central Nat'l Bank, 216 Ill. App. 3d 279, 576 N.E.2d 1055 (1991), the contract incorporated an earlier version of A201. The contractor was justified in failing to proceed until a written change order acknowledging additional compensation was received, because the requested change involved a major change in the scope of the project.

In Framingham Heavy Equip. Co., Inc. v. John T. Callahan & Sons, Inc., 61 Mass. App. Ct. 171, 807 N.E. 2d 851 (2004), the subcontractor was not required to continue working without being paid. The subcontractor was entitled to payment under a Construction Change Directive without waiting for a Change Order and, not having been paid, was permitted to stop work.

emergency endangering life or property arising under Section 10.4. (**4.3.5; modified reference**)

The contractor must give written notice before doing the work for which it believes it is entitled to an extra.[420]

The only exception is for an emergency when life or property is threatened. One such case involved emergency reroofing by a contractor at the verbal direction of the architect.[421] The court held that the contractor was entitled to additional compensation even though the claim was not timely filed because the matter was an emergency and fell under the provisions of this paragraph. Otherwise, if the contractor proceeds without written notice, it is not entitled to any extra compensation.[422]

---

[420] Ida Grove Roofing & Improvement, Inc., v. City of Storm Lake, 378 N.W.2d 313 (Iowa Ct. App. 1985); Moore Constr. v. Clarksville Dep't of Elec., 707 S.W.2d 1 (Tenn. Ct. App. 1985). In Whitfield Constr. Co. v. Commercial Dev. Corp., 392 F. Supp. 982 (D. V.I. 1975), the court stated:

As a general rule, the provision in a private building or construction contract that alterations or extras must be ordered in writing is valid and binding upon the parties, and therefore, so long as such a provision remains in effect no recovery can be had for alterations or extras done without a written order in compliance therewith.

In Castle Concrete Co. v. Fleetwood Assocs., 13 Ill. 2d 289, 268 N.E.2d 474 (1971), the court stated the elements that a contractor must show before he is entitled to an extra: (a) the work was outside the scope of the contract; (b) the extra items were ordered by the owner; (c) the owner agreed to pay extra, either by words or conduct; (d) the extras were not furnished by the contractor as his voluntary act; and (e) the extra items were not rendered necessary by any fault of the contractor. In Chiappisi v. Granger Contracting Co., 352 Mass. 174, 223 N.E.2d 924 (1967), the subcontractor scaled the architect's drawing and miscalculated the amount of spray insulation required. He asked for an extra at the conclusion of the work. The court held that the extra was not permitted because the subcontractor failed to follow the claims procedures. Notice was timely in Board of Educ. v. Tracy Trombley Constr. Co., 122 A.D. 421, 505 N.Y.S.2d 233 (1986). The contractor became aware of an error on July 6 and requested a change order from the architect on July 11. On August 22, the contractor sent a verified notice of claim to the owner. The letter to the architect constituted sufficient notice. Stelko Elec., Inc. v. Taylor Cmty. Schs. Bldg. Corp., 826 N.E.2d 152 (Ind. Ct. App. 2005) (claim not valid where contractor failed to give notice). In American Nat'l Elec. Corp. v. Poythress Commercial Contractors, Inc., 604 S.E.2d 315 (N.C. Ct. App. 2004), the subcontract incorporated A201. The subcontractor was aware that its work was being delayed in April 2000, but did not notify the general contractor of its delay claim in writing until a letter dated 20 September 2000. This lack of timely notice defeated the claim.

[421] Pioneer Roofing Co. v. Mardian Constr. Co., 152 Ariz. 455, 733 P.2d 652 (1986).

[422] In United States v. Centex Constr. Co., 638 F. Supp. 411 (W.D. Va. 1985), a subcontractor brought an action against the contractor for additional work. The contract had this provision:

Notice of any damage or additional cost which subcontractor alleges the contractor . . . [has] caused or [is] causing it by [its] act or omission shall be filed in writing with the Contractor . . . within seven days from commencement of such alleged damage or additional cost . . . No claims for such damage shall be valid unless the Subcontractor complies with all the requirements of this paragraph.

The subcontractor argued that it was under economic duress. This argument was rejected. The claim was denied by the court for failing to comply with the notice requirement.

The purpose of this is to protect the owner from unfair cost increases. For instance, if the architect issues a change order for a minor change, believing that it does not change the contract sum or contract time (the only change the architect can make without the owner's approval), the contractor must notify the parties if that change would affect the cost or time. Then, the change can be reconsidered and reevaluated before it takes effect.

Note that if the owner terminates the contract, for whatever reason, the contractor must follow the procedures, including the dispute resolution procedures, unless the owner and contractor can come to terms on their own.

15.1.5 Claims for Additional Time

15.1.5.1 If the Contractor wishes to make a Claim for an increase in the Contract Time, written notice as provided herein shall be given. The Contractor's Claim shall include an estimate of cost and of probable effect of delay on progress of the Work. In the case of a continuing delay only one Claim is necessary. (**4.3.7.1; no change**)

The 21-day time limit applies here. A continuing delay might be a strike or action of a governmental body.

15.1.5.2 If adverse weather conditions are the basis for a Claim for additional time, such Claim shall be documented by data substantiating that weather conditions were abnormal for the period of time, could not have been reasonably anticipated and had an adverse effect on the scheduled construction. (**4.3.7.2; no change**)

Written documentation is required under this provision.[423] The contractor must prove that the specific weather conditions delayed the work. Unusually cold temperatures may not cause a delay if the building is enclosed and the contractor should have provided for temporary heat.

15.1.6 Claims for Consequential Damages. The Contractor and Owner waive Claims against each other for consequential damages arising out of or relating to this Contract. This mutual waiver includes: (**4.3.10; no change**)

    .1 damages incurred by the Owner for rental expenses, for losses of use, income, profit, financing, business and reputation, and for loss of management or employee productivity or of the services of such persons; and (**4.3.10.1; no change**)

    .2 damages incurred by the Contractor for principal office expenses including the compensation of personnel stationed there, for losses of financing,

---

[423] For a case that turned on severe weather, *see* Roger Johnson Constr. Co. v. Bossier City, 330 So. 2d 338 (La. Ct. App. 1976); McDevitt & Street Co., v. Marriott Corp., 713 F. Supp. 906 (E.D. Va. 1989).

    In Bellim Constr. Co. v. Flagler County, 622 So. 2d 21 (Fla. Dist. Ct. App. 1993), the architect first approved a six-day extension of time requested by the contractor. When the owner withheld money from the contractor, the contractor sued. The architect then rescinded his earlier approval, citing lack of proof as to weather delays and triggering a liquidated damages provision. The court found this unreasonable and stated that the architect had waived the required documentation.

business and reputation, and for loss of profit except anticipated profit arising directly from the Work. **(4.3.10.2; no change)**

This mutual waiver is applicable, without limitation, to all consequential damages due to either party's termination in accordance with Article 14. Nothing contained in this Section 15.1.6 shall be deemed to preclude an award of liquidated damages, when applicable, in accordance with the requirements of the Contract Documents. **(4.3.10; change reference; delete "direct")**

This section is a waiver of certain damages known as consequential damages.[424] This type of damage is different than direct damages, although both must be foreseeable to be recoverable.[425] Direct damages flow directly from the contract, or the breach of that contract. For instance, if the contractor fails to complete the project, the owner can recover the cost to complete as direct damages.[426] If either the owner or the contractor is damaged, they waive any recovery for the damages specified in ¶¶ 15.1.6.1 and 15.1.6.2. Note that this list is not exclusive, so that if other damages are incurred that are deemed by a court to be consequential damages,[427] those would also be waived. This is a way of limiting damages, and courts

---

[424] *Black's Law Dictionary* (8th Ed., 2004) defines *consequential damages* as "Losses that do not flow directly and immediately from an injurious act but that result indirectly from the act."

[425] Wright Schuchart, Inc., v. Cooper Indus., 40 F.3d 1247 (9th Cir. 1994) (and, consequential damages are always necessary to make a buyer whole).

[426] Northwestern Eng'g v. Ellerman, 23 N.W.2d 273 (S.D. 1946). *See, Restatement (Second), Contracts* § 347(a) (1981), which lists direct damages as "the loss that may fairly and reasonably be considered as arising naturally."

[427] Section 2-715 of the Uniform Commercial Code defines *consequential damages* as:

(a) any loss resulting from general or particular requirements and needs of which the seller at the time of contracting had reason to know and which could not reasonably be prevented by cover or otherwise; and

(b) injury to person or property proximately resulting from any breach of warranty.

   Thus, if there is an injury to the owner's property caused by the breach of any warranty by the contractor, this provision may waive recovery for such damages.

   The court in Arthur Andersen & Co., v. Perry Equipment Corp., 945 S.W.2d 812 (Tex., 1997), compared direct damages versus consequential damages:

Actual damages are those damages recoverable under common law . . . At common law, actual damages are either "direct" or "consequential" . . . Direct damages are the necessary and usual result of the defendant's wrongful act; they flow naturally and necessarily from the wrong . . . Direct damages compensate the plaintiff for the loss that is conclusively presumed to have been foreseen by the defendant from his wrongful act . . . Consequential damages, on the other hand, result naturally, but not necessarily, from the defendant's wrongful acts . . . Under common law, consequential damages need not be the usual result of the wrong, but must be foreseeable . . . and must be directly traceable to the wrongful act and result from it.

   In Pacific Coast Title Ins. Co., v. Hartford Accident & Indem. Co., 7 Utah 2d 377; 325 P.2d 906 (Utah 1958), the Utah Supreme Court discussed whether attorneys fees expended in an action against a third party were recoverable as consequential damages of an original breach. The contractor had failed to pay its subcontractors, laborers, and material suppliers. As a result,

have generally upheld such provisions.[428] The provision that the waiver is applicable to termination is not a limitation, because "arising out of or relating to" is considered to be very broadly inclusive. See the discussion as to broad form arbitration clauses at ¶ 15.4.1. If the contract documents contain a liquidated damages clause, ¶ 15.1.6 does not waive them. One question that may arise is to what extent delay damages are waived by this provision. That question is not addressed by ¶ 15.1.6, so the parties may consider adding a clause to cover this issue.

The impetus for this waiver of consequential damages was the *Perini* case.[429] In that case, Perini was the construction manager for the renovation of a casino in Atlantic City, New Jersey. Perini's duties were to coordinate between the owner and architect, supervise the trade contractors, and establish a guaranteed maximum price for the project. Perini's fee was to be $600,000. Construction was to be completed by May 31, 1984, but certain parts of the work were not completed until September 1, 1984. The casino area and food court were timely completed, but the new entrance and luxury hotel suites were delayed. After the owner terminated Perini for cause, Perini filed for arbitration. The owner counterclaimed for its lost profits. The arbitration award of some $14.5 million in favor of the owner survived appeal. These types of lost profits are consequential damages that the revised AIA language is intended to eliminate.

Owners might try to get around this limitation by drafting a liquidated damages clause that actually includes anticipated consequential damages.[430] Liquidated

---

they filed liens against the homes under construction. Pacific, the title insurer, defended against the liens and then filed suit against the contractor's performance bond to recover the attorneys fees and costs incurred. The court acknowledged that, generally, attorneys fees are not recoverable absent statutory or contractual authorization, but nevertheless awarded them to Pacific as consequential damages. The issue of foreseeability was the prime determinant, with the court finding that it was foreseeable that the contractor's failure to pay would bring about the series of events that occurred, including the filing of liens and Pacific's retention of attorneys to defend.

In EBWS, LLC v. Britly Corp., 928 A.2d 497 (Vt. 2007), the court compared direct damages with consequential damages. Direct damages are for "losses that naturally and usually flow from the breach itself, and it is not necessary that the parties actually considered these damages." "In comparison, special or consequential damages must pass the tests of causation, certainty and foreseeability, and, in addition, be reasonably supposed to have been in the contemplation of both parties at the time they made the contract."

Consequential damages can include water damage to a rug resulting from a leak (Hartzell v. Justus Co., 693 F.2d 770 (8th Cir., 1992)); idle time, interest or finance charges, loss of use of goods, overhead, and labor and equipment rental (Wright Schuchart, Inc., v. Cooper Indus., 40 F.3d 1247 (9th Cir. 1994)).

[428] *See, e.g.,* Adams Lab., Inc. v. Jacobs Eng'g Co., 486 F. Supp. 383 (N.D. Ill. 1980). Note that parties sometimes attempt to circumvent this limitation by asking for recovery in tort. In Lincoln Pulp & Paper Co. v. Dravo Corp., 436 F. Supp. 262 (D. Me. 1977), the owner was permitted tort recovery against the engineer despite such language.

[429] Perini Corp. v. Greate Bay Hotel & Casino, Inc., 129 N.J. 479, 610 A.2d 364 (1992).

[430] In City of Milford v. Coppola Constr., 2004 WL 3090680 (Conn. Super. Dec. 1, 2004), the city hired the contractor to lift several houses. The contractor brought specialized equipment to the site to perform the work, but the city needed to correct the plans to obtain the required permits, resulting in a delay of seven months during which the equipment lay idle. The contractor's claims

damages are a fixed sum that is set out in the contract[431] and replaces direct damages.[432] The contractor, of course, would have no corresponding provision allowing for its consequential damages in the guise of liquidated damages. Owners of projects with a high potential of lost income in the event of a project delay will want to delete this entire waiver, with the contractors then taking on an extremely high-risk project. If the owner does eliminate this waiver, ¶ 3.18.1 should be modified to include indemnification for loss of use, which was removed in the 1997 language because of this waiver, as well as the language of ¶ 11.3.3, relating to loss of use insurance.

Note that the provision found in ¶ 14.4.3, allowing "reasonable overhead and profit on the Work not executed," appears to be in some conflict with the waiver of consequential damages provision found here, even though this waiver provision states that it applies to Article 14. Paragraph 15.1.6.2 permits the contractor to recover "anticipated profit arising directly from the Work" as an exception to the waiver. Thus, the contractor's anticipated profit is permitted by both provisions. However, ¶ 15.1.6.2 implies that the contractor's overhead expenses are waived, while ¶ 14.4.3 permits recovery of overhead for the uncompleted work. These apparent conflicts could be harmonized by allowing the contractor to recover its anticipated overhead costs for the uncompleted work as an exception to the waiver of consequential damages, but only in the situation where the owner terminates for convenience.

---

were arbitrated, with the arbitrator awarding the contractor damages. The city then brought an action to modify and correct the award on the ground that idle equipment and materials were not matters submitted to the arbitrator. The court examined the AIA contract, including the waiver of consequential damages provision, and held that the arbitrator's award should stand. These provisions "indicate that not all types of damages are amendable to arbitration." Since the parties had waived consequential damages, the arbitrator could not award such damages. The issue for the court was "whether the idle equipment and materials were consequential damages, and therefore not arbitrable, or whether they were other than consequential damages, such as liquidated direct damages, and therefore properly arbitrable." The last AIA provision [from the 1997 version of this provision] that "*[n]othing contained in this Subparagraph 4.3.10 shall be deemed to preclude an award of liquidated direct damages. . . .* " means that such damages could be awarded by the arbitrator. "In order to qualify as consequential damages, therefore, the idle equipment and materials would have to have been special, unusual, indirect and not ordinarily foreseeable to the city." The court found that the loss from this contractor's idle equipment in this case arose in the usual course of the performance of this contract, was foreseeable, and was the direct result of the city's breach.

[431] United States v. United Eng'g & Constr. Co., 234 U.S. 236 (1914) (this is "a fixed sum as liquidated damages, not wholly disproportionate to the loss for each day's delay . . .").

[432] Wise v. U.S., 249 U.S. 361 (1919) ("It is obvious that the extent of the loss which would result to the government from delay in performance must be uncertain and difficult to determine and it is clear that the amount stipulated for is not excessive, having regard, to the amount of money which the government would have invested in the buildings at the time when such delay would occur, to the expense of securing or continuing in other buildings during such delay, and to the confusion which must necessarily result").

§ 4.86   Initial Decision: ¶ 15.2

15.2 INITIAL DECISION

15.2.1 Claims, excluding those arising under Sections 10.3, 10.4, 11.3.9, and 11.3.10, shall be referred to the Initial Decision Maker for initial decision. The Architect will serve as the Initial Decision Maker, unless otherwise indicated in the Agreement. Except for those Claims excluded by this Section 15.2.1, an initial decision shall be required as a condition precedent to mediation of any Claim arising prior to the date final payment is due, unless 30 days have passed after the Claim has been referred to the Initial Decision Maker with no decision having been rendered. Unless the Initial Decision Maker and all affected parties agree, the Initial Decision Maker will not decide disputes between the Contractor and persons or entities other than the Owner. (**4.4.1; substantial modifications**)

This provides for a person to serve as an Initial Decision Maker. Under prior versions of A201, the architect served this purpose. In this version, the architect still serves this purpose unless the owner and contractor specify another person to fulfill this function.

If there is a claim by either the owner or contractor, a decision by this person will be a condition precedent to any mediation.[433] In other words, without a decision by

---

[433] In Beaver Construction Co. v. Lakehouse, 742 So. 2d 159 (Ala. 1999), the owner argued that the word "litigation" in the 1997 version of this provision allows a party to opt for litigation and avoid an arbitration demand by the other party. The court rejected this argument and ordered arbitration.

The architect's decision was a condition precedent to arbitration in State Street Bridgeport, LP v. HRH/Atlas Construction, Inc., 2002 WL 377542 (Conn. Super. Jan. 25, 2002). However, in Zandri Constr. Corp. v. Wolfe, 737 N.Y.S.2d 400, 291 A.D.2d 625 (2002), the failure of the contractor to establish the date when final payment was due precluded a claim that the owner was barred from suing the contractor for failing to comply with this provision that all claims arising prior to the date final payment was due must be presented to the architect. In Manalili v. Commercial Mowing and Grading, 442 So. 2d 411 (Fla. App. 2d Dist. 1983), the owners were entitled to submit their dispute to arbitration despite the fact that they failed to first present the dispute to an architect where no architect was involved in the project. Instead, an engineer was presented with the dispute and the engineer found that the contractor had not satisfactorily performed. *See also* Auchter Co. v. Zagloul, 949 So. 2d 1189 (Fla. App. 1st Dist. 2007).

In Anagnostopoulos v. Union Turnpike Mgmt. Corp., 300 A.D.2d 393, 751 N.Y.S.2d 762 (2002), the contractor agreed to perform demolition and excavation work and construct a new house for the owner. The owner terminated the contractor during construction due to the contractor's failure to comply with the plans and specifications. The contractor filed a demand for arbitration and the owners moved to stay the arbitration because the contractor had not fulfilled this provision's condition precedent of submitting its claim to the architect. The court agreed and stayed the arbitration without prejudice until ripe, pursuant to Section 14.2 of A201. The owners then hired a substitute contractor to complete the house. After the house was completed, the contractor tried to revive its former claim, but the AAA refused due to the prior stay. Once again, the contractor submitted the matter to the court and the owner objected based on the contractor's failure to timely submit the issue to the architect. The trial court granted the motion to compel arbitration on the grounds of equity. The appellate court reversed:

A court has the jurisdiction to determine whether contractual conditions precedent to arbitration have been fulfilled. . . . In this case, the Supreme Court should have denied the motion to compel arbitration based upon the respondent's failure to fulfill in a timely fashion the contractually mandated condition precedent. . . . Consensual arbitration is a creature of contract. . . . The Supreme Court had no authority to substitute its notions of equity for what the parties have agreed to in their contract.

In Fe-Ri Constr., Inc. v. Intelligroup, Inc., 218 F. Supp. 2d 168 (D.P.R. 2002), the parties did not contest that there was a debt. The contractor sued to collect its fee. The owner moved to dismiss on the basis that the conditions precedent, namely submission to the architect followed by mediation and/or arbitration, were not met. Thus, the case was dismissed without prejudice. The argument that this was not a "claim" but a breach of contract was rejected.

In Cameo Homes, Inc. v. Kraus-Anderson Constr. Co., 2003 WL 22867640 (D. Minn. Dec. 3, 2003), the court held that summary judgment in favor of the owner was appropriate where the contractor failed to provide the architect with written notice of its claims, which was a condition precedent to the litigation.

Whether the condition precedent of submitting the matter to the architect was met is for the arbitrator to determine. *See* dissent in Peterson Constr., Inc. v. Sungate Dev., L.L.C., 2003 WL 22480613 (Tex. App.-Corpus Christi, Oct. 30, 2003).

In Tekmen & Co., v. Southern Builders, Inc., 2005 WL 1249035 (Del. Super. May 25, 2005), the owner contracted to have a hotel built. Following substantial completion, the owner began to observe leaks and other defects. The contractor made numerous trips back to make repairs. Finally, after making final payment, the owner brought a court action against the contractor. The contractor moved to dismiss for lack of jurisdiction, citing the arbitration clause. The owner never submitted its claim to the architect. The court held that the AIA language, "prior to the date final payment is due," can be reasonably interpreted to mean that a party seeking to arbitrate must first submit the claim to the architect only with regard to claims arising before final payment is due. However, the duty to submit the claim to arbitration did not end upon final payment. "Where submission of the claims to the architect is a condition precedent to arbitration, a party's duty to arbitrate is not extinguished merely because the condition precedent is finite in time." The court held that arbitration was required, and the arbitrator was to determine whether submission to the arbitration was a required condition precedent.

However, in Aberdeen Golf & Country Club v. Bliss Constr., Inc., 932 So. 2d 235 (Fla. App. 4th Dist. 2005), the appellate court affirmed a trial court's finding of waiver of the arbitration provision. When mold was discovered, the contractor gave notice to the architect and made a claim for extras. The architect confirmed the presence of mold and agreed that it had caused delays that could affect the price. Rather than initiate mediation, the owner refused to pay the contractor's draw request and terminated the contract. Several weeks later, the owner wrote to the contractor that it wanted arbitration, but apparently never initiated arbitration with the AAA. The court analyzed the ADR provisions in the AIA agreement, including this paragraph, finding that arbitration was not set forth as a precondition to litigation and that the ADR system was meant to function in place of the courts while progress was being made on the contract. "If either party terminated, all bets would be off and either could have its day in court." The ADR provision failed of its essential purpose. *Aberdeen* was cited in Auchter Co. v. Zagloul, 949 So. 2d 1189 (Fla. App. 1st Dist. 2007), where the court disagreed with that interpretation of the AIA language. Auchter stated that the arbitration provision of the AIA documents survives termination of the contract so long as the dispute concerns matters arising under the contract.

In Commonwealth Constr. Co. v. Cornerstone Fellowship Baptist Church, Inc., 2006 WL 2567916 (De. Super. Aug. 31, 2006), the owner's failure to submit its claims to the architect resulted in a waiver of the claims.

In Lopez v. 14th St. Dev., LLC, 835 N.Y.S.2d 186 (App. Div. 1st Dep't 2007), interpreting the 1997 version of A201, the owner brought an action for indemnification against the contractor after being sued by an injured employee of the contractor. The contractor sought to stay the action pending arbitration. The court reversed, finding that the indemnity claim arose long after work

the Initial Decision Maker, no party can proceed to the next step of the dispute resolution process, which is mediation, nor to the final step of litigation or arbitration. The exception to this is, if a claim has been submitted to the Initial Decision Maker and there is no decision rendered, after 30 calendar days, either party to the claim may go to the next step, namely, mediation. If a claim is not referred to the Initial Decision Maker, it cannot proceed unless both the owner and contractor waive this provision.

A Maryland court held that the earlier version of this paragraph (AIA Document A201, ¶ 2.2.9 (1976)) did not represent an agreement to arbitrate when the contract lacked the normal arbitration clause.[434] The parties can alter this agreement to make the architect's (or Initial Decision Maker's) decision final and binding on the parties.[435] The owner may also want to change the time periods, particularly on smaller projects.

---

was completed and final payment was made. Finding that the AIA document was "not a model of clarity," the court reviewed the provisions that require the architect to review claims as a condition precedent to arbitration. There was, however, no further provision in the document subjecting claims that arise after completion of the work to arbitration.

[434] Joseph F. Trionfo & Sons v. Ernest B. LaRosa, 38 Md. App. 598, 381 A.2d 727 (1978). This provision was an administrative provision and did not evidence an intent to arbitrate. The contract language was also read as not intending that a dispute over payment be submitted to the architect after the work has been completed.

If the arbitration provision of ¶ 4.5.1 is deleted, this reference to arbitration would have no effect, and the parties' only recourse would be litigation. Glenn H. Johnson Constr. Co. v. Board of Educ., 245 Ill. App. 3d 18, 614 N.E.2d 208 (1993).

[435] John W. Johnson, Inc. v. J.A. Jones Constr. Co., 369 F. Supp. 484 (E.D. Va. 1973). However, in H.R.H. Prince Ltc. Faisal M. Saud v. Batson-Cook Co., 161 Ga. App. 219, 291 S.E.2d 249 (1982), the court held that the architect was not an arbitrator (which may have been unenforceable under Georgia law). The architect is the owner's representative during the construction, and submittal of claims to the architect is merely a preliminary stage to arbitration.

In Regional Sch. Dist. No. 9 v. Wayne Constr. Co., 2000 Conn. Super. LEXIS 1285 (May 18, 2000), using the 1987 version of A201, this paragraph was amended to read as follows:

All claims, requests for extras, extensions of time, or interpretations of the Contract drawings shall be submitted to the Architect in writing. The Architects' decision and interpretation on such matters shall be binding and final, and shall be accepted by the Contractor and Owner in all cases. The Architects' decision in such instances shall not be subject to any subsequent review in any court or legal proceeding. In the event there is a conflict of this provision and any other provision of the Contract as to the effect of the Architect's decision, this provision shall prevail.

The architect refused to certify that the contractor had achieved substantial completion and the contractor's payment request was unilaterally adjusted. The contractor was then declared in default. The contractor demanded arbitration which was objected to by the owner based on this amended language which purportedly gave the architect exclusive and final authority over the contractor's claims. The court held that the agreement, when viewed as a whole, shows an unambiguous agreement by the parties to an unrestricted submission to arbitration. The quoted language merely means that a decision by the architect on technical matters, such as claims or requests for extras, extensions of time, or interpretations of the contract drawings, is final and binding on the parties. It does not mean that the architect's decisions are binding on the arbitrators

If the claim relates to hazardous materials, emergencies, or insurance claims, it need not be referred initially to the Initial Decision Maker or architect. If the parties cannot agree on a claim related to any of these matters, including hazardous materials claims, they must proceed immediately to mediation and, thereafter, to litigation or arbitration. Note that this person's role as initial arbitrator is limited to disputes between the owner and contractor.

15.2.2 The Initial Decision Maker will review Claims and within ten days of the receipt of a Claim take one or more of the following actions: (1) request additional supporting data from the claimant or a response with supporting data from the other party, (2) reject the Claim in whole or in part, (3) approve the Claim, (4) suggest a compromise, or (5) advise the parties that the Initial Decision Maker is unable to resolve the Claim if the Initial Decision Maker lacks sufficient information to evaluate the merits of the Claim or if the Initial Decision Maker concludes that, in the Initial Decision Maker's sole discretion, it would be inappropriate for the Initial Decision Maker to resolve the Claim. (**4.4.2; substantial revisions**)

If none of the five conditions stated is present, the parties cannot litigate or take a claim to arbitration absent a decision of the Initial Decision Maker.[436] If, however, more than 30 days pass without a decision, the parties can proceed to mediation.

This time limit starts upon receipt of the written claim. A request for additional data extends the time until the additional data are received. If the claim is for any substantial amount, the surety (if any) should be notified (assuming the claim is against the contractor). The Initial Decision Maker also has the option of not deciding a claim if it would be "inappropriate." This might occur if the architect

---

as well or that the architect has the final word on any claim or controversy arising out of or related to the contract. Otherwise, there would be no need for arbitrators to render a full and final settlement of any claim or controversy arising out of or related to the contract, as provided for in the arbitration clause of A201 (which had not been modified).

In J.R.J. Constr., Inc. v. D.A. Sullivan & Sons, Inc., 2006 WL 2848567 (Mass. Super. Aug. 30, 2006), the architect exceeded his authority by issuing decisions that constituted a "rewriting of the contract."

[436] Board of Educ. v. Hatzel & Buehler, Inc., 156 A.D.2d 684, 549 N.Y.S.2d 447 (1989). One court found that, under the 1970 version, no specific penalty or sanction is provided for failing to present such a claim to the architect under this provision. Stauffer Constr. Co. v. Board of Educ., 54 Md. App. 658, 460 A.2d 609 (1983).

In Marsh v. Loffler Hous. Corp., 102 Md. App. 116, 648 A.2d 1081 (1994), the contractor could file for arbitration without first presenting the claim to the architect if the claim related to a mechanic's lien.

In Fru-Con Constr. Co. v. Southwestern Redevelopment Corp., 908 S.W.2d 741 (Mo. Ct. App. 1995), the contract was amended to allow arbitration only for claims that were less than $200,000. The contractor submitted a claim consisting of numerous claims, and the architect determined that only one of these claims exceeded the limit. The contractor filed suit, and the owner sought to arbitrate, asking the court to stay the lawsuit pending the arbitration. The trial court declined to stay the proceeding. The appellate court reversed, saying that the architect had the power to make the initial determination under the contract. Because the architect had determined that all the claims except one were subject to arbitration, the court proceeding should have been stayed.

is the Initial Decision Maker and a substantial error in the drawings created the claim. If a decision is made, there must be some specificity.[437]

One question is whether the five actions in ¶ 15.2.2 are all considered "decisions." Clearly, 2, 3, and 5 are decisions. However, the first and fourth items may or may not be considered decisions that permit the next step in the process to proceed. However, ¶ 15.2.4 has additional time limits regarding the first item.

> *15.2.3 In evaluating Claims, the Initial Decision Maker may, but shall not be obligated to, consult with or seek information from either party or from persons with special knowledge or expertise who may assist the Initial Decision Maker in rendering a decision. The Initial Decision Maker may request the Owner to authorize retention of such persons at the Owner's expense.* **(4.4.3; change architect to Initial Decision Maker)**

This provision allows the Initial Decision Maker to consult with experts on a particular topic. Normally, arbitrators should not obtain evidence or interview persons outside an arbitration. The Initial Decision Maker, although an arbitrator with respect to claims, has broader powers to discuss the claim with other persons by virtue of this agreed-upon provision.

> *15.2.4 If the Initial Decision Maker requests a party to provide a response to a Claim or to furnish additional supporting data, such party shall respond, within ten days after receipt of such request, and shall either (1) provide a response on the requested supporting data, (2) advise the Initial Decision Maker when the response or supporting data will be furnished or (3) advise the Initial Decision Maker that no supporting data will be furnished. Upon receipt of the response or supporting data, if any, the Initial Decision Maker will either reject or approve the Claim in whole or in part.* **(4.4.4; change architect to Initial Decision Maker)**

The Initial Decision Maker can ask the other party to the claim to provide a response to the claim. Alternatively, or in addition to that request, the party making the claim can be asked for more information to support the claim. The claimant who receives this request for more information then can make one of three responses: (1) provide the additional data or other information that was requested; (2) advise the Initial Decision Maker that it will take some time to assemble such information or data, with a target date for that response; or (3) refuse to provide any further data or information. The Initial Decision Maker then has the option of rejecting or approving the claim in its entirety or in part. If the answer is a request for additional time under (2), the Initial Decision Maker can either make a decision or wait until the requested information is provided before making a decision.

---

[437] In General Ins. Co. v. K. Capolino Constr. Corp., 983 F. Supp. 403 (S.D.N.Y. Oct. 7, 1997), the court stated that the architect should have specified the work that the contractor was required to perform under the architect's interpretation of the contract. The architect had informed the contractor that he was to "perform all aspects of your contract in accordance with the specifications, not in accordance with your desires." This was not sufficient.

*15.2.5 The Initial Decision Maker will render an initial decision approving or reject-
ing the Claim, or indicating that the Initial Decision Maker is unable to resolve the
Claim. This initial decision shall (1) be in writing; (2) state the reasons therefor; and
(3) notify the parties and the Architect, if the Architect is not serving as the Initial
Decision Maker, of any change in the Contract Sum or Contract Time or both. The
initial decision shall be final and binding on the parties but subject to mediation and,
if the parties fail to resolve their dispute through mediation, to binding dispute
resolution.* (**4.4.5; substantial modifications**)

The Initial Decision Maker must render the decision in writing. The decision
must be explained and must specify if the contract sum, contract time, or both are to
be modified. Unless one of the parties then submits the matter to mediation and, if
mediation is unsuccessful, to either arbitration or litigation, this decision becomes
final and becomes a change order. Under ¶ 15.2.6.1, the party in whose favor this
decision is rendered may issue, within 30 days of the date of the decision, a demand
that the other party invoke mediation within 60 days of the decision or forfeit the
right to challenge the decision.[438] Then, if mediation is initiated and mediation is
not successful, either arbitration or litigation will follow. This process is a way to
finalize these decisions and not leave them out there until the end of the project. It
will likely become standard practice for the winning party in these decisions to

---

[438] In Ingrassia Constr. Co., Inc. v. Vernon Township Bd. of Educ., 784 A.2d 73 (N.J. Super. 2001),
the owner terminated the contractor. Pursuant to ¶ 4.4.6 of A201-1997, the architect certified that
sufficient cause existed to terminate. However, the architect was licensed in Canada, but not
anywhere in the United States. That made the certification invalid. The court then examined the
consequences that flow from a termination based on a defective architect's certificate. It held that
an invalid certificate deprives the owner of any finality, presumption of correctness, or obligation
of judicial deference that would otherwise attach to a proper architect's certification. This then
left the parties to their common-law causes of action for breach of contract. In this case, under an
earlier version of A201, the arbitration provisions were stricken from the contract. This meant that
a proper architect's certificate would not be subject to arbitration, but would be "final and binding
on the parties." The court stated that "it appears that if the architect's certificate is given in good
faith, without gross error, and complies with the procedural and substantive requirements of the
contract, it is conclusive with respect to the facts stated therein." The contractor argued that the
architect's certification is a condition precedent to the owner's right to terminate the contractor.
The court disagreed with this assertion. The absence of this certificate may be a condition
precedent to the owner's right to terminate under the contract, but not "to the owner's right to
exercise its common-law right of termination subject to the normal and traditional burden of proof
of material breach from which it would be largely exempted by a proper certificate."
    In Beers Constr. Co. v. Pikeville United Methodist Hosp., 2005 Fed. Appx. 0325N, 2005 WL
977264 (6th Cir. Apr. 28, 2005), the parties changed a provision in the agreement to read that the
architect's decision "shall be final and binding on the parties but subject to legal proceedings. . . ." The owner argued that the architect's decision was final and binding. The court,
however, held that the contract contemplated that legal proceedings could be undertaken by
any party not satisfied by the architect's determination.
    In J. Caldarera & Co, Inc. v. State of Louisiana, 2006 WL 3813721 (La. App. 1st Cir. Dec. 28,
2006), the contract between the owner and contractor deleted the arbitration section. However, the
agreement also provided that claims are to be initially reviewed by the architect, and the architect's final decision is "subject to arbitration," possibly pursuant to this provision of A201.
Because of this, the court found that the contract requires arbitration.

issue a 30-day letter. Failure to do so will result in the claim remaining open and a party aggregating a number of these claims at the end of the project for mediation and then litigation or arbitration.

> *15.2.6 Either party may file for mediation of an initial decision at any time, subject to the terms of Section 15.2.6.1.* (**new**)

> *15.2.6.1 Either party may, within 30 days from the date of an initial decision, demand in writing that the other party file for mediation within 60 days of the initial decision. If such a demand is made and the party receiving the demand fails to file for mediation within the time required, then both parties waive their rights to mediate or pursue binding dispute resolution proceedings with respect to the initial decision.* (**new**)

These are new provisions for 2007. Paragraph 15.2.6 states that either party has an unlimited amount of time, subject only to that jurisdiction's statute of limitations, within which to demand mediation[439] and thereafter pursue the dispute resolution procedures of the agreement. However, if either party files a written demand with the other side in relation to a decision made by the Initial Decision Maker, and that demand is made within 30 calendar days from the date of that decision (not the date it was received, but the date on that decision), then the other side must file for mediation within 60 days of the initial decision. If the other side fails to timely file for mediation, then the initial decision becomes final and not subject to appeal.

Note that this provision is similar to ¶ 4.4.6 of AIA A201-1997.[440] Under that prior version, the architect could make a decision on a matter and issue the decision in writing, along with specific language that the decision was final but subject to mediation and arbitration and that, unless a demand for arbitration was filed within 30 days of that decision, it would become final.

Under the current language, the specific language notifying the parties of their options is not required. If the dissatisfied party fails to initiate mediation within the stated 60 days, that decision becomes automatically final. This is likely to be a trap for parties who are not in the habit of reading their agreements.

> *15.2.7 In the event of a Claim against the Contractor, the Owner may, but is not obligated to, notify the surety, if any, of the nature and amount of the Claim. If the*

---

[439] If a party waits until just before the applicable statute of limitations runs to file a demand for mediation, it is quite likely that, if the mediation is unsuccessful, the arbitration or litigation will be filed after the statute of limitations expires. Therefore, parties would be advised to file for mediation several months prior to such time to avoid this problem.

[440] **4.4.6** When a written decision of the Architect states that (1) the decision is final but subject to mediation and arbitration and (2) a demand for arbitration of a Claim covered by such decision must be made within 30 days after the date on which the party making the demand receives the final written decision, then failure to demand arbitration within said 30 days' period shall result in the Architect's decision becoming final and binding upon the Owner and Contractor. If the Architect renders a decision after arbitration proceedings have been initiated, such decision may be entered as evidence, but shall not supersede arbitration proceedings unless the decision is acceptable to all parties concerned.

Claim relates to a possibility of a Contractor's default, the Owner may, but is not obligated to, notify the surety and request the surety's assistance in resolving the controversy. **(4.4.7; change architect to owner)**

Normally, the owner would want to notify the surety if there is any significant claim against the contractor. This provision assures that failing to do so will not invalidate the surety agreement. This provision does not require written notice to the surety.[441]

15.2.8 If a Claim relates to or is the subject of a mechanic's lien, the party asserting such Claim may proceed in accordance with applicable law to comply with the lien notice or filing deadlines. **(4.4.8; delete last clause of prior paragraph)**

This provision clarifies what has been the practice regarding mechanics' lien claims. The party asserting the lien is permitted to record or otherwise perfect its mechanics' lien. It cannot, however, proceed to litigation, except possibly to file a suit and then have that suit stayed pending the outcome of the mediation.[442] In some states, and under some circumstances, failure to timely file a lawsuit could result in the loss of lien rights. In that situation, the filing of a lawsuit and a subsequent stay of that suit would comply with this provision. Rather, the dispute resolution procedure must be followed. If the party were not permitted to do this, the strict laws regulating these liens might invalidate the lien unless the lien is promptly perfected.

§ 4.87   Mediation: ¶ 15.3

15.3 MEDIATION

15.3.1 Claims, disputes, or other matters in controversy arising out of or related to the Contract, except those waived as provided for in Sections 9.10.4, 9.10.5, and 15.1.6 shall, be subject to mediation as a condition precedent to binding dispute resolution. **(4.5.1; modifications)**

Mediation is recognized as an extremely effective dispute resolution process. Although essentially a voluntary procedure, mediation is now a precondition to arbitration or litigation. If one party refuses to participate in the mediation, the other party could presumably ask the arbitrator or court for a summary finding in its favor based on a breach of this provision. Note that claims that are waived,

---

[441] RLI Ins. Co. v. St. Patrick's Home for the Infirm and Aged, 452 F. Supp. 2d 484 (S.D. N.Y. 2006).

[442] In Dominion Consulting & Mgmt., Inc. v. Davis, 2004 WL 288545 (Va. Cir. Ct. Jan. 13, 2004), this provision was initially held to apply so as to permit a party to proceed with its mechanics' lien claim. Upon reconsideration, however, it was shown that the contract actually incorporated the 1987 version of A201, which did not contain this exception for mechanics' lien claims. Thus, arbitration on all issues was permitted to proceed.

such as consequential damages, are not subject to mediation, arbitration, or litigation.

> *15.3.2 The parties shall endeavor to resolve their Claims by mediation which, unless the parties mutually agree otherwise, shall be administered by the American Arbitration Association in accordance with its Construction Industry Mediation Procedures in effect on the date of the Agreement. A request for mediation shall be made in writing, delivered to the other party to the Contract, and filed with the person or entity administering the mediation. The request may be made concurrently with the filing of binding dispute resolution proceedings but, in such event, mediation shall proceed in advance of binding dispute resolution proceedings, which shall be stayed pending mediation for a period of 60 days from the date of filing, unless stayed for a longer period by agreement of the parties or court order. If an arbitration is stayed pursuant to this Section 15.3.2, the parties may nonetheless proceed to the selection of the arbitrator(s) and agree upon a schedule for later proceedings.* (**4.5.2; modifications**)

Mediation must be held prior to arbitration or litigation. According to one court,[443] the 60-day provision means that a demand for mediation stays any legal or equitable proceedings for 60 days. In that case, when one party filed a request for mediation, the other party took the position that mediation was not a reasonable alternative at the time. When litigation was initiated, there was no request for a stay of the legal action pending mediation. This constituted a waiver of the mediation requirement.

> *15.3.3 The parties shall share the mediator's fee and any filing fees equally. The mediation shall be held in the place where the Project is located, unless another location is mutually agreed upon. Agreements reached in mediation shall be enforceable as settlement agreements in any court having jurisdiction thereof.* (**4.5.3; no change**)

If the parties reach a settlement (the goal of mediation), that settlement may be enforced in a court without the need for arbitration or litigation, just like any other settlement agreement.

§ 4.88 Arbitration: ¶ 15.4

See the discussion about arbitration under ¶ 8.3 of AIA Document B101 in **Chapter 2.** All the AIA documents provide for arbitration through the American Arbitration Association, although beginning with the 2007 version of the documents, arbitration must be opted-in—it is not the default.[444] As an alternative, the owner and architect may consider giving the architect full and final authority to arbitrate.

---

[443] LBL Skysystems (USA), Inc. v. APG-America, Inc., 2005 WL 2140240 (E.D. Pa. Aug. 31, 2005).

[444] However, the electronic version of the AIA documents published in November, 2007, automatically checked "arbitration," effectively making that the default for users of the electronic documents.

A New York case included the following language in the general conditions that gave the architect such authority:

> To resolve all disputes and to prevent litigation the parties to this Contract authorize the Architect to decide all questions of any nature whatsoever arising out of, under or in connection with, or in any way related to or on account of, this Contract (including claims in the nature of breach of Contract or fraud or misrepresentation before or subsequent to acceptance of the Contractor's Proposal and claims of a type which are barred by the provisions of the Contract), and his decisions shall be conclusive, final and binding on the parties. His decision may be based on such assistance as he may find desirable, including advice of engineering specialists. The effect of his decision shall not be impaired or waived by any negotiations or settlement offers in connection with the question decided, whether or not he participated therein himself, or by any prior decision of others, which prior decisions shall be deemed subject to review, or by any termination or cancellation of this Contract.[445]

An interesting Missouri case involved a hospital project in which the owner had "prequalified" several bidders.[446] The contract was awarded to the second lowest bidder, and the low bidder demanded arbitration. The owner sought a court order staying the arbitration. The contractor's argument that, by custom and practice, a prequalified bidder must be awarded a contract if it submits the lowest bid, was rejected. The court held that this contractor was not entitled to arbitrate the matter because it was not the "Contractor" as defined in the contract documents.

15.4.1 *If the parties have selected arbitration as the method for binding dispute resolution in the Agreement, any Claim subject to, but not resolved by, mediation shall be subject to arbitration, which, unless the parties mutually agree otherwise, shall be administered by the American Arbitration Association in accordance with its Construction Industry Arbitration Rules in effect on the date of the Agreement. A demand for arbitration shall be made in writing, delivered to the other party to the Contract, and filed with the person or entity administering the arbitration. The party filing a notice of demand for arbitration must assert in the demand all Claims then known to that party on which arbitration is permitted to be demanded.* **(4.6.1; 4.6.5—last sentence; substantial modifications)**

---

[445] Maross Constr. v. Central Reg'l Transp., 66 N.Y.2d 341, 488 N.E.2d 67, 69, 497 N.Y.S.2d 321 (1985).

In Schaper & Assocs. v. Soleimanzadeh, 87 Md. App. 555, 590 A.2d 583 (1991), the architect was the arbiter under this clause: "It is understood by all that in the event of disputes between the Owner and the Contractor all disputes will be decided by Arif Hodzic as an arbitrator. . . . Arbitrator's decision will be final. Owner or contractor will not be allowed to take this decision to any higher court." The court upheld the award on technical grounds, even though the contractor received no notice of the arbitration until the award was given. The trial court said of the "half baked arbitration" that "they contracted for that and therefore they were bound by whatever rules were applicable to it[,] as dumb as they might be." Before drafting a new provision, parties should consult a competent attorney.

See also M&L Bldg. Corp. v. Housing Auth., 35 Conn. App. 379, 646 A.2d 244 (1994).

[446] St. Luke's Hosp. v. Midwest Mech. Contractors, Inc., 681 S.W.2d 482 (Mo. Ct. App. 1984).

If mediation is unsuccessful, arbitration may be initiated, so long as the parties have selected arbitration as the dispute resolution mechanism. Otherwise, the parties would initiate litigation following an unsuccessful mediation. The following discussion assumes that the parties have selected arbitration as their dispute resolution procedure. If they have not, this entire Section 15.4 can be ignored and the normal litigation process will apply.

Some parties may want to change "shall be subject" to "may be subject" in the first sentence. This effectively deletes arbitration from the Agreement and would create a conflict if arbitration is selected in the Agreement. If arbitration is eliminated from the Agreement, the parties can always decide later to arbitrate, but it cannot be forced. If a party wants to arbitrate possible future conflicts, this wording should not be changed, and arbitration must be selected in the Agreement.[447]

The parties must attempt to mediate the dispute prior to arbitration. This is a precondition to arbitration. What happens if the party seeking to arbitrate refuses to mediate? Because this is a precondition, presumably any arbitration without prior mediation would be invalid. If the party opposing the arbitration refuses to mediate, the claimant should still try to mediate, possibly obtaining a ruling by the mediator that mediation is not viable under the circumstances or that the other party refused to participate. This can be used to have the litigation dismissed on the basis that a condition precedent to litigation was not satisfied. Note that matters that have been contractually waived, such as consequential damages, are not subject to mediation or arbitration.

---

[447] In Ringwelski v. Pederson, 919 P.2d 957 (Colo. Ct. App. 1996), the contract contained the following arbitration provision:

> Any disagreement arising out of this contract or from the breach thereof shall be submitted to arbitration, and judgment upon the award rendered may be entered in the court of the forum, state or federal, having jurisdiction.
>
> It is mutually agreed that the decision of the arbitrators shall be a condition precedent to any right of legal action that either party may have against the other. The arbitration shall be held under the Standard Form of Arbitration Procedure of the American Institute of Architects or under the Rules of the American Arbitration Association.

After a dispute developed, the parties went through arbitration and an award was made. Thereafter, the losing plaintiffs filed a court action with the same allegations as in the prior arbitration. The defendant moved to confirm the award and dismiss the court action. The trial court refused to do so, finding that the first part of the second paragraph of the arbitration clause indicated an intent that the arbitration be nonbinding. It also found persuasive the use of the word "may" in the first paragraph. The appellate court reversed, finding that such an interpretation would severely limit the applicability of the first sentence's provision for entry of judgment on an award. The court went on to state that it construed this language to mean that the prevailing party has the option to obtain a judgment for the purpose of enforcement or collection if necessary.

In Joseph Francese Inc. v. Enlarged City Sch. Dist. of Troy, 263 A.D.2d 582, 693 N.Y.S.2d 280 (App. Div. 1999), the primary arbitration provision in the owner-architect agreement was stricken, although there remained a reference to arbitration in the claims section of the contract. When a dispute arose, the contractor demanded arbitration. Defendant then moved to stay the arbitration, and the trial court granted the stay. The appellate court confirmed. The contractor then

In a Virginia case, an award was entered in an arbitration over defective design and construction of a waterfront bulkhead.[448] The award required additional work on the bulkhead, but because of another controversy the work was never done. A storm then caused further damage, and the owner filed a second demand for arbitration, asking that the previous arbitration panel be reassembled. Following that arbitration, the contractor moved to vacate the award. The issue was whether the first arbitration was final so as to preclude a second arbitration. The court held that the language of this AIA provision reflects the parties' understanding that the arbitration process would end with the first arbitration award. The court rejected the owner's argument that the failure to comply with the first arbitration award constituted a breach of the construction contract, thus triggering the arbitration provision.

Disputes pertaining to the performance of the construction contract, even if they arise after the warranty has expired, are subject to arbitration.[449] Disputes relating

---

filed the instant action, and the defendant moved for summary judgment on the basis that the action was filed after the one-year statute of limitations had expired, thereby barring the action. Both the trial court and the appellate court agreed. The filing of the arbitration demand did not toll the statute of limitations because there was no good faith basis for the contractor to believe that arbitration was an available forum. The undeleted references to arbitration were rendered meaningless by virtue of the fact that the entire arbitration section had been deleted. The court stated that "clearly, the deletion of section 4.5 'Arbitration'—the procedural heart and soul of the contract—left all other references to arbitration unsupported."

In Thermal C/M Serv., Inc. v. Penn Maid Dairy Prods., 831 A.2d 1189 (Pa. Super. 2003), the Addendum to the contract documents included a provision that removed the architect's role:

[I]t is understood and agreed between the parties hereto that there is no "Architect" who is a participant with respect to the administration of the Contract. To the extent the word "Architect" appears in any provision in the Contract (including the General Conditions), other than with respect to the design of the Project (and the modification thereof) such provision concerning the authority of the Architect is of no force or effect; such provisions are hereby explicitly amended to reserve such authority to Owner, the Contractor, or both the Owner and Contractor, as appropriate and by mutual agreement and resolution between the parties. Accordingly, Article 4.2 of the General Conditions is hereby deleted in its entirety.

The court refused to order arbitration, although apparently for a different reason.

In Cumberland Cas. & Sur. Co. v. Lamar Sch. Dist., 2004 WL 2294409 (Ark. App. Oct. 13, 2004), the supplementary conditions read as follows: "In subparagraph 4.5.1: Change the words 'shall be subject to arbitration upon the written demand of either party' to read: 'Shall be subject to arbitration if both parties agree to arbitrate.' " The general conditions form used was AIA Document A201-1997. Unfortunately, subparagraph 4.5.1 of the 1997 A201 does not contain this language. Apparently, the architect had mistakenly used the 1987 version of A201 for the modifications, thereby making the supplementary conditions ambiguous. The owner presented an affidavit of the architect stating that the architect had intended to modify the contract to eliminate arbitration. This was of no avail, however, and the court ordered arbitration.

[448] Waterfront Marine Constr. Co. v. North End 49ers, 251 Va. 417, 468 S.E.2d 894 (1996).

[449] K.L. House Constr. Co. v. City of Albuquerque, 91 N.M. 492, 576 P.2d 752 (1978). A dispute arose between the owner and contractor regarding a roof installation. After the expiration of the warranty period, the owner sought to have the entire roof replaced and demanded arbitration. The contractor resisted arbitration, but the court held that the broad AIA arbitration provision applied even after the expiration of the warranty period.

to rescission of the contract itself are subject to arbitration when the alleged basis for the rescission does not include any allegation that the arbitration clause itself was secured by fraud.[450]

The arbitration clause is effective, even after termination of the contract.[451] Note that certain claims are waived by the owner if the owner asserts them after final payment is made. Paragraph 9.10.4 lists the exceptions to this rule. One court held that claims relating to aesthetic effect may, at least in part, be subject to arbitration.[452]

In Lost Creek Mun. Util. Dist. v. Travis Indus. Painters, Inc., 827 S.W.2d 103 (Tex. App. 1992), the owner had hired a contractor to paint the interior of a water reservoir. After the contractor had received final payment, problems were discovered with the paint and the owner demanded that the contractor correct its work. The contractor performed the corrective work under protest and later learned that the defects were probably not its fault. The contractor then sought arbitration and the owner sought to enjoin the arbitration on the grounds that the warranty work was not subject to arbitration. The court rejected this argument, stating that any doubts as to the scope of coverage of an arbitration clause must be resolved in favor of arbitration.

[450] In Ronbeck Constr. Co. v. Savanna Club Corp., 592 So. 2d 344 (Fla. 1992), the owner sought rescission against the construction contractor, as well as breach of contract, damages for fraud, conversion, civil theft, and conspiracy. The parties had used the 1970 AIA Document A107 which contained the standard arbitration clause. Because all the claims arose out of that contract, the court ordered arbitration.

[451] In re Costa & Head (Atrium), Ltd. v. Duncan, Inc., 486 So. 2d 1272 (Ala. 1986); Shamokin Area Sch. Auth. v. Farfield Co., 308 Pa Super. 271, 454 A.2d 126 (1982); Chester City Auth. v. Aberthaw Constr. Co., 460 Pa. 342, 333 A.2d 758 (1975).

In Canon Sch. Dist. No. 50 v. W.E.S. Constr. Co., 177 Ariz. 526, 869 P.2d 500 (1994), the owner sought to have the arbitration clause declared invalid on the basis that a state statute provided an exclusive remedy for dispute resolution. The court, after an extensive discussion of statutory construction, held that the arbitration provision applied.

In Judelson v. Christopher O'Connor, Inc., No. 95 CV 0371181, 1995 Conn. Super. LEXIS 1375 (May 2, 1995), a state statute required the contractor to be licensed as a home improvement contractor. Because the contractor was not properly licensed, the arbitration provision was not enforceable by the contractor. In McCarl's, Inc. v. Beaver Falls Mun. Auth., 847 A.2d 180 (Pa. Cmwlth. 2004), when a dispute arose, the parties entered into a letter settlement agreement and work continued. Subsequently, the owner refused to make further payments to the contractor and the contractor filed suit. The owner demanded arbitration. The contractor argued that the letter agreement superseded and was substituted for the original agreement that contained the arbitration clause. The court held for arbitration. The original contract stated that all modifications issued subsequent to the contract were to be considered part of the contract. In order to cancel the arbitration provisions of the original contract, the letter agreement must expressly cancel or otherwise nullify the arbitration provisions.

In Warwick Township Water & Sewer Auth. v. Boucher & James, Inc., 2004 WL 557597 (Pa. Super. Mar. 23, 2004), the trial court refused to compel arbitration, on the ground that the arbitration provision, based in part on a provision similar to ¶ 4.4.1 of A201, applied only to disputes arising during construction. Arbitration also applied to a count for negligence.

In BFN-Greeley, LLC v. Adair Group, Inc., 141 P.3d 937 (Colo. App. 2006), the court held that arbitrators have authority to decide post-construction claims relating to time extensions and change orders.

[452] Norton Heights Fire Dep't v. Vuono-Lione, Inc., 168 Conn. 276, 362 A.2d 982 (1976). The court said that the architect still had the authority to make the decision as to what is "artistic effect."

This "arising out of or relating to" language (see the definition of a "Claim" at ¶ 15.1.1) has been held to be a "broad form" of arbitration agreement as opposed to language such as "arising in connection with."[453] The use of such a broad form assures that virtually any dispute related to the construction project will be subject to arbitration. An owner's argument that the inclusion of this arbitration agreement indicated a waiver by the architect of the right to a mechanics' lien was rejected by the court.[454]

When a contractor filed bankruptcy, the bankruptcy judge had authority to deny arbitration because the rights of other parties who were not subject to the arbitration would be affected.[455]

In May Constr. Co. v. Benton Sch. Dist. No. 8, 320 Ark. 147, 895 S.W.2d 521 (1995), the court denied arbitration on the basis of this provision. The contractor had requested the use of a substitute for certain curing materials to be used as a sealer on the concrete floors. The contractor represented that the proposed substitute met or exceeded the quality levels of the original materials. After the application of the substitute, the finish on the concrete floor began experiencing gross and unsightly scuff marks from student traffic, to the extent that the floors' appearance became totally unacceptable because the sealer was magnifying all traffic in an unsightly manner. Further, the floor became sticky in some areas. The owner sued, and the contractor sought arbitration. The owner objected on the basis that the claim related only to the aesthetic appearance of the floor, and the court agreed. The contractor's argument, that the claim was really one for breach of contract, was rejected.

[453] Village of Westfield v. Loitz Bros. Constr. Co., 165 Ill. App. 3d 338, 519 N.E.2d 37 (1988) (also holding that right to arbitrate could be assigned despite a nonassignability clause in contract). *See also* County of Middlesex v. Gevyn Constr. Corp., 450 F.2d 53 (1st Cir. 1971) (general contractor filed arbitration against municipality for benefit of a subcontractor; appellate court overturned the lower court stay of arbitration based on broad form of arbitration clause); Robert Lamb Hart Planners & Architects v. Evergreen, Ltd., 787 F. Supp. 753 (S.D. Ohio 1992) (*followed Village of Westfield);* Sears, Roebuck & Co. v. Glenwal Co., 325 F. Supp. 86 (S.D.N.Y. 1970) (subsoil conditions, delays, and requests for extra work are arbitrable); Ozdeger v. Altay, 66 Ill. App. 3d 629, 384 N.E.2d 82 (1978); Roosevelt Univ. v. Mayfair Constr. Co., 28 Ill. App. 3d 1045, 331 N.E.2d 1045, 835 (1975); Sisters of St. John v. Geraghty Constr., 67 N.Y.2d 997, 494 N.E.2d 102, 502 N.Y.S.2d 997 (1986) ("disputes relating to extra work allegedly authorized and required for execution and completion of respondent's renovation of the convent arise out of or relate to the Contract Documents and thus fall generally within the arbitration clause").

Crown Oil & Wax Co. v. Glen Constr. Co., 320 Md. 546, 578 A.2d 1184 (1990), involved the attempt by the assignee of the owner to enforce the arbitration provision against the contractor. The court held that this broad arbitration clause required the parties to arbitrate, despite the language in ¶ 13.2.1.

In Bishop Contracting Co. v. Center Bros., 213 Ga. App. 804, 445 S.E.2d 780 (1994), the arbitration clause applied to extra work by a subcontractor. A claim for defamation was subject to arbitration in Metzler Contracting Co., LLC v. Stephens, 2007 WL 1977732 (D. Haw., July 3, 2007).

For cases involving an arbitration provision that states: "all disputes arising in connection with this contract," *see* Harrison F. Blades, Inc. v. Jarman Mem'l Hosp. Bldg. F, 109 Ill. App. 2d 224, 248 N.E.2d 289 (1969); Silver Cross Hosp. v. S.N. Nielsen Co., 8 Ill. App. 3d 1000, 291 N.E.2d 247 (1972).

The broad arbitration provision applies to indemnification provisions in a construction contract, such as ¶ 3.18. Contract Constr., Inc. v. Power Tech. Ctr. Ltd. P'ship, 100 Md. App. 173, 640 A.2d 251 (1994).

See also McCarney v. Nearing, Staats, Prelogar & Jones, 866 S.W.2d 881 (Mo. Ct. App. 1993).

[454] Buckminster v. Acadia Village Resort, 565 A.2d 313 (Me. 1989).

[455] In re F&T Contractors, Inc., 649 F.2d 1229 (6th Cir. 1981).

Because the General Conditions (AIA Document A201) are usually incorporated into the subcontracts between the general contractor and the subcontractors, the provisions of the General Conditions apply to the relationship between the general contractor and the subcontractors in those cases. In a South Carolina case, the subcontractor sued the general contractor for breach of contract.[456] The general contractor sought to stay the action and bring an arbitration proceeding under this provision of the General Conditions. The appellate court held that because this and other provisions of the General Conditions were incorporated by reference into the subcontract, the subcontractor was bound to arbitrate.

Changes and change orders are also subject to arbitration.[457] In a New York case, the contractor never signed the AIA contract that incorporated the arbitration clause by reference.[458] The court found that the arbitration agreement was still valid, holding that the state arbitration statute required merely a written agreement

---

[456] Godwin v. Stanley Smith & Sons, 386 S.E.2d 464 (S.C. 1989). However, the court refused to order arbitration in Wooster Assocs. v. Walter Jones Constr. Co., No. C.A. No. 2900, 1994 Ohio App. LEXIS 5911 (Dec. 21, 1994). The contractor claimed a right to arbitration, stating that one of the exhibits to his contract contained an index to drawings, which in turn contained outline specifications, which in turn contained a provision incorporating AIA Document A201, which in turn contained the arbitration provision. The court found this too far removed.

In AJM Packaging Corp. v. Crossland Constr. Co., 962 S.W.2d 906 (Mo. Ct. App. 1998), the court denied arbitration between the contractor and the owner. A101 incorporated the 1987 version of A201 by reference. The specifications, on the other hand, incorporated the 1976 version of A201. Finally, another part of the specifications deleted all references to arbitration. Based on this, the court found that the contractor, who wanted to arbitrate, had failed to prove that there was a valid arbitration provision.

In Albertson's, Inc. v. PDG, Inc., 2002 WL 31298862 (N.D. Tex., Oct. 8, 2002), the owner-contractor agreement deleted the arbitration provisions, but the subcontracts did not. The owner sued the general contractor and the general filed a third-party action against the sub. The sub moved to compel arbitration. The general argued that the reference in the subcontract requiring that arbitration "be conducted in the same manner and under the same procedures" as provided in the general contract meant that the sub could not compel arbitration. The court found that the arbitration clause could be enforced. The reference to the general contract is required only to determine which procedures will govern the arbitration process. It does not control whether or not disputes will be arbitrated.

In Robert J. Denley Co., Inc., v. Neal Smith Const. Co., Inc., 2007 WL 1153121 (Tenn. Ct. App., April 19, 2007), the court compelled arbitration where A101 incorporated A201. The bonding company had standing to demand arbitration and there was no waiver by filing an answer.

[457] In C&M Ventures, Inc. v. Wolf, 587 So. 2d 512 (Fla. 1991), the appellate court overturned the lower court's ruling that only the original contract was subject to arbitration, and not the claimed changes. Paragraph 1.1.1 defines Contract Documents to include modifications to the original contract. *Accord* Granger N., Inc. v. Cianchette, 572 A.2d 136 (Me. 1990). In that case, the arbitrator had the authority to award the cost of unsigned change orders to the contractor, even though the requirements of the General Conditions that such change orders must be signed were not met. The parties can modify such a requirement.

In Bishop Contracting Co. v. Center Bros., 213 Ga. App. 804, 445 S.E.2d 780 (1994), the court found that the contract provided for extra work or changes in the work, so that those items were subject to arbitration.

[458] Liberty Mgmt. v. Fifth Ave. & Sixty-Sixth St. Corp., 208 A.D.2d 73, 620 N.Y.S.2d 827 (1995).

to arbitrate. The statute does not require that the writing be signed so long as there is other proof that the parties actually agreed on it. Further, change orders are also subject to arbitration.

In an Alabama case, the court reviewed this provision in a dispute over whether the federal or state arbitration act applied.[459] Because the construction involved

---

[459] Maxus, Inc. v. Sciacca, 598 So. 2d 1376 (Ala. 1992). *See also* Lost Creek Mun. Util. Dist. v. Travis Indus. Painters, Inc., 827 S.W.2d 103 (Tex. App. 1992) (federal law applied). In Osteen v. T.E. Cuttino Constr. Co., 434 S.E.2d 281 (S.C. 1993), the court found that the Federal Arbitration Act applied when the arbitration clause of the standard AIA language did not comply with a state law that required that an arbitration clause in a contract be set forth in bold type. The trial court had ruled that state law applied because of the choice of law provision (¶ 13.1.1) in finding that the arbitration clause was invalid. The Supreme Court of South Carolina reversed, finding that the contract should be read so as to uphold arbitration if interstate commerce was involved (this was a single family home). One issue that sometimes arises is whether statutory claims are subject to arbitration. In United States ex rel. Tanner v. Daco Constr., Inc., 38 F. Supp. 1299 (N.D. Okla. 1999), the court held that, but for the insertion of a special provision in the contract that specifically exempted claims under federal law, a Miller Act claim would have been arbitrable. While this case was undoubtedly correct in its interpretation of the current state of the law, it also demonstrates the expansive reading of arbitration law that courts, primarily at the federal level but also in many state courts, are taking. It seems likely that as attorneys become more creative in utilizing arbitration to obtain results not possible in court, the pendulum will reverse, and courts may, in the not too distant future, again review arbitration decisions more carefully.

Blanton v. Stathos, 351 S.C. 534, 570 S.E.2d 565 (2002), deals with the South Carolina Arbitration Act, which requires that contracts with an arbitration provision contain a prominent notice on the first page of the contract, which the AIA forms do not include. If this is not complied with, then the arbitration provision is not enforceable. The court in *Blanton* held that the FAA applied.

In Dunn Indus. Group, Inc. v. City of Sugar Creek, 112 S.W.3d 421 (Mo. 2003), the court found that the FAA applied, rejecting a claim that the state's equitable mechanic's lien statutes place exclusive jurisdiction for the claims in the case in the state court, not arbitration. The Supremacy Clause defeats such an argument.

In Huntsville Utilities v. Consolidated Constr. Co., 2003 WL 21205396 (Ala. May 23, 2003), the Supreme Court of Alabama continued that state's vehement stance against arbitration. After a dispute arose and suit was filed, various parties moved to stay the proceedings and compel arbitration. The trial court denied the motion and the state Supreme Court affirmed. The issue centered on whether the FAA applied or not. The Supreme Court stated that the party moving for arbitration had the burden of proving the existence of a contract containing a written arbitration clause and relating to a transaction that substantially affects interstate commerce. The case of Sisters of the Visitation v. Cochran Plastering Co., 775 So. 2d 759 (Ala. 2000), was cited as precedent and identified a number of factors for a court to determine whether the FAA applied. The Supreme Court said the movant did not meet that burden despite the fact that one party was incorporated in Delaware with its sole office in Alabama and some 499 invoices evidenced out-of-state material purchases.

Shortly after *Huntsville* was decided, the United States Supreme Court over turned *Sisters of the Visitation* in The Citizens Bank v. Alafabco, Inc. et al., 123 S. Ct. 2037 (2003), finding that the Alabama court interpreted the FAA too narrowly, and further finding that the five-part test in *Sisters of the Visitation* is improper where the FAA is involved. This case should sound the death knell for attempts to avoid the broad reach of the FAA in construction cases. There is virtually no construction project in the United States that would not fall within the reach of the FAA. Apparently the Alabama Supreme Court finally got the message in Huntsville Utils. v. Consolidated Constr. Co., 2003 WL 22064079 (Ala. Sept. 5, 2003).

interstate commerce, the court held that the Federal Arbitration Act applied, not the Alabama arbitration statute. (The interstate commerce in this construction project included shipping of materials by common or other carriers across state lines, as well as ordering, following up, and paying for materials by using the United States mails, telephones, and interstate financial transaction settlement procedures and institutions. Also, escrow funds were held, paid out, and invested by a national banking association using the United States mails, telephones, and interstate financial transaction settlement procedures and institutions.) This is an interesting case, as there is probably no construction project that does not involve some interstate commerce, but most cases are determined under state arbitration acts.

In an Illinois case, the parties deleted this standard arbitration clause by the Additional Supplementary Conditions.[460] When a dispute developed between the parties as to whether arbitration should take place, the contractor argued that other provisions of A201 referring to arbitration were not deleted from the contract and that arbitration was appropriate. The court rejected this argument on the basis that the clear intent of the parties was that disputes should be litigated. Further, under AAA rules, the parties to a contract must specifically authorize the AAA to

---

In Mpact Constr. Group, LLC v. Superior Concrete Constructors, Inc., 785 N.E.2d 632 (Ind. App. 2003), the court found that the FAA applied. "Even though a matter falls within the scope of the FAA, courts generally apply state law to the issue of whether the parties agreed to arbitrate their claims. . . . The FAA preempts state law, however, to the extent that they conflict." Here, the contract between the owner and general contractor incorporated A201, and arbitration was required between those parties. A second issue was whether certain subcontractors were subject to arbitration. The court held that they were not, apparently because the subcontract agreements were not standard AIA forms. The court, in a footnote, stated that "we note the possibility that this outcome was forecast by the General Instructions to AIA Form A201 (General Conditions), which cautions the practitioner against the use of non-AIA forms to avoid inconsistency in language and intent."

In Foodbrands Supply Chain Services, Inc. v. Terracon Inc., 2003 WL 23484633 (D. Kan. Dec. 8, 2003), the court held that the FAA applied because the contract involved interstate commerce. Under Kansas law, the matter would likely have been litigated in court because it involved a multiplicity of parties.

In an interesting case, Peterson Constr., Inc. v. Sungate Dev., L.L.C., 2003 WL 22480613 (Tex. App.—Corpus Christi, Oct. 30, 2003), the court held that the contract, which included the A201-1997, was governed by the FAA and not the Texas Arbitration Act. When the trial judge refused to compel arbitration, the contractor filed an appeal. The court held that such an order was not subject to interlocutory appeal, although an interlocutory appeal would have been available under the Texas Act. The dissent's analysis provides a good overview of this area of law.

In Construction Services, Inc. v. Regency Hosp. Co., 2006 WL 3247322 (S.D. Miss. Nov. 8, 2006), the court ordered arbitration and dismissed the case based on the AIA language under the FAA.

[460] Glenn H. Johnson Constr. Co. v. Board of Educ., 245 Ill. App. 3d 18, 614 N.E.2d 208 (1993). In Weitz Co. v. Shoreline Care Ltd. P'ship, 39 Conn. App. 641, 666 A.2d 835 (1995), the project consisted of two phases. The supplementary conditions had deleted the arbitration provision from the contract for phase one work, although phase two work was subject to arbitration. When a dispute arose, the contractor sued the owner, who sought arbitration. The court enjoined the owner from arbitration on the basis that the dispute pertained to phase one work, and there was no agreement to arbitrate that part of the work.

administer an arbitration, or must specify that they agree to arbitration under its rules, before the AAA may assume jurisdiction of a dispute. Deleting ¶ 4.5.1 (A201-1987 version) took out the only paragraph that gave the AAA jurisdiction, because ¶ 4.5.2 (A201-1987) is effective only if ¶ 4.5.1 (A201-1987) is not deleted. The court held that all of the other paragraphs of A201 that discuss arbitration depend on this paragraph. If this paragraph is deleted, those other references to arbitration are without force. If the parties use the AIA 2007 family of documents, such issues should not occur. If, on the other hand, some non-AIA documents are used, or if the drafters improperly modify the AIA documents, such problems are possible.

Sometimes there is a question as to whether an AIA contract or the arbitration provisions are incorporated into the contract. This often arises when there are contradictory provisions in various documents that comprise the contract.[461]

The final sentence states that if any known claim is not included in the arbitration demand by a party, that claim is waived. This would also apply to counterclaims in arbitration. If a party wants to assert a new claim or counterclaim after the arbitration has been initiated, it is up to the arbitrator as to whether the amendment will be permitted. This is in accordance with the AAA rules. Based on this provision, it is very unlikely that an arbitrator who refused to allow a late claim would be overturned. Arbitrators are usually quite liberal about permitting amendments to claims and counterclaims. However, after an award has been made, this provision would bar any further arbitration or litigation related to the project.

15.4.1.1 *A demand for arbitration shall be made no earlier than concurrently with the filing of a request for mediation, but in no event shall it be made after the date when the institution of legal or equitable proceedings based on the Claim would be barred by the applicable statute of limitations. For statute of limitations purposes, receipt of a written demand for arbitration by the person or entity administering the*

---

[461] In Cleveland Jet Ctr., Inc. v. Structural Sales Corp., 1995 Ohio App. LEXIS 4113 (Sept. 22, 1995), the court found that A201 was not incorporated because the parties elected to use the 1978 edition of A111 instead of the newer version, which expressly incorporated A201.

In DJM Constr., Inc. v. Rust Eng'g Co., 1996 U.S. Dist. LEXIS 5455 (N.D. Ill. Apr. 24, 1996), the court denied a subcontractor's demand for arbitration. An attachment to the contract provided that A401 and A201 were to be part of the contract. However, the contract also incorporated the owner-contractor agreement, which required litigation of all disputes. The court found that this was a more specific clause that took precedence over the standard general arbitration language in the AIA documents. In RTKL Assocs., Inc. v. Baltimore County, 147 Md. App. 647, 810 A.2d 512 (2002), the architect's proposal read as follows: "RTKL fees are based upon the Detailed Scope of Services and the Standard Form of Agreement between Client and Architect 1987 Edition, AIA Document B-141. All Client-generated contracts will require review and acceptance by RTKL's legal counsel before any work may proceed." This proposal was incorporated into the non-AIA contract drafted by the owner and signed by the parties. The disputes provision in that contract was stricken. When a dispute arose, the architect wanted to compel arbitration, arguing that, by striking the disputes clause in the owner's form, the disputes provision in B141 (now B101) should apply. The court rejected this and held that there was no valid arbitration clause.

arbitration shall constitute the institution of legal or equitable proceedings based on the Claim. **(4.6.3; substantial modifications)**

The arbitration must be commenced within the specified times, but no later than the statute of limitations established by the appropriate state law governing the project. Note that ¶ 13.7.1 places a ten-year outside time limit from the time of substantial completion within which to initiate arbitration. Prior versions of A201 required that the time must be "reasonable," and one court has held that this means 30 days.[462] Under Paragraph 15.4.1.1, arbitration can be initiated at any time within the stated limits, with an outside time limit of ten years after substantial completion. One court[463] has held that if several claims are being presented, some time-barred and others not, arbitration may proceed.

---

[462] Lyons v. Krathen, 368 So. 2d 906 (Fla. Dist. Ct. App. 1979) (construing the 1970 version of A201). Based on a clause similar to ¶ 4.4.6 (A201-1997) an arbitration filed nine months after first written notice of a dispute and four months after the decision of the architect was not timely. The decision did not state whether the architect's decision stated that it was final, and it can be assumed that it did not.

    However, in Riggi v. Wade Lupe Constr. Co., 176 A.D.2d 1177, 575 N.Y.S.2d 613 (App. Div. 1991), a delay of six months and participation in litigation did not constitute either a waiver or an untimely demand for arbitration. The architect had made a determination that the owner had cause to terminate the contractor, but this determination did not include the 30-day language of ¶ 4.4.6 (A201-1997) (not in the 2007 version of A201). Thus, six months was reasonable.

    In Miller Bldg. Corp. v. Coastline Assocs. Ltd. P'ship, 411 S.E.2d 420 (N.C. 1992), the court held that the arbitration demand was timely, because only two months had passed from the time the other party filed suit to collect unpaid fees.

    This clause was examined in 200 Levee Drive Assocs. v. Bor-Son Bldg., 441 N.W.2d 560 (Minn. Ct. App. 1989), which found that the trial court should determine whether the arbitration was filed within the statute of limitations.

    In Des Moines Asphalt & Paving Co. v. Colcon Indus. Corp., 500 N.W.2d 70 (Iowa 1993), the owner timely filed for arbitration. Only two months elapsed between the time the owner was served with a foreclosure notice and the time it filed its motion to compel arbitration, and the trial court should have compelled the arbitration. The court further held that it was for the arbitrator to determine whether the owner had demanded arbitration within a reasonable time. In Robbinsdale Pub. Sch., Indep. Sch. Dist. No. 281 v. Haymaker Constr. Inc., 1997 Minn. App. LEXIS 1155 (Oct. 14, 1997), the court held that the arbitrator and not the court should determine whether the claim for arbitration was timely filed.

[463] Office of Irwin G. Cantor, P.C. v. Swanke Hayden Connell & Partners, 186 A.D.2d 71, 588 N.Y.S.2d 19, 20 (1992). The owner obtained a large arbitration award against the architect for the Trump Tower. The architect then sought arbitration against the engineer. The engineer argued that the six-year statute of limitations had run and arbitration could not be initiated. The court held, however, that one of the claims, for indemnification, did not accrue until payment by the architect. Because that claim was not time-barred, all matters should be presented to the arbitrator. According to the court, "Where several theories of liability are predicated upon the same facts and some are timely whereas others are not, the judiciary will not separate between the different types of claims but will allow them all to proceed to arbitration."

    In Belfatto & Pavarini v. Providence Rest Nursing Home, Inc., 582 N.Y.S.2d 744 (App. Div. 1992), the court found that certain causes of action based on breach of contract and architectural malpractice were barred by the statute of limitations. Those matters could not be arbitrated. However, the court allowed arbitration of an indemnification claim, because the statute of limitations had not yet run on that claim.

15.4.2 The award rendered by the arbitrator or arbitrators shall be final, and judgment may be entered upon it in accordance with applicable law in any court having jurisdiction thereof. **(4.6.6)**

The award is normally converted into a standard judgment that is enforceable by the courts by normal means, including attaching the assets of the losing party. Arbitrator's awards are difficult to overturn. A Louisiana case involved a cost overrun when an architect designed a house for the owner.[464] There had been discussion that the house was to cost $60,000. The lowest bid was $110,270. The architect then prepared a second set of plans, and bids came in at $79,240. The owner then told the architect that the project was postponed and the architect's fees would not be paid. The architect commenced arbitration and obtained an award based on the contractual percentage of the $110,270 bid. The owner appealed the award. The court held that unless a party can show that the arbitrator was somehow prejudiced or the award was procured by undue means (these were statutory grounds in the Louisiana Arbitration Act) that the amount of the award could not be challenged.

In a Texas case, the contractor argued that ¶ 9.7, providing that "if the Owner does not pay the Contractor within seven days after the date established in the Contract Documents the amount . . . awarded by arbitration (now "binding dispute resolution"), then the Contractor may . . . stop the work," allowed it to stop the work when the owner failed to pay an arbitration award within seven days.[465]

---

[464] Firmin v. Garber, 353 So. 2d 975 (La. 1977).

However, in Tretina Printing, Inc. v. Fitzpatrick & Assocs., 262 N.J. Super. 45, 619 A.2d 1037 (1993), the court held that an arbitrator may not make an award that is wholly bereft of evidential support. The arbitrator had itemized his award, allowing the court to review the basis for each line item. The arbitrator had ignored the guaranteed maximum price in the contract, as well as other provisions. Most arbitrators will not give reasons or itemize awards, in order to avoid having the award overturned by a court.

In Forge Square Assocs. Ltd. P'ship v. Constr. Servs., 43 Conn. Supp. 32, 638 A.2d 654, 655 (1993), following an arbitration award in favor of the contractor, the owner sought to vacate the award. Apparently the architect had certified that grounds existed for termination of the contractor (*see, e.g.,* ¶ 2.6.18, 2.6.19 of B141-1997 and ¶ 14.2.2 of A201). The owner argued that this constituted a manifest disregard for the law and so exceeded the arbitrators' powers that the award should be vacated. The court disagreed, stating the rule that "in order to prevail on this application to vacate, the plaintiff must demonstrate that the award reflects an egregious or patently irrational rejection of clearly controlling legal principles." Because the award did not give specific reasons, the court would have had to speculate on the specific reasons for the award.

In RaDec Constr., Inc. v. School Dist. No. 17, 248 Neb. 338, 535 N.W.2d 408 (1995), the court found that the architect's determination of the amount owed by a contractor for a change was patently erroneous and therefore legally equivalent to bad faith, so the court could reverse the architect's determination. In Lauro v. Visnapuu, 351 S.C. 507, 570 S.E.2d 551 (Ct. App. 2002), the trial court vacated the arbitrator's award on the basis of a "manifest disregard for the law." The appellate court reversed. Although an award may be vacated for this reason, this non-statutory ground requires something more than a mere error of law, or failure on the part of the arbitrator to understand or apply the law. "An erroneous application of the law, however, does not constitute manifest disregard."

[465] Snyder v. Eanes Indep. Sch. Dist., 860 S.W.2d 692 (Tex. App. 1993).

The court stated that state statute permitted the owner to challenge the award within a 90-day period. Paragraph 15.4.2 provides that arbitration awards are to be subject to applicable law. Therefore, the owner's failure to pay the arbitration award within seven days did not violate a covenant to pay an award within a specified time.

In an unusual case, the contractor agreed to construct an addition to the defendant's house.[466] When a dispute arose, the parties arbitrated. The award provided that the contractor would obtain the necessary approvals from the local building department, including a certificate of occupancy. The award further provided that when the documents were received by the contractor, they would be turned over to the owner, who would then pay the contractor $101,000 as full and final payment. Neither party sought to confirm or vacate the arbitration award within the statutory time. The owner then thwarted the contractor's performance. The municipality refused to issue the occupancy certificate until an additional smoke detector was installed. The owner refused to allow the contractor to install the smoke detector and did not pay the money to the contractor. The contractor then brought a court action to either enforce the award or obtain a new arbitration. The owner defended on the grounds that the contractor had failed to confirm the award. The appellate court found that the statutory procedures are not the exclusive means for judicial enforcement of an arbitration award. It confirmed the award, also finding that the effect of the award was to preclude issues of fact and law from further litigation.

In a New York case, this provision was stricken from the form contract, an A401 subcontract.[467] The issue was whether striking this language made the arbitration

---

[466] Spearhead Constr. Corp. v. Bianco, 39 Conn. App. 122, 665 A.2d 86 (1995).

[467] St. Lawrence Explosives Corp. v. Worthy Bros. Pipeline Corp., 916 F. Supp. 187 (N.D.N.Y. 1996).
In Ringwelski v. Pederson, 919 P.2d 957 (Colo. Ct. App. 1996), the contract contained the following arbitration provision:

Any disagreement arising out of this contract or from the breach thereof shall be submitted to arbitration, and judgment upon the award rendered may be entered in the court of the forum, state or federal, having jurisdiction.
It is mutually agreed that the decision of the arbitrators shall be a condition precedent to any right of legal action that either party may have against the other. The arbitration shall be held under the Standard Form of Arbitration Procedure of the American Institute of Architects or under the Rules of the American Arbitration Association.

When a dispute developed, the parties went through arbitration and an award was made. The losing plaintiffs then filed a court action with the same allegations as in the arbitration. The defendant moved to confirm the arbitration award and dismiss the court action. The trial court refused to do so, finding that the first part of the second paragraph of the arbitration clause indicated an intent that the arbitration be nonbinding. It also found persuasive the use of the word "may" in the first paragraph. The appellate court reversed, finding that such an interpretation would severely limit the applicability of the first sentence's providing for entry of judgment on an award. The court went on to state that it construed this language to mean that the prevailing party has the option to obtain a judgment for the purpose of enforcement or collection if necessary. In New Concept Constr. Co., Inc. v. Kirbyville Consolidated Indep. Sch. Dist., 119 S.W.3d 468 (Tex. App. 2003), the contract incorporated the general conditions and included this provision:

9.1 All matters relating to the validity, performance, interpretation of [sic] construction of the contract documents or breach thereof shall be governed by and construed in accordance

provision nonbinding. Based on the liberal federal policy in favor of arbitration, and based on ¶ 6.1 (4.6.2 in the 1997 A201, 15.4.1 here) of the agreement (which incorporated the rules of the American Arbitration Association, which in turn provide that judgment may be entered on the arbitration award), the court held that arbitration was binding.

15.4.3 The foregoing agreement to arbitrate and other agreements to arbitrate with an additional person or entity duly consented to by parties to the Agreement shall be specifically enforceable under applicable law in any court having jurisdiction thereof. (**4.6.4; substantial revisions**)

Prior versions of this document, as well as other AIA documents, prohibited joinder of parties. This meant that an owner who had a construction problem that involved both design and construction issues, could not force both the contractor and the architect into one single arbitration. This provided a tactical advantage to architects and contractors. Under the 2007 documents, if the parties select arbitration as the dispute resolution method, joinder is permitted (see below).

§ 4.89   Consolidation or Joinder: ¶ 15.4.4

15.4.4 CONSOLIDATION OR JOINDER

15.4.4.1 Either party, at its sole discretion, may consolidate an arbitration conducted under this Agreement with any other arbitration to which it is a party provided that (1) the arbitration agreement governing the other arbitration permits consolidation, (2) the arbitrations to be consolidated substantially involve common questions of law or fact, and (3) the arbitrations employ materially similar procedural rules and methods for selecting arbitrator(s). (**4.6.4; substantial modifications—allowed joinder of other parties**)

Prior versions of A201 stated that no person who is not a party to the agreement may be joined in the arbitration, except by agreement,[468] or by

---

with the laws of the state of Texas. The Contractor shall not institute any action of [sic] proceeding in any way relating to this agreement against the Owner except in a court of competent jurisdiction in the County in which the work was performed.

The court went to great pains to harmonize this provision with the arbitration clauses in A201: "We interpret the forum selection clause and the arbitration provision together to mean that the contractor must file any court proceeding not precluded by the arbitration provision in the County in which the work was performed—for example, an action to enforce arbitration and render judgment on an arbitration award, or an action where arbitration has been waived." The dissent provided a more cogent argument against arbitration against the "misguided slavish adherence to the 'arbitrate at every opportunity' mentality."

[468] The architect is not a party to A201, and joinder cannot be forced by the contractor. James Stewart Polshek v. Bergen Iron Works, 142 N.J. Super. 516, 362 A.2d 63 (1976). A surety is only an incidental beneficiary of the contract and cannot enforce the arbitration agreement. Aetna Cas. & Sur. Co. v. Jelac Corp., 505 So. 2d 37 (Fla. Dist. Ct. App. 1987).

In Buck Run Baptist Church, Inc. v. Cumberland Surety Ins. Co., 983 S.W.2d 501 (Ky. 1998), the issue was whether the surety could compel arbitration against the owner. The general contractor was hired by the owner to construct a new worship center. A payment and performance bond was obtained from the surety. The bond stated that the AIA contract was made part of the bond. When the general contractor was unable to perform, the owner terminated the contractor, and the surety retained another contractor to complete the project. When a dispute arose about the performance of the substitute contractor, the owner refused to pay the remaining contract balance. The surety sought to arbitrate the dispute, and the owner refused to arbitrate.

The owner argued that the performance bond at issue was an insurance contract. Under Kentucky law, insurance contracts are exempt from arbitration. The court found that the surety stood in the shoes of the original contractor when it undertook to complete the construction. The dispute involved a construction contract, and not the applicability of the Kentucky insurance exemption. The court found it notable that the owner had the advice of a professional architect, who had recommended termination of the original construction contract. This was not a contract of adhesion, and the parties were not free to repudiate promises to arbitrate.

The court also looked at the differences between an insurance policy and a surety bond such as the one here. An insurance policy is a contract of indemnity whereby the insurer agrees to indemnify the insured for any loss resulting from a specific event. The insurer undertakes the obligation based on an evaluation of the market's risks and losses, and these are actuarially predicted. A surety bond, on the other hand, is written based on an evaluation of a particular contractor and the contractor's capacity to perform a given contract. A surety's relationship to its principal is more like that of a creditor-debtor relationship than that of a traditional insurer-insured.

In Dunn Indus. Group, Inc. v. City of Sugar Creek, 2002 WL 31548615 (Mo. App. Nov. 19, 2002), the court distinguished the surety cases from the instant case, where the parent company guaranteed performance of the contractor's work. The guarantee here did not incorporate by reference the construction contract. Mere reference to the construction contract is insufficient to establish that the guarantor bound itself to the arbitration provision in the construction contract.

Another Florida court held the opposite, Henderson Inv. Corp. v. International Fidelity Ins. Co., 575 So. 2d 770 (Fla. 1991), stating that "it would not be third party beneficiary status which would give a surety the right to invoke an arbitration clause, but rather that surety's status as the real party in interest/joint and several obligor, where it has expressly incorporated the terms of a construction contract in its performance bond."

In Windowmaster Corp. v. B.G. Danis Co., 511 F. Supp. 157 (S.D. Ohio 1981), the general contractor sought to enforce arbitration not only against its subcontractor but also against the sub's surety. The court held that the surety could not be forced to arbitrate, despite the presence of a subcontract bond.

Acknowledging that a majority of jurisdictions allow arbitration against a surety, the court in Hartford Accident v. Scarlett Harbor Assocs. Ltd. P'ship, 109 Md. App. 217, 674 A.2d 106 (1996), *aff'd,* 346 Md. 122, 695 A.2d 153 (1997), held that the surety was not bound to arbitrate. Although the bond incorporated the AIA contract that contained the arbitration clause, it did not mandate arbitration against the surety, according to the court. The arbitration language itself only requires the owner and contractor to arbitrate and does not expressly include the surety. Further, the bond indicated an intent to litigate by virtue of the standard language that requires initiation of any suit within two years. In Liberty Mut. Ins. Co. v. Mandaree Pub. Sch. Dist. 36, 459 F. Supp. 2d 866 (D.N.D. 2006), the surety initially consented to arbitration with the owner. However, the owner then unilaterally hired a contractor to perform remedial work and the surety took the position that this constituted a forfeiture of all rights under the performance bond. The surety then revoked its consent to arbitration. The court ruled against the owner on the motion to compel arbitration. For a contrary holding, *see* Higley S., Inc. v. Park Shore Dev. Co., 494 So. 2d 227 (Fla. Dist. Ct. App. 1986).

In City of Piqua v. Ohio Farmers Ins. Co., 84 Ohio App. 3d 619, 617 N.E.2d 780 (1992), the owner sought arbitration against the surety, and the surety appealed. The appellate court held that the surety was bound by the arbitration provision in the contract.

In the case of Stallings & Sons, Inc. v. Sherlock, Smith & Adams, Inc., 670 So. 2d 861 (Ala. 1995), the owner-architect agreement was not an AIA agreement and did not contain an arbitration agreement. The owner-contractor agreement, however, incorporated A201. The contractor sued the architect for suppressing information about the project and dereliction of duties. The trial court stayed that action pending arbitration of the matter. The contractor appealed on the basis that there was no arbitration agreement between it and the architect. The supreme court of Alabama agreed with the contractor, finding that not only was there no agreement to arbitration, but also the arbitration provision of A201 (¶ 4.5.5) specifically prohibited joinder.

In Curtis G. Testerman Co. v. Buck, 340 Md. 569, 667 A.2d 649 (1995), the owner and contractor were involved in a dispute. The contractor sought arbitration and moved to dismiss the owner's complaint against him personally. The court refused to dismiss him personally. The contract had the name "Curtis G. Testerman, Inc." instead of the actual name of the corporation, "Curtis G. Testerman Company." The appellate court held that this was not fatal and that the owner understood that he was contracting with a corporation. The individual should have been dismissed and was not personally liable under the contract, and he could not be compelled to arbitrate. The nonconsolidation language was one reason for not joining the individual.

An interesting issue that may arise is the extent to which a guarantor of a construction contract is subject to arbitration. This was the issue in Scinto v. Sosin, 721 A.2d 552 (Conn. App. Ct. 1998). Barbara Scinto was a guarantor of, but not a signatory to, the contract. Because the guarantee did not refer to the arbitration clause, and the arbitration clause did not refer to the guarantee, the court had no difficulty finding that she was not subject to the arbitration provision. The court also found that Robert Scinto, who had signed the contract both as guarantor and in his individual capacity, could not be forced to arbitrate based merely on the guarantee.

In Callahan & Assocs. v. Honorable Patrick A. Clark, 1996 Tex. App. LEXIS 5527 (Dec. 12, 1996), the appellate court upheld this clause and refused to allow consolidation.

In Maxum Founds., Inc. v. Salus Corp., 817 F.2d 1086 (4th Cir. 1987), the parties had a modified clause that read: "No arbitration shall include by consolidation, joinder or in any other manner, parties other than the Owner, the Contractor and any other persons substantially involved in a common question of fact or law, whose presence is required if complete relief is to be accorded in the arbitration." The court found that this provision permitted consolidation of arbitration. Sometimes an argument is made that a nonparty to the agreement is a third-party beneficiary of the contract and should be permitted to arbitrate. This was rejected in Duchess of Dixwell Avenue, Inc. v. The Neri Corp., 1999 Conn. Super. LEXIS 2077 (Aug. 4, 1999). There, a contractor was hired to repair a culvert carrying a stream beneath land that contained plaintiff, a restaurant. The contract did not list the restaurant as an "Owner." When a dispute arose and the contractor demanded arbitration, the restaurant sought to file a counterclaim in the arbitration for the contractor's delay and interference with its business during construction, claiming that it was a third-party beneficiary of the contract. Finding that the contractor did not even know the name of the restaurant corporation until the arbitration and that there was no evidence that the contractor intended any entity other than the named "Owner" to enforce the contract or benefit from the contractor's services, the court held that the restaurant was not a third-party beneficiary and could not participate in the arbitration.

In A.S.W. Allstate Painting v. Lexington Ins. Co., 94 F. Supp. 2d 782 (W.D. Tex. 2000), a fire caused substantial damage to a building under construction. The owner's insurer paid the owner and put the contractor on notice that it had a claim. The contractor brought a declaratory action against the insurer to bar any subrogation claim. The insurer then sought to compel arbitration of the claim. The court held that there was no agreement to arbitrate with the insurer by virtue of this provision and the anti-subrogation provision (¶ 11.4.7 of A201).

In Chain Elec. Co. v. National Fire Ins. Co., 2006 WL 2973044 (S.D. Miss. Oct. 16, 2006), a contractor obtained a performance and payment bond from a surety. When a dispute arose with a subcontractor, the subcontractor filed suit against both the contractor and the surety. The matter was arbitrated and an award was issued. This case discusses the law concerning a surety not a party to the litigation.

waiver.[469] Thus, a contractor may have a claim not only against the owner for extras due to delay or additional work, but also against the architect. Because of that nonconsolidation language, the contractor could not pursue a single arbitration action against both the owner and the architect unless all parties agree to it.[470] If the

---

[469] The parties can waive the nonconsolidation provision, just like any other provision of the contract. Thus, when a case is consolidated without objection by a party, that party cannot later complain that it was improperly consolidated contrary to the express provision of this paragraph.

In Ure v. Wangler Constr. Co., 232 Ill. App. 3d 492, 597 N.E.2d 759 (1992), the owner had entered into separate contracts with an architect and a contractor. Each contract had language similar to the AIA nonconsolidation provision. After the building was substantially built, disputes arose among the three parties, and the owner filed separate demands for arbitration with the AAA. The AAA assigned a single case number to the demands and treated them as a consolidated file. No objection to consolidation was raised by any party. Over the course of several months, a number of administrative matters concerning the arbitration occurred. Letters indicating the consolidation were sent to the parties but no objection was made. Finally, at the first evidentiary hearing, the contractor objected to the consolidation for the first time, during opening statements. The arbitrator consulted with the tribunal administrator from the AAA and ruled that there had been a waiver of the nonconsolidation provision. The hearings continued, and an award was entered in favor of the owners. Wangler filed a petition to vacate the award, alleging that the arbitrator exceeded his authority by allowing the consolidation in the face of a contractual provision prohibiting consolidation.

The trial court confirmed the arbitration award and, on appeal, the appellate court found that there had been a waiver. "A contractual right with respect to arbitration can be waived as can any other contract right." The court also found that the arbitrator had the power to determine whether a waiver had occurred. "Where arbitrability is not contested, questions concerning timeliness and waiver in asserting contractual rights are for arbitrators to decide, rather than courts."

[470] *But see* Episcopal Hous. Corp. v. Federal Ins. Co., 273 S.C. 181, 255 S.E.2d 451 (1979). In Gavlik Constr. Co. v. H.F. Campbell Co., 526 F.2d 777 (3d Cir. 1975), the court construed the 1970 version of A201, which did not have nonconsolidation language, and held that the contractor could join the owner when the subcontractor commenced arbitration proceedings against the contractor over a payment dispute. But in Del E. Webb Constr. v. Richardson Hosp. Auth., 823 F.2d 145 (5th Cir. 1987), the court construed a more recent version of A201 that included the present nonconsolidation language and found that the contractor could not force consolidation.

The subcontractor cannot be named as an additional party to an arbitration between the architect and owner because the language in the agreement restricts the arbitration to matters arising out of "this agreement." Cumberland-Perry v. Bogar & Bink, 261 Pa. Super. 350, 396 A.2d 433 (1978).

In Garden Grove Cmty. Church v. Pittsburgh-Des Moines Steel Co., 140 Cal. App. 3d 251, 191 Cal. Rptr. 15 (1983), the owner had the following provision in its contract with the builder:

All claims, disputes . . . arising out of, or relating to, this Contract or the breach thereof shall be settled by arbitration . . . provided, however, that Owner shall not be obligated to arbitrate any such claim, dispute, or other matter, if Owner, in order to fully protect its interests, desires in good faith to bring in or make a party to any such claim, dispute, or other matter, the Construction Manager, the Architect, or any other third party who has not agreed to participate in and be bound by the same arbitration proceeding.

The owner had standard AIA language in the contract with the architect. The court held that this escape clause relieved the owner from the duty to arbitrate if all the necessary parties were not joined in the arbitration proceeding.

In Raffa Assocs. v. Boca Raton Resort & Club, 616 So. 2d 1096 (Fla. Dist. Ct. App. 1993), the contractor sought arbitration against the owner and the architect after an injured worker sued the

contractor had elected to pursue both actions, it would probably have initiated an arbitration action and a court action. It was then possible that the resolution of the first-decided action would affect the later action, based on the legal doctrines of res judicata or collateral estoppel.

This has now changed with this consolidation language. At the owner's or contractor's option, arbitrations between the two parties to this document—the owner and the contractor—can be consolidated with other arbitrations to which the owner and/or contractor is a party, such as with the architect or with a subcontractor. The arbitration provision in those contracts must closely parallel the one in the A201. This will be the case if the contractor has used standard AIA documents or if the A201 is incorporated into those agreements. The issues must be the same or substantially similar.

When there is a possibility of litigation following an arbitration proceeding, a court reporter would be a wise investment. This typically occurs when other parties are involved in the project, but not in the arbitration proceeding, such as a dispute among the owner, architect, and contractors. However, a court reporter is usually not possible if the party who needs the record is not a party to the arbitration, because arbitrations are not public. Also, such a record may not be conclusive as to what issues were actually decided. The finding of the arbitrator is often extremely succinct, affording little insight other than the outcome. Arbitrators are not usually required to give reasons to support or clarify their award[471] unless the parties agree to have the arbitrators do so.

15.4.4.2 *Either party, at its sole discretion, may include by joinder persons or entities substantially involved in a common question of law or fact whose presence is required if complete relief is to be accorded in arbitration, provided that the party sought to be joined consents in writing to such joinder. Consent to arbitration involving an additional person or entity shall not constitute consent to arbitration of any claim, dispute or other matter in question not described in the written consent.* **(new)**

Either the owner or the contractor may bring in any other party to an arbitration between the owner and contractor, if there is a common question of law or fact and the presence of that third party is required to grant "complete relief." The American Arbitration Association has a rule that covers the situation where one or more parties in two different arbitrations that are both administered under AAA rules object to consolidation of the arbitrations.[472] If the parties have agreed to the

---

owner and the architect for negligence. The owner and the architect then filed third-party complaints against the contractor. The contractor could force the owner to arbitrate its third-party complaint, but this provision barred arbitration against the architect.

[471] Carteret County v. United Contractors, Inc., 462 S.E.2d 816 (N.C. Ct. App. 1995).

[472] Rule R-7, Consolidation or Joinder, states:

If the parties' agreement or the law provides for consolidation or joinder of related arbitrations, all involved parties will endeavor to agree on a process to effectuate the consolidation or joinder.

AAA rules, then it will be an arbitrator, and not a court, that determines whether particular arbitrations are subject to joinder.

15.4.4.3 The Owner and Contractor grant to any person or entity made a party to an arbitration conducted under this Section 15.4, whether by joinder or consolidation, the same rights of joinder and consolidation as the Owner and Contractor under this Agreement. **(4.6.4; substantial revisions)**

This provision allows joined parties to the arbitration to bring in others who are "substantially involved in a common question of law or fact" and who are necessary parties. This particular provision does not require that the arbitration agreements be substantially similar to this document, but it does require that the agreements between these other parties allow for arbitration or that these parties agree to arbitrate.

If they are unable to agree, the Association shall directly appoint a single arbitrator for the limited purpose of deciding whether related arbitrations should be consolidated or joined and, if so, establishing a fair and appropriate process for consolidation or joinder. The AAA may take reasonable administrative action to accomplish the consolidation or joinder as directed by the arbitrator.

CHAPTER 5

ALTERNATIVE LANGUAGE FOR AIA DOCUMENT A201

§ 5.1   Introduction

The language in this chapter is offered as possible alternative language to AIA Document A201. The user is cautioned that some of these provisions are mutually incompatible, and each paragraph and each sentence should be carefully reviewed with legal counsel before use. Note further that some of these changes require changes to other documents.

The AIA has also made these documents available in electronic format, so that changes can be made by computer directly on the document. This results in a single document incorporating all changes. The next-best method of amending A201 is to attach separate written amendments that refer back to A201. This is normally done in supplementary general conditions. An alternative method is to graphically delete material and insert new material in the margins of the standard form. However, this method may create ambiguity, and it may result in the stricken material's being rendered illegible, and is the least preferred method.

The following introductory paragraph may be added at the beginning of the Supplementary Conditions, if that method is utilized:

The Supplementary Conditions contain modifications and additions to the General Conditions of the Contract for Construction, AIA Document A201, 2007 Edition. Where any part of the AIA General Conditions is modified or voided by the Supplementary Conditions, the unaltered portions shall remain in effect. The paragraph numbering system of AIA Document A-201, 2007 Edition is continued in the Supplementary Conditions.

§ 5.2   Basic Definitions: ¶ 1.1

The bidding documents could be included in the contract documents by striking the last sentence of ¶ 1.1.1 and adding the following:

> Bidding Documents, including but not limited to advertisement or invitation to bid, Instructions to Bidders, other information furnished by the Owner in anticipation of receiving bids or proposals, the Contractor's Bid, and addenda or portions of addenda relating to any bidding documents.

Another:

> The Contract Documents shall include the bidding documents as listed in the Instructions to Bidders and any alterations made thereto by addenda. In the event of a conflict or contradiction within the Contract Documents and for the resolution of same, the following order of hierarchy shall prevail:
>
> 1) Contract; 2) Addenda; 3) Supplementary General Conditions; 4) General Conditions; 5) Specifications; 6) Drawings; 7) Instructions To Bidders; 8) Invitation To Bid; 9) Sample Forms.

Another alternative is to add, in ¶ 1.1.1, first sentence, after the word "Specifications":

> Bidding Documents, including without limitations the Advertisement or Invitation to Bid, the Instructions to Bidders, the Bid Proposal Form, other information furnished by the Owner in anticipation of receiving bids or proposals, the Contractor's Bid, and addenda or portions thereof relating to the Bidding Document.

Delete the last sentence of this subparagraph.

The owner may want to insert the following provision at the end of this paragraph:

> **1.1.1.1** The Contractor acknowledges and warrants that it has closely examined all the Contract Documents, that they are suitable and sufficient to enable the Contractor to complete the Work in a timely manner for the Contract Sum, and that they include all work, whether or not shown or described, which reasonably may be inferred to be required or useful for the completion of the Work in full compliance with all applicable codes, laws, ordinances, and regulations.

Additional language under ¶ 1.1.3 the owner could use to further define the work is:

> **1.1.3.1** The Work shall include the obligation of the Contractor to visit the site of the project before submitting a proposal. Such site visit shall be for the purpose of familiarizing the Contractor with the conditions as they exist and the character of the operations to be carried on under the Contract Documents,

including all existing site conditions, access to the site, physical characteristics of the site and surrounding areas.

The following language could be added to clarify that the owner and architect have no responsibility for nonwork items:

1.1.3.2 Nothing in these General Conditions shall be interpreted as imposing on either the Owner or Architect, or their respective agents, employees, officers, directors, or consultants, any duty, obligation, or authority with respect to any items that are not intended to be incorporated into the completed project, or that do not comprise the Work, including but not limited to the following: shoring, scaffolding, hoists, weatherproofing, or any temporary facility or activity, because these are the sole responsibility of the Contractor.

Another:

1.1.3 The term "Work" shall mean supervision, labor, equipment, tools, material, supplies, incidentals operations and activities required by the Contract Documents or reasonably inferable by Contractor therefrom as necessary to produce the results intended by the Contract Documents, in a safe, expeditious, orderly, and workmanlike manner, and in the best manner known to each respective trade.

If shop drawings are to be included as part of the contract documents, modify ¶ 1.1.5 to include "shop drawings following approval by the Architect," and also modify ¶ 3.12.4. The following additional provisions could be added to the definitions section:

1.1.9 ADDENDA

Addenda are written or graphic instruments issued by the Architect prior to receipt of the bids which modify or interpret the Bidding Documents by additions, deletions, clarifications, or corrections.

1.1.10 BID

A Bid is a complete and properly signed proposal to do the Work for the sums stipulated therein, submitted in accordance with the Bidding Documents.

Acceptance of the Bid by the Owner will result in a contract between the Owner and Contractor.

1.1.11 BASE BID

The Base Bid is the sum stated in the Bid for which the Bidder offers to perform the Work described in the Bidding Documents as the base, to which Work may be added or from which Work may be deleted for sums stated in Alternate Bids.

1.1.12 ALTERNATE BID

An Alternate Bid (or Alternate) is the amount stated in the Bid to be added or deducted from the amount of the Base Bid if the corresponding change in the Work, as described in the Bidding Documents, is accepted.

1.1.13 BIDDER

A Bidder is a person or entity who submits a Bid.

1.1.14 SUB-BIDDER

A Sub-bidder is a Person or entity who submits a Bid to a Bidder for materials, equipment, or labor for a portion of the Work.

1.1.15 FINAL COMPLETION

Final Completion is achieved at the time that final inspection has been performed by the Architect and the final Certificate for Payment issued by the Architect to the Owner.

1.1.16 NOT IN CONTRACT (NIC)

Work not included in the work of this Contractor.

1.1.17 EQUAL TO (OR APPROVED EQUAL)

Products by manufacturers other than those specified in the Contract Documents which the Contractor may submit for substitution as equal to those specified in the Contract Documents and which may be incorporated in the Work after review and acceptance by the Architect of the information about such products and acceptance by the Owner.

Another version of this:

1.1.17.1 Whenever in the Contract Documents any article, appliance, device, or material is designated by the name of a manufacturer, vendor, or by any proprietary or trade name, the words "or approved equal," shall automatically follow and shall be implied unless specifically indicated otherwise. The standard products of manufacturers other than those specified will be accepted when, prior to the ordering or use thereof, it is proven to the satisfaction of the Owner and the Architect they are equal in design, appearance, spare parts availability, strength, durability, usefulness, serviceability, operation cost, maintenance cost, and convenience for the purpose intended. Any general listings of approved manufacturers in any Contract Document shall be for informational purposes only and it shall be the Contractor's sole responsibility to ensure that any proposed "or equal" complies with the requirements of the Contract Documents.

1.1.17.2 The Contractor shall submit to Owner and Architect a written and full description of the proposed "or equal" including all supporting data, including technical information, catalog cuts, warranties, test results, installation instructions, operating procedures, and similar information demonstrating that the proposed "or equal" strictly complies with the Contract Documents. The Architect or Owner shall take appropriate action with respect to the submission of a proposed "or equal" item. If Contractor fails to submit proposed "or equals" as set forth herein, it shall waive any right to supply such items. The Contract Sum and Contract Time shall not be adjusted as a result of any failure by Contractor to submit proposed "or equals" as provided for herein. All documents submitted in connection with preparing an "or equal" shall be clearly and obviously marked as a proposed "or equal" submission.

1.1.17.3 No approvals or action taken by the Architect or Owner shall relieve Contractor from its obligation to ensure that an "or equal" article, appliance, devise or material strictly complies with the requirements of the Contract Documents. Contractor shall not propose "or equal" items in connection with Shop Drawings or other Submittals, and Contractor acknowledges and agrees that no approvals or action taken by the Architect or Owner with respect to Shop Drawings or other Submittals shall constitute approval of any "or equal" item or relieve Contractor from its sole and exclusive responsibility. Any changes required in the details and dimensions indicated in the Contract Documents for the incorporation or installation of any "or equal" item supplied by the Contractor shall be properly made and approved by the Architect at the expense of the Contractor. No "or equal" items will be permitted for components of or extensions to existing systems when, in the opinion of the Architect, the named manufacturer must be provided in order to ensure compatibility with the existing systems, including, but not limited to, mechanical systems, electrical systems, fire alarms, smoke detectors, etc. No action will be taken by the Architect with respect to proposed "or equal" items prior to receipt of bids, unless otherwise noted in the Special Conditions.

1.1.18 FORCE MAJEURE

An act of God, fire, tornado, hurricane, flood, earthquake, explosion, war on American soil, civil disturbance, labor strikes, and similar unavoidable circumstances beyond Contractor's control, not caused by the negligent act or omission of Contractor or breach of this Agreement, its Subcontractors, or anyone else for whom Contractor is responsible, and not caused by Contractor's breach of a project labor or a "no strike" agreement.

Another Force Majeure clause:

The time for Completion of Construction shall be extended for the period of any reasonable delay which is due exclusively to causes beyond the control and without the fault of [Hutton], including Acts of God, fires, floods and acts or omissions of the [City] with respect to matters for which the [City] is solely responsible: Provided, however, that no extension of time for completion shall be granted [Hutton] unless within ten (10) days after the happening of any event relied upon by [Hutton] for such an extension of time [Hutton] shall have made a request therefor in writing to the [City], and provided further that no delay in such time of completion or in the progress of the work which results from any of the above causes or from any changes in construction which may be made pursuant to Subsection "d" of this Section 1 [which allowed the City to modify the construction plans and provide extensions therefor] shall result in any liability on the part of the [City]. Time extensions due to weather will be considered only when [Hutton] is on site.[1]

1.1.19 HAZARDOUS MATERIALS

Includes friable asbestos or asbestos-containing materials, polychlorinated biphenyls (PCBs), petroleum products, natural gas, source material, special nuclear materials, and by-product materials regulated under the Atomic Energy Act (42 U.S.C. §§ 2011, *et seq.*), pesticides regulated under the Federal Insecticide, Fungicide and Rodenticide Act (7 U.S.C. §§ 136, *et seq.*), and any hazardous waste, toxic or dangerous substance or related material, including any substance defined, determined or identified as "hazardous waste," or "toxic substance," or "contaminant" (or comparable term) in any Environmental Law.

1.1.20 PUNCH LIST

A written list of items that are minor, uncompleted, or unacceptable items of the Work that do not interfere with the Owner's use and occupancy of any part of the Project.

1.1.20 PUNCH LIST

"Punch List" means the list of items, prepared in connection with the inspection of the Project by the Owner and/or Architect in connection with Substantial Completion of the Work or a portion of the Work, which the Owner or Architect has designated as remaining to be performed, completed or corrected before the Work will be accepted by the Owner.

1.1.21 KNOWLEDGE

The terms "knowledge," "recognize" and "discover," their respective derivatives and similar terms in the Contract Documents, as used in reference to the Contractor, shall be interpreted to mean that which the Contractor knows or should know, recognizes or should recognize and discovers or should discover in exercising the care, skill, and diligence of a diligent and prudent contractor familiar with the work. Analogously, the expression "reasonably inferable" and similar terms in the Contract Documents shall be interpreted to mean reasonably inferable by a diligent and prudent contractor familiar with the work and in accordance with the highest standards in the contracting profession.

---

[1] *See* Hutton Contracting Co., v. City of Coffeyville, 487 F.3d 772 (10th Cir. 2007).

These are modifications that could be adapted for ¶ 1.1.2:

1.1.2.1 Contractor acknowledges that, except only for those representations, statements or promises expressly contained in this Agreement and any exhibits attached to it and incorporated by reference in it, no representation, statement or promise, oral or in writing, of any kind whatsoever, by the Owner, its officers, directors, agents or employees, has induced Contractor to enter into this Agreement or has been relied upon by Contractor, including any with reference to: (i) the meaning, correctness, suitability or completeness of any provisions or requirements of this Agreement; (ii) the nature of the Work to be performed; (iii) the nature, quantity, quality or volume of any materials, equipment, labor and other facilities needed for the performance of this Agreement; (iv) the General Conditions which may in any way affect this Agreement or its performance; (v) the compensation provisions of this Agreement; or (vi) any other matters, whether similar to or different from those referred to in (i) through (v) immediately above, affecting or having any connection with this Agreement, its negotiation, any discussions of its performance or those employed or connected or concerned with it.

1.1.2.2 Contractor acknowledges that Contractor was given ample opportunity and time and was requested by the Owner to review thoroughly all documents, including the Drawings and Specifications and Addenda, forming this Agreement before signing this Agreement in order that it might request inclusion in this Agreement of any statement, representation, promise or provision that it desired or on which it wished to place reliance. Contractor did so review those documents, and either every such statement, representation, promise or provision has been included in this Agreement or else, if omitted, Contractor relinquishes the benefit of any such omitted statement, representation, promise or provision and is willing to perform this Agreement in its entirety without claiming reliance on it or making any other claim on account of its omission.

1.1.2.3 This Agreement was the result of negotiations between the Owner and Contractor, and has been reviewed by the Owner, Contractor and their respective counsel. Accordingly, this Agreement shall be deemed to be the product of both parties and no ambiguity shall be construed in favor of or against either party.

§ 5.3   Execution, Correlation, and Intent: ¶ 1.2

A possible additional paragraph under ¶ 1.2.1 to establish an order of precedence for the contract documents is the following:

1.2.3.1 In the event of conflicting provisions, specifications [or the drawings] will take precedence over the drawings [or the specifications]. Notwithstanding the foregoing, the more specific provision will take precedence over the less specific; the more stringent will take precedence over the less stringent; the more expensive item will take precedence over the less expensive. On all drawings, figures take precedence over scaled dimensions. Scaling of dimensions, if done, is done at the Contractor's own risk.

Another provision:

1.2.3.1 In the event of conflicting provisions among the Contract Documents that were not called to the Owner's or Architect's attention prior to award of the Contract, the Architect shall determine which of the conflicting requirements shall govern, generally taking as a guideline the more stringent requirement or more expensive material, unless, in the opinion of the Architect, another requirement is more appropriate. The Architect's decision shall be final in such case, and the Architect's decision shall not be further reviewable by arbitration or by litigation. [Alternatively to this last sentence: The Architect's decision shall be final unless the Contractor shall file a Demand for Arbitration [or lawsuit] within seven (7) calendar days of receiving the Architect's written decision in this matter.]

Another:

1.2.1 The Specifications are that portion of the Contract Documents consisting of the written requirements for materials, equipment, construction system, standards and workmanship and performance of related services for the Work identified in the Contract for Construction. Specifications are separated into divisions for convenience of reference only. Organization of the Specifications into divisions, sections and articles, and arrangement of Drawings shall not control the Contractor in dividing the Work among Subcontractors or in establishing the extent of Work to be performed by any trade. Such separation will not operate to make the Owner or the Architect an arbiter of labor disputes or work agreements.

1.2.2 The drawings herein referred to, consist of drawings prepared by the Architect and are enumerated in the Contract Documents.

1.2.3 Drawings are intended to show general arrangements, design, and dimensions of work and are partly diagrammatic. Dimensions shall not be determined by scale or rule. If figured dimensions are lacking, they shall be supplied by the Architect on the Contractor's written request to the Owner.

1.2.4 The intent of the Contract Documents is to include all items necessary for the proper execution and completion of the Work by the Contractor. The Contract Documents are complimentary, and what is required by one shall be as binding as if required by all; performance by the Contractor shall be required only to the extent consistent with the Contract Documents and reasonably inferable from them as being necessary to produce the intended results.

1.2.5 In the event of inconsistencies within or between parts of the Contract Documents, or between the Contract Documents and applicable standards, codes and ordinances, the Contractor shall (1) provide the better quality or greater quantity of Work or (2) comply with the more stringent requirement; either or both in accordance with the Owner's and/or Architect's interpretation. On the Drawings, given dimensions shall take precedence over scaled measurements, and large scale drawings over small scale drawings. Before ordering any materials or doing any Work, the Contractor and each Subcontractor shall verify measurements at the Work site and shall be responsible for the correctness of such measurements. Any difference which may be found shall be submitted to the Owner's Representative and Architect for resolution before proceeding with the Work. If a minor change in the Work is found necessary due to actual field conditions, the Contractor shall submit detailed drawings of such departure for the approval by the Owner and Architect before making the change.

The following provisions may be added to shift responsibility to the contractor regarding discrepancies among the drawings and specifications:

1.2.4 If work is required in a manner to make it impossible to produce work of the quality required by or reasonably inferred from the Contract Documents, or should discrepancies appear among the Contract Documents, the Contractor shall request in writing an interpretation from the Architect before proceeding with the work. If the Contractor fails to make such request, no excuse will thereafter be entertained for failure to carry out work in the required manner or provide required guarantees, warranties, or bonds, and the Contractor shall not be entitled to any change in the Contract Sum or the Contract Time on account of such failure.

1.2.5 Should conflict occur in or between Drawings and Specifications, the Contractor is deemed to have included the better quality and larger quantity of work in the Bid.

§ 5.4   Ownership and Use of Drawings, Specifications, and Other Instruments of Service: ¶ 1.5

The general contractor faces potential liability if the plans and specifications infringe on the copyright of some other party. Often, the contractor will have no way of verifying whether or not these plans and specifications are original, or whether some other architect may file a copyright infringement claim against the owner and contractor. Such a filing will result in substantial costs to the contractor and will usually delay the project, resulting in additional costs. The contractor may want to add language similar to the following:

1.5.3 The Owner warrants to the Contractor that the Owner has the necessary licenses in order to grant to the Contractor a license sufficient to permit the Contractor to complete the Project. Owner agrees to indemnify Contractor from any and all losses, claims and damages, including attorneys' fees and costs, incurred by the Contractor relative to any suit alleging any breach of copyright brought by any party. Further, Contractor shall be entitled to additional compensation if the Project is delayed or stopped as a result of the filing of any action, in court or otherwise, by any person alleging copyright infringement relative to the Project, unless due to the fault of Contractor. This warranty shall survive the termination of this Agreement.

If the owner has obtained the copyright (ownership of the actual copyright, not merely a license) from the design professionals, change ¶ 1.5 as follows:

1.5.1 The Drawings, Specifications and other documents, including those in electronic form, prepared by the Architect and the Architect's consultants are Instruments of Service and are for use solely with respect to this Project. The Owner shall retain all common-law, statutory and other reserved rights, including copyrights, in accordance with the Owner/Architect Agreement. The Contractor, Subcontractors, Sub-subcontractors, and material or equipment suppliers shall not own or claim a copyright in the Instruments of Service. Submittal or distribution to meet official regulatory requirements or for other purposes in connection with this Project is not to be construed as a waiver of the Owner's reserved rights.

1.5.2 The Contractor, Subcontractors, Sub-subcontractors and material or equipment suppliers are authorized to use and reproduce applicable portions of the Instruments of Service furnished to them but solely for use in the execution of their Work. All copies made under this authorization shall bear the statutory copyright notice, if any, shown on the Instruments of Service. The Contractor, Subcontractors, Sub-subcontractors, and material or equipment suppliers may not use the Instruments of Service on other projects without the specific written consent of the Owner.

§ 5.5   Transmittal of Digital Data: ¶ 1.6

The parties can reference the recent AIA document E201-2007, Digital Data Protocol Exhibit:

1.6.1 AIA Document E201-2007 is hereby made a part of the Contract Documents.

§ 5.6   Owner—General: ¶ 2.1

Many owners will want to delete ¶ 2.1.2 so as to avoid having to give information to the contractor that would enable the contractor to file a lien against the property. Another option is to modify that paragraph:

2.1.2 As a condition precedent to the recording or perfection of a mechanic's lien, Contractor shall give Owner seven (7) days' prior written notice and opportunity to cure. Said notice shall include any and all documents that Contractor intends to record as well as a statement specifying which invoices have not been timely paid.

§ 5.7   Information and Services Required of the Owner: ¶ 2.2

This provision shifts the burden of surveys and other information to the contractor. If this provision is included, ¶ 3.7.3 should be replaced:

2.2.3 The Owner shall not be responsible for furnishing surveys or other information as to the physical characteristics, legal limitations, and utility locations for the site of the Project. However, the Owner shall furnish to the Contractor a legal description of the Project site, which shall not be a Contract Document. The Contractor shall confirm the location of each utility. The Contractor represents that it has inspected the site and is satisfied as to the condition thereof, including, without limitation, all structural, surface, and subsurface conditions thereof. The Contractor shall undertake to make such further investigations, including soil borings and otherwise, to determine the subsurface conditions. The Owner may, at the Contractor's request, make available to the Contractor such soil borings, tests, and reports as it may have available, but in no event shall such information constitute any warranty or representation by the Owner as to the accuracy of such information, and the Contractor shall not be entitled to rely on the accuracy of such information furnished by the Owner. The Contractor shall make no claims for any subsurface conditions.

Another provision:

2.2.3.1 The furnishing of such information by the Owner shall not relieve the Contractor from its duties under the Contract Documents, specifically as to inspection of the site and the Contract Documents. The Owner shall not be required to furnish the Contractor with any information as to subsurface conditions. If the Owner or the Architect has made any investigations of subsurface conditions, such investigations were made solely for the information of the Owner and Architect and not for the Contractor's information. No such information shall be construed to be a part of the Contract Documents. The Contractor acknowledges that, if the Owner or Architect furnishes any such information to the Contractor, no waiver of the foregoing shall be implied, and the Contractor shall not be entitled to rely on such information but rather shall conduct its own investigation of such subsurface conditions. Further, no warranty of the accuracy of any such information shall be implied. The Contractor warrants that it is experienced in the type of Work undertaken pursuant to this contract and has the necessary expertise to form its own conclusions as to the necessity for conducting investigations of a type and nature as is calculated by Contractor to provide it with the necessary information so as to properly carry out the Work hereunder. If the Contractor discovers conditions that vary from those that it anticipated, whether such anticipation was reasonable or not, the Contractor's sole remedy against the Owner will be an extension of the Contract

Time, but in no event will such condition entitle the Contractor to an increase in the Contract Sum.

A possible clause to add to ¶ 2.2.4 is as follows:

The Contractor shall, within 21 days of receipt of any information furnished by the Owner pursuant to this Paragraph, verify and confirm the accuracy of information so furnished. In the event of any inaccuracies, the Contractor shall promptly notify the Owner, who shall correct any such inaccuracy. Failure to notify the Owner within the said 21 days shall act to bar any claims by the Contractor arising from the inaccuracy of any such information.

The following is a more generous provision for the contractor regarding the number of drawings to be furnished:

2.2.5 The Owner will furnish to the Contractor ten (10) sets of Drawings and Specifications at no cost. The Contractor may obtain additional copies at cost of reproduction.

Another clause is similar to the AIA provision. Note that this language refers to the Project Manual, a concept that was deleted from the 2007 version of A201. If this language is used, a definition from the prior version of A201 should be included:

2.2.5 The Owner shall furnish the Contractor with one set of reproducible Drawings and one Project Manual. The Contractor is given permission to make copies of these documents for use on this Project. All costs of reproduction are the responsibility of the Contractor.

Another version of this section:

2.2 INFORMATION AND SERVICES REQUIRED OF OWNER

2.2.1 Owner may furnish surveys describing physical characteristics, legal limitations, and utility locations for the site of the Project, and a legal description of the site. The furnishing of these surveys and the legal description of the site is for the convenience of Contractor only and shall not relieve Contractor from its duties under the Contract Documents in general. Neither Owner nor the Architect shall be required to furnish Contractor with any information concerning subsurface characteristics or concealed conditions of the areas where the Work is to be performed. If Owner or Architect has made investigations of subsurface characteristics or concealed conditions of the areas where the Work is to be performed, such investigations, if any, were made solely for the purposes of Owner's study and Architect's design. Neither such investigations nor the records thereof are a part of the Contract between Owner and Contractor. To the extent such investigations or the records thereof are made available to Contractor by Owner or Architect, such information is furnished solely for the convenience of Contractor. Neither Owner nor Architect assumes any responsibility whatsoever with respect to the sufficiency or accuracy of the investigations thus made, the records thereof, or of the interpretations set forth therein or made by Owner or Architect in its use thereof, and there is no warranty or guaranty, either express or implied, that the conditions indicated by such investigations or records thereof are representative of those existing throughout the areas where the Work is to be performed, or any part thereof, or that unforeseen developments may not occur, or that materials other than or in proportions different from those indicated may not be encountered. Contractor shall undertake such further investigations and studies as might be necessary or useful to determine subsurface characteristics and patent and concealed conditions. In connection with the foregoing, Contractor shall be solely responsible for locating and shall locate prior to performing any Work all utility lines, telephone lines and cables, sewer lines, water pipes, gas lines, electrical lines, including, without limitation, all buried pipelines and buried telephone cables, and shall perform the Work in such a manner so as to avoid damaging any such lines, cables, pipes, and pipelines.

2.2.2 Except for those permits and fees which are the responsibility of Contractor under the Contract Documents, Owner shall secure and pay for necessary approvals, easements, assessments, and charges required for construction, use, or occupancy of permanent structures or for permanent changes in existing facilities. Such approvals and the like shall be provided by Owner within a time and in a manner so as to avoid any unreasonable delays in the Work or Progress Schedule and shall include only such approvals for permanent facilities which are necessary to perform the Work as set forth in the Contract Documents.

2.2.3 Upon the written request of Contractor, information or services required to be furnished by Owner shall be furnished by Owner with reasonable promptness to avoid unreasonable delay in the orderly progress of the Work.

2.2.4 After execution of the Contract for Construction and prior to commencement of the Work, Contractor will be furnished, without charge, one complete set of reproducible Drawings, one Project Manual, and one copy of each Addendum free of charge.

Another version:

2.2.3 The Owner may provide, at its own cost and expense for Contractor's information and use at Contractor's risk, the following information:

2.2.3.1 Surveys describing the property, boundaries, topography and reference points for use during construction, including existing service and utility lines;

2.2.3.2 Geotechnical studies describing subsurface conditions, and other surveys describing other physical conditions at the Site;

2.2.3.3 Temporary and permanent easements, zoning and other requirements and encumbrances affecting land use, or necessary to permit the proper design and construction of the Project;

2.2.3.4 To the extent available, as-built and record drawings of any existing structures at the Site; and

2.2.3.5 To the extent available, environmental studies, reports and impact statements describing the environmental conditions, including Hazardous Conditions, in existence at the Site.

2.2.4 The information identified in Section 2.2.3, above, is not warranted or represented by the Owner to be accurate. The Contractor will not be entitled to rely on it and if the Contractor does rely on such information, then Contractor does so at its own risk. When such information is provided by the Owner, the Contractor acknowledges that the Owner has not verified such information. Site plans prepared by Owner's design professionals or others are based on surveys performed by consultants, and have not been verified by the Owner. Site plans do not constitute any representation by the Owner to the Contractor of Site boundaries or other characteristics.

Another provision:

2.2.4 Information or services required of the Owner by the Contract Documents shall be furnished by the Owner with reasonable promptness. Any other information or services relevant to the Contractor's performance of the Work under the Owner's control shall be furnished by the Owner after receipt from the Contractor of a written request for such information or services. Neither the Owner nor the Architect shall be required to conduct investigations or to furnish the Contractor with any information concerning subsurface conditions or other conditions regarding the area where the Work is to be performed beyond that which is provided in the Contract Documents. The Contractor shall not be entitled to rely on the accuracy of any information or services furnished pursuant to this Paragraph that is not shown on the Contract Documents.

§ 5.8   Owner's Right to Stop the Work: ¶ 2.3

An alternative provision:

2.3.1 If Contractor fails to correct Work that is not in accordance with the requirements of the Contract Documents as required by Section 12.2; or fails to carry out the Work in accordance with the Contract Documents; or fails or refuses to provide a sufficient amount of properly supervised and coordinated labor, materials, or equipment so as to be able to complete the Work within the Contract Time; or fails to remove and discharge (within ten (10) days) any lien filed upon Owner's or Landlord's property by anyone claiming by, through, or under Contractor; or disregards the instructions of Architect or Owner when based on the requirements of the Contract Documents, Owner may order Contractor to stop the Work, or any portion thereof, until the cause for such order has been eliminated; provided, however, the right of Owner to stop the Work shall not give rise to a duty on the part of Owner to exercise this right for the benefit of

Contractor or any other person or entity, and any delay resulting from such work stoppage shall not extend any Milestone Date identified in the Contract for Construction or the required dates of Substantial or Final Completion.

Another alternative provides as follows:

2.3.1 If the Contractor fails to correct Work that is not in accordance with the requirements of the Contract Documents as required by Paragraph 12.2 or fails to carry out Work in accordance with the Contract Documents, the Owner may issue a written order to the Contractor to stop the Work, or any portion thereof, until the cause for such order has been eliminated; however, the right of the Owner to stop the Work shall not give rise to a duty on the part of the Owner to exercise this right for the benefit of the Contractor or any other person or entity, except to the extent required by Subparagraph 6.1.3, nor shall the exercise of the Owner's right hereunder give rise to any claim by the Contractor for additions to the Contract Sum or Contract Time.

Another provision:

2.3.2 If, after consultation with the Architect, suspension of the Work is warranted by reason of unforeseen conditions which may adversely affect the quality of the Work if such Work were continued, the Owner may suspend the Work by written notice to the Contractor. In such event, the Contract Time shall be adjusted accordingly, and the Guaranteed Maximum Price shall be adjusted by Change Order to the extent, if any, that additional costs are incurred by reason of such suspension. If the Contractor, in its reasonable judgment, believes that a suspension is warranted by reason of unforeseen circumstances which may adversely affect the quality of the Work if the Work were continued, the Contractor shall immediately notify the Owner and the Architect of such belief.

§ 5.9   Owner's Right to Carry Out the Work: ¶ 2.4

An alternative paragraph to the notice provisions of ¶ 2.4.1 is as follows:

2.4.1 If the Contractor defaults or neglects to carry out the Work in accordance with the Contract Documents and fails within a ten-day period after receipt of written notice from the Owner to commence and continue correction of such default or neglect with diligence and promptness, the Owner may, after such ten-day period and without prejudice to other remedies the Owner may have, immediately correct such deficiencies. In such case, an appropriate Change Order shall be issued, deducting from payments then or thereafter due the Contractor the cost of correcting such deficiencies, including compensation for the Architect's additional services and expenses made necessary by such default, neglect, or failure, including, without limitation, the Owner's reasonable attorneys' fees. Such action by the Owner and amounts charged to the Contractor are both subject to the prior approval of the Architect. Such Change

Order shall be deemed to have been executed by the Contractor, whether or not actually signed by the Contractor, unless the Change Order is shown to have been prepared in bad faith by the Owner and Architect. If payments then or thereafter due the Contractor are not sufficient to cover such amounts, the Contractor shall pay the difference to the Owner.

Add the following new ¶ 2.5, including 2.5.1:

2.5 ADDITIONAL RIGHTS

2.5.1 The rights stated in Article 2 shall be in addition and not in limitation of any other rights of the Owner granted in the Contract Documents or at law or in equity.

Another, more onerous, alternative provision:

2.4.1 If the Contractor fails to carry out the Work in accordance with the Contract Documents and fails within a two-business day period after written notice from the Owner or Architect to eliminate (or commence to eliminate and thereafter work diligently to eliminate) such failure, the Owner may, regardless of whether an Event of Default has occurred and without prejudice to other remedies Owner may have, correct such deficiencies. In such case, an offset may be made, deducting from payments then or thereafter due Contractor the cost of correcting such deficiencies, including compensation for Owner's and Architect's additional services and expenses made necessary by such default, neglect, or failure. If payments then or thereafter due Contractor are not sufficient to cover such amounts, Contractor shall promptly pay the difference to Owner. The correction of such deficiencies by Owner or by others shall not relieve Contractor of any obligation or liability for the Work and shall not operate to waive any right or claim of Owner.

§ 5.10   Miscellaneous Owner's Provisions: ¶ 2.5

The following additional provision could be inserted as ¶ 2.5 when the owner wants the right to audit the contractor's books, especially in cost-plus contracts or in case of shared savings:

2.5 OWNER'S RIGHT TO AUDIT

2.5.1 The Contractor shall keep full and accurate records of all costs incurred and items billed in connection with the performance of the Work, which records shall be open to audit by the Owner or its authorized representatives during performance of the Work and until three years after Final Payment. In addition, the Contractor shall make it a condition of all subcontracts relating to the Work that any and all Subcontractors will keep accurate records of costs incurred and items billed in connection with their work and that such records shall be open to audit by the Owner or its authorized representatives during performance of the Work and until two years after its completion.

Another provision:

2.5 OWNER'S RIGHT TO PERSONNEL

2.5.1 The Owner reserves the right to have the Contractor and/or subcontractors remove person(s) and/or personnel from any and all work on the project <u>with cause but without cost</u> to the Owner. Such requests from the Owner may be made verbally or in writing and may be done directly or indirectly through the Architect/Engineer or on-site representative. Cause may be, but not limited to, any of the following: incompetence, poor workmanship, poor scheduling abilities, poor coordination, disruptive to the facility or others, poor management, cause delay or delays, disruptive to the project, will not strictly adhere to facility procedures and project requirements either willfully or unknowingly, insubordination, drug/alcohol use, possession of contraband, belligerent acts or actions, etc. The Contractor shall provide replacement person(s) and/or personnel acceptable to the Owner at no cost to the Owner.

2.5.2 Any issue or circumstance relating to or resulting out of this clause shall not be construed or interpreted to be interference with or impacting upon the Contractor's responsibilities and liabilities under the Contract Documents.

2.5.3 Person(s) and/or personnel who do not perform in accordance with the Contract Documents, shall be deemed to have provided the Owner with cause to have such persons removed from any and all involvement in the Work.

2.5.4 The Contractor agrees to indemnify and hold harmless the Owner from any and all causes of action, demands, claims, damages, awards, attorneys' fees, and other costs brought against the Owner and/or Architect/Engineer by any and all person(s) or personnel.

§ 5.11   Contractor—General: ¶ 3.1

An additional provision:

3.1.1.1 The Contractor warrants to Owner that it is currently licensed in the jurisdiction where the Project is located to perform the work required by the Contract Documents. If, at any time, the Contractor's license lapses, Contractor shall make every effort to immediately reinstate all required licenses, and Contractor shall be responsible to the Owner for all delay and other costs incurred by Owner as a result of such lapse, including any additional costs of hiring substitute contractors. This provision shall apply notwithstanding the provisions of Section 15.1.6.

Another provision to be added to ¶ 3.1.3:

Quality Control (i.e. ensuring compliance with the Contract Documents) is the responsibility of the Contractor. Testing, observations and/or inspections performed or provided by the Owner are for Quality Assurance (i.e. confirming compliance with the Contract Documents) purposes and are for the benefit of the Owner.

§ 5.12   Review of Contract Documents and Field Conditions by Contractor: ¶ 3.2

Stronger language could be substituted for ¶ 3.2.1 to assure that the contractor has carefully studied the project and project site:

3.2.1 Execution of the Contract by the Contractor is a representation that the Contractor has carefully examined the Contract Documents and the site, and represents that the Contractor is thoroughly familiar with the nature and location of the Work, the site, the specific conditions under which the Work is to be performed, and all matters that may in any way affect the Work or its performance. The Contractor further represents that as a result of such examinations and investigations, the Contractor thoroughly understands the Contract Documents and their intent and purpose, and is familiar with all applicable codes, ordinances, laws, regulations, and rules as they apply to the Work, and that the Contractor will abide by same. Claims for additional time or additional compensation as a result of the Contractor's failure to follow the foregoing procedure and to familiarize itself with all local conditions and the Contract Documents will not be permitted.

Another provision:

3.2.1 Execution of the Contract by the Contractor is a representation that the Contract Documents are complete, are sufficient to have enabled the Contractor to determine the cost of the Work shown therein, and that the Contract Documents are sufficient to enable it to construct the work shown therein, and otherwise to fulfill all its obligations hereunder, including but not limited to Contractor's obligation to construct the Work for an amount not in excess of the Contract Sum on or before the date of Substantial Completion established in the Agreement. The Contractor further acknowledges and declares that it has visited and examined the site, examined all conditions affecting the Work of every kind and nature, and is fully familiar with all of such conditions. In connection therewith, the Contractor specifically represents and warrants to the Owner that it has, by careful examination, satisfied himself as to: (a) the nature, location, and character of the site, including, without limitation, the surface and subsurface conditions of the site and all structures and obstructions pertaining thereto, both natural and manmade, and all surface and subsurface water conditions of the site and the surrounding area; (b) the nature and character of the area in which the Project is located, including, without limitation, its climatic conditions, available labor supply and labor costs, and available equipment supply and equipment costs; and (c) the quality and quantity of all materials, supplies, tools, equipment, labor, and services of any kind necessary to complete the Work within the cost and timeframe required by the Contract Documents and otherwise in accordance with the Contract Documents. In connection with the foregoing, after having carefully examined all Contract Documents, and having visited the site, the Contractor acknowledges and declares that it has no knowledge of any discrepancies, omissions, ambiguities, or conflicts in and among said Contract Documents and that if it becomes aware of any such discrepancies, omissions, ambiguities, or conflicts, it will promptly notify Owner and Architect of such fact.

Another provision:

3.2.1.1 Each bidder shall visit the site of the work and thoroughly inform itself relative to construction hazards and procedure, labor, and all other conditions and factors, local and otherwise, that would affect the prosecution and completion of the work and the cost thereof. It must be understood and agreed that all such factors have been properly investigated and considered in the preparation of every proposal submitted, as there will be no subsequent financial adjustment, to any contract awarded thereunder, which is based on the lack of such prior information or its effect on the cost of the work.[2]

Another provision:

3.2.1.1 The Contractor acknowledges that it has taken all measures reasonably necessary to verify and ascertain the nature and location of the Work, and that it has investigated and satisfied itself as to all general and local conditions that may affect the Work or its cost, including but not limited to (a) conditions relating to transportation, handling, storage and disposal of materials and equipment; (b) availability of labor, power, water and other utilities, and roads; (c) uncertainties of weather, tides, river stages, and similar physical characteristics of the site and its surroundings; (d) conditions of the soil and grounds; and (e) character of equipment and other facilities required relative to the Work, both prior to commencement of the Work at the site and during the performance of the Work. Contractor further acknowledges that it has fully satisfied itself as to the nature, character, quality and quantity of surface and subsurface conditions, materials or obstacles to be encountered to the extent that such information is reasonably ascertainable from an inspection of the site, from information available from the local municipality and other public bodies, and from the Contract Documents. Any failure on the part of the Contractor to take the actions described and acknowledged herein shall not relieve the Contractor from responsibility for estimating properly the difficulty and cost of properly performing the Work in a timely manner and without additional cost to Owner.

---

[2] *See* Green Constr. Co. v. Kansas Power & Light Co., 1 F.3d 1005 (10th Cir. 1993).

3.2.1.2 The Owner assumes no responsibility for any conclusions or interpretations made by the Contractor based on the information made available by the Owner, unless such information is included in the Contract Documents.

Another:

3.2.1 Execution of the Contract by the Contractor is a representation that the Contractor has visited the site, become familiar with local conditions under which the Work is to be performed and correlated personal observations with requirements of the Contract Documents. Contractor represents that it has performed its own investigation and examination of the Work site and its surroundings and satisfied itself before entering into this Contract as to:

3.2.1.1 conditions bearing upon transportation, disposal, handling, and storage of materials;

3.2.1.2 the availability of labor, materials, equipment, water, electrical power, utilities and roads;

3.2.1.3 uncertainties of weather, river stages, flooding and similar characteristics of the site;

3.2.1.4 conditions bearing upon security and protection of material, equipment, and Work in progress;

3.2.1.5 the form and nature of the Work site, including the surface and subsurface conditions;

3.2.1.6 the extent and nature of Work and materials necessary for the execution of the Work and the remedying of any defects therein; and

3.2.1.7 the means of access to the site and the accommodations it may require and, in general, shall be deemed to have obtained all information as to risks, contingencies and other circumstances.

The following language could be added to ¶ 3.2.3:

The accuracy of grades, elevations, dimensions, or locations of existing conditions is not guaranteed by the Architect or Owner, and the Contractor is responsible for verifying same.

Here is a possible additional clause under ¶ 3.2:

3.2.5 Should any words or numbers that are necessary to a clear understanding of the Work be illegible or omitted, or should an error, or discrepancy occur in any of the Contract Documents, the Contractor shall immediately notify the Architect of such omission, error, or discrepancy, and the Contractor shall not proceed with that portion of the Work until clarification is received. If the

Contractor proceeds without so notifying the Architect, the Contractor shall be responsible for the cost of correcting same, including any resulting damage.

Another alternate provision:

3.2.1 The Contractor shall examine the site of the Work and adjacent premises and the various means of approach to the site, and shall make all necessary investigations in order to inform itself thoroughly as to the character and magnitude of all work involved in the complete execution of the work shown in the Contract Documents. The Contractor shall further inform itself as to the facilities for delivering, handling, and installing the construction plant and other equipment and the conditions and difficulties that will be encountered in the performance of the Work. No plea of ignorance of conditions that exist or that may hereafter exist, or of the difficulties that will be encountered in the performance of the Work, as a result of failure to make necessary examinations and investigations, will be accepted as sufficient excuse for any failure or omission on the part of the Contractor to fulfill in every detail all the requirements of the Contract Documents, or will be accepted as a basis for any claim whatsoever for extra compensation or for an extension of time.

Another provision:

3.2.1 In addition to and not in derogation of the Contractor's duties under these General Conditions, the Contractor shall take field measurements and verify field conditions and shall carefully compare such field measurements and conditions and other information known to the Contractor with the Contract Documents before commencing Work. Errors, inconsistencies, or omissions discovered or which reasonably should have been discovered shall be reported to the Architect and Owner at once. The Contractor shall be responsible for any errors, inconsistencies, or omissions that are not discovered but that reasonably should have been discovered by a prudent contractor performing the services to be provided hereunder.

A provision that places greater responsibility on the contractor regarding codes and ordinances is as follows:

3.2.2 Any design errors or omissions noted by the Contractor during this review shall be reported promptly to the Architect. The Contractor shall also report any nonconformity with any applicable law, statute, ordinance, building code, rule, or regulation that the Contractor knows or reasonably should have known.

Another provision:

3.2.2 Since the Contract Documents are complementary, before starting each portion of the Work, the Contractor shall carefully study and compare the various Drawings and other Contract Documents relative to that portion of the Work, as well as the information furnished by the Owner pursuant to Section 2.2.3, shall take field measurements of any existing conditions related to that portion of the Work and shall observe any conditions at the site affecting it.

These obligations are for the purpose of facilitating construction by the Contractor and are not for the purpose of discovering errors, omissions, or inconsistencies in the Contract Documents; however, any errors, inconsistencies or omissions discovered by the Contractor shall be reported promptly to the Architect Owner and Architect, as a request for information in such form as the Architect Owner may require. The Contractor hereby specifically acknowledges that the Contract Documents including the Drawings, the Specifications, and all Addenda are sufficient to enable it to commence construction of the Work outlined therein. The Contractor further acknowledges that it has visited the site, examined all identifiable conditions affecting the Work, is fully familiar with all of the conditions thereon and affecting the same.

§ 5.13 Supervision and Construction Procedures: ¶ 3.3

The AIA language unfortunately includes provisions suggesting that the Contract Documents might include specific instructions concerning the means, methods and techniques of construction. This is almost always a bad idea and could lead to misunderstandings by a court in the event of litigation over construction accidents. The following is a revised version that eliminates this language:

3.3.1 The Contractor shall supervise and direct the Work, using the Contractor's best skill and attention. The Contractor shall be solely responsible for, and have control over, construction means, methods, techniques, sequences and procedures and for coordinating all portions of the Work under the Contract.

The following language added to the end of ¶ 3.3.1 places a greater burden on the contractor:

The Contractor shall engage workers who are skilled in performing the Work, and all Work shall be performed with care and skill and in a good workmanlike manner under the full-time supervision of an approved engineer or foreman. The Contractor shall be liable for all property damage, including repairs and replacements of the Work and economic losses, which proximately result from the breach of this duty. The Contractor shall advise the Architect (a) if a specified product deviates from good construction practices; (b) if following the Specifications will affect any warranties; or (c) any objections which the Contractor may have to the Specifications. Nothing contained in Section 1.1.3 shall alter the responsibilities established in this Section.

Paragraph 3.3.2 could be made somewhat more inclusive by adding the word "any" before "other persons" and adding "or with any subcontractor, sub-subcontractor, supplier or similar party in privity with the Contractor" at the end of the paragraph. The owner may want to add the following language to the end of ¶ 3.3.2:

and for any damages, losses, costs and expenses, including but not limited to attorneys' fees, resulting from such acts and omissions.

The following are additional clauses that may be added after ¶ 3.3.3:

3.3.4 The Contractor shall inspect all materials delivered to the premises and shall reject any materials that will not conform with the Contract Documents when properly installed.

3.3.5 Immediately upon commencement of the Work, the Contractor shall retain the services of a licensed surveyor to verify the boundaries of the site and establish benchmarks for the Work. All discrepancies shall be immediately brought to the attention of the Architect.

3.3.6 The Contractor shall be responsible for and coordinate any and all inspections required by any governmental body that has jurisdiction over the Project. Failure to obtain any permits, licenses, or other approvals because of the failure of the Contractor to conform to this requirement shall not extend the Contract Time, and the Contractor shall not be entitled to an increase in the Contract Sum therefor.

Another provision:

3.3.4 The Contractor shall assign to this Project a competent, technically trained office project manager who shall handle all office functions including checking, approving, and coordinating shop drawings and approving purchases and disbursements, pay-out requests, and correspondence.

Another provision:

3.3.1 Contractor acknowledges that timely completion of the Work in accordance with the terms of said Documents is of crucial importance to Owner. Contractor shall provide the best skill and judgment of its officers and employees and shall cooperate with Owner and Architect to further the interests of Owner and to bring about timely completion of the Work. Contractor shall furnish sufficient business administration and superintendence and provide at all times an adequate supply of labor and materials to secure execution of the Work in the best and soundest way and in the most expeditious and economical manner consistent with the interests of Owner. In the event of delays and/or unforeseen events, whether or not the same should entitle Contractor to an adjustment in the Contract Sum and/or Contract Time pursuant to Articles 7 and 8 hereof, Contractor shall use diligent efforts to maintain scheduled completion dates. Such efforts shall include rephasing events, decreasing overly conservative durations on subsequent events, increasing activity overlap, and using float on noncritical events. The float available in the Progress Schedule shall be used by Owner and Contractor whenever possible to offset the impact of delays. Contractor shall be responsible for coordinating its Work with the Work of any other contractors and/or activities at the job site.

3.3.2 Contractor shall supervise and direct the Work, using Contractor's best skill and attention. Contractor shall be solely responsible for and have control over construction means, methods, techniques, sequences, and procedures and for coordinating all portions of the Work in accordance with the requirements of the Contract Documents.

3.3.3 Contractor shall be responsible to Owner for acts and omissions of Contractor's employees, Subcontractors and their agents and employees, and other persons performing portions of the Work under contract or other arrangement with Contractor.

3.3.4 Contractor shall give notices and comply with applicable laws, ordinances, rules, regulations, and lawful orders of public authorities bearing on the Work, including those with respect to the safety of persons and property and their protection from damages, injury, or loss. Contractor shall promptly remedy damage and loss to property at the site caused in whole or in part by Contractor, its Subcontractor, or anyone directly or indirectly employed by any of them or by anyone for whose acts they may be liable, except for damage or loss attributable solely to acts or omissions of Owner or Architect or by anyone for whose acts either of them may be liable and not attributable to the fault or negligence of Contractor, its Subcontractor, or anyone directly or indirectly employed by them. The foregoing obligations of the Contractor are in addition to Contractor's obligations under other provisions hereunder.

3.3.5 Contractor shall not be relieved of obligations to perform the Work in accordance with the Contract Documents either by activities or duties of the Architect under the Contract for Construction or by tests, inspections, or approvals required or performed by persons other than Contractor.

3.3.6 Contractor shall be responsible for inspection of portions of Work already performed under the Contract for Construction to determine that such portions are in proper condition to receive subsequent Work.

3.3.7 Contractor has the responsibility to ensure that all material suppliers and Subcontractors, their agents, and employees adhere to the Contract Documents, and that they order materials on time, taking into account the current market and delivery conditions, and that they provide materials on time. Contractor shall coordinate its Work with that of all others on the Project, including coordinating deliveries, storage, installations, and use of construction utilities. Contractor shall be responsible for the space requirements, location, and routing of its equipment. In areas and locations where the proper and most effective space requirements, locations, and routing cannot be made as indicated, Contractor shall meet with all others involved before installation to plan the most effective and efficient method of overall installation.

3.3.8 Contractor shall establish and maintain bench marks and all other grades, lines, and levels necessary for the Work; report errors or inconsistencies to Owner and Architect before commencing Work; and, if applicable, review the placement of the building(s) and permanent facilities on the site with Owner and Architect after all lines are staked out and before foundation

Work is started. Contractor shall provide access to the Work for Owner, the Architect, other persons designated by Owner, and governmental inspectors. Any encroachments made by Contractor or its Subcontractors on adjacent properties caused by construction as revealed by an improvements survey, except for encroachments arising from errors or omissions not reasonably discoverable by Contractor in the Contract Documents, shall be the sole responsibility of Contractor, and Contractor shall correct such encroachments within thirty (30) days of the improvement survey (or as soon thereafter as reasonably possible), at Contractor's sole cost and expense, either by the removal of the encroachment (and subsequent reconstruction on the Project site) or agreement with the adjacent property Owner(s) (in form and substance satisfactory to Owner in its sole discretion) allowing the encroachments to remain.

3.3.9 Contractor shall verify at the Work site the measurements indicated on the Drawings and Specifications and shall establish correctly the lines, levels, and positions for the Work, and be responsible for their accuracy and proper correlation with control lines, monuments, and data, as established by surveys furnished by Owner. Work shall be erected square, plumb, level, true to line and grade, in the exact plane and to the correct elevation and/or sloped to drain as indicated. To ensure the proper execution of its subsequent Work, Contractor shall measure all work already in place (including but not limited to utilities and grades installed or prepared by others) and shall at once report to Architect and Owner any discrepancy between said work and the Drawings and Specifications for the Work.

3.3.10 Any discrepancy or omission in the dimensions or elevations shown on the Drawings and Specifications or found in previous work which may prevent accurate layout or construction of the Work, shall immediately be reported by Contractor to Owner and Architect. If Contractor performs, permits, or causes performance of any Work when Contractor knows or reasonably should have known that such discrepancy or omission exists, without first obtaining further instruction from Architect or Owner, Contractor shall bear any and all costs arising therefrom including, without limitation, the costs of correction thereof without increase or adjustment in the Contract Sum. Omissions from the Drawings or Specifications, or the mis-description of details of Work which are reasonably inferable in order to carry out the intent of the Drawings and Specifications, or which are customarily performed, shall not relieve Contractor from performing such omitted or mis-described details of the Work, and they shall be performed as if fully and correctly set forth and described in the Drawings and Specifications, at no additional cost to Owner.

The following provision could be added:

3.3.4 It is understood and agreed that the relationship between the Owner and Contractor is that of independent contractor, and that nothing contained in the Contract Documents shall be construed to make the Contractor the agent, servant, or employee of the Owner or to create any joint venture, partnership, or other association between the Owner and Contractor other than that of independent contractor.

Another provision reads as follows:

3.3.1.1 In the event the Contractor determines that any unsafe conditions exist pursuant to ¶ 3.3.1, the Contractor may suspend performance in the affected areas until the unsafe conditions are remedied, and the Contractor will be entitled to an equitable adjustment in the Contract Sum and Contract Time.

§ 5.14 Labor and Materials: ¶ 3.4

An alternate provision:

3.4.1 Unless otherwise provided in the Contract Documents, Contractor shall provide and pay for labor, materials, equipment, tools, construction equipment and machinery, water, heat, utilities, telephone, transportation, and other facilities and services necessary for proper execution and completion of the Works, whether temporary or permanent and whether or not incorporated or to be incorporated in the Work.

3.4.2 Contractor shall enforce strict discipline and good order among Contractor's employees and other persons carrying out the Work. Contractor shall not permit employment of unfit persons or persons not skilled in tasks assigned to them. Contractor shall also be responsible for labor peace on the Project and shall at all times use its best efforts and judgment as an experienced contractor to adopt and implement policies and practices designed to avoid work stoppages, slowdowns, disputes, or strikes where reasonably possible and practical under the circumstances and shall at all times maintain Project-wide labor harmony. Except as specifically provided in this Agreement, Contractor shall not be entitled to any adjustment in the Contract Sum or the Contract Time and shall be liable to Owner for all damages suffered by Owner occurring as a result of work stoppages, slowdowns, disputes, or strikes by the work force of or provided by Contractor or its Subcontractors.

3.4.3 Unless otherwise specifically provided in the Contract Documents, all equipment, material, and articles incorporated in the Work shall be new and of the most suitable grades for the intended purpose. Materials shall conform to manufacturer's standards in effect at the date of execution of the Contract for Construction and shall be installed in strict accordance with manufacturer's latest directions. Contractor shall, if required by Owner or Architect, furnish satisfactory evidence as to the kind and quality of any materials. All packaged materials shall be shipped to the site in the original containers clearly labeled, and delivery slips shall be submitted with bulk materials, identifying thereon the source and warranting quality and compliance with Contract Documents.

3.4.4 When the Contract Documents require the Work, or any part of same, to be above the standards required by applicable laws, ordinances, rules, and regulations, and other statutory provisions pertaining to the Work, such Work shall be performed and completed by Contractor in accordance with the Contract Documents. When the Contract Documents describe the Work in general terms, but not in complete detail, Contractor understands and acknowledges only the best general practice is to be employed. Any design detail furnished by Contractor shall be in conformance with applicable laws and shall be sufficient for the purposes intended. Contractor shall closely inspect all materials as delivered and all Work as performed and shall promptly reject and return all substandard materials and redo all substandard Work without awaiting Architect's inspection and rejection thereof.

§ 5.15   Warranty: ¶ 3.5

An alternate provision:

3.5.1 The Contractor warrants to the Owner and Architect that materials and equipment furnished under the Contract shall be in first-class condition and new unless otherwise required or permitted by the Contract Documents, that the Work will be free from defects not inherent in the quality required or permitted, and that the Work will conform to the requirements of the Contract Documents. The Contractor further warrants that all workmanship shall be of the highest quality and in full conformance with the Contract Documents, and that all labor shall be performed by persons well qualified in their respective trades. Work not conforming to these requirements, including substitutions not properly approved and authorized, may be considered defective. The Contractor's warranty excludes remedy for damage or defect caused by abuse, modifications not executed by the Contractor, improper or insufficient maintenance, improper operation, or normal wear and tear and normal usage. If required by the Architect or Owner, the Contractor shall furnish satisfactory evidence as to the kind and quality of materials and equipment.

The following provision adds protection for the owner when the contractor is providing equipment:

3.5.2 The Contractor shall furnish maintenance and 24-hour call-back service for the equipment provided by it for a period of three (3) months after final completion and acceptance of the work. This work shall include all necessary adjustments, greasing, oiling, cleaning, supplies, and parts to keep the equipment in proper operation except such parts made necessary by misuse, accidents, or negligence not caused by the Contractor or any of its Subcontractors.

§ 5.16   Taxes: ¶ 3.6

The following language favorable to the owner could be added at the end of ¶ 3.6:

including but not limited to all sales taxes, use taxes, occupational taxes, excise taxes, Social Security benefits, unemployment compensation taxes, or similar levies on all materials, labor, tools, and equipment furnished under this

Agreement, as required by the statutes applicable to the Project, including all federal, state and local taxes.

Other provisions that could be inserted after ¶ 3.6:

3.6.2 The Contractor's fee includes, and the Contractor shall be solely responsible for paying, any and all taxes, excises, duties, and assessments ("taxes") arising out of the Contractor's performance of the Work in any manner levied, assessed, or imposed by any government or agency having jurisdiction.

3.6.3 The Contractor shall promptly pay and discharge when due, unless the validity or application to the Work is being contested in good faith, any and all taxes, together with any interest and penalties, the responsibility and liability for which is assumed by the Contractor pursuant to the preceding paragraph. If any such taxes are levied, assessed, or imposed on the Owner, the Owner shall notify the Contractor and the Contractor shall promptly pay and discharge the taxes, but upon the written request and at the expense of the Contractor, the Owner shall assist the Contractor in contesting the validity or application of such taxes. If the Owner receives a refund of all or any part of any taxes (including a refund of interest or penalties), the amount refunded to the Owner shall promptly be remitted to the Contractor, less any expenses of the Owner associated with contesting the taxes not previously reimbursed by the Contractor to the Owner.

A provision that can be used where the Project is owned by a tax-exempt body:

3.6.2 The Owner, a public body, is exempt from all applicable federal, state, and local sales tax. Retail sales tax shall not be included in the contract amount.

Another provision:

3.6.1 Contractor shall file or cause to be filed on a timely basis (including lawful extensions of time to file) all tax returns that are or shall be required to be filed by it with respect to its performance of the Work pursuant to the laws, regulations, or administrative requirements of each governmental body with taxing power over its performance of the Work, and it shall pay all taxes required to be paid, including all taxes and other assessments on income, unemployment, occupation, payroll, excise, property, transfer, real property transfer or gains taxes, withholding, sales, use, value added, franchise, gross receipts taxes, license, recording, documentation and registration fees, and customs duties, imposed by any governmental body, including interest and penalties on taxes; provided, however, that Contractor shall not be responsible for taxes, charges, assessments, or fees relating to Owner's business or ownership of the Project site. Contractor warrants that adequate provision has been made in the Contract Sum for all such taxes for which it is responsible here-under. Contractor shall be responsible for and shall indemnify, defend, protect, and hold harmless Owner against any penalties or interest assessed as a result of underpayments or late payments of such taxes. If Contractor does not have a sales tax identification number for the state in which the Project is located, it will obtain such number prior to receiving any payments from Owner.

§ 5.17   Permits, Fees, Notices, and Compliance
with Laws: ¶ 3.7

If any changes are made here, they must be coordinated with ¶ 2.2.2. To indicate in a more positive manner that the contractor is to obtain the permits, the owner may want to delete in ¶ 3.7.1 the words, "Unless otherwise provided in the Contract Documents,".

The following clause could be inserted after ¶ 3.7.2:

3.7.2.1 If the Contractor fails to give such notices, it shall be liable for and shall indemnify and hold harmless the Owner and the Architect and their respective employees, officers, and agents against any resulting fines, penalties, judgments, or damages, including reasonable attorneys' fees, imposed on or incurred by the parties indemnified hereunder. Further, and notwithstanding any other provision in the Contract Documents, including Section 15.1.6, Contractor shall be responsible for all damages, including consequential damages, resulting from such failure.

In ¶ 3.7.3, the owner may want to replace "costs" with "costs, damages, losses, expenses of every kind, including, but not limited to, reasonable attorneys' fees."

The owner may want to include this additional paragraph after ¶ 3.7.1:

3.7.1.1 Copies of any and all permits, licenses, and certificates shall be delivered to the Architect and Owner as soon as they are obtained. Along with the request for final payment, the Contractor shall deliver the originals of such permits, licenses, and certificates to the Architect.

Another provision:

3.7.1 Contractor shall secure and pay for construction permits, fees, licenses, and engineering and inspection charges required by any governmental authority or other person or entity having jurisdiction over the Work. Said permits shall include, without limitation, both temporary and permanent permits, building permits, certificates of occupancy, curb-breaking permits, highway entrance permits, water permits, and all similar permits or certificates. Any construction and/or seismic equipment tax shall also be included. Contractor shall secure and pay all fees and charges for utility connections and installations. Owner and/or Architect will deliver original drawings and specifications to the Building Department and pay the initial plan check fee. If any governmental authority or other person or entity having jurisdiction requires special bonding of Contractor or the posting of any security for any part of the Work (not including Contractor's performance bond), Contractor shall pay the premiums for such bond or post such security.

Another provision:

3.7.1 Unless otherwise provided in the Contract Documents, the Contractor shall secure and pay for the building permit and other permits and governmental fees, licenses and inspections, if required, necessary for proper execution and completion of the Work which are customarily secured after execution of the Contract and which are legally required when bids are received or negotiations concluded. The Contractor shall procure all certificates of inspection, use and occupancy and all permits and licenses, pay all charges and fees and give all notices necessary and incidental to the due and lawful prosecution of the Work. Certificates of inspection, use, and occupancy shall be delivered to the Owner upon completion.

§ 5.18   Concealed or Unknown Conditions: ¶ 3.7.4

Another version:

3.7.4 Concealed or Unknown Conditions. If the Contractor encounters conditions at the site that are (1) subsurface or otherwise concealed physical conditions that differ materially from those indicated in the Contract Documents or (2) unknown physical conditions of an unusual nature, that differ materially from those ordinarily found to exist and generally recognized as inherent in construction activities of the character provided for in the Contract Documents, the Contractor shall promptly provide notice to the Owner and the Architect before conditions are disturbed and in no event later than ten days after first observance of the conditions. Failure to do so shall be deemed a waiver of all claims related to the condition. The Architect will promptly investigate such conditions and, if the Architect determines that they differ materially and cause an increase or decrease in the Contractor's cost of, or time required for, performance of any part of the Work, will recommend an equitable adjustment in the Contract Sum or Contract Time, or both. The Owner and Contractor shall thereafter negotiate in good faith regarding the Architect's recommendation and execute a Change Order if agreement is reached. If the Architect determines that the conditions at the site are not materially different from those indicated in the Contract Documents and that no change in the terms of the Contract is justified, the Architect shall promptly notify the Owner and Contractor in writing, stating the reasons. If either party disputes the Architect's determination or recommendation, or if no agreement between the Owner and Architect is reached, that party may proceed as provided in Article 15.

Another version, less favorable to the contractor:

3.7.4 Concealed or Unknown Conditions. If the Contractor encounters conditions at the site that are subsurface or otherwise concealed physical conditions that differ materially from those indicated in the Contract, the Contractor shall promptly provide notice to the Owner and the Architect before conditions are disturbed and in no event later than ten days after first observance of the conditions. Failure to do so shall be deemed a waiver of all claims related to the condition. The Architect will promptly investigate such conditions and, if the Architect determines that they differ materially and cause an increase or decrease in the Contractor's cost of, or time required for, performance of any part of the Work, will recommend an equitable adjustment in the Contract Sum or Contract Time, or both. The burden shall be on the Contractor to substantiate any such increase in cost or time, and failure to do so shall result in a rejection of such claim. If the Architect determines that the conditions at the site are not materially different from those indicated in the Contract Documents and that no change in the terms of the Contract is justified, the Architect shall promptly notify the Owner and Contractor in writing, stating the reasons. If either party disputes the Architect's determination or recommendation, or if no agreement between the Owner and Architect is reached, that party may proceed as provided in Article 15.

The new provision found in ¶ 3.7.5 could be modified to make it more contractor-friendly:

3.7.5 If, in the course of the Work, the Contractor encounters human remains or recognizes the existence of burial markers, archaeological sites or wetlands not indicated in the Contract Documents, the Contractor shall immediately suspend any operations that would affect them and shall notify the Owner and Architect. Upon receipt of such notice, the Owner shall promptly take any action necessary to obtain governmental authorization required to resume the operations. The Contractor shall continue to suspend such operations until otherwise instructed by the Owner but shall continue with all other operations that do not affect those remains or features. The Owner shall be responsible to the Contractor for all costs and extensions of time associated therewith pursuant to Article 15, including lost profits and other consequential damages.

§ 5.19   Allowances: ¶ 3.8

A different version:

3.8.1 The Contract sum shall include all allowances specifically set forth in the Contract Documents. The Contractor shall furnish and install the allowance items and incorporate same into the Work.

3.8.2 The Contract Sum shall include the Contractor's cost of installation, taxes, delivery to the site, and all equipment related to such items. The cost of each allowance item shall be the actual cost to the Contractor, giving credit to the Owner for any trade discount.

3.8.3 If the actual cost of any allowance item is less than the allowance, the Contract Sum shall be decreased accordingly, evidenced by a Change Order, for such difference. If the actual cost of any allowance item is greater than the allowance, the Contract Sum shall be increased accordingly, evidenced by a Change Order, for such difference.

3.8.4 Allowance items shall be promptly selected by the Owner and/or Architect. If there is a delay in such selection, the Contractor shall be entitled to a Change Order to reflect any delay in time and an increase in the Contract Sum if such delay results in any additional costs to the Contractor.

§ 5.20   Superintendent: ¶ 3.9

Another provision:

3.9.1 Prior to starting Work, Contractor shall designate the Project Manager, Superintendent, and other key individuals who shall be assigned to the Project through and including Final Completion. The Superintendent shall be in attendance at the Project site throughout the Work, including completion of the punchlist. The Superintendent shall be approved by Owner in its sole discretion. Said representatives shall be qualified in the type of Work to be undertaken and shall not be changed during the course of construction without the prior consent of Owner. Should a representative leave Contractor's employ, Contractor shall promptly designate a new representative. Owner shall have the right, at any time, to direct a change in Contractor's representatives if their performance is unsatisfactory. In the event of such demand, Contractor shall, within seven (7) days after notification thereof, replace said individual(s) with an individual satisfactory to Owner, in Owner's sole discretion. If said replacement is disapproved, the Contract may, at Owner's option, be terminated for cause. The Superintendent shall represent Contractor, and communications given to the Superintendent shall be as binding as if given to Contractor. Owner shall have no obligation to direct or monitor Contractor's employees.

Here is another provision regarding key personnel:

3.9.1 The Contractor will, immediately upon receiving a fully executed copy of the Agreement, assign and maintain during the term of the Agreement and any extension of it, an adequate staff of competent personnel, including but not limited to, a Project Manager and Superintendent, who are fully equipped, licensed as appropriate, available as needed, qualified and assigned to perform the Work. Contractor will include among its staff such Key Personnel (including a Project Manager and a Superintendent) and positions as identified in Exhibit B.

3.9.1.1 Upon award of the Agreement, the Contractor shall submit a project staff organizational chart that includes the names and resumes of employees proposed for key positions for the Project. The name and title of each individual indicated on the project staff organizational chart ("Key Personnel") shall be set forth in Exhibit B. All Key Personnel must have experience on projects of a like size and scope, regarding their responsibility, on this Project and must be approved by the Owner.

3.9.1.2 If any Key Personnel furnished by the Contractor for the Project in accordance herewith should be unable to continue in the performance of assigned duties for reasons due to death, disability or termination, the Contractor will promptly notify the Architect and Owner explaining the circumstances. Changes in assignment of Key Personnel due to commitments not related to this Agreement are prohibited without the Owner's written approval.

3.9.1.3 On request by the Owner, the Contractor will furnish to the Architect and Owner within seven (7) Days the name of the person substituting for the individual unable to continue, together with any information the Owner may require to judge the experience and competence of the substitute person. Upon approval by the Owner, such substitute person will be assigned to the Project and if the Owner rejects the substitute, the Contractor will have seven (7) Days thereafter to submit a second substitute person. Such process will be repeated until a proposed replacement has been approved by the Owner.

3.9.1.4 In the event that, in the opinion of the Owner, the performance of personnel, including but not limited to Key Personnel, of the Contractor assigned to this Project is at an unacceptable level, such personnel will cease to be assigned to this Project, will return to the Contractor, and the Contractor will furnish to the Owner the name of a substitute person or persons in accordance with the previous paragraph.

Additional provisions that could be inserted under ¶ 3.9 are as follows:

3.9.4 As directed by the Architect, there is to be held at the Architect's field office a meeting of the representatives of the various trades engaged about the Work, for furthering the progress of the Work and giving clarifications by the Architect and instructions by the Owner. If the Contractor's representatives fail to attend or to execute the instructions given them, they shall on request of the Owner be dismissed from the Work and other representatives must be immediately substituted.[3]

3.9.3 The Contractor shall not employ a proposed superintendent to whom the Owner or Architect has made reasonable and timely objection. The Contractor shall not change the superintendent without the prior written consent of the Owner, which consent shall not be unreasonably withheld. The superintendent shall be present at the Project until substantial [or final] completion. At the Owner's request, the Contractor shall assign a different superintendent to the Project.

3.9.3 The Contractor's superintendent and necessary assistants shall be acceptable to the Owner. The Contractor shall notify the Owner in writing of any proposed change in such personnel, including the reason therefor, prior to making any change. Such personnel shall not be changed except with the consent of the Owner, unless such personnel cease to be in the employ of the Contractor.

---

[3] Adapted from E.C. Ernst, Inc. v. General Motors Corp., 482 F.2d 1047 (5th Cir. 1973), *aff'd after remand*, 537 F.2d 105 (5th Cir. 1976).

Another provision:

3.9.1 The Contractor shall employ competent personnel, supervisors, project managers, project engineers, and all others who shall be assigned to the Work throughout its duration whether at the site or not. Contractor's personnel extends to those employed by the Contractor whether at the site or not. The Owner shall have right to review and approve or reject all replacement of Contractor's personnel. All personnel assigned by the Contractor to the Work shall possess the requisite experience, skills, abilities, knowledge and integrity to perform the Work. The superintendent and others as assigned shall be in attendance at the Project site during the performance of the Work. The superintendent shall represent the Contractor. All communications given to the Contractor's personnel such as the project manager or the superintendent, whether verbal, electronic or written, shall be as binding as if given to the Contractor. It is the Contractor's responsibility to appropriately staff, manage, supervise and direct the Work which is inclusive of the performance, acts, and actions of his personnel or subcontractors. As such, the Contractor further agrees to indemnify and hold harmless the Owner and the Architect, and to protect and defend both from and against all claims, attorneys' fees, demands, causes of action of any kind or character, including the cost of defense thereof, arising in favor of or against the Owner, Architect, Contractor, their agents, employees, or any third parties on account of the performance, behavior, acts or actions of the Contractor's personnel or subcontractors.

§ 5.21   Contractor's Construction Schedules: ¶ 3.10

The owner may want to add at the end of ¶ 3.10.1:

The construction schedule shall not be changed without the written consent of the Owner and Architect.

Another provision adds to the end of Subparagraph 3.10.1 the following:

The Owner's or Architect's silence as to a submitted schedule that exceeds time limits current under the Contract Documents shall not relieve the Contractor of its obligation to meet those time limits, nor shall it make the Owner or Architect liable for any of Contractor's damages incurred as a result of increased construction time or not meeting those time limits. Similarly, the Owner's or Architect's silence as to a Contractor's schedule showing performance in advance of such time limits shall not create or infer any rights in favor of the Contractor for performance in advance of such time limits.

Another version:

3.10.2 The Contractor shall prepare a submittal schedule, promptly after being awarded the Contract and thereafter as necessary to maintain a current submittal and shall submit the schedule(s) for the Architect's review. The Architect's review shall not be unreasonably delayed or withheld. The submittal schedule shall 1) be coordinated with the Contractor's construction schedule, and 2) allows the Architect reasonable time to review submittals. If the Contractor fails to submit a submittal schedule, the Contractor shall not be entitled to any increase in Contract Sum or extension of Contract Time based on the time required for review of submittals. Neither the Contractor's preparation nor the Architect's receipt or review shall modify the Contractor's responsibility to make required submittals or to do so in a timely manner to provide for review in accordance with paragraph 4.2.7 as modified herein.

Another provision:

3.10.4 Should the Contractor fail to comply with the progress schedule or, in the Owner's opinion, otherwise fail, refuse, or neglect to supply a sufficient amount of labor or material in the prosecution of the Work, Owner shall have the right to (1) direct Contractor to furnish such additional labor and/or materials as may, in the Owner's opinion, be required to comply with the progress schedule or otherwise diligently prosecute the work, or (2) furnish such additional labor and/or materials as may be required to comply with said schedule. Any costs incurred by Owner pursuant to the exercise of its rights under this paragraph shall be borne by the Contractor and shall not increase the Contract Sum.[4]

Another provision:

3.10.1 Contractor, promptly after being awarded the Contract for Construction and before commencing Work, shall prepare and submit for Owner's and Architect's review and approval a preliminary schedule for the Work. Within ten (10) days following Owner's Notice to Proceed, Contractor shall provide to Owner and Architect a schedule of performance of the Work, showing timely completion of the Work and timely achievement of each Milestone Date as required by the Contract for Construction and meeting all other requirements of this Section 3.10 (the "Progress Schedule"). Upon receipt of Contractor's proposed Progress Schedule, Owner may accept the proposed Progress Schedule as submitted or reject it, noting deficiencies. If such schedule is rejected, the deficiencies noted shall be corrected and a new proposed Progress Schedule shall be submitted within ten (10) days. In any case, a complete Progress Schedule must be approved by Owner prior to any payments' being made.

3.10.2 The Progress Schedule shall be in the form of a network using critical path methodology (CPM), clearly showing construction activities, dependencies, and durations. The critical path activities shall be highlighted, float time for noncritical activities shall be shown, and the start and stop stated for each activity shall be listed. Longer-duration activities shall be broken into subactivities when the Work can be completed in phases (i.e., south half, north half, etc.). Contractor will be allowed flexibility in schedule, logic, and content; however, the Progress Schedule must be broken down by all trades, indicating ordering, delivery, and Milestone Dates, and the following activities must be included in all cases, if covered by the Scope of Work: (a) award of Contract;

---

[4] *See* Marriott Corp. v. Dasta Constr. Co., 26 F.3d 1057, 1065 (11th Cir. 1994).

(b) site delivery and mobilization; (c) demolition; (d) pour foundations; (e) underground utilities; (f) pour slabs phase; (g) exterior walls phase; (h) columns; (i) floor and roof structure phase; (j) roof decking; (k) roofing (drying); (l) HVAC duct work; (m) fire sprinkler piping; (n) interior stud walls phase; (o) drywall; (p) lath and plaster phase; (q) painting phase; (r) ordering and delivery of long-lead materials; (s) completion of any parking structures; and (t) Milestone Dates. For all long-lead materials and for the purchase of any materials or equipment with a cost of $5,000 or more, the Progress Schedule shall include a Material Purchase Log, indicating the item of material or equipment, the quantity required, the estimated lead time, and, to the extent known, Contractor's purchase order number, the date ordered, the scheduled delivery date, and the actual or committed delivery date.

3.10.3 Contractor shall perform the Work in accordance with the Progress Schedule as well as within the Milestone Dates and completion dates specified in the Contract for Construction. The times set forth in the Contract for Construction for all Milestone Dates and the time of completion must govern, and the Progress Schedule must be adjusted to meet these dates. Contractor shall maintain such Progress Schedule on a current basis in accordance with the provisions of this Section 3.10 and shall keep proper records to substantiate actual activity durations and completion dates.

3.10.4 Contractor shall submit to Owner monthly with each application for payment, and at such additional times as may be required by Owner, three copies of a Monthly Status Report in such form as Owner reasonably requests. Each Monthly Status Report shall concisely but completely describe, in narrative form, the then current status of the Work including, without limitation:

3.10.4.1 A review of actual progress during the month in comparison to the Progress Schedule and, if actual progress is behind schedule, discussion of any "workaround" or "catch-up plan" that Contractor has employed or will employ to recover the original Progress Schedule;

3.10.4.2 A concise statement of the outlook for meeting future Progress Schedule dates, and the reasons for any change in outlook from the previous report;

3.10.4.3 A concise statement of significant progress on major items of Work during the report period, and progress photographs as necessary to document the current status of the Work;

3.10.4.4 A review of any significant technical problems encountered during the months and the resolution or plan for resolution of such problems;

3.10.4.5 An explanation of any corrective action taken or proposed;

3.10.4.6 A complete review of the status of Change Orders, including a review of any changes in the critical path of the construction Progress Schedule which result from Change Orders approved by Owner during the month, as well as a review of the schedule impact of Change Order requests then pending;

3.10.4.7 A summary of any Claims anticipated by Contractor with respect to the Work, including the anticipated cost and schedule impacts of any such Claims;

3.10.4.8 A cumulative summary of the number of days of, and the extent to which the progress of the Work was delayed by, any of the causes for which Contractor could be entitled to an extension of the Contract Time;

3.10.4.9 A marked copy of the current Progress Schedule showing the status of each element of the Work; and

3.10.4.10 An updated Material Purchase Log.

3.10.5 Contractor shall submit to Owner monthly with each application for payment, and at such additional times as may be required by Owner, for Owner's review and approval, three copies of an updated Progress Schedule meeting all the requirements of this Section 3.10, including:

3.10.5.1 Actual versus estimated percent completion for each activity and Project total;

3.10.5.2 Actual versus estimated work in place for each activity and Project total;

3.10.5.3 Actual versus estimated manpower for each activity and Project total;

3.10.5.4 Actual versus estimated cash flow; and

3.10.5.5 Any change in the critical path.

3.10.6 If the progress of the Work is behind the Progress Schedule to such an extent that Owner reasonably determines that Contractor will be unable to meet any of the critical path dates set forth in the Progress Schedule, including without limitation any Milestone Date, Owner may direct Contractor to accelerate its work, at Contractor's own cost, without any adjustment to the Contract Sum. Such acceleration may include employing such additional forces or paying such additional overtime wages as may be required to place the progress of the Work in conformity with the Progress Schedule and to assure timely substantial completion of the Work and achievement of all Milestone Dates.

3.10.7 In addition, if the progress of the Work is behind the Progress Schedule to such an extent that Owner reasonably determines that Contractor will be unable to meet any of the critical path dates set forth in the Progress Schedule, including without limitation any Milestone Date, or Contractor fails to take prompt and adequate corrective action to Owner's satisfaction to bring the progress of the Work in compliance with the Progress Schedule, Owner may, in addition to any other right or remedy provided herein, proceed as provided in Sections 2.3 or 2.4.

3.10.8 Whenever significant changes to the Project occur, such as added or deleted activities, they must be reflected on a revised Progress Schedule to be submitted to Owner for its review and approval.

3.10.9 Contractor shall be responsible on a daily basis to maintain all information which affects the length of specific activities on the Progress Schedule, times when Contractor will perform specific jobs, and other data relevant to the Progress Schedule as required by the Architect or Owner. Contractor shall make available at any time such information for review by the Architect or Owner.

3.10.10 Contractor shall prepare and keep current, for the Architect's review, a schedule of submittals which is coordinated with Contractor's Progress Schedule, and allow the Architect reasonable time to review submittals.

Other provisions:

3.10.1 The Contractor, promptly after being awarded the Contract, shall prepare and submit for the Owner's and Architect's information a Contractor's construction schedule for the Work. The schedule shall not exceed time limits current under the Contract Documents, shall be revised at appropriate intervals as required by the conditions of the Work and Project, shall be related to the entire Project to the extent required by the Contract Documents, and shall provide for expeditious and practicable execution of the Work. The Contractor warrants to the Owner and Architect that the Work can be completed according to the schedule and that the schedule is accurate at the time of creation of same. This schedule shall:

3.10.1.1 indicate the dates for the start and completion of the various elements of the Work, and shall be revised at the time of issuance of any Change Order affecting the Contract Time, or at such other intervals as required by conditions of the Work;

3.10.1.2 provide a graphic representation of activities and events that will occur during performance of the Work in sufficient detail, and as acceptable to the Owner and Architect, to show the sequencing of the various trades for each part of the Work. However, review of such schedules shall not constitute the undertaking of or responsibility for construction means, methods, sequences or techniques;

3.10.1.3 identify each phase of construction and occupancy; and

3.10.1.4 set forth dates that are critical in ensuring the timely and orderly completion of the Work in accordance with the requirements of the Contract Documents as amended from time to time.

3.10.3.1 If the Contractor submits a schedule, or otherwise expresses an intent to achieve Substantial or Final Completion of the Work or any portion thereof, prior to the date any such completion is required by the Contract Documents, no liability on the part of the Owner or Architect shall be created thereby. The Contractor shall not be entitled to any adjustment in the Contract Sum or Contract Time for failure to achieve such early completion dates.

Another provision:

The Contractor shall prepare and submit to Owner a monthly summary report in a form and with sufficient detail approved by Owner. At a minimum the report shall specify whether the Project is on schedule and, if not, the reasons therefore and the Contractor's recommendations as to how to meet the schedule. Such monthly report shall be an updated current Project schedule and shall list the status of all change requests or modifications. The Contractor shall hold weekly progress meetings at the Project Site or at such other times and places approved by Owner. Progress of the Work shall be reported in detail with reference to construction schedules. Each Subcontractor, if requested by Owner or Contractor, shall have present a competent representative to report the condition of its work and to receive information or instructions.

§ 5.22 Documents and Samples at the Site: ¶ 3.11

The owner may want to insert the following words at the end of ¶ 3.11.1: "Delivery of such items shall be a condition precedent before final payment is made."

In jurisdictions in which the contractor has a set of approved permit drawings, the following provision can be inserted after ¶ 3.11.1:

3.11.2 The Contractor shall maintain all approved permit drawings and other documents at the site, so as to make them accessible to inspectors, the Architect, and the Owner at all times that the Work is in progress. Such documents shall be delivered to the Architect before final payment.

§ 5.23 Shop Drawings, Product Data, and Samples: ¶ 3.12

Sometimes the parties want the shop drawings to be contract documents. Such a change may be made by substituting this language for ¶ 3.12.4:

3.12.4 Shop Drawings, Product Data, Samples, and similar submittals shall become Contract Documents only after written acceptance of each such item by the Architect and Owner. The purpose of their submittal is to demonstrate for those portions of the Work for which submittals are required the way the

Contractor proposes to conform to the information given and the design concept expressed in the Contract Documents. Review by the Architect is subject to the limitations of Subparagraph 4.2.7. In the event of any discrepancy between any such submittal and the other Contract Documents, the other Contract Documents shall control, unless the Contractor has specifically noted the discrepancy and that discrepancy is approved in writing by the Architect and Owner.

The following language could replace 3.12.10 to expand on this section:

3.12.10 When professional certification of performance criteria of materials, systems, or equipment is required by the Contract Documents, the Contractor shall provide the person or party providing the certification with full information on the relevant performance requirements and on the conditions under which the materials, systems, or equipment will be expected to operate at the project site. The certification shall be based on performance under the operating conditions at the project site. The Architect shall be entitled to rely on the accuracy and completeness of such certifications.

Another provision:

3.12.7 The Contractor shall perform no portion of the Work for which the Contract Documents require submittal and review of Shop Drawings, Product Data, Samples or similar submittals until the respective submittal has been approved by the Architect. Shop Drawings for Architectural, Structural, Mechanical and Electrical Work shall be submitted to the Architect for confirmation that these documents are consistent with the requirements of the Contract Documents, with the Architect to retain the approved Shop Drawings for delivery to Owner at the completion of the Project. The Contractor shall assemble for the Architect's approval and for transmittal to the Owner three complete copies in loose leaf binders of all operating and maintenance data from all manufacturers whose equipment is installed in the Work. Upon request from Contractor, Architect shall prepare a detailed list of operations and maintenance submittal requirements.

Another:

Add the following to the end of paragraph 3.12.7: "Should the Contractor, Subcontractors or Sub-subcontractors install, construct, erect or perform any portion of the Work without approval of any requisite submittal, the Contractor shall bear the costs, responsibility, and delay for removal, replacement, and/or correction of any and all items, material, and /or labor."

§ 5.24   Use of Site: ¶ 3.13

Another provision:

3.13.1 The Contractor shall confine operations at the site to areas permitted by applicable laws, statutes, ordinances, codes, rules and regulations, and lawful orders of public authorities and the Contract Documents and shall not unreasonably encumber the site with materials or equipment. Failure to comply with such laws, statutes, ordinances, codes, rules and regulations, and lawful orders that result in any delay to the Project or other cost to the Owner shall be the responsibility of the Contractor and shall promptly be remedied by the Contractor, with any cost borne by the Owner to be deducted from the Contract Sum. To the extent there is any delay resulting therefrom, the Owner shall be entitled to damages resulting from such delay, notwithstanding the provisions of Section 15.1.6.

3.13.2 The Contractor shall limit operations and storage of material to the area within the Work limit lines shown on Drawings, except as necessary to connect to existing utilities, shall not encroach on neighboring property, and shall exercise caution to prevent damage to existing structures.

3.13.3 Only materials and equipment which are to be used directly in the Work shall be brought to and stored on the site by the Contractor. After equipment is no longer required for the Work, it shall be promptly removed from the site. Protection of construction materials and equipment stored at the site from weather, theft, damage and all other adversity is solely the responsibility of the Contractor.

3.13.4 No project signs shall be erected without the written approval of the Owner.

3.13.5 The Contractor shall ensure that the Work is at all times performed in a manner that affords reasonable access, both vehicular and pedestrian, to the site of the Work and all adjacent areas. The Work shall be performed, to the fullest extent reasonably possible, in such a manner that public areas adjacent to the site of the Work shall be free from all debris, building materials and equipment likely to cause hazardous conditions. Without limitation of any other provision of the Contract Documents, Contractor shall not interfere with the occupancy or beneficial use of (1) any areas and buildings adjacent to the site of the Work or (2) the Work in the event of partial occupancy. Contractor shall assume full responsibility for any damage to the property comprising the Work or to the owner or occupant of any adjacent land or areas resulting from the performance of the Work.

3.13.6 The Contractor shall not permit any workers to use any existing facilities at the site, including, without limitation, lavatories, toilets, entrances, and parking areas other than those designated by Owner. The Contractor, Subcontractors of any tier, suppliers and employees shall comply with instructions or regulations of the Owner governing access to, operation of, and conduct while in or on the premises and shall perform all Work required under the Contract Documents in such a manner as not to unreasonably interrupt or interfere with the conduct of Owner's operations.

3.13.7 The Contractor and each Subcontractor of any tier shall have its name, acceptable abbreviation or recognizable logo and the name of the city and state of the mailing address of the principal office of the company, on each motor vehicle and motorized self-propelled piece of equipment which is used in connection with the Work. The signs are required on such vehicles during the time the Contractor is working on the Project.

3.13.8 The Contractor shall keep the site of the Work and surrounding areas free from accumulation of waste materials, rubbish, debris, and dirt resulting from the Work and shall clean the site and surrounding areas as requested by the Architect and/or the Owner, including mowing of grass greater than 6 inches high. The Contractor shall be responsible for the cost of cleanup and removal of debris from premises. The building and premises shall be kept clean, safe, in a workmanlike manner, and in compliance with OSHA standards at all times. At completion of the Work, the Contractor shall remove from and about the site tools, construction equipment, machinery, fencing, and surplus materials. Further, at the completion of the work, all dirt, stains, and smudges shall be removed from every part of the building, all glass in doors and windows shall be washed, and entire Work shall be left broom clean in a finished state ready for occupancy. The Contractor shall advise its Subcontractors of any tier of this provision, and the Contractor shall be fully responsible for leaving the premises in a finished state ready for use to the satisfaction of the Owner's Representative. If the Contractor fails to comply with the provisions of this paragraph, the Owner may do so and the cost thereof shall be charged to the Contractor.

§ 5.25   Cutting and Patching: ¶ 3.14

The following provision could be added:

3.14.3 The Contractor shall locate, protect, and save from injury utilities of all kinds, either above or below grade, inside or outside of any structure, found in the areas affected by its work. Contractor shall be responsible for all damage caused to such utility by the operation of equipment or delivery of materials or as the direct or indirect result of any of its work and shall repair all such damage at its expense and as a part of the work included in the Contract Documents. The Contractor shall not be entitled to any increase in the Contract Sum or the Contract Time on account of such damage to any utility.

Another:

3.14.3 If the Work involves renovation and/or alteration of existing improvements, Contractor acknowledges that cutting and patching of the Work is essential for the Work to be successfully completed. Contractor shall perform any cutting, altering, patching, and/or fitting of the Work necessary for the Work and the existing improvements to be fully integrated and to present the visual appearance of an entire, completed, and unified project. In performing any Work which requires cutting or patching, Contractor shall use its best efforts to protect and preserve the visual appearance and aesthetics of the Work to the reasonable satisfaction of both the Owner and Architect.

§ 5.26   Cleaning Up: ¶ 3.15

The following provision could be substituted:

3.15.1 The Contractor shall keep the site of the Work and adjacent premises as free from material, debris and rubbish as is practicable, and shall remove same from any portion of the site, if, in the opinion of the Owner's Project Manager, such material, debris, or rubbish constitutes a nuisance or is objectionable in any way to the Owner. The Contractor further agrees to remove all machinery, materials, implements, barricades, staging, falsework, debris, and rubbish connected with or caused by the Work immediately upon the completion of the Work, and to clean all structures and work under the Contract Documents to the satisfaction of the Owner's Project Manager, and to leave the premises in perfect condition insofar as affected by the Work hereunder. Contractor acknowledges that Owner and/or other tenants in the building, if any, will continue to occupy and must maintain continuous operations in the building in which the Work is located. It is critical that these operations shall not suffer any significant interference, including, without limitation, any interruption in utilities or unreasonable noise, dust, odor, or vibration. Contractor shall perform the Work and limit its use of the Project site in such a manner as to minimize any interference with occupancies and operations in the building and in accordance with applicable building rules and regulations. Without limiting the generality of the foregoing, at no additional cost to Owner, Contractor shall provide and apply dust control at all times, including holidays and weekends, as required, to prevent the spread of dust and to avoid the creation of a nuisance at the Work site or in the surrounding area as a result of construction activities. Dust control shall be by sprinkler water or other approved means, except that no chemicals, oil, or similar palliatives shall be used. Quantities and equipment shall be sufficient to control dust effectively. When weather conditions warrant, sprinkling equipment shall be on hand and immediately available at all times. Architect or Owner shall have the authority to order dust control Work whenever required in its opinion; however, dust control shall be effectively maintained at all times whether or not specifically ordered by Architect or Owner. Contractor shall also take proper measures, at no additional cost to Owner, to prevent tracking mud onto public streets or roads or property of third persons. Such measures shall include but are not limited to covering muddy areas on site with clean, dry sand. All ingress/egress from the site shall be maintained in a dry condition, and any mud tracked onto public streets or roads, or other areas of the building, or property of third persons shall be immediately removed and the affected area cleaned. Architect or Owner may order such Work at any time that conditions warrant. Contractor shall be liable for all costs, liability, and expense, including but not limited to court costs, for all claims related to dust or windblown materials attributed to the Work hereunder.

3.15.2 Contractor shall be held responsible for all daily clean-up of construction materials and debris and building dust control. Clean-up shall include removal of materials and debris from the building and placement in a debris box or other proper disposal. Special consideration is required for the immediate removal and/or protection of material or debris which poses a hazard to Owner's customers, employees, fixtures, and floor coverings (i.e., hazardous materials, broken glass, sawdust, materials that pose a tripping hazard, etc.), including utilization of protective coverings for newly installed floor covering and fixtures. Certain construction activities, including but not limited to drywall sanding, spray painting, sawing, etc., create dust which must be controlled to protect Owner's customers, employees, equipment, and merchandise. Contractor shall take steps as necessary to control the dust created by these operations, including but not limited to Visqueen, ventilation, or a solid construction barrier. The clean-up of a given area is a rigid requirement:

3.15.2.1 Contractor is to ensure that it has an adequate labor force to maintain a clean work area at all times during the Work;

3.15.2.2 If for any reason Contractor does not clean up its work, it will receive a written notice from Owner's Project Manager before Owner takes over the Work and back-charges Contractor for the cost incurred;

3.15.2.3 Any damage done by Contractor shall be its responsibility to report and rectify as soon as possible; and

3.15.2.4 Clean-up shall include removal by Contractor of all excess materials, such as unused paint and solvents, etc., upon completion of the Work.

3.15.3 Upon completion of the Work, Contractor shall provide final clean-up of all surfaces, including, without limitation, tile, glass, storefronts, carpet, wall finishes, equipment, pavers, sidewalks, and streets.

Another:

3.15.3 The Contractor shall pump, bail, or otherwise keep any general excavations free of water. The Contractor shall keep all areas free of water before, during and after concrete placement.

§ 5.27 Access to Work: ¶ 3.16

Another provision:

3.16.1 The Architect and the Owner and their respective representatives shall, at all times, have access to the Work; and the Contractor shall provide proper and safe facilities for such access.

3.16.2 When requested by the Owner or Architect, the Contractor, at no extra charge, shall provide scaffolds or ladders in place as may be required by the Architect or the Owner for examination of Work in progress or completed.

§ 5.28 Royalties, Patents, and Copyrights: ¶ 3.17

Contractors may be unwittingly named in a copyright suit brought by a prior architect. This language may help:

> **13.17.2** Owner warrants to Contractor that the Instruments of Service provided to the Contractor are free of any claim by any third party, and that the Owner either owns the copyright or a license to use the Instruments of Service for the intended purpose herein. In the event that any claim of copyright infringement is brought against the Contractor other than contemplated by Section 13.17.1, including but not limited to a claim by the Architect or the Architect's consultants, Owner will defend and hold harmless the Contractor and its subcontractors from any such claim, including costs and expenses, including, but not limited to attorneys fees. The Contractor shall be entitled to retain counsel of its own choice, all costs of which are to be reimbursed by Owner.

§ 5.29 Indemnification: ¶ 3.18

Substitute language for ¶ 3.18.1 more favorable to the contractor is as follows:

> **3.18.1** To the fullest extent permitted by law, the Contractor shall indemnify and hold harmless the Owner and agents and employees of the Owner from and against claims, damages, losses, and expenses arising out of or resulting from performance of the Work, provided that such claim, damage, loss, or expense is attributable to bodily injury, sickness, disease, or death, or to injury to or destruction of tangible property (other than the Work itself), but only to the extent caused by negligent acts or omissions of the Contractor, a Subcontractor, anyone directly or indirectly employed by them, or anyone for whose acts they may be liable, unless such claims, damages, losses, and expenses are caused in whole or in part by the Owner, the Architect, Architect's consultants, the Owner's own forces or other contractors employed by the Owner, or any employee of any of them. Such obligation shall not be construed to negate, abridge, or reduce other rights or obligations of indemnity which would otherwise exist as to a party or person described in this Paragraph 3.18.

The owner may want to amend the coverage in ¶ 3.18.1 to include officers and directors of the owner.

When the project involves multiple owners, such as a condominium or townhouse project, the owner may want to insert the following provision after ¶ 3.18.1:

> **3.18.1.1** Pursuant to this Section 3.18, the definition of "Owner" shall include each owner of each townhome within the townhouse project, and shall include each beneficiary of any land trust, and each co-tenant, partner, or other person or entity that owns any portion of any such unit.

The owner may want to insert the following provisions in those states where they are applicable:

3.18.3 The obligations set forth in this section shall, but not by way of limitation, specifically include all claims arising or alleged to arise under the [state] Structural Work Act [insert other applicable act].

3.18.4 The purchase of insurance by the Contractor with respect to the obligations required herein shall in no event be construed as fulfillment or discharge of such obligations.

3.18.5 None of the foregoing provisions shall deprive the Owner or the Architect of any action, right, or remedy otherwise available to them or any of them at common law.

3.18.6 If any party is requested but refuses to honor the indemnity obligations hereunder, then the party indemnifying shall, in addition to all other obligations, pay the cost of bringing any such action, including attorneys' fees, to the party requesting indemnity.

If concerned with labor problems, an owner may wish to add the following language:

3.18.7 The Contractor agrees to indemnify, defend, and hold harmless the Owner from and against any and all administrative and judicial actions (including reasonable attorneys' fees related to any such actions) and judgments incurred by the Owner in connection with any labor-related activity arising from the Contractor's performance of the Work. As used in these Contract Documents, "labor-related activity" includes but is not limited to strikes, walkouts, informational or organizational picketing, use of placards, distribution of hand-outs, leaflets, or other similar acts at or in the vicinity of the Project or in the vicinity of any other facility where the Owner conducts business. The Owner shall advise the Contractor if any labor-related activity occurs, and the Contractor shall arrange for the legal representation necessary to protect the Owner's interest, provided such representation is approved by the Owner in advance.

To avoid the possibility that a court may invalidate this provision under a state anti-indemnification statute,[5] the following amendment to ¶ 3.18 may be considered:

This indemnification shall not be construed to indemnify any indemnitee from its own negligence. To the extent any party indemnified hereunder is negligent,

---

[5] In Herington v. J.S. Alberici Constr. Co., 266 Ill. App. 3d 489, 639 N.E.2d 907 (1994), the subcontract had the following provision:

(c) For all Subcontract work performed in Illinois or performed by Illinois Subcontractors, the following term is in effect:

this provision shall not apply to such party, but it shall continue to be effective as to all other parties not so negligent.

Another provision that might be added to protect the owner and architect from subcontractor claims is the following:

3.18.8 The Contractor will hold harmless the Owner, the Architect, Architect's Consultants, and their agents and employees from all liability, loss, or expense, including but not limited to attorneys' fees, arising out of claims by subcontractors or suppliers of any material or equipment for installation or incorporation in the Work, including any items especially designed or fabricated for the Work or for tools or equipment rented or leased for the Work.

The following provision relates to ¶ 3.18.2; however, it should be placed in a different part of the Supplementary Conditions to avoid voiding it in those states where a court strictly construes an anti-indemnification statute. The effect of this provision is to reinforce the waiver of workers' compensation liability that is available in some jurisdictions:

16.1 To the fullest extent permitted by law, the Contractor waives any liability cap imposed by law provided to an employer who has paid any injured employee's workers' compensation benefits. The Contractor shall require similar waivers from each Subcontractor and their respective Sub-subcontractors, so that every entity providing labor for the Project shall be subject to this provision. The intent of this provision is to place liability for workers' injuries on the party responsible for the safety of such workers and not to avoid any limitation on such liability by virtue of operation of law. It is further the intent

---

Subcontractor hereby assumes the entire liability for its own negligence and the negligence of its own employees; in addition, Subcontractor hereby assumes the entire liability arising from any alleged violation of the Structural Work Act (Chapter 48, Sections 60-69 Ill. Revised Statutes) that Subcontractor knew of or by the exercise of ordinary care should have known of. Subcontractor agrees to indemnify and save harmless Contractor and its agents, servants and employees, from and against all loss, expense, damage or injury, including legal fees, that Contractor may sustain as a result of any claims predicated or said allegations of Subcontractor's own negligence or on Subcontractor's alleged violation of the Structural Work Act as above set forth. This provision shall specifically not require Subcontractor to indemnify Contractor from Contractor's own alleged negligence in violation of Chapter 29, Section 61 of the Illinois Revised Statutes. In the event claim of any such loss, expense, damage or injury, as above defined and limited, is made against Contractor, its agents, servants or employees, Contractor may: (1) withhold from any payment due or hereafter becoming due to Subcontractor under the terms of this contract, an amount sufficient in Contractor's judgment to protect and indemnify Contractor from all such claims, expenses, legal fees, loss, damage or injury as above defined and limited; or (2) require Subcontractor to furnish a surety bond in such amount so determined; or (3) require Subcontractor to provide suitable indemnity acceptable to Contractor.

The court held that this clause did not violate public policy and was effective to waive workers' compensation limits under Illinois law.

of the parties that this provision is not to be construed as an indemnification provision.

This is a form of indemnification that is being used in Illinois. It can be modified for other jurisdictions by replacing the citations with appropriate ones for the jurisdiction:

1. The Contractor agrees to protect, defend, indemnify, and hold the Owner, its officers, directors, agents, and employees (hereafter "the Indemnified Parties"), free and harmless from and against any and all claims, damages, demands, injury or death, arising out of or being in any way connected with the Contractor's performance under this Agreement. The indemnification provided herein will be effective to the maximum extent permitted by applicable law. This indemnity extends to all legal costs, including without limitation, the following: attorneys' fees, costs, liens, judgments, settlements, penalties, professional fees, or other expenses incurred by the Owner, including but not limited to, fines and penalties imposed by public entities and the reasonable settlement of such claims. This indemnification is not limited by any amount of insurance required under this Agreement. Further, the indemnity contained in this section will survive the expiration or termination of this Agreement. Only to the extent necessary to prevent any indemnity under this Agreement from being void under 740 ILCS 35/1, "The Construction Contract Indemnification for Negligence Act," this indemnity agreement shall not require an Indemnified Party to be indemnified against that party's own negligence. To the extent permissible by law, Contractor waives any limits to the amount of its obligations to indemnify, defend, or contribute to any sums due under any claim by an employee of Contractor that may be subject to the Workers Compensation Act, 820 ILCS 305/1 *et seq.*, or any other law or judicial decision (e.g., *Kotecki v. Cyclops Welding Corp.*, 146 Ill. 2d 155 (1991)).

2. The Contractor will be solely responsible for the defense of any and all claims, demands, or suits against the Indemnified Parties, including without limitation, claims by an employee, subcontractor, agents, or servants of Contractor even though the claimant may allege that the Indemnified Parties were in charge of the Work or allege negligence on the part of the Indemnified Parties. Owner will have the right, at its sole option, to participate in the defense of any such suit, without relieving the Contractor of its obligations hereunder.

3. "Injury" or "damage" as these words are used in this section will be construed to include, but will not be limited to, injury or damage consequent upon the failure of use or misuse by Contractor, its subcontractors, agents, servants, or employees, of any scaffolding, hoists, cranes, stays, ladders, supports, rigging, blocking or any and all other kinds of items of equipment, whether or not the same be owned, furnished, or loaned by the indemnified Parties.

§ 5.30   Publicity: ¶ 3.19

The owner may want to insert the following clause to limit the contractor's right to use the project for publicity purposes:

3.19 PUBLICITY

3.19.1 The Contractor shall not divulge information concerning this project to anyone (including, without limitation, information in applications for permits, variances, etc.) without the Owner's prior written consent. The Contractor shall obtain a similar agreement from firms, subcontractors, suppliers, and others employed by the Contractor. The Owner reserves the right to release all information as well as to time its release, form, and content. This requirement shall survive the expiration of the contract.

§ 5.31   Security: ¶ 3.20

On some projects, the owner may want particular security precautions. The following provision is for a remodeling project, but it can be adapted to various requirements:

3.20 SECURITY

3.20.1 Only previously authorized personnel will be permitted on the construction site. The Contractor shall, before the commencement of the Work, submit to the Construction Coordinator the names of all personnel either directly employed by the Contractor or in the employ of any Subcontractor who will be present on the site.

3.20.1.1 All construction personnel will be required to register with the Owner's security forces and/or personnel and wear badges as furnished by the Owner. Personnel who do not display badge identification will be removed from the site until properly registered and/or wearing badges. All badges shall be maintained and controlled by the Contractor and shall be returned to the Owner upon completion of the Work.

3.20.1.2 If additional or special personnel are needed for the efficient completion of the Work, then the Contractor shall submit a list of names of all such additional personnel at least 48 hours before their appearance on the site.

3.20.1.3 Site-parked mobile equipment and operable machinery and hazardous parts of the new construction subject to mischief shall be kept locked or otherwise made inoperable whenever left unattended.

§ 5.32   Contractor's Qualifications: ¶ 3.21

The owner may want to add this language:

3.21 CONTRACTOR'S QUALIFICATIONS

3.21.1 The Contractor represents and warrants the following to the Owner (in addition to any other representation and warranty given by the Contractor to the Owner) as an inducement to the Owner to enter into the Owner-Contractor Agreement, which representations and warranties shall survive the execution of the Contract Documents and final completion of the work and final payment therefor:

3.21.1.1 The Contractor is financially solvent, able to pay its debts as they mature, and possessed of sufficient working capital to complete the Work and perform its obligations under the Contract Documents in an efficient and capable manner;

3.21.1.2 The Contractor is able to furnish the tools, materials, supplies, equipment, and labor required to complete the Work and perform its obligations under the Contract Documents, and has sufficient experience and competence to do so;

3.21.1.3 The Contractor is authorized to do business in the state where the Project is located and is properly licensed by all necessary governmental, public, and other authorities having jurisdiction over the Contractor and the project;

3.21.1.4 The person(s) executing the Owner-Contractor Agreement is properly authorized to do so;

3.21.1.5 The Contractor has visited the site and has become familiar with the Contract Documents and the conditions at the site; has correlated the Contract Documents with the site conditions and with all applicable codes, ordinances, regulations, laws, and decrees; and knows of no reason why the Work cannot be performed exactly as shown on the Contract Documents, unless previously stated otherwise in writing to the Owner and Architect.

§ 5.33   Lender's Architect: ¶ 3.22

If the lender will have an inspecting architect, the following provision may be appropriate:

3.22 LENDER'S ARCHITECT

3.22.1 If the Owner's lender requires the services of an inspecting architect or other representative, the Owner may require the concurrence of such inspecting architect or lender's representative in each instance where the approval of the Architect herein is required by any provision of the Contract Documents. The Contractor shall fully cooperate with such inspecting architect or representative.

§ 5.34   Contractor's Subordination: ¶ 3.23

The Owner may want to add the following:

3.23 Contractor shall and hereby does subordinate any and all liens, rights, and interests (whether choate or inchoate and including, without limitations, all mechanics' and materialmen's liens under the applicable laws of the state in which the Project is located, whether contractual, statutory, or constitutional) owned, claimed, or held or to be owned, claimed, or held by Contractor in and to any part of the Work or the property on which the Work is performed, to the liens securing payments of sums now or hereinafter borrowed by Owner in connection with the development, design, and/or construction of the Project and to all liens, rights, and interests of Owner. Contractor shall promptly execute such further and additional evidence of the subordination of liens, rights, and interest of Owner or Owner's interim or permanent lenders. The subordination of lien is made in consideration of and as an inducement to the execution and delivery of the Contract for Construction and shall be applicable despite any dispute between the parties hereto or any others, or any default by Owner under the Contract or otherwise.

§ 5.35   Record Drawings: ¶ 3.24

3.24.1 The Contractor shall maintain a set of Record Drawings on site in good condition and shall use colored pencils to mark up said set with "record information" in a legible manner to show: (1) bidding addendums, (2) executed change orders, (3) deviations from the Drawings made during construction; (4) details in the Work not previously shown; (5) changes to existing conditions or existing conditions found to differ from those shown on any existing drawings; (6) the actual installed position of equipment, piping, conduits, light switches, electric fixtures, circuiting, ducts, dampers, access panels, control valves, drains, openings, and stub-outs; and (7) such other information as either Owner or Architect may reasonably request. The prints for Record Drawing use will be a set of "blue line" prints provided by Architect to Contractor at the start of construction. Upon Substantial Completion of the Work, Contractor shall deliver all Record Drawings to Owner and Architect for approval. If not approved, Contractor shall make the revisions requested by Architect or Owner. Final payment and any retainage shall not be due and owing to Contractor until the final Record Drawings marked by Contractor as required above are delivered to Owner.

3.24.2 Operating Instructions and Service Manuals

3.24.2.1 The Contractor shall submit four (4) volumes of operating instructions and service manuals to the Architect at the time of Substantial Completion, or as soon thereafter as practicable. Submission of all of the following shall be a condition precedent to Final Payment to the Contractor. The operating instructions and service manuals shall contain:

.1 Start-up and Shutdown Procedures: Provide a step-by-step write up of all major equipment. When manufacturer's printed start-up, trouble shooting and shut-down procedures are available, they may be incorporated into the operating manual for reference.

.2 Operating Instructions: Written operating instructions shall be included for the efficient and safe operation of all equipment.

.3 Equipment List: List of all major equipment as installed shall include model number, capacities, flow rate, and name-plate data.

.4 Service Instructions: The Contractor shall be required to provide the following information for all pieces of equipment.

   (a) Recommended spare parts including catalog number and name of local suppliers or factory representative.

   (b) Belt sizes, types, and lengths.

   (c) Wiring diagrams.

.5 Manufacturer's Certificate of Warranty:

Manufacturer's certificates of warranty shall be obtained for all major equipment. Warranty shall be obtained for at least one year from the date of Substantial Completion. Where longer period is required by the Contract Documents, the longer period shall govern.

.6 Parts catalogs: For each piece of equipment furnished, a parts catalog or similar document shall be provided which identifies the components by number for replacement ordering.

3.24.3 Submission

.1 Manuals shall be bound into volumes of standard 8 1/2″ x 11″ hard binders. Large drawings too bulky to be folded into 8 1/2″ x 11″ shall be separately bound or folded and in brown envelopes, cross-referenced and indexed with the manuals.

.2 The manuals shall identify the Owner's project name, project number, and include the name and address of the Contractor and major Subcontractors of any tier who were involved with the activity described in that particular manual.

§ 5.36   Architect—General: ¶ 4.1

This section must be coordinated with the owner/architect agreement, particularly if changes have been made to the standard form AIA agreement. If changes are made here, they would not affect the architect's duties to the owner unless the owner/architect agreement contains those same provisions.

§ 5.37   Administration of the Contract: ¶ 4.2

Another provision:

4.2.1 The Architect will provide administration of the Contract as described in the Contract Documents, and will be an Owner's representative (1) during construction, (2) until final payment is due, and (3) with the Owner's concurrence, from time to time during the one-year period for correction of Work described in Section 12.2. Notwithstanding these provisions, no act or omission of the Architect shall be considered a waiver of any of the Owner's rights and interests. The Architect will have authority to act on behalf of the Owner only to the extent provided in the Contract Documents, unless otherwise modified in writing in accordance with other provisions of the Contract.

If the owner does not want the architect to actually reject work without first consulting with the owner, the first sentence of ¶ 4.2.6 could be replaced with the following:

The Architect shall notify the Owner whenever the Architect believes that any Work does not conform to the Contract Documents, whereupon the Owner may reject or accept same.

The following provision could be added to provide additional clarification that the architect is not in charge of the work. Note that, despite this language, a court may still find that the owner is "in charge" of the work.

The Architect will have the right to recommend to the Owner rejection of Work which does not conform to the Contract Documents, but Architect will not have the authority to stop the Work. Neither the Owner nor the Architect is in charge of the work or responsible for the use of supporting devices in connection with the Contract. The Contractor is solely and exclusively responsible for and in charge of site safety.

The following provision can be inserted after ¶ 4.2.7 to make sure that the contractor does not simply trace the architect's drawings. If electronic copies of documents are furnished to the contractor, this clause should be modified accordingly:

4.2.7.1 The Contractor shall not submit any shop drawing that is merely a tracing or other copy of any of the Contract Documents. Each shop drawing must be prepared by the Contractor or a subcontractor or supplier of the Contractor. The Architect shall have the authority to reject any shop drawing that violates this provision, and no extension of the Contract Time shall be given on account of such rejection.

The following provision expands on the architect's interpretation duties:

4.2.11 The Architect will interpret and decide matters concerning performance under, and requirements of, the Contract Documents on written request of either the Owner or Contractor. Upon receipt of such request, the Architect shall promptly notify the non-requesting party in writing of such request. The Architect's response to such requests will be made in writing within any time limits agreed upon or otherwise with reasonable promptness. If no agreement is made concerning the time within which interpretations required of the Architect shall be furnished in compliance with this Section 4.2, then delay shall not be recognized on account of failure by the Architect to furnish such interpretations until 15 days after written request is made for them. Both the Contractor and Owner shall proceed diligently with their respective duties under the Contract Documents pending the Architect's interpretation, and thereafter in conformance with such interpretation, but subject to other provisions contained herein.

§ 5.38 Definitions: ¶ 5.1

Another version:

5.1.1 Subcontractor and Lower-tier Subcontractor. A Subcontractor is a person or organization who has a contract with the Contractor to perform any of the Work. The term "Subcontractor" is referred to throughout the Contract Documents as if singular in number and means a Subcontractor or its authorized representative. The term "Subcontractor" also is applicable to those furnishing materials to be incorporated in the Work whether work performed is at the Owner's site or off site, or both. A lower-tier Subcontractor is a person or organization who has a contract with a Subcontractor or another lower-tier Subcontractor to perform any of the Work at the site. Nothing contained in the Contract Documents shall create contractual relationships between the Owner or the Architect and any Subcontractor or lower-tier Subcontractor of any tier.

§ 5.39   Award of Subcontracts and Other Contracts for Portions of the Work: ¶ 5.2

An alternative clause to ¶ 5.2.1 that gives time limits for submittals by the contractor after an award is:

5.2.1 Within ten (10) days after execution of the Owner-Contractor Agreement, the Contractor shall submit to the Architect an accurate itemized labor and material cost schedule showing all subcontractors' names, addresses, telephone numbers, nature of work, and subcontract costs. Unless the Architect or Owner objects to any one or more subcontractors or material suppliers within seven (7) days after receipt of such list, it shall be deemed accepted.

The following paragraph could be added to obtain additional information from the contractor:

5.2.1.1 In addition to information which may be required prior to the execution of the Contract, not later than thirty (30) days after execution of the Contract, the Contractor shall furnish to the Owner through the Architect the names of persons or entities proposed as manufacturers for each of the products identified in the Specifications and/or Drawings and, where applicable, the name of the installing Subcontractor.

The owner may want to obtain copies of the subcontracts. This clause could be added after ¶ 5.2.4 to obtain them:

5.2.5 Upon request, the Contractor shall provide to the Owner an executed copy of all subcontracts, purchase orders, and other agreements relating to the Work.

Another provision:

5.2.4 The Contractor shall not change a Subcontractor, person or entity previously selected if the Owner or Architect makes reasonable objection to such substitute. The Owner may require the Contractor to change any Subcontractor previously approved and, if at such time the Contractor is not in default hereunder, the Contract Sum shall be increased or decreased by the difference in cost occasioned by such change and/or the Contract Time equitably adjusted to the extent occasioned by such change.

§ 5.40   Subcontractual Relations: ¶ 5.3

The following additional provisions regarding subcontractors could be added to provide more protection for the owner and architect:

5.3.1 By written agreement consistent with the provisions hereof, Contractor shall require each Subcontractor, to the extent of the Work to be performed by the Subcontractor, to be bound to Contractor and, to the extent herein provided, directly to Owner, by the terms of the Contract Documents which shall be incorporated by reference in such Subcontract Agreement. Each Subcontractor shall assume toward Contractor all the obligations and responsibilities which Contractor, by these Contract Documents, assumes toward Owner and Architect. In addition, each Subcontractor shall waive any rights it may have against Owner for damage by fire or other perils covered by property insurance maintained by the Subcontractor or required to be maintained by the Subcontractor or under the terms of the Contract Documents. Each Subcontract agreement shall preserve and protect the rights of Owner and Architect under the Contract Documents with respect to the Work to be performed by the Subcontractor so that subcontracting thereof will not prejudice such rights, and each Subcontract agreement shall allow to the Subcontractor, unless specifically provided otherwise in the Subcontract agreement, the benefit of all rights, remedies, and redress against Contractor that Contractor, by the Contract Documents, has against Owner. Contractor shall require each Subcontractor to enter into similar agreements with Sub-subcontractors. Contractor shall make available to each proposed Subcontractor, prior to the execution of the Subcontract agreement, copies of the Contract Documents to which the Subcontractor will be bound, and, upon written request of the Subcontractor, identify to the Subcontractor terms and conditions of the proposed Subcontract agreement which may be at variance with the Contract Documents. Subcontractors shall similarly make copies of applicable portions of such Contract Documents available to their respective proposed Sub-subcontractors.

5.3.2 Without limitation on the generality of the foregoing, each Subcontract agreement and each Sub-subcontract agreement shall include, and shall be deemed to include, the following provisions:

5.3.2.1 An agreement that the Owner is a third-party beneficiary of the Subcontract (or Sub-subcontract), entitled to enforce any rights thereunder for its benefit, and that the Owner shall have the same rights and remedies against the Subcontractor (or Sub-subcontractor) as the Contractor (or Subcontractor) has, including but not limited to the right to be compensated for any loss, expense, or damage of any nature whatsoever incurred by the Owner resulting from any breach of representations and warranties, expressed or implied, if any, arising out of the agreement and any error, omission, or negligence of the Subcontractor (or Sub-subcontractor) in the performance of any of its obligations under the agreement;

5.3.2.2 A requirement that the Subcontractor (or Sub-subcontractor) promptly disclose to the Contractor (or Subcontractor) any defect, omission, error, or deficiency in the Contract Documents or in the Work of which it has knowledge;

5.3.2.3 A provision requiring the Subcontractor (or Sub-subcontractor) to maintain worker's compensation insurance as required by the laws of the state of _____(), and employer's liability, commercial general liability, comprehensive automobile liability, and excess liability insurance with limits and terms as provided in Paragraph [11.1] or as approved in writing by Owner. These policies shall provide that they may not be altered or canceled except with thirty (30) days' advance written notice to Owner, by registered mail. The Subcontractor (or Sub-subcontractor) shall provide a copy of the applicable insurance policies to Owner upon request;

5.3.2.4 A provision that the agreement shall be terminable upon seven (7) days' written notice by the Contractor (or Subcontractor), or, if the Subcontract (or Sub-subcontract) has been assigned to Owner, by Owner;

5.3.2.5 A provision that neither the Contractor nor the Subcontractor (or the Sub-subcontractor) shall have the right to require arbitrations of any disputes in those cases in which the Owner is a party or in which the outcome could affect the Contract Sum or the Contract Time, except at the sole election of the Owner;

5.3.2.6 A provision that the Owner and its authorized representatives shall have the right to conduct audits or other examinations of the Subcontractor's (or Sub-subcontractor's) books and records relating to Claims and any Change Orders in which the price is determined on the basis of actual costs incurred;

5.3.2.7 A provision requiring the Subcontractor (or Sub-subcontractor) to submit certificates and waivers of liens for work completed by it and its Sub-subcontractors as a condition to the disbursement of the progress payment next due and owing; and

5.3.2.8 A provision requiring submission to the Contractor (or Subcontractor, as the case may be) of applications for payment in a form approved by the Owner, together with clearly defined invoices and billings supporting all such applications.

5.3.3 The Contractor shall be responsible for any and all Subcontractors working under it and shall carry insurance for all Subcontractors or ensure that they are carrying it themselves so as to relieve the Owner, Architect, and Architect's Consultants of any and all liability.

5.3.4 The Owner or Architect assumes no responsibility for the overlapping or omission of parts of the work by various Subcontractors in their Contracts with the Contractor, because this is solely the Contractor's responsibility.

Additional provisions:

5.3.2 Without limitation on the foregoing, each subcontract agreement and each sub-subcontract agreement shall include, and shall be deemed to include, the following:

5.3.2.1 An agreement that the Owner is a third-party beneficiary of the subcontract or sub-subcontract, entitled to enforce any rights thereunder for its benefit, and that the Owner shall have the same rights and remedies against the Subcontractor (or Sub-subcontractor) as the Contractor (or Subcontractor) has, including, but not limited to the right to be compensated for any loss, expense, or damage of any nature whatsoever incurred by the Owner resulting from any breach of contract, warranty, whether express or implied, arising out of the agreement and any error, omission or negligence of the Subcontractor or sub-contractor in the performance of any obligation under the agreement; and

5.3.2.2 A requirement that the Subcontractor or Sub-subcontractor promptly disclose to the Contractor or Subcontractor any defect, omission, error or deficiency in the Contract Documents or in the Work of which it has, or should have had, knowledge.

5.3.2.3 On request of the Owner or Architect, the Contractor shall promptly furnish true and correct copies of all applicable contracts, work orders, requisitions, purchase orders and other documents relating to the Work.

5.3.2 Any part of the work performed for the Contractor by a Subcontractor shall be pursuant to a written Subcontract between the Contractor and such Subcontractor. Each such Subcontract shall:

.1 require that such Work be performed in accordance with the requirements of the Contract Documents;

.2 waive all rights the contracting parties may have against one another or that the Subcontractor may have against the Owner for damages caused by fire or other perils covered by the property insurance described in the Contract Documents (except for the amount of the deductibles, the right to recover the same is not waived); and

.3 require the Subcontractor to carry and maintain liability insurance in accordance with the Contract Documents.

§ 5.41   Contingent Assignment of Subcontracts: ¶ 5.4

The owner may want to add the following after ¶ 5.4.1.2:

5.4.1.3 The Contractor shall promptly submit to the Owner a copy of each sub-contract upon execution of same. Each subcontract shall contain a contingent assignment of the subcontract to the Owner, consistent with this section.

§ 5.42   Owner Payments to Subcontractors: ¶ 5.5

The owner may want to insert the following provisions following ¶ 5.4.3:

5.5 OWNER PAYMENTS TO SUBCONTRACTORS

5.5.1 In the event of any default hereunder by the Contractor, or in the event the Owner or Architect fails to approve any Application for Payment that is not the fault of a Subcontractor, the Owner may make direct payment to the Subcontractor, less appropriate retainage. In that event, the amount so paid the Subcontractor shall be deducted from the payment to the Contractor.

5.5.2 Nothing contained herein shall create any obligation on the part of the Owner to make any payments to any Subcontractor, and no payment by the Owner to any Subcontractor shall create any obligation to make any further payments to any Subcontractor.

§ 5.43   Changes: ¶¶ 7.1 through 7.4

The following provision is an attempt to limit verbal changes:

7.1.1 Changes in the Work or omissions of Work previously ordered may be accomplished after execution of the Contract for Construction, and without invalidating the Contract Documents, only by Change Order, Field Order, or Bulletin, subject to the limitations stated in this Article 7 and elsewhere in the Contract Documents. No Change Order, Field Order, or Bulletin shall be effective unless issued and executed as herein specified. The Contractor specifically agrees that if it proceeds on an oral order to change the Work, it shall waive any claim for additional compensation for such work and the Contractor shall not be excused from compliance with the Contract Documents. The requirements set forth in this Article 7 are the essence of the Contract Documents. Accordingly, no course of conduct or dealings between the parties, no oral, express, or implied acceptance of alterations or additions to the Work, and no claim that the Owner has been unjustly enriched by any alteration or addition to the Work shall be the basis for any claim to an increase in the Contract Sum or Contract Time.

The following additional language requires the contractor to submit more information regarding changes:

7.1.4 For any changes in the Work involving a request by the Contractor of more than a three (3) calendar day extension of time, the Contractor shall submit critical path schedules showing the original schedule and the impact of the proposed change justifying the requested extension of time. The Owner may at its option refuse the extension of time and have the Contractor perform the work within the original schedule, provided all reasonable costs for completing the work, including overtime and acceleration costs, are included in the Change Order.

The following also relates to an extension of time:

7.1.5 If the Owner requests from the Contractor a proposal for additional work which involves additional cost and time, at the Owner's option the Owner may extend the completion date for the additional work only without extending the Project completion date, provided costs are included in the Change Order to compensate the Contractor for reasonable remobilization costs.

The following is a detailed set of change provisions that could be used on some projects:

7.2 BULLETINS AND FIELD ORDERS

7.2.1 A Bulletin is a written order directed to Contractor and issued by the Architect directing a minor change, or making a clarification in the Work, or requesting information from Contractor about the Work. A Bulletin is to be used for minor changes or clarifications which the Architect believes will have no impact on the Contract Sum or Contract Time.

7.2.2 A Field Order is a written order directed to Contractor and signed by Owner directing a change, or making a clarification in the Work, or requesting information from Contractor about the Work. A Field Order signed by Contractor indicates receipt of the Field Order.

7.2.3 A Field Order or Bulletin shall not be recognized as having any impact upon the Contract Sum or the Contract Time and Contractor shall have no Claim therefor unless it shall, in no event later than five (5) days from the date such direction or order was given, submit to Owner for Owner's approval its estimates of any adjustment in the Contract Sum or Contract Time to which Contractor believes it is entitled hereunder as a result of the change in the Work described in the Field Order or Bulletin, including sufficient detail to allow Owner to evaluate the price. Information furnished by Contractor must include quantities, unit prices, labor rates and hours, productivity factors, markups, and cumulative effect or such other information as may be reasonably requested by Owner.

7.2.4 Upon receipt of a Bulletin or Field Order, Contractor shall promptly proceed with the Work involved, or as otherwise directed by the Field Order or Bulletin.

7.3 PRICING CHANGES

7.3.1 Changes in the Work or omissions of Work previously ordered may be accomplished after execution of the Contract for Construction, and without invalidating the Contract Documents, only by Change Order, Bulletin, or Field Order, subject to the limitations stated in this Article 7 and elsewhere in the Contract Documents.

7.3.2 When submitting its change proposal, Contractor shall include and set forth, in clear and precise detail, breakdowns of labor and materials for all trades involved and the estimated impact on the Progress Schedule. If requested by Owner, Contractor shall furnish spreadsheets from which the breakdowns were prepared, plus spreadsheets, if requested, of any Subcontractors.

7.3.3 When directions for a change are given, Contractor may be directed by Field Order to initiate Work immediately, prior to pricing, or to submit a price prior to initiating Work.

7.3.4 Except for Guaranteed Maximum Price Contracts, under which Change Order pricing shall be based on the GMP method described below, Owner shall establish one of the following methods for final Change Order pricing:

STIPULATED: Fixed price based on agreed estimate.

GUARANTEED MAXIMUM PRICE (GMP): Guaranteed maximum price for the Change Order based on agreed estimate. Final price to be established after actual costs are accounted for. Final price would then be the lower of the Guaranteed Maximum Price or the total reimbursable costs.

TIME AND MATERIAL (T&M): Final price to be based on reimbursable cost of Work.

7.3.5 In the event that (i) Owner approves a Change Order to be priced on a time and material basis pursuant to Paragraph 7.3.4 hereof, (ii) Owner issues a Field Order to proceed with a change in the Work, Contractor timely notifies Owner of Contractor's belief that it is entitled to an adjustment in the Contract Sum or the Contract Time as a result of such change, and Owner and Contractor cannot agree on the amount of such adjustment, (iii) Contractor is entitled to compensation as the result of a concealed condition pursuant to Paragraph 4.3.9 hereof, or (iv) the Contract Sum includes Allowances pursuant to Paragraph 3.7, subject to the limitations of those sections, the amount of Contractor's allowable costs shall be governed by this Paragraph 7.3.5. Allowable costs include and are strictly limited to the following:

7.3.5.1 The actual and reasonable cost of additional materials and equipment, required as a result of such change, purchased by the Contractor (or any Subcontractor) and used in the Work, including sales taxes, freight, and delivery changes. Owner reserves the right to approve materials and sources of supply of materials furnished by the Contractor, or if necessary to facilitate the progress of the work in order to furnish the materials to the Contractor.

7.3.5.2 Rental charges at not more than the prevailing rental rate for the additional or reduced time that machinery or construction equipment (excluding hand tools and small tools customarily owned by construction workers or contractors) is required as a result of such change. The prevailing rate may be determined by reference to the most recent edition of the "Compilation of Nationally Averaged Rental Rates" of the Equipment Bluebook or Associates Equipment Distributors, or similar publication.

7.3.5.3 Material, construction equipment rental, and other costs shall be substantiated by vendors' invoices. In no event shall the cost of such items exceed the average current wholesale prices at which the items are available in the quantities required, delivered to the site, less applicable cash or trade discounts.

7.3.5.4 Labor costs, including welfare and fringe benefits, shall be the actual labor cost required as a result of such change in the Work, including no more than one (1) working foreman, but not including any supervisory or administrative personnel. Labor costs shall be substantiated by time cards (signed by the Superintendent daily) and labor summaries.

7.3.5.5 The cost of Work performed by Subcontractors or Sub-subcontractors based on the actual direct costs of material, labor, and construction equipment determined pursuant to the provisions of this Paragraph.

7.3.5.6 For deleted Work otherwise required to be performed hereunder, the Contract Sum shall be reduced by an amount equal to the net savings to the Contractor and all Subcontractors and Sub-subcontractors on account of the deleted work for material, labor, and construction equipment, plus the percentage fee thereof calculated in accordance with Paragraph 7.3.11 hereof.

7.3.5.7 If the Contract for Construction is written on the basis of a Guaranteed Maximum Price, no cost shall be considered an Allowable Cost pursuant to this Paragraph 2.3.5 which would not be considered a Cost of the Work pursuant to the Contract for Construction.

7.3.6 If Contractor does not submit preliminary and final estimates within the time periods set forth in Paragraph 7.4.2, and/or costs within ten (10) days of completing the Work, Owner may direct the Architect or a qualified consultant to develop an estimate which at Owner's sole discretion shall be used for Change Order pricing. If Contractor does not agree with the Change Order Pricing, it must assume the burden of establishing new Change Order pricing at no cost to Owner. Specifically, Contractor may meet with the Architect, Owner's consultants, or another mutually agreeable consultant, but Contractor must pay all fees associated with this review at no cost to Owner.

7.3.7 When submitting estimates and costs, Contractor must summarize and certify its submittal on Owner's Change Order Recap form.

7.3.8 Contractor shall separately price and account for each Bulletin and Field Order. Contractor shall not summarize more than one Bulletin or Field Order per change estimate.

7.3.9 Each Change Order shall be executed without regard to other Change Orders, and cumulative totals should not be used. Approved Change Orders should be listed on the Application for Payment each month.

7.3.10 Failure of Contractor and Owner to agree on an adjustment of the Contract Sum or extension of Time for performance under this Contract shall not excuse Contractor from proceeding with the prosecution and performance of the Work as changed. Owner shall have the right within its sole discretion to require Contractor to commence performance of changes to the Work. If Contractor and Owner cannot agree on an appropriate stipulated sum or guaranteed maximum price for any change, the adjustment to the Contract Sum, if any, shall, subject to the provisions of this Article 7, be determined by Owner on the basis of the allowable costs incurred as a result of such change as set forth in Section 7.3.5 plus a fee as set forth in Section 7.3.11.

7.3.11 In addition to the allowable costs incurred for a change, the General Contractor (and any Subcontractors) is entitled to a fee. Such fee shall be included in any agreed-upon stipulated sum or guaranteed maximum price for the Change Order and shall be added to the allowable costs for Change Orders priced on a time-and-material basis or as to which no agreement is reached. The additional fee shall be Contractor's (and Subcontractor's) sole reimbursement for overhead, profit, and any other cost (including insurance required in this Contract) not specifically reimbursable under the Contract. For Work to be omitted, no such fee shall be credited, and if the Work consists of both extra and omitted items within one Bulletin, any fee hereunder shall be computed only on the excess (if any) of the cost of the extra Work over the cost of the omitted Work. No such fee shall be applied to gross receipts or use tax amounts or bond premiums paid by Contractor. The following fees shall apply to Contractor and all Subcontractors and represent the maximum fees chargeable on each Contract tier, except for Contractors whose Contract for Construction is based on a Guaranteed Maximum Price, who shall receive mark-ups for all categories on a percentage basis equal to Contractor's Fee stated in the Contract for Construction.

7.3.11.1 For Work to be performed by the Subcontractors, such Subcontractors, cumulative of all tiers, shall be permitted an additional fee of fifteen percent (15%) of the allowable costs, as set forth in Subparagraph 7.3.5, incurred by such Subcontractors;

7.3.11.2 For Work to be performed by the Contractor's own forces, the Contractor shall be permitted an additional fee in the amount of fifteen percent (15%) of the allowable costs, as set forth in Subparagraph 7.3.5., incurred directly by the Contractor;

7.3.11.3 For Work to be performed by the Subcontractors, the Contractor shall be permitted an additional fee in the amount of five percent (5%) of the allowable costs, as set forth in Subparagraph 7.3.5, incurred by Subcontractor's, but not including the allowance for Subcontractors' fee provided in Subparagraph 7.3.11.A above.

7.3.12 Changes in the Work shall be performed under applicable provisions of the Contract Documents, and Contractor shall proceed promptly, unless otherwise provided in the Bulletin, Field Order, or Change Order.

7.3.13 If unit prices are stated in the Contract Documents or subsequently agreed upon, and if quantities originally contemplated are so changed in a proposed Change Order Bulletin or Field Order that application of such unit prices to quantities of Work proposed will cause substantial inequity to Owner or Contractor, the applicable unit prices shall be equitably adjusted.

7.4 PROCESSING CHANGE ORDERS

7.4.1 If a change in the Work is to be ordered, a Field Order or Bulletin shall be issued by Owner to Contractor describing the change. When time does not permit the processing of a Change Order in advance of commencing the change in the Work, upon receipt of a written authorization from Owner, Contractor shall proceed with a change in the Work, and Contractor shall concurrently proceed with submission of a change estimate as provided in Paragraph 7.2.3.

7.4.2 Within five (5) days of its receipt of a Field Order or Bulletin, Contractor shall provide a preliminary estimate of any change in Contract Sum or Contract Time associated with the change described in the Field Order or Bulletin. Within ten (10) days following receipt of a Field Order or Bulletin, Contractor shall submit a change estimate to Owner, setting forth any requested adjustment in the Contract Sum or the Contract Time and including an itemization of all costs of material and labor with extensions listing quantities and total costs and a substantiation of any Claim for an extension of the Contract Time. If no change estimate is submitted by Contractor within such period, it shall be conclusively presumed that the change described in the Field Order or Bulletin does not call for any Work that will result in an increase in the Contract Sum or the Contract Time, and such change shall be performed by Contractor without any such increase. If Contractor is unable to submit the above information within the time limit, it shall notify Owner in writing, setting forth for Owner's written approval a date by which Contractor will submit the information as well as a schedule for the performance of the Work for which a change estimate will be forthcoming.

7.4.3 If Owner accepts a change estimate submitted by Contractor, Contractor shall prepare a Change Order that is based on such change estimate for review by Architect and execution by Contractor and Owner, and the Contract Sum and the Contract Time shall be adjusted as provided in the approved Change Order.

7.4.4 Nothing contained herein shall limit the right of Owner to order changes in Work by Change Orders that have not been signed by Contractor, and Contractor shall promptly perform all Work required under the Contract Documents or a Change Order despite its refusal to accept or execute the Change Order.

7.4.5 No Change in the Work shall be the basis of an addition to the Contract Sum or a change in the Contract Time unless and until such change has been authorized by a Change Order executed and issued in accordance with the Contract Documents. Changes in the Work may be made without notice to Contractor's sureties, and absence of such notice shall not relieve such sureties of any of their obligations to Owner.

In a Connecticut case,[6] the contract modified the standard AIA language to make the contractor responsible for all change orders unless the owner agreed

---

[6] Naek Constr. Co., Inc., v. PAG Charles Street Ltd. P'ship, 2004 WL 2757623 (Conn. Super. Nov. 3, 2004).

to pay. Although the contractor performed more than $1 million of extra work, the court held that it was not entitled to be paid for such work. Here is the modified language:

> Contractor understands and acknowledges that it is solely responsible for any excess over the Contract Sum, except as may be specifically provided in writing by the Owner and CHFA, as well as, at the option of the Owner, the City of New Haven and/or the Project Investors. Contractor and Owner acknowledge that no additional funds shall be provided to the Contractor for added costs during the construction period. Accordingly, all Change Orders, whether necessitated by circumstances such as (including, but not limited to), unforseen conditions, changes in building code requirements or any government official's interpretation of such code requirements or other modifications to the specifications, plans or drawings, shall be "no additional cost" Change Orders. Such Change Orders shall be prepared and executed solely for the expressed purpose of providing documentation for changes made in the scope of work and for the preparation of "as-built" plans, drawings and surveys. All Change Orders must be approved in writing in advance of any change in the scope of work by the Owner and CHFA and, at the option of the Owner, by the City of New Haven and/or Project Investors. Any additional cost caused by any approved Change Order shall be the sole responsibility of the Contractor.

The court rejected the argument that there was a cardinal change.

§ 5.44 Agreed Overhead and Profit Rates: ¶ 7.5

The following additional language may be modified to suit a particular situation:

7.5 AGREED OVERHEAD AND PROFIT RATES

7.5.1 For any adjustments to the Contract Sum based on other than the unit price method, overhead and profit combined shall be calculated at the following percentages of the cost attributable to the change in the Work:

.1 For the Contractor for Work performed by the Contractor's own forces, ten percent (10%) of the cost;

.2 For the Contractor, for Work performed by the Contractor's Subcontractors, five percent (5%) of the amount due the Subcontractor;

.3 For each Subcontractor or Sub-subcontractor involved, for Work performed by that Subcontractor's or Sub-subcontractor's own forces, ten percent (10%) of the cost;

.4 For each Subcontractor, for Work performed by the Subcontractor's Sub-subcontractors, five percent (5%) of the amount due the Sub-subcontractor;

.5 Costs to which overhead and profit is to be applied shall be determined in accordance with Sub-subparagraphs 7.3.6.1 through 7.3.6.5;

.6 When both additions and credits are involved in any one change, the allowance for overhead and profit shall be figured on the basis of the net increase, if any;

.7 To facilitate checking of quotations for extras or credits, all proposals shall be accompanied by a complete itemization of costs including labor, material, and subcontracts. When major cost items are Subcontracts, they shall be itemized also.

§ 5.45   Definitions: ¶ 8.1

The parties can modify ¶ 8.1.3 to provide for the owner's occupancy before final completion:

8.1.3 The date of Substantial Completion is the date certified by the Architect in accordance with Paragraph 9.8, or the date when the Owner occupies or uses the Work or a designated portion thereof, whichever comes first.

Additional language that may be inserted after ¶ 8.1.3 is:

8.1.3.1 In the absence of a certificate establishing the date of Substantial Completion, the date of the Architect's final Certificate for Payment or the final payment by the Owner to the Contractor, whichever occurs first, shall be considered the date of Substantial Completion.

§ 5.46   Progress and Completion: ¶ 8.2

This provision may be added to give the owner additional leverage to require the contractor to start the work on time:

8.2.4 Failure by the Contractor to commence actual physical work on the Project within seven (7) days from the date set forth in the Notice to Proceed shall entitle the Owner to consider the Contractor in default of this Contract. Thereafter, and without further notice, the Owner may revoke the Notice to Proceed, declare the Contractor in default, and terminate the Contract. Thereafter, the Owner may retain the services of another contractor to perform the Work and the Contractor shall be liable to the Owner for the cost necessary to perform the work in excess of the Cost of the Work.

§ 5.47   Delay and Extensions of Time: ¶ 8.3

The owner may want to add this provision to avoid the possibility that an ordinary act by the owner results in a time extension for the contractor:

> In Paragraph 8.3.1, add the word "wrongful" before the word "act" in the first sentence.

Liquidated damages provisions are sometimes inserted after ¶ 8.3.1. One such liquidated damages provision is the following:

> **8.3.1.1** Failure to Complete the Work on Time. It is mutually agreed by and between the parties hereto that time shall be an essential part of this contract and that if the Contractor fails to complete its contract within the time specified and agreed upon, the Owner will be damaged thereby; and because the amount of said damages, inclusive of expenses for inspection, superintendence, and necessary traveling expenses, is difficult if not impossible to definitely ascertain and prove, it is hereby agreed that the amount of such damages shall be the appropriate sum set forth below in the Schedule of Liquidated Damages as liquidated damages for every working day's delay in finishing the work in excess of the number of working days prescribed; and the Contractor hereby agrees that said sum shall be deducted from monies due the Contractor under the contract or, if no money is due the Contractor, the Contractor hereby agrees to pay to the Owner as liquidated damages, and not by way of penalty, such total sum as shall be due for such delay, computed aforesaid.[7]

Another:

> The time of the Completion of Construction of the Project is of the essence of the Contract. Should [Hutton] neglect, refuse or fail to complete the construction within the time herein agreed upon, after giving effect to extensions of time, if any, herein provided, then, in that event and in view of the difficulty of estimating with exactness damages caused by such delay, the [City] shall have the right to deduct from and retain out of such monies which may be then due, or which may become due and payable to [Hutton], the sum of FIVE HUNDRED DOLLARS ($500.00) per day for each and every day that such construction is delayed on its completion beyond the specified time, as liquidated damages and not as a penalty.[8]

Another liquidated damages provision:

> **.1   General.** Liquidated damages are applicable when specified in Section _____ of the Project Manual.

---

[7] Adapted from Associated Eng'rs & Contractors, Inc. v. State, 58 Haw. 187, 567 P.2d 397 (1977). A contractor was delayed in completion of the work because of weather. The contract required that certain work could not be performed unless minimum temperatures were present, and the contractor argued that it was impossible to perform the work within the required time period. The court held that he must pay the liquidated damages.

[8] *See* Hutton Contracting Co., v. City of Coffeyville, 487 F.3d 772 (10th Cir. 2007).

.2 Not a Penalty. Failure on the part of the Contractor to complete the work within the contract time including such extensions thereof as approved by Owner, will result in added expense, loss and damage to Owner. Liquidated damages are established because such added expense, loss and damage are not reasonably ascertainable and not as a penalty to the Contractor.

.3 Amount. When incorporated, the amount of liquidated damages is specified in Section _____ of the Project Manual and represents a fair and reasonable amount for compensation caused by delay.

.4 Computation. The Contractor shall pay to Owner as liquidated damages the stated sum for each calendar day completion is delayed beyond the contract time as adjusted for any extensions approved by Owner.

.5 Determining Completion. The Contractor will be deemed to have satisfied the requirements for completion upon substantial completion of all work required by the Contractor for purposes of computing liquidated damages.

.6 Non-waiver. The following acts shall not constitute a waiver of the Contractor's obligation to pay liquidated damages:

    A. acceptance of or payment for any portion of the work;

    B. substantial completion of a portion of the work or occupancy by Owner; or

    C. Owner's requiring or allowing the Contractor to complete the work.

.7 Additional Costs/Claims of Other Parties. Owner's right to recover liquidated damages is in addition to and not a substitute for any right of recovery for additional costs incurred to complete the work, should the Contractor fail to do so. Nor shall Owner's right to recover liquidated damages be a substitution for or bar to recovery of any additional compensation Owner may be obliged to pay the Architect or contractors for other work on the project caused by Contractor's delay or other failure to perform.

.8 Other Rights and Remedies. The rights and remedies of Owner herein provided are in addition to any other rights and remedies provided under the contract or by operation of law.

The contractor may wish to include more specific language related to some of these contingencies. The following is referred to as a *strike clause:*

8.3.1.2 The Contractor shall not be liable for any loss, damage, or delay caused by strikes, lockouts, fire, explosion, theft, floods, riot, civil commotion, war, malicious mischief, act of God, or by any cause beyond the Contractor's reasonable control, and in any event, the Contractor shall not be liable for consequential damages.[9]

---

[9] Adapted from Curtis Elevator Co. v. Hampshire House, Inc., 142 N.J. Super. 537, 362 A.2d 73 (1976).

To avoid the possibility that a court may interpret a "delay" claim as being different from other types of claims, the following clarifying language may be added at the end of ¶ 8.3.2:

> Claims for delay shall be made in accordance with applicable provisions of Article 15.

Another provision:

8.3.1 The time for completion of the Work specified in the Contract Documents is an essential part of this contract. If the Contractor finds it impossible to complete the Work on or before the time for completion specified in the Contract Documents, then, not less than ten (10) days prior to the expiration of the stipulated time for completion, Contractor may make written request to the Owner for an extension of time. Contractor shall set forth fully in its request the reasons which it believes justify the granting of this request. If the Owner finds that the Work was delayed because of conditions beyond the control of the Contractor or that the quantities of work done or to be done are in excess of the estimated quantities by an amount sufficient to warrant additional time, the Owner may grant an extension of time for completion to such date as appears reasonable and proper, which extension of time must be approved by the surety. The extended time for completion shall then be considered as in full force and effect the same as if it were the original time for completion.

8.3.2 Should the Contractor fail to complete the Work within the time agreed upon in the contract or within such extra time as may have been allowed by extensions, there may be deducted from any moneys due or that may become due the Contractor the sum set forth below for each and every calendar day, exclusive of Sundays and legal holidays, that the Work shall remain uncompleted. This sum shall be considered and treated not as a penalty but as liquidated damage due the Owner from the Contractor by reason of inconvenience to the Owner, added cost of engineering and supervision, maintenance of detours, and other items that have caused an expenditure of funds resulting from Contractor's failure to complete the Work within the time specified in the contract.

8.3.3 Permitting the Contractor to continue and finish the Work or any part of it after the time fixed for its completion or after the date to which the time for completion may have been extended shall in no way operate as a waiver on the part of the Owner or any of its rights under the Contract Documents.

Another provision:

8.3.1 Date of Commencement and Time of Completion. Contractor agrees that it will begin work immediately upon receipt of notice to proceed from the Owner, and that it will diligently proceed with said work such that the same shall be completed with 180 days from the date of such notice to proceed.

8.3.2 Best Efforts. The Contractor acknowledges that the services to be performed are essential to the effective operation of the Owner, and that, therefore, the Contractor will exercise its best efforts to complete the services called for under this Agreement in the minimum time possible, and within the time specified in such work orders as may be issued by the Owner to the Contractor. In the event that the Contractor for good cause shown cannot complete the services for a particular task or phase within the time agreed to, the Contractor shall make a written request to the Owner, in accordance with Paragraph 8.3.4 below.

8.3.3 Notice of Conditions Causing Delay.

8.3.3.1 Within five (5) working days after the commencement of any condition which is causing or may cause delay in completion, the Contractor must notify the Owner in writing of the effect, if any, of such condition upon the time progress schedule, and must state why and in what respects, if any, the condition is causing or may cause such delay.

8.3.3.2 Failure to strictly comply with this requirement may, in the discretion of the Owner, be deemed sufficient cause to deny any extension of time on account of delay in completion arising out of or resulting from any change, extra work, suspension, or other condition.

8.3.4 Extension of Time.

8.3.4.1 Any extension or extensions of time for the completion of the work may be granted by the Owner subject to the provisions of this section, but only upon written application therefor by the Contractor to the Owner.

8.3.4.2 An application for an extension of time must set forth in detail the source and nature of each alleged cause of delay in the completion of the work, the date upon which each such cause of delay began, ended, or will end, and the number of days' delay attributable to each of such causes. It must be submitted prior to completion of the work.

8.3.4.3 If such an application is made, the Contractor shall be entitled to an extension of time for delay and completion of the work caused solely: (a) by the acts or omissions of the Owner, its officers, agents, or employees; or (b) by unforeseeable supervening conditions entirely beyond the control of either party hereto (such as, but not limited to, acts of God or the public enemy, war or other national emergency making performance temporarily impossible or illegal, or strikes or labor disputes).

8.3.4.4 The Contractor shall, however, be entitled to an extension of time for such causes only for the number of calendar days of delay which the Owner may determine to result solely from such causes, and then only if the Contractor shall have strictly complied with all the requirements of this Section. The Owner shall make such determination within thirty (30) calendar days after receipt of the Contractor's application for an extension of time; provided, however, said application complies with the requirements of this Paragraph.

8.3.4.5 The Contractor shall not be entitled to receive a separate extension of time for each one of several causes of delay operating concurrently, but, if at all, only for the actual period of delay in completion of the work as determined by the Owner, regardless of the number of causes contributing to produce such delay. If one of several causes of delay operating concurrently results from any act, fault, or omission of the Contractor or of its Subcontractor, if any, and would of itself (regardless of the concurrent causes) have delayed the work, no extension of time will be allowed for the period of delay resulting from such act, fault, or omission.

8.3.4.6 The granting of an application for an extension of time for causes of delay other than those herein referred to shall be entirely within the discretion of the Owner.

8.3.5 Delay Claims. Contractor represents and warrants that the provisions herein contained for extension of time are fair and adequate and that Contractor has had an opportunity to make provision for any and all delays within the contemplation of the parties. Accordingly, it is understood and agreed that Contractor shall not have or assert any claim for damages or prosecute any suit, action, cause of action, arbitration claim, or other proceeding against the Owner for such damages arising from any delay or hindrance in the completion of the work called for in this Agreement caused by any act or omission on the part of the Owner, its agents, servants, and employees, or otherwise.

Another provision:

8.3.6 Delays and Extensions of Time

8.3.6.1 If (a) Work on the critical path as shown on the Progress Schedule is delayed at any time by (i) an act or neglect of Owner, any employee of Owner, or any separate contractor employed by Owner, (ii) changes ordered in the Work in accordance with the provisions of Article 7, (iii) unusually severe weather conditions which preclude the safe performance of the Work, (iv) war or national conflicts or priorities arising therefrom, (v) fires beyond the reasonable control of Contractor, (vi) floods beyond the reasonable control of Contractor, (vii) earthquakes, (viii) off-site or area-wide labor disputes which are beyond the reasonable control of Contractor, or (ix) civil disturbances, and for no other cause or causes; (b) Contractor would otherwise have been able to perform its obligations timely under this Agreement but for such delay; (c) Contractor has taken reasonable precautions to foresee, prevent, and mitigate the effects of delays resulting from such causes; and (d) the Contractor has given written notice as required by Paragraph 8.3.2, then subject to the provisions of this Section 8.3, the Contract Time shall be appropriately extended by Change Order by the number of working days of delay on the critical path of the Work actually and directly caused by such occurrence. Contractor shall provide a critical path analysis of such delay claim which clearly identifies the effect of such delay on any critical path activities. Such extension of Contract Time shall be net of any delays caused by or a result of the fault or negligence of Contractor or which are otherwise the responsibility of Contractor or its agents or Subcontractors, and shall also be net of any contingency or "float" time allowance included in the Progress Schedule. Owner may, at its option, authorize extra Work in order to accelerate the Progress Schedule and minimize or eliminate the impact of the delay. No extension shall be made or allowed nor shall such extra Work be authorized unless Contractor makes a written request therefor within five (5) calendar days after the first occurrence of the delay. Any claims by Contractor relating to time shall be made promptly in accordance with applicable provisions of Article 7, otherwise they shall be deemed waived. Whenever Contractor knows or reasonably suspects that any actual or potential labor dispute is delaying or threatens to delay the timely performance of the work, Contractor shall immediately give notice thereof, including all relevant information with respect thereto, to Architect and Owner. In the case of a continuing cause of delay, only one request shall be necessary, which request shall affirmatively state that the delay is a continuing one and the reasons therefor. All delay requests hereunder shall describe the nature of the delay and estimate its probable effect on the progress of the Work. The effect of any delay shall also be shown on the latest Progress Schedule. All extensions of Time for completion of the Work or requests for acceleration on account of a delay permitted hereunder must be evidenced by a written Change Order approved by Owner. Contractor shall, in the event of any occurrence likely to cause a delay, cooperate in good faith with Owner to minimize and mitigate the impact of any such occurrence and do all things reasonable under the circumstances to achieve scheduled completion dates. Contractor shall advise and consult with Owner in connection with any delay and its effect on the Progress Schedule and shall take such action on Owner's behalf with respect thereto as Owner may request in accordance with the terms and conditions of this Contract.

8.3.6.2 Except as herein provided, Contractor shall not be entitled to any increase in the Contract Sum nor any monetary payment, reimbursement, or compensation over and above the Contract Sum for any delay in the commencement, prosecution, hindrance, or obstruction in the performance of the Work; loss of productivity or other similar claims (collectively referred to in this Section 8.3.6 as "delays"), or any loss, cost, damage, or expense of any kind, including but not limited to consequential damages, lost opportunity costs, impact damages, or other similar remuneration, which may arise out of or be caused by any delay in the Work from any cause or any extensions of the Contract Time hereunder, whether or not such delays are foreseeable. Contractor expressly waives any right to claim such loss, cost, damage, or expense on account thereof. Owner's exercise of any of its rights under the Agreement, including without limitation its rights under Article 7, regardless of the extent or number of such Changes, or Owner's exercise of any of its remedies to suspend or stop the Work, or Owner's right to require the correction or re-execution of any defective Work, shall not under any circumstances be construed as intentional interference with or delay of Contractor's performance of the Work.

8.3.6.3 If adverse weather conditions are the basis for a Claim for additional Time, such Claim shall be documented by data to substantiate that weather conditions were abnormal for the period of time and could not have been reasonably anticipated, and that weather conditions had a direct and adverse effect on scheduled Critical Path activities. Requests for extensions of Contract Time because of adverse weather conditions shall include U.S. Weather Bureau Climatological Reports for the months involved plus a report indicating the average precipitation, temperature, etc., for the past ten (10) years from the nearest reporting station. The ten-year average will be the basis for determining the number of adverse weather days and their effect on construction which Contractor would normally expect to encounter. Extensions of Time may be requested for any month of construction for days actually lost because of adverse weather in excess of the normally expected lost time; provided, however, if Owner determines that the seasonal average of adverse weather days during construction is less than what would be normally expected, no Change Order shall be issued and the request for extension of Time shall be denied. Any extension of Time under this Paragraph 8.3.6.3 shall be net of any delays caused by or resulting from the causes other than weather and shall also be net of any contingency or float time allowance in the Progress Schedule.

8.3.7 Acceleration: In the event of an excusable delay which extends scheduled completion dates despite diligent efforts by Contractor, and in lieu of granting an extension of Time, Owner, at its discretion, may direct Contractor to accelerate its performance to meet the Progress Schedule, in which case Owner shall issue a Change Order to increase the Contract Sum to include the additional cost of the Work, if any, reasonably incurred by Contractor to meet the Progress Schedule. Upon request, Contractor shall provide Owner with the options available for acceleration, including the costs and impact on the Progress Schedule. In presenting costs, Contractor shall credit Owner for those costs which would not be incurred as a result of Owner's willingness to invest extra funds to compress the Progress Schedule. Owner shall only be responsible for the actual premium costs of acceleration specifically authorized in advance for a critical activity in order to offset an excused delay. Owner shall not be responsible for premium costs which do not accelerate Critical Path activities.

8.3.8 Owner's Remedy for Delay: Owner may seek recovery for actual damages suffered as a result of delays caused by Contractor or its agents or subcontractors for failing to meet the following: (1) scheduled Substantial Completion date for any portion of the Work, (2) scheduled occupancy date for any portion of the Work, (3) scheduled Substantial Completion date for the entire Work, and (4) scheduled occupancy date for the entire Work. The dates referenced herein shall be subject to adjustment as provided in the Contract Documents. Such damages shall be documented in a deductive Change Order.

In order to ensure that the contractor will obtain an extension of time in the event hazardous materials are discovered (**see ¶** 10.3 of A201), the following language could be added in the list of reasons for an extension: "discovery of any hazardous materials."

§ 5.48   Contract Sum: ¶ 9.1

An alternative clause:

9.1.1 The Contract Sum is stated in the Agreement and is the maximum amount payable by the Owner to the Contractor for performance of the Work under the Contract Documents. The Contract Sum may only be increased pursuant to a Change Order signed by the Owner. Completion of the Work is a condition precedent to the Owner's obligation to pay the full Contract Sum.

§ 5.49   Schedule of Values: ¶ 9.2

A possible additional clause favorable to the owner could be inserted after ¶ 9.2.1:

9.2.2 The Schedule of Values shall state the names of all Subcontractors, Sub-subcontractors, and material suppliers and the amounts to become due each. It shall state the value of work to be completed by the Contractor's own forces. At the direction of the Architect, it shall include quantities, if applicable. The Contractor's overhead and profit shall each be carried as separate items. The total for all items shall aggregate the Contract Sum.

Another provision:

9.2.1 Within ten (10) days after the Contract for Construction is awarded, the Contractor shall submit to the Owner and the Architect a complete itemized Schedule of Values that includes all the line items on the Invitation to Bid. The Schedule of Values shall contain separate line items to include a list of values showing all principal trades and allocating values as the Work will be per-formed. The Contractor's profit, fees, taxes, overhead, and General Conditions must be itemized separately and not be prorated across other categories. The Owner shall have the right to request cost breakdowns in other formats for tax and accounting purposes. The Schedule of Values shall reflect accurate cost breakdowns and be supported by evidence of correctness as the Architect may direct or as required by the Owner, including copies of contracts with sub-contractors and material suppliers. This Schedule of Values, when approved by the Architect and the Owner, shall be used to monitor the progress of the Work and as a basis for Certificates for Payment. Each item shall show its total Scheduled Value, value of previous applications, value of the application, percentage completed, value completed, and value yet to be completed. All blanks and columns must be filled in, including every percentage-complete figure. The costs used in compiling the Schedule shall not be construed as fixing a basis for the costs of changes in the Work. The Contractor is to add approved Change Orders to the Schedule of Values monthly. Approval of the Schedule of Values from time to time shall not constitute evidence of a change in the Contract Sum.

Another:

9.2.1 Within fifteen (15) days after receipt of the Notice to Proceed, the Contractor shall submit to the Owner and Architect a schedule of values allocated to various portions of the Work, prepared in such form and supported by such data to substantiate its accuracy as the Owner or Architect may require. This schedule, unless objected to by the Owner or Architect, shall be used as a basis for reviewing the Contractor's Applications for Payment. The values set forth in such schedule shall not be used in any manner as fixing a basis for additions to or deletions from the Contract Sum.

9.2.2 The progress and payment schedule of values shall show the following:

9.2.2.1 The proposed schedule for tasks identified in the Contractor's Progress and Payment Schedule in bar chart form.

9.2.2.2 Important milestones which may impact the progress schedule (such as the anticipated delivery of structural steel, completion of rough-in, completion of utility relocations, etc.).

9.2.2.3 Rate of progress proposed by the Contractor in terms of cumulative percent complete, shown as an "S" curve superimposed on the bar chart schedule.

9.2.2.4 Anticipated monthly payments by the Owner based on the rate of progress proposed by the Contractor.

9.2.2.5 The dates shown for Tasks on the Contractor's progress and payment schedule shall agree with the start and finish dates provided on the Contractor's Schedule of Values.

§ 5.50   **Applications for Payment: ¶ 9.3**

The following clauses may be inserted at the end of ¶ 9.3.1 to assure that the contractor will provide waivers of lien:

The Contractor shall furnish the Owner, with each Application for Payment, a notarized Contractor's affidavit in a form satisfactory to the Owner's title company, along with proper waivers of lien and other supporting documentation sufficient to satisfy said title company that payment is properly due the Contractor.

Another version of a retention clause:

9.3.1.3 The Architect will authorize ninety percent (90%) of the amount due the Contractor on account of progress payments so long as there are no outstanding liens or claims and so long as in the opinion of the Owner the previous work has been done properly and is on schedule for completion of the construction and the unpaid balance is sufficient to complete the unfinished work. No interest shall be paid on retention.

The Contractor shall submit each Application for Payment on the standard Sworn Contractor's Statement available from the [name] Company. The Application for Payment shall list each item shown on the approved Schedule of Values and shall be properly executed and sworn under oath, as required by the Owner's title company. Each Application for Payment shall be accompanied by notarized Partial or Final Waivers of Lien on forms acceptable by the Owner's title company, one waiver for each separate payment requested and shown for that Application.

A clause that allows for retention could be added after ¶ 9.3.1.2:

9.3.1.3 Until the Work is 50 percent (50%) complete, the Owner shall pay 90 percent (90%) of the amount due the Contractor on account of progress payments. At the time the Work is 50 percent (50%) complete and thereafter, the Architect will authorize remaining partial payments to be paid in full. Notwithstanding the foregoing, if the Architect determines that the Contractor is not reasonably performing the Work, either by failure to reasonably follow the schedule or to adequately perform the work (all to be determined in the reasonable judgment of the Architect), the Owner shall continue to make progress payments at the rate of 90 percent (90%) of the amount due the Contractor for each payment.

To assure payment for materials stored off-site, the contractor may want to change the second sentence of ¶ 9.3.2 to read, "Payment shall be made . . ."
The following is an example of a "no-lien" clause that could be added after ¶ 9.3.3. Note that no-lien provisions are not legal in all jurisdictions.

9.3.3.1 The Contractor and all Subcontractors and material suppliers hereby waive all mechanic's liens and material supplier's liens. The Contractor shall hold the Owner harmless from any costs and damages, including but not limited to attorneys' fees, incurred by the Owner as a result of the filing of any mechanic's liens or material supplier's liens, or any litigation or arbitration arising out of such liens. Notwithstanding this provision, the Contractor shall, along with each Application for Payment, furnish verified written statements as required by [state statute], and appropriate waivers of lien from all Subcontractors and material suppliers that furnish labor or materials to the project.[10]

Another example of a no-lien clause is:

9.3.3.1 The Contractor, for itself and for its Subcontractors, laborers, material suppliers, and all others directly or indirectly acting for, through, or under it or

---

[10] Adapted from Luczak Bros. v. Generes, 116 Ill. App. 3d 286, 451 N.E.2d 1267 (1983).

any of them, covenants and agrees that no mechanic's liens or claims will be filed or maintained against the Project, the Premises, or any part thereof, or any interest therein or any improvements thereon, or the Owner, or against any monies due or to become due from the Owner to the Contractor, for or on account of any work, labor, services, materials, equipment, or other items performed or furnished for or in connection with the Work, and the Contractor for itself and its Subcontractors, laborers, and material suppliers, and all others above mentioned does hereby expressly waive, release, and relinquish all rights to file or maintain such liens and claims, and agrees further that this waiver of the right to file or maintain mechanic's liens and claims shall be an independent covenant and shall apply as well to work, labor, and services performed and materials, equipment, and other items furnished under any Change Order or supplemental agreement for extra or additional work in connection with the Project as to the original Work covered by the Contract Documents. If any Subcontractor, laborer, or material supplier of the Contractor or any other person directly or indirectly acting for, through, or under it or any of them files or maintains a mechanic's lien or claim as aforesaid, the Contractor agrees to cause such liens and claims to be satisfied, removed, or discharged at its own expense by bond, payment, or otherwise within ten (10) days from the date of the filing thereof, and upon its failure so to do, the Owner shall have the right, in addition to all other rights and remedies provided under the Contract Documents or by law, to cause such liens or claims to be satisfied, removed, or discharged by whatever means the Owner chooses, at the entire cost and expense of the Contractor, such cost and expense to include reasonable attorneys' fees and disbursements. The Contractor agrees to indemnify, protect, and save harmless the Owner from and against any and all such liens and claims and actions brought or judgments rendered thereon, and from and against any and all loss, damages, liability, costs, and expenses, including reasonable attorneys' fees and disbursements, which the Owner may sustain or incur in connection therewith.

Another version:

9.3.1 The Contractor shall submit to the Owner an itemized Application for Payment for operations completed in accordance with the Schedule of Values. The Application for Payment shall be on a form reasonably acceptable to the Owner. Such application shall be supported by such data to substantiate the Contractor's right to payment as the Owner or the Architect may require, such as copies of requisitions from Subcontractors and material suppliers, and shall reflect retainage. Any allowance included in the Application for Payment shall be separately itemized with supporting data attached. The Contractor shall furnish with each Application for Payment:

9.3.1.1 Cost data to support the application, including, without limitation, copies of Subcontractors' pay requests that accurately reflect current percentage of completion on a line-item basis, verified invoices, and labor sheets;

9.3.1.2 A conditional waiver and release of lien for itself and for each Subcontractor or material supplier who furnished labor, equipment, materials, or services to the Project during the period covered by the Application for Payment, in the form set forth by applicable law, as required to assure an effective waiver of mechanics' liens and stop notices under applicable law;

9.3.1.3 An unconditional waiver and release of lien (excluding any retention) for itself and for each Subcontractor or material supplier who furnished labor, equipment, materials, or services to the Project prior to the period covered by the Application for Payment in the form set forth in applicable law, as required to assure an effective waiver of mechanics' liens and stop notices under applicable law; and

9.3.1.4 The Monthly Progress Report and Updated Project Schedule as set forth herein.

9.3.2.1 Subject to the approval of the Owner's lender for the Project, if any, payments shall be made on account of materials and equipment delivered and suitably stored at the site for subsequent incorporation in the Work within thirty (30) days after such approved Application for Payment or such longer period as may be approved in advance by the Owner and Owner's lender or specifically authorized by the Contract Documents. If approved in advance by the Owner and Owner's lender, if any, payment may similarly be made for materials and equipment suitably stored off the site at a location agreed upon in writing. Payment for materials and equipment stored on or off the site shall be conditioned upon compliance by the Contractor with procedures satisfactory to the Owner and Owner's lender, if any, to establish the Owner's title to such materials and equipment or otherwise protect the Owner's and its lender's interest, and shall include applicable insurance, storage, transportation to the site, as well as all other costs associated with getting the materials and equipment from the place of storage to its final destination at the site, for such materials and equipment stored off the site. The Contractor warrants and agrees that title to all Work will pass to the Owner either by incorporation in the construction or upon the receipt of payment therefor by the Contractor, whichever occurs first, free and clear of all liens, claims, security interests, or encumbrances whatsoever, that the vesting of such title shall not impose any obligations on the Owner or relieve the Contractor of any of its obligations under the Contract for Construction, that the Contractor shall remain responsible for damage to or loss of the Work, whether completed or under construction, until responsibility for the Work has been accepted by the Owner in the manner set forth in the Contract Documents, and that no Work covered by an Application for Payment will have been acquired by the Contractor or by any other person performing Work at the site or furnishing materials and equipment for the Project, subject to an agreement under which an interest therein or an encumbrance thereon is retained by the seller or any other party or otherwise imposed by the Contractor or such other party.

9.3.2.2 When an Application for Payment includes materials or equipment stored off the project site or stored on the Project site but not incorporated in the Work for which no previous payment has been requested, a complete description of such materials or equipment shall be attached to the application. Suitable storage which is off the Project site shall be a bonded warehouse or appropriate storage facility approved by the Owner and Owner's lender, with the stored materials or equipment properly tagged and identifiable for this Project and properly segregated from other materials. The materials or equipment shall not have anything attached suggesting that such material or equipment is subject to any ownership interest or security other than that of the Owner or the Contractor. The Owner's written approval shall be obtained before the use of an off-site storage is made. Such approval may be withheld in the Owner's sole discretion.

In states such as Illinois, which have a statutory payment procedure, the provisions of A201 should be modified. Here is one possibility, which also includes a provision requiring the contractor to list all pending claims and changes:

9.3.1 At least ten days before the date established for each progress payment, the Contractor shall submit to the Architect a sworn statement, in form and substance satisfactory to the Owner and the Title Company and in compliance with Section 5 of the Illinois Mechanics Lien Act (770 ILCS 60/5). This statement ("Section 5 Affidavit") shall be an affidavit, notarized, setting forth (a) the names and addresses of all parties hired by the Contractor to furnish labor and materials for the Project, (b) their respective trade, (c) the total amount of their contract, (d) the amount previously actually paid to such party, (e) the amount of the current payment requested, if any, and (f) the balance due or to become due, if any, for each such party. In addition, the Contractor shall furnish such other supporting data as may reasonably be required by the Owner or Architect, such as copies of requisitions from Subcontractors and material suppliers, and reflecting retainage if provided for in the Contract Documents. The Contractor's sworn statements shall be accompanied by such statements, waivers and affidavits sufficient to reasonably indicate to the Owner the proper payment of all lienable services, labor and materials. In the event that any part of the funds to pay the Contractor are from a lender, such statements, waivers and affidavits shall also be sufficient to induce the title insurance company, if any, to issue its mechanics lien coverage to such lender(s). Lien waivers shall be on forms customarily used by Chicago Title Insurance Company. All affidavits from any tier subcontractor shall comply with the requirements for a Section 5 Affidavit set forth above.

9.3.1.1 In the event the Owner receives a notice from a Subcontractor pursuant to Section 24 of the Illinois Mechanics Lien Act (770 ILCS 60/24), notwithstanding anything contained to the contrary in the Contract Documents, the Owner may elect to retain sufficient funds from the Contractor to pay such Subcontractor giving notice and to make those payments pursuant to Section 27 of the Illinois Mechanics Lien Act (770 ILCS 60/27).

9.3.1.2 As provided in Subparagraph 7.3.8, the Contractor's Section 5 Affidavit may include requests for payment on account of changes in the Work which have been properly authorized by Construction Change Directives, or by interim determinations of the Architect, but not yet included in Change Orders.

9.3.1.3 The Contractor's Section 5 Affidavit may not include requests for payment for portions of the Work for which the Contractor does not intend to pay to a Subcontractor or material supplier, unless such Work has been performed by others whom the Contractor intends to pay.

9.3.1.4 The Contractor shall also submit, contemporaneously with the submittal of each Section 5 Affidavit, a statement of unapproved and pending Claims and Change Orders ("Pending Claims Affidavit"). The Pending Claims Affidavit shall be sworn under oath and shall represent the Contractor's best estimate of the value of each and every Claim and pending Change Order not fully approved. Each such item shall be separately listed, with a date of initiation, description, and value. To the extent any such item relates to any Subcontractor and/or material supplier, identify each, along with an estimated value for each trade. If the Contractor fails to list any Claim within 21 days of the date the Contractor first became aware of such Claim, such Claim will be deemed waived. The Owner shall not be required to make payments on account of any item listed on the Pending Claims Affidavit unless and until any such pending Claim is resolved as set forth herein and unless and until any such pending Change Order is fully approved.

§ 5.51 Decisions to Withhold Certification: ¶ 9.5

The following language could be added in ¶ 9.5.1.2 to help the contractor avoid over-withholding due to questionable claims by subcontractors:

.2 third-party claims filed or reasonable evidence indicating probable filing of such claims unless security acceptable to the Owner is provided by the Contractor. However, such third-party claims must arise out of the work of this Contractor and must not overlap with the claims of any other entity, and the owner must have a good-faith basis for such withholding.

Some alternative language favorable to the owner to cover situations in which the contractor fails to pay the subcontractors could be added after ¶ 9.5.1.3:

9.5.1.3.1 If any claim or lien is made or filed with or against the Owner, the Project, or the Premises by any person claiming that the Contractor or any Subcontractor or other person under it has failed to make payment for any labor, services, materials, equipment, taxes, or other items or obligations furnished or incurred for or in connection with the Work, or if at any time there shall be evidence of such nonpayment or of any claim or lien for which, if established, the Owner might become liable and which is chargeable to the Contractor, or if the Contractor or any Subcontractor or other person under it causes damage to the Work or to any other work on the Project, or if the Contractor fails to perform or is otherwise in default under any of the terms or provisions of the Contract Documents, the Architect shall withhold certification, and the Owner shall have the right to retain from any payment then due or thereafter to become due an amount which the Architect shall deem sufficient to (1) satisfy, discharge, and/or defend against any such claim or lien or any action which may be brought or judgment which may be recovered thereon, (2) make good any such nonpayment, damage, failure, or default, and (3) compensate the Owner for and indemnify it against any and all losses, liability, damages, costs, and expenses, including reasonable attorneys' fees and disbursements, which may be sustained or incurred by the Owner in connection therewith. The Owner shall have the right to apply and charge against the Contractor so much of the amount retained as may be required for the foregoing purposes. If such amount is insufficient therefor, the Contractor shall be liable for the difference and pay the same to the Owner.

In ¶ 9.5.1.6, the owner may want to change the word "and" after "Contract Time" to "and/or" in order to withhold certification if the contract time is exceeded, regardless of whether enough money is available.

In ¶ 9.5.1.7, the owner may want to delete the word "repeated."

An additional ground for withholding certification could be added after ¶ 9.5.1.7:

.8 rejection of the Work or any part of the Work by any governmental authority having jurisdiction over the Project, or by the Owner's lender.

.9 failure to comply with applicable Laws.

The owner may want to add this language at the end of ¶ 9.5.2:

The Owner shall not be deemed to be in default of the Contract by reason of withholding payment while any of the above grounds remain uncured.

The owner may want to add these provisions after ¶ 9.5.3:

9.5.4 If the Contractor disputes any determination of the Architect with respect to any Application for Payment, the Contractor shall nevertheless continue to diligently prosecute the Work.

9.5.5 No interest shall accrue to the Contractor on account of payments withheld hereunder. The Architect's determination regarding issuance of, withholding of, or adjusting certification of payments shall be subject to arbitration [or, "final and binding and not subject to arbitration"], but shall not subject the Architect to any liability from the Contractor, Surety, Owner, or any other party whatsoever.

Additional provisions:

9.5.4 If Contractor disputes any determination by the Owner with regard to any payment, Contractor shall nevertheless expeditiously continue to prosecute the Work. The Owner shall not withhold approval without reasonable justification.

9.5.5 If there exists a bona fide dispute over an amount to be paid to Contractor, Owner shall not be deemed to be in breach of this Contract by reason of the withholding of the portion of the payment in dispute pursuant to any provision of the Contract Documents.

§ 5.52 Progress Payments: ¶ 9.6

To avoid any confusion over whether the architect can communicate directly with the subcontractors concerning payment, the following could be added to ¶ 9.6.3:

> Notwithstanding paragraph 4.2.4, the Architect and Subcontractor may communicate directly on the matters covered in this Subparagraph. The Architect and Owner are hereby also authorized to communicate directly with any Subcontractor or material supplier regarding any issue of payment, liens, or quality of work or materials.

If the owner wants to retain the option of making direct payments to the subcontractors, state lien laws should be reviewed before inserting this provision after ¶ 9.6.7:

> **9.6.8** Notwithstanding any other provision to the contrary, the Owner reserves the right to make payment directly to any Subcontractor of the Contractor (or jointly to the Contractor and Subcontractor) in such amounts as the Owner determines to protect the Owner's interest and the Owner's property from a lien or asserted lien or other claim, and the amount owed the Contractor shall be reduced by the amount of any such payment by the Owner. Exercise of this option shall not create any claims or rights by any Subcontractor or other party against the Owner or the Owner's funds. This right may also be exercised through the Owner's title company making any such payments.

Other provisions:

> **9.6.1** After the Contractor has issued an Application for Payment and the Owner has approved it, the Owner shall make payment within thirty (30) days of such approval. The Owner may refuse to make payment on any Application for Payment for any default of the Contract, including but not limited to those defaults set forth in Paragraph 9.5.1. The Owner shall not be deemed in default by reason of withholding payment while any such defaults remain uncured.

> **9.6.2** Amounts paid to the Contractor on account of Work performed by Subcontractors shall be held in trust by the Contractor for the benefit of such Subcontractors. The Contractor shall promptly pay each Subcontractor, upon receipt of payment from the Owner, out of the amount paid to the Contractor on account of such Subcontractor's portion of the Work, the amount to which said Subcontractor is entitled, reflecting percentages actually retained from payments to the Contractor on account of such Subcontractor's portion of the Work. The Contractor shall, by appropriate agreement with each Subcontractor, require each Subcontractor to make payments to Sub-subcontractors in a similar manner.

> **9.6.7** The Owner may, at its election and at any time, make payments jointly to the order of the Contractor and to any Subcontractor or supplier.

There is often an argument as to whether the language in ¶ 9.6.2 is a conditional payment clause. As written, it can easily be construed to be such a clause. If it is, the subcontractors will be paid only if the contractor is paid. This may cause the subcontractors to inflate their prices to the general contractor to account for this risk. The owner may want to make the following modification to give the subcontractors greater assurances that this provision is not a conditional payment clause:

9.6.2 The Contractor shall promptly pay each Subcontractor, upon receipt of payment from the Owner, out of the amount paid to the Contractor on account of such Subcontractor's portion of the Work, the amount to which said Subcontractor is entitled, reflecting percentages actually retained from payments to the Contractor on account of such Subcontractor's portion of the Work. This provision is not to be construed as a "conditional payment" provision. In the event that payment to the Contractor is delayed without fault of the Subcontractor, payment to the Subcontractor shall be made within a reasonable time for work properly performed by the Subcontractor. The Contractor shall, by appropriate agreement with each Subcontractor, require each Subcontractor to make payments to Sub-subcontractors in similar manner.

§ 5.53   Failure of Payment: ¶ 9.7

This is a provision favorable to the owner that has been used by a number of owners, but has a totally different context than the AIA language:

9.7.1 If the Owner is entitled to reimbursement or payment from the Contractor under or pursuant to the Contract Documents, such payment by Contractor shall be made promptly upon demand by the Owner. Notwithstanding anything contained in the Contract Documents to the contrary, if the Contractor fails to promptly make any payment due the Owner, or the Owner incurs any costs and expenses to cure any default of the Contractor or to correct defective Work, the Owner shall have an absolute right to offset such amount against the Contract Sum and may, in the Owner's sole discretion, elect either to: (1) deduct an amount equal to that to which the Owner is entitled from any payment then or thereafter due the Contractor from the Owner, or (2) issue a written notice to the Contractor reducing the Contract Sum by an amount equal to that to which the Owner is entitled.

§ 5.54   Substantial Completion: ¶ 9.8

The owner may want to provide for an occupancy permit by inserting the following language at the end of ¶ 9.8.1:

and when all required occupancy permits, if any, have been issued and copies of same have been delivered to the Owner.

In order to limit the number of inspections, the architect may want to add the following clause to ¶ 9.8.2:

The Architect will make only two (2) such inspections to determine substantial completion. If these inspections determine that the work is not substantially complete, either because of major items not completed or an excessive number of punchlist items, successive inspections requested by the Contractor shall be charged to the Contractor at a rate of $400.00 per person per half day.

Another way to accomplish this is to add the following to ¶ 9.8.3:

The Contractor shall ensure the project is substantially complete prior to requesting any inspection by the Architect so that no more than one (1) inspection is necessary to determine Substantial Completion. If the Contractor does not perform adequate inspections to develop a comprehensive list as required in Paragraph 9.8.2 and does not complete or correct such items upon discovery or notification, the Contractor shall be responsible and pay for the costs of the Architect's additional inspections to determine Substantial Completion.

Other alternate provisions:

9.8.1 Substantial Completion is the stage in the progress of the Work when the Work or designated portion thereof (which the Owner agrees to accept separately) is sufficiently complete in accordance with the Contract Documents so the Owner can legally and practicably occupy and utilize the Work for its intended use.

9.8.1.1 The Work will not be considered suitable for Substantial Completion review until all Project systems included in the Work are operational as designed and scheduled, all designated or required governmental inspections and certifications, including certificates of occupancy, have been made and posted, designated instruction of the Owner's personnel in the operation of systems has been completed, and all final finishes within the Contract Documents are in place. In general, the only remaining Work shall be minor in nature, so that the Owner or Owner's tenants could occupy the building on that date and the completion of the Work by the Contractor would not materially interfere with or hamper the normal business operations of the Owner or Owner's tenants (or those who claim by, through, or under Owner). As a further condition of Substantial Completion acceptance, the Contractor shall certify that all remaining Work will be completed within thirty (30) consecutive calendar days or as agreed upon following the Date of Substantial Completion.

9.8.2 When the Contractor considers that the Work or a portion thereof which the Owner agrees to accept separately is substantially complete, the Contractor shall prepare and submit to the Architect a comprehensive list of items to be completed or corrected. The Contractor shall proceed promptly to complete and correct items on the list. Failure to include an item on such list does not alter the responsibility of the Contractor to complete all Work in accordance with the

Contract Documents. Upon receipt of the Contractor's written list and written notification that the Work is ready for inspection, the Architect will make an inspection to determine whether the Work (or designated portion thereof which Owner agrees to accept separately) is substantially complete. The Architect will make such inspection within ten (10) working days and issue an Inspection Report ("Punchlist") within five (5) working days of the inspection. If the Architect's inspection discloses any incomplete item, whether or not it is included on the Contractor's list, the Contractor shall, before issuance of the Certificate of Substantial Completion, complete or correct such item upon notification by the Architect (unless such item is minor in nature). The Contractor shall then submit a written request for another inspection by the Architect to determine Substantial Completion. The Architect will make a "back-check" inspection to verify completion of Punchlist Work. If the Work is still not complete (except for items minor in nature) after the back-check inspection, the Architect shall revise the Punchlist to indicate the status of uncompleted Work. The updated Punchlist will then be issued within five (5) working days. When the Work (or designated portion thereof which the Owner agrees to accept separately) is substantially complete, the Architect will prepare a Certificate of Substantial Completion which, if approved by the Owner, shall establish the date of Substantial Completion, shall establish responsibilities of the Owner and the Contractor for security, maintenance, heat, utilities, damage to the Work and insurance, and shall fix the time within which Contractor shall finish all items on the list accompanying the Certificate. The Certificate of Substantial Completion shall be submitted to the Owner and the Contractor for their written acceptance of responsibilities assigned to them in such Certificate.

9.8.3 Upon Substantial Completion of the Work (or designated portion thereof which the Owner agrees to accept separately) and upon application by the Contractor, certification by the Architect, and approval by the Owner, the Owner shall make payment reflecting the adjustment in retainage, if any, for such Work or portion thereof as provided in the Contract Documents. The Contractor shall reimburse the Owner for the extra architectural service charges, travel costs, and other expenses associated with all inspections after the first back-check inspection.

9.8.4 The Owner shall not be required to accept, unless specifically agreed in writing, Substantial Completion when it occurs before the expiration of the Contract Time.

Another:

9.8.4 When the Work or designated portion thereof is substantially complete, the Architect will prepare a Certificate of Substantial Completion which shall establish the date of Substantial Completion and which shall establish responsibilities of the Owner and Contractor for security, maintenance, heat, utilities, damage to the Work and insurance. After issuance of the Certificate of Substantial Completion, the Contractor shall finish and complete all remaining items within thirty (30) calendar days of the date on the Certificate. The Architect shall identify and fix the time for completion of specific items which may be excluded from the thirty (30) calendar day time limit. Failure to complete any items within the specified time frames may be deemed by the Owner as default of the contract on the part of the Contractor.

§ 5.55 Partial Occupancy or Use: ¶ 9.9

The owner may want to add this no-damage-for-delay clause after ¶ 9.9.1:

9.9.1.1 It is expressly understood and agreed that the Contractor shall not be entitled to any damages or compensation from the Owner, or be reimbursed for any losses, on account of any delay or delays resulting from the Owner's partial occupancy of the Work. Upon the Contractor's request for a Change Order (for time only), the Architect shall determine the number of days the Contractor has been delayed and shall promptly issue a Change Order to the Contractor for the appropriate number of days' delay.

Alternative provisions:

9.9.1 The Owner and its lessees and separate contractors may occupy or use any completed or partially completed portion of the Work at any stage of construction regardless of whether the Contract Time has expired (hereinafter sometimes referred to as "Partial Occupancy"). Such Partial Occupancy may commence whether or not the applicable portion of Work is substantially complete.

9.9.2 In the event of Partial Occupancy, the Contractor shall promptly secure endorsement from its insurance carrier(s), consent from its surety(ies), if any, and consent from public authorities that have jurisdiction over the Work to permit Partial Occupancy.

9.9.3 In the event of Partial Occupancy before Substantial Completion as provided above, the Contractor shall cooperate with the Owner in making available for the Owner's use and benefit such building services as heating, ventilating, cooling, water, lighting, telephone, elevators, and security for the portion or portions to be occupied, and if the Work required to furnish such services is not entirely completed at the time the Owner desires to occupy the aforesaid portion or portions, the Contractor shall make every reasonable effort to complete such Work or make temporary provisions for such Work as soon as possible so that the aforementioned building services may be put into operation and use. In the event of Partial Occupancy prior to Substantial Completion, provided the Contractor has met all Milestone Dates set forth in the Contract for Construction, mutually acceptable arrangements shall be made between the Owner and the Contractor in respect to the operation and cost of necessary security, maintenance, and utilities, including heating, ventilating, cooling, water, lighting, telephone services, and elevators. The Owner shall assume proportionate and reasonable responsibility for the cost of the above services, reduced by any savings to the Contractor for such services realized by reason of Partial Occupancy. Further, mutually acceptable arrangements shall be made between the Owner and the Contractor in respect to insurance and damage to the Work. The Contractor's acceptance of arrangements proposed by the Owner in respect to such matters shall not be unreasonably withheld, delayed, or conditioned.

9.9.4 In each instance, when the Owner elects to exercise its right of Partial Occupancy as described herein, the Owner will give the Contractor and Architect advance written notice of its election to take the portion or portions involved, and immediately prior to Partial Occupancy, the Owner, Contractor, and Architect shall jointly inspect the area to be occupied or portion of the Work to be used to determine and record the conditions of the same.

9.9.5 It shall be understood, however, that Partial Occupancy shall not: (1) constitute final acceptance of any Work; (2) relieve the Contractor for responsibility for loss or damage because of or arising out of defects in or malfunctioning of any Work, material, or equipment, nor from any other unfulfilled obligations or responsibilities under the Contract Documents; or (3) commence any warranty period under the Contract Documents, provided that the Contractor shall not be liable for ordinary wear and tear resulting from such Partial Occupancy.

9.9.6 Subject to the terms and conditions provided herein, if Contractor claims that delay or additional cost is involved because of Partial Occupancy by Owner, Contractor shall make such claim as provided elsewhere in the Contract Documents.

§ 5.56   Final Completion and Final Payment: ¶ 9.10

In order to limit the number of final inspections, the architect may want to add the following clause to ¶ 9.10.1:

The Architect will make only two (2) such inspections to determine final completion. If these inspections determine that the Work is not finally complete, successive inspections requested by the Contractor shall be charged to the Contractor at a rate of $400.00 per person per half day.

An additional clause that could be inserted in ¶ 9.10.2 is:

Should there prove to be any such claim, obligation, or lien after final payment is made, the Contractor shall refund to the Owner all monies that the Owner shall pay in satisfying, discharging, or defending against any such claim, obligation, or lien or any action brought or judgment recovered thereon and all costs and expenses, including reasonable attorneys' fees and other costs of such defense, incurred in connection therewith.

The owner may want to substitute the following sentence in place of ¶ 9.10.4 to indicate that the owner does not waive any claims by making final payment to the contractor:

9.10.4 The making of final payment shall not constitute a waiver of any claims by the Owner.

The following provision makes it clear that the contractor is responsible for the architect's fees resulting from delay, in this case 60 days after substantial completion:

9.10.6 In addition to any other damages, failure of the Contractor to achieve final completion within sixty (60) days after the specified date of Substantial Completion, subject to authorized extensions, will result in the Contractor's being responsible for excess Architect's fees. Excess Architect's fees include the cost of all necessary Architect's services, as determined by the Owner and Architect, incurred after sixty (60) days beyond the date of Substantial Completion. Excess Architect's fees will be deducted from the amount due the Contractor.

Additional alternate provisions:

9.10.1 Upon receipt of written notice that the Work is ready for final inspection and acceptance and upon receipt of a final Application for Payment, but in no event later than 30 days after the date of Substantial Completion, the Architect will promptly make such inspection and issue a Final Inspection Report within five (5) working days of the inspection.

9.10.2 The Contractor shall diligently inspect and supervise the Work and correct Work that does not conform to the requirements of the Contract Documents with promptness and diligence. If the Work is still not complete at the time of the Final Inspection, the Architect will make another "back-check" inspection upon the Contractor's written notice or no later than ten (10) working days after the Contractor's receipt of the Final Inspection Report. The Contractor will reimburse the Owner for the extra architectural service charges, travel costs, and other expenses associated with all inspections after the first back-check inspection, including the inspection for the Final Inspection Report. If at any time the Architect's or Owner's inspections indicate that the Contractor has not promptly or diligently pursued corrective Work, the Owner may, at its option, take over and correct Work forty-eight (48) hours after the Contractor's receipt of written notice of such failure to comply with the Contract Terms. The Owner may offset any funds due the Contractor for any corrective Work performed. If remaining funds due the Contractor are inadequate to reimburse the Owner, the Contractor shall, promptly upon demand, pay such amounts to the Owner.

9.10.3 Upon receipt of written notice that the Work is ready for final inspection and acceptance and upon receipt of a final Application for Payment clearly identified by the Contractor as for final payment, the Architect will promptly make such inspection and, when the Architect finds the Work acceptable under the Contract Documents fully performed, the Architect will promptly issue a final Certificate for Payment stating that, to the best of the Architect's knowledge, information, and belief, and on the basis of the Architect's observations and inspections, the Work has been completed in accordance with terms and conditions of the Contract Documents and that the entire balance found to be due Contractor and noted in said final Certificate is due and payable. The Architect's final Certificate for Payment will constitute a further representation that conditions precedent listed in Paragraph 9.10.4 have been fulfilled.

9.10.4 Neither final payment nor any remaining retained percentage shall become due until the Contractor submits to the Architect and Owner: (1) an affidavit that payrolls, bills for materials and equipment, and other indebtedness connected with the Work for which the Owner or Owner's property might be responsible or encumbered (less amounts withheld by the Owner) have been paid or otherwise satisfied; (2) a certificate evidencing that insurance required by the Contract Documents to remain in full force after final payment is currently in effect and will not be materially reduced or canceled or allowed to expire until at least thirty (30) days' prior written notice has been given to Owner; (3) a written statement satisfactory to the Owner that the insurance will cover the period required by the Contract Documents; (4) consent of surety, if any, to final payment; (5) if required by the Owner, other data establishing payment or satisfaction of obligations, such as receipts, releases and waivers of liens, claims, security interests, or encumbrances arising out of the Contract for Construction, to the extent and in such form as may be designated by the Owner and Owner's lender; (6) certification by the Contractor that (i) all Work has been completed in accordance with the Contract Documents, (ii) the final Application for Payment includes all claims of the Contractor against the Owner arising in connection with the Project and constitutes a waiver and release of any and all claims not presented in that application, except for claims arising out of third party actions, cross-claims and counterclaims, and (iii) the Record Drawings maintained by the Contractor pursuant to the Contract Documents and delivered to the Owner or Architect are complete and accurate in all respects; and (7) evidence of compliance with all requirements of the Contract Documents, such as notices, certificates, affidavits, or other requirements to complete obligations under the Contract Documents, including but not limited to (i) instruction of the Owner's representatives in the operation of mechanical, electrical, plumbing, and other systems; (ii) delivery of keys to the Owner with keying schedule (master, submaster, and special keys); (iii) delivery to the Owner of Contractor's warranties as set forth in the Contract Documents and each written warranty and assignment thereof prepared in duplicate, certificates of inspections, and bonds for the Architect's review and delivery to the Owner; (iv) delivery to the Owner of printed or typewritten operating, servicing, maintenance, and cleaning instructions for all Work (parts lists and special tools for mechanical and electrical work) in approved form; (v) delivery to the Owner of the Record Drawings; (vi) delivery to the Owner of a Final Waiver and Release of Liens covering all Work for itself and for each Subcontractor, vendor, and material supplier who furnished labor, materials, and services to the Work, executed by an authorized officer and duly notarized; (vii) delivery to the Owner of final waivers of lien from each subcontractor and material supplier who furnished labor, materials, and services to the Work, executed by their respective officers and duly notarized; and (viii) delivery of sales and use tax certificate number of the Contractor. In addition to the foregoing, all other submissions required by other Articles and Paragraphs of the Specifications and other Contract Documents shall be submitted to the Owner before approval of final payment. If a Subcontractor refuses to furnish a release or waiver required by the Owner and Owner's lender (if any), the Contractor must furnish a bond satisfactory to the Owner and Owner's lender (if any) to indemnify the Owner and Owner's lender (if any) against such lien. If such lien remains unsatisfied after payments are made, the Contractor shall refund to the Owner all money that the Owner may be compelled to pay in discharging such lien, including all costs and attorneys fees.

Another provision:

9.10.5 Contractor shall complete Punch List work within ninety (90) days after Substantial Completion, except for Punch List work requiring long-lead time for ordering materials. Contractor shall remove, within thirty (30) days after the date of Substantial Completion, all construction equipment and material not needed for the completion of Punch List work ("Final Completion"). Regardless of the actual date of Final Completion, Contractor shall remove all trailers, scaffolding, tools and other equipment from the Site as directed by the Architect or Owner.

Another provision:

9.10.1.1 Final Completion shall be achieved no later than thirty (30) days after Substantial Completion unless modified by a Change Order. Failure of the Contractor to achieve Final Completion pursuant to this Paragraph shall be considered a material breach of the Contract.

9.10.1.2 The Contractor shall notify the Architect and Owner in writing of the date when the Work has reached or will reach Final Completion and will be ready for final inspection. This notification shall be given at least ten (10) days prior to said date. If, in the opinion of the Architect, the Work is not complete, the Contractor shall diligently pursue completion of any unfinished work and shall reimburse the Owner for the cost of subsequent inspections by the Architect, or such costs may be deducted from any sums due the Contractor.

9.10.1.3 Approval of any Work pursuant to any inspection hereunder shall not act to release the Contractor from its responsibilities under the Contract.

9.10.4 A reasonable sum may be withheld until the Contractor delivers to the Owner record drawings and other items required pursuant to Section 3.11.1, the instructions and maintenance manuals required to be furnished pursuant to Section 3.12.7, and a final statement of the cost of the Work broken down according to the budget and in a form which has been approved by the Owner's lender has been furnished.

§ 5.57   Safety Precautions and Programs: ¶ 10.1

The following provision must be tailored to each state. It replaces Paragraph 10.1.1 with the following in an attempt to limit safety-related actions brought by workers and others:

10.1.1 The Contractor expressly agrees that it is in charge of and in control of the Work and that it shall have sole exclusive responsibility to comply with the requirements of the Structural Work Act. Neither the Owner nor the Architect is in charge of the Work or in control of the execution of the Work. The obligation of the Contractor under this Section 10.1.1 shall be construed to include but not be limited to injury or damage because the Contractor, its agents, and employees failed to use or misused any scaffold, hoist, crane, stay, ladder, support, or other mechanical contrivance erected or constructed by any person, or any or all other kinds of equipment, whether or not owned or furnished by the Contractor under the requirements of the [state] Structural Work Act, [citation]. The Contractor expressly agrees that it is exclusively responsible for compliance with OSHA and local regulations for construction and that it is the "employer" within the meaning of those regulations. Any provision in the Contract Documents in conflict with this paragraph shall be null and void. It is the express intent of the parties that this provision be given broad and liberal construction to effectuate the intent of the parties that the Contractor, and not the Architect or Owner, is in charge of the Work.

§ 5.58   Safety of Persons and Property: ¶ 10.2

The following provision could be inserted after ¶ 10.2.2 to provide additional protection for the owner:

10.2.2.1 If the Contractor fails to give such notices or fails to comply with such laws, ordinances, rules, regulations, and lawful orders, it shall be liable for and shall indemnify and hold harmless the Owner and the Architect and their respective employees, officers, and agents, against any resulting fines, penalties, judgments, or damages, including reasonable attorneys' fees, imposed on or incurred by the parties indemnified hereunder.

The owner may want to insert this additional paragraph after ¶ 10.2.7:

10.2.8 The Contractor shall promptly report to the Architect and Owner in writing all accidents arising out of or in connection with the Work that cause death, personal injury, or property damage. The report shall give full details, including statements of witnesses, hospital reports, and other information in the possession of the Contractor. In addition, in the event of any serious injury or damage, the Contractor shall immediately notify the Owner and Architect by telephone of such accident.

Other provisions:

10.2.1 The Contractor shall comply with all applicable safety laws, rules, regulations, or standards and shall take all necessary precautions for the safety of and shall provide all necessary protection to prevent damage, injury, or loss to:

.1 All persons involved in or affected by the Project;

.2 The Work and materials and equipment to be incorporated therein, whether in storage on or off the site, under the care, custody, or control of the Contractor or Contractor's Subcontractors or Sub-subcontractors; and

.3 Other property at the site or adjacent thereto, such as trees, shrubs, lawns, walks, pavements, roadways, structures, and utilities not designated for removal, relocation, or replacement in the course of construction.

10.2.2 The Contractor shall give notices and comply with applicable laws, ordinances, rules, regulations, and lawful orders of public authorities bearing on safety of persons or property or their protection from damage, injury, or loss. The Contractor shall provide all facilities and shall follow all procedures required by the Occupational Safety and Health Act (OSHA), including but not limited to providing and posting all required posters and notices, and shall otherwise be responsible for all other mandatory safety laws.

10.2.3 The Contractor shall erect and maintain, as required by existing conditions and performance of the Contract, all necessary safeguards for safety and protection, including posting danger signs and other warnings against hazards, promulgating safety regulations, and notifying the owners and users of adjacent sites and utilities.

10.2.4 When the use or storage of explosives, combustibles, or other hazardous materials or equipment or unusual methods is necessary for execution of the Work, the Contractor shall exercise utmost care and carry on such activities under supervision of properly qualified personnel.

10.2.5 The Contractor shall promptly remedy at its sole cost and expense any damage and/or loss (other than damage or loss insured under the Owner's all-risk builder's risk insurance, subject to the Contractor's liability to pay any deductible amount) to property referred to in Paragraphs 10.2.1.2 and 10.2.1.3, caused in whole or in part by the Contractor, a Subcontractor, a Sub-subcontractor, or anyone directly or indirectly employed by any of them, or by anyone for whose acts they may be liable and for which the Contractor is responsible, except damage or loss arising solely from the wanton and willful negligence or the malicious acts or omissions of the Owner. The obligations of the Contractor under this indemnification shall not extend to the liability of the Architect, its agents, or employees arising out of (1) the preparation or approval of maps, Drawings, opinions, reports, surveys, Change Orders, designs, or Specifications, or (2) the giving of or the failure to give directions or instructions by the

Architect, its agents, or employees, provided such giving or failure to give directions or instructions by the Architect is the primary cause of the injury or damage. This indemnification is in addition to such other indemnifications that are part of the Contract Documents.

10.2.6 The Contractor shall designate a responsible member of the Contractor's organization at the site whose duty shall be the prevention of accidents. This person shall be the Contractor's Superintendent unless otherwise designated by the Contractor in writing to the Owner and Architect.

10.2.7 The Contractor shall not impose or permit loading upon any part of the Work, construction site, or upon or adjacent to the Work site, in excess of safe limits, or permit loading that will result in stress or damage to the structural, architectural, mechanical, electrical, or other components of the Work. Design of all temporary construction, equipment, and appliances used in construction of the work and not a permanent part thereof, including, but not limited to, hoisting equipment, cribbing, shoring, and temporary bracing of structural steel, is the sole responsibility of the Contractor. The Contractor shall immediately request the Architect's instructions if there is any question as to the need for bracing or shoring of a building component. In the event that such bracing or shoring relates to construction means, methods, or techniques for which the Contractor is solely responsible, then the Contractor shall retain such licensed architects or engineers as may be required to render competent advice as to the matters raised, and who will certify that such bracing or shoring will comply with applicable codes and requirements, and the Architect and Owner will be entitled to rely on such professional certification. The Architect's instructions, if any, shall be binding, and any required shoring or bracing shall be promptly installed, but such instructions, if any, shall not confer any obligation or liability on the Architect beyond that conferred by the Architect's contract with the Owner. All such items shall conform with the requirements of governing codes and all laws, ordinances, rules, regulations, and orders of all governmental authorities or other entities having jurisdiction. The Contractor shall take special precaution, such as shoring of masonry walls and temporary tie bracing of structural steel work, to prevent possible wind damage during construction of the Work. The installation of such bracing or shoring shall not damage or cause damage to Work in place or installed by others. The Contractor must promptly repair any damage that does occur, at no cost to the Owner, or, at the Owner's option, shall pay to the Owner the full costs of such repairs.

10.2.8 Upon commencement of installation of the Owner's furniture, equipment, or finish work, or delivery of the Owner's property to the Work site for storage and protection thereon, the Contractor shall supply continuous security guard services to the Work site when no construction activity or personnel are on the Work site, including evenings and nights, weekends and holidays, until the Owner's property is secured, completely enclosed and lockable, so as to prevent unauthorized entry thereon.

10.2.9 The Contractor shall, at all times, provide and maintain all necessary and proper safeguards in and around the Construction Work, in order to protect all persons working, entering, or visiting in or near the Project, and to protect from theft and vandalism all materials, equipment, tools, and personal property of every description located at the Property or stored for use on the Project site.

10.2.10 The Contractor shall not cause or permit any "Hazardous Materials" (as defined herein) to be brought upon, kept, or used in or about the Project site except to the extent such Hazardous Materials: (i) are necessary for the prosecution of the Work; (ii) are required pursuant to the Contract Documents; and (iii) have been approved in writing by Owner. Any Hazardous Materials allowed to be used on the Project site shall be used, stored, and disposed of in compliance with all applicable laws relating to such Hazardous Materials. Any unused or surplus Hazardous Materials, as well as any other Hazardous Materials that have been placed, released, or discharged on the Project site by the Contractor or any of its employees, agents, suppliers, or Subcontractors, shall be removed from the Project site at the earlier of: (i) the completion of the Work requiring the use of such Hazardous Materials; (ii) the completion of the Work as a whole; or (iii) within twenty-four (24) hours following the Owner's demand for such removal. Such removal shall be undertaken by the Contractor at its sole cost and expense and shall be performed in accordance with all applicable laws. Any damage to the Work, the Project site, or any adjacent property resulting from the improper use of or any discharge or release of Hazardous Materials shall be remedied by the Contractor at its sole cost and expense and in compliance with all applicable laws. The Contractor shall immediately notify the Owner of any release or discharge of any Hazardous Materials on the Project site. The Contractor shall provide the Owner with copies of all warning labels on products which the Contractor or any of its Subcontractors will be using in connection with the Work, and the Contractor shall be responsible for making any and all disclosures required under applicable "Community Right-to-Know" or similar laws. The Contractor shall not clean or service any tools, equipment, vehicles, materials, or other items in such a manner as to cause a violation of any laws or regulations relating to Hazardous Materials. All residue and waste materials resulting from any such cleaning or servicing shall be collected and removed from the Project site in accordance with all applicable laws and regulations. The Contractor shall immediately notify the Owner of any citations, orders, or warnings issued to or received by the Contractor, or of which the Contractor otherwise becomes aware, which relate to any Hazardous Materials on the Project site. Without limiting any other indemnification provisions pursuant to law or specified in this Agreement, the Contractor shall indemnify, defend (at the Contractor's sole cost, and with legal counsel approved by Owner), and hold the Owner and Architect harmless from and against any and all claims, demands, losses, damages, disbursements, liabilities, obligations, fines, penalties, costs, and expenses for removing or remedying the effect of any Hazardous Materials on, under, from, or about the Project site, arising out of or relating to, directly or indirectly, the Contractor's failure to comply with any of the requirements herein. As used herein, the term "Hazardous Materials" means any hazardous or toxic substances, materials, and wastes listed in the United States Department of Transportation Hazardous Materials Table, or listed by the Environmental Protection Agency as hazardous substances, and any substances, materials, or wastes that are or become regulated under federal, state, or local law.

Another provision:

10.2.9 A decision by the Owner to supplement or duplicate the security measures otherwise required to be performed or provided by the General Contractor in accordance with these Contract Documents shall not in any way relieve the General Contractor of its responsibility to safeguard and secure all Work and the Work Area or impose any liability or responsibility on the Owner for damages or loss caused by trespass or theft.

Another:

10.2.3.1 During the performance of the Work, the Contractor shall be responsible for providing and maintaining warning signs, lights, signal devices, barricades, guard rails, fences, and other devices appropriately located on site which shall give proper and understandable warning to all persons of danger of entry onto land, structure, or equipment.

10.2.3.2 The Contractor shall maintain at his own cost and expense, adequate, safe and sufficient walkways, platforms, scaffolds, ladders, hoists and all necessary, proper, and adequate equipment, apparatus, and appliances useful in carrying on the Work and which are necessary to make the place of Work safe and free from avoidable danger, and as may be required by safety provisions of applicable laws, ordinances, rules, regulations and building and construction codes.

10.2.3.3 The Contractor shall be responsible for all shoring required to protect the Work or adjacent property and shall pay for any damage caused by failure to shore or by improper shoring or by failure to give proper notice. Shoring shall be removed only after completion of permanent supports.

Additional language:

10.2.4.1 When use or storage of explosives or other hazardous materials or equipment or unusual methods are necessary, the Contractor shall give the Owner reasonable notice.

10.2.4.2 Contractor is solely responsible for the proper use or storage of explosives or other hazardous materials or equipment or unusual methods.

10.2.8 When required by the Contract Documents, or reasonably inferable therefrom, the Contractor shall shore up, brace, underpin and protect foundations and other portions of the existing structures which are in any way affected by the Work. The Contractor, before commencement of any part of the Work, shall give notice to the Owner, who will notify adjoining landowners or other parties.

§ 5.59   Hazardous Materials: ¶ 10.3

The contractor may want to expand the list of hazardous materials to be more inclusive by adding the following provision:

> **10.3.4** The definition of "hazardous materials" shall include, but not be limited to, all of the materials and substances listed as being hazardous from time to time by the United States Environmental Protection Agency and the corresponding agency in the state where the project is located. The Owner shall promptly furnish to the Contractor any environmental report in the possession of the Owner or its agents.

§ 5.60   Emergencies: ¶ 10.4

An alternate provision:

> **10.4.1** In an emergency affecting the safety of persons or property, Contractor shall act to prevent or minimize damage, injury, or loss. Contractor shall promptly notify the Owner and Architect, which notice may be oral followed by written confirmation, of the occurrence of such an emergency and Contractor's action.

§ 5.61   Contractor's Liability Insurance: ¶ 11.1

In ¶ 11.1.1, the owner may want to insert the words "acceptable to the Owner" after the word "companies" in the first sentence. Another addition is the insertion of "Owner, Architect, and such other parties as may be designated by the Owner" after the words "protect the Contractor" in the first sentence.

The following provision could be added after ¶ 11.1.3 to include the owner and architect as additional insureds on the contractor's policies, but consult an insurance advisor before modifying any portion of Article 11:

> **11.1.3.1** The policies and the certificates required herein shall name the Owner and Architect as additional insureds and shall be subject to the approval of the Owner and Architect. The Contractor shall furnish the Owner and Architect copies of any endorsements that are subsequently issued amending coverage or limits.

The following is a hold-harmless agreement that is currently in use and includes an obligation to purchase insurance:

> WHEREAS ABC Co. has erected or may cause to be erected scaffolds, platforms, ramps, ladders, passageways, and other equipment (the "Scaffolding") to be used in connection with its work at [name of job], located at _____(2), and

WHEREAS [name of contractor] (the "Contractor") has a contract and/or a purchase order to perform work at said location and desires to use the Scaffolding in performing such work,

NOW THEREFORE, in consideration of the undertakings of the Contractor hereinafter stated, and for other good and valuable consideration, ABC Co. hereby consents to the Contractor's use of the Scaffolding at such times and under such circumstances as are consistent with ABC Co. operations, on the following terms and conditions:

(1) The Contractor agrees to indemnify and hold harmless ABC Co. from any and all claims, losses, damages, or expenses of any kind, including all legal expenses and attorney's fees, arising out of the maintenance or use of the Scaffolding by the Contractor, its agents, servants, employees, or licensees, even though such claims may be without any foundation whatsoever, including but not limited to claims for personal injuries, death, or property damage, or claims arising out of any statute, rule, or regulation of any governmental authority having jurisdiction over the Project.

(2) The Contractor further agrees (i) to obtain insurance with a reliable insurance company authorized to do business in the state of _____() and acceptable to ABC Co., covering the liability assumed in paragraph (1) above, (ii) to obtain insurance from such an insurance company providing Contractor with workers' compensation and public liability insurance for the Contractor's work on the location, and (iii) to furnish ABC Co. with certificates of insurance evidencing the foregoing coverages.

(3) ABC Co. does not guarantee or make any representations concerning the safety, suitability, or legal sufficiency of any of the Scaffolding and does not agree to make available or put in place any particular equipment for the use of the Contractor, such use to be solely at the convenience of ABC Co. ALL WARRANTIES OF ANY SORT, INCLUDING FITNESS FOR PURPOSE, ARE SPECIFICALLY EXCLUDED.

Additional language:

11.1.3.1 Certificates of Insurance shall be in a form acceptable to the Owner and shall be filed with the Owner prior to commencement of the Work. In addition to Certificates of Insurance, the Contractor shall furnish a written endorsement to the Contractor's general liability insurance policy that names the Owner and Architect as additional insureds. The endorsement shall provide that the Contractor's liability insurance policy shall be primary, and that any liability insurance of the Owner or Architect shall be secondary and noncontributory.

11.1.4 The aggregate limits of Insurance required by the Contract Documents shall apply, in total, to this Contract only. This shall be indicated on the insurance certificate or an attached policy amendment.

11.1.5 The insurance policies and Certificates of Insurance required by the Contract shall contain a provision that no material alteration, cancellation, nonrenewal, or expiration of the coverage contained in such policy or evidenced by such Certificates of Insurance shall have effect unless the Owner and/or Architect, as the case may be, have been given at least thirty (30) days' prior written notice. The Contractor shall provide a minimum of thirty (30) days' written notice to the Owner and Architect of any proposed reduction of coverage limits, including every coverage limit identified in the Contract Documents, or any substitution of insurance carriers.

11.1.6 In no event shall any failure of the Owner or Architect to receive certified copies or Certificates of Insurance required pursuant to the Contract Documents or to demand receipt of such documents prior to the Contractor's commencing the Work be construed as a waiver of the Contractor's obligations to obtain insurance pursuant to this Article 11. The obligation to procure and maintain any insurance required by the Contract Documents is a separate responsibility of the Contractor and independent of the duty to furnish a certified copy or certificate of such insurance policies.

Additional provisions:

11.1.1.5 claims for damages, other than to the Work itself, because of injury to or destruction of tangible property, including explosion, collapse, underground hazards and damage to underground utilities and loss of use resulting therefrom;

11.1.1.9 claims under any scaffolding, structural work or safe place law, or any law with respect to protection of adjacent landowners.

Additional provisions:

11.1.3 All insurance coverages procured by Contractor shall be provided by agencies and insurance companies acceptable to and approved by Owner. Any insurance coverage shall be provided by insurance companies that are duly licensed to conduct business in the state where the Project is located as an admitted carrier. The form and content of all insurance coverage provided by Contractor are subject to the approval of Owner. All required insurance coverages shall be obtained and paid for by Contractor. Any approval of the form, content or insurance company by Owner shall not relieve the Contractor from the obligation to provide the coverages required herein.

11.1.4 All insurance coverage procured by the Contractor shall be provided by insurance companies having policyholder ratings no lower than "A-" and financial ratings not lower than "XI" in the Best's Insurance Guide, latest edition in effect as of the date of the Contract, and subsequently in effect at the time of renewal of any policies required by the Contract Documents. Insurance coverages required hereunder shall not be subject to a deductible amount on a per-claim basis of more than $10,000.00 and shall not be subject to a per-occurrence deductible of more than $25,000.00. Insurance procured by Contractor covering the additional insureds shall be primary insurance and any insurance maintained by Owner shall be excess insurance.

11.1.5 All insurance required hereunder shall provide that the insurer's cost of providing the insureds a defense and appeal, including attorneys' fees, shall be supplementary and shall not be included as part of the policy limits but shall remain the insurer's separate responsibility. Contractor shall cause its insurance carriers to waive all rights of subrogation against the Owner and its officers, directors, partners, employees and agents.

11.1.6 The Contractor shall furnish the Owner with certificates, policies or binders which indicate the Contractor and Owner are covered by the required insurance showing the type, amount, class of operations covered, effective dates and date of expiration of policies. Such certificates, policies or binders shall be submitted to Owner prior to the start of the Work. All certificates, policies and binders shall be executed by a duly authorized agent of each of the applicable insurance carriers and shall contain the statement that: "The insurance covered by this certificate will not be canceled or altered except after sixty (60) days' written notice has been received by Owner." All certificates, policies and binders shall be in a form acceptable to the Owner. Contractor shall provide certified copies of all insurance policies required above within ten (10) days of Owner's written request for said copies.

11.1.7 With respect to all insurance coverages required to remain in force and affect after final payment, Contractor shall provide Owner additional certificates, policies and binders evidencing continuation of such insurance coverages along with Contractor's application for final payment and shall provide certificates, policies and binders thereafter as requested by Owner.

11.1.8 The maintenance in full current force and effect of such forms and amounts of insurance and bonds required by the Contract Documents shall be a condition precedent to Contractor's exercise or enforcement of any rights under the Contract Documents.

11.1.9 Failure of Owner to demand certificates, policies and binders evidencing insurance coverages required by the Contract Documents, approval by Owner of such certificates, policies and binders or failure of Owner to identify a deficiency from evidence that is provided by Contractor shall not be construed as a waiver of Contractor's obligations to maintain the insurance required by the Contract Documents.

11.1.10 The Owner shall have the right to terminate the Contract for cause if Contractor fails to maintain the insurance required by the Contract Documents.

11.1.11 If Contractor fails to maintain the insurance required by the Contract Document, Owner shall have the right, but not the obligation, to purchase said insurance at Contractor's expense. If Owner is damaged by Contractor's failure to maintain the insurance required by the Contract Documents, Contractor shall bear all reasonable costs properly attributable to such failure.

11.1.12 By requiring the insurance set forth herein and in the Contract Documents, Owner does not represent or warrant that coverage and limits will necessarily be adequate to protect Contractor, and such coverages and limits shall not be deemed as a limitation on Contractor's liability under the indemnities granted to Owner in the Contract Documents.

11.1.13 If Contractor's liability policies do not contain a standard separation of insureds provision, such policies shall be endorsed to provide cross-liability coverage.

11.1.14 If any part of the Work hereunder is to be subcontracted, the Contractor shall: (1) cover any and all Subcontractors in its insurance policies; (2) require each Subcontractor to secure insurance which will protect said Subcontractor and supplier against all applicable hazards or risks of loss designated in accordance with the provisions hereunder; and (3) require each Subcontractor or supplier to assist in every manner possible in the reporting and investigation of any accident, and upon request, to cooperate with any insurance carrier in the handling of any claim by securing and giving evidence and obtaining the attendance of witnesses as required by any claim or suit.

§ 5.62   Owner's Liability Insurance: ¶ 11.2

Another version:

11.2.1 The Owner shall be responsible for and may at its option, maintain such insurance or self insurance as will protect it from its contingent liability to others for damages because of bodily injury, including death, which may arise from operations under this Contract, and any other liability for damages which the Contractor is required to insure under any provision of this Contract. This insurance is not for the benefit of the Contractor.

A version that shifts responsibility:

11.2.1 The Contractor shall purchase and maintain such insurance as will protect the Owner from his contingent liability to others for damages because of bodily injury, including death, and property damage, which may arise from operations under this Contract and other liability for damages which the Contractor is required to insure under any provision of this Contract. Certificate of this insurance will be filed with the Owner.

§ 5.63 Property Insurance: ¶ 11.3

The current language of ¶ 11.3.1 may create an ambiguity if the "Project" is significantly larger than the "Work." Under this provision, the owner is required to purchase builder's risk insurance in the value "for the entire Project at the site." If the "Work" is a small part of the "Project" and the remainder of the Project is completed and the contractor damages some of this completed work that had been performed by others, there is a chance that a builder's risk policy may not pay for this damage, although a different policy may cover such damage. As with all other provisions involving insurance, a competent insurance advisor should be consulted. Here is a modified version of the AIA language:

11.3.1 Unless otherwise provided, the Owner shall purchase and maintain, in a company or companies lawfully authorized to do business in the jurisdiction in which the Project is located, property insurance written on a builder's risk "all-risk" or equivalent policy form in the amount of the cost of the Work, as modified from time to time, on a replacement cost basis without optional deductibles. Such property insurance shall be maintained, unless otherwise provided in the Contract Documents or otherwise agreed in writing by all persons and entities who are beneficiaries of such insurance, until final payment has been made as provided in Section 9.10 or until no person or entity other than the Owner has an insurable interest in the property required by this Section 11.3 to be covered, whichever is later. This insurance shall include interests of the Owner, the Contractor, Subcontractors and Sub-subcontractors in the Project.

Here are some alternative provisions that shift the burden for property insurance to the contractor. These provisions should be carefully tailored to the project's specific requirements, with input from the owner's insurance advisers.

11.3 PROPERTY INSURANCE

11.3.1 The Contractor shall purchase and maintain, in a company or companies lawfully authorized to do business in the jurisdiction in which the Project is located, property insurance written on a builder's risk "all-risk" or equivalent policy form in the amount of the initial Contract Sum, plus the value of subsequent Contract modifications and cost of materials supplied or installed by others, comprising the total value for the entire Project at the site on a replacement cost basis without optional deductibles. Such property insurance shall be maintained, unless otherwise provided in the Contract Documents or otherwise agreed in writing by all persons and entities who are beneficiaries of such insurance, until final payment has been made as provided in Paragraph 9.10 or until no person or entity other than the Owner has an insurable interest in the property required by this Paragraph 11.3 to be covered, whichever is later. This insurance shall include interests of the Owner, the Contractor, Subcontractors, and Sub-subcontractors in the Project. In addition, the Owner may, but is not required to, obtain additional insurance coverage that may, in part, duplicate the coverage required of the Contractor. In such case, the Contractor's insurance shall be considered the primary policy.

11.3.1.1 Property insurance shall be on an "all-risk" or equivalent policy form and shall include, without limitation, insurance against the perils of fire (with extended coverage) and physical loss or damage including, without duplication of coverage, theft, vandalism, malicious mischief, collapse, earthquake, flood, windstorm, falsework, testing and startup, temporary buildings and debris removal, including demolition occasioned by enforcement of any applicable legal requirements, and shall cover reasonable compensation for the Architect's and the Contractor's services and expenses required as a result of such insured loss.

11.3.1.2 If the Contractor fails to purchase such property insurance required by the Contract and with all of the coverages in the amount described above, the Owner may do so and deduct the cost of such insurance from the amounts due the Contractor. If the Owner is damaged by the failure or neglect of the Contractor to purchase or maintain insurance as described above, then the Contractor shall bear all reasonable costs properly attributable thereto.

11.3.1.3 If the property insurance requires deductibles, the Contractor shall pay costs not covered because of such deductibles.

11.3.1.4 This property insurance shall cover portions of the Work stored off the site and also portions of the Work in transit.

11.3.1.5 Partial occupancy or use in accordance with Paragraph 9.9 shall not commence until the insurance company or companies providing property insurance have consented to such partial occupancy or use by endorsement or otherwise. The Owner and the Contractor shall take reasonable steps to obtain consent of the insurance company or companies and shall, without mutual written consent, take no action with respect to partial occupancy or use that would cause cancellation, lapse, or reduction of insurance.

11.3.2 Boiler and Machinery Insurance. The Contractor shall purchase and maintain boiler and machinery insurance required by the Contract Documents or by law, which shall specifically cover such insured objects during installation and until final acceptance by the Owner; this insurance shall include interests of the Owner, Contractor, Subcontractors, and Sub-subcontractors in the Work, and the Owner and Contractor shall be named insureds.

11.3.3 Loss of Use Insurance. The Owner, at the Owner's option, may purchase and maintain such insurance as will insure the Owner against loss of use of the Owner's property due to fire or other hazards, however caused. To the extent not otherwise covered by the Contractor's insurance, the Contractor shall be responsible to the Owner for loss of use of the Owner's property, including consequential losses due to fire or other hazards, however caused.

11.3.4 If the Contractor requests in writing that insurance for risks other than those described herein or other special causes of loss be included in any property insurance policy obtained by the Owner, the Owner shall, if possible, include such insurance, and the cost thereof shall be charged to the Contractor by appropriate Change Order.

11.3.5 If during the Project construction period the Owner insures properties, real or personal or both, at or adjacent to the site by property insurance under policies separate from those insuring the Project, or if after final payment property insurance is to be provided on the completed Project through a policy or policies other than those insuring the Project during the construction period, the Owner shall waive all rights in accordance with the terms of Subparagraph 11.3.7 for damages caused by fire or other causes of loss covered by this separate property insurance. All separate policies shall provide this waiver of subrogation by endorsement or otherwise.

11.3.6 Before an exposure to loss may occur, the Contractor shall file with the Owner a copy of each policy that includes insurance coverages required by this Paragraph 11.3. Each policy shall contain all generally applicable conditions, definitions, exclusions, and endorsements related to this Project. Each policy shall contain a provision that the policy will not be canceled or allowed to expire, and that its limits will not be reduced, until at least 30 days' prior written notice has been given to the Owner.

11.3.7 <u>**Waivers of Subrogation.**</u> The Owner and Contractor waive all rights against (1) each other and any of their subcontractors, sub-subcontractors, agents, and employees, each of the other, and (2) the Architect, Architect's consultants, separate contractors described in Article 6, if any, and any of their subcontractors, sub-subcontractors, agents, and employees, for damages caused by fire or other causes of loss to the extent covered by property insurance obtained pursuant to this Paragraph 11.3 or other property insurance applicable to the Work, except such rights as they have to proceeds of such insurance held by the Owner as fiduciary. The Owner or Contractor, as appropriate, shall require of the Architect, Architect's consultants, separate contractors described in Article 6, if any, and the subcontractors, sub-subcontractors, agents, and employees of any of them, by appropriate agreements, written where legally required for validity, similar waivers each in favor of other parties enumerated herein. The policies shall provide such waivers of subrogation by endorsement or otherwise. A waiver of subrogation shall be effective as to a person or entity even though that person or entity would otherwise have a duty of indemnification, contractual or otherwise, did not pay the insurance premium directly or indirectly, and whether or not the person or entity had an insurable interest in the property damaged.

11.3.8 A loss insured under the Owner's property insurance may, at the Owner's option, be adjusted by the Owner as fiduciary and made payable to the Owner as fiduciary for the insureds, as their interests may appear, subject to requirements of any applicable mortgagee clause and of Subparagraph 11.3.10. The Contractor shall pay the Subcontractors their just shares of insurance proceeds received by the Contractor, and by appropriate agreements, written where legally required for validity, shall require Subcontractors to make payments to their Sub-subcontractors in similar manner.

11.3.9 The Owner shall not have any responsibility to any party other than the Contractor with respect to proceeds of any insurance policy. After a loss, the Owner may choose to terminate the Contract for convenience as described herein, or continue with the construction of the Project. If after such loss no other special agreement is made and unless the Owner terminates the Contract for convenience, replacement of damaged property shall be performed by the Contractor after notification of a Change in the Work in accordance with Article 7.

Additional provisions:

11.3.11 The Contractor shall secure, pay for and maintain whatever Fire or Extended Coverage Insurance the Contractor may deem necessary to protect the Contractor against loss of owned or rented capital equipment and tools, including any tools owned by mechanics, and any tools, equipment, scaffolding, staging, towers and forms owned or rented by the Contractor. The requirements to secure and maintain such insurance are solely for the benefit of the Contractor. Failure of the Contractor to secure such insurance or to maintain adequate levels of coverage shall not obligate the Owner, the Architect, or the Architect's consultants or their agents and employees for any losses of owned or rented equipment.

§ 5.64   Performance Bond and Payment Bond: ¶ 11.4

The following language could be substituted for ¶ 11.4:

11.4 PERFORMANCE AND PAYMENT BOND

11.4.1 The Contractor shall furnish to the Owner and keep in force during the term of the Contract performance and labor and material payment bonds, guaranteeing that the Contractor will perform its obligations under the Contract and will pay for all labor and materials furnished for the Work. Such bonds shall be issued in a form and by a Surety reasonably acceptable to the Owner, shall be submitted to the Owner for approval as to form, shall name the Owner and its lender as obligees, and shall be in an amount equal to at least 100% of the Contract Sum (as the same may be adjusted from time to time pursuant to the Contract). The Contractor shall deliver the executed, approved bonds to the Owner within seven (7) days after execution of this Agreement.

11.4.2 Subcontractors and Sub-subcontractors having a contract with a value in excess of $ _____ (2) shall also obtain and provide performance and labor and material payment bonds, issued in an amount and form and by a surety reasonably acceptable to the Owner and naming the Owner and its lender as obligees. The Owner shall from time to time and at any time have the right to increase or decrease the value of subcontracts requiring such bonds.

11.4.3 The costs of all bonds furnished hereunder shall be included in the Contract Sum.

11.4.4 The Owner shall have the right to waive any bonds required to be provided hereunder, in which event the amount of the premium of any such waived bond shall be deducted from the Contract Sum by appropriate Change Order.

11.4.5 Upon the request of any person or entity appearing to be a potential beneficiary of bonds covering payment of obligations arising under the Contract, the Contractor shall promptly furnish a copy of the bonds or shall permit a copy to be made.

11.4.6 If any Surety hereunder makes any assignment for the benefit of creditors, or commits any act of bankruptcy, or is declared bankrupt, or files a voluntary petition for bankruptcy, or in the reasonable opinion of the Owner is insolvent, the Contractor shall immediately furnish and maintain another Surety satisfactory to the Owner.

11.4.7 If the Owner or Contractor is damaged by the failure of the other to purchase or maintain any insurance or bond required by these Contract Documents, without the written consent of the other, then the party failing to so purchase or maintain such insurance or bonds shall pay all costs incurred by the other party, including but not limited to reasonable attorneys' fees.

Another provision:

11.4.1 Before commencing its Work, the Contractor shall obtain and furnish to the Owner a Performance Bond and Labor and Material Payment Bond in the amount of 100% of the applicable Contract Sum, conditioned upon the performance by the Contractor of all undertakings, covenants, terms, conditions, and agreements of the Contract Documents, and upon the prompt payment by the Contractor to all persons supplying labor and materials in the performance of the Work provided by the Contract Documents, and in a form acceptable to Owner. The Contractor shall deliver its required bonds not later than the date of execution of the Contract and deliver the required Subcontractor bonds to the Owner not later than the date of execution of the subcontract with any such Subcontractor, or if the Work is commenced prior thereto in response to a notice to proceed, the Contractor shall, prior to commencement of the Work, submit evidence satisfactory to Owner that such bonds will be issued.

11.4.2 The bonds shall be executed by the Contractor and a corporate bonding company licensed to transact such business in the state in which the Project is located and named on the current list of Surety Companies Acceptable on Federal Bonds, published in the Treasury Department's Circular Number 570. If at any time a surety on any such bond is declared a bankrupt or loses its right to do business in the state in which the Work is to be performed, or is removed from the list of surety companies accepted on Federal bonds, the Contractor shall, within ten (10) days after notice from the Owner to do so, substitute an acceptable bond (or bonds) in such form and sum and signed by such other surety or sureties as may be satisfactory to the Owner. No further payments from the Owner shall be deemed due nor shall be made until the new surety or sureties shall have furnished an acceptable bond to the Owner. The bonds furnished hereunder shall name as obligees the Owner, Owner's partners and affiliates, any lender(s) of Owner secured in whole or in part by a lien on the Project, and the title insurance company(ies) that has (have) issued title policies to Owner or its lender(s), and the bonds shall be automatically increased in the amount of any additive Change Orders. The bond(s) shall have affixed to it (them) a certified and current copy of a power of attorney for the attorney-in-fact who executes the bonds on behalf of the surety.

11.4.3 The premium for bonds required above shall be paid by the Owner unless such premium is included in the Contract Documents as the obligation of the Contractor.

11.4.4 Upon the request of any person or entity appearing to be a potential beneficiary of bonds covering payment of obligations arising under the Contract Documents, the Contractor shall promptly furnish a copy of the bonds or shall permit a copy to be made.

§ 5.65   Acceptance of Nonconforming Work: ¶ 12.3

Another version:

12.3.1 If the Owner prefers to accept Work which is not in accordance with the requirements of the Contract Documents, the Owner may do so instead of requiring its removal and correction, in which case the Contract Sum will be adjusted as appropriate and equitable. Such adjustment shall be effected whether or not final payment has been made. Upon request of the Owner or Architect, the Contractor shall provide all requested information concerning such Work, including costs of materials and labor, necessary to evaluate the cost of such nonconforming Work.

§ 5.66   Governing Law: ¶ 13.1

Another provision:

13.1.1 This Contract shall be interpreted, construed, enforced and regulated under and by the laws of the State of _____. Whenever possible, each provision of this Contract shall be interpreted in a manner as to be effective and valid under applicable law. If, however, any provision of this Contract, or a portion thereof, is prohibited by law or found invalid under any law, only such provision or portion thereof shall be ineffective, without invalidating or affecting the remaining provisions of this Contract or valid portions of such provision, which are hereby deemed severable. Contractor and Owner further agree that in the event any provision of this Contract, or a portion thereof, is prohibited by law or found invalid under any law, this Contract shall be reformed to replace such prohibited or invalid provision or portion thereof with a valid and enforceable provision which comes as close as possible to expressing the intention of the prohibited or invalid provision.

A typical forum selection clause could be added after ¶ 13.1.1:

13.1.2 The parties agree that any litigation arising out of this contract will be brought in the [name of court]. If jurisdiction is refused by such court for any reason, such case shall be brought in the [name of alternative court]. Any such action shall be heard by the court without a jury.[11]

Another version:

13.1.2 Contractor and Owner each agree that the State of Illinois, Circuit Court of Cook County shall have exclusive jurisdiction to resolve all Claims and any issue and disputes between Contractor and Owner. Contractor agrees that it shall not file any petition, complaint, lawsuit or legal proceeding against Owner in any other court other than the Circuit Court, Cook County, Illinois.

Here is a provision that could be added here or as its own section. Note that if arbitration is to be kept in the agreement, this language should be modified:

13.1.3 All judicial proceedings brought with respect to this Agreement may be brought in (1) any court of the State of Illinois of competent jurisdiction and (2) any Federal court of competent jurisdiction having *situs* within the boundaries of the Federal court district of the Northern District of Illinois, and by execution and delivery of this Agreement, the Contractor accepts, and shall cause its Subcontractors to accept, generally and unconditionally, the exclusive jurisdiction of the aforesaid courts, and irrevocably agrees to be bound by any final judgment rendered thereby from which no appeal has been taken or is available. The Contractor designates and appoints its Project Superintendent as its agent in Chicago, Illinois, to receive on its behalf service of

---

[11] In Bryant Elec. Co. v. City of Fredericksburg, 762 F.2d 1192 (4th Cir. 1985), the court found that the following forum selection clause was valid:

> All claims, disputes and other matters in question between OWNER and CONTRACTOR arising out of, or relating to the Contract Documents or the breach thereof, except for claims which have been waived by the making or acceptance of final payment . . . shall be decided by the circuit Court of the City of Fredericksburg.

The federal court declined jurisdiction based on this clause.

all process in any such proceedings in such court (which representative will be available to receive such service at all times), such service being hereby acknowledged by such representative to be effective and binding service in every respect. Said agent may be changed only upon the giving of written notice by the Contractor to the Owner of the name and address of a new Agent for Service of Process who works within the geographical boundaries of Cook County and is employed by the Contractor. The Contractor irrevocably waives and shall cause its Subcontractors to irrevocably waive any objection (including without limitation any objection of the laying of venue or based on the grounds of forum non conveniens) which it may now or hereafter have to the bringing of any action or proceeding with respect to this Agreement in the jurisdiction set forth above. Nothing herein will affect the right to serve process in any other manner permitted by law or will limit the right of the Owner to bring proceedings against the Contractor in the courts of any other jurisdiction.

§ 5.67 Successors and Assigns: ¶ 13.2

If the contractor wants to prohibit any assignment of the contract, the standard provision must be modified. One possibility is as follows:

13.2.1 The Owner and Contractor respectively bind themselves, their partners, successors, and legal representatives to the other party hereto and its partners, successors, and legal representatives in respect to covenants, agreements, and obligations contained in the Contract Documents. Neither party to the Contract shall assign the Contract in whole or in part without the written consent of the other, and any such attempted assignment shall be null and void. No party hereto shall have any contractual obligations under this Contract to any other party.

Another provision:

13.2.1 The Contractor hereby binds itself, its partners, successors, assigns and legal representatives to the Owner in respect to covenants, agreements and obligations contained in the Contract Documents. Contractor shall not assign the Contract or proceeds hereof without written consent of the Owner. If Contractor attempts to make such an assignment without such consent, it shall be void and confer no rights on third parties, and Contractor shall nevertheless remain legally responsible for all obligations under the Contract.

The form language requires the lender to assume the obligations of the contractor, meaning that the lender would have to pay the contractor all unpaid invoices for the period prior to the time the lender took over the project. Here is a provision that should be more acceptable to lenders:

13.2.2 The Owner may, without consent of the Contractor, assign the Contract to an institutional Lender providing construction financing for the Project. In such event, the Lender shall assume the Owner's rights under the Contract Documents, but the Lender shall not be required to assume any liabilities or obligations of the Owner incurred prior to the Lender's assumption of rights. The Contractor shall execute all consents reasonably required to facilitate such assignment. In the event the Lender assumes the rights of the Owner, the Contractor shall continue performance of its work hereunder and shall not stop work because of any breach of contract of the Owner occurring prior to such assumption by the Lender.

§ 5.68   Written Notice: ¶ 13.3

The following is an example of a notice provision that may be added after ¶ 13.3.1:

13.3.2 Any written notice required hereunder must be served as follows:

If to the Owner:

John Owner
111 S. Michigan
Chicago, IL 60603

with a copy to:

Fred Attorney
1 N. LaSalle St.
Chicago, IL 60602

The following language can be added to ¶ 13.3.2 above:

Notice may also be made by facsimile transmission. In such case, notice will be deemed received when the transmission is made. The party making such facsimile transmission shall also forward a copy of such notice by regular mail.

Another provision:

Where in any of the Contract Documents there is any provision in respect to the giving of any notice, such notice shall be deemed to have been given:

As to the Owner, when written notice shall be delivered to the Project Manager of the Owner, or shall have been placed in the United States mails addressed to the Director of Facilities of Owner at the place where bids or proposals for the contract were opened;

As to the Contractor, when a written notice shall be delivered to the chief representative of the Contractor at the site of the Project, or shall have been placed in the United States mails addressed to the Contractor at the place stated in the Bid Proposal as to the address of Contractor's permanent place of business;

As to the surety on the performance bond, when a written notice shall have been placed in the United States mails addressed to the surety at the home office of such surety or to its agent or agents who executed such performance bond in behalf of such surety.

§ 5.69   Interest: ¶ 13.6

Language favorable to an owner could be substituted in ¶ 13.6.1:

13.6.1 No payment due or unpaid shall bear any interest charges.

§ 5.70   Time Limits on Claims: ¶ 13.7

This section was largely eliminated in the 2007 version of A201. While eliminating this language is very favorable to owners, contractors may want to incorporate the old AIA language back into this document:

13.7 COMMENCEMENT OF STATUTORY LIMITATION PERIOD

13.7.1 As between the Owner and Contractor:

.1 Before Substantial Completion. As to acts or failures to act occurring prior to the relevant date of Substantial Completion, any applicable statute of limitations shall commence to run and any alleged cause of action shall be deemed to have accrued in any and all events not later than such date of Substantial Completion;[12]

---

[12] If the event occurs before *substantial completion* (the date shown by the architect on the certificate of substantial completion), the statute of limitations starts to run no later than that date, but if a party can demonstrate an earlier date, that will be the effective date.

   In American Prod. Co. v. Reynolds & Stone, 1998 Tex. App. LEXIS 7387 (Nov. 30, 1998), the owner added a new building and modified an existing structure at its facilities. The architect certified Phase I of the project as substantially complete on October 1, 1984. On October 7, 1994, the roof on the addition collapsed, apparently due to the omission of certain specified anchor bolts. Texas has a 10-year statute of repose (a statute of repose is similar to a statute of limitations in that it imposes an absolute time limit in which to bring an action). Because more than 10 years had elapsed from the date of substantial completion, the court dismissed the action. The owner argued that substantial completion must be measured from the date the entire project is substantially complete, not just one phase. Because the construction contract expressly provided for phased construction, it was proper to have separate dates of substantial completion. Even though the architect had inadvertently left the date on the Certificate of Substantial Completion blank, there was other evidence indicating the correct date, and the architect filed an affidavit attesting to the correct date. The architect was assigned exclusive authority to determine the substantial completion date by the contract, and the court would not disturb that determination.

In Louisville/Jefferson County Metro Gov't v. HNTB Corp., 2007 WL 1100743 (W.D.Ky., April 11, 2007), it was argued that there had not been substantial completion so as to trigger the running of the statute of limitations. The project was for a baseball field, and the plaintiff argued that the project had failed to meet the standards of the governing body for baseball. The court rejected this argument on the basis that baseball games were being played at the facility without apparent problems, so that it was sufficiently complete for the owner to occupy and use the facility.

In Harbor Court Assocs. v. Leo A. Daly Co., 179 F.3d 147 (4th Cir. 1999), construction on a project—a combination office tower, hotel, and garage located in Baltimore—started in mid-1984, and a final Certificate of Completion was issued on September 11, 1987. Other than some minor chipping and cracking of the outer brick veneer, no problems appeared until April 1996, when a 15-square-foot section of brick suddenly and without warning exploded off the face of the structure. The consulting engineers concluded that the structure suffered from fundamental and latent defects in design and construction.

In the resulting lawsuit by the owner, the defendant architect moved for summary judgment based on the AIA language and the fact that Maryland had a three-year statute of limitations. The plaintiff argued that the "discovery rule" should apply, whereby the statute of limitations does not start to run until a defect is discovered. The federal court, however, held that the parties can contractually set their own rule that departs from the discovery rule, and that the AIA clause did just that. If the discovery rule had applied, the statute of limitations would not even start to run until the owner either knew or should have known that there was a problem. In this case, that would probably have meant that the three-year period would have started in April of 1996, thus permitting the architect to be sued. Instead, the court found that the three-year period started on the date of substantial completion and ended three years later, in September of 1990.

In Northridge Homes, Inc. v. John W. French & Assocs., Inc., 1999 Mass. Super. LEXIS 435 (Nov. 15, 1999), the owner sued the architect after severe leaks were found in the roofs of the townhouses designed by the defendant. The architect provided construction administration for the project and issued a Certificate of Substantial Completion effective February 17, 1993. Massachusetts has a three-year statute of limitations. Suit was filed on April 30, 1997.

Starting in 1994, and continuing into 1995 and 1996, the architect met with the owner for the purpose of resolving the roof leak problems. After the lawsuit was filed, the architect moved for summary judgment based on the AIA language that requires that suit be filed within the applicable three-year period following the date of substantial completion. The court found that, for actions that occurred before the date of substantial completion, the time had expired. The court permitted the lawsuit to proceed to give the owner a chance to demonstrate that the architect misrepresented facts, thereby misleading the owner into waiting to file suit. This is referred to as the doctrine of equitable estoppel, and places a substantial burden of proof on the owner. If the owner can prove that the architect misrepresented facts with the intent to mislead the owner, then the limitations period is extended.

One argument of the owner was that the Certificate of Substantial Completion was never signed by the owner. Therefore, it should be disregarded. The court rejected this contention, based on the language found in AIA Document A201 (now §§ 9.8.4 and 9.8.5). The court held that, based on these provisions, by failing to sign the Certificate, the owner does not negate the date of Substantial Completion. This language places the responsibility for determining the date of Substantial Completion entirely on the architect, and does not require the owner's signature to set that date. At most, a failure to sign reflects the owner's refusal to accept the responsibilities assigned under the Certificate.

In another case, College of Notre Dame of Maryland v. Morabito Consultants, 132 Md. App. 158, 752 A.2d 265 (2000), an owner sued the structural engineer hired by the architect for breach of

.2 Between Substantial Completion and Final Certificate for Payment. As to acts or failures to act occurring subsequent to the relevant date of Substantial Completion and prior to issuance of the final Certificate for Payment, any applicable statute of limitations shall commence to run and any alleged cause of action shall be deemed to have accrued in any and all events not later than the date of issuance of the final Certificate for Payment; and

.3 After Final Certificate for Payment. As to acts or failures to act occurring after the relevant date of issuance of the final Certificate for Payment, any applicable statute of limitations shall commence to run and any alleged cause of action shall be deemed to have accrued in any and all events not later than the date of any act or failure to act by the Contractor pursuant to any Warranty provided under Paragraph 3.5, the date of any correction of the Work or failure to correct the Work by the Contractor under Paragraph 12.2, or the date of actual commission of any other act or failure to perform any duty or obligation by the Contractor or Owner, whichever occurs last.

This clause generally provides for all other acts that are not covered above. The parties can also fix limitations periods shorter than in the statutes.

In one case, a court stated that when

the contract requires the architect to conduct inspections to determine completion dates and issue a final certificate, issuance of that certificate represents a significant contractual right of the owner and concomitant obligation of the architect. The architect's issuance of the certificate marks the completion of its performance and the point when the Statute of Limitations starts to run for a breach of its contractual undertaking. However, that same result does not obtain when the owner itself

---

contract and negligence. The engineer had been hired to inspect a building to be renovated in 1991. In 1997, significant movement in the building was detected and a subsequent engineer concluded that the first engineer had failed to properly calculate the loads. In December 1998, the owner filed suit and the engineer moved to dismiss on the grounds that its contract contained the limitations provisions found above. In fact, both the owner-architect agreement and the architect-consultant agreement contained this language.

The owner argued that the accrual clause is like an exculpatory clause and should be read more stringently. The court held that this is not an exculpatory clause because it does not relieve any party from liability. Instead, it alters the time for accrual of a cause of action from what the law would otherwise impose. Such contractual modifications are generally not disfavored in law. Here, the Maryland three-year statute of limitations barred the action.

See also Town of Hopedale v. Alderman & MacNeish, Inc., 2001 Mass. Super. LEXIS 162, 2001 WL 544024 (Mass. Super. Ct. Mar. 12, 2001); Gustine Uniontown Assocs., Ltd. v. Anthony Crane Rental, Inc., 786 A.2d 246 (Pa. 2002) (court determined which state statute of limitations period applied and then found that suit had been filed within that time following the date of substantial completion). *Gustine* was followed in Trinity Church v. Lawson-Bell, 925 A.2d 710 (N.J. Super. 2007).

In Williamson Pounders Architects v. Tunica County, Mississippi, 2007 WL 2903216 (N.D. Miss., Sept. 28, 2007), the issue was whether the complaint was timely filed. The owner argued that the architect was aware of a claim prior to substantial completion, so that the cause of action accrued on that date. The court agreed with the architect, finding that the contract set the accrual date as the date of substantial completion, making the action timely.

controls issuance of the final certificate. The final certificate in such circumstances may indicate the owner's acceptance of the work for purposes of contractual guarantees or equitable price adjustments, but it does not represent completion of the contractual obligations of the architect or general contractor for purposes of triggering the Statute of Limitations.[13]

In a New York case, an owner sued a contractor based on a defective roof.[14] The contractor had agreed to furnish a two-year guarantee but did not furnish any written guarantee. The court held that the statute of limitations commenced at the time of completion of the project when the guarantee should have been provided, and that there was no guarantee.

In order to fix limitation periods shorter than the law provides, the parties can use the following language:

13.7.1 The parties agree that no action hereunder may be brought by either party more than one year after the date of the final Certificate for Payment, or the date of the Contractor's last substantial work, whichever is earlier, except in the case of an action for nonpayment of any amounts due under the Contract Documents, in which case no action may be brought more than two years after payment is due.

Here is another provision:

13.7.1 As between the Owner and Contractor, any applicable statute of limitations shall commence to run and any alleged cause of action shall be deemed to have accrued (notwithstanding any knowledge or lack of knowledge on the part of either the Owner or the Contractor) in any and all events not later than the

---

[13] Penn York Constr. Corp. v. Peter Bratti Assocs., 60 N.Y.2d 987, 459 N.E.2d 486, 487, 471 N.Y.S.2d 261 (1983).

[14] Board of Educ. v. Thompson Constr. Corp., 111 A.D.2d 497, 488 N.Y.S.2d 880 (1985). As to the architect, based on the *continuous treatment doctrine*, Borgia v. City of N.Y., 12 N.Y.2d 151, 187 N.E.2d 777 237 N.Y.S.2d 319 (1962), the statute of limitations did not start to run until the confidential professional relationship existed between the owner and architect. The doctrine is also called the *continuing relationship doctrine*. *See* Northern Mont. Hosp. v. Knight, 811 P.2d 1276 (Mont. 1991).

In Russo Farms, Inc. v. Vineland Bd. of Educ., 675 A.2d 1077, 1091, 1092 (N.J. 1996), farmers sued for damages caused by improper siting and construction of a school across from their property and by an inadequate drainage system. Among the defendants were the architect and contractor, who were sued in tort. They defended based on the running of the statute of limitations. The plaintiffs alleged a continuing tort theory. In rejecting this argument, the court held that the architect and contractor had no control over the school property after the end of construction:

> It is only when the new injury results from a new breach of duty that a new cause of action accrues. "For there to be a continuing tort there must be a continuing duty." . . . "That they never corrected the problem does not render the tort continuing." Their "mere failure to right a wrong and make plaintiff whole cannot be a continuing wrong which tolls the statute of limitations, for that is the purpose of any lawsuit and the exception would obliterate the rule." . . . Thus, the new damages are simply continuing damages, not a new tort. As a result, the only tort alleged is negligence, and that tort accrued in 1980–81 when the injury first occurred.

earlier of the date of Substantial Completion or the date of the issuance of a Certificate of Occupancy by the governmental entity charged with issuing such certificates.

13.7.2 In the event neither a Certificate of Occupancy nor a Certificate of Substantial Completion is issued, Substantial Completion shall be deemed to have occurred no later than the date when the Owner has tendered payment to the Contractor in excess of 95% of the Contract Sum.

Another provision:

13.7.1 As between the Owner and Contractor, any applicable statute of limitations shall commence to run and any alleged cause of action shall be deemed to have accrued in any and all events as provided by applicable law.

§ 5.71 Confidentiality: ¶ 13.8

At this point, the owner may want to insert a clause restricting the contractor's ability to communicate confidential information:

13.8 CONFIDENTIALITY

13.8.1 The Contractor acknowledges that certain of the Owner's valuable, confidential, and proprietary information may come into the Contractor's possession. Accordingly, the Contractor agrees to hold all information it obtains from or about the Owner in strictest confidence, not to use such information other than for the performance of the services, and to cause any of its employees, subcontractors, or consultants to whom such information is transmitted to be bound to the same obligation of confidentiality to which the Contractor is bound. The Contractor shall not communicate the Owner's information in any form to any third party without the Owner's prior written consent. In the event of any violation of this provision, the Owner shall be entitled to preliminary and permanent injunctive relief as well as an equitable accounting of all profits or benefits arising out of such violation, which remedy shall be in addition to any other rights or remedies to which the Owner may be entitled.

13.8.2 If Contractor is presented with a request for documents by any entity, governmental or administrative agency or with a *subpoena duces tecum* regarding any records, data or documents which may be in Contractor's possession by reason of this Agreement, Contractor must immediately give notice to the Owner with the understanding that the Owner will have the opportunity to contest such process by any means available to it before the records or documents are submitted to a court or other third party. Contractor, however, is not obligated to withhold the delivery beyond the time ordered by the court or administrative agency, unless the *subpoena* or request is quashed or the time to produce is otherwise extended.

§ 5.72 Preconstruction Conference: ¶ 13.9

A preconstruction conference is an excellent way to start a project. Such a conference includes all the major parties to the work and should be held shortly after award of the contract. The following clause can be used for this purpose:

13.9 PRECONSTRUCTION CONFERENCE

13.9.1 Within 14 days after execution of the Agreement, a conference will be held for review and acceptance of the schedule referred to in Section 3.10, to establish procedures for handling shop drawings and other submittals, for processing Applications for Payment, and to establish a general working relationship among the parties to the Work. The conference will be held at a time and place to be determined by the Owner. The conference will be attended by a representative of the Contractor, the major Subcontractors and material suppliers, the Architect, and the Owner.

§ 5.73 Contractor's Responsibility for Additional Architectural Fees: ¶ 13.10

To assure that the contractor does not cause additional architect's fees for the owner, the following clause can be used:

13.10 CONTRACTOR'S RESPONSIBILITY FOR ADDITIONAL ARCHITECTURAL FEES

13.10.1 If more than two submittals are required for any shop drawing or other submittal, the Contractor shall be liable for any Architect's fees incurred as the result of such submittals. If the Work is not complete after submittal of the Contractor's written notice pursuant to Paragraph 9.10.1, the Contractor shall be liable for any additional Architect's fees incurred for any inspection following the initial inspection after receipt of such notice. If the Contractor defaults and causes the Architect to provide additional services, the Contractor shall be responsible for same. If the Contractor submits an extensive number of claims and the majority of such claims are rejected, the Contractor shall be responsible for any additional Architect's fees for any such rejected claims. Any funds due under this paragraph shall be deducted by the Owner from the amounts due the Contractor for such additional Architect's fees and paid directly to the Architect.

§ 5.74 Miscellaneous Provisions

13.11 Contractor's Authority To Enter Into This Agreement. Execution of this Agreement by the Contractor is authorized and signature(s) of each person signing on behalf of the Contractor have been made with complete and full authority to commit the Contractor to all terms and conditions of this Agreement, including each and every representation, certification, and warranty contained herein, attached hereto and collectively incorporated by reference herein, or as may be required by the terms and conditions hereof. If other than a sole proprietorship, Contractor must provide satisfactory evidence that the execution of the Agreement is authorized in accordance with the business entity's rules and procedures in accordance with applicable laws and regulations.

§ 5.75 Termination by the Contractor: ¶ 14.1

The owner may want to change these paragraphs as follows:

14.1.1 The Contractor may terminate the Contract if the Work is stopped for a period of 90 consecutive days through no act or fault of the Contractor or a Subcontractor, Sub-subcontractor or their agents other persons or entities performing portions of the Work under direct or indirect contract with the Contractor, for any of the following reasons:

.1 issuance of an order of a court or other public authority having jurisdiction which requires all Work to be stopped; or

.2 an act of government, such as a declaration of national emergency which requires all Work to be stopped.

14.1.2 If one of the reasons described in Section 14.1.1 exists, the Contractor may, upon fourteen days' written notice to the Owner and Architect, terminate the Contract and recover from the Owner payment for Work properly executed.

§ 5.76 Termination by the Owner for Cause: ¶ 14.2

Because it is questionable whether the owner is entitled to attorneys' fees and other costs associated with a contractor's default, the owner may want to insert a clause such as the following after ¶ 14.2.4:

14.2.4.1 The costs of finishing the Work include, without limitation, all reasonable attorneys' fees, additional title costs, insurance, additional interest because of any delay in completing the Work, and all other direct and indirect and consequential costs incurred by the Owner by reason of the termination of the Contractor as stated herein.

A different version of a clause providing for termination by the owner follows. Note that bankruptcy laws may have an effect on this provision. Delete Subparagraph 14.2.1 in its entirety and replace it with the following:

14.2.1 If the Contractor shall institute proceedings or consent to proceedings requesting relief or arrangement under the Federal Bankruptcy Act or any similar or applicable federal or state law; or if a petition under any federal or state bankruptcy or insolvency law is filed against the Contractor and such petition is not dismissed within sixty (60) days from date of said filing; or if the Contractor admits in writing its inability to pay its debts generally as they become due; or if Contractor makes a general assignment for the benefit of its creditors; or if a receiver, liquidator, trustee, or assignee is appointed on account of Contractor's bankruptcy or insolvency; or if a receiver of all or any substantial portion of the Contractor's properties is appointed; or if the Contractor abandons the Work; or if Contractor fails, except in cases for which extension of time is provided, to prosecute promptly and diligently the Work or to supply enough properly skilled workers or proper materials for the Work; or if Contractor submits an Application for Payment, sworn statement, waiver of lien, affidavit, or document of any nature whatsoever which is intentionally falsified; or if Contractor fails to make prompt payment to Subcontractors or for materials or labor or otherwise breaches its obligations under any subcontract with a Subcontractor; or if a mechanic's or materialman's lien or notice of lien is filed against any part of the Work or the site of the Project and not promptly bonded or insured over by the Contractor in a manner satisfactory to the Owner; or if the Contractor disregards any laws, statutes, ordinances, rules, regulations, or orders of any governmental body or public or quasi-public authority having jurisdiction of the Work or the site of the Project; or if Contractor otherwise violates any provision of the Contract Documents, then the Owner, without prejudice to any right or remedy available to the Owner under the Contract Documents or at law or in equity, may, after giving seven (7) days' written notice to the Contractor and the surety under the Performance Bond and under the Labor and Material Payment Bond described in Paragraph 11.4, terminate the employment of the Contractor. If requested by the Owner, the Contractor shall remove any part or all of its equipment, machinery, and supplies from the site of the Project within seven (7) days from the date of such request, and in the event of the Contractor's failure to do so, the Owner shall have the right to remove or store such equipment, machinery, and supplies at the Contractor's expense. In case of such termination, the Contractor shall not be entitled to receive any further payment for Work performed by the Contractor through the date of termination. The Owner's right to terminate the Owner-Contractor Agreement pursuant to this Subparagraph 14.2.1 shall be in addition to and not in limitation of any rights or remedies existing hereunder or pursuant hereto or at law or in equity.

A further provision:

14.2.4 If the unpaid balance of the Contract Sum exceeds all costs to the Owner of completing the Work, then the Contractor shall be paid for all Work performed by the Contractor to the date of termination. If such costs to the Owner of completing the Work exceed such unpaid balance, the Contractor shall pay the difference to the Owner immediately upon the Owner's demand. The costs to the Owner of completing the Work shall include but not be limited to the cost of any additional architectural, managerial, and administrative services required thereby, any costs incurred in retaining another contractor or other subcontractors, any additional interest or fees which the Owner must pay by reason of delay in completion of the Work, attorney's fees and expenses, and any other damages, costs, and expenses the Owner may incur by reason of completing the Work or any delay thereof. The amount, if any, to be paid to the Contractor shall be certified by the Architect, upon application, in the manner provided in Paragraph 9.4, and this obligation for payment shall survive the termination of the Contract.

In ¶ 14.2.2.3, the owner is required to furnish to the contractor, upon the contractor's written request, a "detailed accounting" of the costs required to finish the project following termination of the contractor for cause. The contractor may want to have the following language inserted at the end of that paragraph:

In the event the Contractor requests such information, the information shall be compiled and furnished in accordance with Generally Accepted Accounting Principles as normally applied in the construction industry in the locale of the Project.

In *Adams Builders & Contractors, Inc. v. York Saturn, Inc.*, 60 Mass. App. Ct. 1101, 798 N.E.2d 586 (2003), the contract included this provision:

The Owner shall be entitled to collect from the Contractor all direct, indirect, and consequential damages suffered by the Owner on account of the Contractor's default, including, without limitation, additional services and expenses of the Architect made necessary thereby and attorneys' fees incurred by the Owner on account of such default.

§ 5.77   Termination by the Owner for Convenience: ¶ 14.4

The following provisions can be substituted for the form provisions:

14.4 TERMINATION BY THE OWNER FOR CONVENIENCE

14.4.1 Notwithstanding any other provision to the contrary in any Agreement or the General Conditions, the Owner reserves the right at any time and in its absolute discretion to terminate the services of the Contractor and the Work by giving written notice to the Contractor. In such event, the Contractor shall be entitled to and the Owner shall reimburse the Contractor for an equitable portion of the Contractor's fee based on the portion of the Work completed before the effective date of termination and for any other costs attributable to such termination.

14.4.1.1 Subject to the approval of HUD, the performance of work under this contract may be terminated by the Owner in accordance with this paragraph in whole, or from time to time in part, whenever the Contracting Officer shall determine that such termination is in the best interest of the Owner. Any such termination shall be effected by delivery to the Contractor of a Notice of Termination, specifying the extent to which the performance of the work under the contract is terminated and the date upon which such termination becomes effective.

14.4.1.2 If the performance of the work is terminated, either in whole or in part, the Owner shall be liable to the Contractor for reasonable and proper costs of termination, which costs shall be paid to the Contractor within 90 days of receipt by the Owner of a properly presented claim setting out in detail: (1) the total cost of the work performed to date of termination less the total amount of contract payments made to the Contractor; (2) the cost (including reasonable profit) of settling and paying claims under subcontracts and material orders for work performed and materials and supplies delivered to the site, payment for which has not been made by the Owner to the Contractor or by the Contractor to the subcontractor or supplier; (3) the cost of preserving and protecting the work already performed until the Owner or assignee takes possession thereof or assumes responsibility therefor; (4) the actual or estimated cost of legal or accounting services reasonably necessary to prepare and present the termination claim to the Owner; and (5) an amount constituting a reasonable profit on the value of the work performed by the Contractor.[15]

§ 5.78   Claims and Disputes: ¶ 15.1

Another provision:

15.1.2 Time Limits on Claims. Claims by either party arising prior to the date Final Payment is due must be initiated within 21 days after occurrence of the event giving rise to such Claim or within 21 days after the claimant first recognizes the condition giving rise to the Claim, whichever is later. Claims must be initiated by written notice to the Architect and the other party. Such notice shall be entitled "Notice of Claim." Each Claim shall be set forth on a separate written notice. Failure to give written notice of a Claim within such time limits shall constitute a waiver of such Claim.

Another:

15.1.2 Notice of Claims. Claims by either the Owner or Contractor must be initiated by written notice to the other party and to the Initial Decision Maker with a copy sent to the Architect, if the Architect is not serving as the Initial Decision Maker. Claims by either party must be initiated within 21 days after occurrence of the event giving rise to such Claim If the Contractor wishes to reserve its rights under this paragraph, written notice concerning any event that may give rise to a claim must be given within 21 days of the event, whether or not any impact in money or time has been determined. The written notice must be a separate and distinct correspondence provided in hardcopy to the Initial Decision Maker, Architect and Owner and must delineate the specific event and outline the causes and

---

[15] Adapted from Linan-Faye Constr. Co. v. Housing Auth., 847 F. Supp. 1191 (D.N.J. 1994).

reasons for a claim or potential claim. Written remarks or notes of a generic nature are invalid. Comments made at progress meetings, walk-throughs, inspections, in e-mails, voice mails, and other such communications do not meet the requirement of providing notice of a claim or potential claim.

The owner may want to use this alternative provision in place of the standard AIA clause in ¶ 15.1.5.2 in order to minimize claims:

15.1.5.2 Extensions of time will not be granted for delays caused by unfavorable weather, unsuitable ground conditions, inadequate construction force, or the failure of the Contractor to place orders for equipment or materials sufficiently in advance to insure delivery when needed.[16]

Additional provisions related to weather claims:

15.1.5.2.1 Inclement or adverse weather shall not be a prima facie reason for the granting of an extension of time, and the Contractor shall make every effort to continue work under prevailing conditions. The Owner may grant an extension of time if an unavoidable delay occurs as a result of inclement/severe/adverse weather and such shall then be classified as a "Delay Day". Any and all delay days granted by the Owner are and shall be non-compensable in any manner or form. The Contractor shall comply with the notice requirements concerning instances of inclement/severe/adverse weather before the Owner will consider a time extension. Each day of inclement/severe/adverse weather shall be considered a separate instance or event and as such, shall be subject to the notice requirement of Section 15.1.2.

15.1.5.2.2 An "inclement", "severe", or "adverse" weather delay day is defined as a day on which the Contractor is prevented by weather or conditions caused by weather resulting immediately therefrom, which directly impact the current controlling critical-path operation or operations, and which prevent the Contractor from proceeding with at least 75% of the normal labor and equipment force engaged on such critical path operation or operations for at least 60% of the total daily time being currently spent on the controlling operation or operations.

15.1.5.2.3 The Contractor shall consider normal/typical/seasonal weather days and conditions caused by normal/typical/seasonal weather days for the location of the Work in the planning and scheduling of the Work to ensure completion within the Contract Time. No time extensions will be granted for the Contractor's failure to consider and account for such weather days and conditions caused by such weather for the Contract Time in which the Work is to be accomplished.

15.1.5.2.4 A "normal", "typical", or "seasonal" weather day shall be defined as weather that can be reasonably anticipated to occur at the location of the Work for each particular month involved in the Contract Time. Each month involved shall not be considered individually as it relates to claims for additional time due

---

[16] *See* Peter Kiewit Sons' Co. v. Iowa S. Utils. Co., 355 F. Supp. 376 (S.D. Iowa 1973).

to inclement/adverse/severe weather but shall consider the entire Contract Time as it compares to normal/typical/seasonal weather that is reasonably anticipated to occur. Normal/typical/seasonal weather days shall be based upon U.S. National Weather Service climatic data for the location of the Work or the nearest location where such data is available.

15.1.5.2.5 The Contractor is solely responsible to document, prepare and present all data and justification for claiming a weather delay day. Any and all claims for weather delay days shall be tied directly to the current critical-path operation or operations on the day of the instance or event which shall be delineated and described on the Critical-Path Schedule and shall be provided with any and all claims. The Contractor is solely responsible to indicate and document why the weather delay day(s) claimed are beyond those weather days which are reasonably anticipated to occur for the Contract Time. Incomplete or inaccurate claims, as determined by the Architect or Owner, may be returned without consideration or comment.

An alternative provision for the waiver of consequential damages is as follows:

15.1.6 Claims for Consequential Damages. The Contractor and Owner waive Claims against each other for consequential damages arising out of or relating to this Contract. This mutual waiver includes, but is not limited to, the following:

1. damages incurred by the Owner for rental expenses, for losses of use, income, profit, financing, business, and reputation, for delay damages of any sort, and for loss of management or employee productivity or of the services of such persons; and

2. damages incurred by the Contractor for principal office expenses, including the compensation of personnel stationed there, for losses of financing, business, and reputation, for delay damages of any sort, and for loss of profit except anticipated profit arising directly from the Work.

This mutual waiver is applicable, without limitation, to all consequential damages due to either party's termination in accordance with Article 14. To the extent liquidated damages are set forth elsewhere in the Contract Documents, such liquidated damages shall apply and supersede, to the extent of any inconsistencies, the provisions of this Section 15.1.6.

Another version:

15.1.6 Claims for Consequential Damages. The Contractor and Owner waive Claims against each other for consequential damages arising out of or relating to this Contract. This mutual waiver includes:

.1 damages incurred by the Owner for rental expenses, for losses of use prior to the date of Substantial Completion as amended from time to time, income, profit, financing, business and reputation, and for loss of management or employee productivity or of the services of such persons, and for attorneys' fees, insurance and interest (excluding post-judgment interest); and

.2 damages incurred by the Contractor for principal office expenses including the compensation of personnel stationed there, rent, utilities and office equipment, for losses of financing, business and reputation, and for loss of profit except anticipated profit arising directly from the Work, and for attorneys' fees, insurance and interest (excluding post-judgment interest). This mutual waiver is applicable, without limitation, to all consequential damages due to either party's termination in accordance with Article 14. Nothing contained in this Section 15.1.6 shall be deemed to preclude an award of liquidated damages, when applicable, in accordance with the requirements of the Contract Documents. This Section 15.1.6 does not apply to Section 3.18.

15.1.7 Waiver of Claims against the Architect. Notwithstanding any other provision to the contrary, the Contractor waives all claims against the Architect and the Architect's consultants for consequential damages listed above.

Another provision:

15.1.6 Contractor agrees that Owner shall not be liable to Contractor for any special, indirect, incidental, or consequential damage whatsoever, whether caused by Owner's negligence, fault, errors or omissions, strict liability, breach of contract, breach of warranty or other cause or causes whatsoever. Such special, indirect, incidental or consequential damages include, but are not limited to loss of profits, loss of savings or revenue, loss of anticipated profits, labor inefficiencies, idle equipment, home office overhead, and similar types of damages.

Another provision:

15.1.7 In the absence of a Change Order in accordance with Section 7.2 or a Construction Change Directive in accordance with Section 7.3 authorizing an addition or alteration, no course of conduct or dealings between the parties, nor express or implied acceptance of alterations or additions to the Work, and no claim that the Owner has been unjustly enriched by any alteration or addition to the Work, whether or not there is in fact any such unjust enrichment, shall be the basis for any claim to an increase in the Contract Sum or change in the Contract Time.

§ 5.79 Initial Decision: ¶ 15.2

An alternative claims procedure is as follows:

15.2 RESOLUTION OF CLAIMS AND DISPUTES

15.2.1 The following Claims procedure shall supersede any other provisions to the contrary and shall be the exclusive procedure for resolution of any and all Claims arising out of or related to the Contract.

15.2.1.1 Claims Covered. All claims that increase or decrease the Contract Sum or the Contract Time are included herein, including Claims asserted, and not waived, after final payment.

15.2.1.2 Notice. The party asserting a Claim must give written notice (effective upon receipt) of the Claim to the Architect no later than twenty-one (21) days after the party becomes aware or should have become aware of the existence of the Claim. The notice must a) give a general description of the nature of the Claim and the date on which the party first became aware of the existence of the Claim, and b) state whether a change in the Contract Cost and/or Contract Time is sought. The amount of such change may be omitted from the notice if the party cannot make an accurate determination of such change. Amendments to the amount of the change sought shall be permitted upon a showing of due diligence and strict compliance with other provisions of the Claims procedure. No amendments shall be permitted if the Claim is not timely made. This Claims procedure must be followed, even if the Claim asserts fault on the part of the Architect. Copies of the notice must be given to the Owner or Contractor, as the case may be.

15.2.1.3 Architect's Response. Upon receipt of a Claim, the Architect shall have ten (10) days to make one of the following responses:

15.2.1.3.1 Approve, Disapprove, or Modify the Claim. Unless the Architect specifically states otherwise, this decision shall be final and binding upon the parties unless one of the parties initiates arbitration within thirty (30) days of receipt of this decision. In the event of a modification, the Architect shall specify the amount of change in Contract Sum or Contract Time permitted. In the event of an approval where the Claim does not specify the change in Contract Sum or Contract Time sought, the claims procedure shall be stayed until the Claimant specifies such change. Thereafter the Architect shall have ten (10) days to reconsider the approval.

15.2.1.3.2 Request Additional Information from the Claimant. In the absence of a written agreement by the parties, the Claimant shall have ten (10) days to submit additional information. Thereafter, the Architect shall have ten (10) days to make one of the responses listed in Subparagraph 15.2.1.3.1 above.

15.2.1.4 Absence of Architect's Response. If the Architect fails to make a response under ¶ 15.2.1.3, or if the position of Architect is vacant, the Claim shall be conclusively deemed rejected unless arbitration is initiated within forty (40) days of receipt of the Notice of Claim by the Architect or other party.

15.2.1.5 Conversion to Change Order. Any Claim that is resolved and not rejected by any of the above procedures shall be considered an approved Change Order following the end of any period in which arbitration may be initiated.

Another alternative:

Neither the Contractor nor the Surety shall be entitled to present any claim or claims to the Owner either during the prosecution of the Work or upon completion of the Contract for additional compensation for performance of any work that was not covered by the approved Drawings, Specifications, and/or Contract, or for any other cause, unless Contractor or Surety shall give the Owner due notice of its intention to present such claim or claims as hereinafter designated.

The written notice, as above required, must have been given to the Owner, with a copy to the Engineer, prior to the time the Contractor shall have performed such work or that portion thereof giving rise to the claim or for additional compensation; or shall have been given within ten (10) days from the date the Contractor was prevented, either directly or indirectly, by the Owner or his authorized representative from performing any work provided by the Contract, or within ten (10) days from the happening of the event, thing, or occurrence giving rise to the alleged claim.[17]

Another provision:

Any questions or disagreements arising as to the true intent of the Specifications or Drawings or the kind and quality of the work required thereby, shall be decided by the Architect, whose interpretations thereof shall be final, conclusive, and binding as to all parties.[18]

A possible form for the owner's use in compliance with ¶ 15.2.6.1 follows:

OWNER'S DEMAND FOR INITIATION OF MEDIATION

Project:

to: _____

          (Contractor)

Re: Initial Decision rendered on (date)

Pursuant to Section 15.2.6.1 of AIA Document A201-2007, the Owner hereby demands that the Contractor file for mediation within sixty (60) days of the initial decision herein. Failure to do so will result in the initial decision being final.

Date:

(Owner's signature)

The 1997 version of A201 contained an important provision that protects the owner at ¶ 4.4.6, in which the architect's decision on a claim becomes final in

---

[17] *See* McKeny Constr. Co. v. Town of Rowlesburg, 187 W. Va. 521, 420 S.E.2d 281 (1992).

[18] Adapted from M&L Bldg. Corp. v. Housing Auth., 35 Conn. App. 379, 646 A.2d 244 (1994).

certain circumstances. An alternative provision that could be used here is as follows:

Every written decision of the Architect (Initial Decision Maker) shall result in the Architect's (Initial Decision Maker's) decision becoming final and binding upon the Owner and Contractor unless either party demands arbitration (mediation) within 21 days after the date of the Architect's (Initial Decision Maker's) decision. Any party seeking to enforce the Architect's (Initial Decision Maker's) written decision may have judgment entered upon it as if the award is an arbitration award and in accordance with the applicable arbitration laws of the state where the project is located and in any court having jurisdiction thereof.

§ 5.80   Mediation: ¶ 15.3

Another provision:

15.3.2 The parties shall endeavor to resolve their Claims by mediation which, unless the parties mutually agree otherwise, shall be administered by the American Arbitration Association in accordance with its Construction Industry Mediation Procedures in effect on the date of the Agreement. A request for mediation shall be made in writing, delivered to the other party to the Contract, and filed with the person or entity administering the mediation. In no event shall any such Claim be subject to binding dispute resolution until the conclusion of the mediation in good faith by the parties. In the event that the Claim is not settled at the mediation, the mediator shall render a statement that the party making the Claim pursued the mediation in good faith, which statement shall be a condition precedent to further dispute resolution.

§ 5.81   Arbitration: ¶ 15.4

The parties may want to provide for arbitration only when the amount in dispute is below a certain amount. The following provision can be substituted for ¶ 15.4 and tailored to specific requirements:

15.4 ARBITRATION

15.4.1 All claims, disputes, and other matters in question involving amounts in dispute of less than $100,000 between any of the Architect, Construction Manager, Owner, Contractors, Surety, Subcontractors, or any material suppliers arising out of or relating to agreements to which two or more of said parties are bound, or the Contract Documents, or the breach thereof, except as provided in Section 4.2.13 with respect to the Architect's decisions on matters relating to artistic effect, shall be decided by arbitration in accordance with the Construction Industry Arbitration Rules of the American Arbitration Association then obtaining, as modified herein, unless the parties mutually agree otherwise. Prior to arbitration, the parties shall endeavor to resolve disputes by mediation in accordance with the provisions of Section 15.3. Claims, disputes, and other matters in question involving amounts in dispute of $100,000 or more may be submitted to arbitration only with the Owner's written consent. The Owner, Architect, Construction Manager, Subcontractors, and material suppliers who have an interest in the dispute shall be joined as parties to the arbitration. The Owner's contract with the Architect and the Contractor's contract with the Subcontractors shall require such joinder. The arbitrator shall have authority to decide all issues between the parties, including but not limited to claims for extras, delay and liquidated damages, matters involving defects in the Work, rights to payment, whether matters decided by the Architect involve artistic effect and whether the necessary procedures for arbitration have been followed. The foregoing agreement to arbitrate and any other agreement to arbitrate with an additional person or persons duly consented to by the parties to the Owner-Contractor, Owner-Architect or Owner-Construction Manager Agreements shall be specifically enforceable under prevailing arbitration law. The award rendered by the arbitrator shall be final, and judgment may be entered upon it in accordance with applicable law in any court having jurisdiction thereof.

15.4.2 Notice of the demand for arbitration shall be filed in writing with the other party to the arbitration and with the American Arbitration Association. The demand for arbitration shall be made within a reasonable time after the claim, dispute, or other matter in question has arisen, and in no event shall it be made after the date when institution of legal or equitable proceedings based on such claim, dispute, or other matter in question would be barred by the applicable statute of limitation.

15.4.3 Unless otherwise agreed in writing, all parties shall carry on the work and perform their duties during any arbitration proceedings, and the Owner shall continue to make payments as required by agreements and the Contract Documents.

15.4.4 If any proceeding is brought to contest the right to arbitrate and it is determined that such right exists, the losing party shall pay all costs and attorneys' fees incurred by the prevailing party.

15.4.5 In addition to the other rules of the American Arbitration Association applicable to any arbitration hereunder, the following shall apply:

15.4.5.1 Promptly upon the filing of the arbitration each party shall be required to set forth in writing and to serve upon each other party a detailed statement of its contentions of fact and law.

15.4.5.2 All parties to the arbitration shall be entitled to the discovery procedures and to the scope of discovery applicable to civil actions under [state] law, including the provisions of the Civil Practice Act and [state] Supreme Court Rules applicable to discovery. Such discovery shall be noticed, sought, and governed by those provisions of [state] law.

15.4.5.3 The arbitration shall be commenced and conducted as expeditiously as possible consistent with affording reasonable discovery as provided herein.

15.4.5.4 These additional rules shall be implemented and applied by the arbitrator(s).

15.4.6 In the event of any litigation or arbitration between the parties hereunder, all attorneys' fees and other costs shall be borne by the party determined to be at fault and, in the event that more than one party is determined to be at fault, shall be allocated equitably by the court or arbitrator.

Note that this language allows consolidation of the various claims (unlike the standard language, which prohibits consolidation) and permits attorneys' fees to be recovered by the prevailing party.

When the parties are in different locations, the following provision fixes the locale of the arbitration. Without such a provision, the first party to file an arbitration request has an advantage in fixing the location of the arbitration hearings. If there is a dispute, the American Arbitration Association has the authority to fix the locale under its rules. This provision avoids this problem:

Any arbitration hearing shall be heard at the offices of the AAA located in Chicago, Illinois. Any party requesting a site visit shall pay all costs of the AAA, including the arbitrator's fees, related to such site visit, unless both parties agree to a site visit.

CHAPTER 6

AIA DOCUMENT B103 STANDARD FORM OF AGREEMENT BETWEEN OWNER AND ARCHITECT FOR A LARGE OR COMPLEX PROJECT

681

§ 6.1   Introduction

AIA Document B103 is a new document based on B141-1997, parts 1 and 2. It is designed to be used for large or complex projects. This document assumes that the owner will retain consultants to provide cost estimating and scheduling services and that the architect will rely on the services of those consultants. If the owner will not be retaining cost and scheduling consultants, it may be more appropriate to use AIA Document B101. This document refers to AIA Document A201-2007, found in **Chapter 4**, and is intended to be used with that document.

To conserve space, reference is made to AIA Document B101, found in **Chapter 2**, whenever the language is the same. Readers should then consult the corresponding provision in that document for comments. At the end of each paragraph of this document is a boldface reference to the 1997 version of B141. This is to assist the reader in relating this document to the corresponding language in that version of B141.

§ 6.2   Prior Editions

The current edition of B103 was issued in November 2007 as part of a major revision of the most significant AIA documents. The prior version of this document is B141-1997. Future revisions are expected every ten years.

CAUTION: The recommended method of amending this document is to use the electronic documents software that is available from the AIA. Using this software, the document is amended on its face, with additions and deletions indicated either in the margin, or by underlines and strikethroughs. Another way is to attach separate written amendments that refer back to B103. This is normally done by filling in Article 12, or additional pages referred to in Article 12. An alternative method is to graphically delete material and insert new material in the margins of the standard form. However, this method may create ambiguity, and it may result in the stricken material's being rendered illegible. It is illegal to make any copies of this document in violation of the AIA copyright or to reproduce this document by computerized means. Refer to the instruction sheet that comes with B102 for additional information about that copyright.

The cited language of this document is as published by AIA in late 2007 in the Electronic Documents. There appear to be some minor inconsistencies between this document and others that are probably attributable to scrivenors' errors. AIA often corrects such errors in intervening years without any notice and without any way to determine exactly when a particular version of a document was actually released.

§ 6.3   Title Page

See the discussion in § 2.3.

§ 6.4   Article 1: Initial Information

ARTICLE 1 INITIAL INFORMATION

1.1 This Agreement is based on the Initial Information set forth in this Section 1.1.

(Note the disposition for the following items by inserting the requested information or a statement such as "not applicable, "unknown at time of execution" or "to be determined later by mutual agreement.") **(1.1.1; minor modifications)**

1.1.1 The Owner's program for the Project:

(Identify documentation or state the manner in which the program will be developed.) **(1.1.2.3; minor modification)**

See the discussion of ¶ A.1.1 of Exhibit A to B101 in § 2.36.

1.1.2 The Project's physical characteristics:

(Identify or describe, if appropriate, size, location, dimensions, or other pertinent information, such as geotechnical reports; site, boundary and topographic surveys; traffic and utility studies; availability of public and private utilities and services; legal description of the site; etc.) **(1.1.2.2; modifications)**

See the discussion of ¶ A.1.2 of Exhibit A to B101 in § 2.36.

1.1.3 The Owner's budget for the Cost of the Work, as defined in Section 6.1:

(Provide total and, if known, a line item breakdown.) **(1.1.2.5; substantial modifications)**

See the discussion of ¶ A.1.3 of Exhibit A to B101 in § **2.36.** The number here may or may not have been created with the help of the owner's Cost Consultant.

1.1.4 The Owner's anticipated design and construction schedule: **(new)**

This section is for the owner and architect to agree to a preliminary schedule for both the architect's services and the construction schedule. It is anticipated that these dates may change, since these dates will be filled in before the project is designed and before a contractor is hired. In addition, there may be many other reasons why these dates cannot be met that are not anticipated at the time this document is prepared. However, these dates do provide a framework for calculating the architect's fee, which is one of the primary reasons for including these dates. If the dates are changed, the architect may be entitled to additional compensation.

.1   Design phase milestone dates, if any:

Here, insert any milestone dates for the design phase of the project.

.2   Commencement of construction:

This is the date that construction is anticipated to begin.

.3   Substantial Completion date or milestone dates:

This is the anticipated date of substantial completion of the project.

.4   Other:

Here, any other milestone dates should be listed. Alternatively, the parties could reference a schedule that has already been developed.

1.1.5 The Owner intends the following procurement or delivery method for the Project:
(Identify method such as competitive bid, negotiated contract or construction management.) **(1.1.2.7; minor modifications)**

See the discussion of ¶ A.1.5 of Exhibit A to B101 in § **2.36.**

1.1.6 The Owner's requirements for accelerated or fast-track scheduling, multiple bid packages, or phased construction are set forth below:
(List number and type of bid/procurement packages.) **(new)**

If the project will utilize fast-track construction procedures[1] or other accelerated methods, requirements for such methods should be indicated here. In such procedures, the architect will be issuing several bid packages at different times, and the

---

[1] For cases involving fast-track construction procedures, *see, e.g.*, HOK Sport, Inc. v. FC Des Moines, 495 F.3d 927 (8th Cir. 2007; Steelcase, Inc., v. U.S., 165 F.3d 28 (Table), (6th Cir.

parties can set forth the anticipated schedule of these packages, and the number of distinct packages anticipated. Under these methods, the architect's fee will likely be substantially higher due to additional work involved in coordination issues, dealing with increased errors and omissions during the construction process, and the increased liability that results from such errors.

1.1.7 Other Project information:

(Identify special characteristics or needs of the Project not provided elsewhere, such as environmentally responsible design or historic preservation requirements.) **(1.1.2.8; modifications)**

See the discussion of ¶ A.1.6 of Exhibit A to B101 in § **2.36.**

1.1.8 The Owner identifies the following representative in accordance with Section 5.4:

(List name, address and other information.) **(1.1.3.1; minor modifications)**

See the discussion of ¶ A.2.1 of Exhibit A to B101 in § **2.36.**

1.1.9 The persons or entities, in addition to the Owner's representative, who are required to review the Architect's submittals to the Owner are as follows:

(List name, address and other information.) **(1.1.3.2; modifications)**

See the discussion of ¶ A.2.2 of Exhibit A to B101 in § **2.36.**

1.1.10 The Owner will retain the following consultants and contractors:

(List name, address and other information.) **(1.1.3.3; modifications)**

See the discussion of ¶ A.2.3 of Exhibit A to B101 in § **2.36.**

   .1   Cost Consultant:

   .2   Scheduling Consultant:

1998) ("The fast-track method enables the design and construction processes to run concurrently, with the design phase of the project always keeping just ahead of the construction phase."); Roberts & Schaefer Co., v. Hardaway Co., 152 F.3d 1283 (11th Cir. 1988) ("In the construction business, this means that construction commences under a schedule of simultaneous design, building, and construction. In other words, in this case, R & S was to begin construction before it completed the design and finalized a set of fully coordinated plans."); Marriott Corp., v. Dasta Const. Co., 26 F.3d 1057 (11th Cir. 1994) ("Under the fast track method, construction on a building begins before a final set of fully coordinated plans is completed. Rather, the architectural plans and specifications are designed and modified as the building's actual construction progresses. The advantage of the fast track method, as opposed to building from plans completed at the outset, is that it enables construction to begin at a much earlier stage in the project. The method's disadvantage results from increased difficulty in scheduling and coordinating the project since the construction progress schedule must be modified to account for constant plan changes.").

    .3 Geotechnical Engineer:

    .4 Civil Engineer:

    .5 Other, if any:

(List any other consultants or contractors retained by the Owner, such as a Project or Program Manager, construction contractor, or construction manager as constructor.) **(new)**

Here, the agreement will list various consultants to be provided by the owner. The ones that need to be included are the first two: Cost Consultant and Scheduling Consultant, otherwise the B101 is probably a more appropriate form. The other types of consultants may need to be retained by the owner, depending on the specific project.

1.1.11 *The Architect identifies the following representative in accordance with Section 2.3:*

(List name, address and other information.) **(1.1.3.4; modifications)**

See the discussion of ¶ A.2.4 of Exhibit A to B101 in § **2.36.**

1.1.12 *The Architect will retain the consultants identified in Sections 1.1.12.1 and 1.1.12.2:*

(List name, address and other information.) **(1.1.3.5; modifications)**

1.1.12.1 *Consultants retained under Basic Services:*

    .1 Structural Engineer:

    .2 Mechanical Engineer:

    .3 Electrical Engineer:

See the discussion of ¶ A.2.5.1 of Exhibit A to B101 in § **2.36.**

1.1.12.2 *Consultants retained under Additional Services:* **(new)**

See the discussion of ¶ A.2.5.2 of Exhibit A to B101 in § **2.36.**

1.1.13 *Other Initial Information on which the Agreement is based:* **(1.1.4; minor modification)**

See the discussion of ¶ A.2.6 of Exhibit A to B101 in § **2.36.**

1.2 *The Owner and Architect may rely on the Initial Information. Both parties, however, recognize that such information may materially change and, in that event, the Owner and the Architect shall appropriately adjust the schedule, the Architect's services and the Architect's compensation.* **(1.1.6; revisions)**

See the discussion of ¶ 1.3 of B101 in § **2.4.**

§ 6.5   Article 2: Architect's Responsibilities

ARTICLE 2 ARCHITECT'S RESPONSIBILITIES

2.1 The Architect shall provide the professional services as set forth in this Agreement. (**1.2.3.1; substantial revisions**)

See the discussion of ¶ 2.1 of B101 in **§ 2.5.**

2.2 The Architect shall perform its services consistent with the professional skill and care ordinarily provided by architects practicing in the same or similar locality under the same or similar circumstances. The Architect shall perform its services as expeditiously as is consistent with such professional skill and care and the orderly progress of the Project. (**1.2.3.2; substantial revisions**)

See the discussion of ¶ 2.2 of B101 in **§ 2.5.**

2.3 The Architect shall identify a representative authorized to act on behalf of the Architect with respect to the Project. (**1.2.3.3; substantial revisions**)

See the discussion of ¶ 2.3 of B101 in **§ 2.5.**

2.4 Except with the Owner's knowledge and consent, the Architect shall not engage in any activity, or accept any employment, interest or contribution that would reasonably appear to compromise the Architect's professional judgment with respect to this Project. (**1.2.3.5; no change**)

See the discussion of ¶ 2.4 of B101 in **§ 2.5.**

2.5 The Architect shall maintain the following insurance for the duration of this Agreement. If any of the requirements set forth below exceed the types and limits the Architect normally maintains, the Owner shall reimburse the Architect for any additional cost. (**new**)

In the following sections, the agreement will list the types of insurance and the limits for each that the architect will maintain for the "duration of the Agreement." Note that this is not the duration of the project. The question then becomes, just what is this duration? Certainly the duration lasts through Basic Services. The architect's Basic Services extend to the date of final completion, although they may be extended to 60 days after Substantial Completion as stated in ¶ 4.3.2.6 (beyond which the services would be Additional Services). Section 4.3.4 lists the anticipated length of the architect's services, although the architect's services may not actually be completed by that time and the architect will be entitled to additional compensation if the services extend past that date. Presumably, then, the duration of the agreement is the earlier of final completion by the contractor or termination of the agreement pursuant to one of the termination provisions.

If any of the following limits exceed the insurance normally carried by the architect, the owner is to reimburse the architect for the additional costs of the excess. For instance, if the architect normally carries $1 million in professional liability insurance and the owner wants $2 million, the owner will reimburse the architect for the additional cost of the excess $1 million at the time each premium becomes due until the termination of the agreement.

2.5.1 Comprehensive General Liability with policy limits of not less than___($___) for each occurrence and in the aggregate for bodily injury and property damage. (**new**)

This is general liability insurance that does not cover professional liability.

2.5.2 Automobile Liability covering owned and rented vehicles operated by the Architect with policy limits of not less than___($___) combined single limit and aggregate for bodily injury and property damage. (**new**)

This is standard automobile liability insurance.

2.5.3 The Architect may use umbrella or excess liability insurance to achieve the required coverage for Comprehensive General Liability and Automobile Liability, provided that such umbrella or excess insurance results in the same type of coverage as required for the individual policies. (**new**)

Sometimes, it is financially advantageous to obtain "umbrella" or "excess" insurance coverage instead of obtaining a higher limit on the primary insurance policies. For instance, the owner may want $5 million in general and automobile liability coverage. If the architect's general liability coverage is $1 million, the architect could increase the general liability policy to $5 million, or obtain umbrella or excess coverage of $4 million.

2.5.4 Workers' Compensation at statutory limits and Employers Liability with a policy limit of not less than___($___). (**new**)

2.5.5 Professional Liability covering the Architect's negligent acts, errors and omissions in its performance of professional services with policy limits of not less than___($___) per claim and in the aggregate. (**new**)

This will generally be the most costly insurance coverage. Increasing this coverage will be far more expensive than increasing general liability or automobile liability.

2.5.6 The Architect shall provide to the Owner certificates of insurance evidencing compliance with the requirements in this Section 2.5. The certificates will show the Owner as an additional insured on the Comprehensive General Liability, Automobile Liability, umbrella or excess policies. (**new**)

The owner can be listed as an additional insured on the specific types of insurance listed. Owners sometimes ask to be named as an additional insured on the architect's professional liability policy, but this is not an option, as the insurance carriers will refuse to permit this. The owner may want to look at a "wrap" policy or similar insurance that insures the entire project team and includes professional liability coverage. That type of insurance will be very costly.

§ 6.6   Article 3: Scope of Architect's Basic Services

ARTICLE 3 SCOPE OF ARCHITECT'S BASIC SERVICES

3.1 The Architect's Basic Services consist of those described in Article 3 and include usual and customary structural, mechanical, and electrical engineering services. Services not set forth in this Article 3 are Additional Services. (**2.4.1; modified**)

See the discussion of ¶ 3.1 of B101 in § **2.6.**

3.1.1 The Architect shall manage the Architect's services, consult with the Owner, research applicable design criteria, attend Project meetings, communicate with members of the Project team and report progress to the Owner. (**2.1.1; substantial revisions**)

See the discussion of ¶ 3.1.1 of B101 in § **2.6.**

3.1.2 The Architect shall coordinate its services with those services provided by the Owner and the Owner's consultants. The Architect shall be entitled to rely on the accuracy and completeness of services and information furnished by the Owner and the Owner's consultants. The Architect shall provide prompt written notice to the Owner if the Architect becomes aware of any error, omission or inconsistency in such services or information. (**1.2.3.2, 2.1.2; substantial revisions**)

See the discussion of ¶ 3.1.2 of B101 in § **2.6.**

3.1.3 As soon as practicable after the date of this Agreement, the Architect shall submit to the Owner and the Scheduling Consultant a schedule of the Architect's services for inclusion in the Project schedule. The schedule of the Architect's services shall include design milestone dates, anticipated dates when cost estimates or design reviews may occur, and allowances for periods of time required (1) for the Owner's review (2) for the performance of the Owner's consultants, and (3) for approval of submissions by authorities having jurisdiction over the Project. (**1.2.3.2, 2.1.2; substantial revisions**)

This paragraph assumes that the owner will be hiring a Scheduling Consultant who will take input from the owner and architect and, later, from the contractor to develop a schedule for the project. This schedule is intended to be continually modified to reflect changes to the project. The architect provides input in terms of design milestone dates, when cost estimating may occur by another of the owner's consultants, when design reviews can occur, and so forth.

3.1.4 Upon the Owner's reasonable request, the Architect shall submit information to the Scheduling Consultant and participate in developing and revising the Project schedule as it relates to the Architect's services. (**new**)

The architect is expected to furnish information to the Scheduling Consultant upon the owner's reasonable request. The architect should, however, reach an agreement with the owner as to what documents should be copied to this consultant on an ongoing basis. The consultant will likely also attend numerous meetings, or at least obtain copies of meeting minutes, in order to keep up to date on changes and events that may impact the project schedule.

3.1.5 Once the Owner and the Architect agree to the time limits established by the Project schedule, the Owner and the Architect shall not exceed them, except for reasonable cause. (**1.2.3.2; modifications**)

Once the Scheduling Consultant has prepared the Project schedule and the owner and architect agree to that schedule, the owner and architect will abide by the time limits set forth in the schedule.

3.1.6 The Architect shall not be responsible for an Owner's directive or substitution made without the Architect's approval. (**new**)

See the discussion of ¶ 3.1.4 of B101 in § **2.6.**

3.1.7 The Architect shall, at appropriate times, contact the governmental authorities required to approve the Construction Documents and the entities providing utility services to the Project. In designing the Project, the Architect shall respond to applicable design requirements imposed by such governmental authorities and by such entities providing utility services. (**1.2.3.6; substantial revisions; added reference to utilities**)

See the discussion of ¶ 3.1.5 of B101 in § **2.6.**

3.1.8 The Architect shall assist the Owner in connection with the Owner's responsibility for filing documents required for the approval of governmental authorities having jurisdiction over the Project. (**2.1.6; minor modification**)

See the discussion of ¶ 3.1.6 of B101 in § **2.6.**

§ 6.7   Schematic Design Phase Services: ¶ 3.2

3.2 SCHEMATIC DESIGN PHASE SERVICES

This phase is actually composed of several sub-phases. First, there is an evaluation of all of the information provided by the owner and obtained by the architect. Next, the architect presents the owner with the results of this evaluation and an agreement as to the requirements of the project is reached. Then, the architect prepares a preliminary design for the owner's approval. Finally, a schematic design is prepared.

Generally, schematic design involves the first sketches of the project in which the design concept is formulated, site conditions are investigated, and other elements affecting the project are initially examined. This phase should end with the owner's approving a set of schematic design drawings and other documents and the architect's proceeding to the design development phase.

3.2.1 The Architect shall review the program and other information furnished by the Owner, and shall review laws, codes, and regulations applicable to the Architect's services. **(1.2.3.6; substantial revisions)**

See the discussion of ¶ 3.2.1 of B101 in **§ 2.7.**

3.2.2 The Architect shall prepare a preliminary evaluation of the Owner's program, schedule, budget for the Cost of the Work, Project site, and the proposed procurement or delivery method and other Initial Information, each in terms of the other, to ascertain the requirements of the Project. The Architect shall notify the Owner of (1) any inconsistencies discovered in the information, and (2) other information or consulting services that may be reasonably needed for the Project. **(2.3.1, 2.3.2, 2.3.3; modifications)**

See the discussion of ¶ 3.2.2 of B101 in **§ 2.7.**

3.2.3 The Architect shall present its preliminary evaluation to the Owner and shall discuss with the Owner alternative approaches to design and construction of the Project, including the feasibility of incorporating environmentally responsible design approaches. The Architect shall reach an understanding with the Owner regarding the requirements of the Project. **(new)**

See the discussion of ¶ 3.2.3 of B101 in **§ 2.7.**

3.2.4 Based on the Project requirements agreed upon with the Owner, the Architect shall prepare and present for the Owner's approval a preliminary design illustrating the scale and relationship of the Project components. **(2.4.2; substantial modifications)**

See the discussion of ¶ 3.2.4 of B101 in **§ 2.7.**

3.2.5 Based on the Owner's approval of the preliminary design, the Architect shall prepare Schematic Design Documents for the Owner's approval. The Schematic Design Documents shall consist of drawings and other documents including a site plan, if appropriate, and preliminary building plans, sections and elevations; and may include some combination of study models, perspective sketches, or digital modeling. Preliminary selections of major building systems and construction materials shall be noted on the drawings or described in writing. **(2.4.2.1; substantial modifications)**

See the discussion of ¶ 3.2.5 of B101 in **§ 2.7.**

3.2.5.1 The Architect shall consider environmentally responsible design alternatives, such as material choices and building orientation, together with other considerations based on program and aesthetics, in developing a design that is consistent with the Owner's program, schedule and budget for the Cost of the Work. The Owner may obtain other environmentally responsible design services under Article 4. **(new)**

See the discussion of ¶ 3.2.5.1 of B101 in **§ 2.7.**

3.2.5.2 The Architect shall consider the value of alternative materials, building systems and equipment, together with other considerations based on program and aesthetics in developing a design for the Project that is consistent with the Owner's schedule and budget for the Cost of the Work. **(2.1.3; revised)**

See the discussion of ¶ 3.1.6 of B101 in **§ 2.7.** This paragraph omits the word "program" (prior to "schedule") from the list of items to be considered by the architect in designing the project. This is probably an oversight, as the word is included in B101, and is stated earlier in the paragraph.

3.2.6 The Architect shall submit the Schematic Design Documents to the Owner and the Cost Consultant. The Architect shall meet with the Cost Consultant to review the Schematic Design Documents. **(new)**

The design is forwarded to both the owner and the Cost Consultant at this stage, and the architect will again meet with the Cost Consultant to review this design and answer questions.

3.2.7 Upon receipt of the Cost Consultant's estimate at the conclusion of the Schematic Design Phase, the Architect shall take action as required under Section 6.4, and request the Owner's approval of the Schematic Design Documents. If revisions to the Schematic Design Documents are required to comply with the Owner's budget for the Cost of the Work at the conclusion of the Schematic Design Phase, the Architect shall incorporate the required revisions in the Design Development Phase. **(new)**

After the Cost Consultant reviews the architect's schematic design drawings, a new estimate for the Cost of the Work will be prepared and forwarded to the architect. Section 6.4 addresses the likely scenario where this new estimate is greater than the owner's budget. The architect will work with the Cost Consultant and the owner to alter the project to bring the cost back in line with the budget, or, if the owner chooses, the budget can be changed so that the design is maintained. The architect will not be entitled to additional fees to change the drawings unless the owner has directed changes that resulted in the higher cost. The architect should have alerted both the owner and the Cost Consultant at the time that the owner's directive was received.

§ 6.8   Design Development Phase Services: ¶ 3.3

3.3 DESIGN DEVELOPMENT PHASE SERVICES

Once the owner has approved the schematic design phase, the project moves into the design development phase. Sometimes, these phases overlap, particularly when there are changes to the project. If Section 1.1.6 indicates that the project will be "fast-track" or other accelerated project type, these phases will certainly overlap to a great degree as the architect will be issuing drawings to the contractor in a piecemeal fashion, with the design for one element not starting until after construction documents are issued for a different element. The possibility of errors or omissions is greatly increased to virtual certainty.

3.3.1 Based on the Owner's approval of the Schematic Design Documents, and on the Owner's authorization of any adjustments in the Project requirements and the budget for the Cost of the Work pursuant to Section 5.3, the Architect shall prepare Design Development Documents for the Owner's approval. The Design Development Documents shall illustrate and describe the development of the approved Schematic Design Documents and shall consist of drawings and other documents including plans, sections, elevations, typical construction details, and diagrammatic layouts of building systems to fix and describe the size and character of the Project as to architectural, structural, mechanical and electrical systems, and such other elements as may be appropriate. The Design Development Documents shall also include outline specifications that identify major materials and systems and establish in general their quality levels. **(2.4.3.1; substantial revisions)**

See the discussion of ¶ 3.3.1 of B101 in **§ 2.8.** The reference to Section 5.3 is not included in B101.

3.3.2 Prior to the conclusion of the Design Development Phase, the Architect shall submit the Design Development documents to the Owner and the Cost Consultant. The Architect shall meet with the Cost Consultant to review the Design Development Documents. **(new)**

The design is forwarded to both the owner and the Cost Consultant at this stage, and the architect will again meet with the Cost Consultant to review this design and answer questions.

3.3.3 Upon receipt of the Cost Consultant's estimate at the conclusion of the Design Development Phase, the Architect shall take action as required under Sections 6.5 and 6.6 and request the Owner's approval of the Design Development Documents. **(new)**

Section 6.5 gives the owner three options if the Cost Consultant determines that the final design development drawings result in a cost greater than the current budget: (1) increase the budget; (2) change the drawings to bring the cost down; or (3) another alternative. If the owner decides to have the architect redraw the drawings to bring the cost down, the architect is required to do this at no additional cost to the owner, but that is the limit of the architect's liability to the owner. Of course, this assumes that the owner did not cause this increase by, for instance, increasing the size or complexity of the project.

§ 6.9   Construction Documents Phase Services: ¶ 3.4

3.4 CONSTRUCTION DOCUMENTS PHASE SERVICES

This is the phase in which the design is translated into the final "blueprints," although real blueprints seldom are used anymore. Drawings and specifications are coordinated to assure that all the various elements, including structural, electrical, and mechanical, fit together in a logical and economical fashion to achieve the design purpose of the project. At the end of this phase, a set of documents, including drawings, specifications, and other documents, is produced from which a general contractor can construct the project (subject to required submissions by the contractor, such as shop drawings).

In a suit to recover his fees, an architect was not required to prove that he delivered the final plans to the owner, because the contract did not require such delivery.[2] Actions by architects to recover fees are rather common. Of course, the architect should keep the owner advised of the progress of the documents and should send check sets to the owner, retaining copies of transmittals in case proof is required later. This will aid in preventing an owner from later claiming that it was unaware of any progress on the project. A difficult situation arises when the owner balks at paying for the cost of printing these check sets, which can be quite expensive. One answer might be to have periodic meetings in the architect's office to review the original documents, then send minutes of the meetings to all parties.

Regular meetings of the project team are important. On a large project, they are vital. Careful minutes, detailing all decisions and the reasons behind these decisions, can come in quite handy at the litigation phase of the project. It also helps the team preserve a businesslike attitude toward the project. Another valuable tool is the telephone memo. Each call of any importance is transcribed immediately afterward. These can be distributed or merely filed with the project file. Many important decisions are made as a result of telephone calls to the owner, contractor, distributer, or manufacturer. It is almost impossible to later remember that the conversation took place at all, much less what was said.

At least one court has found that an architect could be liable to a contractor for negligent misrepresentation.[3] The architect was aware that certain subsurface debris was located on a site but failed to disclose this on his drawings. In reliance on the drawings, the contractor submitted a low bid. After the debris was discovered by the contractor, additional work was required to remove it, causing extra expense to the contractor.

---

[2] Michalowski v. Richter Spring Corp., 112 Ill. App. 2d 451, 251 N.E.2d 299 (1969).

[3] Gulf Contracting v. Bibb County, 795 F.2d 980 (11th Cir. 1986). Lack of privity did not deter the court. Privity is not required to support an action for negligent misrepresentation, although liability is limited to a foreseeable person for whom the information was intended, such as the contractor. A number of cases have stated that the architect can be liable to a contractor for negligence. *See, e.g.*, Huber, Hunt & Nichols, Inc. v. Moore, 67 Cal. App. 3d 278, 136 Cal. Rptr. 603 (1977).

In Clevecon, Inc. v. Northeast Ohio Reg'l Sewer Dist., 90 Ohio App. 3d 215, 628 N.E.2d 143 (1993), the general contractor sued the architect for malpractice to recover delay damages. The jury found the architect negligent in the preparation and drafting of the plans and specifications and in the administration of the project. It also found that the architect had breached warranties in the preparation and drafting of the plans and specifications. In affirming the trial court's judgment, the appellate court found that the architect had exercised a substantial amount of control over the project, giving orders to the contractor during the construction.

3.4.1 Based on the Owner's approval of the Design Development Documents, and on the Owner's authorization of any adjustments in the Project requirements and the budget for the Cost of the Work, the Architect shall prepare Construction Documents for the Owner's approval. The Construction Documents shall illustrate and describe the further development of the approved Design Development Documents and shall consist of Drawings and Specifications setting forth in detail the quality levels of materials and systems and other requirements for the construction of the Work. The Owner and Architect acknowledge that in order to construct the Work the Contractor will provide additional information, including Shop Drawings, Product Data, Samples and other similar submittals, which the Architect shall review in accordance with Section 3.6.4. **(2.4.4.1; substantial revisions)**

See the discussion of ¶ 3.4.1 of B101 in **§ 2.9.**

3.4.2 The Architect shall incorporate into the Construction Documents the design requirements of governmental authorities having jurisdiction over the Project. **(1.2.3.6; revisions)**

See the discussion of ¶ 3.4.2 of B101 in **§ 2.9.**

3.4.3 During the development of the Construction Documents, the Architect shall assist the Owner in the development and preparation of (1) bidding and procurement information that describes the time, place and conditions of bidding, including bidding or proposal forms; (2) the form of agreement between the Owner and Contractor; and (3) the Conditions of the Contract for Construction (General, Supplementary and other Conditions). The Architect shall also compile a project manual that includes the Conditions of the Contract for Construction and Specifications and may include bidding requirements and sample forms. **(2.4.4.2; revisions)**

See the discussion of ¶ 3.4.3 of B101 in **§ 2.9.**

3.4.4 Prior to the conclusion of the Construction Documents Phase, the Architect shall submit the Construction Documents to the Owner and the Cost Consultant. The Architect shall meet with the Cost Consultant to review the Construction Documents. **(new)**

Again, the architect will forward the construction documents to both the owner and the Cost Consultant at this stage, and the architect will meet with the Cost Consultant to review this design and answer questions. Typically, there are several submissions of the construction documents to the owner and Cost Consultant at various stages of development of these drawings during this phase. This allows for adjustments to the project within this phase.

3.4.5 Upon receipt of the Cost Consultant's estimate at the conclusion of the Construction Documents Phase, the Architect shall take action as required under Section 6.7 and request the Owner's approval of the Construction Documents. **(new)**

If the Cost Consultant determines that the cost exceeds the current budget, the owner can direct the architect to make further revisions, but the architect will be compensated for this work as an additional service unless the architect made modifications to the drawings without authorization. The reason for the additional fee to the architect is that such an increase in cost would not be the architect's fault if the architect merely took the last approved drawings from the design development set and finalized them. Presumably, the fault would lie with the Cost Consultant or there might have been a substantial unforeseen increase in costs of labor or materials.

§ 6.10   Bidding or Negotiation Phase Services: ¶ 3.5

3.5 BIDDING OR NEGOTIATION PHASE SERVICES

3.5.1 GENERAL

The Architect shall assist the Owner in establishing a list of prospective contractors. Following the Owner's approval of the Construction Documents, the Architect shall assist the Owner in (1) obtaining either competitive bids or negotiated proposals; (2) confirming responsiveness of bids or proposals; (3) determining the successful bid or proposal, if any; and, (4) awarding and preparing contracts for construction. **(2.5.1, 2.5.2, 2.5.3; substantial revisions)**

See the discussion of ¶ 3.5.1 of B101 in **§ 2.10.**

3.5.2 COMPETITIVE BIDDING

3.5.2.1 Bidding Documents shall consist of bidding requirements and proposed Contract Documents. **(2.5.4.1; minor revisions)**

See the discussion of ¶ 3.5.2.1 of B101 in **§ 2.11.**

3.5.2.2 The Architect shall assist the Owner in bidding the Project by

> *.1   facilitating the reproduction of Bidding Documents for distribution to prospective bidders,*

This is frequently handled by the architect. The owner pays for the costs of reproduction, messenger services, and so on. As part of the basic services, the architect "facilitates" the distribution of the bidding documents, answers questions, and otherwise assists in the bidding process. This document assumes that the owner and the owner's other consultants will take a larger role in this process than under B101.

> *.2   participating in a pre-bid conference for prospective bidders, and*

Pre-bid conferences are extremely valuable to assure that all bidders have the same understanding of the project. They give each bidder an opportunity to ask questions of the architect and the consultants (who should also attend this meeting). The meeting is usually conducted at the job site. Here, the architect is "participating" in this conference, which is organized by other consultants to the owner, unlike under B101, where the architect organizes this conference.

> *.3   preparing responses to questions from prospective bidders and providing clarifications and interpretations of the Bidding Documents in the form of addenda.*

During the bidding process, the bidders often find discrepancies in the drawings and specifications. These are then called to the architect's attention (**see** AIA Document A201, ¶ 3.2.3), and the architect must then issue an addendum that clarifies or corrects the problem. If the problem is a major one, the bidding period may be extended to allow for reissue of the documents.

3.5.2.3 *The Architect shall consider requests for substitutions, if the Bidding Documents permit substitutions, and shall prepare and distribute addenda identifying approved substitutions to all prospective bidders.* (**2.5.4.4; minor modifications**)

See the discussion of ¶ 3.5.2.3 of B101 in **§ 2.11.**

3.5.3 NEGOTIATED PROPOSALS

Instead of bidding the project, the owner may elect to negotiate directly with one or more pre-selected general contractors. This section deals with that alternative. Note that there are a number of subtle differences between the corresponding provisions of this document, B103, and B101, the more traditional arrangement.

3.5.3.1 *Proposal Documents shall consist of proposal requirements, and proposed Contract Documents.*

See the discussion of ¶ 3.5.3.1 of B101 in **§ 2.12.**

3.5.3.2 *The Architect shall assist the Owner in obtaining proposals by*

> *.1   facilitating the reproduction of Proposal Documents for distribution to prospective contractors, and requesting their return upon completion of the negotiation process;*

Here, the architect "facilitates" the reproduction of the documents, as opposed to B101, where the architect "procures." The responsibility is thus on the owner to procure these documents.

> *.2   participating in selection interviews with prospective contractors; and*

Here, the architect merely "participates," while under B101, the architect also "organizes" this process.

> *.3   participating in negotiations with prospective contractors.* (**2.5.5; modifications**)

Again, under this document, the architect "participates" while under B101, the architect also prepares a report of the results of the negotiations.

3.5.3.3 The Architect shall consider requests for substitutions, if the Proposal Documents permit substitutions, and shall prepare and distribute addenda identifying approved substitutions to all prospective contractors. **(2.5.5.4; minor modification)**

See the discussion of ¶ 3.5.3.3 of B101 in **§ 2.12.**

§ 6.11   Construction Phase Services: ¶ 3.6

3.6 CONSTRUCTION PHASE SERVICES

3.6.1 GENERAL

3.6.1.1 The Architect shall provide administration of the Contract between the Owner and the Contractor as set forth below and in AIA Document A201™–2007, General Conditions of the Contract for Construction. If the Owner and Contractor modify AIA Document A201–2007, those modifications shall not affect the Architect's services under this Agreement unless the Owner and the Architect amend this Agreement. **(2.6.1.1; substantial modifications)**

See the discussion of ¶ 3.6.1.1 of B101 in **§ 2.13.**

3.6.1.2 The Architect shall advise and consult with the Owner during the Construction Phase Services. The Architect shall have authority to act on behalf of the Owner only to the extent provided in this Agreement. The Architect shall not have control over, charge of, or responsibility for the construction means, methods, techniques, sequences or procedures, or for safety precautions and programs in connection with the Work, nor shall the Architect be responsible for the Contractor's failure to perform the Work in accordance with the requirements of the Contract Documents. The Architect shall be responsible for the Architect's negligent acts or omissions, but shall not have control over or charge of, and shall not be responsible for, acts or omissions of the Contractor or of any other persons or entities performing portions of the Work. **(2.6.1.3, 2.6.2.1; substantial modifications)**

See the discussion of ¶ 3.6.1.2 of B101 in **§ 2.13.**

3.6.1.3 Subject to Section 4.3, the Architect's responsibility to provide Construction Phase Services commences with the award of the Contract for Construction and terminates on the date the Architect issues the final Certificate for Payment. **(2.6.1.2; modifications; deleted reference to entitlement to additional fees after 60 days after Substantial Completion)**

See the discussion of ¶ 3.6.1.3 of B101 in **§ 2.13.**

§ 6.12   Evaluations of the Work: ¶ 3.6.2

3.6.2 EVALUATIONS OF THE WORK

3.6.2.1 The Architect shall visit the site at intervals appropriate to the stage of construction, or as otherwise required in Section 4.3.3, to become generally familiar with the progress and quality of the portion of the Work completed, and to determine, in general, if the Work observed is being performed in a manner indicating that the Work, when fully completed, will be in accordance with the Contract Documents. However, the Architect shall not be required to make exhaustive or continuous on-site inspections to check the quality or quantity of the Work. On the basis of the site visits, the Architect shall keep the Owner reasonably informed about the progress and quality of the portion of the Work completed, and report to the Owner (1) known deviations from the Contract Documents and from the most recent construction schedule, and (2) defects and deficiencies observed in the Work. **(2.6.2.1, 2.6.2.2; substantial modifications)**

See the discussion of ¶ 3.6.2.1 of B101 in **§ 2.14.** This paragraph omits the reference to the construction schedule "submitted by the Contractor," since the owner's scheduling consultant will be preparing schedules. The documents are not clear as to the degree the owner's scheduling consultant has duties to advise the owner and architect as to deviations by the contractor from the latest approved schedule. Presumably, with the presence of such a consultant, the architect's duty to monitor adherence to such a schedule is lessened.

3.6.2.2 The Architect has the authority to reject Work that does not conform to the Contract Documents. Whenever the Architect considers it necessary or advisable, the Architect shall have the authority to require inspection or testing of the Work in accordance with the provisions of the Contract Documents, whether or not such Work is fabricated, installed or completed. However, neither this authority of the Architect nor a decision made in good faith either to exercise or not to exercise such authority shall give rise to a duty or responsibility of the Architect to the Contractor, Subcontractors, material and equipment suppliers, their agents or employees or other persons or entities performing portions of the Work. **(2.6.2.5; minor modification)**

See the discussion of ¶ 3.6.2.2 of B101 in **§ 2.14.**

3.6.2.3 The Architect shall interpret and decide matters concerning performance under, and requirements of, the Contract Documents on written request of either the Owner or Contractor. The Architect's response to such requests shall be made in writing within any time limits agreed upon or otherwise with reasonable promptness. **(2.6.1.7; minor modification)**

See the discussion of ¶ 3.6.2.3 of B101 in **§ 2.14.**

3.6.2.4 Interpretations and decisions of the Architect shall be consistent with the intent of and reasonably inferable from the Contract Documents and shall be in writing or in the form of drawings. When making such interpretations and decisions, the Architect shall endeavor to secure faithful performance by both Owner and Contractor, shall not show partiality to either, and shall not be liable for results of interpretations or decisions rendered in good faith. The Architect's decisions on matters relating to aesthetic effect shall be final if consistent with the intent expressed in the Contract Documents. **(2.6.1.8, 2.6.1.9; minor modifications)**

See the discussion of ¶ 3.6.2.4 of B101 in **§ 2.14.**

3.6.2.5 Unless the Owner and Contractor designate another person to serve as an Initial Decision Maker, as that term is defined in AIA Document A201–2007, the Architect shall render initial decisions on Claims between the Owner and Contractor as provided in the Contract Documents. **(new as to Initial Decision Maker, 2.6.1.9)**

See the discussion of ¶ 3.6.2.5 of B101 in **§ 2.14.**

§ 6.13   Certificates for Payment to Contractor: ¶ 3.6.3

3.6.3 CERTIFICATES FOR PAYMENT TO CONTRACTOR

3.6.3.1 The Architect shall review and certify the amounts due the Contractor and shall issue certificates in such amounts. The Architect's certification for payment shall constitute a representation to the Owner, based on the Architect's evaluation of the Work as provided in Section 3.6.2 and on the data comprising the Contractor's Application for Payment, that, to the best of the Architect's knowledge, information and belief, the Work has progressed to the point indicated and that the quality of the Work is in accordance with the Contract Documents. The foregoing representations are subject (1) to an evaluation of the Work for conformance with the Contract Documents upon Substantial Completion, (2) to results of subsequent tests and inspections, (3) to correction of minor deviations from the Contract Documents prior to completion, and (4) to specific qualifications expressed by the Architect. **(2.6.3.1; modified)**

See the discussion of ¶ 3.6.3.1 of B101 in **§ 2.15.**

3.6.3.2 The issuance of a Certificate for Payment shall not be a representation that the Architect has (1) made exhaustive or continuous on-site inspections to check the quality or quantity of the Work, (2) reviewed construction means, methods, techniques, sequences or procedures, (3) reviewed copies of requisitions received from Subcontractors and material suppliers and other data requested by the Owner to substantiate the Contractor's right to payment, or (4) ascertained how or for what purpose the Contractor has used money previously paid on account of the Contract Sum. **(2.6.3.2; no change)**

See the discussion of ¶ 3.6.3.2 of B101 in **§ 2.15.**

3.6.3.3 The Architect shall maintain a record of the Applications and Certificates for Payment. **(2.6.3.3; minor modifications)**

See the discussion of ¶ 3.6.3.3 of B101 in **§ 2.15.**

§ 6.14   Submittals: ¶ 3.6.4

3.6.4 SUBMITTALS

3.6.4.1 The Architect shall review the Contractor's submittal schedule and shall not unreasonably delay or withhold approval. The Architect's action in reviewing submittals shall be taken in accordance with the approved submittal schedule or, in the absence of an approved submittal schedule, with reasonable promptness while allowing sufficient time in the Architect's professional judgment to permit adequate review. **(2.6.4.1; substantial modifications)**

See the discussion of ¶ 3.6.4.1 of B101 in **§ 2.16.**

3.6.4.2 In accordance with the Architect-approved submittal schedule, the Architect shall review and approve or take other appropriate action upon the Contractor's submittals such as Shop Drawings, Product Data and Samples, but only for the limited purpose of checking for conformance with information given and the design concept expressed in the Contract Documents. Review of such submittals is not for the purpose of determining the accuracy and completeness of other information such as dimensions, quantities, and installation or performance of equipment or systems, which are the Contractor's responsibility. The Architect's review shall not constitute approval of safety precautions or, unless otherwise specifically stated by the Architect, of any construction means, methods, techniques, sequences or procedures. The Architect's approval of a specific item shall not indicate approval of an assembly of which the item is a component. **(2.6.4.1; substantial modifications)**

See the discussion of ¶ 3.6.4.2 of B101 in **§ 2.16.**

3.6.4.3 If the Contract Documents specifically require the Contractor to provide professional design services or certifications by a design professional related to systems, materials or equipment, the Architect shall specify the appropriate performance and design criteria that such services must satisfy. The Architect shall review shop drawings and other submittals related to the Work designed or certified by the design professional retained by the Contractor that bear such professional's seal and signature when submitted to the Architect. The Architect shall be entitled to rely upon the adequacy, accuracy and completeness of the services, certifications and approvals performed or provided by such design professionals. **(2.6.4.3; minor modifications)**

See the discussion of ¶ 3.6.4.3 of B101 in **§ 2.16.**

3.6.4.4 Subject to the provisions of Section 4.3, the Architect shall review and respond to requests for information about the Contract Documents. The Architect shall set forth in the Contract Documents the requirements for requests for information. Requests for information shall include, at a minimum, a detailed written statement that indicates the specific Drawings or Specifications in need of clarification and the nature of the clarification requested. The Architect's response to such requests shall be made in writing within any time limits agreed upon, or otherwise with reasonable promptness. If appropriate, the Architect shall prepare and issue supplemental Drawings and Specifications in response to requests for information. **(2.6.1.5, 2.6.1.6; modified)**

See the discussion of ¶ 3.6.4.4 of B101 in **§ 2.16.**

3.6.4.5 The Architect shall maintain a record of submittals and copies of submittals supplied by the Contractor in accordance with the requirements of the Contract Documents. **(2.6.4.2; no change)**

See the discussion of ¶ 3.6.4.5 of B101 in **§ 2.16.**

§ 6.15   Changes in the Work: ¶ 3.6.5

3.6.5 CHANGES IN THE WORK

3.6.5.1 The Architect may authorize minor changes in the Work that are consistent with the intent of the Contract Documents and do not involve an adjustment in the Contract Sum or an extension of the Contract Time. Subject to the provisions of Section 4.3, the Architect shall prepare Change Orders and Construction Change Directives for the Owner's approval and execution in accordance with the Contract Documents. **(2.6.5.1; substantial modifications)**

See the discussion of ¶ 3.6.5.1 of B101 in **§ 2.16.**

3.6.5.2 The Architect shall maintain records relative to changes in the Work. **(2.6.5.4; no change)**

See the discussion of ¶ 3.6.5.2 of B101 in **§ 2.16.**

§ 6.16   Project Completion: ¶ 3.6.6

3.6.6 PROJECT COMPLETION

3.6.6.1 The Architect shall conduct inspections to determine the date or dates of Substantial Completion and the date of final completion; issue Certificates of Substantial Completion; receive from the Contractor and forward to the Owner, for the Owner's review and records, written warranties and related documents required by the Contract Documents and assembled by the Contractor; and issue a final Certificate for Payment based upon a final inspection indicating the Work complies with the requirements of the Contract Documents. **(2.6.6.1; minor modifications)**

See the discussion of ¶ 3.6.6.1 of B101 in § **2.18.**

***3.6.6.2** The Architect's inspections shall be conducted with the Owner to check conformance of the Work with the requirements of the Contract Documents and to verify the accuracy and completeness of the list submitted by the Contractor of Work to be completed or corrected.* (**2.6.6.2; minor modification**)

See the discussion of ¶ 3.6.6.2 of B101 in § **2.18.**

***3.6.6.3** When the Work is found to be substantially complete, the Architect shall inform the Owner about the balance of the Contract Sum remaining to be paid the Contractor, including the amount to be retained from the Contract Sum, if any, for final completion or correction of the Work.* (**2.6.6.3; minor modifications**)

See the discussion of ¶ 3.6.6.3 of B101 in § **2.18.**

***3.6.6.4** The Architect shall forward to the Owner the following information received from the Contractor: (1) consent of surety or sureties, if any, to reduction in or partial release of retainage or the making of final payment; (2) affidavits, receipts, releases and waivers of liens or bonds indemnifying the Owner against liens; and (3) any other documentation required of the Contractor under the Contract Documents.* (**2.6.6.4; added (3); minor modifications**)

See the discussion of ¶ 3.6.6.4 of B101 in § **2.18.**

***3.6.6.5** Upon request of the Owner, and prior to the expiration of one year from the date of Substantial Completion, the Architect shall, without additional compensation, conduct a meeting with the Owner to review the facility operations and performance.* (**2.7.2; minor modifications**)

See the discussion of ¶ 3.6.6.5 of B101 in § **2.18.**

§ 6.17   Article 4: Additional Services

ARTICLE 4 ADDITIONAL SERVICES

***4.1** Additional Services listed below are not included in Basic Services but may be required for the Project. The Architect shall provide the listed Additional Services only if specifically designated in the table below as the Architect's responsibility, and the Owner shall compensate the Architect as provided in Section 11.2.*

(Designate the Additional Services the Architect shall provide in the second column of the table below. In the third column indicate whether the service description is located in Section 4.2 or in an attached exhibit. If in an exhibit, identify the exhibit.) (**2.8.3; substantial revisions**)

See the discussion of ¶ 4.1 of B101 in § **2.19.**

Services	Responsibility (*Architect, Owner or Not Provided*)	Location of Service Description (*Section 4.2 below or in an exhibit attached to this document and identified below*)
4.1.1 Programming		
4.1.2 Multiple preliminary designs		
4.1.3 Measured drawings		
4.1.4 Existing facilities surveys		
4.1.5 Site Evaluation and Planning (B203™–2007)		
4.1.6 Building information modeling		
4.1.7 Civil engineering		
4.1.8 Landscape design		
4.1.9 Architectural Interior Design (B252™–2007)		
4.1.10 Value Analysis (B204™–2007)		
4.1.11 Detailed cost estimating		
4.1.12 On-site project representation		
4.1.13 Conformed construction documents		
4.1.14 As-designed record drawings		
4.1.15 As-constructed record drawings		
4.1.16 Post occupancy evaluation		
4.1.17 Facility Support Services (B210™–2007)		
4.1.18 Tenant-related services		
4.1.19 Coordination of Owner's consultants		
4.1.20 Telecommunications/data design		
4.1.21 Security Evaluation and Planning (B206™–2007)		
4.1.22 Commissioning (B211™–2007)		
4.1.23 Extensive environmentally responsible design		

Services	Responsibility *(Architect, Owner or Not Provided)*	Location of Service Description *(Section 4.2 below or in an exhibit attached to this document and identified below)*
4.1.24 LEED® Certification (B214™–2007)		
4.1.25 Historic Preservation (B205™–2007)		
4.1.26 Furniture, Finishings, and Equipment Design (B253™–2007)		

See the discussion of this table in B101 in **§ 2.19.** This table omits line 4.25 of B101, Fast-track design services, as such services are contemplated to be normal in a complex project for which this document is designed. Section 1.1.6 calls for stating the owner's requirements for accelerated or fast-track scheduling and the like.

4.2 Insert a description of each Additional Service designated in Section 4.1 as the Architect's responsibility, if not further described in an exhibit attached to this document. **(new)**

See the discussion of ¶ 4.2 of B101 in **§ 2.20.**

4.3 Additional Services may be provided after execution of this Agreement, without invalidating the Agreement. Except for services required due to the fault of the Architect, any Additional Services provided in accordance with this Section 4.3 shall entitle the Architect to compensation pursuant to Section 11.3 and an appropriate adjustment in the Architect's schedule. **(1.3.3.1; extensive modifications)**

See the discussion of ¶ 4.3 of B101 in **§ 2.20.**

4.3.1 Upon recognizing the need to perform the following Additional Services, the Architect shall notify the Owner with reasonable promptness and explain the facts and circumstances giving rise to the need. The Architect shall not proceed to provide the following services until the Architect receives the Owner's written authorization: **(1.3.3.2, 2.8.2; substantial revisions)**

See the discussion of ¶ 4.3.1 of B101 in **§ 2.20.**

    .1 Services necessitated by a change in the Initial Information, previous instructions or approvals given by the Owner, or a material change in the Project including, but not limited to, size, quality, complexity, the Owner's schedule or budget for Cost of the Work, or procurement or delivery method, or bid packages in addition to those listed in Section 1.1.6; **(1.3.3.2.1, 1.3.3.2.4; modifications)**

See the discussion of ¶ 4.3.1.1 of B101 in § **2.20.** The last clause, related to bid packages in addition to those listed in Section 1.1.6, is not included in B101. If the owner decides to have the architect prepare more bid packages than are listed in Section 1.1.6, the architect will be entitled to additional fees. If no number of bid packages is listed in that section, one package is assumed. Alternatively, if a fast-track process is listed, and the owner changes that procedure, the architect may be entitled to additional fees.

> **.2** *Services necessitated by the Owner's request for extensive environmentally responsible design alternatives, such as unique system designs, in-depth material research, energy modeling, or LEED® certification;* **(new)**

See the discussion of ¶ 4.3.1.2 of B101 in § **2.20.**

> **.3** *Changing or editing previously prepared Instruments of Service necessitated by the enactment or revision of codes, laws or regulations or official interpretations;* **(new)**

See the discussion of ¶ 4.3.1.3 of B101 in § **2.20.**

> **.4** *Services necessitated by decisions of the Owner not rendered in a timely manner or any other failure of performance on the part of the Owner or the Owner's consultants or contractors;* **(1.3.3.2.3, 1.3.3.2.5; modifications)**

See the discussion of ¶ 4.3.1.4 of B101 in § **2.20.**

> **.5** *Preparing digital data for transmission to the Owner's consultants and contractors, or to other Owner authorized recipients;* **(new)**

See the discussion of ¶ 4.3.1.5 of B101 in § **2.20.**

> **.6** *Preparation of design and documentation for alternate bid or proposal requests proposed by the Owner;* **(new)**

See the discussion of ¶ 4.3.1.6 of B101 in § **2.20.**

> **.7** *Preparation for, and attendance at, a public presentation, meeting or hearing;* **(1.3.3.2.6; minor modifications)**

See the discussion of ¶ 4.3.1.7 of B101 in § **2.20.**

> **.8** *Preparation for, and attendance at a dispute resolution proceeding or legal proceeding, except where the Architect is party thereto;* **(1.3.3.2.6; minor modifications)**

See the discussion of ¶ 4.3.1.8 of B101 in § **2.20.**

> **.9** *Evaluation of the qualifications of bidders or persons providing proposals;* **(new)**

See the discussion of ¶ 4.3.1.9 of B101 in **§ 2.20.**

> *.10 Consultation concerning replacement of Work resulting from fire or other cause during construction; or* **(2.8.2.4; modifications)**

See the discussion of ¶ 4.3.1.10 of B101 in **§ 2.20.**

> *.11 Assistance to the Initial Decision Maker, if other than the Architect.* **(new)**

See the discussion of ¶ 4.3.1.11 of B101 in **§ 2.20.**

4.3.2 *To avoid delay in the Construction Phase, the Architect shall provide the following Additional Services, notify the Owner with reasonable promptness, and explain the facts and circumstances giving rise to the need. If the Owner subsequently determines that all or parts of those services are not required, the Owner shall give prompt written notice to the Architect, and the Owner shall have no further obligation to compensate the Architect for those services:* **(2.8.2; modifications)**

See the discussion of ¶ 4.3.2 of B101 in **§ 2.20.**

> *.1 Reviewing a Contractor's submittal out of sequence from the submittal schedule agreed to by the Architect;* **(2.8.2.1; minor modifications)**

See the discussion of ¶ 4.3.2.1 of B101 in **§ 2.20.**

> *.2 Responding to the Contractor's requests for information that are not prepared in accordance with the Contract Documents or where such information is available to the Contractor from a careful study and comparison of the Contract Documents, field conditions, other Owner-provided information, Contractor-prepared coordination drawings, or prior Project correspondence or documentation;* **(2.8.2.2; modifications)**

See the discussion of ¶ 4.3.2.2 of B101 in **§ 2.20.**

> *.3 Preparing Change Orders, and Construction Change Directives that require evaluation of Contractor's proposals and supporting data, or the preparation or revision of Instruments of Service;* **(2.8.2.3; modifications)**

See the discussion of ¶ 4.3.2.3 of B101 in **§ 2.20.**

> *.4 Evaluating an extensive number of Claims as the Initial Decision Maker;* **(2.8.2.5; add Initial Decision Maker)**

See the discussion of ¶ 4.3.2.4 of B101 in **§ 2.20.**

> *.5 Evaluating substitutions proposed by the Owner or Contractor and making subsequent revisions to Instruments of Service resulting therefrom; or* **(2.8.2.6; modifications)**

See the discussion of ¶ 4.3.2.5 of B101 in § **2.20.**

> *.6 To the extent the Architect's Basic Services are affected, providing Construction Phase Services 60 days after (1) the date of Substantial Completion of the Work or (2) the anticipated date of Substantial Completion, identified in Initial Information, whichever is earlier.* (**2.8.2.8; modifications**)

See the discussion of ¶ 4.3.2.6 of B101 in § **2.20.**

4.3.3 The Architect shall provide Construction Phase Services exceeding the limits set forth below as Additional Services. When the limits below are reached, the Architect shall notify the Owner: (**2.8.1; modifications**)

See the discussion of ¶ 4.3.3 of B101 in § **2.20.**

> *.1 ( ) reviews of each Shop Drawing, Product Data item, sample and similar submittals of the Contractor* (**2.8.1.1; delete "up to"**)

See the discussion of ¶ 4.3.3.1 of B101 in § **2.20.**

> *.2 ( ) visits to the site by the Architect over the duration of the Project during construction* (**2.8.1.2; delete "up to"**)

See the discussion of ¶ 4.3.3.2 of B101 in § **2.20.**

> *.3 ( ) inspections for any portion of the Work to determine whether such portion of the Work is substantially complete in accordance with the requirements of the Contract Documents* (**2.8.1.3; delete "up to"**)

See the discussion of ¶ 4.3.3.3 of B101 in § **2.20.**

> *.4 ( ) inspections for any portion of the Work to determine final completion* (**2.8.1.4; delete "up to"**)

See the discussion of ¶ 4.3.3.4 of B101 in § **2.20.**

4.3.4 If the services covered by this Agreement have not been completed within ( ) months of the date of this Agreement, through no fault of the Architect, extension of the Architect's services beyond that time shall be compensated as Additional Services. (**1.5.9; revise reference to Additional Services**)

See the discussion of ¶ 4.3.4 of B101 in § **2.20.**

§ 6.18   Article 5: Owner's Responsibilities

ARTICLE 5 OWNER'S RESPONSIBILITIES

5.1 Unless otherwise provided for under this Agreement, the Owner shall provide information in a timely manner regarding requirements for and limitations on the

Project, including a written program which shall set forth the Owner's objectives, schedule, constraints and criteria, including space requirements and relationships, flexibility, expandability, special equipment, systems and site requirements. Within 15 days after receipt of a written request from the Architect, the Owner shall furnish the requested information as necessary and relevant for the Architect to evaluate, give notice of or enforce lien rights. (**1.2.2.1; modifications**)

See the discussion of ¶ 5.1 of B101 in § **2.21.**

5.2 The Owner shall furnish the services of a Scheduling Consultant that shall be responsible for creating the overall Project schedule. The Owner shall adjust the Project schedule, if necessary, as the Project proceeds. (**new**)

This Scheduling Consultant will work with the owner, architect, and contractor to create and revise the project schedule. See ¶ 3.1.3, where the architect will submit to the owner's Scheduling Consultant a schedule of the architect's services for inclusion into the project schedule. The architect and owner should also reach some agreement as to the extent that this Scheduling Consultant will be responsible for monitoring the progress of the work of the contractor for deviations from the schedule. Will the architect be entitled to rely on this consultant's evaluation of the progress of the work?

5.3 The Owner shall establish and periodically update the Owner's budget for the Project, including (1) the budget for the Cost of the Work as defined in Section 6.1; (2) the Owner's other costs; and, (3) reasonable contingencies related to all of these costs. The Owner shall furnish the services of a Cost Consultant that shall be responsible for preparing all estimates of the Cost of the Work. If the Owner significantly increases or decreases the Owner's budget for the Cost of the Work, the Owner shall notify the Architect. The Owner and the Architect shall thereafter agree to a corresponding change in the budget for the Cost of the Work or in the Project's scope and quality. (**1.2.2.2; revisions; add reference to Cost Consultant**)

It is the owner's duty to initially establish a budget. Thereafter, the owner, with the assistance of the owner's Cost Consultant, is under a continuing obligation to update the project budget and cannot "significantly" increase or decrease the budget for the part of the work that affects the architect without the architect's agreement. This means that if the budget is changed by more than a minor amount, the architect may have to change the design and/or drawings and specifications, resulting in additional compensation for the architect. The failure by the owner to perform these updates, if required, may constitute a breach of the agreement if the architect is affected thereby.

The architect is also entitled to rely on the Cost Consultant's estimates of the cost of the work. During the entire design process, the architect will furnish this consultant with the drawings and specifications as they progress and will rely on feedback from this consultant as to whether the project conforms to the current budget. Presumably, this consultant is better able to determine such costs than the architect. If the bids then come in too high, the architect will not be responsible and will be entitled to additional compensation if the owner decides that the project must be redrawn to bring the costs down.

5.3.1 The Owner acknowledges that accelerated, phased or fast-track scheduling provides a benefit, but also carries with it associated risks. Such risks include the Owner incurring costs for the Architect to coordinate and redesign portions of the Project affected by procuring or installing elements of the Project prior to the completion of all relevant Construction Documents, and costs for the Contractor to remove and replace previously installed Work. If the Owner selects accelerated, phased or fast-track scheduling, the Owner agrees to include in the budget for the Project sufficient contingencies to cover such costs. (**new**)

Under this agreement, the owner may choose to use a fast-track type construction process or other non-traditional procedure. Such a process inherently will result in more changes and additional costs, necessitating a larger contingency. Of course, the owner is also able to save considerable time using such a process, offsetting the additional construction costs. This provision is an acknowledgement by the owner of such additional costs and an agreement to budget for additional contingencies. The owner will not be able to make a claim against the architect for additional costs due to uncoordinated drawings and specifications, or for errors, unless there are an inordinate number of such mistakes by the architect.

5.4 The Owner shall identify a representative authorized to act on the Owner's behalf with respect to the Project. The Owner shall render decisions and approve the Architect's submittals in a timely manner in order to avoid unreasonable delay in the orderly and sequential progress of the Architect's services. (**1.2.2.3; revisions**)

See the discussion of ¶ 5.3 of B101 in § **2.21.**

5.5 The Owner shall furnish surveys to describe physical characteristics, legal limitations and utility locations for the site of the Project, and a written legal description of the site. The surveys and legal information shall include, as applicable, grades and lines of streets, alleys, pavements and adjoining property and structures; designated wetlands; adjacent drainage; rights-of-way, restrictions, easements, encroachments, zoning, deed restrictions, boundaries and contours of the site; locations, dimensions and necessary data with respect to existing buildings, other improvements and trees; and information concerning available utility services and lines, both public and private, above and below grade, including inverts and depths. All the information on the survey shall be referenced to a Project benchmark. (**2.2.1.2; added "wetlands"**)

See the discussion of ¶ 5.4 of B101 in § **2.21.**

5.6 The Owner shall furnish services of geotechnical engineers, which may include but are not limited to test borings, test pits, determinations of soil bearing values, percolation tests, evaluations of hazardous materials, seismic evaluation, ground corrosion tests and resistivity tests, including necessary operations for anticipating subsoil conditions, with written reports and appropriate recommendations. (**2.2.1.3; minor modifications**)

See the discussion of ¶ 5.5 of B101 in § **2.21.**

5.7 The Owner shall coordinate the services of its own consultants with those services provided by the Architect. Upon the Architect's request, the Owner shall furnish copies of the scope of services in the contracts between the Owner and the Owner's consultants. The Owner shall furnish the services of consultants other than those designated in this Agreement, or authorize the Architect to furnish them as an Additional Service, when the Architect requests such services and demonstrates that they are reasonably required by the scope of the Project. The Owner shall require that its consultants maintain professional liability insurance and other liability insurance as appropriate to the services provided. **(1.2.2.4; substantial modifications; added reference to "other liability insurance")**

See the discussion of ¶ 5.6 of B101 in § **2.21.** Note the reference to "other liability insurance" required to be maintained by the owner's consultants. This reference is not present in B101. Presumably this was inserted in this document because large or complex projects will have more consultants hired by the owner than normal projects. At the very least, this document requires that the owner hire cost and scheduling consultants on whom the owner and architect will rely.

5.8 The Owner shall furnish tests, inspections and reports required by law or the Contract Documents, such as structural, mechanical, and chemical tests, tests for air and water pollution, and tests for hazardous materials. **(1.2.2.5; minor revisions)**

See the discussion of ¶ 5.7 of B101 in § **2.21.**

5.9 The Owner shall furnish all legal, insurance and accounting services, including auditing services, that may be reasonably necessary at any time for the Project to meet the Owner's needs and interests. **(1.2.2.6; no change)**

See the discussion of ¶ 5.8 of B101 in § **2.21.**

5.10 The Owner shall provide prompt written notice to the Architect if the Owner becomes aware of any fault or defect in the Project, including errors, omissions or inconsistencies in the Architect's Instruments of Service. **(1.2.2.7; minor modification)**

See the discussion of ¶ 5.9 of B101 in § **2.21.**

5.11 Except as otherwise provided in this Agreement, or when direct communications have been specially authorized, the Owner shall endeavor to communicate with the Contractor and the Architect's consultants through the Architect about matters arising out of or relating to the Contract Documents. The Owner shall promptly notify the Architect of any direct communications that may affect the Architect's services. **(2.6.2.4; minor modifications)**

See the discussion of ¶ 5.10 of B101 in § **2.21.**

5.12 Before executing the Contract for Construction, the Owner shall coordinate the Architect's duties and responsibilities set forth in the Contract for Construction with the Architect's services set forth in this Agreement. The Owner shall provide the Architect a copy of the executed agreement between the Owner and Contractor, including the General Conditions of the Contract for Construction. **(new)**

See the discussion of ¶ 5.11 of B101 in **§ 2.21.**

5.13 The Owner shall provide the Architect access to the Project site prior to commencement of the Work and shall obligate the Contractor to provide the Architect access to the Work wherever it is in preparation or progress. **(2.6.2.3; modifications)**

See the discussion of ¶ 5.12 of B101 in **§ 2.21.**

§ 6.19   Article 6: Cost of the Work

ARTICLE 6 COST OF THE WORK

6.1 For purposes of this Agreement, the Cost of the Work shall be the total cost to the Owner to construct all elements of the Project designed or specified by the Architect and shall include contractors' general conditions costs, overhead and profit. The Cost of the Work does not include the compensation of the Architect, the costs of the land, rights-of-way, financing, contingencies for changes in the Work or other costs that are the responsibility of the Owner. **(1.3.1.1, 1.3.1.2, 1.3.1.3; substantial modifications)**

See the discussion of ¶ 6.1 of B101 in **§ 2.22.**

6.2 The Owner's budget for the Cost of the Work is provided in Initial Information, and may be adjusted throughout the Project as required under Sections 5.3 and 6.4. Evaluations of the Owner's budget for the Cost of the Work represent the Architect's judgment as a design professional.

Under this agreement, as opposed to the B101, the architect has a much lesser duty to the owner related to the cost of the project. Instead, the owner and architect are relying on the owner's Cost Consultant to maintain the budget and give the architect feedback and advice as to whether the drawings and specifications, as they are being developed, are within the budget. Here, the architect is to provide an initial evaluation of the budget at the outset of the project. The standard of the architect's "judgment as a design professional" is actually a low standard and the owner will not be able to place much reliance on this evaluation by the architect.

6.3 The Owner shall require the Cost Consultant to include appropriate contingencies for design, bidding or negotiating, price escalation, and market conditions in estimates of the Cost of the Work. The Architect shall be entitled to rely on the accuracy and completeness of estimates of the Cost of the Work the Cost Consultant prepares as the Architect progresses with its Basic Services. The Architect shall prepare, as an Additional Service, revisions to the Drawings, Specifications or other documents required due to the Cost Consultant's inaccuracies or incompleteness in preparing cost estimates. The Architect may review the Cost Consultant's estimates solely for the Architect's guidance in completion of its services, however, the Architect shall report to the Owner any material inaccuracies and inconsistencies noted during any such review. **(new)**

The Cost Consultant will be working with the architect throughout the project, but particularly during the design phases up to the point where the contract for construction is signed. The budget prepared by this consultant should include the listed contingencies and take into account the state of the economy and likely changes in prices of materials and labor during the construction phase. The goal during the design of the project is to have this consultant work closely with the architect so that the architect can design the project as close to the owner's budget as possible. If the architect proposes some design element that may be too costly, this consultant should advise the architect of the cost impact of that element, permitting the architect to timely revise either that element, or something else on the project so as to maintain the overall budget. If it turns out that the Cost Consultant has made errors that result in a design that is too costly for the owner and the owner wants the architect to revise the drawings to decrease the cost, the architect will be entitled to be reimbursed for the cost of making such revisions, as the architect was relying on the Cost Consultant.

6.4 *If, prior to the conclusion of the Design Development Phase, the Cost Consultant's estimate of the Cost of the Work exceeds the Owner's budget for the Cost of the Work, the Architect, in consultation with the Cost Consultant, shall make appropriate recommendations to the Owner to adjust the Project's size, quality or budget, and the Owner shall cooperate with the Architect in making such adjustments.* **(2.1.7.1, substantial modifications; add role of Cost Consultant)**

If the anticipated cost of the project exceeds the updated budget prior to the end of the Design Development Phase, the project needs to be brought back into line. While the architect is still working on the Design Development Phase, making design changes will be relatively easy, and the architect will be expected to make changes at no additional cost. Of course, this supposes that the owner and Cost Consultant are acting reasonably and making few changes. If, on the other hand, changes are being arbitrarily made by the owner, or the Cost Consultant is not performing properly, the architect will be entitled to additional compensation.

The following section gives various options for accomplishing this if this occurs after the completion of this phase and before the start of the construction documents phase, including increasing the budget. If this occurs prior to this stage of the project, the architect will work with the owner and Cost Consultant to modify the design and the documents to bring the cost back down to the budget amount.

6.5 *If the estimate of the Cost of the Work at the conclusion of the Design Development Phase exceeds the Owner's budget for the Cost of the Work, the Owner shall*

(2.1.7.5; change the stage where this occurs from the end of the Construction Documents Phase to Design Development Phase)

> .1 *give written approval of an increase in the budget for the Cost of the Work;* **(2.1.7.5.1; no change)**
>
> .2 *in consultation with the Architect, revise the Project program, scope, or quality as required to reduce the Cost of the Work; or* **(2.1.7.5.4; minor modifications)**
>
> .3 *implement any other mutually acceptable alternative.* **(new)**

If, at the end of the Design Development Phase, the Cost Consultant determines that the project's project cost exceeds the current budget, the owner can implement one of the three choices listed above. AIA Document B101 has two more alternatives. First, that of terminating the contract for convenience per Paragraph 9.5. Presumably, that would still be an option whether or not it is set forth here. The second option found in B101 but not here is to "authorize rebidding or renegotiating of the Project within a reasonable time." Again, the owner could still opt for this under Section 6.5.3.

> **6.6** *If the Owner chooses to proceed under Section 6.5.2, the Architect, without additional compensation, shall incorporate the required modifications in the Construction Documents Phase as necessary to comply with the Owner's budget for the Cost of the Work at the conclusion of the Design Development Phase Services, or the budget as adjusted under Section 6.5.1. The Architect's modification of the Construction Documents shall be the limit of the Architect's responsibility as a Basic Service under this Article 6.* **(2.1.7.6; modifications)**

If the owner decides not to increase the budget, but chooses the second option of changing the scope of the project to bring the costs down, the architect will make the necessary revisions to the design by incorporating the changes into the "Construction Documents Phase." In most cases involving large or complex projects, there will not be a single set of construction documents. Instead, the project will be fast-tracked or separate bid packages will be prepared. Thus, this wording means that the architect will incorporate changes into certain of the construction documents prepared in the future, and the architect will not be required to redraw the design development drawings, although the architect may have to prepare some sketches to illustrate proposed changes. Doing this will be the limit of the architect's liability for cost increases.

> **6.7** *After incorporation of modifications under Section 6.6, the Architect shall, as an Additional Service, make any required revisions to the Drawings, Specifications or other documents necessitated by subsequent cost estimates that exceed the Owner's budget for the Cost of the Work, except when the excess is due to changes initiated by the Architect in scope, basic systems, or the kinds and quality of materials, finishes or equipment.* **(new)**

If further changes are necessary to bring the cost of the project within the budget, the architect will be entitled to make such changes as an additional service and for an additional fee, unless the architect is the cause of the cost overrun. The architect can avoid any liability here by running any proposed changes by the Cost Consultant so that the architect will not make any revision without cost input from this consultant.

§ 6.20   Article 7: Copyrights and Licenses

ARTICLE 7 COPYRIGHTS AND LICENSES

This section is entitled "Copyrights and Licenses" (a more accurate description of what this section is about) while the equivalent section in B101 is entitled "Instruments of Service." The actual language of the following provisions is the same in both documents.

__7.1__ The Architect and the Owner warrant that in transmitting Instruments of Service, or any other information, the transmitting party is the copyright owner of such information or has permission from the copyright owner to transmit such information for its use on the Project. If the Owner and Architect intend to transmit Instruments of Service or any other information or documentation in digital form, they shall endeavor to establish necessary protocols governing such transmissions. **(new)**

See the discussion of ¶ 7.1 of B101 in **§ 2.23.**

__7.2__ The Architect and the Architect's consultants shall be deemed the authors and owners of their respective Instruments of Service, including the Drawings and Specifications, and shall retain all common law, statutory and other reserved rights, including copyrights. Submission or distribution of Instruments of Service to meet official regulatory requirements or for similar purposes in connection with the Project is not to be construed as publication in derogation of the reserved rights of the Architect and the Architect's consultants. **(1.3.2.1; 1.3.2.3; add reference to consultants and electronic form)**

See the discussion of ¶ 7.2 of B101 in **§ 2.23.**

__7.3__ Upon execution of this Agreement, the Architect grants to the Owner a nonexclusive license to use the Architect's Instruments of Service solely and exclusively for purposes of constructing, using, maintaining, altering and adding to the Project, provided that the Owner substantially performs its obligations, including prompt payment of all sums when due, under this Agreement. The Architect shall obtain similar nonexclusive licenses from the Architect's consultants consistent with this Agreement. The license granted under this section permits the Owner to authorize the Contractor, Subcontractors, Sub-subcontractors, and material or equipment suppliers, as well as the Owner's consultants and separate contractors, to reproduce applicable portions of the Instruments of Service solely and exclusively for use in performing services or construction for the Project. If the Architect rightfully terminates this Agreement for cause as provided in Section 9.4, the license granted in this Section 7.3 shall terminate. **(1.3.2.2; 1.3.2.3; substantial modifications)**

See the discussion of ¶ 7.3 of B101 in § **2.23.**

7.3.1 In the event the Owner uses the Instruments of Service without retaining the authors of the Instruments of Service, the Owner releases the Architect and Architect's consultant(s) from all claims and causes of action arising from such uses. The Owner, to the extent permitted by law, further agrees to indemnify and hold harmless the Architect and its consultants from all costs and expenses, including the cost of defense, related to claims and causes of action asserted by any third person or entity to the extent such costs and expenses arise from the Owner's use of the Instruments of Service under this Section 7.3.1. The terms of this Section 7.3.1 shall not apply if the Owner rightfully terminates this Agreement for cause under Section 9.4. (**1.3.2.2; revisions**)

See the discussion of ¶ 7.3.1 of B101 in § **2.23.**

7.4 Except for the licenses granted in this Article 7, no other license or right shall be deemed granted or implied under this Agreement. The Owner shall not assign, delegate, sublicense, pledge or otherwise transfer any license granted herein to another party without the prior written agreement of the Architect. Any unauthorized use of the Instruments of Service shall be at the Owner's sole risk and without liability to the Architect and the Architect's consultants. (**1.3.2.2; substantial modifications**)

See the discussion of ¶ 7.4 of B101 in § **2.23.**

§ 6.21   Article 8: Claims and Disputes

ARTICLE 8 CLAIMS AND DISPUTES

8.1 GENERAL

8.1.1 The Owner and Architect shall commence all claims and causes of action, whether in contract, tort, or otherwise, against the other arising out of or related to this Agreement in accordance with the requirements of the method of binding dispute resolution selected in this Agreement within the period specified by applicable law, but in any case not more than 10 years after the date of Substantial Completion of the Work. The Owner and Architect waive all claims and causes of action not commenced in accordance with this Section 8.1.1. (**1.3.7.3; completely rewritten; established ten-year outside limit on all claims**)

See the discussion of ¶ 8.1.1 of B101 in § **2.24.**

8.1.2 To the extent damages are covered by property insurance, the Owner and Architect waive all rights against each other and against the contractors, consultants, agents and employees of the other for damages, except such rights as they may have to the proceeds of such insurance as set forth in AIA Document A201–2007, General Conditions of the Contract for Construction. The Owner or the Architect, as appropriate, shall require of the contractors, consultants, agents and employees of any of them similar waivers in favor of the other parties enumerated herein. (**1.3.7.4; deleted limitation of time to "during construction"; changed reference**)

See the discussion of ¶ 8.1.2 of B101 in **§ 2.24.**

8.1.3 The Architect shall indemnify and hold the Owner and the Owner's officers and employees harmless from and against damages, losses and judgments arising from claims by third parties, including reasonable attorneys' fees and expenses recoverable under applicable law, but only to the extent they are caused by the negligent acts or omissions of the Architect, its employees and its consultants in the performance of professional services under this Agreement. The Architect's duty to indemnify the Owner under this provision shall be limited to the available proceeds of insurance coverage. **(new)**

This indemnification and limitation of liability clause does not appear in B101. This is a limitation of liability provision[4] whereby the maximum recovery of the owner against the architect, due to the negligence of the architect or the architect's employees and consultants, is the "available proceeds of insurance coverage."

---

[4] In W. William Graham, Inc. v. City of Cave City, 289 Ark. 104, 709 S.W.2d 94 (1986), the court construed the following clause:

> The Owner agrees to limit the Engineer's liability to the Owner and to all Construction Contractors and Subcontractors on the Project, due to the Engineer's professional negligent acts, errors or omissions, such that the total aggregate liability of the Engineer to those named shall not exceed Fifty Thousand Dollars ($50,000.00) or the Engineer's total fee for services rendered on this project, whichever is greater.

The court found that the clause did not cover actions for breach of contract and the damages that flowed from such a breach. Therefore, the limit does not apply.

California has allowed such a limitation of liability when the parties have had the opportunity to accept, reject, or modify the provision. Markborough Cal. v. Superior Court, 227 Cal. App. 3d 705, 277 Cal. Rptr. 919 (1991).

In Leon's Bakery, Inc. v. Grinnell Corp., 990 F.2d 44 (2d Cir. 1993), a business whose property was damaged by a fire brought an action against the company that manufactured and installed the fire protection system. The claim was barred by a limitation of liability provision that stated as follows:

LIMITATIONS OF LIABILITY

> In no event shall Seller be liable for special or consequential damages and Seller's liability on any claim whether or not based in contract or in tort or occasioned by Seller's active or passive negligence for loss or liability arising out of or connected with this contract, or any obligation resulting therefore, or from the manufacture, fabrication, sale, delivery, installation, or use of any materials covered by this contract, shall be limited to that set forth in the paragraph entitled "Warranty."

WARRANTY

> Seller agrees that for a period of one (1) year after completion of said installation it will, at its expense, repair or replace any defective materials or workmanship supplied or performed by Seller. Upon completion of the installation, the system will be turned over to the Purchaser fully inspected, tested and in operative condition. As it is thereafter the responsibility of the Purchaser to maintain it in operative condition, it is understood that the Seller does not guarantee the operation of the system. Seller further warrants the products of other manufacturers supplied hereunder, to the extent of the warranty of the respective manufacturer. ALL OTHER EXPRESS OR IMPLIED WARRANTIES OF MERCHANTABILITY OR FITNESS OR OTHERWISE ARE HEREBY EXCLUDED.

At Section 2.5.5, the parties can state the amount of such insurance that the architect will carry during the duration of this agreement. Note that, if the owner

---

In Georgetown Steel Corp. v. Union Carbide Corp., 806 F. Supp. 74 (D.S.C. 1992), the engineer had the following limitation of liability provision in its contract:

WARRANTY AND LIMITATION OF LIABILITY

The only warranty or guarantee made by Law Engineering Testing Company in connection with the services performed hereunder, is that we will use that degree of care and skill ordinarily exercised under similar conditions by reputable members of our profession practicing in the same or similar locality. No other warranty, expressed or implied is made or intended by our proposal for consulting services or by our furnishing oral or written reports.

Our liability for any damage on account of any error, omission, or other professional negligence will be limited to a sum not to exceed $50,000 or our fee, whichever is greater. In the event the client does not wish to limit our professional liability to this sum, we agree to waive this limitation upon receiving client's request, and agreement by the client to pay additional consideration of 4% of our total fee or $200.00 whichever is greater.

In Valhal Corp. v. Sullivan Assocs., Inc., 44 F.3d 195, 205-06 (3d Cir. 1995), the architect had the following provision in the contract:

The OWNER agrees to limit the Design Professional's liability to the OWNER and to all construction Contractors and Subcontractors on the project, due to the Design Professional's professional negligent acts, errors or omissions, such that the total aggregate liability of each Design Professional shall not exceed $50,000 or the Design Professional's total fee for services rendered on this project. Should the OWNER find the above terms unacceptable, an equitable surcharge to absorb the Architect's increase in insurance premiums will be negotiated.

In Pitts v. Watkins, 905 So. 2d 553 (Miss. 2005), involving a home inspection agreement, the Mississippi Supreme Court held the limitation of liability provision and arbitration provision unconscionable. The agreement required arbitration of any claim by the homeowner, but the inspector could bring a court action. The inspector's liability was capped at his fee—$265. "The limitation of liability clause, when paired with the arbitration clause, effectively denies the plaintiff of an adequate remedy and is further evidence of substantive unconscionability."

In Bolingbrook Hotel Corp., Inc. v. Linday, Pope, Brayfield & Assocs., Inc., 2005 WL 1226058 (N.D. Ill. Apr. 20, 2005), the owner sued the architect for improperly designing the plumbing pipes, resulting in inadequate water pressure. The contract contained this indemnification clause:

In any project there will be ambiguities, inconsistencies, errors, and omissions in the Construction Documents. These conditions may result in changes to the Contract Work, Amount, and Time. To the extent of a Contingency Reserve of 5% of the Construction Cost, the Owner agrees to indemnify and hold the Architect harmless with respect to all claims, awards, or legal actions arising out of any suit concerning the aforementioned changes.

The architect argued that this provision limits claims against it by the owner. The owner argued that this language did not apply to design flaws, but rather to errors such as typos and other minor items. The court agreed with the owner, holding that the language indemnifies the architect to the extent that errors in the construction documents that result in changes to the contract work, amount, or time cause damages greater than 5 percent of the construction cost. Thus, it only covered errors that lead to changes to the contract work, amount, or time, none of which applied to the errors alleged by the complaint. This indemnification provision did not protect the architect.

The court found that this clause was not against public policy, stating that an architectural firm and real estate developer have attempted to allocate risks between themselves in such a way that neither is relieved from liability for its own negligence. We see no reason to hold that the policy enunciated in section 491 [68 Pa. Cons. Stat. Ann. § 491] precludes them from doing so.

Because the limit of the architect's liability was $50,000, the federal court lacked diversity jurisdiction. 28 U.S.C. § 1332(a) requires that the amount exceed $50,000. The *Valhal* case was followed in Marbro, Inc. v. Borough of Tinton Falls, 688 A.2d 159 (N.J. Super. Ct. Law Div. 1996).

requires the architect to maintain professional liability insurance in amounts greater than what the architect normally carries, the owner will pay for the excess, but the architect can revert to normal limits after the project ends. If a claim is made after that time, such as in the event of a latent defect,[5] there may not be sufficient insurance to cover the owner's needs. Owners also must realize that the architect's insurance coverage is likely a declining balance coverage, so that the available proceeds are reduced by things like attorneys' and expert fees expended on the case, as well as other claims made during the applicable policy period.[6]

8.1.4 The Architect and Owner waive consequential damages for claims, disputes or other matters in question arising out of or relating to this Agreement. This mutual waiver is applicable, without limitation, to all consequential damages due to either party's termination of this Agreement, except as specifically provided in Section 9.7. **(1.3.6; minor modifications)**

See the discussion of ¶ 8.1.3 of B101 in **§ 2.24.**

§ 6.22   Mediation: ¶ 8.2

8.2 MEDIATION

8.2.1 Any claim, dispute or other matter in question arising out of or related to this Agreement shall be subject to mediation as a condition precedent to binding dispute resolution. If such matter relates to or is the subject of a lien arising out of the Architect's services, the Architect may proceed in accordance with applicable law to comply with the lien notice or filing deadlines prior to resolution of the matter by mediation or by binding dispute resolution. **(1.3.4.1; modified to add binding dispute resolution process)**

See the discussion of ¶ 8.2.1 of B101 in **§ 2.25.**

8.2.2 The Owner and Architect shall endeavor to resolve claims, disputes and other matters in question between them by mediation which, unless the parties mutually agree otherwise, shall be administered by the American Arbitration Association in accordance with its Construction Industry Mediation Procedures in effect on the date of the Agreement. A request for mediation shall be made in writing, delivered to the other party to the Agreement, and filed with the person or entity administering the

---

[5] For instance, if a leak is discovered several years after completion, the cause may be a design defect. The architect would then be sued by the owner, with the architect carrying a different amount of insurance coverage than what was carried during the actual project.

[6] As an example, assume that the owner sues the architect for a significant claim. The architect has a $1 million professional liability policy for the claim. A different owner sues the same architect during the applicable period for an unrelated matter and the insurance carrier expends $250,000 successfully defending that other claim. If the attorneys fees are $500,000 and other costs are $100,000, there will only be $150,000 in "available proceeds" to pay this owner, which is the limit of the architect's liability under this provision.

mediation. The request may be made concurrently with the filing of a complaint or other appropriate demand for binding dispute resolution but, in such event, mediation shall proceed in advance of binding dispute resolution proceedings, which shall be stayed pending mediation for a period of 60 days from the date of filing, unless stayed for a longer period by agreement of the parties or court order. If an arbitration proceeding is stayed pursuant to this section, the parties may nonetheless proceed to the selection of the arbitrator(s) and agree upon a schedule for later proceedings. **(1.3.4.2; substantial modifications)**

See the discussion of ¶ 8.2.2 of B101 in § **2.25.**

8.2.3 The parties shall share the mediator's fee and any filing fees equally. The mediation shall be held in the place where the Project is located, unless another location is mutually agreed upon. Agreements reached in mediation shall be enforceable as settlement agreements in any court having jurisdiction thereof. **(1.3.4.3; no change)**

See the discussion of ¶ 8.2.3 of B101 in § **2.25.**

8.2.4 If the parties do not resolve a dispute through mediation pursuant to this Section 8.2, the method of binding dispute resolution shall be the following:

(Check the appropriate box. If the Owner and Architect do not select a method of binding dispute resolution below, or do not subsequently agree in writing to a binding dispute resolution method other than litigation, the dispute will be resolved in a court of competent jurisdiction.) **(new)**

☐ *Arbitration pursuant to Section 8.3 of this Agreement*
or

☐ *Litigation in a court of competent jurisdiction*
or

☐ *Other (Specify)* **(new)**

See the discussion of ¶ 8.2.4 of B101 in § **2.25.**

§ 6.23   Arbitration: ¶ 8.3

8.3 ARBITRATION

See the discussion of ¶ 8.3 of B101 in § **2.26.**

8.3.1 If the parties have selected arbitration as the method for binding dispute resolution in this Agreement any claim, dispute or other matter in question arising out of or related to this Agreement subject to, but not resolved by, mediation shall be subject to arbitration which, unless the parties mutually agree otherwise, shall be administered by the American Arbitration Association in accordance with its Construction Industry Arbitration Rules in effect on the date of the Agreement. A demand for arbitration shall be made in writing, delivered to the other party to this Agreement, and filed with the person or entity administering the arbitration. (**1.3.5.3, 1.3.5.2; modified to reflect modifications to binding dispute resolution procedures; minor modifications**)

See the discussion of ¶ 8.3.1 of B101 in § **2.26.**

8.3.1.1 A demand for arbitration shall be made no earlier than concurrently with the filing of a request for mediation, but in no event shall it be made after the date when the institution of legal or equitable proceedings based on the claim, dispute or other matter in question would be barred by the applicable statute of limitations. For statute of limitations purposes, receipt of a written demand for arbitration by the person or entity administering the arbitration shall constitute the institution of legal or equitable proceedings based on the claim, dispute or other matter in question. (**1.3.5.3; added last sentence; minor modifications**)

See the discussion of ¶ 8.3.1.1 of B101 in § **2.26.**

8.3.2 The foregoing agreement to arbitrate and other agreements to arbitrate with an additional person or entity duly consented to by parties to this Agreement shall be specifically enforceable in accordance with applicable law in any court having jurisdiction thereof. (**1.3.5.4-last sentence; no change**)

See the discussion of ¶ 8.3.2 of B101 in § **2.26.**

8.3.3 The award rendered by the arbitrator(s) shall be final, and judgment may be entered upon it in accordance with applicable law in any court having jurisdiction thereof. (**1.3.5.5; no change**)

See the discussion of ¶ 8.3.3 of B101 in § **2.26.**

§ 6.24   Consolidation or Joinder: ¶ 8.3.4

8.3.4 CONSOLIDATION OR JOINDER

8.3.4.1 Either party, at its sole discretion, may consolidate an arbitration conducted under this Agreement with any other arbitration to which it is a party provided that (1) the arbitration agreement governing the other arbitration permits consolidation; (2) the arbitrations to be consolidated substantially involve common questions of law or fact; and (3) the arbitrations employ materially similar procedural rules and methods for selecting arbitrator(s). (**1.3.5.4; enabled joinder instead of prohibiting it**)

8.3.4.2 Either party, at its sole discretion, may include by joinder persons or entities substantially involved in a common question of law or fact whose presence is required if complete relief is to be accorded in arbitration, provided that the party sought to be joined consents in writing to such joinder. Consent to arbitration involving an additional person or entity shall not constitute consent to arbitration of any claim, dispute or other matter in question not described in the written consent. **(1.3.5.4; enabled joinder instead of prohibiting it)**

8.3.4.3 The Owner and Architect grant to any person or entity made a party to an arbitration conducted under this Section 8.3, whether by joinder or consolidation, the same rights of joinder and consolidation as the Owner and Architect under this Agreement. **(1.3.5.4; enabled joinder instead of prohibiting it)**

See the discussion of ¶ 8.3.4 of B101 in § **2.27.**

§ 6.25   Article 9: Termination or Suspension

ARTICLE 9 TERMINATION OR SUSPENSION

9.1 If the Owner fails to make payments to the Architect in accordance with this Agreement, such failure shall be considered substantial nonperformance and cause for termination or, at the Architect's option, cause for suspension of performance of services under this Agreement. If the Architect elects to suspend services, the Architect shall give seven days' written notice to the Owner before suspending services. In the event of a suspension of services, the Architect shall have no liability to the Owner for delay or damage caused the Owner because of such suspension of services. Before resuming services, the Architect shall be paid all sums due prior to suspension and any expenses incurred in the interruption and resumption of the Architect's services. The Architect's fees for the remaining services and the time schedules shall be equitably adjusted. **(1.3.8.1; minor modifications)**

See the discussion of ¶ 9.1 of B101 in § **2.28.**

9.2 If the Owner suspends the Project, the Architect shall be compensated for services performed prior to notice of such suspension. When the Project is resumed, the Architect shall be compensated for expenses incurred in the interruption and resumption of the Architect's services. The Architect's fees for the remaining services and the time schedules shall be equitably adjusted. **(1.3.8.2; delete reference to 30 days)**

See the discussion of ¶ 9.2 of B101 in § **2.28.**

9.3 If the Owner suspends the Project for more than 90 cumulative days for reasons other than the fault of the Architect, the Architect may terminate this Agreement by giving not less than seven days' written notice. **(1.3.8.3; change consecutive to cumulative days; minor modification)**

See the discussion of ¶ 9.3 of B101 in § **2.28.**

9.4 Either party may terminate this Agreement upon not less than seven days' written notice should the other party fail substantially to perform in accordance with the terms of this Agreement through no fault of the party initiating the termination. **(1.3.8.4; minor modifications)**

See the discussion of ¶ 9.4 of B101 in **§ 2.28.**

9.5 The Owner may terminate this Agreement upon not less than seven days' written notice to the Architect for the Owner's convenience and without cause. (**1.3.8.5; minor modification**)

See the discussion of ¶ 9.5 of B101 in **§ 2.28.**

9.6 In the event of termination not the fault of the Architect, the Architect shall be compensated for services performed prior to termination, together with Reimbursable Expenses then due and all Termination Expenses as defined in Section 9.7. (**1.3.8.6; change reference**)

See the discussion of ¶ 9.6 of B101 in **§ 2.28.**

9.7 Termination Expenses are in addition to compensation for the Architect's services and include expenses directly attributable to termination for which the Architect is not otherwise compensated, plus an amount for the Architect's anticipated profit on the value of the services not performed by the Architect. (**1.3.8.7; minor modifications**)

See the discussion of ¶ 9.7 of B101 in **§ 2.28.**

9.8 The Owner's rights to use the Architect's Instruments of Service in the event of a termination of this Agreement are set forth in Article 7 and Section 11.9. (**new**)

See the discussion of ¶ 9.8 of B101 in **§ 2.28.**

§ 6.26   Article 10: Miscellaneous Provisions

ARTICLE 10 MISCELLANEOUS PROVISIONS

10.1 This Agreement shall be governed by the law of the place where the Project is located, except that if the parties have selected arbitration as the method of binding dispute resolution, the Federal Arbitration Act shall govern Section 8.3. (**1.3.7.1; modifications**)

See the discussion of ¶ 10.1 of B101 in **§ 2.29.**

10.2 Terms in this Agreement shall have the same meaning as those in AIA Document A201–2007, General Conditions of the Contract for Construction. (**1.3.7.2; modified reference**)

See the discussion of ¶ 10.2 of B101 in **§ 2.29.**

10.3 The Owner and Architect, respectively, bind themselves, their agents, successors, assigns and legal representatives to this Agreement. Neither the Owner nor the Architect shall assign this Agreement without the written consent of the other, except that the Owner may assign this Agreement to a lender providing financing for the Project if the lender agrees to assume the Owner's rights and obligations under this Agreement. **(1.3.7.9; modifications)**

See the discussion of ¶ 10.3 of B101 in § **2.29.**

10.4 *If the Owner requests the Architect to execute certificates, the proposed language of such certificates shall be submitted to the Architect for review at least 14 days prior to the requested dates of execution. If the Owner requests the Architect to execute consents reasonably required to facilitate assignment to a lender, the Architect shall execute all such consents that are consistent with this Agreement, provided the proposed consent is submitted to the Architect for review at least 14 days prior to execution. The Architect shall not be required to execute certificates or consents that would require knowledge, services or responsibilities beyond the scope of this Agreement.* **(1.3.7.8; modifications)**

See the discussion of ¶ 10.4 of B101 in § **2.29.**

10.5 *Nothing contained in this Agreement shall create a contractual relationship with or a cause of action in favor of a third party against either the Owner or Architect.* **(1.3.7.5; no change)**

See the discussion of ¶ 10.5 of B101 in § **2.29.**

10.6 *Unless otherwise required in this Agreement, the Architect shall have no responsibility for the discovery, presence, handling, removal or disposal of, or exposure of persons to, hazardous materials or toxic substances in any form at the Project site.* **(1.3.7.6; modifications)**

See the discussion of ¶ 10.6 of B101 in § **2.29.**

10.7 *The Architect shall have the right to include photographic or artistic representations of the design of the Project among the Architect's promotional and professional materials. The Architect shall be given reasonable access to the completed Project to make such representations. However, the Architect's materials shall not include the Owner's confidential or proprietary information if the Owner has previously advised the Architect in writing of the specific information considered by the Owner to be confidential or proprietary. The Owner shall provide professional credit for the Architect in the Owner's promotional materials for the Project.* **(1.3.7.7; no change)**

See the discussion of ¶ 10.7 of B101 in § **2.29.**

10.8 *If the Architect or Owner receives information specifically designated by the other party as "confidential" or "business proprietary," the receiving party shall keep such information strictly confidential and shall not disclose it to any other person except to (1) its employees, (2) those who need to know the content of such information in order to perform services or construction solely and exclusively* *for the Project, or (3) its consultants and contractors whose contracts include similar restrictions on the use of confidential information.* **(1.2.3.4; substantial modifications)**

See the discussion of ¶ 10.8 of B101 in § **2.29.**

§ 6.27   Article 11: Compensation

ARTICLE 11 COMPENSATION

11.1 For the Architect's Basic Services described under Article 3, the Owner shall compensate the Architect as follows:

(Insert amount of, or basis for, compensation.) **(1.5.1; change reference; minor modifications)**

See the discussion of ¶ 11.1 of B101 in § **2.30.**

11.2 For Additional Services designated in Section 4.1, the Owner shall compensate the Architect as follows:

(Insert amount of, or basis for, compensation. If necessary, list specific services to which particular methods of compensation apply.) **(new)**

See the discussion of ¶ 11.2 of B101 in § **2.30.**

11.3 For Additional Services that may arise during the course of the Project, including those under Section 4.3, the Owner shall compensate the Architect as follows:

(Insert amount of, or basis for, compensation.) **(1.5.2; modifications)**

See the discussion of ¶ 11.3 of B101 in § **2.30.**

11.4 Compensation for Additional Services of the Architect's consultants when not included in Sections 11.2 or 11.3, shall be the amount invoiced to the Architect plus ( ), or as otherwise stated below: **(1.5.3; modifications)**

See the discussion of ¶ 11.4 of B101 in § **2.30.**

11.5 Where compensation for Basic Services is based on a stipulated sum or percentage of the Cost of the Work, the compensation for each phase of services shall be as follows:

Schematic Design Phase:	*percent (%)*
Design Development Phase:	*percent (%)*
Construction Documents Phase:	*percent (%)*
Bidding or Negotiation Phase:	*percent (%)*
Construction Phase:	*percent (%)*
Total Basic Compensation:	*one hundred percent (100%)*

See the discussion of ¶ 11.5 of B101 in § **2.30.**

The Owner acknowledges that with an accelerated Project delivery or multiple bid package process, the Architect may be providing its services in multiple Phases simultaneously. Therefore, the Architect shall be permitted to invoice monthly in proportion to services performed in each Phase of Services, as appropriate. **(new)**

With large or complex projects, "fast-track" construction procedures are often used, as are multiple bid packages. These methods often save the owner considerable time in the overall project schedule, but make it much more likely that there will be coordination and other errors in the drawings and specifications. Change orders and increased and unforeseen costs to the owner are, therefore, much more likely than on a traditional design-bid-build project. This provision allows the architect to bill monthly for several phases at once. Section 1.1.6 requires the parties to discuss fast-track and other procedures that may impact this.

11.6 When compensation is based on a percentage of the Cost of the Work and any portions of the Project are deleted or otherwise not constructed, compensation for those portions of the Project shall be payable to the extent services are performed on those portions, in accordance with the schedule set forth in Section 11.5 based on (1) the lowest bona fide bid or negotiated proposal, or (2) if no such bid or proposal is received, the most recent estimate of the Cost of the Work for such portions of the Project. The Architect shall be entitled to compensation in accordance with this Agreement for all services performed whether or not the Construction Phase is commenced. **(new)**

See the discussion of ¶ 11.6 of B101 in § **2.30.**

11.7 The hourly billing rates for services of the Architect and the Architect's consultants, if any, are set forth below. The rates shall be adjusted in accordance with the Architect's and Architect's consultants' normal review practices.

(If applicable, attach an exhibit of hourly billing rates or insert them below.)

Employee or Category   **Rate (1.5.6; modified)**

See the discussion of ¶ 11.7 of B101 in § **2.30.**

§ 6.28   Compensation for Reimbursable Expenses: ¶ 11.8

11.8 COMPENSATION FOR REIMBURSABLE EXPENSES

11.8.1 Reimbursable Expenses are in addition to compensation for Basic and Additional Services and include expenses incurred by the Architect and the Architect's consultants directly related to the Project, as follows: **(1.3.9.2; modifications)**

See the discussion of ¶ 11.8.1 of B101 in § **2.31.**

    .1   *Transportation and authorized out-of-town travel and subsistence;* **(1.3.9.2.1; modifications)**

See the discussion of ¶ 11.8.1.1 of B101 in **§ 2.31.**

    .2   *Long distance services, dedicated data and communication services, tele-conferences, Project Web sites, and extranets;* **(1.3.9.2.1; substantially expanded definition)**

See the discussion of ¶ 11.8.1.2 of B101 in **§ 2.31.**

    .3   *Fees paid for securing approval of authorities having jurisdiction over the Project;* **(1.3.9.2.2; no change)**

See the discussion of ¶ 11.8.1.3 of B101 in **§ 2.31.**

    .4   *Printing, reproductions, plots, standard form documents;* **(1.3.9.2.3; minor modifications)**

See the discussion of ¶ 11.8.1.4 of B101 in **§ 2.31.**

    .5   *Postage, handling and delivery;* **(1.3.9.2.3; minor modifications)**

See the discussion of ¶ 11.8.1.5 of B101 in **§ 2.31.**

    .6   *Expense of overtime work requiring higher than regular rates, if authorized in advance by the Owner;* **(1.3.9.2.4; no change)**

See the discussion of ¶ 11.8.1.6 of B101 in **§ 2.31.**

    .7   *Renderings, models, mock-ups, professional photography, and presentation materials requested by the Owner;* **(1.3.9.2.5; added photography and presentation materials)**

See the discussion of ¶ 11.8.1.7 of B101 in **§ 2.31.**

    .8   *Architect's Consultant's expense of professional liability insurance dedicated exclusively to this Project, or the expense of additional insurance coverage or limits if the Owner requests such insurance in excess of that normally carried by the Architect's consultants;* **(1.3.9.2.6; modifications)**

See the discussion of ¶ 11.8.1.8 of B101 in **§ 2.31.**

    .9   *All taxes levied on professional services and on reimbursable expenses;* **(new)**

See the discussion of ¶ 11.8.1.9 of B101 in **§ 2.31.**

    .10 Site office expenses; and **(new)**

See the discussion of ¶ 11.8.1.10 of B101 in **§ 2.31.**

.11 *Other similar Project-related expenditures.* (**1.3.9.2.8; no change**)

See the discussion of ¶ 11.8.1.11 of B101 in **§ 2.31.**

11.8.2 For Reimbursable Expenses the compensation shall be the expenses incurred by the Architect and the Architect's consultants plus (   ) of the expenses incurred. (**1.5.4; modifications**)

See the discussion of ¶ 11.8.2 of B101 in **§ 2.31.**

§ 6.29   Compensation for Use of Architect's Instruments of Service: ¶ 11.9

11.9 COMPENSATION FOR USE OF ARCHITECT'S INSTRUMENTS OF SERVICE

If the Owner terminates the Architect for its convenience under Section 9.5, or the Architect terminates this Agreement under Section 9.3, the Owner shall pay a licensing fee as compensation for the Owner's continued use of the Architect's Instruments of Service solely for purposes of completing, using and maintaining the Project as follows: (**new**)

See the discussion of ¶ 11.9 of B101 in **§ 2.32.**

§ 6.30   Payments to the Architect: ¶ 11.10

11.10 PAYMENTS TO THE ARCHITECT

11.10.1 An initial payment of__(__$) shall be made upon execution of this Agreement and is the minimum payment under this Agreement. It shall be credited to the Owner's account in the final invoice. (**1.5.7; modifications**)

See the discussion of ¶ 11.10.1 of B101 in **§ 2.33.**

11.10.2 Unless otherwise agreed, payments for services shall be made monthly in proportion to services performed. Payments are due and payable upon presentation of the Architect's invoice. Amounts unpaid   (   ) days after the invoice date shall bear interest at the rate entered below, or in the absence thereof at the legal rate prevailing from time to time at the principal place of business of the Architect.

(Insert rate of monthly or annual interest agreed upon.) (**1.5.7, 1.5.8; modifications**)

See the discussion of ¶ 11.10.2 of B101 in **§ 2.33.**

11.10.3 The Owner shall not withhold amounts from the Architect's compensation to impose a penalty or liquidated damages on the Architect, or to offset sums requested by or paid to contractors for the cost of changes in the Work unless the Architect agrees or has been found liable for the amounts in a binding dispute resolution proceeding. **(1.3.9.1; substantial modifications)**

See the discussion of ¶ 11.10.3 of B101 in **§ 2.33.**

11.10.4 Records of Reimbursable Expenses, expenses pertaining to Additional Services, and services performed on the basis of hourly rates shall be available to the Owner at mutually convenient times. **(1.3.9.3; modifications)**

See the discussion of ¶ 11.10.4 of B101 in **§ 2.33.**

§ 6.31   Article 12: Special Terms and Conditions

ARTICLE 12 SPECIAL TERMS AND CONDITIONS

12.1 Special terms and conditions that modify this Agreement are as follows: **(1.4.2; No change)**

See the discussion of Article 12 of B101 in **§ 2.34.**

§ 6.32   Article 13: Scope of the Agreement

ARTICLE 13 SCOPE OF THE AGREEMENT

13.1 This Agreement represents the entire and integrated agreement between the Owner and the Architect and supersedes all prior negotiations, representations or agreements, either written or oral. This Agreement may be amended only by written instrument signed by both Owner and Architect. **(1.4.1; Deleted last sentence of prior paragraph)**

See the discussion of ¶ 13.1 of B101 in **§ 2.35.**

13.2 This Agreement is comprised of the following documents listed below: **(1.4.1; modifications)**

> *.1   AIA Document B103™–2007, Standard Form Agreement Between Owner and Architect* **(1.4.1.1; add reference to Digital Data; modifications)**

> *.2   AIA Document E201™–2007, Digital Data Protocol Exhibit, if completed, or the following:* **(new)**

See the discussion of ¶ 13.2.2 of B101 in **§ 2.35.**

> *.3   Other documents:*

(List other documents, if any, including additional scopes of service forming part of the Agreement.) **(1.4.1.3; modifications)**

Here, any other documents that form part of the contract should be listed. The scopes of service referred to might include various scopes of service documents published by the AIA. B101 includes an Exhibit A, Initial Information. In this document, this information is included in Article 1.

This Agreement entered into as of the day and year first written above.

CHAPTER 7

ALTERNATIVE LANGUAGE FOR AIA DOCUMENT B103

§ 7.1   Introduction

The language in this chapter is offered as possible alternative language to AIA Document B103. Much of the language in B103 is identical to the language in B101. Therefore, this chapter only covers the language of B103 that is different than that in B101. Users should consult **Chapter 3** for those paragraphs that are the same as in B101.

The user is cautioned that some of these provisions are mutually incompatible, and each paragraph and each sentence should be carefully reviewed with legal counsel before use. Note further that some of these changes require changes to other documents. Finally, please note that no recommendations are being made or implied by this alternative language. Much of it is more favorable to the owner than the standard AIA language, while some is more favorable to the architect. Each party should consult with an attorney familiar with construction law in the appropriate jurisdiction before making any changes to the standard document, as such changes will often result in serious consequences to the parties. If in doubt, consideration should be given to using the AIA language as-is.

The recommended method of amending B103 is to use the AIA's Electronic Documents software and make the changes directly to the body of the document. Alternatively, attach separate written amendments that refer back to B103. This is normally done in riders to the agreement, or inserting language in Article 12. Another method is to graphically delete material and insert new material in the margins of the standard form. However, this method may create ambiguity, and it may result in rendering the stricken material illegible.

§ 7.2   Schematic Design Phase Services: ¶ 3.2

B103 calls for the owner to hire a Cost Consultant. Here are some alternate provisions for the Schematic Design Phase:

> **3.2.6** The Architect shall submit the Schematic Design Documents to the Owner and the Cost Consultant. The Architect shall meet with the Cost Consultant to review the Schematic Design Documents in furtherance of the Cost Consultant's duties to the Owner to analyze the probable costs of constructing the various elements of the design of the Project. The Architect is entitled to rely on information and advice of the Cost Consultant relative to potential costs of various design alternatives.

> **3.2.7** Upon receipt of the Cost Consultant's estimate at the conclusion of the Schematic Design Phase, the Architect shall take action as required under Section 6.4, and request the Owner's approval of the Schematic Design Documents. If revisions to the Schematic Design Documents are required to comply with the Owner's budget for the Cost of the Work at the conclusion of the Schematic Design Phase, the Architect shall incorporate the required revisions in the Design Development Phase. If the Owner directs the Architect to make revisions to the Schematic Design Phase Documents as a result of any input from the Cost Consultant, the Architect shall be entitled to additional compensation for such revisions.

Similar alternative language, but more favorable to the owner:

> **3.2.6** The Architect shall submit the Schematic Design Documents to the Owner and the Cost Consultant. The Architect shall meet with the Cost Consultant to review the Schematic Design Documents. The Architect shall exercise the Architect's own independent judgment relative to potential costs of various design alternatives, and advise the Owner promptly if the Architect disagrees with any opinion or assessment of the Cost Consultant.

> **3.2.7** Upon receipt of the Cost Consultant's estimate at the conclusion of the Schematic Design Phase, the Architect shall take action as required under Section 6.4, and request the Owner's approval of the Schematic Design Documents. If revisions to the Schematic Design Documents are required to comply with the Owner's budget for the Cost of the Work at the conclusion of the Schematic Design Phase, the Architect shall incorporate the required revisions in the Design Development Phase. To the extent any revisions are required as a result of any error, omission, or failure by the Architect to conform to the Cost Consultant's advice, the Architect shall make such revisions at no cost to the Owner.

§ 7.3   Design Development Phase Services: ¶ 3.3

Alternative language more favorable to the architect:

3.3.2 Prior to the conclusion of the Design Development Phase, the Architect shall submit the Design Development documents to the Owner and the Cost Consultant. The Architect shall meet with the Cost Consultant to review the Design Development Documents in furtherance of the Cost Consultant's duties to the Owner to analyze the probable costs of constructing the various elements of the design of the Project. The Architect is entitled to rely on information and advice of the Cost Consultant relative to potential costs of various design alternatives.

3.3.3 Upon receipt of the Cost Consultant's estimate at the conclusion of the Design Development Phase, the Architect shall take action as required under Sections 6.5 and 6.6 and request the Owner's approval of the Design Development Documents. If the Owner directs the Architect to make revisions to the Design Development Phase Documents as a result of any input from the Cost Consultant, the Architect shall be entitled to additional compensation for such revisions as Additional Services.

§ 7.4   Construction Documents Phase Services: ¶ 3.4

Alternative language more favorable to the architect:

3.4.4 Prior to the conclusion of the Construction Documents Phase, the Architect shall submit the Construction Documents to the Owner and the Cost Consultant. The Architect shall meet with the Cost Consultant to review the Construction Documents. The Architect shall be entitled to rely on the services provided by the Cost Consultant.

§ 7.5   Article 5: Owner's Responsibilities

Alternative language more favorable to the architect:

5.2 The Owner shall furnish the services of a Scheduling Consultant that shall be responsible for creating the overall Project schedule, and shall be responsible for monitoring the activities of the Contractor for conformance to the schedule, as well as advising the Owner and Architect from time to time of any deviations by the Contractor from such schedule, as amended. The Owner shall adjust the Project schedule, if necessary, as the Project proceeds.

5.3 The Owner shall establish and periodically update the Owner's budget for the Project, including (1) the budget for the Cost of the Work as defined in

Section 6.1; (2) the Owner's other costs; and, (3) reasonable contingencies related to all of these costs. The Owner shall furnish the services of a Cost Consultant that shall be responsible for (1) preparing all estimates of the Cost of the Work, (2) assisting the Architect with all facets of budgeting and estimating for the Project, (3) analyzing the Architect's drawings, specifications and other materials on an ongoing basis in order to advise the Architect whether the design is in conformity with the budget. If the Owner significantly increases or decreases the Owner's budget for the Cost of the Work, the Owner shall notify the Architect. The Owner and the Architect shall thereafter agree to a corresponding change in the budget for the Cost of the Work or in the Project's scope and quality.

5.3.1 The Owner acknowledges that accelerated, phased or fast-track scheduling provides a benefit, but also carries with it associated risks. Such risks include the Owner incurring costs for the Architect to coordinate and redesign portions of the Project affected by procuring or installing elements of the Project prior to the completion of all relevant Construction Documents, and costs for the Contractor to remove and replace previously installed Work. If the Owner selects accelerated, phased or fast-track scheduling, the Owner agrees to include in the budget for the Project sufficient contingencies to cover such costs, and, in such event, the Owner waives any claim that Owner may have against the Architect and the Architect's consultants relating to delay, increased costs of the Project, errors or omissions by the Architect or the Architect's consultants related to such accelerated, phased or fast-track scheduling.

5.3.1.1 In further recognition of the risks inherent with accelerated, phased and fast-track scheduling, the Owner waives any and all claims against the Architect to the extent that the Cost of the Work exceeds the Owner's budget at the time of execution of the Owner-Contractor Agreement by more than twenty-five (25%) percent, not including Owner-requested changes that result in increased costs, and such increase is the result of errors or omissions by the Architect and/or the Architect's consultants.

§ 7.6   Article 6: Cost of the Work

Alternate language more favorable to the architect:

6.1 For purposes of this Agreement, the Cost of the Work shall be the greater of the total cost to the Owner to construct all elements of the Project designed or specified by the Architect, or the fair market value of all such elements; and shall include contractors' general conditions costs, overhead and profit. The Cost of the Work does not include the compensation of the Architect, the costs of the land, rights-of-way, financing, contingencies for changes in the Work or other costs that are the responsibility of the Owner.

6.2 The Owner's budget for the Cost of the Work is provided in Initial Information, and may be adjusted throughout the Project as required under Sections 5.3

and 6.4. Evaluations of the Owner's budget for the Cost of the Work represent the Architect's judgment as a design professional, but to the extent that the Owner's Cost Consultant has evaluated the Owner's budget or provided any input towards that budget, the Architect shall have no liability to the Owner, it being understood that both the Owner and the Architect are relying on the Owner's Cost Consultant for all cost information.

6.3 The Owner shall require the Cost Consultant to include appropriate contingencies for design, bidding or negotiating, price escalation, and market conditions in estimates of the Cost of the Work. The Architect shall be entitled to rely on the accuracy and completeness of estimates of the Cost of the Work the Cost Consultant prepares as the Architect progresses with its Basic Services, as well as the advice of the Cost Consultant during all phases of the Architect's work on the Project. The Architect shall prepare, as an Additional Service, revisions to the Drawings, Specifications or other documents required due to the Cost Consultant's inaccuracies or incompleteness in preparing cost estimates, as well as any misleading or inaccurate information or advice given by the Cost Consultant. The Architect may review the Cost Consultant's estimates solely for the Architect's guidance in completion of its services, however, the Architect shall report to the Owner any material inaccuracies and inconsistencies noted during any such review. The Architect shall not be liable to the Owner or the Cost Consultant for the results of such reports to the Owner.

6.4 If, prior to the conclusion of the Design Development Phase, the Cost Consultant's estimate of the Cost of the Work exceeds the Owner's budget for the Cost of the Work, the Architect, in consultation with the Cost Consultant, shall make appropriate recommendations to the Owner to adjust the Project's size, quality or budget, and the Owner shall cooperate with the Architect in making such adjustments.

6.5 If the estimate of the Cost of the Work at the conclusion of the Design Development Phase exceeds the Owner's budget for the Cost of the Work, the Owner shall

.1 give written approval of an increase in the budget for the Cost of the Work;

.2 in consultation with the Architect and Cost Consultant, revise the Project program, scope, or quality as required to reduce the Cost of the Work; or

.3 implement any other mutually acceptable alternative.

6.6 If the Owner chooses to proceed under Section 6.5.2, the Architect, as an Additional Service, shall incorporate the required modifications in the Construction Documents Phase as necessary to comply with the Owner's budget for the Cost of the Work at the conclusion of the Design Development Phase Services, or the budget as adjusted under Section 6.5.1. The Architect shall have no liability to the Owner for any damages resulting from cost overruns or any variance from the Owner's budget, the parties having acknowledged that the

Architect is relying on the Cost Consultant throughout the design process for cost information.

6.7 After incorporation of modifications under Section 6.6, the Architect shall, as an Additional Service, make any required revisions to the Drawings, Specifications or other documents necessitated by subsequent cost estimates that exceed the Owner's budget for the Cost of the Work.

§ 7.7   Claims and Disputes: ¶ 8.1

A version more architect-friendly:

8.1.1 The Owner and Architect shall commence all claims and causes of action, whether in contract, tort, or otherwise, against the other arising out of or related to this Agreement in accordance with the requirements of the method of binding dispute resolution selected in this Agreement within the period specified by applicable law, but in any case not more than four years after the date of Substantial Completion of the Work. The Owner and Architect waive all claims and causes of action not commenced in accordance with this Section 8.1.1.

Another provision that clarifies that the architect's limit of liability to the owner is not just limited to this single paragraph:

8.1.3 The Architect shall indemnify and hold the Owner and the Owner's officers and employees harmless from and against damages, losses and judgments arising from claims by third parties, including reasonable attorneys' fees and expenses recoverable under applicable law, but only to the extent they are caused by the negligent acts or omissions of the Architect, its employees and its consultants in the performance of professional services under this Agreement. The Architect's duty to indemnify the Owner under this provision, as well as the Architect's liability to the Owner pursuant to any other provision of this Agreement, statute or otherwise, shall be limited to the available proceeds of insurance coverage.

CHAPTER 8

AIA DOCUMENT B104 STANDARD FORM OF AGREEMENT BETWEEN OWNER AND ARCHITECT FOR A PROJECT OF LIMITED SCOPE

§ 8.1   Introduction

AIA Document B104 is a new document based on B101-2007, and is an abbreviated version of that document. It is also similar to B151-1997. It is designed to be used for projects of limited scope. This document refers to AIA Document A107-2007 (which contains general conditions similar to A201-2007), and is intended to be used with that document.

To conserve space, reference is made to AIA Document B101, found in **Chapter 2,** whenever the language is the same. Readers should then consult the corresponding provision in that document for comments. At the end of each paragraph of this document is a boldface reference to the 1997 version of B141, or to the 1997 version of B151 indicated thus: **B151-2.2.5**. This is to assist the reader in relating this document to the corresponding language in that version of B141.

§ 8.2   Prior Editions

The current edition of B104 was issued in November 2007 as part of a major revision of the most significant AIA documents. An extremely important point to remember is that documents from different series cannot be mixed. Thus, the 2007 B104 cannot be used on the same project as the 1997 versions of other AIA documents, such as A201.[1] The text in this chapter uses the most recent version of B104. However, the referenced court opinions are from earlier versions of other AIA documents. Because use of the current B104 started in November 2007, it should be expected that no court opinions relating to this edition will appear for at least a year following release. The language of many parts of related or similar documents, however, is close enough to provide insights into judicial interpretation of certain sections of B104. Relevant cases should be closely read to determine if they apply to specific situations.

Future revisions to this document are expected every 10 years. At the end of each section of AIA text there is a reference to the prior version of the B141-1997 language in bold text.

---

[1] An example of problems cause by mixing documents is found in Eis Group/Cornwall Hill v. Rinaldi Constr., 154 A.D.2d 429, 546 N.Y.S.2d 105 (1989), in which the owner-contractor agreement was a 1987 AIA form that incorporated the 1987 version of A201. However, the 1976 version of A201 was actually attached to the contract, along with various modifications which included striking the arbitration clause. The court found an ambiguity (obviously!) and declined to order arbitration.

CAUTION: The recommended method of amending this document is to use the electronic documents software that is available from the AIA. Using this software, the document is amended on its face, with additions and deletions indicated either in the margin, or by underlines and strikethroughs. Another way is to attach separate written amendments that refer back to B104. This is normally done by filling in Article 12, or additional pages referred to in Article 13. An alternative method is to graphically delete material and insert new material in the margins of the standard form. However, this method may create ambiguity, and it may result in the stricken material's being rendered illegible. It is illegal to make any copies of this document in violation of the AIA copyright or to reproduce this document by computerized means. Refer to the instruction sheet that comes with B104 for additional information about that copyright.

The cited language of this document is as published by AIA in late 2007 in the Electronic Documents. There appear to be some minor inconsistencies between this document and others that are probably attributable to scrivenors' errors. AIA often corrects such errors in intervening years without any notice and without any way to determine exactly when a particular version of a document was actually released.

§ 8.3   Title Page

See the discussion in **§ 2.3.**

§ 8.4   Article 1: Initial Information

ARTICLE 1 INITIAL INFORMATION

1.1 This Agreement is based on the Initial Information set forth below:

(State below details of the Project's site and program, Owner's contractors and consultants, Architect's consultants, Owner's budget for the Cost of the Work, and other information relevant to the Project.) (**1.1.1; modifications**)

Here, the parties should identify as much initial information as possible. The parties should consider using Exhibit A to AIA Document B101 for this purpose and attach that document as an exhibit here or listing that document in Section 13.2.

1.2 The Owner and Architect may rely on the Initial Information. Both parties, however, recognize that such information may materially change and, in that event, the Owner and the Architect shall appropriately adjust the schedule, the Architect's services and the Architect's compensation. (**1.1.6; revisions**)

See the discussion of ¶ 1.3 of B101 in **§ 2.4.**

§ 8.5   Article 2: Architect's Responsibilities

ARTICLE 2 ARCHITECT'S RESPONSIBILITIES

The Architect shall provide the professional services set forth in this Agreement consistent with the professional skill and care ordinarily provided by architects practicing in the same or similar locality under the same or similar circumstances. The Architect shall perform its services as expeditiously as is consistent with such professional skill and care and the orderly progress of the Project. (**1.2.3.1, 1.2.3.2; substantial revisions**)

See the discussion of ¶¶ 2.1 and 2.2 of B101 in § **2.5.**

§ 8.6   Article 3: Scope of Architect's Basic Services

ARTICLE 3 SCOPE OF ARCHITECT'S BASIC SERVICES

3.1 The Architect's Basic Services consist of those described in Article 3 and include usual and customary structural, mechanical, and electrical engineering services. (**2.4.1; modified**)

See the discussion of ¶ 3.1 of B101 in § **2.6.** This paragraph omits the language of Section 3.1 of B101, stating that "Services not set forth in Article 3 are Additional Services." This omitted language is actually redundant, since the definition here would exclude any services not set forth in Article 3 or the "usual and customary" services specified.

3.1.1 The Architect shall be entitled to rely on (1) the accuracy and completeness of the information furnished by the Owner and (2) the Owner's approvals. The Architect shall provide prompt written notice to the Owner if the Architect becomes aware of any error, omission or inconsistency in such services or information. (**1.2.3.7; revisions**)

See the discussion of ¶ 3.1.2 of B101 in § **2.6.**

3.1.2 As soon as practicable after the date of this Agreement, the Architect shall submit for the Owner's approval a schedule for the performance of the Architect's services. Once approved by the Owner, time limits established by the schedule shall not, except for reasonable cause, be exceeded by the Architect or Owner. With the Owner's approval, the Architect shall adjust the schedule, if necessary, as the Project proceeds until the commencement of construction. (**1.2.3.2; substantial revisions**)

B101, at Section 3.1.3, calls for a combined schedule, with information about both the work of the architect as well as that of the contractor. This is in keeping with the fact that this document does not contain a section where the proposed dates of commencement of construction and substantial completion are indicated, such as at Section 1.2 of B101. This schedule is only for the architect's work.

3.1.3 The Architect shall assist the Owner in connection with the Owner's respon-sibility for filing documents required for the approval of governmental authorities having jurisdiction over the Project. **(2.1.6; no change)**

See the discussion of ¶ 3.1.6 of B101 in **§ 2.6.**

§ 8.7   Design Phase Services: ¶ 3.2

3.2 DESIGN PHASE SERVICES

3.2.1 The Architect shall review the program and other information furnished by the Owner, and shall review laws, codes, and regulations applicable to the Architect's services. **(1.2.3.6; substantial revisions)**

See the discussion of ¶ 3.2.1 of B101 in **§ 2.7.**

3.2.2 The Architect shall discuss with the Owner the Owner's program, schedule, budget for the Cost of the Work, Project site, and alternative approaches to design and construction of the Project, including the feasibility of incorporating environ-mentally responsible design approaches. The Architect shall reach an understanding with the Owner regarding the Project requirements. **(2.3.1, 2.3.2, 2.3.3; substantial modifications)**

See the discussion of ¶ 3.2.2 of B101 in **§ 2.7.** This paragraph, unlike B101, requires the architect to "discuss" with the owner the program and other informa-tion. B101, on the other hand, requires the architect to prepare an evaluation of these items. This is something that would require substantially more work and effort.

3.2.3 The Architect shall consider the relative value of alternative materials, build-ing systems and equipment, together with other considerations based on program and aesthetics in developing a design for the Project that is consistent with the Owner's schedule and budget for the Cost of the Work. **(B151-2.2.3; modifications)**

Under this provision, the architect is to develop a design that takes into account the program and budget of the owner, as well as the proposed schedule, program, and so forth. This is less specific than the corresponding provisions of B101.

3.2.4 Based on the Project requirements, the Architect shall prepare Design Docu-ments for the Owner's approval consisting of drawings and other documents appro-priate for the Project and the Architect shall prepare and submit to the Owner a preliminary estimate of the Cost of the Work. **(new)**

Unlike the B101, which has two design phases: Schematic Design and Design Development, this contract only requires one design phase. The design drawings will be submitted to the owner for approval, along with the architect's estimate of the construction cost based on that set of design drawings.

3.2.5 *The Architect shall submit to the Owner an estimate of the Cost of the Work prepared in accordance with Section 6.3.* **(B151-2.2.5; modifications)**

See the discussion of ¶ 3.2.6 of B101 in § **2.7.**

3.2.6 *The Architect shall submit the Design Documents to the Owner, and request the Owner's approval.* **(new)**

It is important to obtain the owner's approval of the design documents, otherwise there may be a dispute as to whether the architect should be paid for any work in the construction documents phase. If the owner is not willing to sign a document confirming approval, the architect should send a letter to the owner advising that the architect considers the design phase completed and requesting an acknowledgment from the owner.[2]

§ 8.8   Construction Documents Phase Services: ¶ 3.3

3.3 CONSTRUCTION DOCUMENTS PHASE SERVICES

3.3.1 *Based on the Owner's approval of the Design Documents, the Architect shall prepare for the Owner's approval Construction Documents consisting of Drawings and Specifications setting forth in detail the requirements for the construction of the Work. The Owner and Architect acknowledge that in order to construct the Work the Contractor will provide additional information, including Shop Drawings, Product Data, Samples and other similar submittals, which the Architect shall review in accordance with Section 3.4.4.* **(2.4.4.1; substantial revisions)**

See the discussion of ¶ 3.4.1 of B101 in § **2.9.**

3.3.2 *The Architect shall incorporate into the Construction Documents the design requirements of governmental authorities having jurisdiction over the Project.* **(1.2.3.6; revisions)**

See the discussion of ¶ 3.4.2 of B101 in § **2.9.**

3.3.3 *The Architect shall update the estimate for the Cost of the Work.* **(2.4.3; revisions)**

See the discussion of ¶ 3.4.4 of B101 in § **2.9.**

3.3.4 *The Architect shall submit the Construction Documents to the Owner, advise the Owner of any adjustments to the estimate of the Cost of the Work, take any action required under Section 6.5, and request the Owner's approval.* **(B151-2.3.2; substantial modifications)**

See the discussion of ¶ 3.4.5 of B101 in § **2.9.**

---

[2] For an example of a form that could be used for this purpose, see ¶ 4.3.1.1 of B101.

3.3.5 The Architect, following the Owner's approval of the Construction Documents and of the latest preliminary estimate of Construction Cost, shall assist the Owner in awarding and preparing contracts for construction. **(B151-2.5; modifications)**

This paragraph wraps up the completion of the construction documents phase along with the bidding and negotiating phase found in the B101. On smaller projects, bidding procedures are often very informal.

§ 8.9   Construction Phase Services: ¶ 3.4

3.4 CONSTRUCTION PHASE SERVICES

3.4.1 GENERAL

3.4.1.1 The Architect shall provide administration of the Contract between the Owner and the Contractor as set forth below and in AIA Document A107™-2007, Standard Form of Agreement Between Owner and Contractor for a Project of Limited Scope. If the Owner and Contractor modify AIA Document A107-2007, those modifications shall not affect the Architect's services under this Agreement unless the Owner and the Architect amend this Agreement. **(2.6.1.1; substantial modifications)**

This clause provides that the architect can assume that the General Conditions found in A107-2007 will apply to the construction. If the owner decides to use other general conditions, the architect may be entitled to additional compensation, because that may affect a number of the documents the architect prepares. There is a potential for conflict in this provision if the owner modifies the General Conditions or uses nonstandard general conditions. Under this language, the architect would presumably be entitled to extra compensation for services that it is asked to perform under nonstandard general conditions. Usually, the architect participates in modifications to the General Conditions (A201, or, in this case, A107) by preparing supplementary general conditions and specifications. Presumably, this participation would constitute a written approval by the architect under this provision in lieu of a written amendment to this agreement.

Of course, it is always a good idea for the architect to immediately notify the owner of any discrepancy between the owner-architect agreement and any other nonstandard agreement.

3.4.1.2 The Architect shall advise and consult with the Owner during the Construction Phase Services. The Architect shall have authority to act on behalf of the Owner only to the extent provided in this Agreement. The Architect shall not have control over, charge of, or responsibility for the construction means, methods, techniques, sequences or procedures, or for safety precautions and programs in connection with the Work, nor shall the Architect be responsible for the Contractor's failure to perform the Work in accordance with the requirements of the Contract Documents. The Architect shall be responsible for the Architect's negligent acts or omissions, but shall not have control over or charge of and shall not be responsible for, acts or omissions of the Contractor or of any other persons or entities performing portions of the Work. **(2.6.1.3, 2.6.2.1; substantial modifications)**

See the discussion of ¶ 3.6.1.2 of B101 in § **2.13.**

3.4.1.3 *Subject to Section 4.2, the Architect's responsibility to provide Construction Phase Services commences with the award of the Contract for Construction and terminates on the date the Architect issues the final Certificate for Payment.* **(2.6.1.2; modifications; deleted reference to entitlement to additional fees after 60 days after Substantial Completion)**

See the discussion of ¶ 3.6.1.3 of B101 in § **2.13.**

3.4.2 EVALUATIONS OF THE WORK

3.4.2.1 *The Architect shall visit the site at intervals appropriate to the stage of construction, or as otherwise required in Section 4.2.1, to become generally familiar with the progress and quality of the portion of the Work completed, and to determine, in general, if the Work observed is being performed in a manner indicating that the Work, when fully completed, will be in accordance with the Contract Documents. However, the Architect shall not be required to make exhaustive or continuous on-site observations to check the quality or quantity of the Work. On the basis of the site visits, the Architect shall keep the Owner reasonably informed about the progress and quality of the portion of the Work completed, and report to the Owner (1) known deviations from the Contract Documents and from the most recent construction schedule submitted by the Contractor, and (2) defects and deficiencies observed in the Work.* **(2.6.2.1, 2.6.2.2; substantial modifications)**

See the discussion of ¶ 3.6.2.1 of B101 in § **2.14.**

3.4.2.2 *The Architect has the authority to reject Work that does not conform to the Contract Documents and has the authority to require inspection or testing of the Work.* **(2.6.2.5; substantial modification)**

See the discussion of ¶ 3.6.2.2 of B101 in § **2.14.** Note that B101 contains language that states that the architect assumes no duty towards the contractor or others working on the project. Section 10.5 of this document performs the same function.

3.4.2.3 *The Architect shall interpret and decide matters concerning performance under, and requirements of, the Contract Documents on written request of either the Owner or Contractor. The Architect's response to such requests shall be made in writing within any time limits agreed upon or otherwise with reasonable promptness.* **(2.6.1.7; minor modification)**

See the discussion of ¶ 3.6.2.3 of B101 in § **2.14.**

3.4.2.4 *When making such interpretations and decisions, the Architect shall endeavor to secure faithful performance by both Owner and Contractor, shall not show partiality to either, and shall not be liable for results of interpretations or* *decisions rendered in good faith.* (**2.6.1.8; modifications; deleted reference to aesthetic decisions**)

See the discussion of ¶ 3.6.2.4 of B101 in § **2.14.** The absence of the language of B101 that "(t)he Architect's decisions on matters relating to aesthetic effect shall be final if consistent with the intent expressed in the Contract Documents," means that all decisions of the architect, including aesthetic decisions, are subject to further dispute resolution between the owner and contractor.

3.4.2.5 The Architect shall render initial decisions on Claims between the Owner and Contractor as provided in the Contract Documents. (**2.6.1.9**)

See the discussion of ¶ 3.6.2.5 of B101 in § **2.14.** Unlike B101, there is no reference here to an Initial Decision Maker, as the architect will perform that role.

3.4.3 CERTIFICATES FOR PAYMENT TO CONTRACTOR

3.4.3.1 The Architect shall review and certify the amounts due the Contractor and shall issue certificates in such amounts. The Architect's certification for payment shall constitute a representation to the Owner, based on the Architect's evaluation of the Work as provided in Section 3.4.2 and on the data comprising the Contractor's Application for Payment, that, to the best of the Architect's knowledge, information and belief, the Work has progressed to the point indicated and that the quality of the Work is in accordance with the Contract Documents. (**2.6.3.1; deleted last sentence of prior section**)

3.4.3.2 The issuance of a Certificate for Payment shall not be a representation that the Architect has (1) made exhaustive or continuous on-site inspections to check the quality or quantity of the Work, (2) reviewed construction means, methods, techniques, sequences or procedures, (3) reviewed copies of requisitions received from Subcontractors and material suppliers and other data requested by the Owner to substantiate the Contractor's right to payment, or (4) ascertained how or for what purpose the Contractor has used money previously paid on account of the Contract Sum. (**2.6.3.2; no change**)

See the discussion of ¶ 3.6.3 of B101 in § **2.15.**

3.4.4 SUBMITTALS

3.4.4.1 The Architect shall review and approve or take other appropriate action upon the Contractor's submittals such as Shop Drawings, Product Data and Samples, but only for the limited purpose of checking for conformance with information given and the design concept expressed in the Contract Documents. Review of such submittals is not for the purpose of determining the accuracy and completeness of other information such as dimensions, quantities, and installation or performance of equipment or systems, which are the Contractor's responsibility. The Architect's review shall not constitute approval of safety precautions or, unless otherwise specifically stated by the Architect, of any construction means, methods, techniques, sequences or procedures. (**2.6.4.1; substantial modifications**)

See the discussion of ¶ 3.6.4.2 of B101 in **§ 2.16.** B101, at ¶ 3.6.4.1, contains a reference to the contractor's submittal schedule, which is absent here. This is because the projects for which this document will be used will be less complex and do not warrant the preparation of a submittal schedule. Also absent is the last sentence of ¶ 3.6.4.2, referring to the architect's approval of a specific item as not indicating approval of the entire assembly. This omission should not be taken as implying the reverse of the B101 language.

3.4.4.2 If the Contract Documents specifically require the Contractor to provide professional design services or certifications by a design professional related to systems, materials or equipment, the Architect shall specify the appropriate performance and design criteria that such services must satisfy. The Architect shall review shop Drawings and other submittals related to the Work designed or certified by the design professional retained by the Contractor that bear such professional's seal and signature when submitted to the Architect. The Architect shall be entitled to rely upon the adequacy, accuracy and completeness of the services, certifications and approvals performed or provided by such design professionals. **(2.6.4.3; minor modifications)**

See the discussion of ¶ 3.6.4.3 of B101 in **§ 2.16.**

3.4.4.3 The Architect shall review and respond to written requests for information about the Contract Documents. The Architect's response to such requests shall be made in writing within any time limits agreed upon, or otherwise with reasonable promptness. **(2.6.1.5; modified)**

In the corresponding section of B101, ¶ 3.6.4.4, there is language that the architect is to set forth requirements in the contract documents for requests for information. Such requirements should make it more difficult for contractors to overwhelm the architect with RFIs that are redundant or unnecessary. On smaller projects, for which this document is designed, that should be less of a concern, hence the elimination of this language.

3.4.5 CHANGES IN THE WORK

The Architect may authorize minor changes in the Work that are consistent with the intent of the Contract Documents and do not involve an adjustment in the Contract Sum or an extension of the Contract Time. Subject to the provisions of Section 4.2.2, the Architect shall prepare Change Orders and Construction Change Directives for the Owner's approval and execution in accordance with the Contract Documents. **(2.6.5.1; substantial modifications)**

See the discussion of ¶ 3.6.5.1 of B101 in **§ 2.17.**

3.4.6 PROJECT COMPLETION

The Architect shall conduct inspections to determine the date or dates of Substantial Completion and the date of final completion; issue Certificates of Substantial Completion; receive from the Contractor and forward to the Owner, for the Owner's review and records, written warranties and related documents required by the

Contract Documents and assembled by the Contractor; and issue a final Certificate for Payment based upon a final inspection indicating the Work complies with the requirements of the Contract Documents. **(2.6.6.1; minor modifications)**

See the discussion of ¶ 3.6.6.1 of B101 in § **2.18.**

§ 8.10   Article 4: Additional Services

ARTICLE 4 ADDITIONAL SERVICES

4.1 Additional Services are not included in Basic Services but may be required for the Project. Such Additional Services may include programming, budget analysis, financial feasibility studies, site analysis and selection, environmental studies, civil engineering, landscape design, telecommunications/data, security, measured drawings of existing conditions, coordination of separate contractors or independent consultants, coordination of construction or project managers, detailed cost estimates, on-site project representation beyond requirements of Section 4.2.1, value analysis, quantity surveys, interior architectural design, planning of tenant or rental spaces, inventories of materials or equipment, preparation of record drawings, commissioning, environmentally responsible design beyond Basic Services, LEED® Certification, fast-track design services, and any other services not otherwise included in this Agreement. **(new)**

(Insert a description of each Additional Service the Architect shall provide, if not further described in an exhibit attached to this document.)

B101, at Article 4, includes a table of possible additional services, with assignments of responsibility for each. The parties may want to consider using that table here as a way to clarify which services are contemplated, and by whom. Without some specificity in the contract, the owner and architect will sometimes have a dispute over whether a particular service was a basic service or an additional service.

4.2 Additional Services may be provided after execution of this Agreement, without invalidating the Agreement. Except for services required due to the fault of the Architect, any Additional Services provided in accordance with this Section 4.2 shall entitle the Architect to compensation pursuant to Section 11.3. **(1.3.3.1; extensive modifications)**

If additional services are required of the architect, that will not invalidate this agreement, unless the changes are so significant as to be considered a *cardinal change*.[3] Note that any service that is due to the fault of the architect will not entitle the architect to any additional compensation. This would include any error or omission in the contract documents.

---

[3] See the discussion of cardinal changes at § 4.42.

***4.2.1** The Architect has included in Basic Services ( ) site visits over the duration of the Project during construction. The Architect shall conduct site visits in excess of that amount as an Additional Service.* (**2.8.1.1; modified**)

The parties will insert the number of anticipated site visits. If, during the project, more site visits are required, perhaps because the contractor has questions or there are more pay requests than anticipated, the architect will be entitled to additional compensation. It would be a good idea to notify the owner prior to such additional site visits to avoid the situation where the owner refuses to pay for these additional services on the grounds that the owner would not have approved them had the owner known about the situation.

***4.2.2** The Architect shall review and evaluate Contractor's proposals, and if necessary, prepare Drawings, Specifications and other documentation and data, and provide any other services made necessary by Change Orders and Construction Change Directives prepared by the Architect as an Additional Service.* (**2.8.2.3; modifications**)

The contractor may make proposals concerning changes to the project. The architect is to evaluate such proposals and, if accepted by the owner, if the proposals require any drawings or other documents, the architect will be entitled to additional fees for preparing such documents. This paragraph is not entirely clear as to whether the architect is entitled to additional fees for evaluating every proposal by the contractor. The logical answer is yes. The owner and architect should contemplate that the contractor will perform the work in exact compliance with the contract documents. If the contractor wants to modify the contract documents for any reason, and provides a proposal for review by the owner and architect, that should result in additional fees to the architect for this additional work, but only if the owner authorizes the architect to make such a review. In many cases, the review will be minor and the architect may waive such additional fees.

***4.2.3** If the services covered by this Agreement have not been completed within ___ (___) months of the date of this Agreement, through no fault of the Architect, extension of the Architect's services beyond that time shall be compensated as Additional Services.* (**1.5.9; revise reference to Additional Services**)

See the discussion of ¶ 4.3.4 of B101 in **§ 2.20.**

§ 8.11 Article 5: Owner's Responsibilities

ARTICLE 5 OWNER'S RESPONSIBILITIES

***5.1** Unless otherwise provided for under this Agreement, the Owner shall provide information in a timely manner regarding requirements for and limitations on the Project, including a written program which shall set forth the Owner's objectives, schedule, constraints and criteria, including space requirements and relationships, flexibility, expandability, special equipment, systems and site requirements. Within*

15 days after receipt of a written request from the Architect, the Owner shall furnish the requested information as necessary and relevant for the Architect to evaluate, give notice of or enforce lien rights. **(1.2.2.1; modifications)**

See the discussion of ¶ 5.1 of B101 in **§ 2.21.**

5.2 The Owner shall establish and periodically update the Owner's budget for the Project, including (1) the budget for the Cost of the Work as defined in Section 6.1; (2) the Owner's other costs; and, (3) reasonable contingencies related to all of these costs. If the Owner significantly increases or decreases the Owner's budget for the Cost of the Work, the Owner shall notify the Architect. The Owner and the Architect shall thereafter agree to a corresponding change in the Project's scope and quality. **(1.2.2.2; revisions)**

See the discussion of ¶ 5.2 of B101 in **§ 2.21.**

5.3 The Owner shall furnish surveys to describe physical characteristics, legal limitations and utility locations for the site of the Project, a written legal description of the site, and services of geotechnical engineers or other consultants when the Architect requests such services and demonstrates that they are reasonably required by the scope of the Project. **(2.2.1.2; 2.2.1.3; modifications)**

See the discussion of ¶¶ 5.4 and 5.5 of B101 in **§ 2.21.** The language of B101 contains more extensive requirements as to the survey, and it also requires the services of a geotechnical engineer. Here, such services will be provided only if requested by the architect and necessary for the project.

5.4 The Owner shall coordinate the services of its own consultants with those services provided by the Architect. Upon the Architect's request, the Owner shall furnish copies of the scope of services in the contracts between the Owner and the Owner's consultants. The Owner shall require that its consultants maintain professional liability insurance as appropriate to the services provided. **(1.2.2.4; substantial modifications)**

See the discussion of ¶ 5.6 of B101 in **§ 2.21.**

5.5 The Owner shall furnish tests, inspections and reports required by law or the Contract Documents, such as structural, mechanical, and chemical tests, tests for air and water pollution, and tests for hazardous materials. **(1.2.2.5; minor revisions)**

See the discussion of ¶ 5.7 of B101 in **§ 2.21.**

5.6 The Owner shall furnish all legal, insurance and accounting services, including auditing services that may be reasonably necessary at any time for the Project to meet the Owner's needs and interests. **(1.2.2.6; no change)**

See the discussion of ¶ 5.8 of B101 in **§ 2.21.**

5.7 The Owner shall provide prompt written notice to the Architect if the Owner becomes aware of any fault or defect in the Project, including errors, omissions or inconsistencies in the Architect's Instruments of Service. **(1.2.2.7; minor modification)**

See the discussion of ¶ 5.9 of B101 in § **2.21.**

5.8 The Owner shall endeavor to communicate with the Contractor through the Architect about matters arising out of or relating to the Contract Documents. **(2.6.2.4; modifications)**

See the discussion of ¶ 5.10 of B101 in § **2.21.**

5.9 The Owner shall provide the Architect access to the Project site prior to commencement of the Work and shall obligate the Contractor to provide the Architect access to the Work wherever it is in preparation or progress. **(2.6.2.3; modifications)**

See the discussion of ¶ 5.12 of B101 in § **2.21.**

§ 8.12 Article 6: Cost of the Work

ARTICLE 6 COST OF THE WORK

6.1 For purposes of this Agreement, the Cost of the Work shall be the total cost to the Owner to construct all elements of the Project designed or specified by the Architect and shall include contractors' general conditions costs, overhead and profit. The Cost of the Work does not include the compensation of the Architect, the costs of the land, rights-of-way, financing, contingencies for changes in the Work or other costs that are the responsibility of the Owner. **(1.3.1.1, 1.3.1.2, 1.3.1.3; substantial modifications)**

See the discussion of ¶ 6.1 of B101 in § **2.22.**

6.2 The Owner's budget for the Cost of the Work is provided in Initial Information, and may be adjusted throughout the Project as required under Sections 5.2, 6.4 and 6.5. Evaluations of the Owner's budget for the Cost of the Work, the preliminary estimate of the Cost of the Work and updated estimates of the Cost of the Work prepared by the Architect, represent the Architect's judgment as a design professional. It is recognized, however, that neither the Architect nor the Owner has control over the cost of labor, materials or equipment; the Contractor's methods of determining bid prices; or competitive bidding, market or negotiating conditions. Accordingly, the Architect cannot and does not warrant or represent that bids or negotiated prices will not vary from the Owner's budget for the Cost of the Work or from any estimate of the Cost of the Work or evaluation prepared or agreed to by the Architect. **(2.1.7.2; minor modifications; add first sentence)**

See the discussion of ¶ 6.2 of B101 in § **2.22.**

6.3 In preparing estimates of the Cost of Work, the Architect shall be permitted to include contingencies for design, bidding and price escalation; to determine what materials, equipment, component systems and types of construction are to be included in the Contract Documents, to make reasonable adjustments in the program and scope of the Project and to include in the Contract Documents alternate bids as may be necessary to adjust the estimated Cost of the Work to meet the Owner's budget for the Cost of the Work. The Architect's estimate of the Cost of the Work shall be based on current area, volume or similar conceptual estimating techniques. If the Owner requests detailed cost estimating services, the Architect shall provide such services as an Additional Service under Article 4. (**2.1.7.3; add last two sentences; minor modifications**)

See the discussion of ¶ 6.3 of B101 in § **2.22.**

6.4 If the bidding has not commenced within 90 days after the Architect submits the Construction Documents to the Owner, through no fault of the Architect, the Owner's budget for the Cost of the Work shall be adjusted to reflect changes in the general level of prices in the applicable construction market. (**2.1.7.4; modifications**)

See the discussion of ¶ 6.4 of B101 in § **2.22.**

6.5 If at any time the Architect's estimate of the Cost of the Work exceeds the Owner's budget for the Cost of the Work, the Architect shall make appropriate recommendations to the Owner to adjust the Project's size, quality or budget for the Cost of the Work, and the Owner shall cooperate with the Architect in making such adjustments. (**2.1.7.1, minor modifications**)

See the discussion of ¶ 6.5 of B101 in § **2.22.**

6.6 If the Owner's current budget for the Cost of the Work at the conclusion of the Construction Documents Phase Services is exceeded by the lowest bona fide bid or negotiated proposal, the Owner shall (**2.1.7.5; add "current"**)

   .1   give written approval of an increase in the budget for the Cost of the Work; (**2.1.7.5.1; no change**)

   .2   authorize rebidding or renegotiating of the Project within a reasonable time; (**2.1.7.5.2; no change**)

   .3   terminate in accordance with Section 9.5; (**2.1.7.5.3; change reference**)

   .4   in consultation with the Architect, revise the Project program, scope, or quality as required to reduce the Cost of the Work; or (**2.1.7.5.4; minor modifications***)

   .5   implement any other mutually acceptable alternative. (**new**)

See the discussion of ¶ 6.6 of B101 in § **2.22.**

6.7 If the Owner chooses to proceed under Section 6.6.4, the Architect, without additional compensation, shall modify the Construction Documents as necessary to comply with the Owner's budget for the Cost of the Work at the conclusion of the Construction Documents Phase Services, or the budget as adjusted under Section

6.6.1. The Architect's modification of the Construction Documents shall be the limit of the Architect's responsibility under this Article 6. **(2.1.7.6; modifications)**

See the discussion of ¶ 6.7 of B101 in **§ 2.22.**

§ 8.13   Article 7: Copyrights and Licenses

ARTICLE 7 COPYRIGHTS AND LICENSES

7.1 The Architect and the Owner warrant that in transmitting Instruments of Service, or any other information, the transmitting party is the copyright owner of such information or has permission from the copyright owner to transmit such information for its use on the Project. If the Owner and Architect intend to transmit Instruments of Service or any other information or documentation in digital form, they shall endeavor to establish necessary protocols governing such transmissions. **(new)**

See the discussion of ¶ 7.1 of B101 in **§ 2.23.**

7.2 The Architect and the Architect's consultants shall be deemed the authors and owners of their respective Instruments of Service, including the Drawings and Specifications, and shall retain all common law, statutory and other reserved rights, including copyrights. Submission or distribution of Instruments of Service to meet official regulatory requirements or for similar purposes in connection with the Project is not to be construed as publication in derogation of the reserved rights of the Architect and the Architect's consultants. **(1.3.2.1; 1.3.2.3; add reference to consultants and electronic form)**

See the discussion of ¶ 7.2 of B101 in **§ 2.23.**

7.3 Upon execution of this Agreement, the Architect grants to the Owner a nonexclusive license to use the Architect's Instruments of Service solely and exclusively for purposes of constructing, using, maintaining, altering and adding to the Project, provided that the Owner substantially performs its obligations, including prompt payment of all sums when due, under this Agreement. The Architect shall obtain similar nonexclusive licenses from the Architect's consultants consistent with this Agreement. The license granted under this section permits the Owner to authorize the Contractor, Subcontractors, Sub-subcontractors, and material or equipment suppliers, as well as the Owner's consultants and separate contractors, to reproduce applicable portions of the Instruments of Service solely and exclusively for use in performing services or construction for the Project. If the Architect rightfully terminates this Agreement for cause as provided in Section 9.4, the license granted in this Section 7.3 shall terminate. **(1.3.2.2; 1.3.2.3; substantial modifications)**

See the discussion of ¶ 7.3 of B101 in **§ 2.23.**

7.3.1 In the event the Owner uses the Instruments of Service without retaining author of the Instruments of Service, the Owner releases the Architect and Architect's consultant(s) from all claims and causes of action arising from such uses. The Owner, to the extent permitted by law, further agrees to indemnify and hold harmless the Architect and its consultants from all costs and expenses, including the cost of defense, related to claims and causes of action asserted by any third person or entity to the extent such costs and expenses arise from the Owner's use of the Instruments of Service under this Section 7.3.1. The terms of this Section 7.3.1 shall not apply if the Owner rightfully terminates this Agreement for cause under Section 9.4. **(1.3.2.2; revisions)**

See the discussion of ¶ 7.3.1 of B101 in **§ 2.23.**

7.4 Except for the licenses granted in this Article 7, no other license or right shall be deemed granted or implied under this Agreement. The Owner shall not assign, delegate, sublicense, pledge or otherwise transfer any license granted herein to another party without the prior written agreement of the Architect. Any unauthorized use of the Instruments of Service shall be at the Owner's sole risk and without liability to the Architect and the Architect's consultants. **(1.3.2.2; substantial modifications)**

See the discussion of ¶ 7.4 of B101 in **§ 2.23.**

§ 8.14   Article 8: Claims and Disputes

ARTICLE 8 CLAIMS AND DISPUTES

8.1 GENERAL

8.1.1 The Owner and Architect shall commence all claims and causes of action, whether in contract, tort, or otherwise, against the other arising out of or related to this Agreement in accordance with the requirements of the method of binding dispute resolution selected in this Agreement within the period specified by applicable law, but in any case not more than 10 years after the date of Substantial Completion of the Work. The Owner and Architect waive all claims and causes of action not commenced in accordance with this Section 8.1.1. **(1.3.7.3; completely rewritten; established ten-year outside limit on all claims)**

See the discussion of ¶ 8.1.1 of B101 in **§ 2.24**.

8.1.2 To the extent damages are covered by property insurance, the Owner and Architect waive all rights against each other and against the contractors, consultants, agents and employees of the other for damages, except such rights as they may have to the proceeds of such insurance as set forth in AIA Document A107™-2007, Standard Form of Agreement Between Owner and Contractor for a Project of Limited Scope. The Owner or the Architect, as appropriate, shall require of the contractors, consultants, agents and employees of any of them similar waivers in favor of the other parties enumerated herein. **(1.3.7.4; deleted limitation of time to "during construction"; changed reference)**

See the discussion of ¶ 8.1.2 of B101 in **§ 2.24.**

8.1.3 The Architect and Owner waive consequential damages for claims, disputes or other matters in question arising out of or relating to this Agreement. This mutual waiver is applicable, without limitation, to all consequential damages due to either party's termination of this Agreement, except as specifically provided in Section 9.6. **(1.3.6; minor modifications)**

See the discussion of ¶ 8.1.3 of B101 in **§ 2.24.**

§ 8.15   Mediation: ¶ 8.2

8.2 MEDIATION

8.2.1 *Any claim, dispute or other matter in question arising out of or related to this Agreement shall be subject to mediation as a condition precedent to binding dispute resolution. If such matter relates to or is the subject of a lien arising out of the Architect's services, the Architect may proceed in accordance with applicable law to comply with the lien notice or filing deadlines prior to resolution of the matter by mediation or by binding dispute resolution.* **(1.3.4.1; modified to add binding dispute resolution process)**

See the discussion of ¶ 8.2.1 of B101 in **§ 2.25.**

8.2.2 *Mediation, unless the parties mutually agree otherwise, shall be administered by the American Arbitration Association in accordance with its Construction Industry Mediation Procedures in effect on the date of the Agreement. The parties shall share the mediator's fee and any filing fees equally. The mediation shall be held in the place where the Project is located, unless another location is mutually agreed upon. Agreements reached in mediation shall be enforceable as settlement agreements in any court having jurisdiction thereof.* **(1.3.4.2; 1.3.4.3; substantial modifications)**

See the discussion of ¶¶ 8.2.2 and 8.2.3 of B101 in **§ 2.25.** The language here does not include the 60-day stay provision found in B101.

8.2.3 *If the parties do not resolve a dispute through mediation pursuant to this Section 8.2, the method of binding dispute resolution shall be the following:*

(Check the appropriate box. If the Owner and Architect do not select a method of binding dispute resolution below, or do not subsequently agree in writing to a binding dispute resolution method other than litigation, the dispute will be resolved in a court of competent jurisdiction.) **(new)**

   ☐ *Arbitration pursuant to Section 8.3 of this Agreement*
   or

   ☐ *Litigation in a court of competent jurisdiction*
   or

   ☐ *Other (Specify)* **(new)**

See the discussion of ¶ 8.2.4 of B101 in **§ 2.25.**

§ 8.16   Arbitration: ¶ 8.3

8.3 ARBITRATION

8.3.1 *If the parties have selected arbitration as the method for binding dispute resolution in this Agreement any claim, dispute or other matter in question arising out of or related to this Agreement subject to, but not resolved by, mediation shall be subject to arbitration which, unless the parties mutually agree otherwise, shall be administered by the American Arbitration Association in accordance with its Construction Industry Arbitration Rules in effect on the date of the Agreement.* **(1.3.5.3, 1.3.5.2; modified to reflect modifications to binding dispute resolution procedures; minor modifications)**

See the discussion of ¶ 8.3.1 of B101 in **§ 2.26.** Note that the last sentence of this paragraph in B101 has been deleted here.

8.3.1.1 *A demand for arbitration shall be made no earlier than concurrently with the filing of a request for mediation, but in no event shall it be made after the date when the institution of legal or equitable proceedings based on the claim, dispute or other matter in question would be barred by the applicable statute of limitations. For statute of limitations purposes, receipt of a written demand for arbitration by the person or entity administering the arbitration shall constitute the institution of legal or equitable proceedings based on the claim, dispute or other matter in question.* **(1.3.5.3; added last sentence; minor modifications)**

See the discussion of ¶ 8.3.1.1 of B101 in **§ 2.26.**

8.3.2 *The foregoing agreement to arbitrate and other agreements to arbitrate with an additional person or entity duly consented to by parties to this Agreement shall be specifically enforceable in accordance with applicable law in any court having jurisdiction thereof.* **(1.3.5.4-last sentence; no change)**

8.3.3 *The award rendered by the arbitrator(s) shall be final, and judgment may be entered upon it in accordance with applicable law in any court having jurisdiction thereof.* **(1.3.5.5; no change)**

See the discussion of ¶ 8.3.3 of B101 in **§ 2.26.**

8.3.4 CONSOLIDATION OR JOINDER

8.3.4.1 *Either party, at its sole discretion, may consolidate an arbitration conducted under this Agreement with any other arbitration to which it is a party provided that (1) the arbitration agreement governing the other arbitration permits consolidation; (2) the arbitrations to be consolidated substantially involve common questions of law or fact; and (3) the arbitrations employ materially similar procedural rules and methods for selecting arbitrator(s).*

8.3.4.2 *Either party, at its sole discretion, may include by joinder persons or entities substantially involved in a common question of law or fact whose presence is required if complete relief is to be accorded in arbitration, provided that the* *party sought to be joined consents in writing to such joinder. Consent to arbitration involving an additional person or entity shall not constitute consent to arbitration of any claim, dispute or other matter in question not described in the written consent.*

8.3.4.3 *The Owner and Architect grant to any person or entity made a party to an arbitration conducted under this Section 8.3, whether by joinder or consolidation, the same rights of joinder and consolidation as the Owner and Architect under this Agreement.* **(1.3.5.4; enabled joinder instead of prohibiting it)**

See the discussion of ¶ 8.3.4 of B101 in **§ 2.27.**

§ 8.17   Article 9: Termination or Suspension

ARTICLE 9 TERMINATION OR SUSPENSION

9.1 *If the Owner fails to make payments to the Architect in accordance with this Agreement, such failure shall be considered substantial nonperformance and cause for termination or, at the Architect's option, cause for suspension of performance of services under this Agreement. If the Architect elects to suspend services, the Architect shall give seven days' written notice to the Owner before suspending services. In the event of a suspension of services, the Architect shall have no liability to the Owner for delay or damage caused the Owner because of such suspension of services. Before resuming services, the Architect shall be paid all sums due prior to suspension and any expenses incurred in the interruption and resumption of the Architect's services. The Architect's fees for the remaining services and the time schedules shall be equitably adjusted.* **(1.3.8.1; minor modifications)**

See the discussion of ¶ 9.1 of B101 in **§ 2.28.**

9.2 *If the Owner suspends the Project, the Architect shall be compensated for services performed prior to notice of such suspension. When the Project is resumed, the Architect shall be compensated for expenses incurred in the interruption and resumption of the Architect's services. The Architect's fees for the remaining services and the time schedules shall be equitably adjusted.*

See the discussion of ¶ 9.2 of B101 in **§ 2.28.**

9.3 *If the Owner suspends the Project for more than 90 cumulative days for reasons other than the fault of the Architect, the Architect may terminate this Agreement by giving not less than seven days' written notice.*

See the discussion of ¶ 9.3 of B101 in **§ 2.28.**

9.4 *Either party may terminate this Agreement upon not less than seven days' written notice should the other party fail substantially to perform in accordance with the terms of this Agreement through no fault of the party initiating the termination.*

See the discussion of ¶ 9.4 of B101 in **§ 2.28.**

9.5 The Owner may terminate this Agreement upon not less than seven days' written notice to the Architect for the Owner's convenience and without cause.

See the discussion of ¶ 9.5 of B101 in § **2.28.**

9.6 In the event of termination not the fault of the Architect, the Architect shall be compensated for services performed prior to termination, together with Reimbursable Expenses then due and all Termination Expenses as defined in Section 9.7.

See the discussion of ¶ 9.6 of B101 in § **2.28.**

9.7 Termination Expenses are in addition to compensation for the Architect's services and include expenses directly attributable to termination for which the Architect is not otherwise compensated, plus an amount for the Architect's anticipated profit on the value of the services not performed by the Architect.

See the discussion of ¶ 9.7 of B101 in § **2.28.**

§ 8.18   Article 10: Miscellaneous Provisions

ARTICLE 10 MISCELLANEOUS PROVISIONS

10.1 This Agreement shall be governed by the law of the place where the Project is located, except that if the parties have selected arbitration as the method of binding dispute resolution, the Federal Arbitration Act shall govern Section 8.3. (**1.3.7.1; modifications**)

See the discussion of ¶ 10.1 of B101 in § **2.29.**

10.2 Terms in this Agreement shall have the same meaning as those in AIA Document A107-2007, Standard Form of Agreement Between Owner and Contractor for a Project of Limited Scope. (**1.3.7.2; modified reference**)

This document anticipates that the general conditions found in A107-2007 will be utilized in the contract between the owner and contractor. If other general conditions are to be used, this reference must be changed.

10.3 The Owner and Architect, respectively, bind themselves, their agents, successors, assigns and legal representatives to this Agreement. Neither the Owner nor the Architect shall assign this Agreement without the written consent of the other, except that the Owner may assign this Agreement to a lender providing financing for the Project if the lender agrees to assume the Owner's rights and obligations under this Agreement. (**1.3.7.9; modifications**)

See the discussion of ¶ 10.3 of B101 in § **2.29.**

10.4 If the Owner requests the Architect to execute certificates or consents, the proposed language of such certificates or consents shall be submitted to the Architect for review at least 14 days prior to the requested dates of execution. The Architect shall not be required to execute certificates or consents that would require knowledge, services or responsibilities beyond the scope of this Agreement. **(1.3.7.8; modifications)**

See the discussion of ¶ 10.4 of B101 in § **2.29.** This provision has the same intent as the corresponding one in B101.

10.5 Nothing contained in this Agreement shall create a contractual relationship with or a cause of action in favor of a third party against either the Owner or Architect. **(1.3.7.5; no change)**

See the discussion of ¶ 10.5 of B101 in § **2.29.**

10.6 The Architect shall have no responsibility for the discovery, presence, handling, removal or disposal of or exposure of persons to hazardous materials or toxic substances in any form at the Project site. **(1.3.7.6; modifications)**

See the discussion of ¶ 10.6 of B101 in § **2.29.** Note that B101 includes "unless otherwise required in this Agreement."

10.7 The Architect shall have the right to include photographic or artistic representations of the design of the Project among the Architect's promotional and professional materials. However, the Architect's materials shall not include information the Owner has identified in writing as confidential or proprietary. **(1.3.7.7; modifications)**

See the discussion of ¶ 10.7 of B101 in § **2.29.** B101 gives the architect the specific right of access to take photographs, while that is omitted here. Also eliminated here is the requirement that the owner give the architect professional credit in the owner's promotional material. This is, presumably, because the projects intended for this document will be smaller projects of lesser value to the architect as promotional material.

§ 8.19   Article 11: Compensation

ARTICLE 11 COMPENSATION

11.1 For the Architect's Basic Services as described under Article 3, the Owner shall compensate the Architect as follows:

(Insert amount of, or basis for, compensation.) **(1.5.1; change reference; minor modifications)**

See the discussion of ¶ 11.1 of B101 in § **2.30.**

11.2 *For Additional Services designated in Section 4.1, the Owner shall compensate the Architect as follows:*

(Insert amount of, or basis for, compensation. If necessary, list specific services to which particular methods of compensation apply.) **(new)**

See the discussion of ¶ 11.2 of B101 in **§ 2.30.**

11.3 *For Additional Services that may arise during the course of the Project, including those under Section 4.2, the Owner shall compensate the Architect as follows:*

(Insert amount of, or basis for, compensation.) **(1.5.2; modifications)**

See the discussion of ¶ 11.3 of B101 in **§ 2.30.**

11.4 *Compensation for Additional Services of the Architect's consultants when not included in Section 11.2 or 11.3, shall be the amount invoiced to the Architect plus (),* *or as otherwise stated below:* **(1.5.3; modifications)**

See the discussion of ¶ 11.4 of B101 in **§ 2.30**.

11.5 *Where compensation for Basic Services is based on a stipulated sum or percentage of the Cost of the Work, the compensation for each phase of services shall be as follows:*

Design Development Phase:	*percent (%)*
Construction Documents Phase:	*percent (%)*
Construction Phase:	*percent (%)*
Total Basic Compensation:	ONE HUNDRED PERCENT (100%)

See the discussion of ¶ 11.5 of B101 in **§ 2.30.** Note that the first phase in this document is designated here as the "Design Development Phase," while in Section 3.2 it is designated as "Design Phase Services." This is likely due to scrivenor's error and may be corrected in future releases of this document.

11.6 *When compensation is based on a percentage of the Cost of the Work and any portions of the Project are deleted or otherwise not constructed, compensation for those portions of the Project shall be payable to the extent services are performed on those portions, in accordance with the schedule set forth in Section 11.5 based on (1) the lowest bona fide bid or negotiated proposal, or (2) if no such bid or proposal is received, the most recent estimate of the Cost of the Work for such portions of the Project. The Architect shall be entitled to compensation in accordance with this Agreement for all services performed whether or not the Construction Phase is commenced.* **(new)**

See the discussion of ¶ 11.6 of B101 in **§ 2.30.**

11.7 The hourly billing rates for services of the Architect and the Architect's consultants, if any, are set forth below. The rates shall be adjusted in accordance with the Architect's and Architect's consultants' normal review practices.

(If applicable, attach an exhibit of hourly billing rates or insert them below.)

Employee or Category   ***Rate*** **(1.5.6; modified)**

See the discussion of ¶ 11.7 of B101 in § **2.7.**

§ 8.20   Compensation for Reimbursable Expenses: ¶ 11.8

11.8 COMPENSATION FOR REIMBURSABLE EXPENSES

11.8.1 *Reimbursable Expenses are in addition to compensation for Basic and Additional Services and include expenses incurred by the Architect and the Architect's consultants directly related to the Project, as follows:* **(1.3.9.2; modifications)**

See the discussion of ¶ 11.8.1 of B101 in § **2.31.**

> *.1 Transportation and authorized out-of-town travel and subsistence;* **(1.3.9.2.1; modifications)**

See the discussion of ¶ 11.8.1.1 of B101 in § **2.31.**

> *.2 Long distance services, dedicated data and communication services, teleconferences, Project Web sites, and extranets;* **(1.3.9.2.1; substantially expanded definition)**

See the discussion of ¶ 11.8.1.2 of B101 in § **2.31.**

> *.3 Fees paid for securing approval of authorities having jurisdiction over the Project;* **(1.3.9.2.2; no change)**

See the discussion of ¶ 11.8.1.3 of B101 in § **2.31.**

> *.4 Printing, reproductions, plots, standard form documents;* **(1.3.9.2.3; minor modifications)**

See the discussion of ¶ 11.8.1.4 of B101 in § **2.31.**

> *.5 Postage, handling and delivery;* **(1.3.9.2.3; minor modifications)**

See the discussion of ¶ 11.8.1.5 of B101 in § **2.31.**

> *.6 Expense of overtime work requiring higher than regular rates if authorized in advance by the Owner;* **(1.3.9.2.4; no change)**

See the discussion of ¶ 11.8.1.6 of B101 in § **2.31.**

    .7   *Renderings, models, mock-ups, professional photography, and presentation materials requested by the Owner;* (**1.3.9.2.5; added photography and presentation materials**)

See the discussion of ¶ 11.8.1.7 of B101 in § **2.31.**

    .8   *Expense of professional liability insurance dedicated exclusively to this Project or the expense of additional insurance coverage or limits requested by the Owner in excess of that normally carried by the Architect and the Architect's consultants;* (**1.3.9.2.6; modifications**)

Many architects today do not carry professional liability insurance of any kind. If the owner insists on that insurance, the cost of coverage in excess of what the architect normally carries is reimbursable. If the architect does not carry any insurance, then this provision should be modified to state that it carries no insurance. The owner should be aware of the policy limits and the length of time this amount of insurance will be carried past the termination of the project. Although most claims are made soon after the end of the project, claims can be made many years after final completion. Each state has different statutes of limitations and repose. The architect should seek legal and insurance counsel whenever an owner requests additional insurance.

    .9   *All taxes levied on professional services and on reimbursable expenses;* (**new**)

See the discussion of ¶ 11.8.1.9 of B101 in § **2.31.**

    .10 *Site office expenses; and* (**new**)

See the discussion of ¶ 11.8.1.10 of B101 in § **2.31.**

    .11 *Other similar Project-related expenditures.* (**1.3.9.2.8; no change**)

See the discussion of ¶ 11.8.1.11 of B101 in § **2.31.**

11.8.2 For Reimbursable Expenses the compensation shall be the expenses incurred by the Architect and the Architect's consultants plus ( ) of the expenses incurred. (**1.5.4; modifications**)

See the discussion of ¶ 11.8.2 of B101 in § **2.31.**

§ 8.21   Compensation for Use of Architect's Instruments of Service: ¶ 11.9

11.9 COMPENSATION FOR USE OF ARCHITECT'S INSTRUMENTS OF SERVICE

If the Owner terminates the Architect for its convenience under Section 9.5, or the Architect terminates this Agreement under Section 9.3, the Owner shall pay a licensing fee as compensation for the Owner's continued use of the Architect's Instruments of Service solely for purposes of completing, using and maintaining the Project as follows:

See the discussion of ¶ 11.9 of B101 in § **2.32.**

§ 8.22 Payments to the Architect: ¶ 11.10

11.10 PAYMENTS TO THE ARCHITECT

11.10.1 *An initial payment of ($) shall be made upon execution of this Agreement and is the minimum payment under this Agreement. It shall be credited to the Owner's account in the final invoice.* (**1.5.7; modifications**)

See the discussion of ¶ 11.10.1 of B101 in § **2.33.**

11.10.2 *Unless otherwise agreed, payments for services shall be made monthly in proportion to services performed. Payments are due and payable upon presentation of the Architect's invoice. Amounts unpaid   (   ) days after the invoice date shall bear interest at the rate entered below, or in the absence thereof at the legal rate prevailing from time to time at the principal place of business of the Architect.*

(Insert rate of monthly or annual interest agreed upon.) (**1.5.7, 1.5.8; modifications**)

See the discussion of ¶ 11.10.2 of B101 in § **2.33.**

11.10.3 *The Owner shall not withhold amounts from the Architect's compensation to impose a penalty or liquidated damages on the Architect, or to off set sums requested by or paid to contractors for the cost of changes in the Work unless the Architect agrees or has been found liable for the amounts in a binding dispute resolution proceeding.* (**1.3.9.1; substantial modifications**)

See the discussion of ¶ 11.10.3 of B101 in § **2.33.**

11.10.4 *Records of Reimbursable Expenses, expenses pertaining to Additional Services, and services performed on the basis of hourly rates shall be available to the Owner at mutually convenient times.* (**1.3.9.3; modifications**)

See the discussion of ¶ 11.10.4 of B101 in § **2.33.**

§ 8.23 Article 12: Special Terms and Conditions

ARTICLE 12 SPECIAL TERMS AND CONDITIONS

Special terms and conditions that modify this Agreement are as follows: (**1.4.2; No change**)

See the discussion of Article 12 of B101 in § **2.34.**

§ 8.24   Article 13: Scope of the Agreement

ARTICLE 13 SCOPE OF THE AGREEMENT

13.1 *This Agreement represents the entire and integrated agreement between the Owner and the Architect and supersedes all prior negotiations, representations or agreements, either written or oral. This Agreement may be amended only by written instrument signed by both Owner and Architect.* (**1.4.1; Deleted last sentence of prior paragraph**)

See the discussion of ¶ 13.1 of B101 in § **2.35.**

13.2 *This Agreement incorporates the following documents listed below:*

(List other documents, if any, including additional scopes of service and AIA Document E201™-2007, Digital Data Protocol Exhibit, if completed, forming part of the Agreement.)

With the widespread use of drawings and specifications in digital form, and contractors and owners requesting access to the architect's digital files, accommodations must be made with regards to digital files. See Section 1.6 of A201, which is a new section in 2007 regarding transmission of data in digital form. AIA Document E201-2007 covers procedures regarding the exchange of digital data.

The architect may also want to incorporate other documents as part of the agreement, such as an earlier proposal. Caution should be taken to make sure that such documents are consistent with the terms of this document. Any document not referenced here is not a part of the contract between the parties pursuant to the integration clause of Section 13.1.

This Agreement entered into as of the day and year first written above.

OWNER   *ARCHITECT*

_____   _____
(Signature)   *(Signature)*

_____   _____
(Printed name and title)   *(Printed name and title)*

CHAPTER 9

ALTERNATIVE LANGUAGE FOR AIA DOCUMENT B104

§ 9.1   Introduction

The language in this chapter is offered as possible alternative language to AIA Document B104. Much of the language in B104 is identical to the language in B101. Therefore, this chapter only covers the language of B104 that is different than that in B101. Users should consult **Chapter 3** for those paragraphs that are the same as in B101.

The user is cautioned that some of these provisions are mutually incompatible, and each paragraph and each sentence should be carefully reviewed with legal counsel before use. Note further that some of these changes require changes to other documents. Finally, please note that no recommendations are being made or implied by this alternative language. Much of it is more favorable to the owner than the standard AIA language, while some is more favorable to the architect. Each party should consult with an attorney familiar with construction law in the appropriate jurisdiction before making any changes to the standard document, as such changes will often result in serious consequences to the parties. If in doubt, consideration should be given to using the AIA language as-is.

The recommended method of amending B104 is to use the AIA's Electronic Documents software and make the changes directly to the body of the document. Alternatively, attach separate written amendments that refer back to B104. This is normally done in riders to the agreement, referenced in Article 13, or inserting language in Article 12. Another method is to graphically delete material and insert new material in the margins of the standard form. However, this method may create ambiguity, and it may result in rendering the stricken material illegible.

§ 9.2   Architect's Schedule: ¶ 3.1.2

Here is an alternate provision that makes the schedule optional:

3.1.2 During the performance of this Agreement, the Architect may submit a schedule to the Owner showing the anticipated dates for performance of the Architect's services. This schedule shall take into account the Owner's proposed dates for the award of the contract for construction. However, the parties recognize that the Architect has no control over matters such as approvals by governmental authorities having jurisdiction over the project and, accordingly, the Architect does not guarantee that such schedule will be met.

§ 9.3   Schematic Design Phase Services: ¶ 3.2.4

Here is a provision that allows the architect to designate how many different designs will be submitted to the owner for approval. Depending on the project and the proposed fee, the architect may want to include one, two, or even three designs as Basic Services, with additional designs as Additional Services:

3.2.4 Based on the Project requirements, the Architect shall prepare Design Documents for the Owner's approval consisting of drawings and other documents appropriate for the Project and the Architect shall prepare and submit to the Owner a preliminary estimate of the Cost of the Work. If the Owner requests that the Architect prepare more than [one] design[s] for the project, such additional designs shall be considered Additional Services.

§ 9.4   Submittals: ¶ 3.4.4

This provision is an alternative to the form language:

3.4.4.1 The Architect shall review and approve or take other appropriate action upon the Contractor's submittals such as Shop Drawings, Product Data and Samples, but only for the limited purpose of checking for conformance with information given and the design concept expressed in the Contract Documents. Review of such submittals is not for the purpose of determining the accuracy and completeness of other information such as dimensions, quantities, and installation or performance of equipment or systems, which are the Contractor's responsibility. The Architect's review shall not constitute approval of safety precautions or of any construction means, methods, techniques, sequences or procedures. The Architect may return unnecessary or unwanted submittals without review. To the extent the Contractor does not promptly and properly correct rejected submittals, the Architect shall be entitled to be paid for repeated reviews of the same submittal as Additional Services.

Requests for Information can be a problem for architects. This modifies the form language to allow the architect to be paid for improper RFIs:

3.4.4.3 The Architect shall review and respond to written requests for information about the Contract Documents. The Architect's response to such requests shall be made in writing within any time limits agreed upon, or otherwise with reasonable promptness. However, to the extent that the information requested by the Contractor would be readily ascertainable from a review of the Contract Documents, or otherwise available to the Contractor or known to a reasonably competent Contractor, the Architect's responses to such unnecessary requests for information shall be compensated as Additional Services.

§ 9.5   Additional Services: Article 4

This language modifies the form language more favorably to the architect:

4.1 Additional Services are not included in Basic Services but may be required for the Project. Such Additional Services may include (by way of example only, and not as a limitation): programming, budget analysis, financial feasibility studies, site analysis and selection, environmental studies, civil engineering, landscape design, telecommunications/data, security, measured drawings of existing conditions, coordination of separate contractors or independent consultants, coordination of construction or project managers, detailed cost estimates, on-site project representation beyond requirements of Section 4.2.1, value analysis, quantity surveys, interior architectural design, planning of tenant or rental spaces, inventories of materials or equipment, preparation of record drawings, commissioning, environmentally responsible design beyond Basic Services, LEED® Certification, fast-track design services, and any other services not otherwise specifically included in this Agreement.

This version is more favorable to the owner:

4.1 Additional Services are not included in Basic Services but may be required for the Project. Such Additional Services may include programming, budget analysis, financial feasibility studies, site analysis and selection, environmental studies, civil engineering, landscape design, telecommunications/data, security, measured drawings of existing conditions, coordination of separate contractors or independent consultants, coordination of construction or project managers, detailed cost estimates, on-site project representation beyond requirements of Section 4.2.1, value analysis, quantity surveys, interior architectural design, planning of tenant or rental spaces, inventories of materials or equipment, preparation of record drawings, commissioning, environmentally responsible design beyond Basic Services, LEED® Certification, fast-track design services, and any other services not otherwise included, or reasonably inferable as Basic Services in this Agreement. Prior to performing any Additional Services, Architect shall obtain the written consent of Owner. Any services performed by the Architect without such prior written consent, except in the case of emergency, shall be considered Basic Services, notwithstanding Section 4.2 hereof.

Paragraph 4.2.2 is not very clear as to the circumstances where the architect is to be paid for additional services. Here are some variations:

4.2.2 The Architect shall review and evaluate Contractor's proposals, and if necessary, prepare Drawings, Specifications and other documentation and data, and provide any other services made necessary by Change Orders and Construction Change Directives prepared by the Architect. Such reviews and evaluations shall be an Additional Service.

Another:

4.2.2 To the extent the Architect has prepared Change Orders and Construction Change Directives as an Additional Service, the Architect shall review and evaluate Contractor's proposals, and if necessary, prepare Drawings, Specifications and other documentation and data, and provide any other services made necessary thereby. Such reviews and evaluations shall [not] be an Additional Service.

§ 9.6   Compensation: Article 11

This provision corrects an internal inconsistency in the document by correctly identifying the first phase:

11.5 Where compensation for Basic Services is based on a stipulated sum or percentage of the Cost of the Work, the compensation for each phase of services shall be as follows:

Design Phase:	percent (%)
Construction Documents Phase:	percent (%)
Construction Phase:	percent (%)
Total Basic Compensation:	

CHAPTER 10

AIA DOCUMENT A101 STANDARD FORM OF AGREEMENT BETWEEN OWNER AND CONTRACTOR[1]

§ 10.1   Title Page

AGREEMENT made as of the _____ day of _____ in the year of _____

The date of the agreement should be as early as possible. This can assist in obtaining lien priority, depending on the circumstances.

BETWEEN the Owner:

The owner must be properly identified.[2] The "owner" might not be the title holder of the property but rather a tenant or a beneficiary of the owner. Care should be taken to identify the actual title holder and, if possible, to list that party and obtain its permission. It might be advisable to notify (by registered or certified

---

[2] In Keller Constr. Co. v. Kashani, 220 Cal. App. 3d 222, 269 Cal. Rptr. 259 (1990), the architect entered into a contract with a limited partnership. Kashani, the sole general partner, signed the agreement on behalf of the partnership. A dispute developed and the partnership filed bankruptcy. The architect sought arbitration against Kashani. The court held that Kashani was bound by the arbitration agreement.

In Cheek v. Uptown Square Wine Merchants, 538 So. 2d 663 (La. Ct. App. 1989), the owner argued that he signed the AIA contract in a representative capacity and not individually. The court held otherwise, based on the plain language of the contract.

In Silver Dollar City v. Kitsmiller Constr., 931 S.W.2d 909 (Mo. Ct. App. 1996), the owner thought it was contracting with a joint venture that consisted of two parties. The first page of the contract named both entities the "Contractor" but the signature page listed only one. There was thus an apparent ambiguity and the contract could have been voided by the owner at that time, particularly because the contractors knew of the mistake. However, by failing to take prompt action to rescind or correct the contract, the court found that the owner had ratified the contract.

mail) this "real" owner that the construction work is under way, so that there can be no objection later that the owner was unaware of the work. Any lien would be filed against the interest of this true owner, even if that owner did not know that any work was going on.

The status of the owner should also be indicated. The following are some examples:

Smith Partners, an Illinois general partnership

Smith Partners, Ltd., an Illinois limited partnership

Smith Corp., an Illinois corporation

John Smith, authorized agent for the Owner, XYZ

Smith Bank, Trustee under Trust No. 111.

In the last example, the party executing the contract is a trustee for the owner. In Illinois, for example, a bank can hold title to a parcel of property as a trustee for the beneficial owner. This is sometimes referred to as a secret land trust because the beneficiary is not disclosed to the public. This beneficiary is the true owner and directs the trustee to deal with the property. Usually, the trustee would execute the owner-contractor agreement, which would include an exculpatory clause saying something to the effect that the parties agree that the trustee is acting merely as a trustee and will not be personally liable in case of any default. In this situation, the contractor may find that it cannot enforce its contract against anyone: the trustee who executed the contract has the exculpatory clause, and the beneficiary (true owner) did not sign the contract. Of course, it might be possible to bring in the beneficiary on some other theories, but it would be much cleaner if the beneficiary also executes the contract as an additional party in interest at the outset.

When the owner is a corporation, the contractor would be wise to try to add an individual as a signatory to the contract. This provides additional protection, but the owner may also insist that the contractor also sign in both a corporate and individual capacity. For instance:

Smith Corp., an Illinois corporation and John Smith, individually.[3]

If the owner is a governmental body, the contractor should be sure that the body passed the proper resolution to enable it to enter into the contract. Without such authority, it may be difficult or impossible for the contractor to collect a fee.[4]

---

[3] *See, e.g.,* Cheek v. Uptown Square Wine Merchants, 538 So. 2d 663 (La. Ct. App. 1989).

[4] County of Stephenson v. Bradley & Bradley, Inc., 2 Ill. App. 3d 421, 275 N.E.2d 675 (1971). In Blue Ridge Sewer Improvement Dist. v. Lowry & Assocs., 149 Ariz. 373, 718 P.2d 1026, 1027 (Ct. App. 1986), an engineer was unable to collect fees. He attempted to collect on the basis of quantum meruit (unjust enrichment or implied contract). The court held that "one who provides services under a contract with a political subdivision may not recover for the value of those services under quantum meruit if the contract was entered into in violation of a state law requiring approval of a

Further, lien laws relating to public work are frequently different from those for private work.

and the Contractor:

Here, insert the name of the contractor and its status, such as "an Illinois corporation." Note that an unlicensed contractor may not be entitled to enforce the contract.[5]

---

majority of property owners in the area affected." *Accord*, Scofield Eng'g Co. v. City of Danville, 126 F.2d 942 (4th Cir. 1942); Galion Iron Works & Mfg. Co. v. City of Georgetown, 322 Ill. App. 498, 54 N.E.2d 601 (1944); Dempsey v. City Univ., 106 A.D.2d 486, 483 N.Y.S.2d 24 (1984); Annotation, *Implied Public Contracts*, 154 A.L.R. 356 (1945).

In Murphy v. City of Brockton, 364 Mass. 377, 305 N.E.2d 103, 105 (1973), the architect was unable to collect more from the city than the amount of the initial appropriation. The court stated that "it is our opinion that the plaintiff was obligated to proceed no further with its work under the contract than was covered by an appropriation," despite the fact that a contract was signed between the parties to cover all architectural services.

[5] Alonzo v. Chifici, 526 So. 2d 237 (La. Ct. App. 1988) (contract void because contractor was required to have license).

In Settimo Assocs. v. Environ Sys., Inc., 17 Cal. Rptr. 2d 757 (Ct. App. 1993), an unsuccessful bidder on a private construction project brought suit against the successful bidder for intentional or negligent interference with prospective economic advantage. At the time of bidding, the low bidder held some but not all licenses required to perform the work. Subsequently, that contractor acquired the necessary licenses. The next lowest bidder argued that it submitted the lowest bid of the bidders that were properly licensed. The court denied relief, stating that a private entity was not required to accept bids only from duly licensed contractors. The court noted that the successful bidder's conduct amounted to a misdemeanor that foreclosed any possibility of its suing to enforce its contract.

In Koehler v. Donnelly, 114 N.M. 363, 839 P.2d 980 (1992), an unlicensed contractor was allowed to enforce its lien because it had substantially complied with the licensing requirements, even though it did not hold a valid license when it entered into the contract. The license inadvertently lapsed without the contractor's knowledge, and the contractor took immediate steps to renew the license after learning of the cancellation. In addition, the contractor was competent and financially responsible during the license lapse.

In Spaw-Glass Constr. Servs. v. Vista de Santa Fe, 114 N.M. 557, 844 P.2d 807 (1992), the contractor's failure to be properly licensed should have been raised in the arbitration hearing as an affirmative defense, not in the court confirmation hearing. The same result was reached in In re Hirsch Constr. Corp., 181 A.D.2d 52, 585 N.Y.S.2d 418 (1992), in which the arbitrator held in favor of the contractor. Later, the owner learned that the contractor was not properly licensed. In the subsequent court proceedings to confirm the arbitration award, the owner brought this to the court's attention and lost. The appellate court affirmed, holding that the lack of license should have been raised in the arbitration and that the award could not be challenged on the basis of newly discovered evidence. The court reviewed the public policy argument that the contractor should not be permitted to profit from its lack of license, but it found that the public policy of upholding arbitration awards was not in conflict with the public policy that contractors should be licensed.

In a Missouri case, Strain-Japan R-16 Sch. Dist. v. Landmark Sys., Inc., 965 S.W.2d 278 (Mo. Ct. App. 1998), the contractor sought to arbitrate a dispute with the owner. The owner sued for an injunction against the arbitration proceeding, based on the fact that the contractor was not licensed as an architectural firm, while Missouri law makes contracts for engineering services rendered by an unlicensed firm unenforceable. The work was for a pre-engineered metal building that was to

The Project is:

Here, insert the name and location of the project.

The Architect is:

Here, insert the name and address of the architect for the project.

In an unusual case,[6] Montgomery owned a building in Fayette, Mississippi, which it leased to Family Dollar Stores. In 2001, Family Dollar hired Bowman to be the general contractor for an expansion of the store. Bowman was not licensed as a general contractor and it had been some 35 years since Bowman had been involved with a project that required removal of a load-bearing wall. Bowman hired Turner to be the subcontractor for certain of the work, including removal of a load-bearing wall. While Turner was working on this wall, the roof of the building partially collapsed, injuring a number of patrons and employees of the store, which was open at the time.

Montgomery sued Family Dollar for breach of contract. The form contract that was used had a space for identifying the architect. Someone had entered the name of Family Dollar into that space. The contract had no description of the architect's

---

conform to preliminary plans furnished by the owner. The contract called for plans by the general contractor and specified the particular manufacturer of the pre-engineered structure. The contractor then arranged for plans to be prepared by licensed engineers, which were then approved by the owner. In the lawsuit, the owner argued that its contract with the general contractor is a contract for architectural or engineering services, since it requires the contractor to provide those services. The court rejected these arguments, finding that if that argument were correct, either the owner would be forced to contract directly with architects or engineers, or all corporations engaged in general contracting would need to register as architects or engineers. The court was not willing to read the statutes in that manner. Furthermore, the court found that the owner contracted not for professional engineering services but for a building conforming to its initial sketches.

In Space Planners Architects, Inc. v. Frontier Town-Missouri, Inc., 107 S.W.3d 398 (Mo. App. 2003), an architect sought to foreclose a mechanic's lien. On a motion for summary judgment, the architect presented an uncertified copy of the required architect's license. This was insufficient for the appellate court, however. That court stated that the architect had to plead and prove as part of its case that it was properly licensed. Failing to provide the court with a certified copy of the license required reversal of the summary judgment in the architect's favor.

Another court held that an owner could raise the lack of proper license as a defense against a contractor in Agway, Inc. v. Williams, 185 A.D.2d 636, 585 N.Y.S.2d 643, 644 (1992). The contractor agreed to construct a dairy facility for the owner, including barns as well as various labor-saving devices. The licensing laws in question did not apply "to farm buildings, including barns, sheds, poultry houses and other buildings used directly and solely for agricultural purposes." The court held that this exception "must be read narrowly and must be restricted only so far as its language fairly warrants." The exception was read to apply only to the barns and not the labor-saving devices.

In Judelson v. Christopher O'Connor, Inc., No. 95 CV 0371181, 1995 Conn. Super. LEXIS 1375 (May 2, 1995), a state statute required the contractor to be licensed as a home improvement contractor. Because the contractor was not properly licensed, the arbitration provision was not enforceable by the contractor.

[6] Family Dollar Stores of Mississippi, Inc. v. Montgomery, 946 So. 2d 426 (Miss. App. 2006).

duties (presumably, not an AIA document). The trial court found in favor of Montgomery on the basis of expert testimony that Family Dollar, as the designated architect, breached its implied duties by failing to advise the contractors concerning the temporary shoring of the structure. Family Dollar had argued that the form contract had been improperly completed and that the entry for an architect should have been left blank. Family Dollar argued that neither itself nor Bowman intended for there to be an architect on the project.

At trial, there was substantial evidence on which Montgomery relied that Family Dollar ignored its responsibilities as an architect. Family Dollar relied on the same evidence to support its position that there was no architect. The trial court found that Family Dollar had breached its duties as an architect and awarded Montgomery more than $211,000. Family Dollar appealed.

The appellate court reversed and remanded, based mainly on the lack of any explanation of the architect's duties in the contract.

The Owner and Contractor agree as follows.

§ 10.2   Article 1: The Contract Documents

ARTICLE 1   THE CONTRACT DOCUMENTS

1.1 The Contract Documents consist of this Agreement, Conditions of the Contract (General, Supplementary and other Conditions), Drawings, Specifications, Addenda issued prior to execution of this Agreement, other documents listed in this Agreement and Modifications issued after execution of this Agreement, all of which form the Contract, and are as fully a part of the Contract as if attached to this Agreement or repeated herein. The Contract represents the entire and integrated agreement between the parties hereto and supersedes prior negotiations, representations or agreements, either written or oral. An enumeration of the Contract Documents, other than a Modification, appears in Article 9. (Art. 1; change reference)

These are the documents that form the contract between the owner and contractor[7] for the construction work involved. The best practice is to specifically

---

[7] In W.F. Constr. Co. v. Kalik, 103 Idaho 713, 652 P.2d 661 (Ct. App. 1982), the parties did not sign the owner-contractor agreement but proceeded with the construction. The court admitted the document into evidence as to the content of the oral agreement between the parties.

In Novel Iron Works, Inc. v. Wexler Constr. Co., 26 Mass. App. Ct. 401, 528 N.E.2d 142 (1988), the court upheld an oral contract when there were substantial dealings between the owner and contractor and two draft contracts had been prepared but not signed.

In Resurgence Prop., Inc. v. W.E. O'Neil Constr. Co., 1995 U.S. Dist. LEXIS 11633 (N.D. Ill. Aug. 14, 1995), the court found that the General Conditions were incorporated into the owner-contractor agreement (A101/CM, 1980 version). The court reviewed several provisions that referred to the General Conditions as evidence of an intent to incorporate. *See also* Atlantic Mut. Ins. Co. v. Metron, 83 F.3d 897 (7th Cir. 1996), in which the trial court was reversed on the issue of whether the general conditions were incorporated into the contract. The document was a standard AIA A101/CM. The boilerplate and instructions to that document indicated that it was to list in Article 9 all of the contract documents that are in existence at the time this document is signed.[8] The owner-architect agreement is not a part of the contract documents. Neither are the bidding documents. Many architects and owners, mostly out of habit, include the bid form and invitation to bid as part of the contract documents. This should be done with care, with a provision for handling inconsistencies between these bid documents and the other contract documents. If the bidding documents are to be included, they must be listed in ¶ 9.1.7.[9]

---

be used along with A201/CM. The federal appeals court found the contract ambiguous because the parties failed to actually list A201 in the place where it should have been listed. The matter was sent back to the district court for a factual determination of whether the general conditions were actually incorporated.

[8] In Jim Carlson Constr., Inc. v. Bailey, 769 S.W.2d 480 (Mo. Ct. App. 1989), the court held that this provision incorporated the General Conditions by reference, although the General Conditions were not specifically listed in the later article that enumerated the contract documents. The 1987 version of A101 has eliminated any confusion about incorporating the General Conditions, but this situation could come up again if Supplementary Conditions are not enumerated in Article 8.

In State v. Hon. John L. Henning, 201 W. Va. 42, 491 S.E.2d 42 (1997), the general conditions were incorporated. The owner testified that he had not read the contract, but his secretary advised him to consult an attorney before signing the contract, which he failed to do. She also recognized that other documents were incorporated into the contract. This was sufficient for the court.

In Atlantic Mut. Ins. Co. v. Metron, 83 F.2d 897 (7th Cir. 1996), the trial court was reversed on the issue of whether the general conditions were incorporated into the contract. The document was a standard AIA A101/CM, 1980 edition. The boilerplate and instructions to that document indicated that it was to be used along with A201/CM. The federal appeals court found the contract ambiguous because the parties failed to actually list A201 in the place where it should have been listed. The matter was sent back to the district court for a factual determination whether the general conditions were actually incorporated. The *Jim Carlson* decision was mentioned as relying on an earlier version of A101 which required that A101 could be used only with the 1976 edition of AIA Document A201. That language was not included in the 1980 edition. The current edition does not have this problem. The court rejected the argument that Article 1 incorporates the general conditions because Article 1 also refers to supplementary and other conditions, but neither existed here. Also, there was no requirement in that version of A101 that the general conditions be those prepared by the AIA.

[9] In Citizens Bank v. Harlie Lynch Constr. Co., 426 So. 2d 52, 53 (Fla. Dist. Ct. App. 1983), the construction company was the low bidder on a proposed bank branch facility. The bid was accepted by the owner's board of directors and the contractor was orally notified. The architect was directed to prepare the owner-contractor agreement, but it was not executed because a new board of directors decided the new building was too expensive and refused to proceed with the contract. The contractor then sued for breach of contract. The "Instructions to Bidders" included a provision for bid security, pledging that the "Bidder will enter into a Contract with the Owner on the terms stated in his bid and will, if required, furnish bonds as described hereunder. . . . Should the Bidder refuse to enter into such Contract or fail to furnish such bonds, if required, the amount of the bid security shall be forfeited to the Owner as liquidated damages, not as penalty." This provision evidenced a clear intention to be bound by the terms of the bid. The contract had been accepted, even though the formal written contract had not been signed. Under the current AIA documents in which "Instructions to Bidders" are not contract documents, the case might have been decided differently.

In Cumberland Cas. & Sur. Co. v. Nkwazi, L.L.C., 2003 WL 21354608 (Tex. App.—Austin, June 12, 2003), the issue was whether A201 was a part of the contract documents. After problems developed between the owner and contractor, the owner declared the contractor in default and filed

The second sentence is an integration clause, meaning that any agreements between the parties prior to the signing of the instant agreement are merged into the present agreement. If those prior agreements are inconsistent with the present agreement, they are void.[10]

---

a claim against the performance bond. The bonding company refused coverage, arguing that its performance was excused because the owner materially altered the bonded contract by not hiring an architect to inspect the contractor's work, leading to a substantial overpayment for work the contractor had either improperly performed or failed to perform. The owner countered that its contract did not require it to hire an architect to inspect the work. The issue hinged on what constituted the contract. According to the owner, the bid proposal was the only contract between the parties. The bid proposal contained the following language:

> Bidder has carefully examined the form of contract, instructions to bidders, profiles, grades, specifications and the plans therein . . . and will do all the work and furnish all the material called for in the contract DRAWINGS and specifications. . . . In the event of the award of a contract to the undersigned, the undersigned will execute same on [an AIA] Standard Form Construction Contract and make bond for the full amount of the contract, to secure proper compliance with the terms and provisions of the contract. . . . The work proposed to be done shall be accepted when fully complied and finished to the entire satisfaction of the Architect and the Owner.

However, no AIA document was ever executed. The bonding company argued that the appropriate document was A101, which incorporated A201, which in turn required that the owner hire an architect to monitor construction. Since there was no evidence of another contract and because the owner never misrepresented that an AIA contract had been signed, the court held against the bonding company.

[10] In American Demolition, Inc. v. Hapeville Hotel Ltd. P'ship, 413 S.E.2d 749 (Ga. 1991), the court stated that such a clause barred a fraud claim.

A Mississippi case, Godfrey, Bassett v. Huntington, 584 So. 2d 1254 (Miss. 1991), held an architect liable to a contractor for misrepresenting the terms of a contract. The court stated that if the contractor's recovery was contingent solely on this provision, he would lose. However, the court went beyond the contract. This case is instructive concerning the duties of the architect towards bidders. The architect had been hired by a school to prepare plans for the renovation of certain areas of the school. When the plans went out for bids, they included a requirement in the specifications that the bidders include a $9,000 contingency in their bids. Any unused portion would be credited to the owner. The architect then issued to all bidders an addendum that deleted the contingency. When the bids were opened, all bids exceeded the budget, and the architect was asked to reduce the scope of the work.

When the new plans were reissued, portions of the old plans and specifications were reused, including the section that required the $9,000 contingency. When one of the bidders noticed this, he called the senior partner at the architectural firm (with whom he was friendly) and asked whether the $9,000 contingency was intentionally included. The architect informed the contractor that an addendum was being mailed out to delete this contingency, just as had been done during the prior bidding. In fact, an addendum was sent out, but it did not delete the $9,000 contingency. The contractor never saw the addendum, but he was the low bidder and was awarded the job. He signed the contract that incorporated the addendum. After the work was started, the contractor learned that the contingency had not been eliminated and that he was expected by the owner to have the contingency in place. At the end of the project, the contractor credited the owner the $9,000 and sued the architect to recover this amount.

The court found that it appeared that all parties had made an honest mistake. However, the contractor relied on the architect's representation that the contingency was not included. In the absence of this misrepresentation, the contractor would have read the addendum and learned

Under a prior version of the AIA documents, this provision prohibited the contractor from using parol evidence to show an oral representation by the owner and architect.[11]

§ 10.3   Article 2: The Work of This Contract

ARTICLE 2   THE WORK OF THIS CONTRACT

2.1 The Contractor shall fully execute the Work described in the Contract Documents, except as specifically indicated in the Contract Documents to be the responsibility of others. **(Art. 2; minor modifications)**

The contractor should list any specific work that is not the responsibility of this contractor. This must be very carefully coordinated with the other contract documents. If not everything that is shown on the drawings and specifications is the Work, the parties should define what the "Work" is very carefully.[12] The "Work" is the object of the particular contract documents, and can be a portion of a larger "Project." The definition of Work is found in ¶ 1.1.3 of AIA Document A201. Scaffolding and similar items are not Work under A201.

In one case, the contractor was found to have contracted to furnish the design for a prestressed concrete system.[13] (The owner wanted to save money by not having the architect do this work.) It was the contractor and not the architect who was responsible for the defective design.[14]

---

the truth before submitting his bid. The contractor's negligence in failing to carefully read the contract documents was less than the architect's negligence in failing to make the planned corrections to the specifications. The contractor's negligence resulted directly from the architect's misrepresentation. The court held that the contractor was entitled to the $9,000 from the architect.

[11] Hercules & Co. v. Shama Rest. Corp., 613 A.2d 916 (D.C. 1992) (arbitration in which the contractor was unhappy with the result).

[12] In Taylor v. Allegretto, 112 N.M. 410, 816 P.2d 479 (1991), the owner claimed the work included construction of the shell of a three-unit medical complex and completion of the interior of Unit 2. The appellate court held, however, that because the "Work" was defined in the Owner-Contractor Agreement as "Unit #2 As per plans and specifications," the contractor was only required to work on Unit 2, not anything else:

   The clause making all contract documents [complementary] can only be reasonably read to require that the documents complement each other insofar as they are consistent. Where the language of the agreement particularly typed-in language on a printed form contract specifically indicates that the parties intended to limit the scope of the work to be performed, it is simply not "consistent" to expand the agreement because the plans as originally drawn included additional work.

[13] Stevens Constr. Corp. v. Carolina Corp., 63 Wis. 2d 342, 217 N.W.2d 291 (1974) (prestressed concrete system was designed by the concrete subcontractor; unfortunately, an error was made and the structure was designed to hold 120 pounds per square foot less than it should have).

[14] However, when the architect specified an improper sealant and the sealant sub-subcontractor called this to the contractor's attention, the contractor was not liable to the sub, because the contractor advised the architect who had the final decision on the use of the sealant. Lawrence Dev. Corp. v.

Because the work is only the specific work for which the contractor is hired, damage to work performed under another portion of the project is not recoverable under the provisions of ¶ 12.2.2 of AIA Document A201, which provides that the contractor must correct work within one year.[15]

§ 10.4   Article 3: Date of Commencement and Substantial Completion

This article sets forth the date of commencement (which does not necessarily have to be the date the agreement is signed) and the date of substantial completion, a critical date for both parties.

§ 10.5   Commencement Date: ¶¶ 3.1 and 3.2

3.1 The date of commencement of the Work shall be the date of this Agreement unless a different date is stated below or provision is made for the date to be fixed in a notice to proceed issued by the Owner. **(3.1; no change)**

If the commencement date is to be any date other than the date of signing this document, that date should be inserted here. It is important to place the correct date on this agreement in order to establish the commencement date. The date of this agreement does not need to be the time it is signed. If the parties agree, it can be the date of the start of the project. **See** ¶ 8.2.2 of AIA Document A201 for additional provisions relating to this paragraph.

Some owners give the contractor a formal notice to proceed. If so, that establishes the formal start of the project and sets the time of completion. If this is not contemplated, the contractor is required to give the owner five days' notice before starting the work.

In a New York case,[16] the issue was whether the AIA contract was enforceable in light of the local Workers' Compensation Law, which provides that the employer's liability is limited to that law, unless a contract for the work is entered into after an accident and the contract is intended to apply retroactively. A worker employed by Bernini was injured on November 13, 1998. He sued the owner, who, in turn, sued Bernini. The contract between the owner and Bernini was signed on December 10,

---

Jobin Waterproofing, Inc., 588 N.Y.S.2d 422 (A..D.. 1992). The general contractor had no control over, or ability to change, the sealant specification. Therefore, it owed no duty of care to the sub.

According to J. Lee Gregory, Inc. v. Scandinavian House, Ltd. P'ship, 209 Ga. App. 285, 433 S.E.2d 687 (1993), the Uniform Commercial Code applied to a contract to furnish and install windows in an apartment house. There was a question as to the formation of the contract. Because the sale of the windows was the predominant purpose of the transaction and the installation of the windows was secondary, the UCC applied.

[15] Idaho State Univ. v. Mitchell, 92 Idaho 724, 552 P.2d 776 (1976).

[16] Pena v. Chateau Woodmere Corp., 759 N.Y.S.2d 451 (A.D. 1st Dep't, 2003).

1998, but the space for the date of commencement on the AIA contract indicated November 10, 1998, three days prior to the accident. Because the contract was intended to apply retroactively, the local law did not prohibit enforcement of the indemnification provisions in the general conditions of the AIA agreement.

If, prior to the commencement of the Work, the Owner requires time to file mortgages, and other security interests, the Owner's time requirement shall be as follows: (3.1; modified)

Here, the owner may want to insert information regarding a delay in the commencement of the work if the owner or lender requires time to file or record various documents. The security interests referred to in ¶ 3.1 relate to the interests of the owner's lenders and others who may want to have a priority over the contractor and others working at the project. If the owner permits the contractor to start construction before a lender's mortgage or other security interest is filed, the owner may be in default of its loan.

3.2 The Contract Time shall be measured from the date of commencement. **(3.2; no change***)*

This merely states that the contract time starts on the date of commencement, whether or not the contractor actually starts on that date. The Contract Time is defined in ¶ 8.1.1 of AIA Document A201. Any changes in the contract time because of change orders or construction change directives would adjust this time.

§ 10.6   Substantial Completion Date: ¶ 3.3

3.3 The Contractor shall achieve Substantial Completion of the entire Work not later than (   ) days from the date of commencement, or as follows:

(Insert number of calendar days. Alternatively, a calendar date may be used when coordinated with the date of commencement. If appropriate, insert requirements for earlier Substantial Completion of certain portions of the Work.) **(3.3; modified parenthetical)**

The number of calendar days from the date of commencement until substantial completion is required should be inserted in the space. Other formulas can be inserted here. A201 defines a "day" as a calendar day at ¶ 8.1.4. If business days are intended, both this provision and the A201 would need to be revised.

Portion of Work   Substantial Completion Date

Here, if different parts of the project will have different substantial completion dates, describe each such different phase or part of the project, along with the corresponding date of anticipated substantial completion.

, subject to adjustments of this Contract Time as provided in the Contract Documents.

(Insert provisions, if any, for liquidated damages relating to failure to achieve Substantial Completion on time or for bonus payments for early completion of the Work.) **(3.1; modified parenthetical)**

Substantial completion is defined in ¶ 9.8 of AIA Document A201. The date of substantial completion is an important date for the project. On that date, the contractor is due payment including any remaining retainage, less the value of punch-list items.[17] At this date, the risk of loss passes to the owner, who must insure the building and take other responsibilities for it (**see** ¶ 9.8.4 of A201, although note that that provision states that the architect will "establish responsibilities of the Owner and Contractor for security, maintenance, heat, utilities, damage to the Work and insurance . . ." Thus it is possible that the contractor will remain responsible for the risk of loss until final completion, but that is not customary.). The contractor's warranties start to run from this date, and liquidated damages, if any, end on this date. Note that this date is different from the date, if any, on which a municipality issues a certificate of occupancy. The project may be substantially complete but a certificate of occupancy may not be granted. It is the architect who determines the date of substantial completion.[18] After this date, the contractor is entitled to its fee, less the value of any uncompleted work.[19]

---

[17] Mayfield v. Swafford, 106 Ill. App. 3d 610, 435 N.E.2d 953 (1983). Failure to perform in a workmanlike manner constitutes a breach of the contract. If the owner receives substantial performance, it must pay the contractor the contract sum less the cost of correction of any deficiencies. *See also* J.R. Sinnott Carpentry, Inc. v. Phillips, 110 Ill. App. 3d 632, 443 N.E.2d 597 (1982); Watson Lumber v. Guennwig, 79 Ill. App. 2d 377, 226 N.E.2d 270 (1967).

Without substantial completion, the contractor can recover only in quantum meruit. Stephenson v. Smith, 337 So. 2d 570 (La. Ct. App. 1976).

[18] A contrary opinion was expressed by the court in Holy Family Catholic Congregation v. Stubenrauch Assocs., 136 Wis. 2d 515, 402 N.W.2d 382 (Ct. App. 1987). The case involved an action by a church against the architect and contractor for a leaky roof. The question turned on the date of substantial completion and whether the statute of limitations had run. The court held that it was up to the court, and not the architect, to determine the date of substantial completion. The architect's certificate may be persuasive, but it is not determinative.

In Allen v. A&W Contractors, Inc., 433 So. 2d 839, 841 (La. Ct. App. 1983), the court found that an arbitrator had the power to fix a date of substantial completion different from the date the architect had determined. The fact that the contract provides that the architect will establish the date of substantial completion is "not sacrosanct if the facts show substantial completion at a date earlier than that certified by the owner's architect."

In Stephenson v. Smith, 337 So. 2d 570 (La. Ct. App. 1976), the court held that because the contractor had left more than 10% of the work undone, he had not substantially complied with the contract.

[19] In Forrester v. Craddock, 51 Wash. 2d 315, 317 P.2d 1077, 1082 (1957) (*citing* White v. Mitchell, 123 Wash. 630, 213 P.10 (1923)), the court stated the rule as follows:

Where the builder has substantially complied with his contract, the measure of damages to the owner would be what it would cost to complete the structure as contemplated by the contract. There is a substantial performance of a contract to construct a building where the variations from the specifications or contract are inadvertent and unimportant and may be remedied at relatively small expense and without material change of the building; but where it is necessary, in order to make the building comply with the contract, that the structure, in whole or in material part, must be changed, or there will be damage to parts of the building, or the expense of such repair will be great, then it cannot be said that there has been a substantial performance of the contract.

Adjustments to the contract time are covered in ¶ 8.3 of AIA Document A201. Some contracts provide for liquidated damages[20] if the project is not completed on time. Any extension of time delays the commencement of liquidated damages.[21]

Accord Mayfield v. Swafford, 106 Ill. App. 3d 610, 435 N.E.2d 953 (1983); J.R. Sinnott Carpentry, Inc. v. Phillips, 110 Ill. App. 3d 632, 443 N.E.2d 597 (1982); Watson Lumber v. Guennwig, 79 Ill. App. 2d 377, 226 N.E.2d 270 (1967); E.B. Ludwig Steel v. Waddell, 534 So. 2d 1364 (La. Ct. App. 1988); Cleveland Neighborhood Health Servs., Inc. v. St. Clair Builders, Inc., 64 Ohio App. 3d 639, 582 N.E.2d 640 (1989) ("Where the party obligated to perform under the contract makes an honest effort to do so, and there is no willful omission on its part, substantial performance is all that is required to entitle the party to payment under the contract"); J&J Elec., Inc. v. Gilbert H. Moen Co., 9 Wash. App. 954, 516 P.2d 217 (1973).

[20] "Liquidated damages clauses are essentially artificial damages agreed to at the time of contracting, and these clauses are enforceable if actual damages are difficult to ascertain, and if the liquidated damages provision is a reasonable estimate of the damages which would actually result from a breach of the contract. However, if both of these criteria are not met, then the liquidated damages clause is unenforceable since, in that situation, the provision would actually be a penalty." Calumet Constr. Corp., v. Metropolitan Sanitary Dist. of Greater Chicago, 533 N.E.2d 453 (Ill. App. 1st Dist. 1988).

[21] J.R. Stevenson Corp. v. County of Westchester, 113 A.D.2d 918, 493 N.Y.S.2d 819 (1985). The contract contained a no-damages-for-delay clause, a hold harmless clause, and a liquidated damages clause fixing damages at $300 per day. The owner filed suit against the contractor and asked for actual damages, which exceeded the $300 per day amount. The court held that "a liquidated damages clause which is reasonable precludes any recovery for actual damages. This is so even though the stipulated sum may be less than the actual damages sustained by the injured party." The court further held that a no-damages-for-delay clause will not bar an action based on delays or obstructions that were not within the contemplation of the parties when the contract was executed, or that resulted from wilful or grossly negligent acts of the other party.

In Twin River Constr. Co. v. Public Water Dist., 653 S.W.2d 682 (Mo. Ct. App. 1983), the court held that an owner could recover both liquidated damages and actual damages, as long as they are not duplicative.

In Gutowski v. Crystal Homes, Inc., 26 Ill. App. 2d 269, 167 N.E.2d 422 (1960), the contract contained a liquidated damage clause limiting recovery to $1,000. The contractor had built the house too close to a lot line, in violation of the zoning ordinance. The owner was awarded damages for the decrease in value of the house that exceeded the liquidated damages clause because "it is inconceivable that this clause was meant to cover every possible default or failure by either party to the agreement."

In Arrowhead, Inc. v. Safeway Stores, Inc., 179 Mont. 510, 587 P.2d 411 (1978), a contractor claimed delays of 144 days. The court found that the owner properly allowed the contractor an additional 51 days before assessing liquidated damages.

The court in Hartford Elec. Applicators, Inc. v. Alden, 169 Conn. 177, 363 A.2d 135 (1975), stated that "the majority of jurisdictions in this country hold that where there are delays attributable to both parties the liquidated damages clause based upon the contract date is abrogated; the contract is to be performed in a reasonable time."

In Hartford Accident & Indem. Co. v. Boise Cascade Corp., 489 F. Supp. 855 (N.D. Ill. 1980), the surety brought an action against the owner for the amount due for completion of the project. After the contractor was terminated, the surety completed the project. The owner refused to pay the surety after completion, alleging that the owner suffered damages because the project was delayed beyond the completion date. The court held that the contract provision as to the completion date did not condition the surety's payment on meeting the date. The court found that the contract could have specified liquidated damages but did not. Therefore, the surety was entitled to full payment.

In the past, courts did not favor liquidated damages clauses.[22] The more modern view, favored by a majority of jurisdictions, is to allow the contracting parties to address damages in the contract unless the damages are obviously unreasonable.[23] If the contract contains a valid liquidated damages clause, the injured party is entitled to the liquidated damages, irrespective of whether or not actual damages were incurred.[24]

Because of the waiver of consequential damages found at ¶ 15.1.6 of A201, the owner will want to be careful about how liquidated damages are listed. If the amount of liquidated damages is too high in proportion to the anticipated loss, the owner risks winding up with no liquidated damages at all.

§ 10.7   Article 4: Contract Sum

The contract sum is specified in Article 4, which includes any alternative work bid by the contractor and accepted by the owner. Any such alternatives are listed in Article 4, as are unit prices for quantities unknown at the time of signing.

§ 10.8   Amount of Sum: ¶ 4.1

4.1 The Owner shall pay the Contractor the Contract Sum in current funds for the Contractor's performance of the Contract. The Contract Sum shall be _____ ($ __),

---

[22] General Ins. Co. v. Commerce Hyatt House, 5 Cal. App. 3d 460, 85 Cal. Rptr. 317 (1970). There, the court stated:

> "It is well established that where the owners seek liquidated damages pursuant to the provisions of a contract, they must show that they have strictly complied with all requisites to the enforcement of that contractual provision. An owner whose acts have contributed substantially to the delayed performance of a construction contract may not recover liquidated damages on the basis of such delay. "Liquidated damages are a penalty not favored in equity and should be enforced only after he who seeks to enforce them has shown that he has strictly complied with the contractual requisite to such enforcement." [Citations omitted]

> In Centex-Rodgers Constr. Co. v. McCann Steel Co., 426 S.E.2d 596 (Ga. Ct. App. 1992), the court found that the general contractor was not limited to the liquidated damages provision against a subcontractor because the sub had breached a material term of its contract. The general could recover its actual damages, which were apparently greater than the liquidated damages amount.

[23] Calumet Constr. Corp. v. Metropolitan Sanitary Dist. of Greater Chicago, 533 N.E.2d 453 (Ill. App. 1st Dist. 1988) ("While liquidated damages provisions have traditionally been subject to a court's scrutiny because of their inherent speculative nature, liquidated damages provisions have been recognized more recently as appropriate in circumstances where the complexity of contractual relationships make damages difficult to determine, since reasonably related, agreed upon liquidated damage amounts are easy to apply in such situations and satisfy the needs of the parties.")

[24] Southwest Eng'g Co. v. U.S., 341 F.2d 998 (8th Cir., 1965) (Government entitled to liquidated damages even if there were no actual damages). In Worthington Corp. v. Consolidated Aluminum Corp, 544 F.2d 227 (5th Cir. 1976), the owner sought some $4 million from the contractor. The contract, however, set the liquidated damages at $500 per day, with a maximum of $100,000. The owner was not entitled to more than this maximum.

subject to additions and deductions as provided in the Contract Documents. **(4.1; no change)**

Contract sum is defined in ¶ 9.1 of AIA Document A201. The additions and deductions referred to in ¶ 4.1 are accomplished through change orders, as defined in Article 7 of A201.[25]

§ 10.9   Alternates: ¶ 4.2

4.2 The Contract Sum is based upon the following alternates, if any, which are described in the Contract Documents and are hereby accepted by the Owner:

(State the numbers or other identification of accepted alternates. If the bidding or proposal documents permit the Owner to accept other alternates subsequent to the execution of this Agreement, attach a schedule of such other alternates showing the amount for each and the date when that amount expires.) **(4.2, modified parenthetical)**

Here, the owner can insert (and accept) alternatives on which the contractor bid. The alternatives must be specifically identified. The contract sum that is inserted in ¶ 4.1 includes these alternatives. If this is left blank, then the owner has not accepted any alternates and no alternates are included in the contract sum.

§ 10.10   Unit Prices: ¶ 4.3

4.3 Unit prices, if any:

(Identify and state the unit price; state quantity limitations, if any, to which the unit price will be applicable.) **(4.3, minor modification)**

Item   Units and Limitations   Price Per Unit

Unit prices are prices per unit, such as cubic foot, that the contractor will charge the owner under specific conditions. If certain conditions are anticipated but quantities are unknown at the time of execution of this agreement, the owner can fix the costs per unit and minimize conflicts over change orders by establishing unit prices

---

[25] *See* Atlanta Econ. Dev. Corp. v. Ruby-Collins, Inc., 425 S.E.2d 673 (Ga. Ct. App. 1992), for a case in which the contract sum was not filled in. The court found that this failure created an ambiguity with respect to a no-damage-for-delay clause in the contract and held that the contractor was entitled to delay damages.

In Bouten Constr. Co. v. M&L Land Co., 125 Idaho 957, 877 P.2d 928 (1994), the contractor entered into a guaranteed maximum price (GMP) contract for a fast-track project. The drawings on which the contractor bid were only design development drawings. The final drawings were for a project that cost more than the contractor had bid. The court rejected the contractor's argument that the project had been converted to a cost plus contract. The court found that the contractor had to prove that the increased costs resulted from changes in the work made by the owner or architect. Proving such changes was more difficult because of the lack of detail in the design development drawings on which the initial GMP was based.

ahead of time. In a Florida case, the construction contract included the furnishing of one ton of asphalt for $180.[26] Subsequently, the owner and contractor signed a written change order for an additional 130 tons of asphalt for the same $180 per ton. When the owner then asked for 381 more tons of asphalt, the contractor requested the same price. The owner refused to pay, but the court found for the contractor because the additional asphalt was of the same nature and character as that provided for in the first change order and was therefore chargeable at the same rate. If the owner had not signed the first change order, the court would have charged the owner with a "reasonable" amount, presumably less than $180 per ton. The first change order, however, established the price for like quantities.

§ 10.11   Allowances: ¶ 4.4

Allowances are set forth in the contract documents by the architect. They generally pertain to items that are undecided at the time the contract documents are prepared. For instance, the contract documents may specify an allowance of $15 per yard for carpeting of a type and color to be selected. If the owner subsequently selects carpeting that costs $20 per yard, the contractor is entitled to a change order for the additional $5 per yard.

> **4.4** *Allowances included in the Contract Sum, if any:*
>
> *(Identify allowance and state exclusions, if any, from the allowance price.)* **(new)**
>
> **Item**   **Price**

Allowances are covered in Section 3.8 of A201. That section provides guidance to how allowances are handled and how the final cost is later derived, based on the actual costs.

§ 10.12   Article 5: Payments

The timing and computation of progress payments to the contractor are detailed in Article 5. Modifications to such payments and retainage reductions are also accounted for.

§ 10.13   Progress Payments: ¶ 5.1.1

> **5.1 PROGRESS PAYMENTS**
>
> **5.1.1** *Based upon Applications for Payment submitted to the Architect by the Contractor and Certificates for Payment issued by the Architect, the Owner shall make progress payments on account of the Contract Sum to the Contractor as provided below and elsewhere in The Contract Documents.* **(5.1.1, no change)**

---

[26] Forest Constr. v. Farrell-Cheek Steel Co., 484 So. 2d 40 (Fla. Dist. Ct. App. 1986).

Applications for payment are covered in ¶ 9.3 of AIA Document A201, and certificates for payment are covered in ¶ 9.4. Other provisions of Article 9 of A201 deal with decisions to withhold certification (¶ 9.5), progress payments (¶ 9.6), and the owner's failure to pay (¶ 9.7). The supplementary general conditions usually provide for additional documents that are required of the contractor when submitting applications for payments. These may include waivers of lien from each subcontractor, supplier, and others that provide labor and materials for the project.

§ 10.14   Payment Intervals: ¶ 5.1.2

5.1.2 The period covered by each Application for Payment shall be one calendar month ending on the last day of the month, or as follows: **(5.1.2, no change)**

If the parties agree that the contractor will be paid other than for a calendar month period, that period must be specified here. Note that this provision applies only to the period for which the contractor is getting paid, not when the contractor is to be paid. For instance, the application for payment may cover work performed in the calendar month of July, but payment may not be due until several weeks later.

§ 10.15   Payment Dates: ¶ 5.1.3

5.1.3 Provided that an Application for Payment is received by the Architect not later than the _____ day of a month, the Owner shall make payment of the certified amount to the Contractor not later than the _____ day of the _____ same month. If an Application for Payment is received by the Architect after the application date fixed above, payment shall be made by the Owner not later than _____ (_____) days after the Architect receives the Application for Payment.

(Federal, state or local laws may require payment within a certain period of time.) **(5.1.3, minor change, add parenthetical)**

This provision determines when the contractor gets paid, depending on when it submits the application for payment. This should carefully be coordinated with the lender, title company, and others that will process payments. Monthly payments are assumed. If the contractor will be paid other than monthly, this paragraph should be revised.

§ 10.16   Schedule of Values: ¶ 5.1.4

5.1.4 Each Application for Payment shall be based on the most recent schedule of values submitted by the Contractor in accordance with the Contract Documents. The schedule of values shall allocate the entire Contract Sum among the various portions of the Work. The schedule of values shall be prepared in such form and supported by such data to substantiate its accuracy as the Architect may require. This schedule, unless objected to by the Architect, shall be used as a basis for reviewing the Contractor's Applications for Payment. **(5.1.4; no change)**

The schedule of values is referred to in ¶ 9.2 of AIA Document A201. It is a breakdown of each trade involved in the construction. The purpose is to aid the architect in determining the percentage of completion of the project at the time of each application for payment. In order to be meaningful, this schedule should give the architect enough detail to evaluate the application. Thus, the architect may require the contractor to break the concrete work down to foundations, floor slabs, and so on.

§ 10.17   Percentage of Completion: ¶ 5.1.5

5.1.5 Applications for Payment shall show the percentage of completion of each portion of the Work as of the end of the period covered by the Application for Payment. **(5.1.5; minor change)**

In this provision, the contractor is required to itemize the percentage of completion, as of the date indicated, by trade or by item shown on the schedule of values.

§ 10.18   Payment Computations: ¶ 5.1.6

5.1.6 Subject to other provisions of the Contract Documents, the amount of each progress payment shall be computed as follows: **(5.1.6; no change**)

5.1.6.1 Take that portion of the Contract Sum properly allocable to completed Work as determined by multiplying the percentage completion of each portion of the Work by the share of the Contract Sum allocated to that portion of the Work in the schedule of values, less retainage of _____ (_____). Pending final determination of cost to the Owner of changes in the Work, amounts not in dispute shall be included as provided in Subparagraph 7.3.9 of AIA Document A201-2007, General Conditions of the Contract for Construction; **(5.1.6.1; change reference)**

Retainage is an amount that is withheld by the owner from the contractor. The purpose of retainage is to give the owner a cushion in case the contractor defaults. It would cost more to bring in a new contractor to finish up the work of a defaulting contractor. The retainage is intended to cover this amount, as well as any miscalculation of the percentage completion.

Under ¶ 5.1.6, each element of the work listed in the schedule of values is assigned a percentage completion. The total value of that item is multiplied by the percentage completion and the retainage is subtracted. For instance, if the item is foundations with a total value (when completed) of $50,000, and the application for payment shows foundations as 80 percent completed, the amount due the contractor for that item is $40,000. If the retainage is ten percent, or $4,000, the contractor is then due $36,000. Any prior payments made to the contractor on account of foundations are subtracted from this amount. For instance, if the contractor had been paid $20,000 for foundations on a prior draw, it is entitled to only $16,000 at this time. This procedure is followed for each line item in arriving at the total due the contractor for that particular payment.

The last sentence of ¶ 5.1.6.1 refers to amounts in dispute when a construction change directive (AIA Document A201, ¶ 7.3) is issued. This occurs when the owner wants a particular change but the parties cannot agree on price. To avoid holding up the project while the parties argue over price, thus increasing the costs even further, a construction change directive is issued. When the parties agree on price and time changes, it becomes a change order (A201, ¶ 7.2). In the meantime, any portion of the price for a construction change directive that is not in dispute can be included in the application for payment.

5.1.6.2 Add that portion of the Contract Sum properly allocable to materials and equipment delivered and suitably stored at the site for subsequent incorporation in the completed construction (or, if approved in advance by the Owner, suitably stored off the site at a location agreed upon in writing), less retainage of _____ *(* _____ *);* **(5.1.6.2; no change)**

Materials stored at the site but not yet incorporated into the work belong to the owner (**see** AIA Document A201, ¶ 9.3.2). If the parties agree, off-site materials suitably stored (presumably in a bonded warehouse) would also belong to the owner upon payment (A201, ¶ 9.3.2). See also A201, ¶ 11.4.1.4 for the owner's insurance coverage for off-site materials.

5.1.6.3 Subtract the aggregate of previous payments made by the Owner; and **(5.1.6.3; no change***)***

5.1.6.4 Subtract amounts, if any, for which the Architect has withheld or nullified a Certificate for Payment as provided in Section 9.5 of AIA Document A201-2007. **(5.1.6.4; modified references)**

Section 9.5 of A201 gives the architect a number of reasons to withhold certification.[27]

§ 10.19   Payment Modifications: ¶ 5.1.7

5.1.7 The progress payment amount determined in accordance with Section 5.1.6 shall be further modified under the following circumstances: **(5.1.7; minor modification)**

---

[27] In School Bd. v. Southeast Roofing, 532 So. 2d 1353 (Fla. Dist. Ct. App. 1988), the court construed an "AIA style construction contract." The architect had reduced the contractor's request, stating that there were deficiencies in the work. The owner refused to pay anything and the contractor filed suit. Unfortunately, the architect's findings were ambiguous, and the case was sent to the lower court for further findings.

Here, insert any additional modifications related to progress payments.

5.1.7.1 Add, upon Substantial Completion of the Work, a sum sufficient to increase the total payments to the full amount of the Contract Sum, less such amounts as the Architect shall determine for incomplete Work, retainage applicable to such work and unsettled claims; and

(Section 9.8.5, of AIA Document A201-2007 requires release of applicable retainage upon Substantial Completion of Work with consent of surety, if any.) **(5.1.7.1; minor modifications)**

At this point, the retainage is usually reduced because the only items left would be minor punchlist items. A typical amount for this paragraph might be 95 percent. The remaining five percent would cover the punchlist items. The architect also has discretion to place a value on the unfinished work (the punchlist items) and include this in the amount withheld from the contractor. The amount withheld should be sufficient to bring in a new contractor to finish in case the prior contractor refuses to complete the work. A new contractor may charge significantly more to complete, because it is not familiar with the project and it is usually reluctant to pick up the pieces of a prior contractor's work.

5.1.7.2 Add, if final completion of the Work is thereafter materially delayed through no fault of the Contractor, any additional amounts payable in accordance with Section 9.10.3 of AIA Document A201-2007. **(5.1.7.2; modified references)**

§ 10.20   Retainage Reductions: ¶ 5.1.8

5.1.8 Reduction or limitation of retainage, if any, shall be as follows:

(If it is intended, prior to Substantial Completion of the entire Work, to reduce or limit the retainage resulting from the percentages inserted in Sections 5.1.6.1 and 5.1.6.2 above, and this is not explained elsewhere in the Contract Documents, insert here provisions for such reduction or limitation.) **(5.1.8; minor change)**

Here, provisions that reduce the retainage at other than substantial completion can be inserted. Some projects reduce or eliminate additional retainage at 50 percent completion.

5.1.9 Except with the Owner's prior approval, the Contractor shall not make advance payments to suppliers for materials or equipment which have not been delivered and stored at the site. **(5.1.9; no change)**

This prohibits the contractor from making any payments to suppliers unless the materials have been delivered to the job site. Presumably, the only penalty for a breach of this provision is that the contractor takes whatever risk there may be in making such payment.

§ 10.21   Final Payment: ¶ 5.2

5.2 FINAL PAYMENT

5.2.1 *Final payment, constituting the entire unpaid balance of the Contract Sum, shall be made by the Owner to the Contractor when:*

5.2.1.1 *the Contractor has fully performed the Contract except for the Contractor's responsibility to correct Work as provided in Section 12.2.2 of AIA Document A201-2007, and to satisfy other requirements, if any, which extend beyond final payment; and*

5.2.1.2 *a final Certificate for Payment has been issued by the Architect.*

5.2.2 *The Owner's final payment to the Contractor shall be made no later than 30 days after the issuance of the Architect's final Certificate for Payment, or as follows:* **(5.2; modified references)**

If nothing is inserted here, the two provisions in the preceding paragraphs must be met before the contractor is entitled to final payment. The owner may, alternatively, insert a date for final payment here, or other provisions for final payment. Note that the work must be fully completed in order for the contractor to be entitled to final payment.[28] The contractor must initiate the architect's final inspection.

---

[28] A Texas case, Ryan v. Thurmond, 481 S.W.2d 199, 202 (Tex. 1972), interpreted the following language (which is similar to AIA language): "Final payment shall be due Thirty days after (1) Substantial Completion of the work provided the work be then (2) fully completed and the contract (3) fully performed." The court held that the contractor was not entitled to payment because the contract was not completed. The contractor had failed to plead substantial performance, and the owner introduced evidence that the contract was not completed.

In Martinson v. Brooks Equip. Leasing, Inc., 36 Wis. 207, 152 N.W.2d 849 (1967), the court construed a similar provision as requiring 100% completion. Substantial completion was not sufficient. The contractor had demanded full payment before completion of the project. The court held that this was a breach of contract.

In Kilianek v. Kim, 192 Ill. App. 3d 139, 548 N.E.2d 598 (1989), the owner sought to overturn an arbitration award in favor of the contractor. The appellate court found that the arbitrator had exceeded his powers by awarding final payment when the architect had not issued a final certificate per this paragraph.

For an interesting case in which the architect had a 50% ownership in the general contracting firm, *see* Professional Builders, Inc. v. Sedan Floral, Inc., 819 P.2d 1254 (Kan. Ct. App. 1991). After the owner lost in an arbitration with the contractor, the owner sought to overturn the arbitration on the grounds of fraud in issuing the equivalent of a final certificate for payment. The court held that this was actually the issue before the arbitrator. The type of fraud necessary to overturn an arbitration award deals with fraud in the arbitration, not that which is totally outside the process of arbitration. The arbitration was upheld.

In In re Modular Structures, Inc., 27 F.3d 72 (3d Cir. 1994), the court held that the general contractor was not entitled to final payment until it had paid its subcontractors, based in part on this provision.

In Johnson City Cent. Sch. Dist. v. Fidelity & Deposit Co. of Md., 641 N.Y.S.2d 426 (A.D. 1996), the dispute centered around timing of filing the suit against the surety. The performance bond had a requirement that suit must be initiated within two years of the date on which final payment under the contract fell due. The architect's final certificate for payment was issued on December 21, 1992. Suit was filed on January 12, 1995. The court looked at this provision of A101 in determining whether the 30-day period meant that the two-year period began on December 21,

Note that even if a final certificate for payment is obtained, the contractor is not relieved from liability for work that does not conform to the contract documents (**see** AIA Document A201, ¶ 9.10.4).

§ 10.22   Article 6: Dispute Resolution

6.1 Initial Decision Maker The Architect will serve as Initial Decision Maker pursuant to Section 15.2 of AIA Document A201-2007, unless the parties appoint below another individual, not a party to this Agreement, to serve as Initial Decision Maker.

(If the parties mutually agree, insert the name, address and other contact information of the Initial Decision Maker, if other than the Architect.) **(new)**

This initial decision maker is a person who will make the initial attempt to resolve disputes. In the absence of naming a person, the arbitrator will fill that role. This provision states that the initial decision maker will be a person not a party to this agreement. Since the architect is not a party to this agreement, the architect could fill that role. Presumably, only the owner and the contractor would be barred from this position unless the parties agreed otherwise. For instance, what if the owner wants to have an employee of the owner as the initial decision maker? **See** also, Section 15.2 of A201-2007 for more about the initial decision maker and the process.

6.2 Binding Dispute Resolution For any Claim subject to, but not resolved by, mediation pursuant to Section 15.3 of AIA Document A201-2007, the method of binding dispute resolution shall be as follows:

(Check the appropriate box. If the Owner and Contractor do not select a method of binding dispute resolution below, or do not subsequently agree in writing to a binding dispute resolution method other than litigation, Claims will be resolved by litigation in a court of competent jurisdiction.) **(new)**

☐ *Arbitration pursuant to Section 15.4 of AIA Document A201-2007*

☐ *Litigation in a court of competent jurisdiction*

☐ *Other (Specify)* **(new)**

For the first time, the 2007 AIA documents have eliminated arbitration as the default mechanism for resolution of disputes, following mediation. The parties

---

1992 or 30 days later. There was some degree of uncertainty as to which construction was more plausible. Holding that there is a general rule of liberal construction in favor of insureds, the court held that the action was not time-barred.

In Humphreys County v. Guy Jones, Jr. Constr. Co., Inc., 2005 WL 646652 (Miss. App., March 22, 2005), the general contractor sued the county for failure to pay. The county argued that the contractor knew, because he had attended a preconstruction meeting, that some of the funding for the project would be coming from federal funds and so might be delayed. The court held that the payment provisions in the AIA document could have been modified by the county, but were not. Therefore, the strict terms of the contract governed, and the county was required to pay the contractor, including interest, costs, and attorneys' fees.

should check one of the three boxes to select arbitration, litigation, or some other dispute resolution method. If all boxes are left blank, or if more than one is checked or it is otherwise unclear as to which method is selected, then litigation is the final process and there will not be arbitration. Of course, mediation is still a condition precedent to any of these processes (**see** Section 15.3 of A201).

§ 10.23   Article 7: Termination or Suspension

Article 7 contains provisions for termination by either the owner or contractor and for suspension by the owner. The actual language is found in A201.

ARTICLE 7 TERMINATION OR SUSPENSION

7.1 The Contract may be terminated by the Owner or the Contractor as provided in Article 14 of AIA Document A201-2007. (**6.1; modified reference**)

The word "may" makes termination optional, according to a Massachusetts court.[29] The parties had amended the A101 to provide that the contract was contingent upon obtaining certain permits and approvals. Incorrectly assuming that the permit process had been completed, the town issued a notice to proceed to the contractor. Upon discovering that construction could not proceed, the town informed the contractor that construction would have to cease. There was a delay of 14 months. Neither side terminated the contract in accordance with A201. The town paid the contractor its mobilization and site work costs. The contractor then sued the town for lost profits pursuant to ¶ 14.4.3. Based on the contingency provision, the court ruled against the contractor.

7.2 The Work may be suspended by the Owner as provided in Article 14 of AIA Document A201-2007. (**7.2; modified reference**)

§ 10.24   Article 8: Miscellaneous Provisions

Besides addressing the integration of the contract documents as a whole and interest rates to be charged on overdue payments to the contractor, the purpose of Article 7 is to provide a place to insert any provisions that are unique for each project.

§ 10.25   Integration of Contract Documents: ¶ 8.1

8.1 Where reference is made in this Agreement to a provision of AIA Document A201-2007 or another Contract Document, the reference refers to that provision

---

[29] BBC Co., Inc. v. Town of Easton, 842 N.E.2d, 2006 WL 266107 (Mass. App. Ct. 2006).

as amended or supplemented by other provisions of the Contract Documents. **(7.1; modified reference)**

This refers to the use of Supplementary Conditions, which are used on almost every project to customize the preprinted A201 General Conditions to the particular project. This provision says that Supplementary Conditions and other related contract documents must be read as being part of the General Conditions, that is, all the contract documents are to be accepted as a whole.

§ 10.26   Interest: ¶ 8.2

8.2 Payments due and unpaid under the Contract shall bear interest from the date payment is due at the rate stated below, or in the absence thereof, at the legal rate prevailing from time to time at the place where the Project is located.

(Insert rate of interest agreed upon, if any.) **(7.2; modified parenthetical)**

In this provision, an interest rate should be inserted. Because many states have usury laws that limit the interest rates that can be charged, legal advice should be obtained before entering an amount here. The legal rate referred to is provided by statute and is usually unrealistically low. One court has held that this is not the maximum rate permitted by law but is rather the statutory rate.[30] If the project is for a single-family home, with the homeowner as owner, a number of state and federal laws may also limit interest rates and may require certain disclosures.

§ 10.27   Representatives: ¶¶ 8.3 through 8.5

8.3 The Owner's representative: **(7.3; minor modification)**

This is the person who speaks for the owner and generally has authority to authorize contract modifications. Normally, this is not the architect.

8.4 The Contractor's representative: **(7.4; minor modification)**

Here, insert the name of the contractor's agent, supervisor, project manager, or other person who will have authority to bind the contractor.

8.5 Neither the Owner's nor the Contractor's representative shall be changed without ten days' written notice to the other party. **(7.5; no change)**

---

[30] Jetty, Inc. v. Hall-McGuff Architects, 595 S.W.2d 918 (Tex. Ct. App. 1980).

In Bolivar Insulation Co. v. R. Logsdon Builders, Inc., 929 S.W.2d 232 (Mo. Ct. App. 1996), the contract had a provision for 18% per year interest. The appellate court reversed the trial court's limitation of interest to the statutory rate of 9%, finding the contractual provision valid.

§ 10.28   Other Provisions: ¶ 8.6

8.6 Other provisions: **(7.6; no change)**

At this point, any other pertinent contract provisions can be inserted.

§ 10.29   Article 9: Enumeration of Contract Documents

This article is meant to precisely define the documents that form the agreement so that there can be no ambiguity. There is space to list each document, as well as any addenda that might have been issued before the agreement is signed.

§ 10.30   Included Documents: ¶ 9.1

9.1 The Contract Documents, except for Modifications issued after execution of this Agreement, are enumerated in the sections below. **(8.1; minor modification)**

Here, list all of the contract documents in general terms that are sufficient to identify each. **See** AIA Document A201, ¶ 1.1.1 for a list of contract documents. Following paragraphs list individual Supplementary and other Conditions of the Contract (¶ 9.1.3), Specifications (¶ 9.1.4), Drawings (¶ 9.1.5), and Addenda (¶ 9.1.6).

9.1.1 The Agreement is this executed AIA Document A101-2007, Standard Form of Agreement Between Owner and Contractor. **(8.1.1; modified reference)**

This refers to this document. This document is then used to refer and incorporate all of the other Contract Documents, such as the A201, the drawings, specifications, and so forth. Thus, the complete listing of all specific Contract Documents occurs in this one document.

9.1.2 The General Conditions are AIA Document A201-2007, General Conditions of the Contract for Construction. **(8.1.2; modified reference)**

It is important not to mix earlier versions of A201 with this agreement in order for the documents to be properly coordinated.[31] In a case that interpreted the 1977 version of A101, the court found that A201 was incorporated into the contract,

---

[31] An example of problems caused by mixing documents is found in Eis Group/Cornwall Hill v. Rinaldi Constr., 154 A.D.2d 429, 546 N.Y.S.2d 105 (1989), in which the owner-contractor agreement was a 1987 AIA form that incorporated the 1987 version of A201. However, the 1976 version of A201 was actually attached to the contract, along with various modifications that included striking the arbitration clause. The court found an ambiguity (obviously!) and declined to order arbitration.

although it was not specifically stated as such.[32] In the 1987 and later versions of A101, A201 is specifically incorporated so that there is no ambiguity. If the parties want to use different General Conditions, ¶ 9.1.2 must be specifically modified.

---

[32] Jim Carlson Constr., Inc. v. Bailey, 769 S.W.2d 480 (Mo. Ct. App. 1989).

    In Walker v. V&V Constr. Co., 28 Mass. App. Ct. 908, 545 N.E.2d 1192 (1989), the court interpreted an AIA A111 contract that did not list A201 as a contract document in the form. The parties had neglected to list A201 as a contract document, but the court held that it was included, even though no copy was ever delivered to the owner. In fact, the parties had erroneously listed the agreement to be document A111a, which was the instruction sheet, and not the form itself.

    See also Cleveland Wrecking Co. v. Central Nat'l Bank, 216 Ill. App. 3d 279, 576 N.E.2d 1055 (1991); Southwest Nat'l Bank v. Simpson & Son, Inc., 14 Kan. App. 2d 763, 799 P.2d 512 (1990) (citing Jim Carlson); R.T. Traynham v. Yeargin Enters., Inc., 403 S.E.2d 329 (S.C. 1991).

    In L.R. Foy Constr. Co. v. Dean L. Dauley & Waldorf Assocs., 547 F. Supp. 166 (D. Kan. 1982), a contractor brought an action to compel arbitration of disputes concerning three construction contracts. The arbitration provision in the A201 form had been deleted from one of the contracts, although the contractor did not notice this when it signed the contract. The contractor argued that, to be valid, any changes in A201 had to be listed in the A101 form. The court rejected this argument. The court also found that, although the two remaining contracts did not specify a location for the arbitration, they specified that arbitration was to be in accordance with AAA rules. Those rules provide that the AAA will determine the location of the arbitration in the event of a dispute among the parties.

    In Welch v. McDougal, 876 S.W.2d 218 (Tex. Ct. App. 1994), A101-1977 was used and failed to list A201 as being the general conditions. The court held that A201 was not incorporated.

    In Cobb County Sch. Dist. v. Mat Factory, Inc., 215 Ga. App. 697, 452 S.E.2d 140, 142 (1994), the issue was whether the contract for construction included reference to a certain California test relating to impact attenuation for a playground surface. The earlier version of this provision provided: "The contract documents, which constitute the entire agreement between the [parties], are enumerated." The California test was not mentioned. The court held that the test was not part of the contract and that the owner could not rely on that test for any fraud in the inducement claim.

    In Atlantic Mut. Ins. Co. v. Metron, 83 F.2d 897 (7th Cir. 1996), the trial court was reversed on the issue of whether the general conditions were incorporated into the contract. The document was a standard AIA A101/CM, 1980 edition. The boilerplate and instructions to that document indicated that it was to be used along with A201/CM. The federal appeals court found the contract ambiguous because the parties failed to actually list A201 in the place where it should have been listed. The matter was sent back to the district court for a factual determination of whether the general conditions were actually incorporated. The *Jim Carlson* decision was mentioned as relying on an earlier version of A101 which required that A101 could be used only with the 1976 edition of AIA Document A201. That language was not included in the 1980 edition. The current edition does not have this problem. The court rejected the argument that Article 1 incorporates the general conditions because Article 1 also refers to supplementary and other conditions, but neither existed here. Also, there was no requirement in that version of A101 that the general conditions be those prepared by the AIA.

    In Heitritter v. Callahan Constr. Co., 670 N.W.2d 430 (Iowa App. 2003), the 1980 A101/CM, which did not incorporate A201, was used. The court found that the preliminary recitals did not become part of the agreement and therefore A201 was not incorporated by reference.

    In 566 New Park Assocs., LLC v. Blardo, 906 A.2d 720 (Conn. App., 2006), the owner entered into a construction contract with a contractor using AIA Document A105. That document incorporates the general conditions, AIA Document A205, by reference, similar to this provision. The trial court inexplicably held that this incorporation clause did not actually incorporate the general conditions. The appellate court ruled that this was error by the trial court.

9.1.3 The Supplementary and other Conditions of the Contract: **(8.1.3; modified)**

Prior versions of this section referred to a "Project Manual," which was defined in the A201. The 2007 version of A201 drops this term, although most architects continue to use a project manual. The project manual, if used, must have a date for identification purposes. See ¶ 1.1.7 of AIA Document A201-1997 for a definition of the Project Manual. In a Michigan case, the project manual stated that A201 was made a part of the specifications.[33] The contractor sought arbitration against the developer based on the arbitration clause of A201. The court denied arbitration because the owner-contractor agreement was not a standard AIA form and did not include an arbitration clause. The court found that the project manual was referred to by the parties not to answer contract issues but rather as a place for the architect's drawings and specifications.

Parties should be cautious when using nonstandard contracts because they are usually not coordinated like the AIA documents. Note that the project manual is not a contract document but it contains parts of the contract documents. If there is no project manual, the documents normally contained in that manual will be separately bound.

Document	*Title*	*Date*	*Pages*

Here, the Supplementary Conditions and other conditions of the contract are enumerated. Each separate part must be identified by document, title, and date. Often, instead of Supplementary Conditions, the A201 is amended on its face. In this listing of documents, it is better to err on the side of too much detail than not enough.

9.1.4 The Specifications:

(Either list the Specifications here or refer to an exhibit attached to this agreement.) **(8.1.4; modified)**

Specifications are defined in ¶ 1.1.6 of AIA Document A201. The specifications are the written documents that provide the contractor with various items of information, such as acceptable manufacturers of components, standards to be adhered to, clean-up, and many other facets of the project that are inappropriate on the drawings. On some smaller projects, the specifications can be found on the drawings. The specifications are usually divided into 16 sections, according to industry practice. Most architects and specifiers follow this practice, and contractors are familiar with it.

The drawings and specifications should be complementary and must be coordinated. These drawings and specifications are the contract documents that tell the contractor how the parts go together. Other contract documents, such as A201, give legal and procedural information.

---

[33] Omega Constr. Co. v. Altman, 147 Mich. App. 649, 382 N.W.2d 839 (1985).

Architects often use specialists to write specifications. Many of these are members of the Construction Specifications Institute, and follow formats designed by that institute or the AIA.

Here, identify each of the sections of the specifications that are included in the project. The specifications index could be attached as an exhibit to this agreement and referenced here. Failing to include this information can result in problems.[34]

9.1.5 *The Drawings:*

(Either list the Drawings here or refer to an exhibit attached to this Agreement.)
(8.1.5; modified)

The Drawings are defined in ¶ 1.1.5 of AIA Document A201. The drawings are sometimes referred to as plans, but they also consist of drawings of details, elevations, and other drawings. They also include drawings that form part of any addenda or other modifications to the contract. Some drawings might even be found in the specifications. Remember to insert the date above.

Here, list all the drawings by drawing number, title, and date of the last revision before execution of the agreement (assuming that last revision was included in the contractor's bid). Further revisions are a change to the contract and usually result in a change order.[35] Revisions before the agreement are not clouded, which is a procedure in which the specific change is marked on the drawing by circling the change and identifying it with a revision number. (The circle looks like a cloud.) Each subsequent change is similarly marked with new revision numbers, and the prior cloud is usually removed. Of course, separate drawings that show only the change can also be used, as can any other method that shows the changes. The important point is that the change is clearly indicated for the parties. Record drawings of each change are kept.

9.1.6 *The Addenda:* **(8.1.6; modified)**

Addenda are changes to the contract documents that are issued before the execution of the owner-contractor agreement. These are issued to clarify the drawings and specifications or to correct omissions. Changes issued after the execution of the agreement are called modifications and include change orders and construction

---

[34] In Trans W. Leasing Corp. v. Corrao Constr. Co., 652 P.2d 1181 (Nev. 1982), the parties had discussed two different sets of roof specifications. The contract failed to identify which one the contractor was to follow, although the contractor did use one set in the installation. When the owner sued the contractor, claiming that the roof was defective and not completed in accordance with the specifications, this became an issue. The court allowed extrinsic evidence as to which set of specifications the parties intended to use. Although the contractor prevailed, much of the controversy might have been avoided by properly filling out this form.

[35] In Kingston Elec., Inc. v. Wal-Mart Prop., Inc., 901 S.W.2d 260 (Mo. Ct. App. 1995), the site lighting drawing was omitted from the list of drawings in ¶ 9.1.5, but the site lighting section of the specifications was listed. This made the contract ambiguous as to whether the site lighting was included in the electrical contractor's contract. The trial court properly considered extrinsic evidence and determined that the site lighting was not included.

change directives. These cannot be included here because they are not known at the time this document is filled out.

Number   *Date*   *Pages*

Here, list each addendum by number and date and number of pages. Addenda normally only carry one date, whereas the drawings may carry many dates, the last of which is the operative date.

Portions of Addenda relating to bidding requirements are not part of the Contract Documents unless the bidding requirements are also enumerated in this Article 9. **(8.1.6; modified reference)**

This is consistent with the principle in the AIA documents (AIA Document A201, ¶ 1.1.1) that bidding documents are procedural and not a part of the contract documents.

9.1.7 Additional documents, if any, forming part of the Contract Documents:

   .1   AIA Document E201-2007, Digital Data Protocol Exhibit, if completed by the parties, or the following: **(new)**

   .2   Other documents, if any, listed below:

(List here any additional documents that are intended to form part of the Contract Documents. AIA Document A201-2007 provides that bidding requirements such as advertisement or invitation to bid, Instructions to Bidders, sample forms and the Contractor's bid are not part of the Contract Documents unless enumerated in this Agreement. They should be listed here only if intended to be part of the Contract Documents.) **(8.1.7; add reference to digital data, change reference)**

With the widespread use of drawings and specifications in digital form, and contractors and owners requesting access to the architect's digital files, accommodations must be made with regards to digital files. Section 1.6 of A201 is a new section in 2007 regarding transmission of data in digital form. AIA Document E201-2007 covers procedures regarding the exchange of digital data. This document is a Contract Document only if it is completed by the parties, although this could occur after the execution of this Agreement.

In the second part of this section, list any other documents that are included as contract documents. Possibilities include shop drawings and submittals, bidding documents, reports, soil borings, and so on.

§ 10.31   Article 10: Insurance and Bonds

10.1 The Contractor shall purchase and maintain insurance and provide bonds as set forth in Article 11 of AIA document A201-2007.

(State bonding requirements, if any, and limits of liability for insurance required in Article 11 of AIA Document A201-2007.) **(new)**

Type of Insurance or Bond   Limit of Liability or Bond Amount ($0.00)

This is a new section for 2007. The specific requirements should be set forth here. It was previously common to place this information in the specifications or supplementary general conditions. Note that only specific requirements for the contractor, but not the owner, are to be listed.

This Agreement is entered into as of the day and year first written above.

In the appropriate spaces, the owner and contractor's representatives must sign and indicate their capacities.

CHAPTER 11

ALTERNATIVE LANGUAGE FOR AIA DOCUMENT A101

§ 11.1   Introduction

The language in this chapter is offered as possible alternative language to AIA Document A101 or as language that can be used to fill in the blanks. The user is cautioned that some of these provisions may be mutually incompatible, and each paragraph and each sentence should be carefully reviewed with legal counsel before using it. Note further that some of these changes require changes to other documents.

CAUTION: The recommended method of amending this document is to use the electronic documents software that is available from the AIA. Using this software, the document is amended on its face, with additions and deletions indicated either in the margin, or by underlines and strikethroughs. Another way is to attach separate written amendments that refer back to A101. An alternative method is to graphically delete material and insert new material in the margins of the standard form. However, this method may create ambiguity, and it may result in the stricken material's being rendered illegible. It is illegal to make any copies of AIA documents in violation of the AIA copyright or to reproduce this document by computerized means. Refer to the instruction sheet that comes with A101 for additional information about that copyright. In some places in A101, blanks are provided in the form to insert the requested information. Wherever possible, the blanks should be filled in, either with information, or with "Not Applicable."

§ 11.2 Substantial Completion Date: ¶ 3.3

A liquidated damages clause could be inserted in ¶ 3.3. One such provision is the following:

3.3.1 FAILURE TO COMPLETE THE WORK ON TIME. It is mutually agreed by and between the parties hereto that time shall be an essential part of this contract and that in case of the Contractor's failure to complete the contract within the time specified and agreed upon, the Owner will be damaged thereby; and because it is difficult to definitely ascertain and prove the amount of said damages, inclusive of expenses for inspection, superintendence, and necessary traveling expenses, it is hereby agreed that the amount of such damages shall be the appropriate sum set forth below in the Schedule of Liquidated Damages as liquidated damages for every working day's delay in finishing the work in excess of the number of working days prescribed; and the Contractor hereby agrees that said sum shall be deducted from monies due the Contractor under the contract or if no money is due the Contractor, the Contractor hereby agrees to pay to the Owner as liquidated damages, and not by way of penalty, such total sum as shall be due for such delay, computed aforesaid.[1]

The following is referred to as a *strike clause*, and could also be included here:

3.3.2 The Contractor shall not be liable for any loss, damage, or delay caused by strikes, lockouts, fire, explosion, theft, floods, riot, civil commotion, war,

---

[1] Adapted from Associated Eng'rs & Contractors, Inc. v. State, 58 Haw. 187, 567 P.2d 397 (1977). A contractor was delayed in completion of the work because of weather. The contract stipulated that certain work could not be performed unless minimum temperatures were present. He argued that it was impossible to perform the work within the required time period. The court held that he must pay the liquidated damages.

In City of Elmira v. Larry Walter, Inc., 76 N.Y.2d 912, 564 N.E.2d 655, 656 (1990), the liquidated damages clause was:

As actual damages for any delay in completing the work . . . are impossible to determine, the Contractors and their Sureties shall be liable for . . . the sum of One Thousand Dollars . . . as fixed, agreed and liquidated damages for each calendar day of delay from the above stipulated completion . . . until such work is satisfactorily completed and accepted.

The contractor defaulted and the owner hired another contractor to complete the work. The appellate court stated that this liquidated damages clause applied only when the completion of the project was delayed, not when the contractor abandons the project. According to the court, the contract language would have to clearly spell out that the liquidated damages clause was also intended to apply to that possibility. This position was rejected by the New Mexico courts in Constr. Contracting & Mgmt., Inc. v. McConnell, 112 N.M. 371, 815 P.2d 1161 (1991); Cincinnati Ins. Co. v. Jasper City Utility Serv. Bd., 2006 WL 2472735 (S.D. Ind. Aug. 24, 2006).

malicious mischief, act of God, or by any cause beyond the Contractor's reasonable control, and in any event, the Contractor shall not be liable for consequential damages.[2]

The following are examples of no damage for delay clauses that favor owners by preventing the contractor from obtaining additional compensation resulting from delays to the project, even when the delay is not the contractor's fault. Note that several states have recently enacted legislation that limits or even voids the effect of such clauses.[3] Legal counsel should be sought to determine the validity of such a clause in any given jurisdiction.

The Owner shall not be liable to the Contractor and/or any Subcontractor for claims or damages of any nature caused by or arising out of delays. The sole remedy against the Owner for delays shall be the allowance of additional time for completion of the Work, the amount of which shall be subject to the claims procedure set forth in the General Conditions.

The Contractor agrees to make no claim for damages for delay in the performance of this Contract occasioned by any act or failure to act of the Owner or any of its officers, directors, employees, architects, or other representatives, or because of any injunctions which may be brought against the Owner or its representatives, and agrees that any such claim shall be fully compensated for by an extension of time to complete performance of the Work as provided herein.

The Contractor shall have no claim against the Owner for an increase in the contract price or a payment or allowance of any kind based on any damage, loss, or additional expense the Contractor may suffer as a result of any delays in prosecuting or completing the work under the Contract, whether such delays are caused by the circumstances set forth in the preceding paragraph or by any other circumstances. It is understood that the Contractor assumes all risks of delays in prosecuting or completing the work under the Contract.

No claim for damages or any claim other than for an extension of time as herein provided shall be made or asserted against the Owner by reason of the delays hereinafter mentioned: [list possible delays].

.1 Should the Contractor be obstructed or delayed in the commencement, prosecution, or completion of the Work because of conditions attributable

---

[2] Adapted from Curtis Elevator Co. v. Hampshire House, Inc., 142 N.J. Super. 537, 362 A.2d 73 (1976).

[3] *E.g.*, Col. Pub. Cont. Code § 7102 (related to public contracts only); Colo. Rev. Stat. § 24-91-103.5; Or. Rev. Stat. § 279.063; Wash. Rev. Code § 4.24.360. *See* Blake Constr. Co., Inc./ Poole & Kent v. Upper Occoquan Sewage Auth., 266 Va. 564, 587 S.E.2d 711 (2003), for a no damages for delay clause held to violate a Virginia state statute.

to the Owner and which by the terms of the Contract may be grounds for an extension of time or money damages, the Contractor shall promptly make claim therefor in writing pursuant to the claims provisions in the General Conditions.

.2 The Owner for just cause shall have the right at any time to delay or suspend the commencement or execution of the whole or any part of the Work without compensation or obligation to the Contractor other than to extend the time for completing the Work for a period equal to that of such time of suspension.

§ 11.3 Alternates: ¶ 4.2

If there are no alternates, the following language could be substituted:

4.2 The Contract Sum does not include any alternates, whether or not any alternates are listed in the Bidding Documents.

If there are alternates, this language could be substituted:

4.2 The Contract Sum stated in Paragraph 4.1 includes the following alternates which shall be considered as part of the Work herein: [list complete description of all accepted alternates, including any deletions from the Work].

§ 11.4 Unit Prices: ¶ 4.3

The following language goes into greater detail:

4.3 In the event of additional work, the following unit prices are in effect:

[list unit prices separately]

In the event that any work is deleted, the unit prices to be applied to deleted work shall be as follows:

[list unit prices for deletions, presumably at a lower rate than for additions]

In the event that particular items are subject to both additions and deletions, the Contractor shall be entitled to use the separate Unit Prices for additions and deletions, as the case may be, for each event that causes such addition or deletion. Any change to the Work that is not specified under the above list for Unit Prices shall be subject to the Change Order procedures in the General Conditions.

§ 11.5   Basis of Payments: ¶ 5.1

The following provisions can be inserted at the end of ¶ 5.1 or in the Supplementary General Conditions:

> The Contractor shall furnish the Owner, with each Application for Payment, a notarized Contractor's affidavit in a form satisfactory to the Owner's title company, along with proper waivers of lien and other supporting documentation sufficient to satisfy said title company that payment is properly due the Contractor.

> The Contractor shall submit each Application for Payment on the standard Sworn Contractor's Statement [available from the Chicago Title Insurance Company]. The Application for Payment shall list each item shown on the approved Schedule of Values, and shall be properly executed and sworn under oath, as required by the Owner's title company. Each Application for Payment shall be accompanied by notarized Partial or Final Waivers of Lien on forms acceptable to the Owner's title company, one Waiver for each separate payment requested and shown for that Application.

§ 11.6   Retainage Reductions: ¶ 5.1.8

A typical clause for reducing retainage is as follows:

> Until the Work is 50 percent complete, the Owner shall pay 90 percent of the amount due the Contractor on account of progress payments. At the time the Work is 50 percent complete and thereafter, the Architect will authorize remaining partial payments to be paid in full. Notwithstanding the foregoing, in the event the Architect determines that the Contractor is not reasonably performing the Work, either by failing to reasonably follow the schedule, or by failing to adequately perform the Work (all to be determined in the reasonable judgment of the Architect), the Owner shall continue to make progress payments at the rate of 90 percent of the amount due the Contractor for each payment.

Another clause reads as follows:

> Until the Work is certified by the Architect to be 50 percent complete, the Owner shall pay 90 percent of the amount due the Contractor on account of progress payments. At the time the Work is so certified to be 50 percent complete and thereafter, the Architect will authorize remaining partial payments to be paid at the rate of 95 percent due the Contractor. Notwithstanding the foregoing, in the event the Architect determines that the Contractor is not reasonably performing the Work, either by failing to reasonably follow the schedule, or by failing to adequately perform the work (all to be determined in the reasonable judgment of the Architect), the Owner shall have the right to make progress payments at the rate of 80 percent of the amount due the Contractor for each payment until such time as the Architect certifies that the Contractor is in full compliance with the schedule and all other conditions of the Contract for Construction.

The Contractor shall have no claim against the Architect, at law or in equity or otherwise, that arises out of the Architect's actions pursuant to this provision.

§ 11.7   Final Payment: ¶ 5.2

The following provision can be inserted at the end of this clause to favor the owner:

In the event the Contractor does not achieve final completion within 60 days after the date of Substantial Completion, allowing for approved extensions of the Contract Time, the Contractor shall not be entitled to any further payment, and the Contractor hereby agrees that such failure to complete the Work within the time set forth above shall constitute a waiver of all claims by the Contractor to any money that may be due. This provision shall not operate as a waiver by the Owner of any claims of any nature against the Contractor arising out of the Contract.

Another provision reads:

The Contractor agrees that the issuance of a Certificate for Payment is a condition precedent to the Contractor's right to payment unless the failure to issue such a Certificate for Payment is done in bad faith. If the position of Architect is vacant, the Owner shall designate another person to act for the purpose of issuing a Certificate for Payment, and such person need not be a licensed architect. If the Owner fails to do so, the provision for the issuance of a Certificate for Payment shall be waived.

§ 11.8   Interest: ¶ 8.2

The following can be inserted in ¶ 8.2, subject to review for compliance with state law:

One percent per month. Payments are due thirty (30) days after the billing date shown on each invoice. All costs of collection, including reasonable attorneys' fees, shall be paid by the Owner.

The last sentence of this clause provides for the payment of attorneys' fees when the owner is in default. The normal American rule is that each party must bear its own attorneys' fees unless a statute or contract specifically provides for them. Most states do not have such statutes for normal architectural projects.[4]

---

[4] However, in the case of frivolous suits, Federal Rule of Civil Procedure 11 and many state statutes, such as Rule 2-611 of the Illinois Code of Civil Procedure, provide for payment of attorneys' fees if a party files an action, motion, or other paper without reasonable basis.

Note that even when the parties to a construction contract have provided for attorneys' fees, the courts will look carefully at the equities of the situation to determine whether such fees will be due. In Willie's Constr. Co. v. Baker, 596 N.E.2d 958 (Ind. Ct. App. 1992), the contractor sued the owner

A provision favorable to the owner is as follows:

No interest shall be due on account of any payment due or unpaid.

Another provision that is more neutral reads:

No interest shall be due on account of any payment due or unpaid, unless the Contractor is found to be owed such payment by any court or arbitrator pursuant to the Contract Documents, in which event interest shall be paid from the date payment was due at the rate of two percent (2%) per annum in excess of the rate of interest announced or published publicly from time to time by [the First National Bank of Chicago], at its principal place of business as its prime or equivalent rate of interest (the "Prime Rate"). The foregoing rate of interest to be charged hereunder shall change automatically without notice and simultaneously with each change in the Prime Rate. A certificate of said Bank as to its Prime Rate in effect on any day shall, for purposes hereof, be conclusive evidence of the Prime Rate in effect on such day.

§ 11.9   Guarantees

The owner may want someone to guarantee the contractor's performance. This is often done when the owner has questions about the contractor's financial ability to perform, or for large projects. One way is to have the contractor obtain a bond. Another is to have a responsible party actually give a personal guarantee. For instance, the owner of the construction company may be asked to guarantee the firm's performance. Here is a form of guarantee that has a provision requiring the guarantor to arbitrate:

Guarantee: For valuable consideration, John Doe and Jane Doe jointly and severally, personally and expressly guarantee the performance of all of the terms and provisions of the Agreement by the Contractor without condition or exception. All of the terms of the Agreement, including, but not limited to, the arbitration provisions, are incorporated into this Guarantee.

Adapted from *Scinto v. Sosin*, 721 A.2d 552 (Conn. App. Ct. 1998).

---

for the balance of the cost of the house. The homeowner counterclaimed. The court found that the contractor had breached the contract and awarded the owner the cost of repairs. The appellate court held that the contractor was not entitled to interest and attorneys' fees on his unpaid bills because the amount awarded to the owners exceeded those costs.

In Marsh v. Loffler Hous. Corp., 102 Md. App. 116, 648 A.2d 1081, 1086 (1994), the contract contained this provision: "In the event that payment under this contract is enforced through legal action, or other collection action, homeowner agrees to pay contractor's costs and attorney's fees related to said action." The court held that attorneys' fees were properly awarded in arbitration.

§ 11.10   Other Provisions: ¶ 8.6

The owner may want the contractor to perform some pre-construction work. These provisions can be adapted to such situations:

8.6.1 Pre-Contract Work performed prior to execution of this Agreement is included as part of the Work of this Agreement, and is comprised of the following:

.1 Contractor's employees' attendance at construction and municipality meetings as specifically requested by Owner with respect to plan design and approval issues or any other purposes as Owner deems necessary;

.2 Preparation of a trade-by-trade cost estimate based on plans;

.3 Preparation of a schedule for the Project, and analysis and updating of the schedule as required;

.4 Contractor shall conduct a value-engineering analysis, including cost, construction feasibility, and considerations relative to labor and material availability;

.5 Contractor shall analyze and make comments and recommendations regarding construction budget;

.6 Contractor shall assist in the selection process and recommend MEP Subcontractors;

.7 Contractor shall assist in the procurement, scheduling, storage and installation of prepurchased long-lead items; and

8.6.2 The Pre-Contract Work Costs are included as part of the Contract Sum stated in Paragraph 4.1 above.

GLOSSARY

AAA   American Arbitration Association. A not-for-profit corporation with offices around the country for the resolution of disputes through mediation and arbitration. AIA contracts call for mediation through the AAA, with arbitration as an optional alternative dispute resolution procedure.

abandonment   A concept related to *cardinal change* (see below). *See L.K. Comstock & Co., Inc., v. Becon Const. Co., Inc.*, 932 F. Supp. 906 (E.D. Ky. 1993). Under this concept, the parties abandon the contract by not following the contractual procedures governing changes to the contract. Some courts distinguish this concept from *cardinal changes. Amelco Electric v. City of Thousand Oaks*, 27 Cal. 4th 228, 38 P.3d 1120 (2002).

addendum (addenda)   A document (drawing or written document) issued by the architect prior to execution of the Owner-Contractor Agreement that modifies or clarifies the bidding documents. The addendum becomes a part of the contract documents. See AIA Document A201, ¶ 1.1.1.

additional services   Services of the architect for which the architect is owed extra fees by the owner. See AIA Document B101, Section 4.

advertisement for bids   A published notice that solicits bids for a project.

affidavit   A written statement made under oath before an officer (notary public) who is authorized to administer oaths.

agency   A relationship in which one party (the agent) acts for another (the principal) with the authority of the principal. The name of the principal may or may not be disclosed.

agent   In law, an agent is a representative of another.

Agreement   In the A201, it refers to the Owner-Contractor agreement, possibly the A101 or similar document. The Agreement would then list all of the other documents that form the Contract between the parties, including the A201.

AIA   American Institute of Architects. 1735 New York Ave. NW., Washington, DC 20006. The largest organization of architects, it is headquartered in Washington, D.C.

alternate bid   A separate amount stated in a bid which, if accepted by the owner, will be added or deducted from the base bid. If accepted, the work that corresponds to the

807

alternate bid becomes part of the agreement between the owner and the contractor. *See FRT Int'l, Inc. v. City of Pasadena,* 62 Cal. Rptr. 2d 1 (1997).

ANSI The American National Standards Institute. 1430 Broadway, New York, NY 10018. A national institute for producing standards used in the construction and other industries. These standards are often incorporated into building codes. It was previously known as the United States of America Standards Institute (USASI) and the American Standards Association (ASA).

anticipatory breach A breach of contract in which one party informs the other of its intent to breach the contract before performance is due. In some cases, this may be cured by repudiating the anticipatory breach and performing under the contract.

Application for payment The contractor's sworn request for payment for the work performed since the previous application, plus materials suitably stored on the site. See AIA Document A201, Section 9.3.

approved equal An alternate material or piece of equipment approved by the architect to be incorporated into the work and substituted in place of that specified or shown on the drawings.

arbitration A method of resolving disputes in which a neutral third party hears evidence and decides the outcome. All states have statutes that provide that proper arbitration awards can be enforced by the courts. To be enforceable, there must be an agreement among the parties to arbitrate the dispute. All AIA contracts have an arbitration clause whereby the parties agree beforehand to arbitrate.

Architect A person who is qualified by education and experience to design structures. No person can be an architect without being licensed in one or more states. Without meeting certain additional requirements, an architect licensed in one state cannot perform architectural services in another state. In the A201, the Architect is the person or firm specifically identified, or its subsequent replacement.

as-built drawings A revised set of drawings that incorporates changes made during the construction process. The architect usually prepares these from marked-up drawings furnished by the contractor at the end of the project.

ASHRAE American Society of Heating, Refrigerating and Air Conditioning Engineers. http://www.ashrae.org. An international organization founded in 1894, that has developed standards for the design of HVAC and related equipment.

assignment A transfer of an interest in something, such as a contract, lease, or title, to another.

ASTM American Society for Testing and Materials. http://www.astm.org. 100 Barr Harbor Dr., West Conshohocken, PA.

attorney-in-fact A person authorized, usually under a power of attorney, to act for another.

Award Often used to refer to the final decision of an arbitrator, which must be confirmed by a court of law to be enforceable. Sometimes refers to the award of a contract to a contractor.

axonometric drawing A method of perspective drawing that shows the plan and partial elevations on the same drawing.

base bid The Contractor's bid for the Work, not including any Alternates.

Basic services The services for which the architect is hired by the owner and for which the architect will be paid compensation. They do not include any Additional Services (AIA Document B101, Section 4) for which the architect is entitled to additional compensation.

beneficial interest The equitable interest in property, rather than the legal interest. This may be under a trust or implied by law. The owner of the beneficial interest gets the benefits and usually has control over the property.

beneficiary One who is entitled to the beneficial ownership of something. Usually applied to a land trust, where the beneficiary is the true owner of the trust property, but the trustee is the legal owner.

bid bond A bond furnished by a bidder. The purpose is to assure that the bidder will enter into a contract with the owner if the owner accepts that bid.

bid form A form to be filled out by a bidder on a project when submitting the bid. This form is usually prepared by the architect and includes information required to evaluate the bid.

bid security A bid bond or other security, such as certified check or cash, to ensure that the contractor will enter into a contract with the owner.

bidding documents Documents, usually prepared by the architect, including advertisement or invitation to bid, Instructions to Bidders, sample forms, and the contractor's bid, which are not part of the contract documents (see AIA Document A201, ¶ 1.1.1), as well as the Contract Documents.

bill of sale A legal instrument that transfers personal, rather than real, property.

building information modeling (BIM) A project database into which the project participants place information. This information can include everything from the drawings, specifications, shop drawings, and so forth, to schedules and cost information.

bulletin A document issued by the architect after execution of the Owner-Contractor Agreement, requesting a proposal for a change in the work. This becomes a part of the contract documents either as a change order or by a construction change directive.

cardinal change   "an alteration in the work so drastic that it effectively requires the contractor to perform duties materially different from those originally bargained for." *Allied Materials & Equip. Co., Inc., v. United States*, 215 Ct. Cl. 406, 569 F.2d 562 (1978).

Certificate for payment   The architect's certification to the owner that the contractor is entitled to payment as shown on the contractor's Request for Payment. See AIA Document A201, ¶ 9.4.

certificate of occupancy   A certificate issued by a municipality stating that a building is ready for occupancy. This may be an indication that the building is substantially complete.

Certificate of substantial completion   The architect's certification to the owner that the work is substantially complete and that the contractor is entitled to payment up to that point. See AIA Document A201, ¶ 9.8.

change in service   Services for which the architect is entitled to additional compensation. Also known as additional services. See AIA Document B101, Section 4.

Change order   A written change to the contract documents and agreed to by the owner, contractor, and architect. It may change the contract sum or contract time, or both. If the contractor does not agree to it, it is a construction change directive. See AIA Document A201, ¶ 7.2.

Claim   A demand or assertion by the owner or contractor, seeking, as a matter of right, an adjustment of the Contract price or time, or other relief with respect to the Contract. Article 15 of A201 governs Claims and the process involved.

class action   A legal action brought on behalf of a number of people (members of a class) and seeking a common remedy.

clerk of the works   A supervisory person, usually full time at the site, who has charge of the work on behalf of the owner. Sometimes employed by the architect. This is not part of the normal architectural services.

code   A comprehensive set of laws that cover a given subject. In architecture, used to refer to building codes.

common law   The modern civil law, with its origins in England. A body of cases that serves as precedent for courts, as distinguished from laws passed by legislatures.

compensatory damages   Damages awarded to cover an actual loss or injury.

completion bond   A bond posted by a contractor to ensure completion of the construction contract. See AIA Document A201, ¶ 11.4.

computer aided design (also known as "**computer assisted drawings**," "**computer aided drafting**," or "**CAD**") Drawings in electronic format such as ".dwg." Used by architects, engineers, and others to create the graphical representations for structures.

condemnation The taking of private property for public use. The United States Constitution requires just compensation for such a taking.

conditions of the contract The portions of the contract documents that deal with the rights and responsibilities of the parties and other procedural matters. Included are the General Conditions (A201), Supplementary Conditions, and other conditions.

condominium A structure that contains two or more spaces that are individually owned. Portions of the structure or surrounding property may be owned in common by all the owners (common area). Regulated by the individual states and some local municipalities.

consequential loss (consequential damages) Damages that are not directly caused by damage to property but may result from it. "Losses that do not flow directly and immediately from an injurious act but that result indirectly from the act." *Black's Law Dictionary* (8th Ed., 2004). For example, damage to goods stored in a building caused by a roof leak. These damages are waived in standard AIA documents (e.g., Section 15.1.6 of A201-2007).

consideration In law, something of value that induces another to enter into a contract. Every contract requires consideration to be enforceable.

construction change directive A change to the contract documents directed to the contractor by the architect and approved by the owner. Once the contractor approves this, it is converted to a change order. See AIA Document A201, ¶ 7.3.

construction manager (CM) The entity who contracts with the owner to manage the construction. This is different from the general contractor, because the CM is an agent of the owner and the owner is liable to the various separate contractors. If the CM is personally liable on the contracts, he is really a general contractor. Note that in a CM situation, the subcontractors may be considered general contractors. *See Sagamore Group, Inc. v. Commissioner of Transportation,* 29 Conn. App. 292, 614 A.2d 1255 (1992) (defines *construction management*).

construction schedule *See* Progress schedule.

consultant One who provides professional services or advice, such as an engineer. See *Best Friends Pet Care, Inc. v. Design Learned, Inc.,* 77 Conn. App. 167, 823 A.2d 329 (2003) (examining "consultants" vs. "contractors").

contingency An event that must occur before a contract is binding. A financing contingency is common in residential sales, where the contract is void if the buyer cannot obtain financing.

contingent fee A fee that is due only upon the happening of the stated contingency. For instance, the architect may prepare drawings contingent on a developer's closing on the property. If no closing occurs, the architect is not paid.

Contract An agreement between two or more parties that creates or modifies a legal relationship. A contract must have consideration. In A201, the Contract is defined in ¶ 1.1.2.

contract administration The duties of the architect during the construction phase. Also known as "construction phase services." See AIA Document B101, Section 3.6.

contract award A notification to a contractor by an owner that the owner intends to enter into a contract for construction with the contractor. Usually follows a bid opening.

Contract documents The documents that form the contract for construction between the owner and contractor. See AIA Document A201, ¶ 1.1.1 for a list.

Contract Sum The price to be paid by the owner to the contractor upon completion of the Work. This can be modified only by a Change Order or the procedure of construction change directive. See AIA Document A101, Art. 4; AIA Document A201, ¶ 9.1.

Contract Time The time allotted for completion of the Work. See AIA Document A101, Art. 3 and A201, Article 8.

Contractor A person or entity who does construction work. A general contractor is hired directly by the owner to perform the work and may, or may not, employ subcontractors to perform portions of the work. A subcontractor is hired by a higher-tiered contractor to perform a portion of the work. In many jurisdictions, a license is required to be a contractor or subcontractor. In A201, the Contractor is the party who is hired by the Owner to perform the Work. See A201, Article 3.

contractor's affidavit Also called the Contractor's Application for Payment. A notarized statement by the contractor listing the various subcontractors and the amounts of work performed by each, as well as payment information. See AIA Document A201, ¶ 9.3.

convey In law, to transfer title to property to someone else.

Cost of the work In A201 this is referred to as the Contract Sum. See ¶ 9.1 of A201.

cost plus A cost plus contract is one where the contractor is entitled to be reimbursed for its costs of time and materials plus a fee. In *Kinetic Sys., Inc. v. Rhode Island Indus. Facilities Corp.,* 2003 WL 22048520 (R.I. Super., Aug. 4, 2003), the project cost went from $30 million to over $90 million, resulting in litigation. The owner attempted to shift the burden of proving the reasonableness of the costs to the contractor. The court held that the contractor was required to prove only the costs themselves, not their reasonableness, absent a showing of fraud or gross negligence.

critical path method (CPM) A scheduling method. Each portion of the process is shown relative to each other part. If a delay of any particular task results in a delay of completion, that task is on the critical path.

CSI Construction Specifications Institute. http://csinet.org. 99 Canal Center Plaza, Suite 300, Alexandria, VA 22314. A national nonprofit organization dedicated to the advancement of construction technology. Includes architects, specifiers, engineers, contractors, and others in the construction industry.

damages The loss suffered by one who has been injured. Actual damages are the cost of repair, injury, and so on. Punitive damages are damages imposed by a court as a means of punishing bad conduct.

date of substantial completion The date shown on the Architect's Certificate of Substantial Completion (see AIA Document A201, ¶ 9.8) when the Work is Substantially Complete. This date is important because it starts the statute of limitations and other periods of importance to the parties.

decree A judgment of a court.

default A failure to perform a legal duty.

default judgment A judgment entered against a party by the court when the party fails to appear at the required time.

defective work Work that does not conform with the contract documents. The architect can reject defective work, but only the owner can accept defective work.

defendant The party against whom a civil or criminal lawsuit is brought.

deponent One who makes a sworn statement, either orally at a deposition, or in writing. If the written statement is an affidavit, the person is the affiant.

deposition Testimony, under oath, of a person taken before an authorized reporter.

design-build A method of building delivery in which the owner contracts with one party who both designs and constructs the building. The design-builder is both the contractor and architect. Typically, the design-builder is a contractor who enters into its own contract with an architect who is then obligated to the design-builder and not the owner, although the architect may still have some liability to the owner.

design development The last stage of design prior to the preparation of the construction documents. This is the stage at which most of the final design decisions are made. See AIA Document B101, ¶ 3.3.

detail   A drawing that shows part of another drawing to explain or clarify some portion of the Work. Details are usually at a larger scale.

direct personnel expense   Cost attributable to salary plus normal benefits. See AIA Document B 141-1997, ¶ 1.3.9.4. This term is not used in the 2007 owner-architect agreements.

direct salary expense   Same as direct personnel expense, except that normal benefits are not included.

discovery rule   The "Discovery Rule" involves defects that are not readily apparent, and must be "discovered" at some later time. Once discovered, the statute of limitations starts to run. In *1000 Virginia Limited P'ship v. Vertecs Corp.,* 127 Wash. App. 899, 112 P.3d 1276 (2005), the court held that "for the discovery rule to apply, the breach must have been undiscovered (latent) and the plaintiff must show he or she reasonably failed to discover it (the 'knew or should have known' test)." The discovery rule originally was judicially created, as opposed to statutes of limitations, which were enacted by legislatures. Subsequently, discovery rules have found their way into laws.

division   Refers to the Specifications, which were divided into 16 divisions, such as concrete, masonry, etc, prior to 2004. At that time, the Construction Specifications Institute (CSI) issued a revised version of MasterFormat™ which is considered the specifications-writing standard for most commercial building design and construction projects in North America.

duplex   A building containing two buildings. Also refers to apartments that are on two floors.

elevation   In architecture, a view of a vertical surface such as an exterior wall. Also, the height above some standard point, such as sea level.

eminent domain   The right of a governmental body to acquire private property for public use, usually through a condemnation proceeding.

encroachment   When a structure such as a building or fence extends onto the property of another.

engineer, civil   A person licensed by the state to practice engineering related to land, drainage, and related matters.

engineer, electrical   A person licensed by the state to practice engineering related to the electrical systems of structures and equipment.

engineer, mechanical   A person licensed by the state to practice engineering related to the mechanical (HVAC) and plumbing systems of a building.

engineer, structural   A person licensed by the state to practice engineering related to the structural systems of buildings and other structures.

estoppel The prevention of one from asserting a legal right because of that person's prior inconsistent position with that assertion.

exemplary damages Damages to punish the offender. Also called punitive damages. Used when there is a deliberate act or gross negligence.

extra Refers to an addition to the contract sum. Extras are valid only as change orders or construction change directives.

fast track A method of building delivery whereby the construction starts before the building is fully designed. The design professionals complete the contract documents in the order in which the contractor needs them. It requires that the contractor be on the building team long before the drawings are completed. It saves the owner time (and therefore money), but it usually results in additional costs as a result of errors, because the documents cannot be as well coordinated as in a traditional project.

fiduciary One who acts in a position of trust to another, usually with regard to financial matters. A fiduciary owes a higher degree of loyalty and dealing. Most states do not consider architects as fiduciaries to their clients.

field order Often used in construction to refer to a written order by an architect or engineer, often written at the jobsite, directing a minor change in the work. A201 does not have a provision for field orders.

field representative The architect's representative at the job site, whose duty is to administer the construction contract.

final acceptance The owner's acceptance of the Work following the architect's final Certificate for Payment.

Final completion The end of construction, as opposed to substantial completion, when some items still remain to be completed. See AIA Document A201, ¶ 9.10. *Decca Design Build, Inc. v. American Auto. Ins. Co.,* 77 P.3d 1251 (Ariz. App. 2003). "Project completion" is completion of all work under the general contract. Final payment is not due until the project is fully complete and the architect has issued a final certificate for payment. See A201, Section 9.10.

final inspection The last of two inspections performed by the architect to determine if the Work is completed and final payment is due the contractor. See AIA Document A201, ¶ 9.10.1.

Final payment The time when a project is fully completed and final payment is due. At this time, all punchlist items are completed and there is nothing left for the contractor to complete. See A201, Section 9.10.

fixed limit of construction cost The agreed-upon maximum cost of the Work under the 1987 version of B141. If there was to be such a limit it had to be agreed upon separately under that version of the document. See AIA Document B141-1987, art. 5. In the 1997 version of B141, the architect was required to conform the design to the budget and the owner can elect several remedies, including having the architect revise the documents for no extra charge, if the lowest bid comes in over the budget. Effectively, then, the 1997 B141 contains a fixed limit of construction cost but the earlier versions did not. Note that the 1997 B151 does not contain such a fixed limit. The 2007 version, AIA Document B101-2007, effectively contains such a fixed limit, although that term is not used.

floor plan   A space layout of a part of a building that shows the sizes of rooms and the function of each. May also show other information for the contractor.

general conditions   Usually refers to AIA Document A201. The part of the contract documents that sets forth the rights and responsibilities of the parties.

general partner   A partner in a partnership with authority to bind the partnership. As opposed to a limited partner who has no management authority at all.

good faith   Done without fraudulent intent.

grade   The angle of slope of land. Also refers to the level or elevation of land.

guarantee   A contract that something or someone will perform as promised. Same as warranty. Distinguished from suretyship, which involves a third party.

guaranteed maximum cost (G-Max)   The amount set forth in the agreement between the owner and contractor as the maximum amount of construction cost. This is frequently accompanied by provisions related to shared cost savings.

Hazardous materials   See A201, Section 10.3. These usually include asbestos, PCBs, petroleum, radioactive materials, and similar materials.

hold harmless agreement   An agreement by which one party agrees to assume the risks of another and pay the costs of the other's defense and other costs. Same as indemnity agreement. *East-Harding, Inc. v. Horace A. Piazza & Assocs.*, 80 Ark. App. 143, 91 S.W.3d 547 (2002) ("Indemnity arises by virtue of a contract and holds one liable for the acts or omissions of another over whom he has no control.").

HVAC   Heating, ventilating and air conditioning. This is usually considered a separate trade, with a single contractor performing this work.

implied warranty   As opposed to an express warranty (which is a warranty explicitly set forth in the agreement). A warranty implied by law, such as an implied warranty of habitability or merchantability.

incumbrance   Any right to or interest in property.

indemnification   *See* hold harmless agreement.

independent contractor   As opposed to an employee. Used to describe a relationship in which the independent contractor is not under the direct control of the party who hired him.

instructions to bidders   The instructions that explain the bidding procedures for the project to the bidders. Not a part of the contract documents under AIA Document A201.

invitation to bid   A solicitation to potential bidders requesting that they submit bids.

isometric drawing   A type of perspective drawing.

job captain   A term used by design firms to denote the person responsible for preparation of the contract documents.

joint venture   A partnership of two or more firms or individuals that is normally limited to one project.

judgment   The decision of a court of law.

jurisdiction   The authority that a court has over a party. State courts usually lack jurisdiction over a party that does not reside or do business in the state. A judgment rendered against a party without jurisdiction is void.

labor and material payment bond   A bond by which a surety guarantees to the owner that the contractor will pay for labor and material used on the project.

latent defect   A defect that is not apparent by normal inspection as is a patent defect.

legal rate of interest   A rate of interest established by law as opposed to established by agreement. Usually less than any prevailing interest rate.

lessee   The party that rents a space.

lessor   The landlord.

letter of intent   A letter that is usually meant to be the prelude to a formal contract. If it contains sufficient information, it may be considered a contract by the courts. Otherwise, it is unenforceable.

lien   A security or incumbrance against a property. This includes mortgages, certain judgments, and mechanics' liens. To be effective, a lien must be filed where title to the property is registered, usually with the county recorder of deeds.

lien waiver Also called waiver of lien. A document by which a party gives up a lien right. Usually used during construction prior to payment to a contractor or subcontractor.

limitation of action A time limit imposed by statute after which rights to bring a lawsuit expire. Different matters have different limitations periods. Also called statute of limitations.

limited partner A partner in a limited partnership with no management authority. As opposed to a general partner.

limited partnership A partnership created under state statute composed of one or more general partners and one or more limited partners. When dealing with limited partnerships, one should deal only with general partners.

liquidated damages The amount of damages specified in a contract that will compensate one party for the breach of contract by the other party. This is a predetermined amount, with no necessary relation to the actual damages. "The term 'liquidated damages' applies to a specific sum of money that has been expressly stipulated by the parties to a contract as the amount of damages to be recovered by one party for a breach of the agreement by the other, whether it exceeds or falls short of the actual damages. *Time Warner Entm't Co., L.P. v. Whiteman*, 802 N.E.2d 886 (Ind., 2004). The enforceability of a liquidated damages clause depends on whether, in the event of a breach, the damages resulting from the breach would have been difficult to ascertain. *Cincinnati Ins. Co. v. Jasper City Utility Serv. Bd.*, 2006 WL 2472735 (S.D. Ind., Aug. 24, 2006).

lis pendens A legal notice recorded with the recorder of deeds against a property that legal proceedings are pending. Anyone taking title to property with a lis pendens recorded takes title subject to the litigation.

load The force exerted on a material. Anticipated loads on buildings and building components are used by the structural engineer in designing the building supporting structure. Wind loads, snow loads, and other types of loads are included in the calculations. Dead loads include the weight of the structure and permanent components of the building. Live loads are the expected weight of people and furnishings. Building codes usually give minimum design loads for various types of occupancies.

low bid The bid that gives the lowest price for the performance of the Work, including selected alternates.

lowest responsible bidder The bidder who submits the lowest bid and is also considered by the owner and architect to be qualified to perform the Work.

lowest responsive bid The bid that is the low bid and complies with all the bidding requirements.

master plan   A zoning plan for a subdivision or larger area. Usually, a comprehensive plan for the growth of a city.

materialman   One who supplies materials for a project. May have mechanics lien rights. *See Vulcraft v. Midtown Bus. Park, Ltd.,* 110 N.M. 761, 800 P.2d 195 (1990) (defines "materialman" as contrasted with "subcontractor"). *Waco Scaffolding Co. v. National Union Fire Ins. Co. of Pittsburgh,* 1999 Ohio App. LEXIS 5058 (Oct. 28, 1999) (discussion of subcontractor and materialmen with respect to insurance coverage).

mechanics' lien   A lien is a right against a property. A mechanic is defined by state law, and usually includes any person who performs services related to the construction of the project, possibly including architects, engineers, and others. Suppliers of materials have rights under mechanic's lien statutes. The lien is normally filed with the recorder of deeds in the county where the project is located.

Milestone   Used in schedules to indicate an important event or date.

Miller Act   40 U.S.C. § 270a-d. An act designed to protect contractors and suppliers on federal government projects. States have adopted similar Little Miller Acts to cover state projects.

misnomer   In law, a mistake in name. Use of the wrong name.

misrepresentation   An untrue statement of fact. Can be negligent or intentional (fraud).

modification   A written amendment to the contract; a change order; a construction change directive; or a written order for a minor change in the Work. A modification changes the contract documents. See AIA Document A201, ¶ 1.1.1.

NCARB   National Council of Architectural Registration Boards. http://www.ncarb.org. 1801 K Street, NW, Suite 1100-K, Washington, D.C. 20006. This national group administers architectural licensing examinations and coordinates qualifications of architects among states. Many architects are registered with NCARB to facilitate licensing in additional states.

negligence   Failure to exercise proper care when one is under a legal duty.

NFPA   National Fire Protection Association. http://www.nfpa.org. 1 Batterymarch Park, Quincy, MA 02169.

NIOSH   National Institute for Occupational Safety and Health. http://www.cdc.gov/niosh/. A governmental organization, part of the CDC, involved in preventing work-related illnesses and injuries. Centers for Disease Control and Prevention, 1600 Clifton Rd., Atlanta, GA 30333.

nominee   One who buys property on behalf of another. Used when the true buyer doesn't want his name revealed.

nonconforming work   Work not in accordance with the contract documents. The architect may reject nonconforming work, but only the owner can accept nonconforming work.

notice to proceed   A notice by the owner to the contractor that directs the contractor to start the Work. It also establishes the date of commencement of the Work.

OSHA   Occupational Safety and Health Administration. http://www.osha.gov. Department of Labor, 200 Constitution Ave. NW, Washington, DC 20210.

outline specifications   Abbreviated specifications used at the schematic design and design development stage.

Owner   One who owns property. In construction contracts, the owner may not be the actual owner, but a lessee, beneficiary, or other person. It is important to verify the status of such an owner. See A201, Article 2.

owner of record   The owner according to the records of the recorder of deeds.

partial occupancy   The use of a portion of the project prior to substantial completion of the whole project. See AIA Document A201, ¶ 9.9.

patent defect   As opposed to a latent defect. A defect that is readily apparent on reasonable inspection. Some courts hold that an owner waives patent defects upon making final payment to a contractor.

payment bond   A bond that assures that the contractor will pay for materials. *See Kammer Asphalt Paving Co. v. East China Township Sch.,* 443 Mich. 176, 504 N.W.2d 635, 637 (1993) (defines payment bond). See AIA Document A201, ¶ 11.4.

penalty clause   A provision in a contract that penalizes the contractor if the Work is completed late. Use of a penalty clause requires a bonus clause in the event the Work is completed early.

perfecting   In law, an instrument such as a lien, deed, or mortgage is perfected when it is recorded. Prior to perfection it is not valid against third parties.

performance bond   A bond posted by a contractor to assure performance of the contract. See AIA Document A201, ¶ 11.4. See *Kammer Asphalt Paving Co. v. East China Township Sch.,* 443 Mich. 176, 504 N.W.2d 635, 637 (1993) (defines performance bond). *Bank of Brewton, Inc. v. International Fidelity Ins. Co.,* 827 So. 2d 747 (Ala. 2002) (A performance bond is an express, written bilateral contract to which the general principles of contract interpretation apply).

plaintiff   The party bringing a lawsuit against the defendant. There are also third-party plaintiffs, which occurs when one of the defendants (the third-party plaintiff) brings in a third-party defendant in an existing lawsuit.

plans   The drawings that an architect prepares. Usually refers to the working drawings. Architects sometimes use this term more narrowly, meaning the actual floor plans, but not elevations or details.

plat   A plan of an area of land dividing it up into lots.

plot plan   A plan of a parcel of property showing where the improvements are to be constructed.

power of attorney   A document that authorizes one to act as attorney-in-fact for another.

prefabrication   The process of assembling components of buildings off-site or not in place. Also called modular construction.

preliminary drawings   Usually refers to the drawings prepared during the design stages of the project.

preliminary estimate of the Cost of the Work   Estimate of the cost of construction prepared by the architect during the design phases of the project. See AIA Document B141-1997, ¶ 2.1.7.1; B101-2007, ¶ 3.2.6.

principal   Among design professionals, refers to an owner, such as a partner or shareholder, of a design firm. In law, this refers to one who hires an agent. A principal can be disclosed (the person who deals with the agent knows the agent is working for someone else) or undisclosed.

priority   In mechanics' lien law, priority establishes the order in which liens are paid off. States differ in the method of establishing priority.

privity   Successive rights or relationships. When one party assigns its contract rights, the new person is in privity with the second party to the contract.

product data   Brochures, schedules, charts, and other diagrams furnished by the contractor for products to be incorporated into the Work. These are usually prepared by the manufacturer of the product.

program   A list of the elements required to be incorporated into the project. A program can be very specific, including required room sizes and equipment specifications. It may be very vague, depending on the owner. It is the responsibility of the owner to provide the architect with the program or to hire the architect to work with the owner to develop the program.

progress payment   Partial payments made by the owner to the contractor during the course of the project. See AIA Document A201, art. 9.

progress schedule (Construction Schedule)   The proposed schedule of the project furnished by the contractor, usually in chart form. Same as construction schedule. See AIA Document A201, ¶ 3.10.1.

project manager Used by design firms to identify the person in charge of the project for the design firm.

project manual The volume usually assembled by the architect, including bidding requirements, sample forms, conditions, and specifications. See 1997 version of AIA Document A201, ¶ 1.1.7 (not included in the 2007 version).

proximate cause The dominant cause of an injury, without which the injury would not have happened.

punchlist A list prepared near the end of a project that shows all work remaining to be performed by the contractor before final completion. This is the contractor's responsibility (see AIA Document A201, ¶ 9.8.2: a *comprehensive list*). Often, the architect and sometimes the owner will each prepare its own punchlist. *See Russo Farms, Inc. v. Vineland Bd. of Educ.*, 675 A.2d 1077 (N.J. 1996) (defines punchlist). A punchlist is a document created in connection with a walk-through by the owner or architect and contractor. The punchlist may include cleanup and touchup work, as well as work that is unacceptable in quality to the owner, contractor, or architect. *Southwest Progressive Enter., Inc. v. R.L. Harkins, Inc.*, 2002 WL 358830 (Tex. App.—El Paso, March 7, 2002).

quantity survey A detailed list of all the materials that form the Work. Used to establish a more accurate cost of the work than is normally done by the architect.

quantum meruit The reasonable value of the work. An implied-in-law contract. This theory of recovery is used when there is no contract or the contract is void.

ratification The act of affirming an act that had no legal effect to begin with. After the affirmance, it is binding on the party that affirmed it. For instance, a principal can affirm an act of an agent that was not authorized, giving it retroactive legal effect.

real property (real estate) Land and property attached to it, such as buildings and fixtures. As opposed to personal property.

record drawings See as-built drawings.

reimbursable expenses Those expenses advanced towards the project for which the design professional will be reimbursed by the owner. See AIA Document B141-1997, ¶ 1.3.9.2; B101-2007, ¶ 11.8.

rendering A drawing of a proposed project, used for presentation purposes.

retainage An amount withheld from the Contractor with each payment. Designed to give the Owner additional assurance that the project will be completed in case of default by the contractor. "[T]he percentage General Contractor and Owner agreed to withhold from each pay request to be paid at the end of the project, or as negotiated through the course of the project. Retainage served to assure General Contractor's completion of the work and to satisfy subcontractors' and suppliers' lien claims." *Cullum Mech. Constr., Inc. v. South Carolina Baptist Hosp.,* 344 S.C. 426, 544 S.E.2d 838 (2001).

Samples A sample of actual products, materials, or work that will be incorporated into the Work. For instance, a brick sample would be an actual brick to be used in the construction. See A201, Sections 3.11 and 3.12. Samples would also include mock-ups of sections of the work.

satisfaction A discharge of an obligation by payment.

scale The amount by which a drawing or detail is reduced. For example, a drawing with a scale of "⅛=1-0" means that each ⅛ inch on the drawing represents one foot in actuality. Also refers to the ruler used by architects. The term *not to scale* means that the drawing itself is not completely accurate and one cannot rely on measuring any portion of that drawing, but written dimensions on such a drawing should be accurate.

scaling The process of thin pieces of hardened concrete breaking away from the concrete mass. *Scaling a drawing* means placing a scale (ruler) on the drawing to determine the dimension of an element. This practice is strongly discouraged.

Schedule of values A list of the various trades, such as concrete, masonry, etc., that will perform work at the project, and the dollar amount of each trade's work. For instance, the contractor may show "concrete—$12,000," or the architect may require greater specificity, such as a breakdown of concrete for footings, floor slabs, etc. With a proper schedule of values, the architect can determine whether the progress of the Work is such as to justify the amounts shown on the contractor's payment application.

Schematic design This is the stage of the project at which the architect sketches the floor plans and elevations to determine the general configuration of the building and to determine whether the budget is reasonably adequate. These drawings are rough and subject to considerable changes in the later stages of the project. See AIA Document B141-1997, ¶ 2.4.2; B101-2007, ¶ 3.2.

seal A device used to imprint an attestation of authenticity. An architect's seal is often required on drawings. Other seals are notary and corporate seals.

sepia A brown drawing used as an original from which prints are made. Usually made from another original.

Shop drawings Drawings, diagrams, schedules, and other data specially prepared for the Work by the contractor, subcontractor, manufacturer, supplier, or distributor that show some portion of the Work. See AIA Document A201, Section 3.12.

site The land on which the project is built.

site development　Refers to the improvements (utilities, grading, etc.) to the site, except for the building itself.

site plan　A drawing that shows the building(s) in outline form on the site, along with the site improvements.

space design　Used to refer to interior design.

Specifications　The written portion of the contract documents giving the contractor information, such as acceptable manufacturers, testing, cleanup, standards, and other items. The specifications are usually bound into a separate book and are divided into sections, usually according to CSI format. See AIA Document A201, ¶ 1.1.6.

statement of probable cost　See preliminary estimate of the Cost of the Work.

statute of limitations　A statutory provision that limits the time in which legal actions may be brought against another party. These time limits vary according to the reason for the claim, and from state to state.

statute of repose　Similar to statute of limitations. The difference is that the time within which to bring an action against a defendant is absolutely cut off, regardless of whether the statute of limitations has run. See AIA Document A201-2007, ¶ 13.7, establishing an absolute 10-year time limit.

stipulated sum contract　"Establishes a fixed price for the work the contractor is to perform. This is the price the owner/developer will pay regardless of the contractor's actual cost of construction." This type of contract places all of the risk of construction on the contractor. *Stelko Elec., Inc. v. Taylor Cmty. Sch. Bldg. Corp.*, 826 N.E.2d 152 (Ind. Ct. App. 2005).

strict liability　Liability without fault. In case of an injury, the defendant is liable, even though he was not negligent.

subcontract　A contract that is not with the owner. Usually refers to the contract between the contractor and a subcontractor.

Subcontractor　A contractor who works for the general contractor. See AIA Document A201, Art. 5. If a contractor works for a subcontractor, it is called a sub-subcontractor. *See Vulcraft v. Midtown Bus. Park, Ltd.*, 110 N.M. 761, 800 P.2d 195 (1990) (defines "materialman" as contrasted with "subcontractor"); *Winter v. Smith*, 914 S.W.2d 527 (Tenn. Ct. App. 1996) (defines subcontractor). *Waco Scaffolding Co. v. National Union Fire Ins. Co. of Pittsburgh*, 1999 Ohio App. LEXIS 5058 (Oct. 28, 1999) (discussion of subcontractor and materialmen with respect to insurance coverage); *Hartford Ins. Co. v. American Automatic Sprinkler*, 201 F.3d 538 (4th Cir. 2000) (distinguishes "contractor" from "subcontractor").

subpoena　A legal procedure used to compel the attendance of a witness or document at a court. Usually refers to the legal document. Also called a writ.

substantial completion   The stage of the project at which the construction is sufficiently complete so that the owner can occupy the building. A punchlist is prepared that details the items remaining before final completion. See AIA Document A201, ¶ 9.8. Similar to substantial performance in law. See *Russo Farms, Inc. v. Vineland Bd. of Educ.*, 675 A.2d 1077 (N.J. 1996) (defines substantial completion).

superintendent   The contractor's agent at the site who is responsible for the execution of the Work and safety. Unless specially designated, the architect does not have a superintendent or any supervisory duties. See AIA Document A201, ¶ 3.9.

supplementary conditions   The portion of the contract documents that supplements the General Conditions. A document that makes A201 specific for the particular job. Instead of Supplementary Conditions, the modern practice is to make revisions directly to A201.

Supplier   A vendor of materials or equipment who does no actual work at the site. This party has a contract with the general contractor or a lower tier subcontractor to furnish materials, supplies, or other materials that are either incorporated into the Work or used in the construction of the Work.

surety   One who voluntarily binds itself to answer for the debt of another, usually for a fee.

surety bond   "A surety bond is a unique form of insurance. While insurance generally insures a principal for his own benefit; [sic] in a surety bond, the principal obtains insurance for the benefit of a third party, known as the obligee. A surety guaranties to the project owner that if the contractor defaults, the project will be completed." *Travelers Cas. & Sur. Co. v. Dormitory Auth. of the State of N.Y.*, 2005 WL 1177715 (S.D.N.Y., May 19, 2005).

survey   A measurement of land, including boundaries and improvements. Also refers to the plan drawn by the surveyor.

syndicate   An association formed for the purpose of participating in a common goal. Syndications are often used for real estate deals. Often refers to the process of limited partnership formation and the raising of equity for a project.

tenant   One who holds an interest in property under a lease or rental agreement. In law, tenancy refers to any possessory interest in property, with or without an agreement.

tender   The offer of money or performance under a contract.

termination expenses   The expenses caused by termination of the architect's contract by the owner. See AIA Document B141-1997, ¶ 1.3.8.7; B101-2007, ¶ 9.7.

third-party beneficiary   One for whom a contract was made, even though that person was not a party to the contract. He might be a direct or indirect beneficiary.

The parties to the contract may not intend that he be a beneficiary. The AIA contracts state that no third-party beneficiaries are intended.

topographic survey   A survey that shows the elevations (heights) of the property.

tort   A private, or civil, wrong. Independent of a contract.

trust   A right of property held by one for the benefit of another.

trustee   One who holds property for the benefit of another.

turnkey   A method of construction delivery in which the owner has minimal input during the process. The owner initially gives a program to the developer, who then contracts with architects and contractors to deliver a fully completed project to the owner. All the owner has to do is ''turn the key'' in the door (and write the check).

undisclosed principal   A principal whose identity is not disclosed by an agent.

Underwriters Laboratories (UL)   http://www.ul.com. 333 Pfingsten Rd., Northbrook, IL 60062. A nonprofit organization that maintains laboratories for the examination and testing of devices, systems, and materials. Materials and devices that meet published standards of performance and manufacture may carry a UL label. The manufacturer must pay UL a fee, and UL may inspect the manufacturing process on an ongoing basis. The lack of a UL label does not necessarily mean that the product is unsafe. Publications listing manufacturers and standards are available from UL.

Uniform Commercial Code (UCC)   A code now adopted in all states that regulates the transfer of personal property and commercial paper, such as checks and notes. May apply to construction, such as the purchase of materials and equipment. *See, e.g., J. Lee Gregory, Inc. v. Scandinavian House, Ltd. P'ship,* 209 Ga. App. 285, 433 S.E.2d 687 (1993) (UCC applied to a contract to furnish and install windows in an apartment house).

Unit price   Work that is paid for on the basis of a price for each piece. For instance, this could be a price per ton of sand, per yard of concrete, and so forth. Often used when exact quantities are not known at bidding time.

usury   An illegal rate of interest on a loan. Many states have laws that regulate interest on certain loans.

vendor   One who sells property, whether real or personal.

void   Having no legal effect. A void contract cannot be enforced. Also, an open or empty space.

voidable   Can be made void by the action of a party, or can be ratified. Not void in and of itself.

waiver The abandonment or relinquishment of a right. A waiver of lien is a written waiver of a right to a mechanics lien.

warranty See guarantee.

Work That part of the project that is being performed by a particular contractor. The contract defines what the Work for that particular contractor is. It may be a part of a larger project. For instance, the Project may be the construction of a university, whereas the Work is the construction of a single university building. It does not include adjacent work of other contractors or items owned by the owner. See AIA Document A201, § 1.1.3.

working drawings Also known as construction drawings. These drawings form part of the contract (the contract documents) between the owner and contractor. These are not the design drawings that lead up to the working drawings. They show the contractor the details of the project, including dimensions and materials. See AIA Document A201, § 1.1.5.

TABLE OF CASES

*Alphabetization is letter-by-letter (e.g., "Hayes Drilling, Inc." precedes "Hay Group, Inc.").
References are to section numbers.*

Case	*Book §*
Gutowski v. Crystal Homes, Inc., 26 Ill. App. 2d 269, 167 N.E.2d 422 (1960)	§§ 4.17, 4.28, 10.6
G-W-L, Inc. v. Robichaux, 643 S.W.2d 392 (Tex. 1982)	§ 4.3
Habitat Architectural Group, P.A. v. Capital Lodging Corp., 28 Fed. Appx. 242, 2002 WL 86682 (4th Cir. Jan. 23, 2002)	§ 2.26
Haemonetics Corp. v. Brophy & Phillips Co., 23 Mass. App. Ct. 254, 501 N.E.2d 524 (1986)	§ 4.65
Hagerman Constr. Corp. v. Long Elec. Co., 741 N.E.2d 390 (Ind. Ct. App. 2000)	§ 4.28
Hagerman Constr., Inc. v. Copeland, 697 N.E.2d 948 (Ind. Ct. App. 1998), *further proceedings,* 1998 Ind. App. LEXIS 2046 (4th Dist. Nov. 25, 1998)	§ 4.28
Hagerstown Elderly Assocs. v. Hagerstown Elderly Bldg. Assocs., 368 Md. 351, 793 A.2d 579 (2002)	§ 4.56
H. & A. Constr. Co., *In re,* 65 B.R. 213 (Bkrtcy. D. Mass. 1986)	§ 4.52
Hanna v. Huer, Johns, Neel, Rivers & Webb, 233 Kan. 206, 662 P.2d 243 (1983)	§§ 4.3, 4.31
Harbor Court Assocs. v. Leo A. Daly Co., 179 F.3d 147 (4th Cir. 1999)	§§ 2.24, 5.70
Harbor Mech., Inc. v. Arizona Elec., 496 F. Supp. 681 (D. Ariz. 1980)	§ 2.3
Harman v. C.E.&M., Inc., 493 N.E.2d 1319 (Ind. Ct. App. 1986)	§ 2.5
Harmon v. Christy Lumber, Inc., 402 N.W.2d 690 (S.D. 1987)	§ 2.1
Harrington v. LaBelle's, Inc., 235 Mont. 80, 765 P.2d 732 (1988)	§ 4.56
Harrington v. McCarthy, 91 Idaho 307, 420 P.2d 790 (1966)	§ 4.40
Harris v. Dyer, 292 Or. 233, 637 P.2d 918 (1981)	§ 4.75
Harris County v. Howard, 494 S.W.2d 250 (Tex. 1973)	§ 2.22
Harris Custom Builders, Inc. v. Hoffmeyer, 834 F. Supp. 256 (N.D. Ill. 1993)	§ 2.23
Harrison F. Blades, Inc. v. Jarman Mem'l Hosp. Bldg. F, 109 Ill. App. 2d 224, 248 N.E.2d 289 (1969)	§§ 2.26, 4.88
Harry Skolnick & Sons v. Heyman, 7 Conn. App. 175, 508 A.2d 64 (1986)	§ 4.31
Hart & Son Hauling, Inc. v. MacHaffie, 706 S.W.2d 586 (Mo. Ct. App. 1986)	§ 4.53
Hartford Accident v. Scarlett Harbor Assocs. Ltd. P'ship, 109 Md. App. 217, 674 A.2d 106 (1996), *aff'd,* 346 Md. 122, 695 A.2d 153 (1997)	§ 4.89
Hartford Accident & Indem. Co. v. Boise Cascade Corp., 489 F. Supp. 855 (N.D. Ill. 1980)	§ 10.6
Hartford Elec. Applicators, Inc. v. Alden, 169 Conn. 177, 363 A.2d 135 (1975)	§§ 4.28, 4.46, 10.6
Hartford Fire Inc. v. Riefolo Constr. Co., 161 N.J. Super. 99, 390 A.2d 1210 (1978)	§§ 4.56, 4.65
Hartzell v. Justus Co., 693 F.2d 770 (8th Cir. 1992)	§ 4.85
Harza Ne., Inc. v. Lehrer McGovern Bovis, Inc., 680 N.Y.S.2d 379 (N.Y. App. Div. 1998)	§ 2.22
Hatzel & Buehler, Inc. v. Orange & Rockland Utils., Inc., 1992 WL 391154 (D. Del. Dec. 14, 1992)	§ 4.3
Haugen v. Raupach, 260 P.2d 340 (Wash. 1953)	§ 4.54

Case	*Book §*
Jacksonville Port Auth. v. Parkhill-Goodloe, 362 So. 2d 1009 (Fla. Dist. Ct. App. 1978)	§§ 4.8, 4.17
Jaeger v. Henningson, Durham & Richardson, Inc., 714 F.2d 773 (8th Cir. 1983)	§ 4.31
Jahncke Serv., Inc. v. Department of Transp., 172 Ga. App. 215, 322 S.E.2d 505 (1984)	§ 4.17
J.A. Jones Constr. Co. v. Greenbrier Shopping Ctr., 332 F. Supp. 1336 (N.D. Ga. 1971)	§ 4.41
Jakubowski v. Alden-Bennett Constr. Co., 763 N.E.2d 790 (Ill. App. Ct. 2002)	§ 4.58
J.A. McDonald, Inc. v. Waste Sys. Int'l, 189 F. Supp. 2d 174 (D. Vt. 2001)	§ 4.81
James A. Cummings, Inc. v. Young, 589 So. 2d 950 (Fla. 1992)	§ 4.31
James J. Gory Mech. Contracting, Inc. v. Philadelphia Hous. Auth., 2001 WL 1736483 (Pa. C.P. July 11, 2001)	§ 2.26
James McKinney & Son, Inc. v. Lake Placid 1980 Olympic Games, Inc., 61 N.Y.2d 836, 462 N.E.2d 137, 473 N.Y.S.2d 960 (1984)	§ 2.29
James Talcott Constr., Inc. v. P&D Land Enters., 141 P.3d 1200 (Mont., 2006)	§§ 4.51, 4.53
J.A. Moore Constr. Co. v. Sussex Assocs. Ltd., 688 F. Supp. 982 (D. Del. 1988)	§ 4.3
J&J Elec., Inc. v. Gilbert H. Moen Co., 9 Wash. App. 954, 516 P.2d 217 (Wash. Ct. App. 1973)	§§ 2.13, 4.31, 4.54, 10.6
Jaroszewicz v. Facilities Dev. Corp., 115 A.D.2d 159, 495 N.Y.S.2d 498 (1985)	§§ 2.13, 2.14, 4.31
J.A. Sullivan Corp. v. Commonwealth, 397 Mass. 789, 494 N.E.2d 374 (1986)	§ 4.31
Javelin Invs., LLC v. McGinnis, 2007 WL 781190 (S.D. Tex. Jan. 23, 2007)	§ 2.23
J. Caldarera & Co. v. Louisiana, 2006 WL 3813721 (La. Ct. App. Dec. 28, 2006)	§§ 2.26, 4.86
J.D. Hedin Constr. Co. v. United States, 347 F.2d 235 (Ct. Cl. 1965)	§ 4.12
Jeffrey A. Grusenmeyer & Assocs. v. Davidson, Smith & Certo Architects, 212 Fed. Appx. 510 2007 WL 62620 (6th Cir. Jan. 8, 2007)	§ 2.23
Jeffrey A. Grusenmeyer & Assocs. v. Davidson, Smith & Certo Architects, 2006 WL 208795 (N.D. Ohio Jan. 25, 2006)	§ 2.23
Jetty, Inc. v. Hall-McGuff Architects, 595 S.W.2d 918 (Tex. App. 1980)	§§ 2.22, 2.33, 10.26
Jewish Bd. of Guardians v. Grumman Allied Indus., 96 A.D.2d 465, 464 N.Y.S.2d 778 (1983)	§§ 2.13, 2.24, 4.31
Jim Arnott, Inc. v. L & E, Inc., 539 P.2d 1333 (Colo. App. 1975)	§ 4.54
Jim Carlson Constr., Inc. v. Bailey, 769 S.W.2d 480 (Mo. Ct. App. 1989)	§§ 10.2, 10.30
J. Lee Gregory, Inc. v. Scandinavian House, Ltd. P'ship, 209 Ga. App. 285, 433 S.E.2d 687 (1993)	§§ 4.3, 4.14, 10.3
J.L. Simmons Co. v. Capital Dev. Bd., 98 Ill. App. 3d 445, 424 N.E.2d 71 (1981)	§ 2.26
J.M. Beeson Co. v. Sartori, 553 So. 2d 180 (Fla. Dist. Ct. App. 1989)	§ 4.54
J.M. Humphries Constr. Co. v. City of Memphis, 623 S.W.2d 276 (Tenn. Ct. App. 1981)	§ 4.40
John G. Danielson, Inc. v. Winchester-Conant Props., Inc., 322 F.3d 26 (1st Cir. 2003)	§ 2.23

Case	*Book §*
Perini Corp. v. Greate Bay Hotel & Casino, Inc., 129 N.J. 479, 610 A.2d 364 (1992)	§§ 4.54, 4.85
Perrit v. Bernhard Mech. Contractors, Inc., 669 So. 2d 599 (La. Ct. App. 1996)	§§ 4.13, 4.58, 4.59
Pertun Constr. Co., U.S. ex rel. v. Harvesters Group, 918 F.2d 915 (11th Cir. 1990)	§ 4.46
Peteet v. Fogarty, 297 S.C. 226, 375 S.E.2d 527 (Ct. App. 1988)	§ 2.22
Peter Kiewit Sons' Co. v. Iowa S. Utils. Co., 355 F. Supp. 376 (S.D. Iowa 1973)	§§ 2.29, 4.46, 5.78
Peterson Constr., Inc. v. Sungate Dev., L.L.C., 2003 WL 22480613 (Tex. App. Oct. 30, 2003)	§§ 2.26, 4.86, 4.88
Pete Wing Contracting, Inc. v. Port Conneaut Investors Ltd. P'ship, 1995 Ohio App. LEXIS 4341 (Sept. 29, 1995)	§§ 2.13, 4.12, 4.54
Philco Corp. v. Automatic Sprinkler Corp. of Am., 337 F.2d 405 (7th Cir. 1964)	§ 4.69
Phoenix Contractors, Inc. v. General Motors Corp., 135 Mich. App. 787, 355 N.W.2d 673 (1984)	§ 4.46
Pickett v. Chamblee Constr. Co., 124 Ga. App. 769, 186 S.E.2d 123 (1971)	§ 4.56
Pierce v. ALSC Architects, P.S., 52 Mont. 93, 890 P.2d 1254 (1995)	§§ 2.5, 4.56
Pierce Assocs. v. Nemours Found., 865 F.2d 530 (3d Cir. 1988)	§ 4.3
Pigott Constr. Int'l, Ltd. v. Rochester Inst. of Tech., 84 A.D. 679, 446 N.Y.S.2d 632 (1981)	§ 4.31
Pinkert v. Olivieri, 2001 WL 641737 (D. Del. May 24, 2001)	§ 4.28
Pinkerton & Laws Co. v. Roadway Express, Inc., 650 F. Supp. 1138 (N.D. Ga. 1986)	§ 4.17
Pioneer Roofing Co. v. Mardian Constr. Co., 152 Ariz. 455, 733 P.2d 652 (1986)	§ 4.85
Pipe Welding Supply v. Haskell, Connor & Frost, 96 A.D.2d 29, 469 N.Y.S.2d 221 (1983), *aff'd,* 61 N.Y.2d 884, 462 N.E.2d 1190, 474 N.Y.S.2d 472 (1984)	§ 2.22
Piqua, City of v. Ohio Farmers Ins. Co., 84 Ohio App. 3d 619, 617 N.E.2d 780 (1992)	§ 4.89
Pitt v. Tyree Org. Ltd., 90 S.W.3d 244 (Tenn. Ct. App. 2002)	§ 4.28
Pitts v. Watkins, 905 So. 2d 553 (Miss. 2005)	§§ 2.24, 2.26, 3.46, 6.21
Pittsburgh, City of v. American Asbestos, 629 A.2d 265 (Pa. Commw. Ct. 1993)	§ 4.28
Pittsburgh Plate Glass Co. v. Kransz, 291 Ill. 84, 125 N.E. 730 (1919)	§ 2.3
Plan Pac., Inc. v. Andelson, 6 Cal. 4th 307, 962 P.2d 158, 24 Cal. Rptr. 2d 597 (1994)	§ 2.26
Point E. Condo. Owner's Assn. v. Cedar House Assoc., 663 N.E.2d 343 (Ohio Ct. App. 1995)	§§ 4.13, 4.35
Polk County v. Widseth.Smith.Nolting, 2004 WL 2940847 (Minn. Ct. App. Dec. 21, 2004)	§ 4.51
Polshek v. Bergen Iron Works, 142 N.J. Super. 516, 362 A.2d 63 (1976)	§§ 4.3, 4.89
Positive Software Solutions v. New Century Mortgage, 476 F.3d 278 (5th Cir. 2007)	§ 2.26
Premier Elec. Constr. Co. v. American Nat'l Bank, 658 N.E.2d 877 (Ill. App. 1st Dist., 1995)	§ 4.35

Case	*Book §*
Resurgence Props., Inc. v. W.E. O'Neil Constr. Co., No. 92C6618, 1995 U.S. Dist. LEXIS 4939 (N.D. Ill. Apr. 14, 1995)	§§ 2.5, 2.15, 2.29, 4.3
Resurgence Props., Inc. v. W.E. O'Neil Constr. Co., 1995 U.S. Dist. LEXIS 11633 (N.D. Ill. Aug. 14, 1995)	§ 10.2
Reynolds v. Long, 115 Ga. App. 182, 154 S.E.2d 299 (1967)	§§ 2.7, 2.28
R.G. Nelson, A.I.A. v. M.L. Steer, 797 P.2d 117 (Idaho 1990)	§§ 2.5, 2.13
R.H. Macy & Co. v. Williams Tile & Terrazzo, 585 F. Supp. 175 (N.D. Ga. 1984)	§§ 2.5, 4.3
Rhoades v. United States, 986 F. Supp. 859 (D. Del. 1997)	§ 4.28
Rian v. Imperial Mun. Serv. Group, Inc., 768 P.2d 1260 (Colo. Ct. App. 1988)	§ 2.13
Richmond v. Grabowski, 781 P.2d 192 (Colo. Ct. App. 1989)	§ 4.65
Richmond Homes Mgmt., Inc. v. Raintree, Inc., 862 F. Supp. 1517 (W.D. Va. 1994)	§ 2.23
Richmond Shopping Ctr. v. Wiley N. Jackson Co., 220 Va. 135, 255 S.E.2d 518 (1979)	§ 4.3
Riggi v. Wade Lupe Constr. Co., 176 A.D.2d 1177, 575 N.Y.S.2d 613 (N.Y. App. Div. 1991)	§ 4.88
Ringwelski v. Pederson, 919 P.2d 957 (Colo. Ct. App. 1996)	§§ 2.26, 4.88
Riviera v. City of Meriden, 2006 WL 1000003 (Conn. Super. Mar. 23, 2006)	§ 4.59
Rivnor Props. v. Herbert O'Donnell, Inc., 633 So. 2d 735 (La. Ct. App. 1994)	§§ 4.13, 4.65
R.J. Griffin & Co. v. Beach Club II Homeowners Ass'n, 384 F.3d 157 (4th Cir. 2004)	§ 2.26
RLI Ins. Co. v. MLK Ave. Redevelopment Corp., 925 So. 2d 914 (Ala. 2005)	§ 4.56
RLI Ins. Co. v. St. Patrick's Home for the Infirm and Aged, 452 F. Supp. 2d 484 (S.D.N.Y. 2006)	§§ 4.52, 4.86
Roadway Package Sys., Inc. v. Kayser, 257 F.3d 207 (3d Cir. 2001)	§§ 2.26, 3.34
Robbinsdale Pub. Sch., Indep. Sch. Dist. No. 281 v. Haymaker Constr. Inc., 1997 Minn. App. LEXIS 1155 (Oct. 14, 1997)	§§ 2.26, 4.88
Robert G. Regan Co. v. Fiocchi, 44 Ill. App. 2d 336, 194 N.E.2d 665 (1963)	§§ 4.4, 4.11
Robert J. Denley Co. v. Neal Smith Constr. Co., 2007 WL 1153121 (Tenn. Ct. App. Apr. 19, 2007)	§§ 2.26, 4.3, 4.88
Robert Lamb Hart Planners & Architects v. Evergreen, Ltd., 787 F. Supp. 753 (S.D. Ohio 1992)	§§ 2.26, 2.27, 4.88
Robert M. Swerdroe, Architect/Planners v. First Am. Inv. Corp., 565 So. 2d 349 (Fla. Dist. Ct. App. 1990)	§ 2.20
Robert P. Willis II, Estate of v. Kiferbaum Constr. Corp., 830 N.E.2d 636 (Ill. Ct. App. 2005)	§ 4.28
Roberts v. Security Trust & Sav. Bank, 196 Cal. 557, 238 P. 673 (1925)	§ 4.31
Roberts & Schaefer Co. v. Hardaway Co., 152 F.3d 1283 (11th Cir. 1988)	§ 6.4
Roberts & Schaefer Co. v. Merit Contracting, Inc., 901 F. Supp. 1349 (N.D. Ill. 1995)	§§ 2.29, 4.72
Robinson v. Powers, 777 S.W.2d 675 (Mo. Ct. App. 1989)	§ 2.9
Rockland, County of v. Primiano Constr. Co., 51 N.Y.2d 1, 409 N.E.2d 951, 431 N.Y.S.2d 478 (1980)	§ 4.31

Case	*Book §*
South Burlington Sch. Dist. v. Calcagni-Frazier-Zajchowski Architects, Inc., 138 Vt. 33, 410 A.2d 1359 (1980)	§§ 2.14, 4.31
Southeastern Sav. & Loan Ass'n v. Rentenbach Constructors, Inc., 907 F.2d 1139 (4th Cir. 1990)	§ 4.49
Southern Md. Hosp. Ctr. v. Edward M. Crough, Inc., 48 Md. App. 401, 427 A.2d 1051 (1981)	§ 4.85
Southern Okla. Health Care Corp. v. JHBR, 900 P.2d 1017 (Okla. 1995)	§ 2.5
South Tippecanoe Sch. Bldg. Corp. v. Shambaugh & Sons, 395 N.E.2d 320 (Ind. Ct. App. 1979)	§ 4.65
South Union v. George Parker & Assocs., 29 Ohio App. 3d 197, 504 N.E.2d 1131 (1985)	§§ 2.18, 4.54
Southwest Eng'g Co. v. United States, 341 F.2d 998 (8th Cir. 1965)	§§ 4.46, 10.6
Southwestern Bell Tel. Co. v. J.A. Tobin Constr., 536 S.W.2d 881 (Mo. 1976)	§ 4.28
Southwest Nat'l Bank v. Simpson & Son, Inc., 14 Kan. App. 2d 763, 799 P.2d 512 (1990)	§ 10.30
Space Planners Architects, Inc. v. Frontier Town-Missouri, Inc., 107 S.W.3d 398 (Mo. App. 2003)	§§ 2.3, 10.1
Spancrete, Inc. v. Ronald E. Frazier & Assocs., 630 So. 2d 1197 (Fla. Dist. Ct. App. 1994)	§§ 2.3, 4.31
Spaw-Glass Constr. Servs. v. Vista de Santa Fe, 114 N.M. 557, 844 P.2d 807 (1992)	§ 10.1
Spearhead Constr. Corp. v. Bianco, 39 Conn. App. 122, 665 A.2d 86 (1995)	§§ 2.26, 4.88
Spearin v. City of N.Y., 160 A.D.2d 263, 553 N.Y.S.2d 372 (1990)	§ 4.46
Spearin; United States v., 248 U.S. 132 (1918)	§§ 4.3, 4.12, 4.17
S.S.D.W. Co. v. Brisk Waterproofing Co., 153 A.D.2d 476, 544 N.Y.S.2d 139 (1989), aff'd, 76 N.Y.2d 228, 556 N.E.2d 1097, 557 N.Y.S.2d 290 (1990)	§§ 4.3, 4.65
Stallings & Sons, Inc. v. Sherlock, Smith & Adams, Inc., 670 So. 2d 861 (Ala. 1995)	§§ 2.27, 4.89
Standard Co. v. Elliott Constr. Co., 359 So. 2d 224 (La. Ct. App. 1978)	§ 2.26
Standard Co. v. Elliott Constr. Co., 363 So. 2d 671 (La. 1978)	§ 2.26
Standhardt v. Flintcote Co., 84 N.M. 796, 508 P.2d 1283 (1973)	§ 2.9
Standley v. Egbert, 267 A.2d 365 (D.C. 1970)	§ 2.28
Stanley v. Chastek, 34 Ill. App. 2d 220, 180 N.E.2d 512 (1962)	§ 2.5
Stanley Consultants v. H. Kalicak Constr. Co., 383 F. Supp. 315 (E.D. Mo. 1974)	§ 2.22
Stark v. Ralph F. Roussey & Assocs., 25 Ill. App. 3d 659, 323 N.E.2d 826 (1975)	§ 2.22
Starks Mech., Inc. v. New Albany-Floyd County Consol. Sch. Corp., 854 N.E.2d 936 (Ind. Ct. App. 2006)	§§ 4.46, 4.85
State v. *See name of opposing party*	
State *ex rel. See name of related party*	
State Farm Fire & Cas. Co. v. White, 777 F. Supp. 952 (N.D. Ga. 1991)	§ 2.23
State Highway Admin. v. Greiner Eng'g Sciences, Inc., 83 Md. App. 621, 577 A.2d 363 (1990)	§§ 2.21, 4.28

Case	*Book §*
Taber Partners I v. Insurance Co. of Am., 875 F. Supp. 81 (D.P.R. 1995)	§§ 4.11, 4.50, 4.52, 4.75
Tamarac Dev. v. Delamater, Freund, 234 Kan. 618, 675 P.2d 361 (1984)	§§ 2.5, 2.14, 4.31
Tanner, United States *ex rel.* v. Daco Constr., Inc., 38 F. Supp. 1299 (N.D. Okla. 1999)	§ 4.88
T.A. Tyre Contractor, Inc. v. Dean, 2005 WL 1953036 (Del. Super. Ct. June 14, 2005)	§ 2.26
Taylor v. Allegretto, 112 N.M. 410, 816 P.2d 479 (1991)	§§ 4.4, 10.3
Taylor v. Cannaday, 230 Mont. 151, 749 P.2d 63 (1988)	§ 2.5
Taylor v. DeLosso, 725 A.2d 51 (N.J. Super. Ct. App. Div. 1999)	§§ 2.21, 4.8
Taylor Pipeline Constr., Inc. v. Directional Rd. Boring, Inc., 438 F. Supp. 2d 696 (E.D. Tex. 2006)	§ 4.52
TC Arrowpoint, L.P. v. Choate Constr. Co., 2006 WL 91767 (W.D.N.C. Jan. 13, 2006)	§§ 2.26, 4.73
Teal Constr. v. Darren Casey Int'l, 46 S.W.2d 417 (Tex. App. 2001)	§ 2.26
Technosteel, LLC v. Beers Constr. Co., 271 F.3d 151 (4th Cir. 2001)	§ 4.35
Teitge v. Remy Constr. Co., 526 N.E.2d 1008 (Ind. Ct. App. 1988)	§ 2.14
Tekmen & Co. v. Southern Builders, Inc., 2005 WL 1249035 (Del. Super. Ct. May 25, 2005)	§ 4.86
Temple Sinai-Suburban Reform Temple v. Richmond, 112 R.I. 234, 308 A.2d 508 (1973)	§ 4.3
Terra Group, Inc. v. Sandefur Mgmt., Inc., 527 So. 2d 849 (Fla. Dist. Ct. App. 1988)	§ 4.66
Texas Bank & Trust Co. v. Campbell Bros., 569 S.W.2d 35 (Tex. 1978)	§ 4.53
Thermal C/M Servs., Inc. v. Penn Maid Dairy Prods., 831 A.2d 1189 (Pa. Super. 2003)	§ 4.88
Thomas Wells & Assocs. v. Cardinal Props., 192 Colo. 197, 557 P.2d 396 (1976)	§ 2.26
3A Indus., Inc. v. Turner Constr. Co., 71 Wash. App. 407, 869 P.2d 65 (1993)	§ 4.35
Three Affiliated Tribes v. Wold Eng'g, 419 N.W.2d 920 (N.D. 1988)	§ 2.5
Tiseo Architects, Inc. v. SSOE, Inc., 431 F. Supp. 2d 735 (E.D. Mich. 2006)	§ 2.23
Tittle v. Giattina, Fisher & Co., Architects, Inc., 597 So. 2d 679 (Ala. 1992)	§ 2.5
Todd Habermann Constr. v. David Epstein, 70 F. Supp. 2d 1170 (D. Colo. 1999)	§ 2.26
Tokio Marine & Fire Ins. Co. v. Employers Ins., 786 F.2d 101 (2d Cir. 1986)	§ 4.65
Tomb & Assocs. v. Wagner, 82 Ohio App. 3d 363, 612 N.E.2d 468 (1992)	§ 4.3
Tonawanda, Town of v. Stapell, Mumm & Beals Corp., 240 A.D. 472, 270 N.Y.S. 377 (1934)	§ 4.56
Touchet Valley Grain Growers, Inc. v. Opp & Seibold Gen. Constr., Inc., 119 Wash. 2d 334, 831 P.2d 724 (1992)	§ 4.65
Town of. *See name of town*	
Townsend v. Muckleshoot Indian Tribe, 2007 WL 316504 (Wash. Ct. App. Feb. 5, 2007)	§§ 4.13, 4.58, 4.59
T-Peg, Inc. v. Isbitski, 2005 WL 768594 (D.N.H. Apr. 6, 2005)	§ 2.23

Case	*Book §*
Weitz Co. v. Shoreline Care Ltd. P'ship, 39 Conn. App. 641, 666 A.2d 835 (1995)	§§ 2.26, 4.88
Welch v. Grant Dev. Co., 120 Misc. 2d 493, 466 N.Y.S.2d 112 (1983)	§§ 2.13, 2.14
Welch v. McDougal, 876 S.W.2d 218 (Tex. App. 1994)	§ 10.30
Wenzel v. Boyles Galvanizing Co., 920 F.2d 778 (11th Cir. 1991)	§§ 2.14, 4.59
Wesleyan Univ. v. Rissil Constr. Assocs., 1 Conn. App. 351, 472 A.2d 23 (1984)	§ 4.3
Westates Constr. Co. v. City of Cheyenne, 775 P.2d 502 (Wyo. 1989)	§ 4.85
West Durham Lumber Co. v. Aetna Cas. & Sur. Co., 184 S.E.2d 399 (N.C. Ct. App. 1971)	§ 4.66
Westerhold v. Carroll, 419 S.W.2d 73 (Mo. 1967)	§§ 2.13, 2.14, 2.15, 4.31
Western Reserve Transit Auth. v. B&B Constr. Co., 1996 Ohio App. LEXIS 143 (Ohio Ct. App. Jan. 16, 1996)	§ 4.17
Western Wash. Corp. of Seventh-Day Adventists v. Ferrellgas, Inc., 7 P.3d 861 (Wash. Ct. App. 2000)	§ 4.65
Westfield, Village of v. Loitz Bros. Constr. Co., 165 Ill. App. 3d 338, 519 N.E.2d 37 (1988)	§§ 4.56, 4.88
Westview Invests., Ltd. v. US Bank Nat'l Ass'n, 138 P.3d 638 (Wash. Ct. App. 2006)	§ 4.52
Westville, Village of v. Loitz Bros. Constr. Co., 165 Ill. App. 3d 338, 519 N.E.2d 37 (1988)	§§ 2.26, 2.29
W.F. Constr. Co. v. Kalik, 103 Idaho 713, 652 P.2d 661 (Ct. App. 1982)	§ 10.2
Whalen v. K-Mart Corp., 519 N.E.2d 991 (Ill. App. Ct. 1988)	§ 4.63
Wheeler & Lewis v. Slifer, 195 Colo. 291, 577 P.2d 1092 (1978)	§§ 2.13, 4.31
Whirlpool Corp. v. Dailey Constr., Inc., 429 S.E.2d 748 (N.C. Ct. App. 1993)	§ 4.49
White v. Mitchell, 123 Wash. 630, 213 P. 10 (1923)	§ 10.6
White Budd Van Ness P'ship v. Major-Gladys Drive Joint Venture, 798 S.W.2d 805 (Tex. 1990)	§ 2.5
Whitehall, City of v. Southern Mech. Contracting, Inc., 269 Ark. 563, 599 S.W.2d 430 (Ct. App. 1980)	§ 4.28
Whitfield Constr. Co. v. Commercial Dev. Corp., 392 F. Supp. 982 (D.V.I. 1975)	§§ 2.18, 2.21, 4.40, 4.85
Whitten Corp. v. Paddock, Inc., 424 F.2d 25 (D. Mass. 1970), *cert. denied,* 421 U.S. 1004, 95 S. Ct. 2407 (1975)	§ 2.9
Whittle v. Pagani Bros. Constr. Co., 422 N.E.2d 779 (Mass. 1981)	§§ 4.28, 4.63
W.H. Lyman Constr. Co. v. Village of Gurnee, 84 Ill. App. 3d 28, 403 N.E.2d 1325 (1980)	§ 4.31
Widett v. United States Fid. & Guar. Co., 815 F.2d 885 (2d Cir. 1987)	§§ 2.29, 4.3, 4.31
Wilco Constr. Co.; State v., 393 So. 2d 885 (La. Ct. App. 1981)	§ 4.56
Wilharm v. M.J. Constr. Co., 1997 Ohio App. LEXIS 591 (Feb. 20, 1997)	§ 2.26
Wilkinson v. Landreneau, 525 So. 2d 617 (La. Ct. App. 1988)	§ 4.12
Wilkinson; State v., 39 P.3d 1131 (Ariz. 2000)	§ 4.11
Willey v. Terry & Wright, Inc., 421 S.W.2d 362 (Ky. Ct. App. 1967)	§ 4.40
Williams & Assocs. v. Ramsey Prod. Corp., 19 N.C. App. 1, 198 S.E.2d 67 (1973)	§§ 2.7, 2.22
Williams & Sons Erectors, Inc. v. South Carolina Steel, 983 F.2d 1176 (2d Cir. 1993)	§ 4.46

INDEX